Maria
Muniagurria

STATISTICAL THEORY

STATISTICAL THEORY

Third Edition

Bernard W. Lindgren
University of Minnesota

MACMILLAN PUBLISHING Co., Inc.
New York
COLLIER MACMILLAN PUBLISHERS
London

Macmillan Publishing Co., Inc.
866 Third Avenue, New York, New York 10022

Collier Macmillan Canada, Ltd.

Library of Congress Cataloging in Publication Data

Lindgren, Bernard William, (date)
Statistical theory.

Bibliography: p.
Includes index.
1. Mathematical statistics. I. Title.
QA276.L546 1976 519.5 74–33099
ISBN 0-02-370830-1 (Hardbound)
ISBN 0-02-979420-X (International Edition)

Printing: 3 4 5 6 7 8 Year: 9 0

PREFACE

Like the earlier editions, this textbook is intended for a year's course in the theory of statistics. No previous knowledge of statistics is assumed, but the material will have special significance for those who have some background in statistical methods. A good command of the first-year calculus course is essential, and multivariable calculus and linear algebra are used in some sections, many of them optional.

The aim of this revision has been to make the material in the book both more accessible to the student and more flexible for the instructor, so that it would be adaptable to courses of various lengths and complexions. There have been significant changes in the order of presentation, and a good bit of the material has been completely rewritten and includes new examples and problems.

Some of the harder material has been put into optional starred sections, and the subsequent development does not depend on them. In particular, the measure-theoretic aspects of probability are optional, as is the material based on linear algebra, including the multivariate normal distribution and the proof of Cochran's theorem.

The most extensive reshuffling has been in the deferring of the decision-theoretic framework until after the special cases of estimation and testing have been studied. This placement eliminates a distinctive (and to some, attractive) feature of the earlier editions, but I have done it for several reasons. First, in following the second edition, students would not reach the garden variety t and F tests until well into the last third of the course, and they would often want this standard material earlier for courses they were taking concurrently. Second, I have been gradually coming to the conclusion that in mathematics generally, the best pedagogical order for the average student—and we have lots of them—is the historical one. We who have "been through it all" appreciate elegance and generality, but we tend to forget that we ourselves did not begin with the ultimate, unified, and general approach in our own learning process. And third, although some might argue that the decision-theoretic approach is basic and all-encompassing, there are other valid points of view that should perhaps not be swept aside by enthusiasm for a nice mathematical structure. Indeed, although statistical mathematics has been extensively developed, the theories of statistical inference are still debated, with no end in sight. I have tried to take this into account, being concerned with the way the mathematical results are used, as much as with proofs of theorems.

Within a course, Chapters 1 through 4 would ordinarily be taken in sequence, but the distribution theory of Section 7.1 could logically be taken up in conjunction with Section 4.3 on sampling distributions. Otherwise, Chapters 5 and 6 should precede the material · that follows. Along with the starred sections, Chapters 8, 9, and 11 may be considered optional—they depend only on Chapters 5 and 6 and may be covered in part or omitted. Chapters 10 and 12, also optional, depend on the material in Chapter 7, and some of Chapter 10 may also be helpful in Chapter 12.

The material is more than ample for a course of three lectures per week. Indeed, it may be that some of the starred material or even one of the later chapters could serve as the basis for term papers for ambitious students. Students should be encouraged to try all of the problems, although not many are likely to conquer them all.

I am grateful to my colleagues in the School of Statistics at the University of Minnesota; they have all, in varying degrees, contributed to this revision—not always knowingly. Several of them read individual chapters. My greatest debt is to Robert Buehler, who read most of the manuscript and whose critical comments resulted in numerous significant improvements.

I am indebted to the editors of *Biometrika*, *The Annals of Mathematical Statistics*, and the *Journal of the American Statistical Association* for permission to adapt tables for use in this book, and to the late Professor Sir Ronald A. Fisher, Cambridge; to Dr. Frank Yates, Rothamsted; and to Oliver and Boyd, Ltd., Edinburgh, Scotland, for permission to reprint a portion of Table 2 from their book *Statistical Tables for Biological, Agricultural, and Medical Research.*

B. W. L.

CONTENTS

STATISTICAL THEORY

1

PROBABILITY MODELS

The term *experiment* will denote doing something or observing something happen under certain conditions, resulting in some final state of affairs or "outcome." The experiment may be physical, chemical, social, industrial, or medical; it may be conducted in the laboratory or in real-life surroundings—in the factory, in the economy, or in the hospital. Occasionally, an experiment is of such a nature that the outcome of the experiment is uniquely defined by a specification of the conditions under which it is performed; indeed, a determinist would insist that this would always be the case if the conditions were completely specified. In practice one finds that experiments are not precisely repeatable, even under supposedly identical conditions. This is the case when there are factors affecting the outcome that the experimenter is not aware of or that he cannot control, and also when factors supposedly under control really are not. The outcome cannot then be predicted from a knowledge of the "conditions" (those taken into account) under which the experiment is performed. One speaks of the experiment as an "experiment involving chance," or simply, as an "experiment of chance." Because of the unpredictability or the element of chance in the experiment, the usual kind of mathematical model involving equations of motion or equations of state (that express physical, chemical, social laws) is inadequate, and a rather new kind of mathematical structure is needed to represent what takes place.

Because the outcome of the experiment is not predictable—is one of many possible outcomes—the model for the experiment should include some kind of list of these possible outcomes. This set of possible outcomes is the "sample space" of the experiment, to be discussed in the first section. The second main ingredient of the mathematical model for an experiment of chance is the notion of probability, which formalizes the common concept that some sets of outcomes may be more or less likely than others. This aspect is taken up in the second section.

Later chapters will then consider the basic problem of statistics—making inferences about certain unknown aspects of the correct model, on the basis of results of one or more trials of the experiment under consideration.

1.1 Sample Spaces

The collection of possible outcomes of an experiment is called a *space*, whose individual points are then the individual outcomes. The word "space" is used here in the mathematical sense of a set of points or objects, rather than in the commonly encountered meaning of "three-dimensional space." The *sample space* of an experiment of chance is simply the set of possible outcomes of the experiment. This is a mathematical concept; operations, relations, and functions will be defined on the points and subsets of this space.

Describing the outcome of an experiment is something that can be done in more than one way. For example, an individual selected by lot from a large group of individuals can be described according to color, sex, or weight, depending upon what is of interest. Thus, the sample space used by one person in constructing a model for an experiment of chance may be different from that used by another. So there is need at the outset for an agreement as to what will be defined and considered as the sample space.

1.1.1 Elementary Outcomes

The term *elementary outcome* will be used to denote one of the various ways in which an experiment can terminate. It is assumed that these ways are defined according to some scheme that is sufficiently detailed to permit them to be used to describe any event likely to be of interest. Because of this dependence on interest, the list of outcomes in one case might be different from the list of outcomes in another case—even though the same physical process is being carried out. When the lists of elementary outcomes are different, the sample spaces are different, and the experiments are thought of as different experiments.

EXAMPLE 1–1. A coin is tossed. If the coin is an ordinary coin, its two faces are different and are referred to as "heads" and "tails" (even though in many instances the tails side actually pictures an eagle). Although it is conceivable that when the coin falls, it may land on edge or roll into a sewer beyond reach, the natural and useful list of outcomes contains just these two: heads up or tails up. These two elementary outcomes make up the sample space, and in this simple experiment there is not much occasion to invent a more complicated model. (One could imagine, however, being interested in the *orientation* of the fallen coin in some reference frame, in which case a more involved model would indeed be required.)

EXAMPLE 1–2. A cube whose six faces are marked with one, two, three, four, five, and six dots, respectively, is called a *die*. When tossed, it falls (on a horizontal surface) with one of the six faces turned up. The up-turned face is usually the outcome of interest, and the six faces (each of which can turn up) are the elementary outcomes. Whether the elementary outcomes are thought of as the faces or as the dot patterns identifying the faces—or indeed as the numbers (1, 2, 3, 4, 5, 6) of dots involved—is a matter of indifference.

EXAMPLE 1–3. Three coins are tossed. The most detailed classification of results would involve knowing what happens with each of the coins. But if one is interested in knowing only the number of heads that show among the three coins, a sample space with the outcomes 0, 1, 2, 3 is adequate.

EXAMPLE 1–4. When a missile is fired at a target on the earth's surface, the result cannot be predicted in terms of known or measurable quantities, because of uncertainties in the propellant, in atmospheric conditions, and in the direction of aiming. The experiment is then best thought of as an experiment of chance. The outcome of the experiment is the landing point of the missile, and the set of all points on the earth's surface (perhaps restricted to those lying within some reasonable distance of the target) would make up the sample space. If a plane surface is assumed, and a rectangular grid is placed over the target area, the landing point's coordinate representation in this grid can be used to identify the outcome. The sample space is then essentially the collection of ordered pairs of numbers—that is, of sets of coordinates (x, y).

EXAMPLE 1–5. Four chips marked, respectively, A, B, C, and D are placed in a container and mixed. If two chips are selected blindly, the outcome cannot be predicted, and a chance model is again appropriate. The elementary outcomes can be thought of as the possible pairs of chips, which can easily be listed: AB, AC, AD, BC, BD, CD. (In writing these down, it is unfortunately necessary—or at least, convenient—to impose an apparent order. That is, in AB, the A is written first and the B second. However, this is nothing more than a quirk of writing things down, and there is no significance in the order. The only question that need be answered in identifying an outcome is this: Which two chips were drawn?)

A related experiment is that in which two chips are drawn *one at a time*, the first not being replaced for the second drawing. In this case, the order *can* be observed and may be significant; to take it into account, a sample space is required in which the elementary outcome AB is distinct from BA. The complete list would be as follows: AB, AC, AD, BC, BD, CD, BA, CA, DA, CB, DB, DC. If the order is not significant, even though the chips are drawn in succession, this 12-point sample space is unnecessarily complicated, but there is no harm in using it.

EXAMPLE 1–6. A certain material is thought of as being composed of tubes or fibers, each assumed to be oriented in a certain direction with respect to some given reference frame. A fiber selected at random will have an orientation that can be any of infinitely many possible orientations of a line in space. These orientations make up the sample space. However, the sample space can be identified with the set of points on a hemisphere of unit radius by identifying a given point on the sphere—or its coordinate representation in some system—with the orientation of the vector from the center of the sphere out to that point. It is immaterial whether the sample space is taken as the set of orientations, the set of points on the hemisphere, or the set of possible coordinate representations of these points.

1.1.2 Composition of Experiments

In many instances, it is easier to analyze a complex experiment by treating it as a combination of two or more simpler experiments, and so it is natural to consider two or more experiments jointly. Suppose, then, that performing an experiment $\mathscr{E}$ is equivalent to first performing an experiment $\mathscr{E}_1$ and then performing an experiment $\mathscr{E}_2$. There is a natural (but not unique) way of itemizing the outcomes of the composite experiment $\mathscr{E}$ in terms of the outcomes of $\mathscr{E}_1$ and of $\mathscr{E}_2$. This way is to associate each elementary outcome of $\mathscr{E}_1$ in turn with each elementary outcome of $\mathscr{E}_2$ to obtain what are then the elementary outcomes of experiment $\mathscr{E}$. Clearly, if the number of elementary outcomes of $\mathscr{E}_1$ is n_1, and the number of elementary outcomes of $\mathscr{E}_2$ is n_2, then the number of elementary outcomes in the composite experiment, counted according to the convention just described, is the product: $n_1 n_2$. (Of course, if either n_1 or n_2 is infinite, there will be infinitely many outcomes in the composite sample space.)

The elementary outcomes of a composite experiment, when specified in this fashion, may turn out to be more "elementary" than needed for some purpose. On the other hand, they will certainly be elementary enough, and it is convenient to have such a standard method of defining a sample space for a composite experiment.

EXAMPLE 1–7. If a coin and a die are tossed together, it does not usually matter whether the coin or die lands first, but the same possible outcomes can be achieved by tossing the coin and then the die. And so the experiment can be thought of as a composite of these simpler experiments (first tossing the coin and then tossing the die). Each of the 2 elementary outcomes in the toss of the coin can be associated with each of the 6 elementary outcomes of the toss of the die to form the following 12 elementary outcomes of the composite experiment:

$$H\text{-}1, H\text{-}2, H\text{-}3, H\text{-}4, H\text{-}5, H\text{-}6; \quad T\text{-}1, T\text{-}2, T\text{-}3, T\text{-}4, T\text{-}5, T\text{-}6.$$

EXAMPLE 1–8. Consider again (as in Example 1–5) the drawing of two chips from a container in which there are four chips marked A, B, C, and D, respectively. The experiment can be performed in two steps: First, one chip is drawn, and then from the remaining chips a second chip is drawn. In the first subexperiment, there are four elementary outcomes; namely, A, B, C, D. In the second subexperiment, the elementary outcomes depend on what was drawn first; but *no matter what is drawn first*, the number of chips available for the second draw is three. Thus, there are $4 \cdot 3 = 12$ elementary outcomes in the composite experiment—just those listed at the end of Example 1–5.

EXAMPLE 1–9. An experiment consists of arranging three objects in a sequence. If the objects are identified as A, B, and C, respectively, the *six* possible arrangements are the elementary outcomes:

$$ABC, \quad ACB, \quad BAC, \quad BCA, \quad CAB, \quad CBA.$$

This count can be obtained without writing out the list by considering the experiment as a composite of three subexperiments. First, select one of the three objects (three outcomes) for the first position of the sequence; second, select one of the remaining two objects (two outcomes) for the second position of the sequence; and third, put the remaining object in the last position (one outcome). The product of the numbers of outcomes in these subexperiments is the desired number of arrangements: $3 \cdot 2 \cdot 1 = 6$.

An obvious generalization of the argument in the last example shows that the number of arrangements of n objects in a sequence is $n! = n(n-1)(n-2)\cdots 3 \cdot 2 \cdot 1$.

If only some, say k, of the n objects are used in an arrangement, then the number of distinct possible arrangements of k objects taken from n is (following the line of reasoning used above)[1]

$$n(n-1)\cdots(n-k+1) \equiv (n)_k.$$

This product of k factors starting with n and decreasing in steps of 1 can be written in terms of factorials:

$$(n)_k = \frac{n!}{(n-k)!}.$$

EXAMPLE 1–10. An assignment of 3 students from a committee of 10 to serve as its chairman, vice-chairman, and secretary is essentially an arrangement of 3 students from the 10. It can be accomplished in 720 ways:

$$(10)_3 = 10 \cdot 9 \cdot 8 = 720.$$

It is also useful to be able to count arrangements (without actually making a list) in cases in which some objects are *alike*. This is not to say that the objects lose their identities; it is just that the identities of "like" objects will be ignored in distinguishing arrangements. The following example will illustrate what is meant.

EXAMPLE 1–11. Four blocks are painted, two red, one blue, and the other yellow. These four blocks can be arranged in 24 ways that are really distinct; but if one chooses to consider one red block as *not* distinguishable from the other, there are 12 possible arrangements:

$$RRYB, \quad RYRB, \quad RYBR, \quad YRRB, \quad YRBR, \quad YBRR,$$
$$RRBY, \quad RBRY, \quad RBYR, \quad BRRY, \quad BRYR, \quad BYRR.$$

These arrangements can be counted without making a list upon realization of the fact that each one corresponds to 2! arrangements in the list of 4! arrangements of four distinct objects. (Putting subscripts on the R's to distinguish them would yield

[1] Another symbol sometimes used for this product is ${}_nP_k$, read "the number of permutations of n things k at a time."

2! arrangements for each one listed; e.g., R_1R_2YB and R_2R_1YB both become $RRYB$ when the subscripts are dropped.) The desired count is, therefore, $4!/2! = 12$.

Again, the generalization is almost obvious. For example, if there are r red objects, b blue objects, and y yellow objects among n objects, and if the remaining objects (if any) are distinct from these and from each other, the number of *color patterns* (which is what is now meant by distinct arrangements) is $n!/(r!\,b!\,y!)$.

A particular instance of some importance is that in which there are just two colors, so that among n objects, k are of one color (or kind) and the remaining $n - k$ are of another color (or kind). The number of distinct arrangements or patterns is

$$\frac{n!}{k!\,(n-k)!}.$$

The importance of this calculation lies in the fact that this count of arrangements is also a count of the number of selections (without regard to order) of k objects from n distinct objects. To arrange the k and $n - k$ objects in a sequence, one can proceed by *selecting* k of the n sequence positions in which to place the objects of one kind; the remaining $n - k$ positions are given to the objects of the other kind. The symbol[2] for the number of distinct combinations or selections of k out of n objects is $\binom{n}{k}$. It is computed as follows:

$$\binom{n}{k} = \frac{n!}{k!\,(n-k)!} = \frac{n(n-1)(n-2)\cdots(n-k+1)}{k(k-1)(k-2)\cdots 3\cdot 2\cdot 1} = \frac{(n)_k}{(k)_k}.$$

This is called a *binomial coefficient*, for reasons that become clear in Problem 1–9. The formula given remains valid for $k = 0$ and $k = n$ if it is understood (as is customary) that $0!$ means 1.

EXAMPLE 1–12. From a container with four chips marked 1, 2, 3, and 4, respectively, two chips are selected. The number of such selections is $\binom{4}{2}$ or 6, which can easily be listed:

1 2, 1 3, 1 4, 2 3, 2 4, 3 4.

Each of these selections corresponds to an arrangement of two A's and two B's:

AABB, *ABAB*, *ABBA*, *BAAB*, *BABA*, *BBAA*.

The numbers 1 3 indicate, for instance, that A's will appear in the first and third positions of the sequence and B's in the other positions: *ABAB*.

[2] Other symbols are sometimes used: ${}_nC_k$, C_k^n.

It should be observed that the binomial coefficient has a symmetry:

$$\binom{n}{k} = \binom{n}{n-k},$$

as is apparent from a comparison of the factorial expressions for the two sides or from the fact that a selection of k automatically selects $n - k$ (to be left behind). That is, selecting k out of n really just separates the n objects into two piles, one with k and one with $n - k$ objects.

Problems

1–1. How many code letters can be made employing four or fewer symbols in sequence for each letter, when each symbol is a dot or a dash? (Examples of such "letters" might be — · ·, — · — —, —, and so forth. Letters in Morse code are formed in this way, so your answer should be at least 26.)

1–2. Compute the following:

(a) $(12)_4$. (b) $\binom{12}{4}$. (c) $\binom{100}{97}$. (d) $\binom{n}{1}$.

1–3. A *function* assigns a value, one of a set of possible values called the *range* of the function, to each point in a set called the *domain* (or domain of definition) of the function. If a domain has exactly m points and the range exactly k values, how many distinct functions can be defined?

1–4. A bowl contains black, white, and red chips. An experiment consists of drawing four chips from the bowl, one at a time, and the outcome is a recorded sequence of observed colors of the chips drawn. How many outcomes are in the sample space of this experiment, assuming that each chip drawn is replaced prior to the next drawing? [Ask yourself: Would it make any difference if the chips drawn were not replaced? For example, consider the cases in which (a) there are only two red chips to begin with, and (b) there are at least four chips of each color.]

1–5. Each symbol in a sequence of 10 symbols is a (+) or a (−).

(a) How many distinct sequences are possible?
(b) How many of these have at least eight +'s in them?
(c) How many sequences contain exactly five +'s and exactly five −'s?
(d) Of the sequences in (c), how many have at least four +'s in a row?

1–6. How many outcomes are there in the sample space of each of the following experiments, using the most detailed classification?

(a) The toss of three ordinary dice.
(b) The toss of n coins.
(c) The selection of 4 persons from a group of 10, without regard to order.
(d) The selection of 4 persons from 10, to serve as president, vice-president, secretary, and treasurer.

1–7. A *standard bridge deck* contains 52 distinct cards, namely, 13 cards of these denominations: 2, 3, . . ., 10, Jack, Queen, King, Ace, in each of four suits: Spades,

Hearts, Diamonds, Clubs. How many elementary outcomes are in the sample space for each of the following experiments?

(a) One card is drawn from the deck.
(b) A "hand" of five cards is dealt from the deck.
(c) A card is drawn from the deck, and only the color and denomination of the card are noted. (Spades and Clubs are black, and Hearts and Diamonds are red.)

1–8. Show that for any positive integer n and any positive integer $j \le n$

$$\binom{n}{j-1} + \binom{n}{j} = \binom{n+1}{j}.$$

1–9. Prove the *binomial theorem*:

$$(x + y)^n = \sum_0^n \binom{n}{k} x^k y^{n-k}, \qquad n = 1, 2, \ldots.$$

[*Hint*: First verify the result for $n = 1$. Then assuming it to be true as written for some given n, show by multiplying through both sides by $(x + y)$ that it remains valid for the next integer $n + 1$.]

1–10. Show the following by expressing each side in terms of factorials:

$$\binom{n}{m}\binom{n-m}{r}\binom{n-m-r}{k-m} = \binom{n}{k+r}\binom{k+r}{k}\binom{k}{m}.$$

Also, interpret each side of the equation as the number of ways of performing a sequence of selections and deduce the equality from this reasoning.

1–11. (a) Show that

$$\sum_0^n \binom{n}{k} = 2^n, \qquad \text{for } n = 1, 2, \ldots.$$

[*Hint*: Set $x = y = 1$ in the result of Problem 1–9.]

(b) Determine the total number of distinct subsets of a set containing n elements (two ways).

1.1.3 Events

Suppose that a card picked at random from a deck of playing cards turns out to be the three of Hearts. This is an elementary outcome, one of the 52 "points" in the sample space of the experiment. However, in the game of "Hearts" one might only be concerned (in the final scoring) about the "heartness" of a card, and ask, "Is it a Heart—or not?" When a Heart *is* picked, he says that the *event* Hearts has happened—whether the card picked is the three, or the Ace, or any other Heart. Thus, there are 13 elementary outcomes that are Hearts, and the appearance of any one of them would cause one to say that the event Hearts has occurred. These 13 outcomes constitute a *subset* of the sample space.

In general, an *event* is a subset of the sample space of an experiment of chance. It may be defined by listing the elementary outcomes that make it up or by giving some property that characterizes or determines that list of outcomes. Events will ordinarily be referred to by names such as $E, F, \ldots$. In specifying an event by making a list of its points, braces will be used, for example: $\{a, e, q, \ldots\}$. In specifying an event by means of a characterizing condition, the notation will also employ braces, as follows:

$$E = \{x \mid x \text{ satisfies condition } E\},$$

which is read "the set of points x such that x satisfies condition E" (whatever condition that may be).

The set of *all* outcomes in a sample space Ω is a particular subset of a sample space and will, therefore, be an event. It is also convenient to define as an event the *empty set*, or set containing no outcomes, denoted by $\varnothing$. This event is defined by any condition that is not satisfied by any of the points in the sample space.

EXAMPLE 1–13. Consider again the drawing of two chips from a container with four chips marked A, B, C, and D, respectively, and assume for the sample space the 12 elementary outcomes in which order is taken into account:

$$\Omega = \{AB, AC, AD, BC, BD, CD, BA, CA, DA, CB, DB, DC\}.$$

(The first letter denotes the first chip drawn and the second letter the second chip drawn.) The condition that the *first chip* is A defines the event

$$E = \{AB, AC, AD\}.$$

The condition that the first chip is A or B and the second is B or D defines the event

$$F = \{AB, AD, BD\}.$$

The condition that one of the chips is A defines the event

$$G = \{AB, AC, AD, BA, CA, DA\}.$$

The condition that both chips be A defines the empty set, since no outcome has this property.

EXAMPLE 1–14. The number of phone calls coming into a certain exchange during a given 5-min period is counted. The possible outcomes of this experiment can be taken to be the nonnegative integers:

$$\Omega = \{0, 1, 2, 3, \ldots\}.$$

An event is then any set of these integers. For instance, the condition that *at least four calls* come in defines the event

$$E = \{4, 5, 6, \ldots\},$$

and the condition that *at most two calls* come in defines the event

$$F = \{0, 1, 2\}.$$

EXAMPLE 1–15. Consider the sample space of Example 1–6, consisting of the possible orientations of the fibers of a material. These orientations can be identified by pairs of colatitude and longitude angles (θ, ϕ), where $0 \leq \theta \leq \pi/2$ and $0 \leq \phi < 2\pi$. Events can then be described by conditions imposed on these coordinate pairs. For instance, the conditions

$$E\colon \theta = \frac{3\pi}{8}, \phi = 0; \qquad F\colon \theta \leq \frac{\pi}{4}; \qquad G\colon 0 \leq \phi < \pi$$

define sets of orientations that are examples of events. If the orientation of a fiber selected at random is determined to be $\theta = \pi/8$, $\phi = \pi/2$, the conditions defining F and G are satisfied, and it is said that F has "happened"—and also that G has happened.

When all of the points in one event E are included in another event F, the one is said to be a *subset* of the other, and one writes $E \subset F$. This would mean that the condition defining F is necessarily true of the outcomes in E, but it may be that some further condition not satisfied by all of the outcomes of F is required to pick out those in the subset E. That is, the statement $E \subset F$ is equivalent to saying, "Condition E implies condition F." Notice that for any event E, $\varnothing \subset E \subset \Omega$.

When the *same* set of outcomes is defined by two superficially different conditions, the events defined by these conditions are really the same event and are said to be equal. That is, $E = F$ means that every outcome in E is also contained in F and conversely, or that both $E \subset F$ and $F \subset E$. To show that $E = F$, then, it suffices to show that any element of E is also in F and that any element of F is also in E.

In case the number of elementary outcomes in a sample space is finite, there are only finitely many distinct events that can be defined. In particular, if the sample space contains N outcomes, there are exactly 2^N subsets or events. For, each of the outcomes either is or is not included in a given subset, so in making up an event one has these two choices for each of the N outcomes; multiplication of the N 2's yields 2^N. (Observe that the events Ω and $\varnothing$ are included in this count—one may use all of the outcomes to make up an event or none of them.) That the number of possible events is 2^N also follows from Problem 1–11(b).

When the number of outcomes in Ω is *not* finite, one runs into difficulties if he admits *every* subset of Ω as an event and then tries to define probabilities consistently for all of them. For this reason, it is often necessary to work with a collection of subsets somewhat smaller than the collection of *all* subsets and to use the term *event* only for one of this smaller collection. This restriction turns out not to be significant in practice, and it will be assumed without comment that subsets arising naturally in the discussion may legitimately be called events. (Optional Section 1.1.6 goes into the matter in more detail.)

1.1.4 Complements, Intersections, and Unions

A person picked from a certain population of people may be male ("male" is an event) or black ("black" is another event). But perhaps there is also reason to consider the condition "black and male," defining an event thought of as a *combination* of the event "black" and the event "male."

Events can be combined or operated on in many ways to form new events. One such operation is that of set *union* or *addition*. A point is in the union of E and F if and only if it lies either in E or in F (or possibly in both). The symbol $+$ and the symbol $\cup$ are both used to denote this operation:

$$E \cup F = E + F = \{E \text{ or } F\} = \{\omega \mid \omega \text{ is in } E \text{ or in } F\}.$$

This operation is both commutative

$$E + F = F + E,$$

and associative,

$$[E + F] + G = E + [F + G],$$

as may readily be established.

The *intersection* or *product* of two events is defined to be that event whose outcomes are those lying in *both* of the events. An intersection is indicated either by ordinary product notation or by the symbol $\cap$:

$$E \cap F = EF = \{E \text{ and } F\} = \{\omega \mid \omega \text{ is in both } E \text{ and } F\}.$$

This operation is also commutative ($EF = FE$) and associative ($[EF]G = E[FG]$). One reason for thinking of set intersection as "multiplication" is that it combines with set union or addition in a distributive law:

$$E(F + G) = EF + EG.$$

However, another distributive law, obtained formally from the above by interchanging indicated unions and intersections, is also valid:

$$E + FG = (E + F)(E + G).$$

Two events are said to be *disjoint* or *mutually exclusive* when their intersection is empty: $EF = \varnothing$. The *difference* $E - F$ of two events is the event consisting of those outcomes lying in E but not in F, that is, outcomes satisfying the condition E but not the condition F. [Notice, however, that $(E - F) + F$ is not necessarily the event E. See Problem 1–13(j) at the end of this section.]

The *complement* of an event E, to be denoted here by E^c (elsewhere by $\bar{E}$ or by CE), is defined to be $E^c = \Omega - E$, or the set of all outcomes in the sample space that are not in E. Observe that $E + E^c = \Omega$, and that $EE^c = \varnothing$. Observe also that a difference can be expressed in terms of the complement: $E - F = EF^c$

EXAMPLE 1–16. A card is drawn from a standard deck of cards (defined in Problem 1–7). The sample space Ω consists of the 52 cards in the deck, any one of which might be drawn. (These are perhaps the most "elementary" outcomes that can be defined here, and again there might be instances in which a coarser classification of results would be adequate.) Consider the following events:

E: The card drawn is a Spade.
F: The card drawn is a Jack.
G: The card drawn is a face card.

(The Jacks, Queens, and Kings are picture faces and are called "face cards.") Then $E + F$, for example, is the set of cards consisting of all of the Spades plus the other three Jacks. The event $E - F$ is the set consisting of all of the Spades except the Jack. The event FG is identical with F. The event $\Omega - E$ is the set consisting of all Hearts, Diamonds, and Clubs. The event $(E + G) - F$ is the set of all Spades, except the Jack, together with the remaining three Queens and three Kings.

EXAMPLE 1–17. A pointer is spun about a pivot, in a horizontal plane, and a scale from 0 to 1 is marked around the circle traced by the tip of the pointer. (The 0 and 1 fall at the same point.) The pointer is spun once and allowed to come to rest. The sample space may be taken to be the set of possible stopping points—essentially, the set of real numbers on the interval from 0 to 1. Consider the following events:

$$E = \{x \mid .5 < x < .8\}.$$
$$F = \{x \mid x > .6\}.$$
$$G = \{x \mid x < .4\}.$$

The event $E + F$ is the set $\{x \mid .5 < x < 1\}$. The intersection EF is the set $\{x \mid .6 < x < .8\}$. The event G^c is $\{x \mid x \geq .4\}$. The difference $E - F$ is $\{x \mid .5 < x \leq .6\}$.

It is helpful to have the schematic picture of these various concepts provided by the "Venn diagram." In such a diagram, the sample space is symbolized by a region in the plane, and its events by subsets of that region. For instance, events E, F, and G might be represented by the points within the circle, rectangle, and triangle, respectively, in Figure 1–1. (The sample space is represented by the points in the larger rectangle.) Figure 1–2 exhibits the events EF, $E + F$, $E - F$, and E^c.

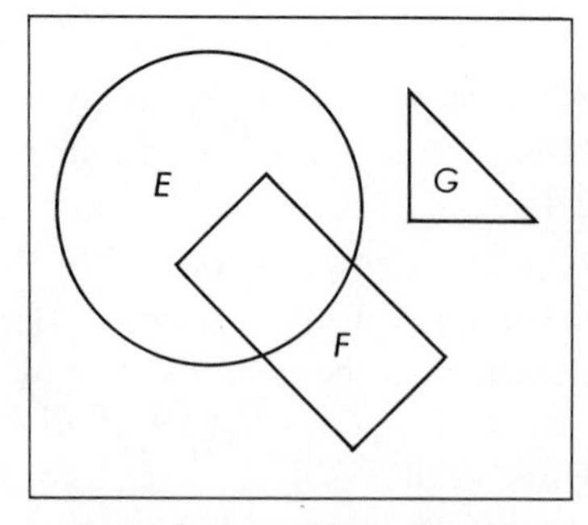

Figure 1–1. A Venn diagram.

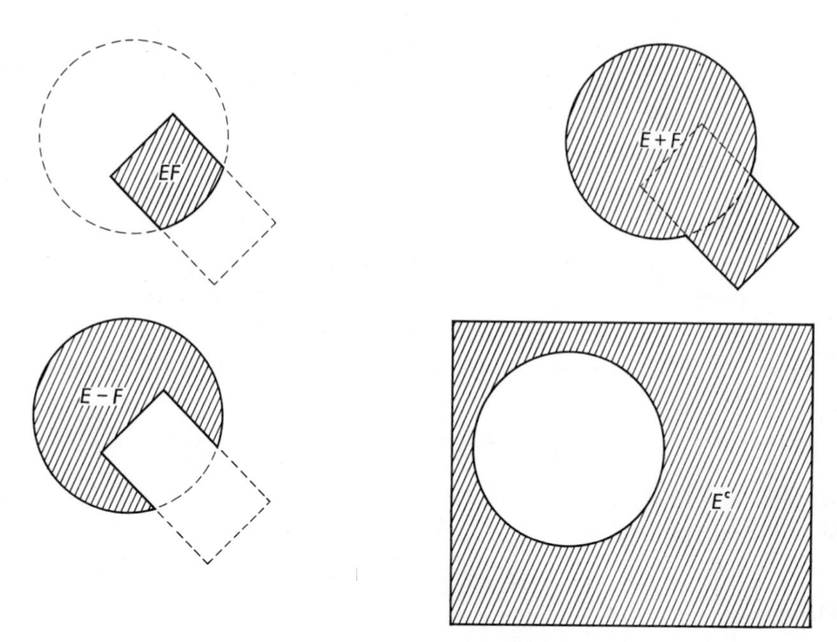

Figure 1–2. Set operations.

Unions and intersections of infinitely many events are easier to define than are sums and products of infinitely many numbers. Even if the number of events is uncountably infinite, the union is the set of all points that are contained in at least one of the sets of the collection. The intersection is the set of points such that each is contained in every set of the collection. The union and intersection of a collection of sets $\{E_\alpha\}$, where the subscript α indexes the sets of the collection, will be denoted by $\bigcup E_\alpha$ and $\bigcap E_\alpha$, respectively. Observe that the definitions apply equally well if the collection happens to be countable or finite, and that in the latter case they coincide with the definitions of union and intersection given earlier for that case.

EXAMPLE 1–18. Let Ω be the set of all real numbers $(-\infty, \infty)$, and let E_n denote the interval $(a - 1/n, a + 1/n)$, for $n = 1, 2, \ldots$. Since the only number in all of these intervals (that is, for each n) is the number a, the intersection of the intervals E_n is the set whose only element is that number:

$$\bigcap E_n = \{a\}.$$

Given any event E, it and its complement E^c constitute a division of the sample space into disjoint sets whose union is the whole space:

$$EE^c = \varnothing \qquad \text{and} \qquad E + E^c = \Omega.$$

Dividing a sample space (or any set, for that matter) into mutually exclusive subsets is called "partitioning." More precisely, a *partition* of Ω is defined to be a collection of events $\{E_\alpha\}$ such that

1. $\bigcup E_\alpha = \Omega$.
2. $E_\alpha E_\beta = \varnothing, \qquad \text{if } \alpha \neq \beta.$

Thus, distinct sets or events in the partition are disjoint, and together the sets of the partition make up the sample space.

An important fact about partitions is that a partition of Ω also partitions any given event in Ω. To be precise, if $\{E_\alpha\}$ is a partition of Ω, then for any event F, the events FE_α and FE_β are disjoint, when $\alpha \neq \beta$, and

$$F = \bigcup FE_\alpha.$$

That is to say, $\{FE_\alpha\}$ is a partition of F. A commonly occurring particular case is that of the partition $\{E, E^c\}$:

$$F = FE + FE^c.$$

Thus, the event F is made up of a part in E and a part not in E, as illustrated in Figure 1–3. Proof of this simpler case proceeds as follows (the general case is quite similar): If ω is an outcome in F, then if ω is in E it is also in FE; if it is also in E^c, then it is in FE^c. In either case, ω is in the union $FE + FE^c$; therefore, the left-hand member is a subset of the right. On the other hand, if ω is an outcome in the union $FE + FE^c$, then it is in FE or it is in FE^c and, in either case, it must be in F. Therefore, the right-hand member of the relation to be proved is a subset of the left. Since each is a subset of the other, the left-hand and right-hand members must be the same set—that is, they are "equal" in the sense of set equality.

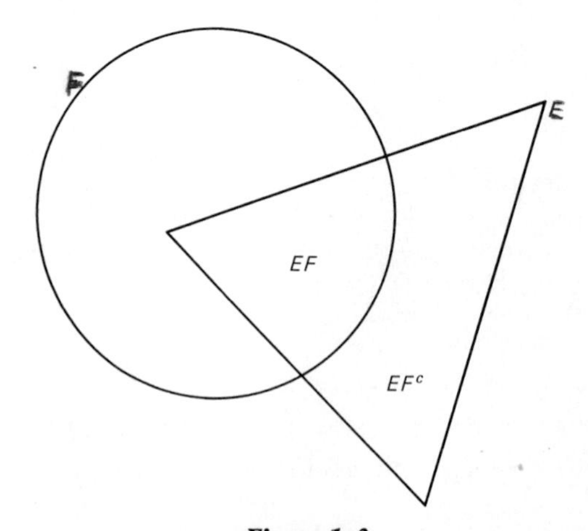

Figure 1–3

Problems

1–12. The sample space for the experiment of tossing two dice and recording the number of points showing on each die contains the 36 elementary outcomes obtained by associating each number of points on one die with each number on the other. In each case, make a list of the outcomes in the event defined by the given condition:

(a) The sum is divisible by 4.
(b) Both numbers are even.
(c) The numbers on the dice are equal.
(d) The numbers differ by at least 4.
(e) The total number of points showing is 10.

1–13. Show the following, for any events E, F, G in a sample space Ω. (Make a Venn diagram in each case, but establish each set equality by showing that each side is contained in the other.)

(a) $\varnothing = \varnothing E$, and $\varnothing + E = E$.
(b) $E = E\Omega$, and $E + \Omega = \Omega$.
(c) $(EF)^c = E^c + F^c$.
(d) $(E + F)^c = E^cF^c$.
(e) $G(F - E) = GF - GE$.
(f) $E(F + G) = EF + EG$.
(g) $EF = E$ if $E \subset F$.
(h) $E + F = E$ if $F \subset E$.
(i) $E + FG = (E + F)(E + G)$.
(j) $(E - F) + F = E + F$.

1–14. Show that for any collection of events $\{E_\alpha\}$,

$$\left(\bigcup E_\alpha\right)^c = \bigcap E_\alpha{}^c \quad \text{and} \quad \left(\bigcap E_\alpha\right)^c = \bigcup E_\alpha{}^c.$$

These relations, which generalize (d) and (c), respectively, of the preceding problem, are sometimes referred to as De Morgan's laws.

1–15. Five cards are selected from a standard bridge deck (Problem 1–7). Let the elementary outcomes of this experiment be all possible selections of 5 cards from the 52. Determine the number of elementary outcomes in each of the following events:

(a) The hand contains exactly one "pair." (A pair consists of two cards of the same denomination; for example, two Kings. The other three cards in the hand are to be of three different denominations.)
(b) The hand contains two pairs of different denominations and a fifth card of still a different denomination.
(c) The hand contains two cards of one denomination and three of another.
(d) The hand contains three cards of one denomination, but the other two do not form a pair.
(e) The hand consists of five cards, all of the same suit.

1–16. The outcomes of a certain experiment are points in a plane, each identified by its rectangular coordinates in a given reference system; that is, ω is (x, y). Sketch the events defined by each of the following conditions:

(a) $x^2 + y^2 = 1$.
(b) $x^2 + y^2 \leq 4$.
(c) $x + y > 3$.
(d) $x - y \leq 2$.
(e) $2x \leq y$.
(f) $x - 3y \leq 6$.
(g) (b) $\cap$ (c).
(h) (e) $\cup$ (f).
(i) (a) $-$ (e).

1–17. A sample space contains just three outcomes, a, b, and c. Determine the eight possible events (specifying each by making a list of its outcomes).

1–18. Consider an *increasing* sequence of events: $E_1 \subset E_2 \subset E_3 \subset \cdots$. Show that (with $E_0 \equiv \varnothing$):

(a) $\bigcup E_i = \bigcup (E_i - E_{i-1})$.
[*Hint*: If ω is in the union $\bigcup E_i$, there is a *first* event in the sequence $E_1, E_2, \ldots,$ which includes it.]

(b) The sets $\{E_i - E_{i-1}\}$ are mutually disjoint.

[These facts can be summarized by saying that the collection $\{E_i - E_{i-1}\}$ is a partition of the union $\bigcup E_i$.]

1.1.5 Functions on a Sample Space

An outcome ω, resulting from the performance of an experiment of chance, may have associated with it one or more numerical characteristics. For example, if the experiment is the drawing of one person from a given group of people (called a *population*), one may be interested in that person's height, his age, his weight, or in all three. Since the person chosen can be any of the outcomes ω, these numerical characteristics of interest are dependent variables—functions of ω:

$$X(\omega) = \text{age of } \omega, \qquad Y(\omega) = \text{weight of } \omega, \qquad \text{etc.}$$

If one has reason to consider simultaneously the age and weight, say, for each ω, the pair $[X(\omega), Y(\omega)]$ defines a vector-valued function on Ω.

A function provides and can be thought of as a *mapping* from the space on which it is defined to the space of its values (from its "domain" to its "range," in mathematical terminology). Each *point* in the sample space Ω has then an *image point* in the set $\mathscr{X}$ of values of the function; and a *set* E of points in Ω has an *image set* of values in this space $\mathscr{X}$ of values of the function. Conversely, given any value x in the value space $\mathscr{X}$ of a function $X(\omega)$, there are certain points in Ω that are assigned that value; these make up what is called the *pre-image* of x. Similarly, a set F of values of a function $X(\omega)$ defines a pre-image set of outcomes ω, namely, the set of all ω's that are assigned values in F by $X(\omega)$.

EXAMPLE 1–19. Let ω denote the six faces of a die, one of which turns up when the die is tossed onto a flat surface. When these faces are marked with dots, as is customary, the function

$$X(\omega) = \text{number of dots on the face } \omega$$

assigns a number to each face. The set ω of values of this function is (for an ordinary die) $\{1, 2, 3, 4, 5, 6\}$. The pre-image of the set $\{x \mid x \leq 4\} = \{1, 2, 3, 4\}$ is the set of faces marked with these numbers of dots: $\{\omega \mid X(\omega) = 1, 2, 3, \text{ or } 4\}$. Instead, if one calls the number of dots on the upturned face the outcome ω, that is, if one takes the sample space to be the set $\{1, 2, 3, 4, 5, 6\}$, then the function $X(\omega) = \omega$ is an example of a function on the sample space. Another function on this sample space might be

$$Y(\omega) = \begin{cases} 1, & \text{if } \omega = 1, 3, \text{ or } 5, \\ 0, & \text{if } \omega = 2, 4, \text{ or } 6. \end{cases}$$

This is an example of an *indicator function* of a set—a function that is defined to have the value 1 for each point of the set and the value 0 for each point not in the set; this particular function "indicates" the set $\{1, 3, 5\}$.

EXAMPLE 1–20. When an electromotive force (voltage) is applied to a known, fixed load, the power consumed is proportional to the square of the voltage. If to determine the power one places a voltmeter across the load, the measured voltage ω can be treated as an outcome of an experiment of chance with sample space $(-\infty, \infty)$, owing to the random inaccuracies introduced in the measurement process. The power W is a function of the outcome ω: $W = k\omega^2$, with values in the interval $[0, \infty)$. Then the set $\{W < k\}$, for instance, has as its pre-image the set $\{-1 < \omega < 1\}$.

EXAMPLE 1–21. Let ω denote the landing point of a projectile aimed at a target point in a plane. The rectangular coordinates $X(\omega)$ and $Y(\omega)$ in a set of perpendicular axes centered at the target point, that is, the coordinate functions in this reference system, can be taken either as single functions of ω or as defining a vector function $[X(\omega), Y(\omega)]$. The value spaces (or range spaces) of the functions $X(\omega)$, $Y(\omega)$ can each be taken to be the whole real line $(-\infty, \infty)$; the value space of the vector $[X(\omega), Y(\omega)]$ is then the whole xy-plane. The set of coordinate pairs

$$E = \{(x, y) \mid x^2 + y^2 \leq 1\}$$

has as its pre-image the set of all landing points within one unit of the target point. The event in $\mathscr{X}$: $\{x \mid x < 0\}$ has as its pre-image the set of landing points to the left of the reference axis from which x is measured: $\{\omega \mid X(\omega) < 0\}$.

Having determined a function $X(\omega)$ to have the value x, say, one may find it necessary to consider yet another function of that value, say $g(x)$. This defines a function of ω as a function of a function:

$$Z(\omega) = g(X(\omega)).$$

Values of this composite function $Z(\omega)$ can be thought of as coming either directly from ω via the function $Z(\omega)$ or indirectly as a value $g(x)$, where x comes from ω as $X(\omega)$. That is, ω is mapped by $X(\omega)$ into a point x, and x is then mapped by $g(x)$ into z:

$$\omega \xrightarrow{X(\omega)} x \xrightarrow{g(x)} z,$$

and this is equivalent to the mapping from ω directly to $g(X(\omega))$:

$$\omega \xrightarrow{Z(\omega)} z.$$

An event E or set of values on the value space of $Z(\omega)$ has a pre-image in the set of values $\mathscr{X}$ of $X(\omega)$, and this in turn has a pre-image set in Ω, which would be the same as the pre-image of E under $Z(\omega)$.

In the case of a vector function $[X_1(\omega), \ldots, X_n(\omega)]$, one might define another vector function $[Y_1(\omega), \ldots, Y_m(\omega)]$ by introducing functions $g_i(x_1, \ldots, x_n)$ of the values of $[X_1(\omega), \ldots, X_n(\omega)]$

$$\begin{aligned} Y_1(\omega) &= g_1[X_1(\omega), \ldots, X_n(\omega)] \\ &\vdots \\ Y_m(\omega) &= g_m[X_1(\omega), \ldots, X_n(\omega)]. \end{aligned}$$

EXAMPLE 1–22. In Examples 1–6 and 1–15, a sample space was taken to be the set of possible orientations of a fiber. For each orientation ω the colatitude and longitude define a vector function of the orientation: $[\theta(\omega), \phi(\omega)]$. The same model might apply in the case of a force of given magnitude K whose direction of application is random, although it would be then more suitable to take the set of *directed* orientations as the sample space. In this case, the coordinate function $\theta(\omega)$ would have values on the interval $[0, \pi]$. (In the identification of orientations with points on a unit sphere, the whole sphere would be used rather than just the upper hemisphere.) The components of force along the direction $\theta = 0$, along the direction $\theta = \pi/2$, $\phi = 0$, and along the direction $\theta = \pi/2$, $\phi = \pi/2$ would be as follows:

$$\begin{aligned} F_1 &= K \cos \theta, \\ F_2 &= K \sin \theta \cos \phi, \\ F_3 &= \mathrm{K} \sin \theta \sin \phi, \end{aligned}$$

thereby defining a vector function $[F_1(\omega), F_2(\omega), F_3(\omega)]$ of the sample point ω as a function of a function.

A function $X(\omega)$ on a sample space Ω induces a partition of the sample space. This partition arises naturally as the collection of sets C_x on which $X(\omega)$ is constant:

$$C_x = \{\omega \mid X(\omega) = x\},$$

where x ranges over the value space of $X(\omega)$. These partition sets are often called *level curves* (or level surfaces) of the function. A common example of such curves is that of the "contour curves" on a map or chart, where each curve indicates a set of points at which altitude—a function of position in the plane—is a given constant.

EXAMPLE 1–23. Suppose that a sample space consists of points in a three-dimensional space, and let $R(\omega)$ denote the distance between the point and a reference origin. Then the condition $R(\omega) = 3$ defines the surface of a sphere of radius 3, and the collection of all sets of the form $\{\omega \mid R(\omega) = \lambda\}$, for λ a nonnegative real number, constitutes a partition of the space. Every point in the sample space lies on one of these spherical surfaces, the surfaces are disjoint, and the union of all such surfaces is the whole space.

A partition determined by a function may also be determined by a different function, if that function is also constant on each set of the partition and has a distinct value on each of these sets. For instance, in Example 1–23, the function $X^2(\omega) + Y^2(\omega)$, where $X(\omega)$ is the horizontal distance and $Y(\omega)$ is the vertical distance to a given set of rectangular coordinate axes, determines the same partition as the function $R(\omega)$, because there is a one-to-one correspondence between values of these two functions.

★1.1.6 Borel Fields

It was mentioned earlier that owing to difficulties encountered in the case of nondiscrete sample spaces, it is often necessary to restrict what are to be called "events" and to work with a family of subsets that is smaller than the collection of all subsets of Ω. Such a family, to be practicable, should have the property that if the basic operations are performed with or on its sets, the results are sets that are again in the family—and so also are "events." That is, if E and F are events, surely $E + F$, EF, and E^c should also be events. A family of sets that includes complements and the union and intersection of any two of its members is called a *field* of sets. A trivial example of a field of sets is the family $\{\varnothing, \Omega\}$, as the reader may easily verify. The family $\{\varnothing, E, E^c, \Omega\}$, where E is any set in Ω, is also a field, containing all complements, unions, and intersections of its members.

If the union and intersection of *two* events are events (i.e., are included in the working family of subsets), then, by induction, so are any *finite* unions and intersections. But when the sample space is not discrete, it may be useful to require *countable* unions and intersections of events to be events.

A *Borel field* of events or sets (also called a σ-field or σ-algebra) is a collection $\mathscr{B}$ of events so constituted as to include along with any event its complement, and also to include the union of any countable number of sets in the collection. Because

$$\bigcap E_n = \left(\bigcup E_n^c\right)^c,$$

it follows that countable intersections of events in a Borel field $\mathscr{B}$ are also contained in $\mathscr{B}$. In particular, the difference $E - F$ or EF^c of two sets E and F in $\mathscr{B}$ is also in $\mathscr{B}$. It it clear, also, that a Borel field always includes Ω and $\varnothing$.

An example of a Borel field is the collection of *all* subsets of a sample space. In the case of a finite sample space this Borel field of all subsets contains 2^N subsets, where N is the number of points in the sample space (see Problem 1–11). A sample space having as many points, say, as the interval of real numbers $[0, 1]$ has altogether too many subsets for this Borel field to be useful in the construction of probability models. In such cases, one works with the "smallest" Borel field containing some simple collection $\mathscr{C}$ of events that are useful—such as the collection of subintervals. By "smallest" Borel field is meant one big enough to contain the given collection but which does not contain a smaller Borel field (i.e., one with fewer members) that also contains the sets of the collection $\mathscr{C}$. The following example illustrates this idea.

EXAMPLE 1–24. A sample space Ω contains four points $\{a, b, c, d\}$. Consider the collection $\mathscr{C}$ consisting of two events: $\{a\}$, $\{c, d\}$. If it is desired to expand this collection to a Borel field, the complements of each event must be included, so $\{b, c, d\}$ and $\{a, b\}$ must be added to the collection. Because unions and intersections must be included, the events $\{a, c, d\}$, $\varnothing$, Ω, and $\{b\}$ must also be added. In summary, then, a Borel field of sets or events that includes the events $\{a\}$ and $\{c, d\}$ consists of the following eight events:

$$\varnothing,\quad \{a\},\quad \{c, d\},\quad \{b, c, d\},\quad \{a, b\},\quad \{a, c, d\},\quad \{b\},\quad \Omega.$$

(The reader should verify that operations of union, intersection, and complementation do not require leaving this collection.) Moreover, each of these events is needed, so there is no smaller Borel field that includes the two given events. This smallest one is called the Borel field *generated* by $\{a\}$ and $\{c\ d\}$. Observe, though, that the collection of all of the $2^4 = 16$ events that can be defined on a sample space with four points is also a Borel field; but it includes the one already defined and is not then the smallest Borel field containing $\{a\}$ and $\{c, d\}$.

EXAMPLE 1–25. Let Ω denote the sample space of the spinning pointer experiment, that is, the interval $0 \leq x < 1$. The collection of events needed to discuss and analyze situations that arise in practice must surely include intervals—sets of the form $(a, b) = \{x \mid a < x < b\}$. The collection of such intervals may seem to be a large collection of sets, but it does not include, for instance, sets containing a single point, or sets consisting of two disjoint intervals. The Borel field generated by the collection of all intervals of the form (a, b) would, however, necessarily include single points (as the intersection of intervals of the form $a - 1/n < x < a + 1/n$ for $n = 1, 2, 3, \ldots$), along with other unions and intersections of intervals. It includes, in particular, closed intervals:

$$\{x \mid a \leq x \leq b\} = \{x \mid a < x < b\} + \{a\} + \{b\},$$

and, similarly, half-open intervals (i.e., including one endpoint but not the other). On the other hand—and this is perhaps the point of concentrating on a Borel field—the Borel field generated by the open subintervals does *not* include *all* subsets;[3] and while it is possible to define probabilities consistently (in a way to be seen in the next section) for the Borel field generated by the open intervals, it happens to be impossible to do so for the set of all subsets of [0, 1].

Problems

1–19. Consider the sample space consisting of the students enrolled in a certain college. Describe the partition sets defined by the following functions.

(a) $X(\omega) =$ age of student ω (at last birthday).

(b) $Y(\omega) = \begin{cases} 1, & \text{if } \omega \text{ is black,} \\ 0, & \text{if } \omega \text{ is not black.} \end{cases}$

[3] To construct a counterexample and show that it is not included in the Borel field generated by open intervals is nontrivial. See P. Halmos, *Measure Theory* (New York: Van Nostrand Reinhold, 1950), p. 69.

1–20. A coin is tossed three times. Make a list of the eight (natural) elementary outcomes in Ω, and for each one compute $X(\omega)$ defined to be the number of heads in ω. Give the partition defined by these functions.

(a) $X(\omega)$.
(b) $[X(\omega) - \frac{3}{2}]^2$.

1–21. A pair of dice is tossed, with outcome $\omega = (x, y)$, where x is the number of points showing on the first, and y on the second die. Let $U(\omega) = |x - y|^2$. What is the pre-image set $\{\omega \mid U(\omega) > 9\}$?

1–22. A sample space contains just the five points a, b, c, d, and e. Consider the function whose values are $X(a) = X(b) = 0$, $X(c) = X(d) = 1$, and $X(e) = 2$. Give the partition induced by $X(\omega)$ and that induced by $Y(\omega) = X^2(\omega)$. [Does $X^2(\omega)$ always induce the same partition as $X(\omega)$?]

1–23. A sample space consists of points in a plane, specified by means of rectangular coordinates relative to a given axis system: $\omega = (x, y)$. Determine the partition induced by $f(\omega) = |x - y|$.

1–24. A sample space consists of the set of points in three dimensions, identified by rectangular coordinates $(x, y, z) = \omega$ in a given reference system. Let $U_1(\omega)$ denote the smallest, $U_2(\omega)$ the second smallest, and $U_3(\omega)$ the largest of the three coordinates of ω. (Ignore the fact that some points ω have coordinates that are not all different.) What is the range space of this vector function $[U_1(\omega), U_2(\omega), U_3(\omega)]$? What are the partition sets of Ω determined by this vector function?

1–25. A cube has its six sides colored red, white, blue, green, yellow, and violet, respectively. Consider the function $X(\omega)$ defined as follows:

$$X(R) = X(W) = 1,$$
$$X(G) = X(B) = 2,$$
$$X(Y) = X(V) = 3.$$

Determine the partition sets of the sample space defined by the function $Y(\omega)$, where Y has values obtained as $y = (x - 2)^2$, for a given value x of $X(\omega)$.

1–26. Let Ω denote the sample space, which is the set of real numbers $(-\infty, \infty)$. Determine the pre-image of the set $\{y \mid y < 1\}$ in the space of values of $Y(\omega) = \omega^2$. If for each y one computes z as

$$z = \frac{y}{1 + y}.$$

What is the range space of the composite function $Z(\omega)$ defined by this relation?

★**1–27.** Show that for any event E, in a sample space Ω, the collection consisting of the events E, E^c, $\varnothing$, and Ω is a Borel field.

★**1–28.** Determine the Borel field generated by the set $\{a, b\}$ in the sample space consisting of the four outcomes a, b, c, d.

★**1–29.** Referring to the sample space of Problem 1–22, show that the Borel field generated by $\{a, b\}$ and $\{e\}$ includes all events of the form $\{\omega \mid X(\omega) \leq \lambda\}$ for various values of λ.

★**1–30.** Consider the sample space of all real numbers.

(a) Is the family of closed intervals a field? Show that any Borel field containing this family would have to include all open intervals.
(b) Show that the Borel field generated by the sets $(-\infty, x]$ is the same as that generated by finite intervals (see Example 1–25).

1.2 Probability

Some dyed-in-the-wool gamblers will lay odds and bet money on almost any experiment whose outcome is uncertain—any experiment of chance. As a matter of fact, scientists, businessmen, physicians, and so forth, who are faced with experiments of chance and *must* deal with them—also lay odds and act according to these odds. That is, people must, either consciously or unconsciously, assess the relative "likelihood" of various outcomes and events in order to interpret and use results of experimentation. *Probability* is a numerical measure of this likelihood whose origin is in a study of repeated trials of an experiment.

In a sequence of trials of a given experiment of chance, the number of times a particular event E occurs is called its *frequency*, and the ratio of that frequency to the number of trials is called its *relative frequency*. It is an observable phenomenon that as one conducts more and more trials, the relative frequency of E tends to stabilize or to approach a limiting proportion. Figure 1–4 illustrates this

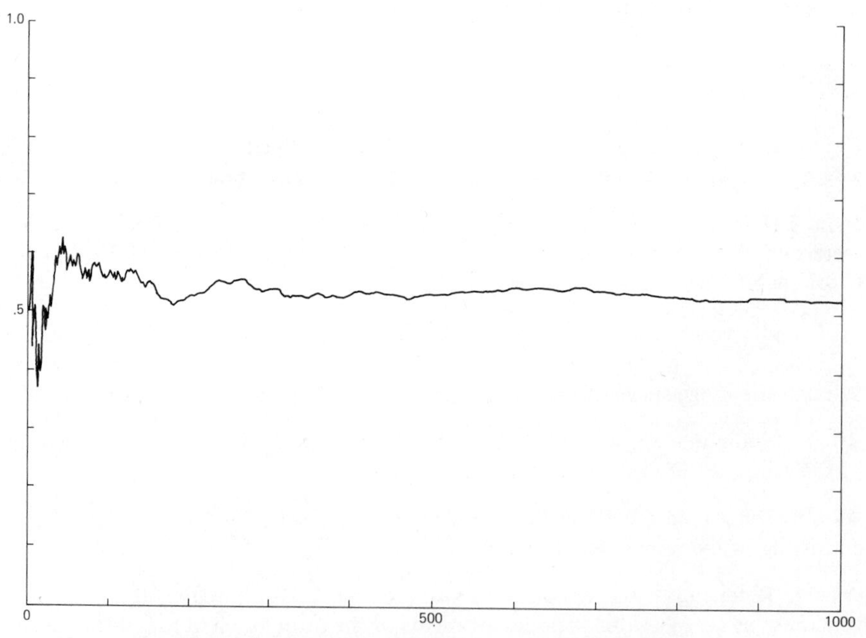

Figure 1–4. Relative frequency—1000 coin tosses.

tendency in a sequence of 1000 tosses of a coin. (Actually, it was a simulated coin "tossed" by a computer.)

If, for example, the relative frequency of E approaches, say, $\frac{2}{3}$ as a limit in repeated trials, this number—called the *probability* of E—is taken to be the basis of belief as to the outcome of a single trial. That is, one lays odds of 2 to 1 on E and calls a bet of \$2 on E against \$1 on E^c a *fair* bet for the outcome of a trial. Conceiving the outcome to be the result of a particular combination of uncontrollable initial conditions and forces, one feels that there are "twice as many" such combinations leading to E as to E^c.

Such feelings often obtain even when an experiment is to be and perhaps *can* be performed only once. Thus, one has certain beliefs and hunches about the passage of a proposed constitutional amendment that he might use to fix his betting odds, just as in the toss of a coin. It is true that in tossing a coin, past experience will condition one's beliefs about chances in a given trial; but for a single trial there is not much difference between the two situations. The main difference, indeed, is that in the case of an amendment, different people will have different beliefs, while there is general agreement about the coin—the chances seem to be an objective thing, an intrinsic property of the coin.

In applying the notion of probability, it may be thought of as subjective or personal—as in the case of the amendment's chances—or as objective, in cases where repeated trials are conceivable. In either situation, the *mathematical model* will be the same kind of structure, and it is this model that is the principal concern here.

EXAMPLE 1–26. A thumbtack tossed on a table will land in one of two positions, with point up or with point down. Repeated tossings of a thumbtack will reveal that the proportion of points up tends toward a limiting value. At any rate, it is assumed that a probability for the event "point up" exists, even though it would be quite hard to guess the correct value of this probability without some experimentation. And even *with* experimentation, one has at best an estimate or approximation of the correct value; different series of trials of the experiment would generally yield different estimates. The complete probability space for this experiment would consist of the two outcomes, "point up" and "point down," and the Borel field that comprises the four events Ω, $\varnothing$, U, and D is certainly adequate (where U and D denote the events containing the single points, respectively, point up and point down). The probabilities for these events are

$$P(\Omega) = 1, \qquad P(\varnothing) = 0, \qquad P(U) = p, \qquad P(D) = 1 - p,$$

where p is a number representing the proportion of U's in an infinitely long series of tosses. Observe that the specification of the value of p would completely define the model.

EXAMPLE 1–27. A man is told by his physician that he has a certain disease. upon inquiry he learns that according to past records, about 85 per cent of patients having this disease have recovered. As viewed by the clinic's statistician, this patient (considered as a trial of an experiment) has a certain chance of recovery which seems to be near 85 per cent. But there is a real question as to whether the patient himself should adopt this as his model—indeed, whether he should consider the whole thing as an experiment of chance. Certainly, the statistician's estimate of the chances of recovery ignores many factors peculiar to the patient, some of which might be known to the patient and so influence his personal probability of recovery. This experiment, as viewed by the patient, would never be repeated, for some aspects of the situation are bound to be different even if he recovers and contracts the disease again. Nevertheless, an assessment (appropriate or not) of his chances of recovery might motivate him to certain action—such as making a will.

EXAMPLE 1–28. An experiment that is difficult to imagine as being repeatable is the observation of some aspect of the weather at a given time. Yet it is becoming common to find weather forecasts given in terms of "chances"—for instance, it might be stated that there is one chance in ten that it will rain the following day. This is equivalent to stating that the probability of the event "rain" for the following day is .1, an item of information that can be quite useful for people making plans dependent on the weather. It could be argued that the experiment is repeatable in the sense that the meteorologist has seen the same patterns before, and found them to result in rain 1 time in 10, but different meteorologists might well view the same data and yet announced different rain probabilities.

1.2.1 Probability Axioms

Whether one adopts the personal or the more traditional, objective attitude in applying probabilities, the mathematical structure of a probability model is the same—the assignment of probabilities to all of the events of interest in the sample space for the experiment in question.

A *probability space* consists of a sample space, a family of subsets called events, and a number called *probability* assigned to each event. Such a space will be denoted $(\Omega, \mathscr{F}, P)$, where Ω is the sample space, $\mathscr{F}$ the family of events, and P the measure of events termed probability. The family $\mathscr{F}$ will have to be, technically, a *Borel field* of subsets (cf. Section 1.1.6); but in many cases it will simply be the collection of all subsets of Ω, and in others it will include all events that occur naturally. It is not absolutely necessary to be concerned about this point in a first exposure.

The probability assigned to events E, written $P(E)$, cannot be completely arbitrary, since it is to be like a *proportion*—a long-run proportion of occurrences of E in a sequence of trails. It has been found that the following axioms lead to a model for probability that embodies all of the properties that intuition would demand for probabilities. In essence, these axioms are basic properties of relative frequencies.

1. $P(\Omega) = 1$.

2. $0 \le P(E) \le 1$, for every event E.
3. $P(\bigcup E_i) = P(E_1) + P(E_2) + \cdots$, for every sequence of disjoint events $E_1, E_2, \ldots$.

Axiom 1 reflects the fact that Ω occurs whenever the experiment is performed and Axiom 2 the fact that a proportion is a number between 0 and 1. Axiom 3 is certainly reasonable for a finite sequence; for instance, if E and F are disjoint, then the proportion of times E occurs plus the proportion of times F occurs equals the proportion of times their union $E + F$ occurs. Axiom 3 is referred to as the axiom of *countable additivity*. [Some prefer the less restrictive axiom of *finite additivity* in which the additive relation of Axiom 3 is required to hold only for finite sequences. See *Theory of Probability* by B. De Finetti (New York: Wiley, 1974)].

EXAMPLE 1–29. The toss of an ordinary die can be represented by a model in which the sample space is the set of integers $\{1, 2, 3, 4, 5, 6\}$, corresponding to the six faces of the die as marked with dots. There are 64 ($=2^6$) possible subsets of the sample space, and one probability model is that in which the probability of any event or subset is taken to be the proportion of outcomes in it, relative to the total number of outcomes:

$$P(E) = \frac{\text{number of outcomes in } E}{6}.$$

This clearly has the property $P(\Omega) = 1$ and $P(\varnothing) = 0$, and Axiom 3 follows from the fact that the number of outcomes in $E + F$ is just the sum of the number in E and the number in F, if E and F are disjoint. Notice that the individual outcomes, each considered as an event whose only element is that outcome, are equally likely.

One easy consequence of the probability axioms is an expression relating the probabilities of an event and of its complement. An event and its complement constitute a partition of the sample space:

$$E + E^c = \Omega \qquad \text{and} \qquad EE^c = \varnothing.$$

Hence, by Axiom 3,

$$P(E + E^c) = P(E) + P(E^c),$$

and by Axiom 1 this probability is 1. Therefore,

$$P(E^c) = 1 - P(E).$$

The significance of this is that it is often more convenient to calculate $P(E^c)$ when one really wants $P(E)$. As an important special case, the probability of the empty set is 0.

$$P(\varnothing) = 1 - P(\Omega) = 0.$$

Another property that can be derived from the axioms is the following. Let $E_1, E_2, \ldots$ be an *ascending* sequence—that is,

$$E_1 \subset E_2 \subset E_3 \cdots.$$

Then

$$P\left(\bigcup_1^\infty E_n\right) = \lim_{n\to\infty} P(E_n), \qquad \text{(ascending sequence).}$$

A similar property deals with *descending* sequences:

$$F_1 \supset F_2 \supset F_3 \supset \cdots,$$

namely,

$$P\left(\bigcap_1^\infty F_n\right) = \lim_{n\to\infty} P(F_n), \qquad \text{(descending sequence).}$$

To establish the first of these relations, let $E_1, E_2, \ldots$ be an ascending sequence and recall that by the result of Problem 1–18, the disjoint sets $\{E_i - E_{i-1}\}$ constitute a partition of $\bigcup E_i$:

$$\bigcup E_n = \bigcup (E_i - E_{i-1}),$$

from which, by Axiom 3, it follows that

$$\begin{aligned} P\left(\bigcup E_n\right) &= \sum_1^\infty P(E_i - E_{i-1}) \\ &= \lim_{n\to\infty} \sum_1^n P(E_i - E_{i-1}). \end{aligned}$$

But the events $E_1, E_2 - E_1, \ldots, E_n - E_{n-1}$ constitute a partition of E_n:

$$E_n = \bigcup_1^n (E_i - E_{i-1})$$

so that (again by Axiom 3),

$$\sum_1^n P(E_i - E_{i-1}) = P(E_n).$$

Substitution of this in the expression for $P(\bigcup E_n)$ above yields the desired result.

The relation for *descending* sequences can be proved similarly, or it can be deduced readily from an application of De Morgan's laws to the ascending sequence of complements together with the result for ascending sequences.

Other useful relations can be derived from the axioms for probability; some of them are given as exercises in Problem 1–31. In particular, this fact is often useful: If $\{E_i\}$ is a countable partition of the sample space, then

$$P(A) = \sum P(AE_i).$$

[This has sometimes been referred to as the "law of total probability." Observe that the partition sets must be at most countable in number in order that the sum be definable. Observe, too, that the sum need only be taken over subscripts i such that the partition set E_i has positive probability; for, if $P(E_i) = 0$, then so also is $P(AE_i) = 0$, and this term can be omitted.]

The axioms for probability and their consequences severely restrict the possibilities in assigning probabilities to events. For instance, if E and F are given probabilities, then $P(E^c)$ and $P(F^c)$ are uniquely determined, as is $P(E + F)$ if E and F are disjoint. On the other hand, by the same token one can construct a probability model by assigning probabilities, not to every single event, but to the events in a class of especially simple events; the probabilities of other useful events may then be derivable from these with the aid of the axioms.

EXAMPLE 1–30. The sample space for a spinning pointer (cf. Example 1–17) is taken to be the interval of real numbers (0, 1), corresponding to a scale from 0 to 1 marked off around the circle traced by the tip of the pointer. It seems consistent with how people bet to assign to any interval of width w a probability proportional to w; that is, doubling the width of an interval doubles the chances of finding that the pointer stops within it:

$$P[(a, b)] = k(b - a).$$

But since $\Omega = (0, 1)$,

$$P[(0, 1)] = 1 = k(1 - 0),$$

and it follows that $k = 1$. The probability of a single value—of an event consisting of a single point—can then be calculated by expressing it as an intersection of successively smaller open intervals. For example,

$$\begin{aligned} P(.2) &= P\left[\bigcap_{n=1}^{\infty}\left(.2 - \frac{1}{n}, .2 + \frac{1}{n}\right)\right] \\ &= \lim_{n\to\infty} P\left(.2 - \frac{1}{n}, .2 + \frac{1}{n}\right) \\ &= \lim_{n\to\infty} \frac{2}{n} = 0. \end{aligned}$$

1.2.2 The Addition Law

A general rule for determining the probability of a union of two events in terms of the probabilities of these events is as follows:

$$P(E + F) = P(E) + P(F) - P(EF).$$

This is reasonable from the point of view that interprets probability as a limiting proportion, since in adding the proportions of a sequence of trials in which E and F occur, respectively, one counts twice those trials in which both occur;

and since he does want to count them *once*, the subtraction of $P(EF)$ is appropriate.

The preceding general addition law is a consequence of Axiom 3, which is a special addition law for disjoint events, as follows. The set $E + F$ can be decomposed into three disjoint parts:

$$E + F = EF + EF^c + E^cF,$$

and then according to Axiom 3,

$$\begin{aligned} P(E + F) &= P(EF) + P(EF^c) + P(E^cF) \\ &= [P(EF) + P(EF^c)] + [P(EF) + P(E^cF)] - P(EF) \\ &= P(E) + P(F) - P(EF), \end{aligned}$$

where the bracketed terms have reduced to $P(E)$ and $P(F)$ by still another application of Axiom 3 and the distributive law.

EXAMPLE 1–31. Let E denote the event that the spinning pointer of the preceding example falls in the interval (0, .5), and F the event that it falls in the interval (.4, .6). According to the probabilities assigned in that example,

$$P(E) = .5, \qquad P(F) = .2, \qquad \text{and} \qquad P(EF) = .1,$$

and then

$$P(E + F) = .5 + .2 - .1 = .6,$$

which is indeed the probability of the union $E + F = (0, .6)$.

★1.2.3 Probability and Borel Sets

Because of the number and diversity of possible events in infinite sample spaces, it is ordinarily not practicable to specify a model by defining a probability for every event. Rather, it is usually the case that some collection of events of simple type will span (in the Borel field they generate) a collection of events that is large enough to include events of interest.

As suggested in Example 1–30. The Borel field generated by the collection of intervals on the real line is a useful class of events. Sets in this class are sometimes called *Borel sets* and are expressible in terms of countable operations on intervals. Similarly, Borel sets in the plane are those generated by rectangles with sides parallel to a given set of coordinate axes.

The point of considering events that are expressible in terms of countable operations on intervals is that a probability measure defined for intervals is automatically extended (by the axioms for probability) to the Borel field generated by them. Thus, if $I_1, I_2, \ldots$ is a descending or ascending sequence, the probability of their intersection or union, respectively, is the limit of $P(I_n)$. If $I_1, I_2, \ldots$ is not ascending, the sequence $I_1, I_1 + I_2, I_1 + I_2 + I_3, \ldots$ *is* and has the same union. But $P(I_1 + I_2 + I_3)$, for example, is computable from the probabilities of intervals by the addition law and its extensions (cf. Problem 1–32). In principle, then, the probabilities of countable unions and intersections of intervals are determined—if not always easy to calculate—by probabilities of intervals.

As seen in Example 1–30, the probabilities of single-point sets are determined from the probabilities of open intervals, and so also are probabilities for half-open and for closed intervals. Indeed, the same Borel field is generated by the family of closed intervals. Furthermore, the same Borel field is generated by the family of semi-infinite intervals of the form $(-\infty, x]$, a fact that will be significant in defining a probability measure by means of a "distribution function," in Section 2.1.5.

1.2.4 The Discrete Case

To construct a probability model in which the sample space has only a finite or countably infinite number of outcomes, it suffices to define a probability as a nonnegative number for each individual outcome in such a way that the total of the assigned probabilities is 1. The probability of a given event then must be, according to Axiom 3, simply the sum of the probabilities assigned to the outcomes that make up the event:

$$P(E) = \sum_{\omega \text{ in } E} P(\omega).$$

(The Borel field of events is here the collection of all possible events.) It is clear, moreover, that $P(\Omega) = 1$; and since the total of the assigned probabilities is 1, any sum of less than all of the assigned probabilities will not exceed 1: $0 \leq P(E) \leq 1$. That is, Axioms 1 and 2 are also satisfied.

EXAMPLE 1–32. One useful model for the experiment of drawing a card from a standard deck is that in which each card has the same chance of being drawn as any other, that is, in which the probability assigned to each card is $\frac{1}{52}$. Since the event "the card is an Ace" includes 4 of the 52 cards, its probability is

$$P(\text{Ace}) = 4 \cdot \tfrac{1}{52} = \tfrac{1}{13}.$$

EXAMPLE 1–33. A coin is tossed until heads first appears. The elementary outcomes of this experiment are sequences each of whose last element is heads, preceded by nothing but tails:

$$H, \quad TH, \quad TTH, \quad TTTH, \quad TTTTH, \quad \text{etc.}$$

One way (seen later to be a natural way) to assign probabilities to these outcomes is according to the formula

$$P(T, T, \ldots, T, H) = \frac{1}{2^k},$$

where k = number of tosses, for $k = 1, 2, 3, \ldots$. According to the formula for the sum of a geometric series, these probabilities total 1:

$$\frac{1}{2} + \frac{1}{4} + \frac{1}{8} + \cdots = \frac{1/2}{1 - 1/2} = 1.$$

The probability that the first heads occurs in three or fewer tosses is

$$\begin{aligned} P(1, 2, \text{ or } 3 \text{ tosses required}) &= P(H) + P(TH) + P(TTH) \\ &= \frac{1}{2} + \frac{1}{4} + \frac{1}{8} = \frac{7}{8}. \end{aligned}$$

The probability that an even number of tosses is required is

$$\begin{aligned} P(2 \text{ or } 4 \text{ or } 6 \text{ or} \cdots \text{tosses required}) &= \frac{1}{4} + \frac{1}{16} + \frac{1}{64} + \cdots \\ &= \frac{1/4}{1 - 1/4} = \frac{1}{3}. \end{aligned}$$

And so on.

Problems

1–31. Show the following relations.
[*Note*: These relations cannot be proved using Venn diagrams, since such a diagram does not represent probability relations. However, sometimes a Venn diagram will suggest how to proceed.]

(a) If $A \subset B$, then $P(A) \leq P(B)$.
(b) If $P(E) = 0$, then $P(EF) = 0$.
(c) $P(E) = P(EF) + P(EF^c)$.
(d) $P(EF) \leq P(E + F) \leq P(E) + P(F)$.
(e) [Extension of (c)] $P(A) = \sum P(AE_n)$, if $\{E_n\}$ is a countable partition of Ω.
(f) Show that if $A \subset B$, then $P(B - A) = P(B) - P(A)$.

1–32. Derive an expression for $P(E + F + G)$ in the general case, in terms of probabilities of E, F, and G and their various intersections.

1–33. Let each of the 52 cards, the elementary outcomes in the sample space for drawing a card from a standard deck, be assigned probability $\frac{1}{52}$ (as in Example 1–32). In this model, determine the probability of each of the following events:

(a) The card is a Spade.
(b) The card is a face card.
(c) The card is a Spade or a face card.
(d) The card is an Ace or a Heart.
(e) The card is not an Ace or a Heart.
(f) The card is black or is an Ace.
(g) The card is red or a face card.

1–34. The outcomes of an experiment are the nine pairs (x, y) where x and y are each one of the integers 1, 2, 3. Let probability $\frac{1}{6}$ be assigned to each of the pairs in which $x = y$ and the probability $\frac{1}{12}$ to each of the other outcomes. Determine the probabilities of each of the following events:

(a) $x = 2$. (b) $y = 1$. (c) $x = 2$ or $y = 1$.
(d) $x + y \leq 4$. (e) $y > 1$. (f) $|x - y| = 1$.

1.2.5 A Priori Models for Finite Spaces

In certain kinds of experiments with finitely many outcomes, the appropriate probability model is strongly suggested by intuition. These are called a priori models, being constructed without actual experimentation, although experience would bear out the usefulness of the model. In these experiments there are reasons, usually geometrical, to believe that all outcomes should be equally likely—have the same chance of occurring. The model that represents this belief assigns equal probabilities to the outcomes; so if there are n outcomes, each is assigned probability $1/n$.

One kind of experiment of this category is that in which a symmetrical object is tossed—a coin, a cube, or some other regular polyhedron. It comes to rest on one of its sides, and this side—or, if more convenient, the side on top (if there is one)—is the outcome. The geometrical symmetry suggests that no one side ought to be weighted more than any other, and one would be greatly surprised to find one side predominating in a long sequence of trials; hence, the outcomes are taken to be equally likely, and each assigned probability $1/n$, where n is the number of sides. The spinning of a roulette wheel and the spinning of the cylinder of a revolver are other experiments of this kind. Incidentally, there is no regular polyhedron having five sides, for example; but an experiment with five equally likely outcomes can be devised. A cylinder, sufficiently long, whose cross section is a regular pentagon, will come to rest (when rolled) with one of its five sides down, and a priori reasoning suggests assigning probability $\frac{1}{5}$ to each of the sides.

Another class of a priori models is that in which the experiment consists of blindly drawing from a finite collection of objects a certain number of them and noting which objects are selected but not taking any order into account. There are $\binom{N}{n}$ distinct selections of n from N objects, and the model suggested by a priori considerations for blind selections assigns equal probabilities to the possible selections, that is, the probability $1/\binom{N}{n}$ to each one. The statement that the selection is made "at random" will be taken to mean that the appropriate model is that in which all possible selections are equally likely. Sometimes the phrase "random selection" is used, but the question "Is this a random selection?" really means "Was the process by which this selection was made one to which the model of equally likely selections applies?"

This last question cannot be answered, strictly speaking. One can mix or shuffle the objects prior to the selection and do the selection in a truly "blind" manner, but aside from exercising precautions such as these, one can never know whether the ideal model is really applicable. Similarly, whether a die and a mechanism by which it is tossed are really represented by the ideal model (in which all sides are equally likely) is something one would never know—although again he could take certain precautions such as using a homogeneous die and giving it a vigorous toss with a substantial spin.

In any model in which the outcomes are equally probable, the probability of an event E is simply the ratio of the number of outcomes in E to the total number of outcomes in the experiment. The problem of computing the probability of an event is then reduced to a problem of counting the number of outcomes for which the condition defining the event is satisfied.

It is to be understood that when questions are asked about probabilities of events in the kinds of experiments discussed here, the a priori or ideal model is assumed or decreed to be the one that governs.

EXAMPLE 1–34. Two chips are selected at random from a container that holds four white and three black chips. What is the probability that both chips are white?

The phrasing of the description of the experiment, that is, "selected at random," means that the $\binom{7}{2}$ possible selections of two chips from seven are to be assigned equal probabilities, $1/\binom{7}{2}$. The desired probability is then the number of selections in which both chips are white, $\binom{4}{2}$, times the probability $1/\binom{7}{2}$:

$$P(\text{both white}) = \frac{\text{number of selections of two white}}{\text{number of selections of two chips}} = \frac{6}{21} = \frac{2}{7}.$$

EXAMPLE 1–35. The 12 faces of a die in the shape of a regular duodecahedron are numbered 1, 2, . . ., 12. What is the probability that an even number turns up when the die is tossed on a table?

Because there are as many even-numbered as odd-numbered faces, and all are equally likely, the probability that an even number turns up is $\frac{1}{2}$:

$$P(\text{even number}) = \frac{\text{number of even faces}}{\text{number of faces}} = \frac{1}{2}.$$

1.2.6 Nondiscrete Cases

Assigning probabilities to individual elementary outcomes, as in the discrete cases above, does not work in the case of a sample space that is uncountably infinite. In most such cases the probabilities assigned to individual outcomes ought to be 0, and it would not be possible to calculate probabilities for events with nonzero probabilities from the zeros assigned to the outcomes. To construct a probability model in such cases, it is necessary to assign probabilities to events rather than to individual points.

When a sample space is nondiscrete, it is usually: (1) a set of ordinary real numbers such as the interval $[0, 1]$ or the whole real line $(-\infty, \infty)$; (2) a set of points in a plane, either a two-dimensional set such as the points within a circle, or some set of points constituting a curve in the plane; (3) a set of points in a three-dimensional space, either a three-dimensional set such as the set of points in a cube, or a two-dimensional set such as a portion of a given surface, or a one-dimensional set of points on a curve; or (4) a set of points in a Euclidean space of dimension higher than three. In such sample spaces it is helpful to have in mind an analogy with distributions of mass.

A distribution of mass implies a function defined on the sets of our three-dimensional space of visualization—for any given region one can define the amount of mass in that region, a nonnegative quantity. Moreover, the amount of mass in a region consisting of two disjoint parts is the sum of the amounts in those parts. The same would be true, then, for a region composed of a finite number of disjoint parts, and could be assumed true for a countable number without offending the intuition. Thus, a given distribution of mass defines a set function having properties like those of probability; that is, the axioms for a probability measure are satisfied, except that the "measure" assigned to the whole space is the total mass, which may not even be finite. If it *is* finite, then the *relative* mass in a region—the fraction of the whole—is a quantity that does satisfy all of the probability axioms. At any rate, a probability distribution can be thought of as having an analog in a distribution of a finite amount of mass, probability corresponding to relative mass. Regions of high probability correspond to regions with a large mass concentration, that is, high density. Lumps or point masses in the distribution correspond to positive probability assigned to those points where the lumps occur. (A distribution having only lumps is then a *discrete* distribution, and is concentrated at most at a countable number of such points.)

Mass distributions (and thus probability) over a sample space of only two dimensions—a surface—would be something like a coat of paint on that surface, spread with variable density, but imagined somehow to be infinitely thin. Distributions on a one-dimensional sample space would be like a coat of paint on a thin wire.

If there is no lumpiness or discreteness, a distribution of mass or probability can be described in terms of a *density* of the distribution. The density at a point is the limiting ratio of the amount of material (mass or probability, as the case may be) contained in a tiny region including the point to the content of that region, as the region approaches 0 in all dimensions. (If the distribution is three-dimensional, the region is a solid region and its content is its volume; if two-dimensional, it is a region on a surface and its content is its area; if one-dimensional, it is a portion of a curve and its content is its length.) A density function has the property that its integral over any region gives the amount of mass or probability in that region:

$$\text{mass in region } E = \int_E (\text{mass density function})\, dR,$$

where E and dR represent a region and element of content (e.g., surface region and element of surface area) in the sample space over which the density is defined.

A mass distribution that has constant density over a certain region E (and is zero outside that region) is said to be *uniform* over E. In such a case, the amount of mass in a subregion is proportional to the content (volume, area, or length) of that subregion, since the constant density factors out of the integral. If a distribution of *probability* is uniform over a region E (and confined to that region), then E must have at most finite content; for the constant value of the density on E would have to be reciprocal of the content of E, inasmuch as the total probability assigned to Ω is 1:

$$\begin{aligned} 1 &= \int_\Omega (\text{probability density})\, dR \\ &= \int_E (\text{probability density})\, dR = \int_E k\, dR = k\,(\text{content of } E). \end{aligned}$$

Notice, too, that if one started by assuming that the probability of each set E is proportional to the content of E, then the density (the limiting ratio as defined above) would turn out to be constant—equal to the constant of proportionality.

In summary then, when a sample space is a portion of the "space" of our experience, a probability measure for sets in the space is a quantity that resembles the geometrical content of sets (in the property of additivity over disjoint sets, and in its nonnegativeness), but which allows for a variable weighting of different portions of the sample space—and even permits weighting individual points with nonzero values (although the geometrical content of individual points is zero).

EXAMPLE 1–36. A consequence of Maxwell's law in physics states that the speed of a molecule of an ideal gas can be thought of as the outcome of an experiment of chance having as its sample space the set of nonnegative real numbers $(0, \infty)$, and having a distribution of probability defined by the density

$$f(v) = (\text{const})v^2 \exp(-hmv^2), \qquad (v > 0),$$

where h and m are physical constants. The "constant" is determined so that the integral of the point density $f(v)$ over the sample space $(0, \infty)$ is unity. The probability that the speed of a molecule falls in a set E (of nonnegative real numbers) is then

$$P(E) = \int_E f(v)\, dv.$$

EXAMPLE 1–37. The probability space postulated for the spinning pointer in Example 1–30 assigned probabilities to intervals proportional to the width of the intervals, in fact, equal to the width. Extending this to more complicated sets, made up from intervals by countable operations, results in a probability measure for sets that is the same as its geometrical measure; the distribution is, therefore, *uniform* on the sample space [0, 1].

EXAMPLE 1–38. The sample space for random orientations (of a fiber in a material, for instance, such as in Example 1–6, 1–15, and 1–22) is essentially the set of points on a unit hemisphere. An event is a set of points on that hemisphere (identified with a corresponding solid angle of orientations), and the ideal model for perfectly random orientations assigns probability to such sets proportional to, their areas on the surface of the unit hemisphere. Since the area of the entire hemisphere is 2π, the probability of an event E is

$$P(E) = \frac{1}{2\pi}\,(\text{area of } E \text{ on hemisphere}).$$

For example, the event $0 \leq \phi \leq \pi/4$ (where ϕ is the longitude coordinate of the orientation) consists of points on $\frac{1}{8}$ of the hemisphere; therefore, it is assigned probability $\frac{1}{8}$.

Problems

• **1–35.** A committee of three is chosen at random from a group of five men and five women. What is the probability that

(a) The committee will consist of all men or all women?
(b) At least one man and at least one woman are on the committee?
(c) At least one man is on the committee?

1–36. The order on a ballot of the names of six candidates for three municipal judgeships is determined in such a way that all orders are equally likely. What is the probability that the names of the three already in office (the incumbents, who are running for reelection) will appear at the head of the list?

1–37. A lot of 20 articles is to be accepted or rejected on the basis of an inspection of 4 drawn at random from the lot. If it is decided to accept the lot when at most 1 of the 4 articles inspected is defective, but otherwise to reject the lot, what is the probability that by following this decision rule the result will be to reject a lot that is only 10 per cent defective?

1–38. The 36 elementary outcomes in the sample space for the experiment of tossing two dice are assigned equal probabilities. Determine the probabilities of the following events (see Problem 1–12):

(a) The sum is divisible by 4.
(b) Both numbers are even.

(c) The numbers on the dice are equal.
(d) The numbers on the dice differ by at least 4.
(e) The total number of points showing is 10.

1–39. Five cards are selected at random from a standard deck of cards. Determine the probabilities of the following events (see Problem 1–15):

(a) The hand contains exactly one pair.
(b) The hand contains two pairs (and an odd card).
(c) The hand contains two cards of one denomination and three of another.
(d) The hand contains three of one denomination, but the other two do not form a pair.
(e) The hand consists of five cards all of the same suit.

1–40. Consider an experiment whose outcomes are positive numbers. Given that the distribution of probability is described by a density function which at any positive number x is e^{-x}, compute the probability that

(a) The outcome exceeds 2.
(b) The outcome is at most 1.
(c) The outcome lies between 1 and 2.

1–41. Consider an experiment whose outcomes are the points in the square bounded by $x = 0$, $y = 0$, $x = 2$, and $y = 2$ in the xy-plane. If probability is distributed uniformly over this sample space, determine probabilities of each of the following events:

(a) $x + y < 2$. (b) $x + y < 1$. (c) $x > \frac{1}{2}$.
(d) $x^2 + y^2 > 1$. (e) $x < 1$ and $y < \frac{3}{2}$. (f) $x > 1$ or $y > 1$.
(g) $|x - y| < 1$. (h) $x - y < 1$. (i) $x - y > 2$.

1–42. Let the sample space of an experiment consist of the points within the unit cube defined by $0 \le x \le 1$, $0 \le y \le 1$, $0 \le z \le 1$. Assuming probability to be uniformly spread throughout the cube, determine the probability of the event consisting of those points (x, y, z) such that $x + y + z \le 1$.

1–43. Refer to the probability space representing random orientations in Example 1–38 and determine the probability of the event $0 \le \theta \le \pi/4$, where θ is the colatitude angle (that is, angular distance from the pole of the hemisphere).

1.3 Dependence and Independence

An aspect of probability theory that perhaps most distinguishes it from measure theory, and from the study of mass distributions, is the concept of independence. Although the concept can be translated into measure theoretic terms, it seems to have no particular significance in any other application of measure theory. It will be helpful, in appreciating the meaning of independence, to introduce it in terms of the notion of conditional probability.

1.3.1 Conditional Probability

It often happens that the sample space for an experiment must be altered to take into account the availability of certain limited information about the outcome of the experiment. Such information may well eliminate certain outcomes as impossible which were otherwise (without the information) possible, and in such a case either the appropriate sample space would omit these impossible outcomes or the probabilities assigned to them would be 0. In the revised model, the probabilities are said to be *conditional* on the occurrence of the event defined by the information. This new model is again a probability space, and the term *conditional* refers only to its origin in a larger probability space.

EXAMPLE 1–39. A standard die is tossed, and before seeing the outcome, one is told (by a bystander who does see the result) that an even number of points is showing. How should he then bet?

Certainly, it would be foolish to place any money on the outcomes 1, 3, or 5—the information that the outcome is an *even* number eliminates these from consideration. The sample space now is effectively just the set of even outcomes $\{2, 4, 6\}$, whose occurrence constitutes the information. But, in addition to betting nothing on the odd outcomes, one would find that his probabilities for the even outcomes are increased over what they are in the standard model for the toss of a die (with no information). For, a priori considerations would again suggest that there is no reason for any one of the outcomes 2, 4, or 6 to predominate, and each would be assigned probability $\frac{1}{3}$. The same conclusion would be reached in considering a long sequence of tosses of the die; in such a sequence the proportions of 1's, 2's, . . ., and 6's would be about equal, and the information that the outcome is even would simply eliminate from the sequence those trials that resulted in an odd number of points—leaving a sequence with equal proportions of 2's, 4's, and 6's.

Conditional probabilities, given an event F (or given that F has occurred), are obtained from the unconditional ones in the original sample space by taking F as a new sample space, and distributing a total probability of 1 over this smaller sample space in such a way that the new probabilities are in the same proportions as the old within F. That is, for any event E contained in an event F of positive probability the conditional probability of E given F, written $P(E \mid F)$, is defined to be

$$P(E \mid F) = \frac{P(E)}{P(F)}, \qquad \text{when } E \subset F.$$

Notice that with this definition (for events in F), conditional probabilities are indeed proportional to the unconditional ones, and further, that the assignment of these conditional probabilities to events satisfies the probability axioms. That is, $P(F \mid F) = 1$, $P(E \mid F) \geq 0$, and the additivity axiom holds because it held in the original space.

Conditioning with the information that F has occurred simply introduces a smaller probability space—and yet it is usually desirable to stay in the framework of the original probability space; thus, the adjective "conditional" is used, to refer to the reduced model. Events in the original sample space define events in the smaller space—for instance, the event G in Ω can be interpreted as a condition that characterizes its outcomes, and imposing this condition in the reduced model amounts to imposing *both* conditions G and F. Thus, the conditional probability of G given F (again assume $P(F) > 0$) is taken to be

$$P(G \mid F) = P(GF \mid F) = \frac{P(GF)}{P(F)}.$$

This definition includes the earlier one, but it can be used without having to check whether or not G is contained in F. The definition is often useful in the form of a *multiplication law*:

$$P(EF) = P(E \mid F)P(F).$$

With this law one can compute the probability of EF from what is sometimes an easier determination of the factors $P(E \mid F)$ and $P(F)$.

EXAMPLE 1–40. A card is drawn at random from a standard deck of cards. (This wording implies that each card is assigned probability $\frac{1}{52}$.) Given that the card drawn is a face card, the probability that it is a Jack would be computed as follows, since a Jack is a face card:

$$P(\text{Jack} \mid \text{face card}) = \frac{P(\text{Jack})}{P(\text{face card})} = \frac{4/52}{12/52} = \frac{1}{3}.$$

But not all Hearts are face cards, so

$$P(\text{Heart} \mid \text{face card}) = \frac{P(\text{Heart and face card})}{P(\text{face card})} = \frac{3/52}{12/52} = \frac{1}{4}.$$

That is, $\frac{1}{3}$ of the face cards are Jacks, and $\frac{1}{4}$ of the face cards are Hearts.

EXAMPLE 1–41. A card is drawn at random from a standard deck, and then a second card is drawn at random from the remaining cards. What is the probability that the first is a Queen and the second is a face card?

Here, although the experiment consists of a sequence of two selections, the wording of the problem implies an assignment of probabilities for the individual draws, namely, on the first draw each of the 52 cards has probability $\frac{1}{52}$ and on the second draw each of the 51 remaining cards has probability $\frac{1}{51}$. Clearly, the probability of a Queen on the first draw is $\frac{4}{52}$; and *given* that a Queen is obtained on the first draw, there are 11 remaining face cards among the 51 remaining cards. Hence,

$$\begin{aligned} &P(\text{Queen on 1st, face card on 2nd}) \\ &\qquad = P(\text{face card on 2nd} \mid \text{Queen on 1st})\, P(\text{Queen on 1st}) \\ &\qquad = \frac{11}{51} \cdot \frac{4}{52} = \frac{11}{663}. \end{aligned}$$

In similar fashion, probabilities for other events in the composite experiment can be determined from the assumption of random selection in the subexperiments.

The technique of the last example can be used to derive a useful result about composite experiments. Suppose that the experiment $\mathscr{E}$ consists of performing an experiment $\mathscr{E}_1$ and then performing an experiment $\mathscr{E}_2$. Suppose, furthermore, that $\mathscr{E}_1$ has m equally likely elementary outcomes, $u_1, \ldots, u_m$, and that $\mathscr{E}_2$ has n equally likely elementary outcomes $v_1, \ldots, v_n$. (These outcomes of $\mathscr{E}_2$ may depend on the result of $\mathscr{E}_1$; but it is assumed that the *number* of available outcomes in $\mathscr{E}_2$ is the same no matter how $\mathscr{E}_1$ turns out, and that these are equally probable.) The elementary outcomes of the composite experiment are the mn distinct pairs (u_j, v_k). The multiplication law is then used to determine probabilities in $\mathscr{E}$:

$$P(u_j, v_k) = P(v_k \mid u_j)P(u_j) = \frac{1}{m} \cdot \frac{1}{n} = \frac{1}{mn}.$$

That is, the elementary outcomes of the composite experiment are equally likely.

EXAMPLE 1–42. A coin is tossed three times. Given that the two elementary outcomes for each toss are equally likely, the eight elementary outcomes

$$HHH, \quad HHT, \quad HTH, \quad THH, \quad TTH, \quad THT, \quad HTT, \quad TTT$$

are equally likely, each having probability $\frac{1}{8}$.

EXAMPLE 1–43. The experiment of selecting one card and then a second card from a standard deck and recording the results so as to keep track of the order has $52 \cdot 51$ elementary outcomes, according to the principle explained in Section 1.1.2. If the outcomes of the first draw are assumed equally likely, and the 51 outcomes of the second draw are also assumed equally likely, then the $52 \cdot 51$ elementary outcomes of the composite experiment are equally likely. Thus, the probability of each elementary outcome or particular sequence of two different cards, such as: Queen of Hearts, 3 of Spades, is 1/2652. The probability of any other event can then be computed from these; for instance, the probability that both cards are Spades is 1/2652 times the *number* of sequences of two distinct cards of which both are Spades:

$$P(\text{both cards Spades}) = \frac{13 \cdot 12}{2652} = \frac{13}{52} \cdot \frac{12}{51}.$$

Writing it in the last form shows that this probability can also be computed as

$$P(\text{both cards Spades}) = P(\text{2nd Spade} \mid \text{1st Spade})\, P(\text{1st Spade}).$$

An extension of the multiplication rule follows from successive applications of the one given, for two events. For instance,

$$\begin{aligned} P(EFG) &= P(E \mid FG)P(FG) \\ &= P(E \mid FG)P(F \mid G)P(G). \end{aligned}$$

Similarly (cf. Problem 1–50),

$$P(EFGH) = P(E \mid FGH)P(F \mid GH)P(G \mid H)P(H).$$

EXAMPLE 1–44. Five chips are drawn, one at a time, from a bowl containing 8 white and 12 black chips. At each drawing the selection is done blindly—at random—from the remaining chips, and none is put back after having been drawn. The colors of the chips drawn are recorded in the sequence in which they are drawn; so the elementary outcomes are sequences of five colors, each being black or white. The probability of a given sequence can be computed using the extended multiplication law, together with the assumption of random selections. For instance, the probability of the sequence (W, B, B, W, B), is computed as follows:

$$\begin{aligned} P(WBBWB) &= P(W)P(B \mid W)P(B \mid WB)P(W \mid WBB)P(B \mid WBBW) \\ &= \frac{8}{20}\cdot\frac{12}{19}\cdot\frac{11}{18}\cdot\frac{7}{17}\cdot\frac{10}{16}. \end{aligned}$$

(The factor $P(W \mid WBB)$, for example, means the conditional probability of white on the fourth draw, given that the first was white and the second and third were black.)

The product can also be obtained in terms of combination symbols as follows:

$$\frac{\binom{8}{2}\binom{12}{3}}{\binom{20}{5}\binom{5}{3}}.$$

Upon noting the origins of these numbers, the "scheme" used to write this down becomes apparent—the 8 and 12 are numbers of white and black chips available, the 2 and 3 are the numbers of white and black chips drawn; and in the denominator $20 = 8 + 12$ and $5 = 2 + 3$ are the number of chips available and the number drawn, respectively. That this is not simply a happenstance can be demonstrated by showing (by induction) that the scheme works generally. A special case of this demonstration will perhaps suffice and suggest the general technique. Assuming that the scheme works for sequences of four, one has

$$P(WBBW) = \frac{\binom{8}{2}\binom{12}{2}}{\binom{20}{4}\binom{4}{2}},$$

and then

$$P(WBBWB) = P(WBBW)P(B \mid WBBW) = \frac{\binom{8}{2}\binom{12}{2}}{\binom{20}{4}\binom{4}{2}}\cdot\frac{10}{16} = \frac{\binom{8}{2}\binom{12}{3}}{\binom{20}{5}\binom{5}{3}}.$$

From this it follows, incidentally, that the probability of a given number of whites among the five draws (disregarding order) can be found by a formula that has another interpretation; for there are precisely $\binom{5}{3}$ sequences in which two are black and three are white and *each* sequence has the same probability:

$$P(2 \text{ whites and } 3 \text{ blacks}) = \binom{5}{3} \cdot \frac{\binom{8}{2}\binom{12}{3}}{\binom{20}{5}\binom{5}{3}} = \frac{\binom{8}{2}\binom{12}{3}}{\binom{20}{5}}.$$

This last expression is the probability that if five beads are *selected at random* from the container with the 8 whites and 12 blacks, there are exactly 2 whites and 3 blacks in the selection—the assumption of a random selection implying that all $\binom{20}{5}$ selections are equally likely. What has been shown in this instance, then, is that if one keeps track only of the numbers of whites and blacks in a selection, performing the selection one at a time at random is equivalent to making a random selection of five. This is actually a general rule (Problem 1–54).

1.3.2 Bayes' Theorem

The multiplication rule can be used in two ways to express the probability of an intersection as a product:

$$P(EF) = P(E \mid F)P(F) = P(F \mid E)P(E),$$

assuming that neither $P(E)$ nor $P(F)$ is 0. Solving the last equality for $P(F \mid E)$ yields

$$P(F \mid E) = \frac{P(E \mid F)P(F)}{P(E)},$$

and provides a means of effectively interchanging the roles of event and condition.

A version of the above relation known as *Bayes' theorem* is obtained when the denominator is rewritten using the law of total probability [Problem 1-31(c)]:

$$P(E) = P(EF) + P(EF^c) = P(E \mid F)P(F) + P(E \mid F^c)P(F^c).$$

Substituting this, one obtains

$$P(F \mid E) = \frac{P(E \mid F)P(F)}{P(E \mid F)P(F) + P(E \mid F^c)P(F^c)}.$$

Similarly, if the sets $A_1, A_2, \ldots$ constitute a partition of Ω, one has the following form of Bayes' theorem:

$$P(A_i \mid E) = \frac{P(E \mid A_i)P(A_i)}{P(E \mid A_1)P(A_1) + P(E \mid A_2)P(A_2) + \cdots}.$$

(Notice that the sum of these conditional probabilities of the partition sets, given E, is 1.)

EXAMPLE 1–45. A certain disease is present in about 1 out of 1000 persons in a given population, and a program of testing is to be carried out using a detection device which gives a positive reading with probability .99 for a diseased person and with probability .05 for a healthy person. It is desired to determine the probability that a person who has a positive reading actually does have the disease.

With obvious notations for diseased, healthy, positive, and negative, the given quantities are as follows:

$$P(D) = .001, \qquad P(+ \mid D) = .99, \qquad P(+ \mid H) = .05.$$

Bayes' theorem permits the computation of $P(D \mid +)$ from these:

$$P(D \mid +) = \frac{P(+ \mid D)P(D)}{P(+)}.$$

The denominator is computed using the law of total probability:

$$P(+) = P(+ \mid D)P(D) + P(+ \mid H)P(H) = .99 \times .001 + .05 \times .999$$
$$= .05094.$$

Then the desired conditional probability is the ratio of the first term to the sum:

$$P(D \mid +) = \frac{.00099}{.05094} = 0.0194,$$

or about one chance in 50. This result may seem a bit odd at first, since the characteristics of the detector as given seem rather good; the explanation is that when a positive reading is obtained, it is much more frequently a machine malfunction than a diseased person which gives the reading. For, even with a perfect detection device, only one in 1000 readings would be positive; whereas with the given device and a completely healthy population, 5 out of every 100 readings would be positive.

Problems

1–44. (a) Determine the probability that a tossed die shows 4, given that the outcome is an even number.
(b) Determine the probability that it shows 6, given that the outcome is divisible by 3.

1–45. A sample space consists of exactly five outcomes, a, b, c, d, and e. Given that $P(\{a, b, c\}) = \frac{1}{2}$ and $P(\{a\}) = \frac{1}{4}$:

(a) Determine the probabilities of all events whose probabilities can be computed from those given.
(b) Compute $P(\{b, c, d\} \mid \{a, b, c\})$.
(c) Compute $P(\{a\} \mid \{a, b, c\})$.

1–46. Six couples gather for three tables of bridge. To determine the starting positions, 12 score cards are passed out at random to the twelve players (one to each), two marked "Table 1, Couple 1," two marked "Table 1, Couple 2," two marked "Table 2, Couple 1," and so on. (At each table, the two having slips marked "Couple 1" will be partners, and those with "Couple 2" slips will be partners.)

(a) What is the probability that Table 1 has all men?
(b) What is the probability that Table 1 has four people who form two married couples (but not necessarily paired that way for the bridge game)?

(c) What is the probability that a certain man draws his own wife as a partner?
(d) What is the probability that a certain man draws his own wife as a partner, given that they draw the same table number?"
(e) What is the probability that all of the men draw their own wives as partners?

1–47. Political prisoners Matthew, Mark, and Luke are told that two of the three are to be executed. Matthew takes the jailer aside and says, "We both know that at least one of the others, Mark or Luke, will be executed; if you know, tell me the name of *one* who will be executed." After mulling this request over, the jailer can see no harm in revealing one name and tells Matthew that Mark will be executed. Thereupon Matthew rejoices because the probability of his execution has just decreased from $\frac{2}{3}$ to $\frac{1}{2}$. Is his elation justified?
[*Hint*: The matter of which prisoner the jailer names if both Mark and Luke are to be executed is crucial. Let p denote the probability that the jailer names Mark, given that both Mark and Luke are to be executed, and use Bayes' theorem to calculate the probability that Matthew escapes execution if the jailer names Mark.]

1–48. Two cards are drawn at random from a standard deck. Determine the probability that

(a) They are both face cards, given that they are of the same suit.
(b) They are both face cards.
(c) Both are Hearts, given that both are red.

1–49. The outcomes of an experiment are real numbers, and the distribution of probability is continuous, having a density function given by $[\pi(1 + \omega^2)]^{-1}$ at the point ω. Determine the probability that the outcome

(a) Exceeds 1, given that it is positive.
(b) Is positive, given that it is on the interval $(-1, 1)$.

1–50. Show the following:

(a) $P(AC \mid B) = P(A \mid BC)P(C \mid B)$.
(b) $P(EFGH) = P(E \mid FGH)P(F \mid GH)P(G \mid H)P(H)$.

1–51. Machines A and B turn out, respectively, 10 and 90 per cent of the total production of a certain type of article. Suppose the probability that machine A turns out a defective article is .01 and that machine B turns out a defective article is .05. What is the probability that an article taken at random from a day's production was made by machine A, given that it is found to be defective?

1–52. Of a large group of college students containing equal numbers of freshmen, sophomores, juniors, and seniors, it is found that 35 per cent of the freshmen, 25 per cent of the sophomores, 20 per cent of the juniors, and 15 per cent of the seniors are girls. What is the probability that a girl picked at random from the group is a freshman?

1–53. Students are classed as either "resident" or "commuter." Of a group of 50 students, 14 are male residents, 20 are female commuters, and 24 are males. What is the probability that a name drawn at random from a list of these students is that of

(a) A female resident?
(b) A resident?
(c) A female, given that the student is a resident?

★**1–54.** A bowl contains M black and $N - M$ white chips; n chips are drawn one at a time at random (without replacing any). Let $\mathscr{P}_n$ denote the proposition that the probability of each sequence of n chip colors consisting of k blacks and $n - k$ whites is given by the formula

$$\frac{\binom{M}{k}\binom{N-M}{n-k}}{\binom{N}{n}\binom{n}{k}}, \qquad \text{for } k = 0, 1, \ldots, n.$$

Show (a) that $\mathscr{P}_1$ is true, and (b) that $\mathscr{P}_n$ implies $\mathscr{P}_{n+1}$, using the multiplication rule illustrated in Example 1–44. By induction, then, $\mathscr{P}_n$ is true for $n = 1, 2, \ldots$, from which follows the general rule given in Example 1–44. [If $a < b$, the symbol $\binom{a}{b}$ is defined to be 0.]

1.3.3 Independent Events

It can happen that a condition does not alter the probability assigned to an event:

$$P(E \mid F) = P(E).$$

That is, the probability that E will occur is the same with as without the information that F has occurred. If this is so and if $P(E) \neq 0$, then (according to Bayes' theorem):

$$P(F \mid E) = \frac{P(F)P(E \mid F)}{P(E)} = P(F),$$

which says that the probability assigned F is the same with as without the information that E has occurred. The multiplication rule in this case assumes the form

$$P(EF) = P(E)P(F),$$

which happens to be true even if E or F has zero probability, since EF is contained both in E and in F. Moreover, if this multiplication relation is true and $P(E) \neq 0$, then

$$P(F \mid E) = \frac{P(EF)}{P(E)} = \frac{P(E)P(F)}{P(E)} = P(F).$$

That is, for events of positive probability, the multiplication rule in the above form is equivalent to the equality of conditional and unconditional probabilities, and either can then be taken to be the definition of *independence* of E and F. Thus, events E and F are said to be *independent* if and only if $P(EF) = P(E)P(F)$. (According to this formulation, an event of zero probability is independent of any other event.)

EXAMPLE 1–46. A card is drawn at random from a standard bridge deck. In Example 1–40 the conditional probability that the card drawn is a Heart given that it is a face card was computed to be

$$P(\text{Heart} \mid \text{face card}) = \frac{3/52}{12/52} = \frac{1}{4}.$$

This is also the probability that the card drawn is a Heart, without the information as to whether or not it is a face card. (One-quarter of all the cards are Hearts, and one-quarter of all the face cards are Hearts.) That is, the events "Heart" and "face card" are independent events. Notice that the multiplication law defining independence holds:

$$\begin{aligned} P(\text{Heart and face card}) &= \frac{3}{52} = \frac{13}{52} \cdot \frac{12}{52} \\ &= P(\text{Heart})P(\text{face card}), \end{aligned}$$

and also that

$$P(\text{face card} \mid \text{Heart}) = \frac{3/52}{13/52} = \frac{3}{13} = P(\text{face card}).$$

(Three-thirteenths of all of the cards are face cards, and three-thirteenths of the Hearts are face cards.)

It should be emphasized that independence is a concept associated with the assignment of probabilities; it is not, therefore, represented in a Venn diagram, which pictures only sets or events. In particular, a student sometimes gets the notion that independence and disjointness are somehow the same thing, whereas disjointness is not defined in terms of probabilities—it is a property that *can* be represented in a Venn diagram. Indeed, disjointness and independence are almost incompatible, as seen in Problem 1–60(c).

The notion of independence can be extended to the case of three events. Events E, F, and G are said to be independent if and only if

(i) They are pairwise independent.
(ii) $P(EFG) = P(E)P(F)P(G)$.

(Pairwise independent, or independent in pairs, means that each combination of two of the three events is a set of independent events, in the sense defined earlier for two events.)

Lest this definition seem unnecessarily complicated, it should be pointed out that (i) does not imply (ii), nor does (ii) imply (i). Example 1–47 will show that pairwise independence does not imply that $P(EFG)$ factors as in (ii); and Problem 1–58 will give an instance in which $P(EFG)$ factors properly, but E, F, and G are not pairwise independent. Neither (i) nor (ii) would then adequately characterize the intuitive notion of independence. For if only the factorization (ii) were assumed, we might not have pairwise independence; and if only pairwise independence were taken as defining independence of E, F, and G, one could not guarantee, for instance, that EF and G are independent. Thus, both (i) and (ii) are needed to define in the model what intuition and practice require for the notion of independence of three events.

Independence of four events could be defined in terms of a factorization of $P(EFGH)$ and a requirement that any three of them are independent, and so on. But the notion can be defined once and for all, for the case of the n events $E_1, E_2, \ldots, E_n$, as follows. The n events $E_1, E_2, \ldots, E_n$ are *independent* if and only if for every subset of k events $E_{i_1}, E_{i_2}, \ldots, E_{i_k}$ the factorization holds:

$$P(E_{i_1}E_{i_2}\cdots E_{i_k}) = P(E_{i_1})P(E_{i_2})\cdots P(E_{i_k}).$$

The implications of this definition seem to include all of what is thought of intuitively as independence.

EXAMPLE 1–47. The probability space for a certain experiment of chance consists of the four equally probable outcomes (1, 0, 0), (0, 1, 0), (0, 0, 1), and (1, 1, 1). Consider the events

E: The first coordinate is 1.
F: The second coordinate is 1.
G: The third coordinate is 1.

Each of these events has probability $\frac{1}{2}$, so

$$P(E)P(F)P(G) = (\tfrac{1}{2})^3 = \tfrac{1}{8}.$$

However, only one of the four outcomes has all three coordinates equal to 1:

$$P(EFG) = P(\{(1, 1, 1)\}) = \tfrac{1}{4}.$$

Thus the three events E, F, and G are *not* independent. But they are pairwise independent; for example,

$$P(EF) = P(\{(1, 1, 1)\}) = \tfrac{1}{4} = P(E)P(F),$$

and similarly,

$$P(FG) = P(F)P(G) \qquad \text{and} \qquad P(EG) = P(E)P(G).$$

To see why one would not want to call E, F, and G independent in this instance, notice that the probability of the event EF depends on whether or not G is known to have occurred:

$$P(EF \mid G) = \frac{P(EFG)}{P(G)} = \frac{1}{2}, \qquad P(EF) = \frac{1}{4}.$$

1.3.4 Independent Experiments

An experiment $\mathscr{E}$, represented by a given probability space, consists of performing first experiment $\mathscr{E}_1$ and then experiment $\mathscr{E}_2$. Outcomes of $\mathscr{E}$ are pairs of outcomes—an outcome of $\mathscr{E}_1$ paired with an outcome of $\mathscr{E}_2$. An event in $\mathscr{E}$ is defined by a condition E_1 on the first element of the pair, and another event in $\mathscr{E}$

is defined by a condition E_2 on the second element of the pair. Experiments $\mathscr{E}_1$ and $\mathscr{E}_2$ are said to be *independent experiments* if each event E_1 relating to $\mathscr{E}_1$ is independent of each event E_2 relating to $\mathscr{E}_2$.

This definition can be used, as its statement implies, to check whether a given probability structure for a composite experiment has the property that the component experiments are independent. But it is more common for the definition to be used the other way—to *construct* a probability space for the composite experiment in order to achieve independence of the component experiments. The probability space can be constructed by using the multiplication rule for independent events to assign probabilities to events of the type E_1E_2, where E_1 and E_2 are conditions only on $\mathscr{E}_1$ and $\mathscr{E}_2$, respectively. (This class of events is not usually large enough to include all events relating to the composite experiment that might be of interest; but, on the other hand, the Borel field generated by this class usually is adequate, and probabilities are extended to this Borel field in the usual manner.)

EXAMPLE 1–48. From a bowl containing four chips numbered 1, 2, 3, 4, a first chip is drawn and then a second, without replacement of the first. The 12 outcomes are as follows:

$$\begin{matrix} 1,2 & 2,1 & 3,1 & 4,1 \\ 1,3 & 2,3 & 3,2 & 4,2 \\ 1,4 & 2,4 & 3,4 & 4,3. \end{matrix}$$

Consider the events

E: The first chip drawn is 1 or 3.
F: The second chip drawn is 1 or 2.

In the model for the composite experiment that assigns equal probabilities ($\frac{1}{12}$) to each outcome, these events have probabilities $P(E) = \frac{1}{2}$ and $P(F) = \frac{1}{2}$, since there are 6 outcomes satisfying condition E and 6 satisfying F. The intersection EF consists of the pairs (1, 2), (3, 1), and (3, 2), so that

$$P(EF) = \tfrac{1}{4} = P(E)P(F).$$

Thus, E and F are independent events. However,

$$P(\text{2nd chip is 3} \mid \text{1st is 3}) = 0 \neq P(\text{2nd chip is 3}),$$

and so the component experiments (drawing the first chip and then drawing the second chip) are *not* independent in this model. All it takes for this conclusion is a *single* instance of a pair of events, one relating to the first experiment and one to the second, that are not independent.

EXAMPLE 1–49. Consider the composite experiment made up of two tosses of a coin (or a toss of two coins). Let the outcomes of the first toss be denoted H and T, and the outcomes of the second h and t. The elementary outcomes of the composite experiment are then the four pairs Hh, Ht, Th, and Tt. For an ideal, balanced, or "fair" coin, one decrees $P(H) = P(T) = \frac{1}{2}$, and $P(h) = P(t) = \frac{1}{2}$. But these probabilities for the component experiment say nothing about probabilities in the composite experiment, without some further assumption about the relationship of the two experiments. One would ordinarily assume that probabilities relating to one coin should not be influenced by information as to how the other one has fallen, and so would assume independence of the tosses. From this assumption and the probabilities for the component experiments, one can construct a model for the composite experiment:

$$P(Hh) = P(H)P(h) = \tfrac{1}{4}, \qquad P(Th) = P(T)P(h) = \tfrac{1}{4},$$
$$P(Ht) = P(H)P(t) = \tfrac{1}{4}, \qquad P(Tt) = P(T)P(t) = \tfrac{1}{4}.$$

And from these one can compute probabilities of events that are not of the type which are the intersections of events in one subexperiment with events in the other. For example,

$$P(\{Ht \text{ or } Th\}) = \tfrac{1}{4} + \tfrac{1}{4} = \tfrac{1}{2}.$$

Although stated above for only two experiments, the notion of independence for any finite number of experiments is similarly defined. It might be noticed that the model has nothing in it that requires that the subexperiments be actually carried out in any particular order, which is as it should be if they are truly independent in the intuitive sense. That is, tossing 10 coins simultaneously is equivalent to tossing them in any given sequence, so long as information about the result for 1 or more of the coins would not alter how one would bet on the remaining coins.

Problems

1–55. A coin and a die are tossed. Assume the 12 elementary outcomes of the composite experiment to be equally likely, and show then that the component experiments (the toss of the coin and the toss of the die) are independent.

1–56. A card is drawn at random from a standard deck. Show that

(a) The events "Heart" and "black" are not independent.
(b) The events "black" and "Ace" are independent.
(c) "Under 10" and "red" are independent.
(d) The three events "red," "face card," and "Heart or Spade," are independent.

1–57. Two chips are drawn at random from four chips numbered 1, 2, 3, 4. List the six (equally likely) possible selections of two. For each outcome, let R denote the magnitude of the difference and S the sum of the numbers on the two chips drawn.

(a) Determine probabilities for the events defined by the condition $R = 2$ and by the condition $S = 5$. Are these independent events?
(b) Are the events $R = 1$ and $S = 5$ independent?

1–58. An experiment can result in one of five outcomes, assigned probabilities as follows: ω_1 with probability $\frac{1}{8}$, ω_2, ω_3, and ω_4 each with probabilities $\frac{3}{16}$, and ω_5 probability $\frac{5}{16}$. Define E, F, and G as follows:

$$E = \{\omega_1, \omega_2, \omega_3\}, \qquad F = \{\omega_1, \omega_2, \omega_4\}, \qquad G = \{\omega_1, \omega_3, \omega_4\},$$

and show that they are not pairwise independent, but that

$$P(EFG) = P(E)P(F)P(G).$$

1–59. Show that when the eight elementary outcomes for the composite experiment of tossing a coin three times are assigned equal probabilities, the three tosses are independent.

1–60. Show the following:

(a) If A and B are independent events, then A and B^c are independent.
(b) If A, B, and C are independent events, then A, B, and C^c are independent.
(c) Disjoint events are not independent unless at least one of them has probability 0.
(d) An event with probability 1 is independent of every other event.

1–61. A die is tossed eight times in succession, and after each toss it is noted whether the result is a six (6) or is not a six (N). Determine the probability:

(a) Of this sequence of results: N, 6, N, 6, N, N, N, 6.
(b) Of this sequence: 6, 6, N, N, N, N, 6, N.
(c) Of any particular sequence in which there are 2 N's and the rest 6's.

1–62. A certain type of device is used to shut off a flow when a container is filled to a certain depth; its "reliability" (probability that it works when it should) is assumed to be .9. A second type of shutoff device is placed "in parallel," that is, so that the flow is shut off if either device works; the reliability of this second device is assumed to be .7. What is the reliability of the combination? What is the probability that when the depth is reached, just one of the devices will work? (Assume that the devices operate independently.)

1–63. A certain device consists of four parts connected so that the device works only if all four parts work, where "works" means operates successfully in a certain mission. If the probability that each part individually works is .9, what is the probability that the device works? (Assume independence.)

1–64. Two teams are to play a series of games for the best four out of seven games; as soon as one team wins four games, the series ends. The probability that team A wins over team B is assumed to be .5 if the game is played on team B's field, and .7 if the game is played on team A's field. The first two games are to be played on team B's field, the next three on team A's field, and the last two (if the series runs that long) on team B's field. What is the probability that team A wins the series on four games? In five games? What is the probability that the series does not run to six games?

2

RANDOM VARIABLES AND THEIR DISTRIBUTIONS

The term *random variable* is an apt one, since it is used to denote a variable that is random—that is, a numerical quantity whose value is determined by an experiment of chance. Other names for the concept are *chance variable* and *stochastic variable*. Because its value is dependent, a random variable is an example of the mathematical idea of *function*; and so the formal definition given in Section 2.1 is in terms of this idea, although the domain of the function is somewhat more general than it is in the usual use of the term.

The adjective "random" in this context does *not* mean (as it does in the case of "random selection") that the appropriate model involves outcomes that are necessarily equally likely. It refers only to the fact that the value of the numerical quantity under consideration cannot be predicted from a knowledge of the experimental conditions. What is different, then, from the general concept of an experiment of chance is that one is concerned with an experiment whose outcomes either are numerical themselves or have numbers assigned to them. The sample space or the induced sample space (in the latter case) is then a space of numbers or a space of vectors of numbers, and the structure of such spaces permits analyses and descriptions that are not possible in the general case.

2.1 Random Variables and Vectors

Consider a probability space $(\Omega, \mathscr{F}, P)$ where, as before, Ω is a sample space, $\mathscr{F}$ a collection of subsets of Ω called events, and $P(E)$ is the probability assigned to the event E.

Definition. *A* random variable *is a measurable function defined on a probability space.*

The condition of measurability[1] is almost always satisfied; indeed, the construction of a *non*measurable function takes considerable ingenuity in continuous

[1] Defined in Section 2.1.2.

cases and some artificiality in discrete cases. Yet it is a condition that seems necessary to make the definition mathematically workable. With this acknowledgment of what is actually needed, it will be assumed in what follows (with impunity, it turns out) that all functions encountered are in fact measurable—without taking the trouble to verify this claim in any case.

To say that a certain physical, biological, or social quantity "is a random variable" means, usually, that the measure of this quantity is numerical and dependent on the outcome of an experiment of chance; but it also carries the implication that the quantity can be represented by the mathematical model of a function defined on an appropriate probability space (although the precise characteristics of this model may be unknown or only partly known).

EXAMPLE 2–1. The resistance of a certain electrical resistor is measured. Because of variations introduced in the process of measurement, repeated measurements of resistance will vary, and the measured resistance is a numerical quantity depending on an experiment of chance—a random variable. Further variation would be introduced if the resistor were taken from a supply of "100-ohm" resistors, whose resistances are in fact not all equal to this nominal value nor even all alike, owing to uncontrolled variables in the process of their manufacture. The "outcome" of the experiment of drawing a resistor and measuring its resistance is perhaps most expediently identified by the reading of the ohmmeter; that is, ω is itself a real number. The simple random variable $X(\omega) = \omega$ is then surely of interest, as is possibly the deviation about the nominal resistance, $Y(\omega) = \omega - 100$.

EXAMPLE 2–2. An individual tosses a coin infinitely often. The sample points or outcomes are then infinite sequences of heads and tails. Some of the random variables that might be of interest are

$X(\omega) =$ number of tosses required to equalize the number of heads and the number of tails for the first time.

$Y(\omega) =$ proportion of the tosses at which the number of heads is ahead of the number of tails.

$Z(\omega) =$ 1 or 0, depending whether the third toss is heads or tails.

Thus, suppose the outcome of the experiment is the sequence

$$H, H, T, H, T, T, T, H, T, \ldots.$$

For this sequence, $X(\omega) = 6$, $Z(\omega) = 0$, and $Y(\omega)$ cannot be determined from only the finite portion of the sequence given.

Although a random variable is a function, it has become common practice to drop the reference to the independent variable ω, and to write just X in place of $X(\omega)$, Y in place of $Y(\omega)$, and so forth. This practice is especially appropriate when ω is a real number and $X(\omega) = \omega$, but it is also generally convenient because the main interest is in the value space of the random variable and in the induced probability distribution in that space.

A *random vector* is a vector-valued function defined on a probability space $(\Omega, \mathscr{F}, P)$ that is measurable with respect to $\mathscr{F}$. But again the verification that a given vector function is actually measurable will always be *assumed* possible, and omitted. As in the case of a single random variable, the simpler notation $(X_1, \ldots, X_n)$ will often be used in place of $[X_1(\omega), \ldots, X_n(\omega)]$.

EXAMPLE 2–3. A person is picked at random from a large population, and his height H, weight W, and age A are determined. Thinking of the people in the population as constituting the sample space and the person picked as an elementary outcome ω, one finds that the triple of measured quantities is a random vector: $[H(\omega), W(\omega), A(\omega)]$, or (H, W, A). Since the population is finite, the collection of all subsets will serve as the Borel field of a probability space, and the phrase "picked at random" calls for the assignment of a probability to each event that is proportional to the number of people in it.

It was seen in Section 1.1.5 that a function on a probability space maps the events of Ω into sets of values in the value space of the function, and conversely, that it determines for each given set of values an event in the sample space—the pre-image of that set. A random variable, being a function on a sample space, does these things; but the sample space has a probability structure, and so the function also provides a natural transfer of probability from Ω to the space of values $\mathscr{X}$.

Because the value space of a random variable $X(\omega)$ is always a set of real numbers, and so is imbedded in the set of *all* real numbers, it will be convenient in the general discussion to take $\mathscr{X}$ to be this set $\mathscr{R}$ of all real numbers, even though some numbers may not actually be assumed as values of $X(\omega)$ in a given case. In this sample space of real numbers, the probability assigned to an event A is just the probability of its pre-image in Ω:

$$P(A) = P[X(\omega) \text{ in } A] = P[\{\omega \mid X(\omega) \text{ in } A\}].$$

This probability distribution in $\mathscr{R}$ is said to be *induced* by the random variable $X(\omega)$ from the distribution in the sample space Ω. Observe, however, that in the special case that sometimes arises, when Ω is itself a set of real numbers and $X(\omega) = \omega$, the induced distribution is essentially the same as the original one.

It should be mentioned that the same distribution in $\mathscr{R}$ can be induced from different sample spaces by correspondingly different functions. For instance, the random variable

$$X(\omega) = \begin{cases} 1, & \text{if a coin falls heads,} \\ 0, & \text{if the coin falls tails,} \end{cases}$$

has precisely the same distribution as the random variable

$$Y(\omega) = \begin{cases} 1, & \text{if a die shows an even number of points,} \\ 0, & \text{if it shows an odd number,} \end{cases}$$

provided that the usual unbiased or fair coin and die are used (that is, equally likely outcomes). Indeed, it may be said that a feature of probability theory that distinguishes it from the theory of measurable functions on measure spaces is that the main interest in probability is only in the distribution in the value space and not in the sample space.

2.1.1 The Distribution Function

To characterize, or to specify or define a distribution in the space $\mathscr{R}$ of values of a random variable $X(\omega)$, it will be seen shortly that a sufficient notion is that of the distribution function:

$$F(\lambda) = P[X(\omega) \leq \lambda] = P(X \leq \lambda).$$

This probability assigned to the set $(-\infty, \lambda]$ is surely a function of λ (recognized in the name *distribution function*), but, of course, it also depends on the nature of the random variable X. If it is necessary to keep that dependence on display, as would be the case when more than one random variable is under discussion, it will be done with a subscript: $F_X(\lambda)$.

One of the inadequacies of traditional functional notation is that the same notation is used for a whole function, $f(x)$, as is used for the specific value of the function corresponding to x. When referring to a whole function (i.e., to the rule for associating values of one variable with those of another), the particular letter employed within the parentheses is not significant: $f(y)$ denotes the same functional relationship as $f(x)$. Perhaps a better notation would be $f(\cdot)$. The x or y used is a "dummy" variable, in that $f(x)$ means that a certain operation is to be performed on *whatever* number x is "plugged" in. Thus, if $F(\lambda)$ is the distribution function of X, then

$$F(\lambda) = P(X \leq \lambda), \qquad F(x) = P(X \leq x), \qquad F(q) = P(X \leq q), \qquad \text{etc.}$$

But the *function* $F(\cdot)$ is the same in each case. A different distribution function results from using a different random variable:

$$F_X(\lambda) = P(X \leq \lambda), \qquad F_Y(\lambda) = P(Y \leq \lambda), \qquad \text{etc.}$$

Here the function $F_X(\cdot)$ defines a relationship different from that defined by $F_Y(\cdot)$ and it makes no difference that the same dummy variable was used in defining both. It is common to find that in referring to a random variable X, the corresponding lower case x is used as a dummy variable (and y for Y, etc.). Indeed, some use the same name for both the random variable and the dummy denoting possible values, a practice that can be convenient when understood, but which leads to such strange statements as this: $P(x \leq x) = F(x)$. The latter practice will be avoided here, but the capital and lower case convention will be used when convenient.

EXAMPLE 2–4. In the model proposed in Examples 1–30 and 1–37 for the stopping position of a spinning pointer, the probability that it stops in a given angle was assumed proportional to the size of the angle. Marking the circle traced by the pointer's tip with a uniform scale from 0 to 1 introduces a random variable $X(\omega)$, which maps the stopping positions ω into the real numbers on the interval [0, 1]. The distribution function of this random variable is

$$F(\lambda) = P(X \leq \lambda) = P(0 \leq X \leq \lambda) = \begin{cases} 0, & \text{if } \lambda < 0, \\ \lambda, & \text{if } 0 \leq \lambda \leq 1, \\ 1, & \text{if } \lambda > 1, \end{cases}$$

The graph of this function is shown in Figure 2–1. Notice that if the pointer's position is identified by the number that marks where it stops on the superimposed scale, then $X(\omega) = \omega$.

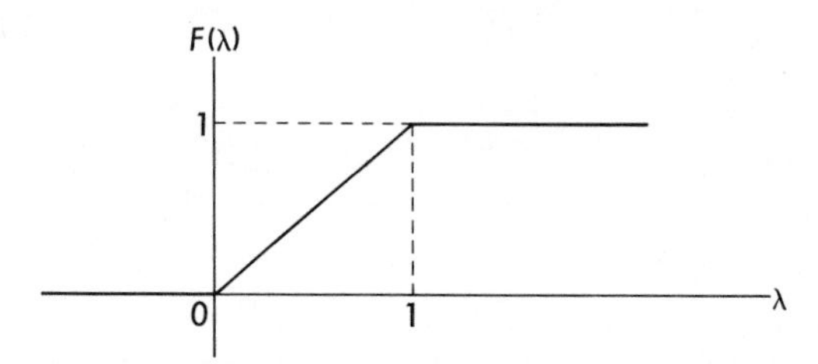

Figure 2–1. Distribution function for the spinning pointer.

EXAMPLE 2–5. Consider again the spinning pointer of the preceding example, and define the following random variable:

$$Y(\omega) = \begin{cases} 0, & \text{if } \omega \text{ is in the left half-circle,} \\ 1, & \text{if } \omega \text{ is in the right half-circle,} \end{cases}$$

where "left" and "right" are measured with respect to some given reference diameter. For this random variable, any interval of Y-values containing just 0 (and not 1) has probability $\frac{1}{2}$, and any interval containing both 0 and 1 has probability 1. That is,

$$F_Y(\lambda) = P(Y \leq \lambda) = \begin{cases} 0, & \text{if } \lambda < 0, \\ \frac{1}{2}, & \text{if } 0 \leq \lambda < 1, \\ 1, & \text{if } \lambda \geq 1. \end{cases}$$

The graph of this function is shown in Figure 2–2. (This distribution in $\mathscr{R}$ is precisely the same as that of the variable defined to be 1 or 0, according to whether a fair coin falls heads or tails.)

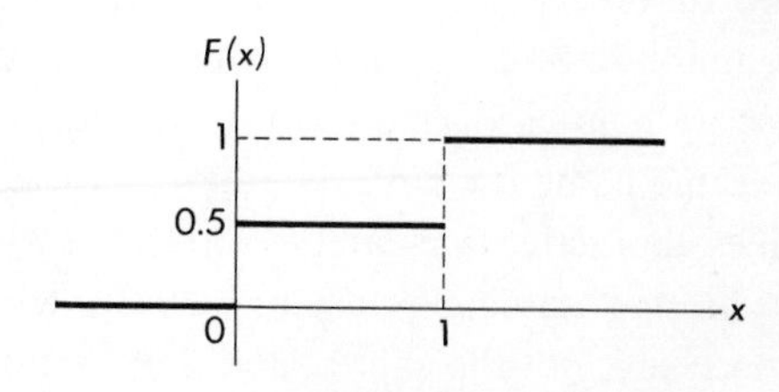

Figure 2–2

An assignment of probability to events can be thought of as a distribution, in the sense that the total probability 1 is distributed or spread out among the points and regions of the sample space, according to relative likelihood of occurrence. A distribution function describes how this probability is distributed by a random variable in the space $\mathscr{R}$ of its values. In particular, it gives the amount assigned to the interval from $-\infty$ up to λ as a function of λ; this amount can be thought of as the amount of probability that would be accumulated in "sweeping up" probability on the axis from the extreme left up to the point λ—if one may take the liberty of sweeping a thing as intangible as probability. The distribution function of a random variable is often called the *cumulative distribution function* or c.d.f.

A c.d.f. not only describes a distribution in terms of accumulated amounts from the left, but also serves as a basis for computing the probability of other events of interest in $\mathscr{R}$. Thus, the probability of the half-open interval $a < x \leq b$ or $(a, b]$ can be computed in terms of $F(\lambda)$ by first applying the addition law to the decomposition

$$(-\infty, b] = (-\infty, a] + (a, b],$$

to obtain

$$F(b) = F(a) + P(a, b],$$

and then transposing for the desired result:

$$P(a < X \leq b) = F(b) - F(a).$$

That is, the probability of an interval open on the left and closed on the right is the difference in the values of $F(\lambda)$ at the two endpoints; or in other words, it is the amount of increase in the function $F(\lambda)$ over the interval.

The probability of a single value, $X = a$, is obtained by expressing the set consisting of that single value as the intersection of a decreasing family $I_1, I_2, \ldots,$ where $I_n = \{x \mid a - 1/n < x \leq a + 1/n\}$. For then

$$\begin{aligned} P(X = a) = P\left(\bigcap I_n\right) &= \lim_{n\to\infty} P(I_n) \\ &= \lim_{n\to\infty}\left[F\left(a + \frac{1}{n}\right) - F\left(a - \frac{1}{n}\right)\right] \\ &= F(a+) - F(a-), \end{aligned}$$

the difference between the limits from the right and left of $F(\lambda)$ at the point $\lambda = a$. This difference is called the *jump* in the function $F(\lambda)$ at $\lambda = a$; if the function $F(\lambda)$ happens to be continuous at $\lambda = a$, the amount of the jump and hence the probability that $X = a$ is 0.

Intervals that are closed, or open, or open on the right and closed on the left can be expressed in terms of single points and intervals that are closed on the right and open on the left:

$$\begin{aligned}(a, b) + \{b\} &= (a, b],\\ [a, b] &= \{a\} + (a, b],\\ [a, b) + \{b\} &= [a, b],\end{aligned}$$

from which

$$\begin{aligned}P(a < X < b) &= F(b) - F(a) - P(X = b),\\ P(a \leq X \leq b) &= F(b) - F(a) + P(X = a),\\ P(a \leq X < b) &= F(b) - F(a) + P(X = a) - P(X = b).\end{aligned}$$

Sets of numbers more complicated than finite combinations of semi-infinite intervals and single points are seldom encountered in practice.

To use the distribution function in such computations, it is necessary to be able to determine its value at a given point. Sometimes, a distribution function can be expressed using simple algebraic or other "elementary" functions, but in many instances it is not so expressible and cannot be evaluated using the familiar tables of logarithms, exponential functions, trigonometric functions, or powers and roots. New tables are then required and have been constructed for many useful distribution functions of "nonelementary" type.

A tabulation of values of a function is often given in terms of a set of conveniently spaced (usually, evenly spaced) values of the independent variable, interpolation being required if a value is needed between tabulated entries. The values of the independent variable used for tabulation are ordinarily taken sufficiently close together for a linear interpolation to be adequate, even though the graph of the tabulated function is not linear.

Another method of tabulation commonly employed for distribution functions is to give values of the independent variable corresponding to equally spaced values of the function. The range of values of a distribution function is [0, 1], and it is this interval that is divided into equal parts. If a distribution function rises steadily from 0 to 1, with no jumps or intervals of constancy, there is a unique number x_p for each p on the interval [0, 1] such that

$$F(x_p) = P(X \leq x_p) = p.$$

The number x_p, a number on the set of values of the random variable X, is that value such that the fraction p of the total probability 1 is assigned to the interval $(-\infty, x_p)$. The number x_p is called a *fractile* or a *quantile* of the distribution. If p is such that $100p = k$, then x_p is called the kth *percentile* of the distribution. Thus, for instance, the probability that X assumes a value less than its 70th percentile is .70. This is illustrated in Figure 2–3. Values x_p are called *deciles*

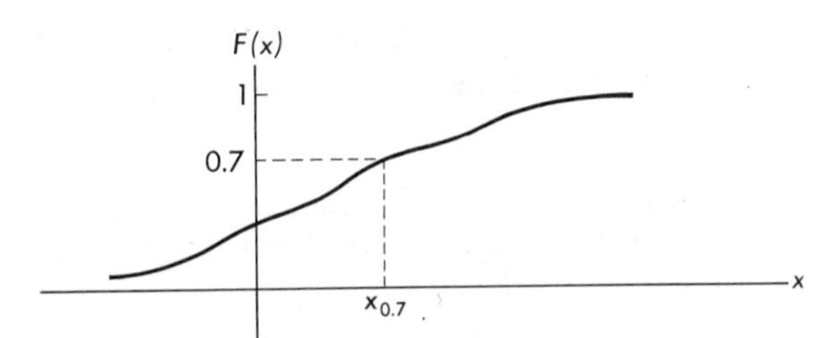

Figure 2–3. Determination of the 70th percentile.

if $10p$ is an integer, and quartiles if $4p$ is an integer. Thus, the first quartile is the same as the 25th percentile, and the third decile is the 30th percentile.

If a distribution function does not increase steadily but has horizontal sections or jumps in its graph, this definition of a percentile is inadequate. But it can be patched in a way that is illustrated for the case of the 50th percentile (or 5th decile or 2nd quartile), called the *median* m of the distribution, as follows:

$$P(X \leq m) \geq \tfrac{1}{2} \quad \text{and} \quad P(X \geq m) \geq \tfrac{1}{2}.$$

That is, the median is any number m satisfying both of these conditions—it is not necessarily unique. Three possible situations are illustrated in Figure 2–4.

★2.1.2 Measurability

A function $X(\omega)$ defined on a sample space is called *measurable* with respect to a Borel field $\mathscr{F}$ if and only if the event $\{\omega \mid X(\omega) \leq \lambda\}$ is in $\mathscr{F}$ for every real λ. The significance of this condition is that when probabilities have been assigned to events in Ω, the event $X(\omega) \leq \lambda$ will *have* a probability, because it is one of the sets in $\mathscr{F}$. This is an important consideration when Ω is not discrete, since there may be sets in Ω that do not have probabilities assigned to them.

When the sample space is discrete, however, there would seldom be any reason to use a Borel field smaller than the field of all subsets of the sample space; and in such a case *every* function on the sample space is automatically measurable.

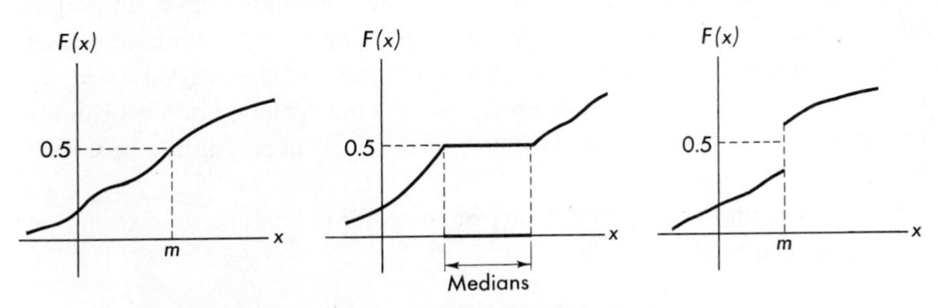

Figure 2–4. Determination of medians in three cases.

In the following example, a nonmeasurable function is constructed on a discrete sample space by using a Borel field smaller than the field of all subsets.

EXAMPLE 2–6. Consider a sample space containing the five elements a, b, c, d, e, and the Borel field generated by the partition of the sample space given by the function $X(\omega)$ defined as follows:

$$X(a) = X(b) = 0, \qquad X(c) = X(d) = 1, \qquad X(e) = 2.$$

Let the probabilities of the events $\{a, b\}$ and $\{e\}$ be $\frac{1}{3}$ and $\frac{1}{6}$, respectively; from these one can easily compute the probabilities of the remaining events of the Borel field:

$$P(\{c, d, e\}) = \tfrac{2}{3}, \qquad P(\{c, d\}) = \tfrac{1}{2}, \qquad P(\{a, b, e\}) = \tfrac{1}{2},$$
$$P(\{a, b, c, d\}) = \tfrac{5}{6}, \qquad P(\{a, b, c, d, e\}) = 1, \qquad P(\varnothing) = 0.$$

With this probability structure on the sample space Ω, the function $X(\omega)$ given above defines a random variable, because $X(\omega)$ is measurable with respect to the given Borel field, as can be readily verified. [For instance, for any λ on the range $0 < \lambda < 1$, the event defined by $X(\omega) \leq \lambda$ is just the event $\{a, b\}$, which is a member of the Borel field.] On the other hand, the function $Y(\omega)$, defined as follows:

$$Y(\omega) = \begin{cases} 0, & \text{for } \omega = a, \\ 1, & \text{for } \omega = b, c, d, \text{ or } e, \end{cases}$$

is *not* measurable. It is the "indicator function" of a set that is not a member of the given Borel field; and the event $Y(\omega) \leq \frac{1}{2}$ (to give one instance, which is enough) is simply $\{a\}$, not a member of the Borel field. As mentioned above, however, if the set of all subsets of $\{a, b, c, d, e\}$ were used as the basic Borel field, then every function on this probability space would be a random variable, including, in particular, the function $Y(\omega)$ just defined.

As mentioned earlier, a random variable $X(\omega)$ provides a natural way of defining probabilities for sets of real numbers. Thus, if $A \subset \mathscr{R}$, one would want

$$P^{\mathscr{R}}(A) = P^{\Omega}(X(\omega) \in A),$$

the superscript serving to emphasize the space in which the event in parentheses lies. (The symbol $\in$ means "is a member of.") Such definition is given a rigorous basis by the assumed measurability of $X(\omega)$, so long as A is a Borel set—a set in the Borel field generated by the collection of semi-infinite intervals $(-\infty, x]$. Indeed, the assumption of measurability permits the definition of a probability space $(\mathscr{R}, \mathscr{B}, P^{\mathscr{R}})$, in which $\mathscr{B}$ is the family of Borel sets of $\mathscr{R}$. This can be seen as follows.

It is convenient to have a temporary notation for the pre-image in Ω of a set in $\mathscr{R}$ as defined by the function $X(\omega)$:

$$E_A = \{\omega \mid X(\omega) \in A\}.$$

Now, when $X(\omega)$ is measurable with respect to $\mathscr{F}$, every Borel set in $\mathscr{R}$ has a pre-image in $\mathscr{F}$. For, pre-images of the complements and unions of sets in $\mathscr{R}$ are the complements and unions (respectively) of the pre-images of these sets:

$$E_{A^c} = \{\omega \mid X(\omega) \text{ in } A^c\} = \{\omega \mid X(\omega) \text{ in } A\}^c = (E_A)^c$$

and

$$E_{\cup A_i} = \left\{\omega \mid X(\omega) \text{ in} \bigcup A_i\right\} = \bigcup \{\omega \mid X(\omega) \text{ in } A_i\} = \bigcup E_{A_i}.$$

Therefore, the construction of a set B in $\mathscr{R}$ by using countable operations of union and complementation has a parallel construction (using these same operations on pre-images) that results in the pre-image of B.

Probability is defined in $\mathscr{R}$, on the events of the Borel field generated by intervals of the form $(-\infty, \lambda]$, as follows:

$$P^{\mathscr{R}}(A) = P^{\Omega}(E_A) = P^{\Omega}(\{\omega \mid X(\omega) \text{ in } A\}),$$

and this probability will usually be denoted simply by $P(X \text{ in } A)$ or $P(A)$. Moreover, if $\mathscr{B}$ denotes the field of Borel sets in $\mathscr{R}$ then $(\mathscr{R}, \mathscr{B}, P^{\mathscr{R}})$ constitutes a probability space. That is, $P^{\mathscr{R}}(A)$ defines a probability measure in the set of real numbers—it is an assignment of probability that satisfies the axioms for a probability space. For, surely,

$$P(X \text{ in } \mathscr{R}) = P(-\infty < X(\omega) < \infty) = P(\Omega) = 1,$$

and if $A_1, A_2, \ldots$ is a sequence of disjoint events in $\mathscr{R}$, then their pre-images are disjoint and

$$P\left(\bigcup A_i\right) = P(E_{\cup A_i}) = P\left(\bigcup E_{A_i}\right) = \sum P(E_{A_i}) = \sum P(A_i).$$

These ideas can be extended to the case of a random vector, the model for studying the "joint" behavior of several random variables. The value space of a vector function with n components is $\mathscr{R}^n$, the space of ordered n-tuples of real numbers. Borel sets in this space are defined as sets in the Borel field generated by sets of the form

$$\{(x_1, \ldots, x_n) \mid x_1 \le \lambda_1, \ldots, x_n \le \lambda_n\},$$

which is a generalization to $\mathscr{R}^n$ of the semi-infinite interval used to obtain Borel sets in $\mathscr{R}^1 = \mathscr{R}$.

Consider now a vector function $[X_1(\omega), \ldots, X_n(\omega)]$ defined on the points ω of a probability space $(\Omega, \mathscr{F}, P)$. In order that Borel sets in $\mathscr{R}^n$ have probabilities induced by this vector function, their pre-images must be in $\mathscr{F}$, the family of sets in Ω that have probabilities assigned to them. They will be in $\mathscr{F}$ if the pre-images of the semi-infinite interval sets, namely,

$$\{\omega \mid X_1(\omega) \le \lambda_1, \ldots, X_n(\omega) \le \lambda_n\},$$

are in $\mathscr{F}$ for every n-tuple of real numbers $(\lambda_1, \ldots, \lambda_n)$. And this in turn will follow if the components $X_i(\omega)$ are random variables—each measurable with respect to $\mathscr{F}$, since the above semi-infinite interval sets are just intersections of events $X_i(\omega) \leq \lambda_i$. Thus, one is led to the following definition.

Definition. *A* random vector *is a vector-valued function on a probability space whose components are random variables.*

The class of available models for vector-valued random phenomena is restricted somewhat by this definition, but as for $n = 1$ the restriction is of little or no practical importance.

As for a random variable on $\mathscr{R}$, a random vector on $\mathscr{R}^n$ induces a probability measure P for the class $\mathscr{B}$ of Borel sets in $\mathscr{R}^n$ such that $(\mathscr{R}^n, \mathscr{B}, P)$ is a probability space.

Problems

2–1. Four coins are tossed. List the 16 outcomes in the sample space that keeps track of the individual coins. For each outcome ω compute $X(\omega)$, the number of heads. Determine the probability assigned to each value of $X(\omega)$ by the assumption that the 16 outcomes are equally likely.

2–2. A dart is thrown at a circular target of radius 1 ft in such a way that the probability of hitting a certain portion of the target is proportional to the area of that region. The dart misses completely one-half of the time. Let a score X be assigned equal to the number of inches from the circumference of the target, except that the score 0 is assigned if the dart misses entirely. Determine

(a) $P(X = 0)$. (b) $P(X > 9)$.
(c) $P(X = 2)$. (d) $P(X < 5)$.
(e) $P(3 < X < 9)$.

2–3. For an infinite sequence of coin tosses let

$$U(\omega) = \text{number of trials needed to get the first heads.}$$
$$V(\omega) = \text{number of trials needed to get two heads.}$$

Determine the following:

(a) $P(U = 2)$. (b) $P(V = 3)$.
(c) $P(U = 4 \text{ and } V = 6)$. (d) $P(U \leq 3 \text{ and } V = 5)$.

2–4. Table I in the Appendix gives values of a certain distribution function, $\Phi(z)$. Plot a sufficient number of points to obtain a smooth graph, and estimate the deciles of the distribution from the graph. (These can be checked with the entries in Table I*a*.)

2–5. Given that a random variable Z has the distribution given in Table I (see Appendix), compute the following:

(a) $\Phi(1.4)$. (b) $\Phi(-2.33)$.

(c) $\Phi(-.436)$. (d) $P(-.3 < Z < 1.4)$.
(e) $P(Z > 2)$. (f) $P(|Z| > 3)$.
(g) $P(|Z - 1| < 2)$. (h) $P(Z^2 < 2)$.

2–6. Toss a fair coin to determine whether to spin a pointer with reward on the scale 0 to 1 (the "uniform" pointer of Example 2–4), or to toss the coin again with reward 0 for tails and 1 for heads. (Assume independent experiments.) Let X denote the reward and determine the distribution function of X.

2–7. A random variable X has the distribution function shown in Figure 2–5. Determine the following:

(a) $P(X = \frac{1}{2})$. (b) $P(X = 1)$.
(c) $P(X < 1)$. (d) $P(X \leq 1)$.
(e) $P(X > 2)$. (f) $P(\frac{1}{2} < X < \frac{5}{2})$.

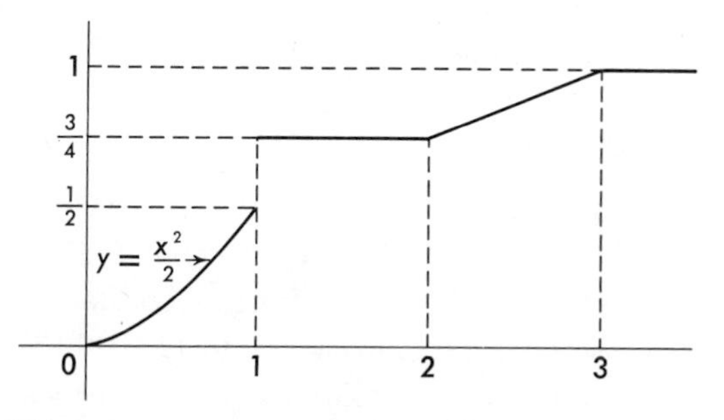

Figure 2–5. Distribution function for Problem 2-7.

2–8. A random variable X has the following distribution function:

$$F(\lambda) = \begin{cases} 1 - .75e^{-\lambda}, & \text{if } \lambda \geq 0, \\ 0, & \text{if } \lambda < 0. \end{cases}$$

Determine the following:

(a) $P(X > 2)$. (b) $P(X \leq 0)$. (c) $P(X = 0)$.

2–9. Determine the probability that a random variable exceeds its 90th percentile, given that it exceeds its 70th percentile. (Assume a steadily increasing c.d.f.)

2–10. Consider the random variables $\theta(\omega)$ and $\phi(\omega)$, as defined in Example 1–38 and earlier examples, the colatitude and longitude angles of a point on the upper hemisphere, with a uniform distribution of probability over this sample space. Determine the distribution functions $F_\theta(\lambda)$ and $F_\phi(\lambda)$.

2–11. Each horizontal line in Table II (Appendix) gives percentiles of a member of a certain class of distribution functions. Plot the distribution function for "5 degrees of freedom" (that is, using the fifth line in the table). From the graph, estimate the following (in which X denotes a random variable with this distribution):

(a) The 40th percentile. (b) $P(X > 8)$.
(c) $P(|X - 5| < 2)$. (d) $P(X < 3)$.

2–12. Each horizontal line in Table III (Appendix) gives certain percentiles of a random variable. Let T have the distribution given opposite 10 "degrees of freedom," where the earlier percentiles are obtained using the fact that this distribution is symmetric about the value 0. [That is, the probability assigned to an interval of negative numbers, $(-a, -b)$ is the same as that assigned to the interval (b, a).] Sketch the distribution function of T and determine the following:

(a) The median of the distribution.
(b) $P(|T| > 3)$.
(c) The 98th percentile of T.
(d) $P(|T - 1| < 1.7)$.

★**2–13.** Consider a sample space with four outcomes: a, b, c, d.

(a) Show that the function $X(\omega)$ defined by

$$X(a) = X(b) = 0, \qquad X(c) = X(d) = 1$$

is measurable with respect to the Borel field generated by the events $\{a, b\}$ and $\{c\}$.

(b) Construct a function $Y(\omega)$ *not* measurable with respect to the Borel field in (a).

2.1.3 Discrete Random Variables

Let $x_1, x_2, x_3, \ldots$ be a sequence of distinct numbers such that there are only finitely many in any finite interval. A random variable X whose distribution function jumps at these values $x_1, x_2, \ldots$, and is constant between adjacent jump points is called *discrete*. As seen in the Section 2.1.1 the amount of a jump in a distribution function is the probability assigned to the value x_i at which it occurs. That is,

$$p_i \equiv [\text{jump at } x_i] = P(X = x_i).$$

And then

$$p_1 + p_2 + \cdots = 1.$$

Thus, the distribution of probability in the space of values of X defines a discrete probability space.

The distribution of a discrete random variable is characterized by the function (or sequence) p_i, sometimes denoted by $p(x_i)$ or $f(x_i)$ and called the probability function of the random variable. In some cases it can be given by a "formula," algebraic or otherwise, but sometimes it is simplest just to make a list of the values and corresponding probabilities in a table:

Values	x_1	x_2	$\ldots$
Probabilities	p_1	p_2	$\ldots$

If the original probability space Ω on which $X(\omega)$ is defined is discrete, then any function $X(\omega)$ defines a *discrete* distribution in the value space, since $X(\omega)$ cannot assume more values than there are ω's. On the other hand, a continuous distribution of probability in Ω can lead to a discrete distribution in the value space of $X(\omega)$, as will be evident in some of the examples to follow.

EXAMPLE 2–7. Three chips are drawn together at random from a bowl containing five chips numbered 1, 2, 3, 4, and 5. There are 10 possible outcomes in this experiment:

$$123, \quad 124, \quad 125, \quad 134, \quad 135, \quad 145, \quad 234, \quad 235, \quad 245, \quad 345,$$

each having probability $\frac{1}{10}$. Let $X(\omega)$ denote the *sum* of the numbers on the chips in outcome ω. The values of X corresponding to the above list of outcomes are, respectively,

$$6, \quad 7, \quad 8, \quad 8, \quad 9, \quad 10, \quad 9, \quad 10, \quad 11, \quad 12.$$

The possible values of $X(\omega)$ and the corresponding probabilities for those values are then as follows:

x_i	6	7	8	9	10	11	12
p_i	.1	.1	.2	.2	.2	.1	.1

The distribution function is that shown in Figure 2–6.

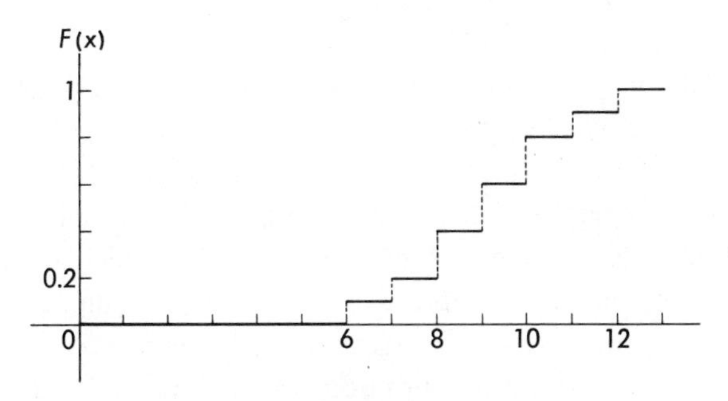

Figure 2–6

EXAMPLE 2–8. In packaging a certain product, it is desirable that at least 8 oz be included in each package. For certain purposes, then, one might be interested only in whether a package contained less than 8 oz or at least 8 oz. Thinking of the possible weights of packages as the points ω in the sample space, introduce the random variable

$$X(\omega) = \begin{cases} 1, & \text{if } \omega < 8 \text{ oz}, \\ 0, & \text{if } \omega \geq 8 \text{ oz}. \end{cases}$$

That is, 1 corresponds to a "defective" package and 0 to a "nondefective" one. The random variable X can take on just the values 0 and 1, and the corresponding probabilities are determined by the distribution of probability in the sample space:

$$p(0) = P^{\Omega}(\omega \geq 8) = P(\text{package weighs 8 oz or more}),$$
$$p(1) = P^{\Omega}(\omega < 8) = P(\text{package weighs less than 8 oz}).$$

EXAMPLE 2–9. The lengths of 10 ft sections of pipe are measured in inches to the nearest inch. Here, even though the length might be thought of as assuming any of a continuum of possible values, the round-off process introduces discreteness into the set of values. The only possible values of the *measured* lengths are

$$\ldots, \quad 118, \quad 119, \quad 120, \quad 121, \quad 122, \quad \ldots,$$

and each value in this list has a corresponding probability—the probability that, in measuring, a length is rounded off to that value. More precise measurements of length would lead to possible values that are more numerous but nonetheless discrete (even should the precision be carried to, say, 10 decimal places). Of course, a discrete model involving lengths to 10 decimal places would be more complicated to deal with than the idealization to a continuous model. In any case, however, the data must be recorded discretely.

2.1.4 Continuous Random Variables

A distribution function that is *continuous*, having no jumps, will define a distribution in the value space of $X(\omega)$ that assigns probability 0 to any single value:

$$P(X = b) = \lim_{n \to \infty} \left[F\left(b + \frac{1}{n}\right) - F\left(b - \frac{1}{n}\right) \right] = 0.$$

However, the class of continuous distribution functions includes many that are not useful in probability theory—those that are nondifferentiable over a large set of points. Models involving a limited number of points of nondifferentiability are useful, as for example the distribution of Example 2–4 (and Figure 2–1). Distribution functions that provide useful models are those which are differentiable everywhere, with the possible exception of at most a finite number of points in any finite interval. In such a case $X(\omega)$ is said to be a *continuous random variable.*[2]

The distribution function of a continuous random variable will then have a derivative except at isolated points, and this derivative is usually denoted by $f(\lambda)$:

$$f(\lambda) = F'(\lambda).$$

Again notice that the dummy variable λ, used here in defining the function $f(\cdot)$, is completely immaterial. That is, $f(x)$, $f(\lambda)$, $f(R)$, and so forth, define the same function.

The derivative function $f(\lambda)$ is called the *density function* of the distribution or of the random variable defining the distribution. If more than one random variable is under consideration, a subscript will be used to distinguish between their density functions, for example, $f_X(\lambda)$. The term *density* is appropriate in accordance with the discussion in Section 1.2.5, as applied to the sample space $\mathscr{R}$, the

[2] The class of distributions included under the heading *continuous* can be (and often is) defined to be just a bit wider than the one given here, namely, as the class of distributions defined by distribution functions that have the property of being recoverable from their derivatives by integration:

$$\int_{-\infty}^{x} F'(\lambda)\, d\lambda = F(x).$$

The functions satisfying the differentiability condition given above do satisfy this condition, but constitute a proper subclass.

set of real numbers. That is, at a point x_0 where the derivative of the distribution function exists,

$$F'(x_0) = f(x_0) = \lim_{h\to 0} \frac{1}{h}\left[F\left(x_0 + \frac{h}{2}\right) - F\left(x_0 - \frac{h}{2}\right)\right]$$
$$= \lim_{h\to 0} \frac{1}{h} P\left(x_0 - \frac{h}{2} < X \le x_0 + \frac{h}{2}\right).$$

This is the limiting value of the ratio of the probability in a small interval about x_0 to the length of the interval; in other words, the density of probability at that point.

EXAMPLE 2–10. In Problem 2–10 the distribution functions of the colatitude angle $\theta(\omega)$ and the longitude angle $\phi(\omega)$ of the random orientation ω were

$$F_\theta(\lambda) = \begin{cases} 0, & \text{for } \lambda < 0, \\ 1 - \cos\lambda, & \text{for } 0 \le \lambda \le \frac{\pi}{2}, \\ 1, & \text{for } \lambda > \frac{\pi}{2} \end{cases}$$

and

$$F_\phi(\lambda) = \begin{cases} 0, & \text{for } \lambda < 0, \\ \frac{\lambda}{2\pi}, & \text{for } 0 \le \lambda \le 2\pi, \\ 1, & \text{for } \lambda > 2\pi. \end{cases}$$

The corresponding density functions are then the derivatives of these (Figure 2–7):

$$f_\theta(\lambda) = \frac{d}{d\lambda}(1 - \cos\lambda) = \sin\lambda, \qquad \text{for } 0 < \lambda < \frac{\pi}{2},$$

$$f_\phi(\lambda) = \frac{d}{d\lambda}\left(\frac{\lambda}{2\pi}\right) = \frac{1}{2\pi}, \qquad \text{for } 0 < \lambda < 2\pi.$$

(The densities are 0 outside the given λ-intervals.)

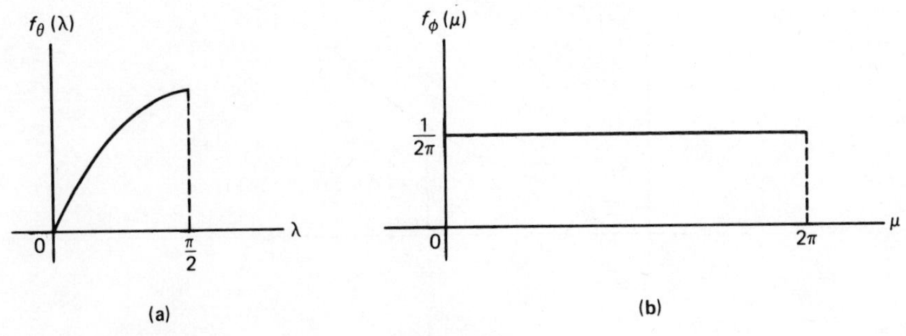

Figure 2–7

A function satisfying the continuity and differentiability conditions imposed here on a distribution function is the integral of its derivative:

$$F(x) = \int_{-\infty}^{x} f(\lambda)\, d\lambda,$$

the lower limit being chosen to satisfy the obvious condition

$$F(-\infty) = P(X \leq -\infty) = 0.$$

From this it follows that

$$P(a < X < b) = F(b) - F(a) = \int_{a}^{b} f(\lambda)\, d\lambda.$$

Notice that whether one or the other endpoint is included is immaterial, since the probability assigned to these individual points is 0 in this continuous case.

A definite integral can be interpreted geometrically as an *area*, namely, the area under the graph of the integrand function that lies above the horizontal axis and between abscissas at the limits of integration. This interpretation can be

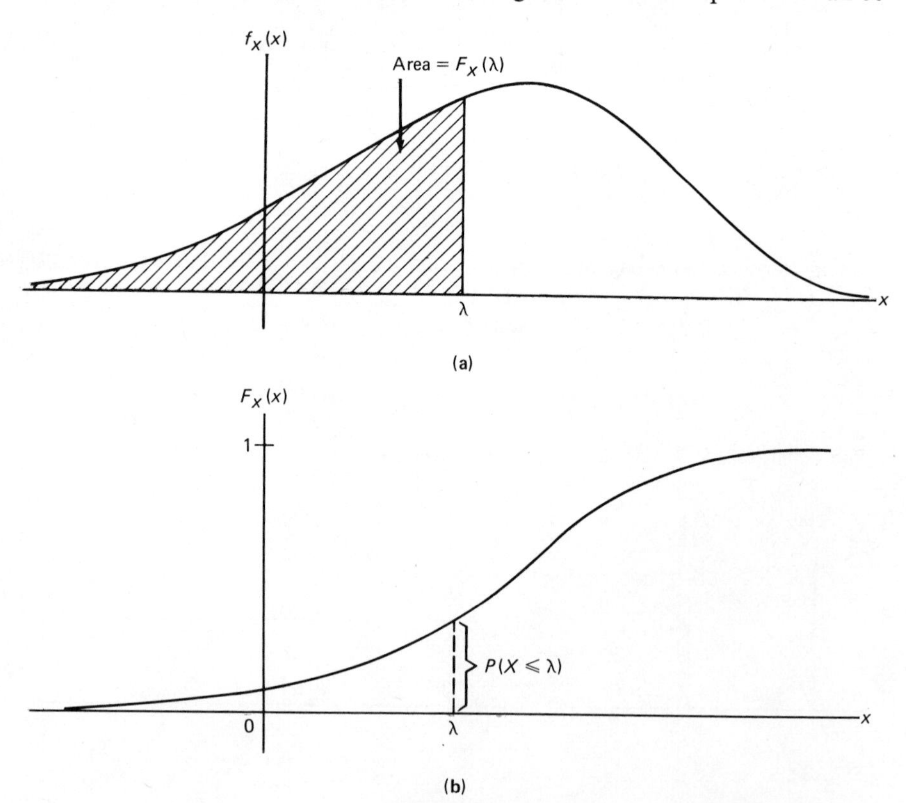

Figure 2–8

applied to probabilities of regions of values of a continuous random variable because such probabilities are definite integrals. As shown in Figure 2–8,

$$F(x) = P(X \leq x) = \int_{-\infty}^{x} f(u)\,du$$
$$= \text{area under the density } f(\cdot) \text{ to the left of the point } x,$$

And again,

$$P(a < X < b) = F(b) - F(a)$$
$$= \int_{a}^{b} f(x)\,dx$$
$$= \text{area under } f(\cdot) \text{ between } x = a \text{ and } x = b,$$

illustrated in Figure 2–9. Both figures show not only the area representation, but also the representation of the probability of an interval as an increment in height on the graph of the c.d.f.

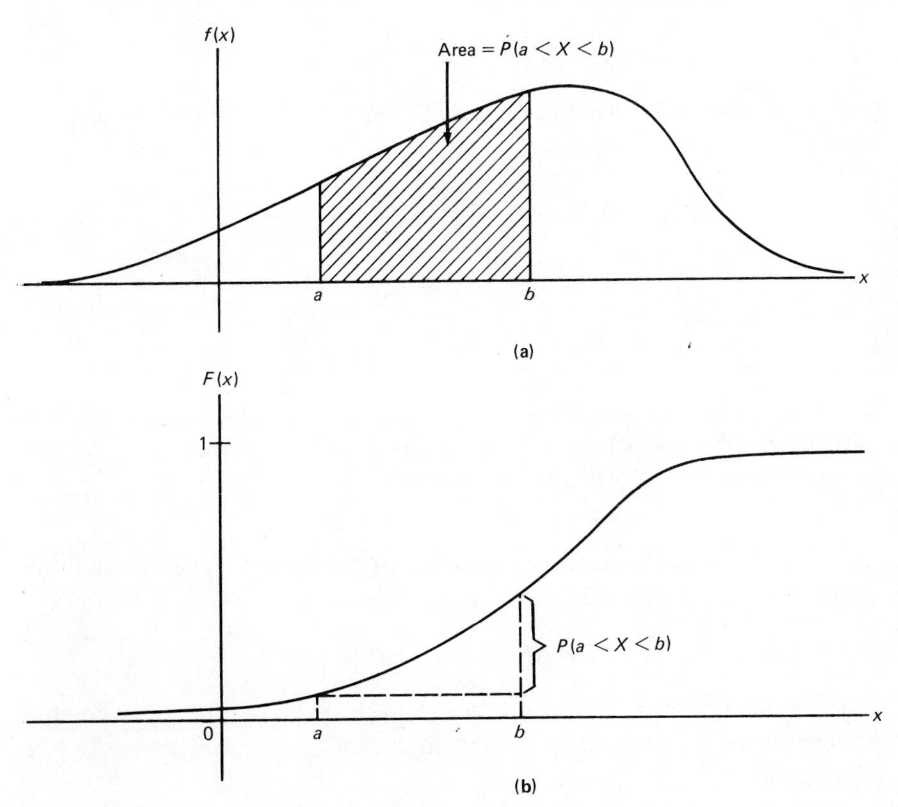

Figure 2–9

A particular case of the calculation of the probability of an interval is the following:

$$1 = P(\mathscr{R}) = P(-\infty < X < \infty) = F(\infty) = \int_{-\infty}^{\infty} f(x)\, dx.$$

That is, the total area under the graph of the density function (and above the x-axis) is 1.

EXAMPLE 2–11. Suppose that a random variable X has the density function

$$f(\lambda) = \sin \lambda, \qquad \text{for } 0 < \lambda < \frac{\pi}{2}$$

as in Example 2–10. The probability of an interval, such as $(\pi/4, \pi/3)$, can be calculated as an integral:

$$\begin{aligned} P\left(\frac{\pi}{4} < X < \frac{\pi}{3}\right) &= \int_{\pi/4}^{\pi/3} \sin \lambda \, d\lambda \\ &= -\cos\frac{\pi}{3} - \left(-\cos\frac{\pi}{4}\right) = .207. \end{aligned}$$

In this case, where $F(x)$ is known (see again Example 2–10), the same probability can be obtained as an increment in F:

$$\begin{aligned} P\left(\frac{\pi}{4} < X < \frac{\pi}{3}\right) &= F\left(\frac{\pi}{3}\right) - F\left(\frac{\pi}{4}\right) \\ &= \left(1 - \cos\frac{\pi}{3}\right) - \left(1 - \cos\frac{\pi}{4}\right) \\ &= -\cos\frac{\pi}{3} - \left(-\cos\frac{\pi}{4}\right). \end{aligned}$$

Notice that precisely the same computation is going on, and that since $F(\lambda)$ is an indefinite integral of $f(\lambda)$, it could have been used (rather than the $-\cos \lambda$ that was used before) in evaluating the definite integral.

A derivative is a *differential coefficient* corresponding to a small change in the independent variable, in the sense that the differential

$$dF(\lambda) = f(\lambda)\, d\lambda$$

is used as an approximation of the change in the value of $F(\lambda)$ corresponding to a change in the independent variable from λ to $\lambda + d\lambda$. This change in $F(\lambda)$ is a probability:

$$P(\lambda < X \leq \lambda + d\lambda) = F(\lambda + d\lambda) - F(\lambda) \doteq f(\lambda)\, d\lambda.$$

This approximation, based on approximating $F(\lambda)$ by the tangent to its graph at λ, can be interpreted in terms of area and the density function graph, as shown in Figure 2–10.

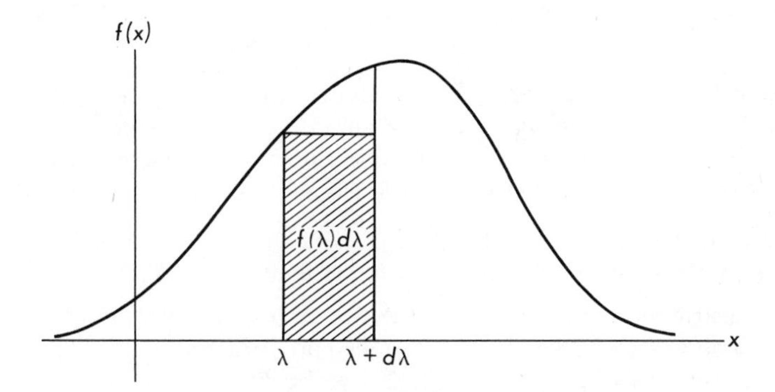

Figure 2–10. The probability element.

It is of interest to note the analogy between certain relations in the discrete case and those in the continuous case:

Discrete	Continuous
$f(x_i) = P(X = x_i)$	$f(x)\,dx \doteq P(x < X < x + dx)$
$F(x) = \sum_{x_i \leq x} f(x_i)$	$F(x) = \int_{-\infty}^{x} f(\lambda)\,d\lambda$
$P(A) = \sum_{x_i \text{ in } A} f(x_i)$	$P(A) = \int_A f(\lambda)\,d\lambda$
$\sum_{\text{all } i} f(x_i) = 1$	$\int_{-\infty}^{\infty} f(x)\,dx = 1$
$f(x_i) = F(x_i) - F(x_{i-1})$	$f(x) = F'(x)$

(In the last relation for the discrete case, it is assumed that the x_i's are arranged in numerical order—that x_{i-1} and x_i are adjacent values of X.)

Problems

2–14. Two articles among 10 articles in a lot are defective. Determine the possible values and the corresponding probabilities for the random variables X, the number of defectives in a random selection of 1 article from the lot, and Y, the number of defective articles in a random selection of 4.

2–15. Five electron tubes in a box are to be tested one at a time until the one defective tube among the five is located. Determine the possible values and the corresponding probabilities for the random variable defined to be the number of tests required. (The answer will depend on whether it is assumed that the tester knows that exactly one defective is present and can omit the fifth test, or is required to carry out the fifth test before he is considered to have "located" the defective.)

2–16. A tester is to identify three cigarettes of three different brands and is told to assign the brand name A to one, the name B to another, and the name C to the third. Suppose that he cannot really tell the difference by smoking the cigarettes and assigns the three names at random. Determine the possible values and corresponding probabilities for the random variable defined to be the number of correct identifications.

2–17. As in Example 2–7, three chips are drawn together at random from a bowl containing five chips numbered 1, 2, 3, 4, and 5. Let Y denote the smallest of the three numbers drawn and R the largest number drawn minus the smallest. Determine the distributions of probability for Y and R. Compute also $P(Y \leq 2)$ and $P(R > 2)$.

2–18. Determine the probability function $f(x_i)$ for the random variable X defined to be the sum of the numbers of points showing in the toss of two dice.

2–19. The distribution function for the value X on the scale [0, 1] at which the spinning pointer stops, when probability is assumed proportional to the angle subtended by a given event, was seen in Example 2–4 to be $F(x) = x$ for $0 < x < 1$, 0 for $x < 0$, and 1 for $x > 1$. Determine the density function of X. Determine also the density function of Y, the value at which the pointer stops when the scale from 0 to 2π is used.

2–20. A random variable X has the distribution function $F(x) = x^2$ for x on $[0, 1]$, 0 for x on $(-\infty, 0)$, and 1 for $x \geq 1$. First, observe that this distribution is of the continuous type and then determine the density function of the distribution.

2–21. A random variable X has the distribution function

$$F(x) = \frac{1}{\pi}\left(\frac{\pi}{2} + \text{Arctan } x\right),$$

where the value of Arctan x is taken to be that on the range from $-\pi/2$ to $\pi/2$. Determine the density function, the 3rd quartile of the distribution, and the probability that $|X| < 1$.

2–22. The random variable Y has density function

$$f(y) = 2e^{-2y}, \qquad \text{for } y > 0.$$

Determine the distribution function, and calculate the probability that $Y > 1$.

2–23. A random variable X has the density function

$$f(x) = \begin{cases} 1 - |x|, & \text{for } |x| < 1, \\ 0, & \text{elsewhere.} \end{cases}$$

(a) Determine and sketch the distribution function of X.
(b) Compute $P(X > 0)$ and $P(|X| > \frac{1}{2})$. Interpret these with respect to both the graph of the density function and graph of the distribution function.

2–24. Let X denote a random variable with density function

$$f(x) = \frac{e^{-x^2}}{\sqrt{\pi}}.$$

Calculate approximately: $P(-.1 < X < .1)$.

2–25. Show that if $f(a - x) = f(a + x)$, so that the graph of $f(x)$ is symmetrical about the ordinate $x = a$, then the median of the distribution with density function $f(x)$ is a.

2.1.5 Specification of a Distribution by Its c.d.f.

Given a probability space $(\Omega, \mathscr{F}, P^{\Omega})$ and a random variable $X(\omega)$ defined on it, the corresponding distribution function $F(x)$ is completely determined as

$$F(x) = P^{\Omega}(X(\omega) \leq x).$$

This function automatically has certain significant properties:

1. $F(-\infty) = 0$.
2. $F(\infty) = 1$.
3. $F(x)$ is nondecreasing in x.
4. $F(x)$ is continuous from the right at each x.

These are not hard to show from the axioms for probability. Proofs of (2) and (3) will be left to Problem 2–30. Given (3), Property (1) is seen as follows: A bounded monotonic function has a limit, which can be evaluated as the limit of a sequence:

$$\begin{aligned} F(-\infty) &= \lim_{n\to\infty} F(-n) = \lim_{n\to\infty} P(X \leq -n) \\ &= P(\bigcap \{X \leq -n\}) = P(\varnothing) = 0. \end{aligned}$$

Property (4) asserts that

$$\lim_{y\to x+} F(y) = F(x).$$

To see this, observe that the limit can be evaluated (because it exists) by taking a decreasing sequence $y_1, y_2, \ldots$ whose limit is x:

$$\begin{aligned} \lim_{n\to\infty} F(y_n) &= \lim_{n\to\infty} P(X \leq y_n) = P(\bigcap \{X \leq y_n\}) \\ &= P(X \leq x) = F(x). \end{aligned}$$

Now, the main interest in a random variable usually has to do with its value space $\mathscr{R}$ and the probability distribution in $\mathscr{R}$ induced by the random variable. [It is shown in Section 2.1.2 that when $X(\omega)$ is defined for ω in a probability space, then $(\mathscr{R}, \mathscr{B}, P^{\mathscr{R}})$ is itself a probability space, where $\mathscr{B}$ is the family of Borel sets and $P^{\mathscr{R}}$ is the induced probability measure in $\mathscr{R}$.] It is often desirable, therefore, in setting up a model for a random phenomenon, to be able to construct the distribution in $\mathscr{R}$ directly—and then to define $X(x) = x$ as the random variable of interest. Indeed, this can be done simply by specifying a distribution function, provided that it satisfies properties (1) through (4) just described.

A function $F(x)$ satisfying properties (1) through (4) can be used to assign probability to intervals:

$$P(a < X \leq b) \equiv F(b) - F(a),$$

and also to more complicated sets by expressing them in terms of combinations of intervals and operations on intervals. [It is shown in Cramér [5], pp. 53–54, that this assignment leads to a proper probability measure in $\mathscr{R}$, that is, one for which $(\mathscr{R}, \mathscr{B}, P^{\mathscr{R}})$ is a probability space.] The term *distribution function* will mean any function satisfying properties (1) through (4), whether it is the starting point in model construction or describes a distribution in $\mathscr{R}$ induced by a random variable on some sample space Ω.

When a model is specified by means of a distribution function, the random variable $X(x) = x$ will be of the continuous type if $F(x)$ is differentiable except possibly for a finite set of points in any finite interval, and will be discrete if $F(x)$ is a function that increases in jumps at a discrete set of x-values.

EXAMPLE 2–12. The function $F(x)$ defined as follows:

$$F(x) = \begin{cases} 0, & \text{for } x < 0, \\ 1 - e^{-x}, & \text{for } x \geq 0 \end{cases}$$

is a nondecreasing function, everywhere continuous (and, therefore, certainly continuous from the right at each x), satisfying the conditions that $F(-\infty) = 0$ and $F(\infty) = 1$. It can, therefore, be used as the basis of a probability model and defines a random variable X having the property that $P(X \leq x) = F(x)$. The variable X is continuous, $F(x)$ being differentiable except at $x = 0$ with derivative

$$f(x) = F'(x) = \begin{cases} 0, & \text{for } x < 0, \\ e^{-x}, & \text{for } x > 0. \end{cases}$$

To avoid the impression that random variables must be either discrete or continuous, the following example is given to illustrate a useful distribution model in which there is a discrete component and a continuous component.

EXAMPLE 2–13. A model describing the life of an electron tube in a certain type of use involves the following distribution function for this life:

$$F(x) = \begin{cases} 1 - R_0 e^{-kx}, & \text{for } x \geq 0, \\ 0, & \text{for } x < 0. \end{cases}$$

This is shown in Figure 2–10. It is evident that there is a jump in the amount $1 - R_0$ at $x = 0$, indicating a probability $1 - R_0$ that a tube taken from the shelf is bad to begin with, or a probability R_0 (possibly less than 1) that the tube is good. The quantity

$$R(x) = P(\text{tube life exceeds } x) = 1 - F(x) = R_0 e^{-kx}$$

is called the *reliability* of the tube in a mission of x time units, and R_0 might then be termed the *initial reliability*.

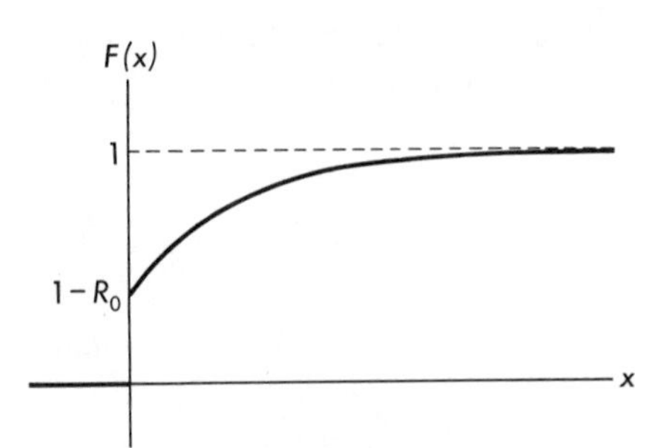

Figure 2–10. Distribution function for tube life in Example 2-13.

The probability model for a continuous distribution can be defined by specifying a density function. When obtained as the derivative of a distribution function defined by a given distribution in $\mathscr{R}$, the density function has these properties:

(i) $f(x) \geq 0$,

(ii) $\int_{-\infty}^{\infty} f(x)\, dx = 1$.

The first of these follows from the fact that the derivative of a nondecreasing function, when it exists, is nonnegative. On the other hand, any function satisfying these conditions and whose integral over the arbitrary interval $(-\infty, x]$ exists can be used to construct a distribution function $F(x)$ as follows:

$$F(x) = \int_{-\infty}^{x} f(u)\, du.$$

Because this satisfies conditions (1) through (4) on p. 71 for distribution functions, it can then be used to construct the distribution in $\mathscr{R}$.

EXAMPLE 2–14. The function $(1 + x^2)^{-1}$ is continuous, nonnegative, and integrable over any interval. The total area under its graph is

$$\int_{-\infty}^{\infty} \frac{dx}{1 + x^2} = \text{Arctan}\, \infty - \text{Arctan}\,(-\infty) = \pi.$$

Hence the function

$$f(x) = \frac{1/\pi}{1 + x^2}$$

can be used as the density of a probability distribution on $(-\infty, \infty)$, the area under its graph being 1. The corresponding distribution function is

$$F(x) = \int_{-\infty}^{x} \frac{1/\pi}{1 + u^2}\, du = \frac{1}{\pi}\left[\text{Arctan}\, x + \frac{\pi}{2}\right],$$

which, of course, increases steadily from 0 at $-\infty$ to 1 at ∞.

A *discrete* distribution in $\mathscr{R}$ can be specified by a sequence of possible values $x_1, x_2, \ldots$ and an assignment of probabilities $p_i = P(\{x_i\})$ having the properties

1. $p_i \geq 0$ for all i.
2. $\sum_i p_i = 1$.

The function $F(x)$ defined from these p_i's as the cumulative sum:

$$F(x) = \sum_{x_i \leq x} p_i,$$

again satisfies the conditions for a distribution function. It is constant between successive values in the ordered list of possible values and jumps an amount p_i at x_i.

EXAMPLE 2–15. The formula

$$\tfrac{1}{2} + (\tfrac{1}{2})^2 + (\tfrac{1}{2})^3 + \cdots = 1$$

follows from the general formula for the sum of a geometric series. Because each term in the sum is nonnegative and their sum is 1, the individual terms can be used as the basis of a probability distribution on any sequence of real numbers.

2.1.6 Functions of Random Variables

A function $g(x)$ defined on the value space of a random variable $X(\omega)$ defines a composite function on Ω:

$$Y(\omega) = g(X(\omega))$$

and so defines a new random variable. The simpler notation $Y = g(X)$ will often be used, the reference to the sample space being dropped but understood. (The question of measurability of Y will be taken up in Section 2.2.5.) The distribution function of Y can be computed from the basic distribution in Ω or from the induced distribution in $\mathscr{R}$:

$$F_Y(y) = P^{\mathscr{R}}(g(X) \leq y) = P^{\Omega}[g(X(\omega)) \leq y].$$

If a model is constructed directly in $\mathscr{R}$ by specification of a c.d.f. $F(x)$ for the random variable $X(x) = x$, then there is no other Ω. The function $g(x)$ is a random variable on the probability space $\mathscr{R}$.

EXAMPLE 2–16. A discrete random variable X has values and probabilities as given in the following table:

x_i	-1	0	1
p_i	.3	.4	.3

The function $g(x) = x^2 - 1$ defines a random variable $Y = X^2 - 1$ whose values are 0 and -1. The value 0 comes from $X = 1$ and also from $X = -1$, and the value -1 from $X = 0$. The probability table for the distribution of Y is then:

y_j	-1	0
$P(Y = y_j)$	.4	.6

EXAMPLE 2–17. Let X have the probability distribution defined by

$$F_X(x) = \begin{cases} 1 - e^{-x}, & \text{for } x \geq 0, \\ 0, & \text{for } x < 0, \end{cases}$$

as in Example 2–12. The function $g(x) = \sqrt{x}$ defines a random variable $Y = \sqrt{X}$, since X is nonnegative with probability 1. [This $g(x)$ does not map negative x's into $\mathscr{R}$, but would only be applied to nonnegative x's. It could be defined arbitrarily for negative x's; and because $P(X < 0) = 0$, this would in no way affect the distribution of Y.] The distribution function of Y is computed as follows:

$$F_Y(y) = P(Y \leq y) = P(\sqrt{X} \leq y) = P(X \leq y^2) = F_X(y^2) = 1 - \exp(-y^2),$$

a computation that is valid for any nonnegative y. If y is a negative number, then $P(Y \leq y) = 0$. The density function of Y can be obtained by differentiation:

$$f_Y(y) = \begin{cases} 2y \exp(-y^2), & \text{for } y > 0, \\ 0, & \text{for } y < 0. \end{cases}$$

EXAMPLE 2–18. Consider the *linear* function $Y = aX + b$, where X is any random variable, and a and b are any constants with $a \neq 0$. Assume first that $a > 0$. Then

$$F_Y(y) = P(aX + b \leq y) = P\left(X \leq \frac{y - b}{a}\right) = F_X\left(\frac{y - b}{a}\right).$$

If X is a continuous random variable, that is, if its distribution function is continuous at every point and differentiable, except perhaps for at most a finite number of points in any finite interval, then the function $F_Y(y)$ has these properties also and Y is continuous. Its density function is obtained by differentiation:

$$f_Y(y) = \frac{d}{dy} F_Y(y) = \frac{d}{dy} F_X\left(\frac{y - b}{a}\right) = \frac{1}{a} f_X\left(\frac{y - b}{a}\right).$$

If a had been assumed negative, the division by a in the early stages would have reversed the direction of the inequality; the result would be the same as for $a > 0$, except that the multiplier would be $-1/a$ in place of $1/a$. The two cases can be summarized in one formula:

$$f_{aX+b}(y) = \frac{1}{|a|} f_X\left(\frac{y - b}{a}\right).$$

The technique employed in this last example can be used more generally. The simplest case is that in which, as in the example, the function $g(x)$ is strictly monotonic, either increasing or decreasing. For definiteness, suppose that $g(x) < g(x')$ whenever $x < x'$, which is what is meant by saying that $g(x)$ is (strictly) monotonically increasing. Suppose further that $g(x)$ is differentiable (and hence continuous). Then there is a unique inverse $x = g^{-1}(y)$ for each y, having the property that $g(g^{-1}(y)) = y$ and $g^{-1}(g(x)) = x$, and

$$P(Y \leq y) = P(g(X) \leq y) = P(X \leq g^{-1}(y)) = F_X(g^{-1}(y)).$$

If X is a continuous random variable, then this distribution function of Y has the properties that make Y a continuous random variable, with density

$$f_Y(y) = \frac{d}{dy} P(Y \leq y) = \frac{d}{dy} F_X(g^{-1}(y)) = f_X(g^{-1}(y)) \frac{d}{dy} g^{-1}(y).$$

The derivative of $g^{-1}(y)$ can be computed as the reciprocal of the derivative of $g(x)$:

$$\frac{d}{dy} g^{-1}(y) = \frac{dx}{dy} = \frac{1}{dy/dx} = \frac{1}{g'(x)} = \frac{1}{g'(g^{-1}(y))}.$$

This may be infinite—$g'(x)$ can vanish at isolated points even when $g(x)$ is strictly monotonic; in this case, the density of Y is infinite, unless f_X is zero and $f(x)/g'(x)$ is an indeterminate form.

The expression for the density of Y can also be derived in terms of probability elements. The differential of $g(x)$ is an approximation of the increment in $g(x)$ corresponding to an increment in x, from x to $x + dx$:

$$\Delta y \doteq dy = g'(x)\, dx.$$

(See Figure 2–11.) The probability assigned to the increment in y is just the probability assigned to dx:

$$f_Y(y)\, dy = P^{\mathscr{Y}}(dy) = P^{\mathscr{X}}(dx) = f_X(x)\, dx$$

$$= \left[f_X(x) \frac{1}{g'(x)} \right] dy.$$

The coefficient of dy is then the density $f_Y(y)$, and, of course, it agrees with the expression obtained previously. [The monotonicity of $g(x)$ was used implicitly in attributing to dy only the probability in the given dx. When the inverse is not single-valued, the probability in dy might come from more than one dx.]

Notice that although the *probability* in dx is the same as in dy, the *density* (probability per unit length) differs by the factor $g'(x)$, which is the ratio of the lengths. Where $g(x)$ is steep, it spreads a given amount of probability on the x-axis over a wider region on the y-axis and produces a thinner distribution there.

It is suggested that rather than memorize the formula for $f_Y(y)$ that holds when Y is a strictly monotonic function of X, whose applicability is rather limited, the reader should be familiar with the *method* of derivation, which is applicable in all cases. Moreover, it would be well to acquire an intuitive feel for what takes place when a random variable undergoes a transformation.

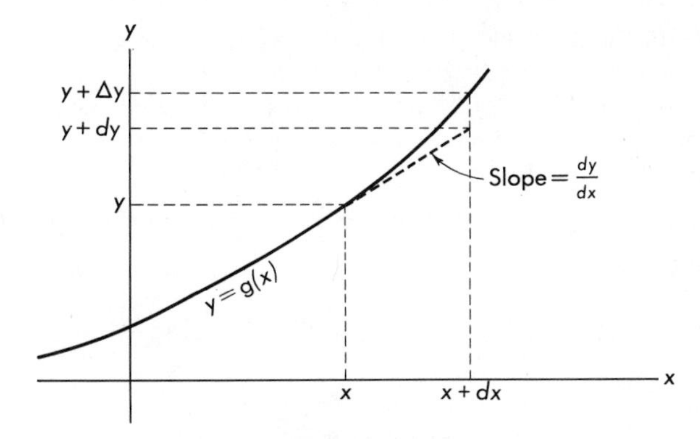

Figure 2–11

EXAMPLE 2–19. Suppose X is a random variable whose distribution is defined by the density in Figure 2–12(a). (A "formula" in several parts could be written for this density, but it is well-defined by the graph.) Suppose further that it is desired to obtain the distribution of $Y = 2(|X| - 1)$. Figure 2–12(b) shows the density of $|X|$, obtained by simply reflecting the negative part of the distribution into the positive part—thereby doubling the density at each positive x. Subtraction of 1 from each possible value of $|X|$ moves the whole distribution to the left 1 unit, as shown in Figure 2–12(c). And, finally, multiplication by 2 stretches the distribution to the range -2 to 2, each value of $|X| - 1$ being doubled; the result is the desired density, shown in Figure 2–12(d).

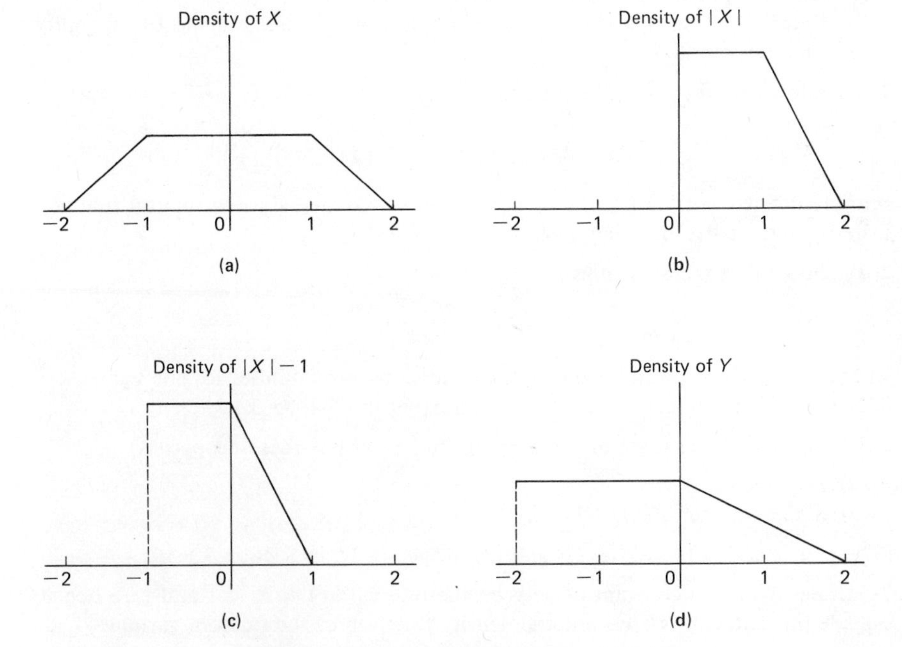

Figure 2–12. Densities for Example 2-19.

EXAMPLE 2–20. Suppose that X has the density $[\pi(1 + x^2)]^{-1}$, and let $Y = X^2$. The function x^2 has two inverses, and so to each element dy on the positive y-axis there correspond two elements dx, one on the positive and one on the negative x-axis, both of which contribute to the density on the y-axis. If y is negative, $P(Y \leq y) = 0$ but for $y \geq 0$,

$$F_Y(y) = P(Y \leq y) = P(X^2 \leq y) = P(-\sqrt{y} \leq X \leq \sqrt{y}) = F_X(\sqrt{y}) - F_X(-\sqrt{y}).$$

Differentiation yields

$$f_Y(y) = \frac{d}{dy} P(Y \leq y) = f_X(\sqrt{y})\frac{1}{2\sqrt{y}} - f_X(-\sqrt{y})\frac{-1}{2\sqrt{y}} = \frac{1/\sqrt{y}}{\pi(1 + y)}, \quad \text{for } y \geq 0.$$

Problems

2–26. In each case, for a suitable choice of constant k, determine whether or not the given function can serve as a density function. For those that can, determine the value that k must have for the given function to be a density:

(a) $f(x) = \begin{cases} kx, & \text{for } 0 < x < 2, \\ 0, & \text{for } x < 0 \text{ or } x > 2. \end{cases}$

(b) $f(x) = \begin{cases} kx(1 - x), & \text{for } 0 < x < 1, \\ 0, & \text{for } x < 0 \text{ or } x > 1. \end{cases}$

(c) ke^{-x}.

(d) $ke^{-|x|}$.

2–27. A discrete random variable X can assume only the value 0, 1, 2, 3, . . .

(a) Can equal probabilities be assigned these values?

(b) Determine k such that $P(X = j) = kp^j$ is a proper assignment of probability, where $0 < p < 1$.

2–28. Show that if $A > 0$, the function

$$F(x) = \frac{1}{2A}(x + A), \qquad \text{for } |x| < A,$$

suitably defined outside the interval $|x| < A$, can serve as a distribution function. Determine the corresponding density function.

2–29. Show that the quantities

$$f(k) = \frac{m^k e^{-m}}{k!},$$

where m is a given positive constant, can serve as probabilities for the values $k = 0, 1, 2, \ldots$. (This model will be discussed further in Chapter 3.)

2–30. Show that the function $F(x) = P[X(\omega) \leq x]$ has these properties:

(a) $F(\infty) = 1$.

(b) If $x < x'$, then $F(x) \leq F(x')$.

[These were called Properties (2) and (3), respectively, in Section 2.1.5.]

2–31. Let X have a constant density on the interval $-1 < x < 1$ and zero density outside that interval. Determine the density function of the random variable

(a) $|X|$. (b) X^2. (c) $(X + 1)/2$.

2–32. A *random number table* is asserted to be a list of outcomes of successive performances of the "spinning pointer" of Example 2–4, the value of X (at each trial) being a random variable with constant density on $0 < x < 1$. Suppose you wanted a similar list of values, but from a variable with constant density on $-1 < x < 1$. How might you use the given table for this purpose?

2–33. Determine the distribution of $Y = X^2 - 7X + 10$, where X denotes the number of points thrown in a toss of a die whose faces are equally likely.

2–34. Let X have the distribution function

$$F(x) = \begin{cases} 0, & x < 0, \\ x^2, & 0 \le x \le 1, \\ 1, & x > 1. \end{cases}$$

Determine the distribution function of $Y = X^2$. (Does this function have a single-valued inverse—and does it make any difference?)

2–35. Let X have a strictly monotonic, continuous distribution function $F(x)$ and define the random variable Y by the relation $Y = F(X)$. (That is, for "g" use the particular function which is the distribution function of X.) Determine the distribution of Y.

2–36. Let X have a strictly monotonic, continuous distribution function $F(x)$ and let $G(x)$ be strictly monotonic, continuous distribution function. Show that $G^{-1}(F(X))$ has $G(x)$ as its distribution function. (G^{-1} denotes the inverse of G.)

2.1.7 Bivariate Distributions

A random vector $(X(\omega), Y(\omega))$ induces a probability distribution in the plane by the definition

$$P(S) = P^{\Omega}[(X(\omega), Y(\omega)) \in S],$$

where S is a region in the plane—in the "value" space $\mathscr{R}_2$ of the random vector. This distribution is characterized by the *distribution function*,

$$\begin{aligned} F(x, y) &\equiv P(X \le x \text{ and } Y \le y) \\ &= P^{\Omega}[X(\omega) \le x \text{ and } Y(\omega) \le y]. \end{aligned}$$

(This is also referred to as the *joint* distribution function.)

EXAMPLE 2–21. Consider again the probability space for random orientations (as in Example 2–10, Problem 2–10, and earlier examples), in which the probability of a region on the surface of the unit hemisphere is proportional to the area of that region. The spherical coordinate functions $\theta(\omega)$ and $\phi(\omega)$, representing colatitude and longitude, respectively, of the point ω, define together a random vector—mapping points on the hemisphere into $\mathscr{R}_2$, the Cartesian plane with coordinates (θ, ϕ). The distribution function of the distribution in this plane is

$$F_{\theta,\phi}(u, v) = P(\theta \le u, \phi \le v).$$

This is the relative area of the portion of the hemisphere bounded by the curves $\phi = 0$, $\phi = v$, and $\theta = u$, as shown in Figure 2–13. Since the area of the entire polar cap bounded by $\theta = u$ is $2\pi(1 - \cos u)$, the portion of it subtended by the angle from $\phi = 0$ to $\phi = v$ has the area $v(1 - \cos u)$. Hence,

$$F_{\theta,\phi}(u, v) = \frac{v}{2\pi}(1 - \cos u), \qquad \text{for } 0 \le v < 2\pi, \quad 0 \le u \le \frac{\pi}{2}.$$

(Similar but simpler computations will yield the correct formulas for other portions of the plane.)

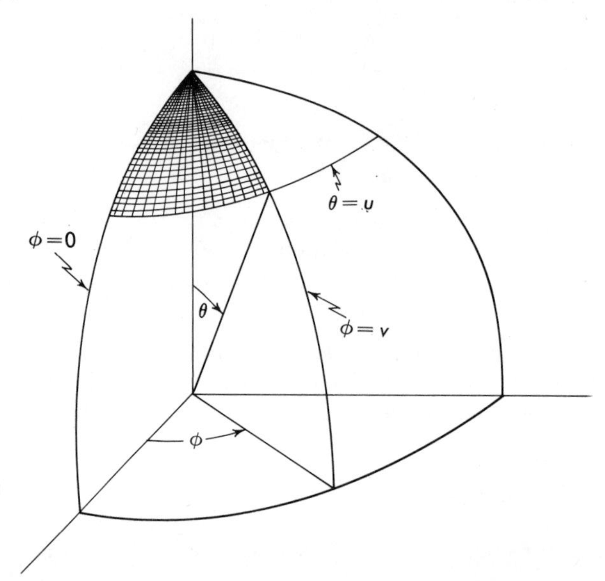

Figure 2–13. Area defining the joint distribution function for (θ, ϕ).

The semi-infinite "intervals" $\{X \le x,\ Y \le y\}$ and the distribution function, which gives their probabilities, are basic because other regions of interest in the plane can be constructed with countable operations on semi-infinite intervals. In particular, a rectangle with sides parallel to the axes (Figure 2–14) is a simple combination of four semi-infinite intervals:

$$\begin{aligned}\{x < X \le x + h, y < Y \le y + k\} &\\ = [\{X \le x + h,\ Y \le y + k\} &- \{X \le x + h,\ Y \le y\}]\\ -[\{X \le x,\ Y \le y + k\} &- \{X \le x,\ Y \le y\}].\end{aligned}$$

And then, upon application of part (*f*) in Problem 1–31,

$$\begin{aligned}&P(x < X \le x + h, y < Y \le y + k)\\ &\quad = F(x + h, y + k) - F(x + h, y) - F(x, y + k) + F(x, y) \equiv \Delta^2 F.\end{aligned}$$

[This is called a *second difference* of the function $F(x, y)$.] More general regions in the plane can be obtained by limiting operations on finite unions of such rectangles (as in the calculation of plane areas by double integrals).

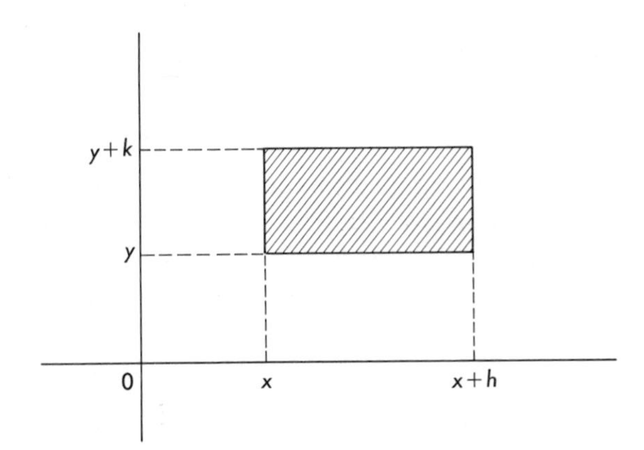

Figure 2–14. Region whose probability is $\Delta^2 F$.

A bivariate distribution function $F(x, y)$ has the following properties:

1. $F(x, \infty)$ and $F(\infty, y)$ are univariate distribution functions, as functions of x and y, respectively.
2. $F(-\infty, y) = F(x, -\infty) = 0$.
3. $\Delta^2 F \geq 0$, for every rectangle with sides parallel to the axes.

(Verification of these properties is called for in Problem 2–39.) The significance of these particular properties is that to define a distribution directly in the plane, it is only necessary to specify a function F having properties (1) through (3) and use F to define probabilities of semi-infinite intervals. It can be shown that a probability measure in the plane can be constructed with this starting point, one that assigns probability to sets or events of practical significance.

[It might be noted that properties (1) and (2) are not enough by themselves, as someone might naïvely think. It can happen that a function satisfies (1) and (2) but does not satisfy property (3); this will be shown in Problem 2–38. But surely property (3) is essential—every rectangle must have a nonnegative probability.]

Two useful types of bivariate distribution stand out: the *discrete* type, with probability concentrated in lumps at isolated points, and the *continuous* type, in which probability is spread over a region and no single points or curves have any positive probability. Again, however, these two types do not include all useful models. For instance, a bivariate distribution might be continuous in one variable and discrete in the other, or be otherwise concentrated along lines or curves in the plane. Also, a distribution can include discrete lumps at isolated points, concentrations along curves, and smears over regions of the plane.

In the case of a discrete distribution it is again convenient to work in terms of a *probability function*:

$$f(x, y) = P(X = x \text{ and } Y = y).$$

This may be specified directly, or it may be defined indirectly by a distribution of probability on an underlying space Ω. The possible outcomes (x, y) are sometimes presented in a rectangular tabulation, all possible values of X along the top margin, say, and all possible values of Y along the left margin; the probability for each pair (x, y) is entered under the x and across from the y. The probabilities must, of course, total 1.

EXAMPLE 2–22. Three balls numbered from 1 to 3 are placed in a container. Two balls are drawn one at a time, at random, without replacement. Let

X = number of the first ball drawn.

Y = number of the second ball drawn.

The outcomes are pairs of numbers (x, y). The possible values of X as well as the possible values of Y are 1, 2, and 3. The probabilities for the various possible combinations are easily computed to be those given in the following table:

Y \ X	1	2	3
1	0	$\frac{1}{6}$	$\frac{1}{6}$
2	$\frac{1}{6}$	0	$\frac{1}{6}$
3	$\frac{1}{6}$	$\frac{1}{6}$	0

The probability of any event concerning (X, Y) can be computed from this table. For instance,

$$P(\text{number drawn first is smaller than number drawn second})$$
$$= P(X < Y) = P[(1, 2), (1, 3), \text{ or } (2, 3)] = \tfrac{1}{6} + \tfrac{1}{6} + \tfrac{1}{6} = \tfrac{1}{2}.$$

A bivariate distribution is said to be of continuous type if its distribution function is continuous and has a second-order, mixed partial derivative function

$$f(x, y) = \frac{\partial^2}{\partial x\, \partial y} F(x, y)$$

from which F can be recovered by integration:

$$F(x, y) = \int_{-\infty}^{x} \int_{-\infty}^{y} f(u, v)\, dv\, du.$$

The derivative function is the limiting value of an increment quotient corresponding to increments Δx and Δy:

$$f(x, y) = \lim \frac{\Delta^2 F}{\Delta x\, \Delta y},$$

as Δx and Δy tend (independently) to 0, where $\Delta^2 F$ is the previously defined second difference of F. This derivative is a *density* function [probability per unit area at the point (x, y)] in the usual sense. It will be referred to as the *joint* density of (X, Y). Like a univariate density, it is a differential coefficient, in the sense that for small increments Δx and Δy,

$$\Delta^2 F \doteq f(x, y)\, \Delta x\, \Delta y,$$

which is the probability element corresponding to those increments.

The probability of a finite plane region S can be calculated, for a continuous distribution with density $f(x, y)$, as follows. Subdividing the region's "shadow" or projection on the x-axis into intervals of width Δx, and on the y-axis into intervals of width Δy, defines a net that divides S into little rectangles of dimensions Δx and Δy (except for irregular pieces along the boundary of S). Then

$$P(S) \doteq \sum \Delta^2 F \doteq \sum f(x, y) \Delta x\, \Delta y,$$

the sums extending over all rectangles inside S. As the subdivisions are made finer, the approximating sums tend to a double integral, which is then the probability of S:

$$P(S) = \int_S \int f(x, y)\, dx\, dy.$$

Regions S that are not finite are handled by truncating and passing to a limit. In particular,

$$P(X \leq x, Y \leq y) = \int_{-\infty}^{y} \int_{-\infty}^{x} f(u, v)\, du\, dv = F(x, y),$$

completing the circle back to the original definition of $F(x, y)$.

EXAMPLE 2–23. The distribution function $F_{\theta,\phi}(u, v)$ of the coordinate variables on the probability space for random orientations (Example 2–21) was found to be

$$F_{\theta,\phi}(u, v) = \begin{cases} 0, & \text{for } v \leq 0 \text{ or } u \leq 0, \\ \dfrac{v}{2\pi}(1 - \cos u), & \text{for } 0 < v < 2\pi,\ 0 < u < \dfrac{\pi}{2}, \\ \dfrac{v}{2\pi}, & \text{for } 0 < v < 2\pi,\ u \geq \dfrac{\pi}{2}, \\ 1 - \cos u, & \text{for } 0 < u < \dfrac{\pi}{2},\ v \geq 2\pi, \\ 1, & \text{for } u \geq \dfrac{\pi}{2},\ v \geq 2\pi. \end{cases}$$

This is differentiable, except at some points along the curves $v = 0$, $u = \pi/2$, and $v = 2\pi$. The second-order, mixed derivative is then the density:

$$f_{\theta,\phi}(u, v) = \begin{cases} \dfrac{1}{2\pi} \sin u, & \text{for } 0 < v < 2\pi,\ 0 < u < \dfrac{\pi}{2}, \\ 0, & \text{elsewhere.} \end{cases}$$

From this, one can calculate such things as

$$P\left(\theta > \frac{\pi}{4}, \frac{\pi}{2} < \phi < \pi\right) = \frac{1}{2\pi} \int_{\pi/2}^{\pi} \int_{\pi/4}^{\pi/2} (\sin u)\, du\, dv$$

$$= \frac{1}{4}\left(\cos\frac{\pi}{4} - \cos\frac{\pi}{2}\right) = \frac{1}{4\sqrt{2}}.$$

[This particular probability could also be computed as

$$F\left(\frac{\pi}{2}, \pi\right) - F\left(\frac{\pi}{2}, \frac{\pi}{2}\right) - F\left(\frac{\pi}{4}, \pi\right) + F\left(\frac{\pi}{4}, \frac{\pi}{2}\right),$$

since the event happens to be a rectangle with sides parallel to the axes in the $\theta\phi$-plane.]

A bivariate density function $f(x, y)$ satisfies the conditions

1. $f(x, y) \geq 0$.
2. $\displaystyle\int_{\mathscr{R}_2}\int f(x, y)\, dx\, dy = 1$.

A function (assumed integrable) that satisfies these conditions can be used to define a bivariate distribution by defining

$$F(x, y) = \int_{-\infty}^{y} \int_{-\infty}^{x} f(u, v)\, du\, dv.$$

In particular, a *uniform* bivariate distribution over a region R in $\mathscr{R}_2$ is defined by $f(x, y) = (\text{area of } R)^{-1}$, since the double integral of this function is 1. In such a case the *probability* of a subregion is just proportional to the area of the subregion.

EXAMPLE 2–24. Suppose that (X, Y) has a distribution which is uniform over a unit circle centered at (0, 0); that is, the density of the distribution is constant over that circle. The constant must be the reciprocal of the area of the circle in order for the volume of the region under the density surface to be 1:

$$f_{X,Y}(x, y) = \begin{cases} \dfrac{1}{\pi}, & \text{for } x^2 + y^2 \leq 1, \\ 0, & \text{elsewhere.} \end{cases}$$

The probability of any plane region is just proportional to the area of that portion of the region that lies inside the unit circle. Thus

$$P\left(X^2 + Y^2 \leq \frac{1}{4}\right) = \frac{\text{area of circle radius } \frac{1}{2}}{\text{area of circle radius } 1} = \frac{1}{4},$$

and

$$P\left(X \le \frac{1}{2}\right) = 1 - \frac{1}{\pi}(\text{area of segment to right of } x = \tfrac{1}{2})$$
$$= 1 - \left(\frac{1}{3} - \frac{\sqrt{3}}{4\pi}\right) \doteq 0.804.$$

A function $g(x, y)$ defines a random variable as a function of the random vector (X, Y):

$$Z(\omega) = g(X(\omega), Y(\omega)).$$

This can be thought of either as the function $Z(\omega)$ on Ω, or if probability is assigned initially in the value space of (X, Y), as a function or a random variable on the probability space $\mathscr{R}_2$. In either case, there is an induced distribution in the space of values of Z, which can be taken to be the set of real numbers $\mathscr{R}$. And this distribution has a distribution function defined in the usual way as $F_Z(z) = P(Z \le z)$.

EXAMPLE 2–25. Let (X, Y) have the joint density function

$$f_{X,Y}(x, y) = \frac{1}{2\pi} \exp\left[-\frac{1}{2}(x^2 + y^2)\right].$$

(It will be verified below that the constant multiplier is chosen so that the total volume under the surface representing this function is 1.) Consider the random variable Z, which is the distance from (0, 0) out to the point (X, Y); that is, $Z^2 = X^2 + Y^2$. The distribution function of Z is computed as follows, for $z > 0$:

$$P(Z \le z) = P(Z^2 \le z^2) = P(X^2 + Y^2 \le z^2)$$
$$= \iint_{x^2+y^2<z^2} \left\{\frac{1}{2\pi} \exp\left[-\frac{1}{2}(x^2 + y^2)\right]\right\} dx\, dy$$
$$= \int_0^{2\pi} \int_0^z \frac{1}{2\pi} \exp\left[-\frac{1}{2}r^2\right] r\, dr\, d\theta = 1 - \exp\left[-\frac{1}{2}z^2\right].$$

(Polar coordinates, $x = r\cos\theta$ and $y = r\sin\theta$, were used in evaluating the double integral.) The density function of Z is $z\exp(-\frac{1}{2}z^2)$ for $z > 0$. Observe that since $P(Z \le \infty) = P(X \le \infty, Y \le \infty) = 1$, the choice of constant multiplier in the given density of (X, Y) has been justified.

The functions $F(x, \infty)$ and $F(\infty, y)$ define distributions in $\mathscr{R}_1$ that are in fact the distributions of X and Y considered separately as single random variables:

$$F(x, \infty) = P(X \le x \text{ and } Y \le \infty) = P(X \le x),$$
$$F(\infty, y) = P(X \le \infty \text{ and } Y \le y) = P(Y \le y).$$

These distributions are called the *marginal distributions* of X and Y; the adjective "marginal" refers only to their origin and does not alter the fact that they are indeed probability distributions in $\mathscr{R}_1$.

In the discrete case, with $p_{ij} = P(X = x_i, Y = y_j)$, the probability that $X = x_i$ is obtained by summing these joint probabilities over all pairs (x_i, y_j) in which the first element is x_i, that is, over all y_j's for fixed x_i:

$$P(X = x_i) = \sum_j P(X = x_i, Y = y_j) = \sum_j p_{ij}.$$

Similarly,

$$P(Y = y_j) = \sum_i p_{ij}.$$

These probabilities are the row and column sums in the rectangular tabulation of the values of p_{ij} described earlier. The writing of these sums in the margins of such an array is the source of the term *marginal distribution.*

EXAMPLE 2–26. As in Example 2–22, let (X, Y) denote the numbers on the first and second balls drawn from a container in which there are three balls numbered 1, 2, 3. The drawing is random, without replacement. The table of probabilities given in that example is reproduced here with the marginal totals also shown:

y_j \ x_i	1	2	3	$f_Y(y_j)$
1	0	$\frac{1}{6}$	$\frac{1}{6}$	$\frac{1}{3}$
2	$\frac{1}{6}$	0	$\frac{1}{6}$	$\frac{1}{3}$
3	$\frac{1}{6}$	$\frac{1}{6}$	0	$\frac{1}{3}$
$f_X(x_i)$	$\frac{1}{3}$	$\frac{1}{3}$	$\frac{1}{3}$	1

(The marginal distributions here are identical, representing the fact that each ball drawn can be 1, 2, or 3, and in each case these results are equally likely. Observe that the probabilities for the second ball drawn are *un*conditional—nothing is assumed known about the result of the first selection.)

In the continuous case, the marginal density function (i.e., the density of the marginal distribution) is obtained by differentiating the marginal distribution function,

$$F_X(x) = F_{X,Y}(x, \infty) = \int_{-\infty}^{x} \int_{-\infty}^{\infty} f_{X,Y}(u, y)\, dy\, du,$$

to obtain

$$f_X(x) = F_X'(x) = \int_{-\infty}^{\infty} f_{X,Y}(x, y)\, dy.$$

That is, to obtain a marginal density one "integrates out" the unwanted variable from the joint density. Interpreted geometrically, the marginal density of X at $x = x_0$ is the area of the cross section at $x = x_0$ of the solid region under the surface that represents the joint density $f_{X,Y}(x, y)$.

EXAMPLE 2–27. Again, let (X, Y) have the distribution given in Example 2–24, with constant density in a unit circle centered at the origin. The density is $1/\pi$ in that circle, and the marginal density of Y at $y = y_0$ can be expressed as an integral:

$$f_Y(y_0) = \int_{-\infty}^{\infty} f_{X,Y}(x, y_0)\, dx = \int_{-a}^{a} \frac{1}{\pi}\, dx$$

or computed as the area of the cross section—a rectangle of height $1/\pi$ and base $2a$, where a is determined (see Figure 2–15) by $a^2 + y_0^2 = 1^2$. Hence, the desired density is

$$f_Y(y_0) = \frac{2}{\pi}(1 - y_0^2)^{1/2}, \qquad \text{for } -1 < y_0 < 1.$$

It is zero outside the interval $[-1, 1]$, and one can easily check to see that the integral of this density function is, as it should be, 1.

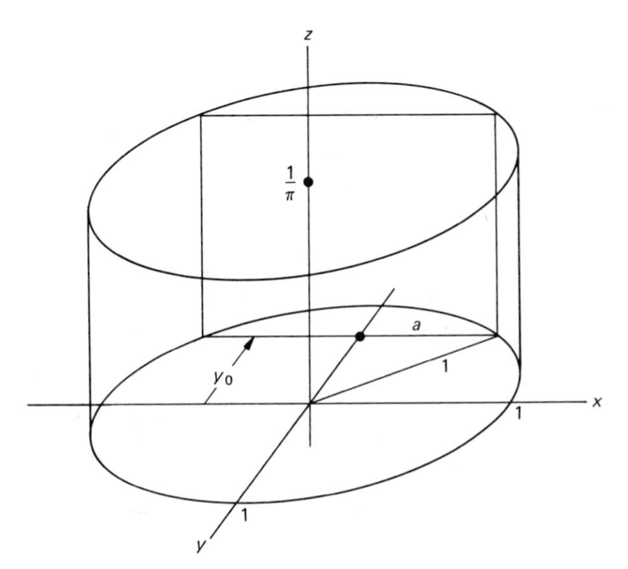

Figure 2–15

Problems

2–37. A random point (X, Y) is distributed uniformly (i.e., with constant density function) on the square whose vertices are $(1, 1)$, $(-1, 1)$, $(1, -1)$, and $(-1, -1)$. Determine the probabilities of these events:

(a) $X^2 + Y^2 < 1$. (b) $2X - Y > 0$.
(c) $|X + Y| < 2$. (d) $|X - Y| < \frac{1}{2}$.

2–38. Show that the following is not a distribution function:

$$F(x, y) = \begin{cases} 1 - e^{-x-y}, & \text{if } x \geq 0 \text{ and } y \geq 0, \\ 0, & \text{otherwise.} \end{cases}$$

[But show, however, that Properties (1) and (2) on p. 81 *are* satisfied.]

2–39. Verify that Properties (1) through (3) on p. 71 are satisfied when $F(x, y)$ is defined as

$$P^{\Omega}(X(\omega) \le x,\ Y(\omega) \le y).$$

2–40. Let X be the number of 5's and Y the number of 6's that turn up in a toss of two standard dice. Construct the table of bivariate probabilities p_{ij} for the distribution of (X, Y); from this table compute the following:

(a) $P(X + Y \ge 1)$.
(b) The probability functions $P(X = x_i)$ and $P(Y = y_j)$.

2–41. Let (X, Y) have the distribution defined by the following table of probabilities:

Y \ X	1	2	3
2	$\frac{1}{12}$	$\frac{1}{6}$	$\frac{1}{12}$
3	$\frac{1}{6}$	0	$\frac{1}{6}$
4	0	$\frac{1}{3}$	0

Determine the following:

(a) The marginal probability distributions.
(b) $P(X = Y)$.
(c) $P(X = 2 \text{ or } Y = 4)$.
(d) $P(X + Y \le 4)$.

2–42. Let (X, Y) have a uniform density over the circle $x^2 + y^2 \le 4$. Determine the following:

(a) $P(Y > kX)$.
(b) $f_X(x)$.
(c) $P(X^2 + Y^2 > 1)$.
(d) The distribution function of $X^2 + Y^2$.
(e) The distribution function of $(X^2 + Y^2)^{1/2}$.

2–43. (a) Show that if $F_1(x)$ and $F_2(x)$ are (univariate) distribution functions, then

$$F(x, y) = F_1(x)F_2(y)$$

is a bivariate distribution function.

(b) Show that if $f_1(x)$ and $f_2(x)$ are univariate densities or probability functions, then

$$f(x, y) = f_1(x)f_2(y)$$

is a bivariate density function or probability function, respectively.

2–44. Let (X, Y) have the distribution defined by the density function

$$f(x, y) = e^{-x-y}, \qquad \text{for } x > 0 \text{ and } y > 0$$

(and zero outside the first quadrant). Determine the following:

(a) The marginal probability distributions.
(b) The joint c.d.f.
(c) $P(X = Y)$.
(d) $P(X + Y \le 4)$.
(e) The distribution function of $Z = X + Y$.

2–45. Let (X, Y) have a constant density on the unit square $0 < x < 1$, $0 < y < 1$. Determine

(a) The marginal distributions of X and Y.
(b) The distribution function of $W = \max(X, Y)$, the larger of X and Y. [*Hint*: $W \leq w$ if and only if both X and Y do not exceed w.]

2–46. Let ω denote a point in a plane, and let probability be distributed uniformly over an equilateral triangle in that plane. Let X, Y, and Z denote the random variables that measure the perpendicular distances from ω to the three sides. It is not hard to see that $X + Y + Z$ is constant so that there are only two "free" variables. Determine the joint distribution of (X, Y).

2.1.8 Conditional Distributions

The information that a given event E of positive probability has occurred alters an initial probability structure; the new probabilities are *conditional* on E. In particular, the new distribution in the value space of a random variable X, given E, is called a *conditional distribution* and is defined by a *conditional c.d.f.*:

$$P(X \leq \lambda \mid E) = \frac{P(X \leq \lambda \text{ and } E)}{P(E)}.$$

The new random variable with this c.d.f. will be denoted $X \mid E$, and the c.d.f. by

$$F_{X|E}(\lambda) = P(X \leq \lambda \mid E).$$

The reduced or restricted measure defined in $\mathscr{R}$ by the c.d.f. of $X \mid E$ is not a new *kind* of measure. The distribution is called "conditional" to emphasize the fact that it came from a previously defined distribution by imposing a condition, the event E.

A conditional distribution, like any distribution in $\mathscr{R}$, may be continuous, in which case its density is

$$f_{X|E}(\lambda) = \frac{d}{d\lambda} F_{X|E}(\lambda).$$

Or it may be discrete, characterized by a (conditional) probability function:

$$f_{X|E}(x) = P(X = x \mid E) = \frac{P(X = x \text{ and } E)}{P(E)}.$$

Or it may be more general—neither discrete nor continuous.

EXAMPLE 2–28. Let Ω consist of the 52 cards in a standard deck, one of which is drawn at random. Let $X(\omega)$ denote the number of points assigned to the card ω as follows:

Ace—4, King—3, Queen—2, Jack—1, other—0.

Let E denote the event that the card drawn is a *Heart or a face card.* Of the 22 equally likely cards in E, 1 is an Ace, 4 are Kings, 4 are Queens, 4 are Jacks, and 9 are other cards. Hence,

$$\begin{aligned}
P(X = 4 \mid E) &= \tfrac{1}{22}\\
P(X = 3 \mid E) &= \tfrac{4}{22}\\
P(X = 2 \mid E) &= \tfrac{4}{22}\\
P(X = 1 \mid E) &= \tfrac{4}{22}\\
P(X = 0 \mid E) &= \tfrac{9}{22}.
\end{aligned}$$

These total 1, and define a distribution of probability on the values 0, 1, 2, 3, 4 somewhat different than the unconditional probabilities (which are $\frac{9}{13}, \frac{1}{13}, \frac{1}{13}, \frac{1}{13}, \frac{1}{13}$, respectively).

EXAMPLE 2–29. Let Ω denote the unit square bounded by (0, 0), (1, 1), (0, 1), and (1, 0) in a given Cartesian coordinate system and assume that the probability of an event in this sample space is proportional to its area. Let $X(\omega)$ be defined to be the x-coordinate of the point ω, and E the event that the x-coordinate exceeds the y-coordinate. Then

$$\begin{aligned}
P(X \le \lambda \mid E) &= \frac{P(X \le \lambda \text{ and } X > Y)}{P(X > Y)}\\
&= \frac{\lambda^2/2}{1/2} = \lambda^2, \qquad \text{for } 0 \le \lambda \le 1.
\end{aligned}$$

The pertinent areas may be determined geometrically from Figure 2–16.

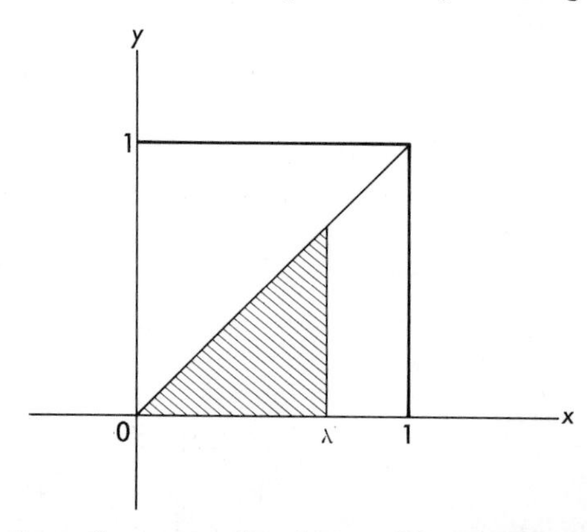

Figure 2–16. Areas for computing the conditional c.d.f. in Example 2-29.

The most commonly used conditional distributions have to do with one of the two variables comprising a random vector, and the condition is then a condition on the value of the other variable. That is, given a random vector $[X(\omega), Y(\omega)]$ one is interested in the conditional distribution of X, given that Y takes on a value in some set. Moreover, that set is usually just a single value. Thus, if (X, Y) has a discrete distribution in the xy-plane, with values (x_i, y_j) for $i = 1, \ldots, m$ and $j = 1, \ldots, n$, the conditional probabilities for X given $Y = y_j$ are computed as follows:

$$f_{X|Y=y_j}(x_i) = P(X = x_i \mid Y = y_j)$$
$$= \frac{P(X = x_i,\ Y = y_j)}{P(Y = y_j)} = \frac{f(x_i, y_j)}{f_Y(y_j)},$$

where $f(x, y)$ is the joint probability function and $f_Y(y)$ is the marginal probability function of Y.

The notation with $X \mid Y = y$ in the subscript is clear but rather cumbersome. A simpler notation will usually be used:

$$f(x \mid y) \equiv f_{X|Y=y}(x),$$

even though it is not so precise. That is, the function $f(y \mid x)$ would mean, in usual mathematical notation, the same function as $f(x \mid y)$ with the variables reversed. Here, however, $f(y \mid x)$ would mean the conditional probability of $Y = y$, given $X = x$, which is a different function of y than $f(x \mid y)$ is of x. In context, this notation is not usually confusing.

Conditioning the random variable X by giving the value of Y does not change the relative odds for those points (x, y) that can still occur. That is, the probability

$$f(x \mid y) = \frac{1}{f_Y(y)} f(x, y)$$

is *proportional* to $f(x, y)$, and the constant of proportionality is just what is needed so that the conditional probabilities add up to 1:

$$\sum_i f(x_i \mid y) = \frac{1}{f_Y(y)} \sum_i f(x_i, y) = \frac{f_Y(y)}{f_Y(y)} = 1.$$

Holding Y fixed at a particular value, $Y = y_j$ amounts to looking at the row of joint probabilities for which $Y = y_j$, when these probabilities $f(x_i, y_j)$ are given in the customary two-way array. The conditional probabilities are then simply these joint probabilities in the row $Y = y_j$ *renormalized to have sum* 1 [through division by the row sum, $f_Y(y_j)$].

Marginal probabilities can be expressed in terms of conditional probabilities. Thus,

$$f_X(x_i) = \sum_j f(x_i, y_j) = \sum_j f(x_i \mid y_j) f_Y(y_j).$$

(This is an instance of what was referred to as the "law of total probability" in Section 1.2.1, inasmuch as the sequence $y_1, y_2, \ldots$ of values of Y constitute a partition of the sample space.) To calculate the *un*conditional probabilities for X, then, one weights each conditional probability with the corresponding probability of the Y-value and sums over all Y-values.

EXAMPLE 2–30. Let (X, Y) have the distribution defined by the table of probabilities $p_{ij} = P(X = x_i \text{ and } Y = y_j)$ as shown. (This is the distribution of Example 2–26, and the marginal probabilities obtained there are also shown.)

y_j \ x_i	1	2	3	
1	0	$\frac{1}{6}$	$\frac{1}{6}$	$\frac{1}{3}$
2	$\frac{1}{6}$	0	$\frac{1}{6}$	$\frac{1}{3}$
3	$\frac{1}{6}$	$\frac{1}{6}$	0	$\frac{1}{3}$
	$\frac{1}{3}$	$\frac{1}{3}$	$\frac{1}{3}$	

The conditional probabilities for X given $Y = 1$ are proportional to those in the first row, being obtained by dividing those entries by their sum:

x_i	1	2	3
$f(x_i \mid 1)$	0	$\frac{1}{2}$	$\frac{1}{2}$

Similarly, the conditional probabilities for Y given $X = 2$ are proportional to those in the second column:

y_j	$f(y_j \mid 2)$
1	$\frac{1}{2}$
2	0
3	$\frac{1}{2}$

The probabilities for $X = 1$, say, given various Y-values are

$$f(1 \mid 1) = 0, \qquad f(1 \mid 2) = \tfrac{1}{2}, \qquad f(1 \mid 3) = \tfrac{1}{2},$$

and their weighted sum gives $P(X = 1)$:

$$f_X(1) = \sum f(1 \mid y_j) f_Y(y_j) = 0 + \tfrac{1}{2} \cdot \tfrac{1}{3} + \tfrac{1}{2} \cdot \tfrac{1}{3} = \tfrac{1}{3}.$$

If Y is continuous, the probability that Y takes on a particular value is 0; an attempt at defining conditional probability as it was defined above in the discrete case yields

$$P(X \leq x \mid Y = y) = \frac{P(X \leq x \text{ and } Y = y)}{P(Y = y)} = \frac{0}{0},$$

which leads nowhere. And yet, since one *can* observe a single value $Y = y$, it is certainly desirable to have a model with conditional probabilities of events given $Y = y$, and to be able to derive these from the model for the random vector (X, Y).

If (X, Y) has a continuous bivariate distribution and a value $Y = y$ is observed, the conditional distribution of X certainly should be continuous along the line of Y-values defined by the given information. Moreover, the density of that distribution should have the property (by analogy with the discrete case) that its weighted integral gives the marginal or unconditional density of X:

$$\int_{-\infty}^{\infty} f(x \mid y) f_Y(y)\, dy = f_X(x),$$

where again a simplified notation has been used for the conditional density:

$$f(x \mid y) \equiv f_{X|Y=y}(x).$$

This desirable property for $f(x \mid y)$ can be reconciled with the definition of a marginal density in terms of the joint density:

$$f_X(x) = \int_{-\infty}^{\infty} f(x, y)\, dy$$

by defining[3]

$$f(x \mid y) \equiv \frac{f(x, y)}{f_Y(y)}.$$

The function $f(x \mid y)$ so defined is (as a function of x) a bona fide density function, being nonnegative with integral

$$\int_{-\infty}^{\infty} f(x \mid y)\, dx = \int_{-\infty}^{\infty} \frac{f(x, y)}{f_Y(y)}\, dx = 1.$$

The integral over a set of X-values defines the conditional probability of that set,

$$P(X \text{ in } A \mid Y = y) = \int_A f(x \mid y)\, dx,$$

and if A is, in particular, the interval $(-\infty, x)$, one obtains the conditional c.d.f.,

$$F(x \mid y) \equiv F_{X|Y=y}(x) = \int_{-\infty}^{x} f(u \mid y)\, du.$$

The definition of conditional density as the joint density divided by the marginal density of the conditioning variable at the given value is intuitively appealing. The geometrical representation of the joint density function is a surface, and the function $f(x \mid y)$, for a given y, varies with x according to the variation along the surface $z = f(x, y)$ in the cross section at the given Y-value. The division by $f_Y(y)$ has the effect of adjusting the heights of this cross-section curve by a constant factor so that the total area under the cross section is 1. This situation is quite analogous to that in the discrete case, in which the conditional probabilities for X given $Y = y_j$ are proportional to the joint probabilities

[3] One might be led to other definitions (as in Problem 2–53), but it can be shown (Lehmann [16], pp. 39 ff.) that the one given is essentially the only one that has the intuitively appealing properties of Problem 2–50.

$f(x_i, y_j)$ in the row of $Y = y_j$, each being divided by the row sum, $f_Y(y_j)$, so that the sum of the conditional probabilities given $Y = y_j$ is equal to 1.

EXAMPLE 2–31. A random vector (X, Y) has a density that is zero outside the triangle with vertices (0, 0), (0, 1), and (1, 0) and is constant within that triangle:

$$f(x, y) = \begin{cases} 2, & \text{if } x + y \leq 1, x \geq 0, \text{ and } y \geq 0, \\ 0, & \text{otherwise.} \end{cases}$$

The marginal density for X is the area under the cross section at $X = x$ as a function of x:

$$f_X(x) = \int_{-\infty}^{\infty} f(x, y)\, dy = \int_0^{1-x} 2\, dy = 2(1 - x), \qquad \text{for } 0 \leq x \leq 1.$$

The conditional density for Y given $X = x$ is then

$$f(y \mid x) = \frac{f(x, y)}{f_X(x)} = \frac{2}{2(1 - x)} = \frac{1}{1 - x}, \qquad 0 < y \leq 1 - x.$$

Its value is 0 outside the indicated range.[4] That is, the density of Y given $X = x$ is constant on the interval $[0, 1 - x]$. This reflects the fact that the cross section of $f(x, y)$ at $X = x$ is of constant height (namely, 2) on this interval. Division by the area of the cross section, namely, by $2(1 - x)$, yields a function whose graph encloses with the horizontal a rectangle of area of 1 (see Figure 2–17).

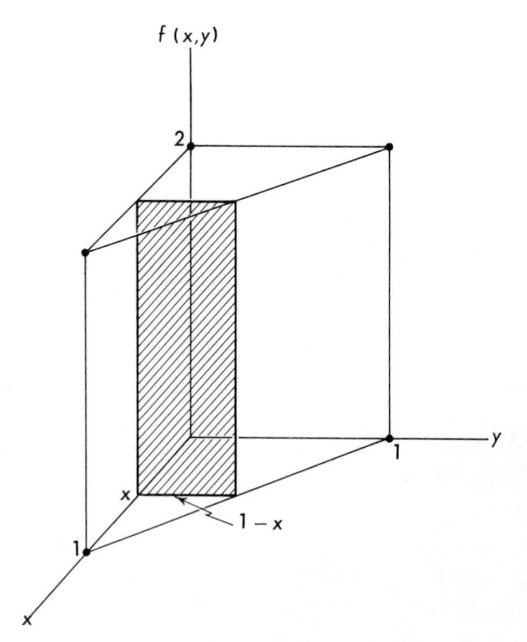

Figure 2–17

[4] It will become increasingly annoying to have to say, repeatedly, "and 0 outside this region." Although it is important to be aware of the complete definition of a density function, we shall henceforth follow the convenient practice of giving its formula only for the region in which it is nonzero, with the understanding that it is "0 outside this region."

EXAMPLE 2–32. Let (X, Y) have the distribution defined by the density function of Problem 2–44, namely,

$$f_{X,Y}(x, y) = e^{-x-y}, \qquad \text{for } x > 0, y > 0.$$

The marginal distribution for Y is the integral of this on x:

$$f_Y(y) = \int_{-\infty}^{\infty} e^{-x-y}\,dx = e^{-y}, \qquad \text{for } y > 0.$$

The conditional density of X given $Y = y$ is then

$$f(x \mid y) = \frac{e^{-x-y}}{e^{-y}} = e^{-x}, \qquad \text{for } x > 0.$$

It will be seen in the next section that the fact that this conditional density of x is independent of y and is indeed equal to the marginal density of X is of particular significance.

2.1.9 Independence

It can happen, as in Example 2–32, that the conditional distribution of one variable in the pair (X, Y) is independent of any condition imposed on the other. The random variables X and Y are then said to be *independent random variables.*

In particular, suppose that for every event A in the value space of X and for each value y of Y,

$$P(X \text{ in } A \mid Y = y) = P(X \text{ in } A).$$

Then for every event B in the value space of Y,

$$\begin{aligned} P(X \text{ in } A \text{ and } Y \text{ in } B) &= \int_B P(X \text{ in } A \mid Y = y) f_Y(y)\,dy \\ &= P(X \text{ in } A) \int_B f_Y(y)\,dy \\ &= P(X \text{ in } A) P(Y \text{ in } B). \end{aligned}$$

(A similar computation with sums yields the same result in the discrete case.) Then the experiment that consists of observing a value of X and the experiment that consists of observing a value of Y are independent experiments, in the terminology of Chapter 1.

If the above factorization is valid for all events A and B, it must hold, in particular, for $\{X \leq x\}$ and $\{Y \leq y\}$:

$$F_{X,Y}(x, y) = F_X(x) F_Y(y).$$

In the discrete case it also holds for the events $\{X = x_i\}$ and $\{Y = y_j\}$:

$$f_{X,Y}(x_i, y_j) = f_X(x_i) f_Y(y_j).$$

In the continuous case, differentiation of the relation for cumulative distribution functions yields the following relation for densities:

$$f_{X,Y}(x, y) = f_X(x)f_Y(y).$$

But then whether f denotes density or probability,

$$f(x \mid y) = \frac{f(x, y)}{f_Y(y)} = \frac{f_X(x)f_Y(y)}{f_Y(y)} = f_X(x),$$

which means that the conditional distribution of X given $Y = y$ is independent of y—the criterion used at the outset to define independence. Therefore, this defining criterion for independence and the various factorization conditions for probabilities, cumulative distribution functions, and density or probability functions are *equivalent*. It does not matter which is taken as defining independence of X and Y.

EXAMPLE 2–33. Let (X, Y) have the joint density e^{-x-y}, for positive x and y, as in Example 2–32. It was seen in that example that

$$f(x \mid y) = e^{-x}, \qquad \text{for } x > 0,$$

which is exactly the marginal density of X. Moreover, the marginal density of Y is e^{-y}, or $y > 0$, so that

$$f(x, y) = e^{-x-y} = e^{-x}e^{-y} = f_X(x)f_Y(y).$$

The component variables X and Y are independent.

EXAMPLE 2–34. Consider the distribution that has a constant density in the triangle bounded by the coordinate axes and by $x + y = 1$ in the plane of values of the random vector (X, Y). This does *not* have the factorization property defining independence of X and Y. The marginal densities are (see Example 2–31)

$$f_X(x) = 2(1 - x), \qquad f_Y(y) = 2(1 - y)$$

for $0 < x < 1$ and $0 < y < 1$, respectively, whereas the joint density is 2 on the given triangle and 0 outside. In this instance, the dependence of X and Y can be deduced without any calculation, if it is noticed that there are events A and B in the interval $[0, 1]$, with positive probabilities, whose intersection is a region in the plane entirely outside the triangle that carries all of the probability. Figure 2–18 illustrates this situation; the shaded region has zero probability, but this zero is not the product of the probabilities of A and B, which are positive.

The reasoning in the last example shows that in general the region in the plane where the joint density or joint probability is not 0 must "factor." That is, it must be the intersection of "cylinder sets" with bases on the coordinate axes. (A cylinder set here is a set in the plane defined by a condition on only one coordinate; its base is the set on the axis of that coordinate defined by the condition. Thus, the events $\{X \text{ in } A\}$ and $\{Y \text{ in } B\}$, when considered as sets in the plane, are cylinder sets with bases A and B, respectively.) If the bases are

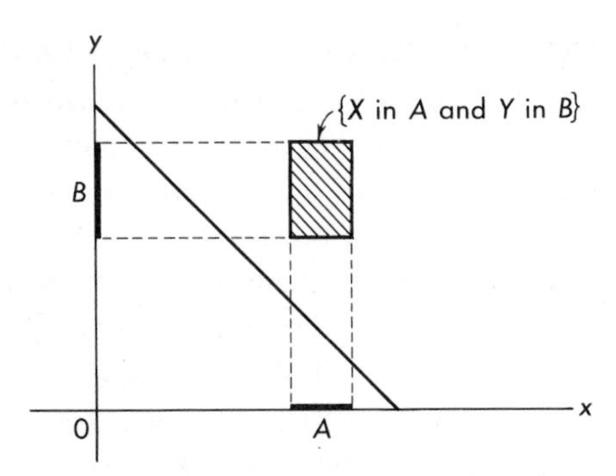

Figure 2–18. Nonfactorization of the support set in Example 2.34.

intervals, the product is an ordinary rectangle with sides parallel to the coordinate axes.

EXAMPLE 2–35. The probability table for (X, Y) given in Example 2–30 and repeated here is another illustration of a situation in which the dependence of X and Y can be seen at a glance.

Y \ X	1	2	3	
1	0	$\frac{1}{6}$	$\frac{1}{6}$	$\frac{1}{3}$
2	$\frac{1}{6}$	0	$\frac{1}{6}$	$\frac{1}{3}$
3	$\frac{1}{6}$	$\frac{1}{6}$	0	$\frac{1}{3}$
	$\frac{1}{3}$	$\frac{1}{3}$	$\frac{1}{3}$	

There are zeros in this table corresponding to combinations of X- and Y-values that have nonzero probabilities, whereas for independence, each entry in the table must be the product of corresponding marginal entries. The following table of probabilities for random variables X' and Y' is one in which the entries are obtained by multiplication of marginal probabilities. The distributions of X' and Y' are identical with the marginal distributions of X and Y above, but here X' and Y' are independent.

Y' \ X'	1	2	3	
1	$\frac{1}{9}$	$\frac{1}{9}$	$\frac{1}{9}$	$\frac{1}{3}$
2	$\frac{1}{9}$	$\frac{1}{9}$	$\frac{1}{9}$	$\frac{1}{3}$
3	$\frac{1}{9}$	$\frac{1}{9}$	$\frac{1}{9}$	$\frac{1}{3}$
	$\frac{1}{3}$	$\frac{1}{3}$	$\frac{1}{3}$	

[It should not be concluded from this example that independence implies equal likelihood of the pairs (x_i, y_j); the pairs are equally likely, simply because the marginal values are equally likely.]

Although not generally the case, it is true for *independent* random variables that the marginal distributions uniquely define a joint distribution. To construct a joint density for a model with independence, it is only necessary to multiply together the marginal densities (or to multiply the marginal c.d.f.'s to obtain the joint c.d.f.). Example 2–35 gives two bivariate distributions having the same marginal distributions, but only one of them has independent marginals.

An important consequence of the independence of two random variables, to be exploited later, is that any functions of the individual variables are independent; that is, if $Z = g(X)$ and $W = h(Y)$ and X and Y are independent, then so also are Z and W independent. For, if A' denotes the set of X-values such that $g(X)$ is in A, and B' denotes the set of Y-values such that $h(Y)$ is in B, then

$$\begin{aligned} P[g(X) \text{ in } A \text{ and } h(Y) \text{ in } B] &= P(X \text{ in } A' \text{ and } Y \text{ in } B') \\ &= P(X \text{ in } A')P(Y \text{ in } B') \\ &= P[g(X) \text{ in } A]P[h(Y) \text{ in } B]. \end{aligned}$$

Problems

2–47. The random vector (X, Y) has a discrete distribution, defined by the accompanying table of probabilities. Determine the following:

$Y \backslash X$	1	2	3
2	$\frac{1}{12}$	$\frac{1}{6}$	$\frac{1}{12}$
3	$\frac{1}{6}$	0	$\frac{1}{6}$
4	0	$\frac{1}{3}$	0

(a) $P(X = 1 \mid X + Y \leq 5)$.
(b) $P(X = 2 \mid Y = 2)$.
(c) $P(Y = 2 \mid X > 1)$.
(d) $f(y \mid 2)$.
(e) $f(x \mid 2)$, $f(x \mid 3)$, and $f(x \mid 4)$; then verify that

$$f_X(x) = \sum_y f(x \mid y) f_Y(y).$$

2–48. Let (X, Y) have a joint density function that is constant over the circle $x^2 + y^2 \leq 4$, and 0 outside that region. Compute the following:

(a) $f(x \mid y)$ for $y = 1$. (b) $P(|X| < 1 \mid Y = .5)$.

2–49. Given the joint density function of (X, Y):

$$f(x, y) = 6(1 - x - y), \quad \text{for} \quad 0 < y < 1 - x, x > 0.$$

Determine the conditional density functions, for X given $Y = y$, and for Y given $X = x$.

2–50. Let (X, Y) have a continuous bivariate distribution, let A be any set of X-values, and B a set of Y-values. Show the following:

(a) $P(A) = \int_{-\infty}^{\infty} P(A \mid Y = y) f_Y(y)\, dy.$

(b) $F_X(x) = \int_{-\infty}^{\infty} F(x \mid y) f_Y(y)\, dy.$

(c) $P(X \text{ in } A \text{ and } Y \text{ in } B) = \int_B P(A \mid Y = y) f_Y(y)\, dy.$

2–51. Show that if $\{E_i\}$ is a partition of the sample space,

$$f_X(x) = \sum f_{X|E_i}(x) P(E_i).$$

2–52. Show that the conditional density of $X \mid Y = y$ can be thought of as a limit of the conditional density of X, given that Y is in a strip $y < Y < y + \Delta y$. That is, show (formally) that

$$\lim_{\Delta y \to 0} f_{X|y<Y<y+\Delta y}(x) = \frac{f(x, y)}{f_Y(y)}.$$

★**2–53.** Suppose that the line $y = y_0$ is approximated by the region between $y = y_0$ and $y = y_0 + \Delta m|x|$ (instead of by the strip used in Problem 2–52). Obtain the conditional density of X, given that (X, Y) is in that region, and show that the limit as $\Delta m \to 0$ is a density function. But then show also that it does *not* have the property that its integral weighted with $f_Y(y)$ produces $f_X(x)$.

2–54. Give a probability table for random variables X and Y that have the same marginals as those in Problem 2–47 but are independent.

2–55. (a) Show, without any computations, that the random variables (X, Y) in each of the Problems 2–47, 2–48, and 2–49 are not independent.
(b) Show that the random variables (θ, ϕ) in Example 2–23 are independent. Show also that the random variables $X = \sin\theta \cos\phi$ and $Y = \sin\theta \sin\phi$ are *not* independent.

2–56. Determine the joint density function of (X, Y), where X and Y are independent and each has a constant density on the interval $[0, 1]$ (and 0 outside).

2–57. Show that if X is discrete with values $x_1, \ldots, x_m$ and Y is discrete with values $y_1, \ldots, y_n$, if the X-values are equally likely and the Y-values are equally likely, and if X and Y are independent, then the mn possible pairs (x_i, y_j) are equally likely.

2–58. Let U be the number of trials needed to get the first heads, and V the number needed to get two heads in repeated tosses of a coin. Show that U and V are dependent.

2.1.10 Multivariate Distributions

A random vector with more than two components defines a distribution in the space of its "values" (ordered triples of real numbers for three, ordered n-tuples of real numbers for n components). This distribution can be discussed in much the same manner as that in which the bivariate distribution of a random vector with two components was given in the preceding sections. The distribution can be characterized by its joint or multivariate *distribution function*:

$$F_{X_1,\ldots,X_n}(x_1, \ldots, x_n) = P(X_1 \le x_1, \ldots, \text{and } X_n \le x_n)$$
$$= P^{\Omega}(X_1(\omega) \le x_1, \ldots, \text{and } X_n(\omega) \le x_n).$$

[As is customary by now, probability is assigned to any set in the space of n-tuples as the probability of the image set in Ω under the vector function defining $(X_1, \ldots, X_n)$.] Marginal distributions of various types are obtained by setting the arguments corresponding to unwanted variables equal to ∞. For instance, the bivariate marginal distribution function of X_1 and X_2 is

$$F_{X_1,X_2}(x_1, x_2) = F_{X_1,\cdots,X_n}(x_1, x_2, \infty, \ldots, \infty).$$

Clearly, then, if n-dimensional Euclidean space itself is the sample space Ω, with $\omega = (x_1, \ldots, x_n)$, and probabilities of events are defined by specifying as a distribution function some function $F(x_1, \ldots, x_n)$, then this function must have the property that an $n - k$ dimensional distribution function is obtained upon setting k of the arguments x_i equal to ∞. It must also have the property that it is zero if $x_i = -\infty$ for any i; and to make sure that the probability of every "rectangle" (rectangular parallelepiped) is nonnegative, its nth order mixed difference must be nonnegative. (This is the quantity that appears in the numerator of the fraction whose limit defines the nth order, mixed partial derivative, when this exists.)

Again, there are two special classes of distributions of importance—continuous and discrete. The continuous distributions are defined by distribution functions that are continuous and have an nth order, mixed partial derivative,

$$f(x_1, \ldots, x_n) = \frac{\partial^n}{\partial x_1 \cdots \partial x_n} F(x_1, \ldots, x_n),$$

called the density function of the distribution, which is sufficiently smooth to permit recovery of the distribution function by integration:

$$F(x_1, \ldots, x_n) = \int_{-\infty}^{x_1} \cdots \int_{-\infty}^{x_n} f(u_1, \ldots, u_n)\, du_n \cdots du_1.$$

The probability of any event S is obtained as the multiple integral over S of the joint density function:

$$P[(X_1, \ldots, X_n) \text{ in } S] = \int_S \cdots \int f(x_1, \ldots, x_n)\, dx_1 \cdots dx_n,$$

and of course the probability of the whole space, $F(\infty, \ldots, \infty)$, is the integral of the density over the whole space and is 1. A distribution can be specified by defining its density to be a particular nonnegative function $f(x_1, \ldots, x_n)$ having the property that its integral over the whole space is 1.

It is helpful to consider the marginal distribution of a subset of variables as the *projection* of the probability or "mass" in a joint distribution of all the variables onto the subspace of values of the subset. For example, given (X, Y, Z), the joint distribution of (X, Y) is the projection onto the xy-plane of the joint distribution of (X, Y, Z) in space. All of the probability in a right cylinder of base A in the xy-plane is collapsed down onto the plane and is the probability of A.

EXAMPLE 2–36. Consider the distribution for (X, Y, Z) defined by the density

$$f(x, y, z) = \frac{3}{4\pi}, \qquad \text{for } x^2 + y^2 + z^2 \leq 1.$$

That is, the density is constant inside a unit sphere, and the value of that constant is the reciprocal of the volume of the sphere [so that the integral of $f(x, y, z)$ over the sphere is 1]. The probability of a subset of this sphere is then proportional to its volume. The distribution of Z is obtained by projecting the probability in the (X, Y, Z)-space onto the Z-axis. Thus, for $|\lambda| < 1$,

$$P(Z \leq \lambda) = \frac{\text{volume of portion of sphere below } Z = \lambda}{\text{volume of sphere}}$$

$$= \frac{3}{4\pi}\int_{-1}^{\lambda} \pi(1 - z^2)\, dz = \frac{1}{4}(2 + 3\lambda - \lambda^3)$$

and the density of Z is, therefore,

$$f_Z(\lambda) = \tfrac{3}{4}(1 - \lambda^2), \qquad \text{for } -1 < \lambda < 1.$$

Notice that the density is large near $z = 0$ and smallest at $z = 1$ and -1, as is to be expected. (It vanishes outside $|z| < 1$.)

Conditional distributions are defined just as in the bivariate case. If the given condition is that certain components of a random vector have specified values, the conditional density (speaking in terms of the continuous case) is the joint density of all the components divided by the joint marginal density of those components whose values are given—with those particular values used wherever the corresponding dummy variables appear.

EXAMPLE 2–37. Consider the uniform distribution in the unit sphere, as introduced in the preceding example. The joint conditional density of X and Y, given $Z = k$, would be

$$f(x, y \mid Z = k) = \frac{f(x, y, k)}{f_Z(k)} = \frac{3/(4\pi)}{3(1 - k^2)/4} = \frac{1}{\pi(1 - k^2)},$$

for $x^2 + y^2 \leq 1 - k^2$. That is, the conditional distribution, given $Z = k$ is uniform within the circle of radius $1 - k^2$. The density of this distribution in the plane $Z = k$ varies just as does the joint density of (X, Y, Z) in that plane, and the division by $f_Z(k)$ furnishes precisely the right normalization to make the integral of the conditional density over the cross section equal to 1.

As in the bivariate case, independence can be defined in any of several ways, according to one of the following criteria:

1. $f(x_1, \ldots, x_n) = f_{X_1}(x_1)f_{X_2}(x_2)\cdots f_{X_n}(x_n)$.
2. $F(x_1, \ldots, x_n) = F_{X_1}(x_1)F_{X_2}(x_2)\cdots F_{X_n}(x_n)$.
3. $P(X_1 \text{ in } A_1, \ldots, \text{and } X_n \text{ in } A_n) = P(X_1 \text{ in } A_1)\cdots P(X_n \text{ in } A_n)$, for every choice of the n events $A_1, \ldots, A_n$.

[The f in (1) denotes the density function in continuous cases or the probability function in discrete cases.] These conditions are equivalent. Notice that independence automatically obtains in the marginal distribution of any subset of $(X_1, \ldots, X_n)$, since it follows from (2) upon replacing the dummy variables corresponding to the remaining variables by ∞, or from (3) by taking the A's corresponding to those remaining variables to be the interval $(-\infty, \infty)$.

The factorization condition for independence can be used to check a given multivariate distribution for independence of the components; but it is more often used to *construct* a multivariate model that is to incorporate an observed or postulated independence by multiplying marginal densities or c.d.f.'s to form the multivariate density or c.d.f.

EXAMPLE 2–38. The model for an experiment of chance is defined by the density function $f(x)$ for the random variable X. Suppose that the experiment is repeated n times in such a way that no performance is influenced by the result of any others. One would then like a model for the results of the successive trials of the experiment. These make up the random vector $(X_1, X_2, \ldots, X_n)$ in which X_i denotes the result of the ith trial of the experiment. The joint density function of the n observations is defined to be the *product* of the marginal distributions; and because each X_i is a replica (in distribution) of the basic experiment, these marginal distributions are identical and defined by the density $f(x)$. Thus,

$$f_{X_1,\ldots,X_n}(x_1, \ldots, x_n) = f_{X_1}(x_1)\cdots f_{X_n}(x_n) = f(x_1)f(x_2)\cdots f(x_n) = \prod_{i=1}^{n} f(x_i).$$

For instance, if

$$f(x) = Ke^{-x^2},$$

then the joint density of n independent observations on this density is

$$\begin{aligned} f_{X_1,\ldots,X_n}(x_1, \ldots, x_n) &= K^n \prod \exp(-x_i^2) \\ &= K^n \exp(-\textstyle\sum x_i^2). \end{aligned}$$

Again, any functions of the individual variables, $g_1(X_1), \ldots, g_n(X_n)$ are independent if $X_1, \ldots, X_n$ are. Moreover, any functions of nonoverlapping subsets of the independent random variables $X_1, \ldots, X_n$ define independent random variables. Thus, for example, if X_1, X_2, X_3, and X_4 are independent, so are $Y = g(X_1, X_4)$ and $Z = h(X_2, X_3)$. For, if A and B denote sets in the value space of Y and of Z, and A' and B' denote their respective pre-images under the functions g and h, then

$$\begin{aligned} P(Y \text{ in } A \text{ and } Z \text{ in } B) &= P[(X_1, X_4) \text{ in } A' \text{ and } (X_2, X_3) \text{ in } B'] \\ &= \iint_{A'}\iint_{B'} f(x_1, x_2, x_3, x_4)\, dx_2\, dx_3\, dx_1\, dx_4 \\ &= \iint_{A'} f_{X_1,X_4}(x_1, x_4)\, dx_1\, dx_4 \cdot \iint_{B'} f_{X_2,X_3}(x_2, x_3)\, dx_2\, dx_3 \\ &= P[(X_1, X_4) \text{ in } A']P[(X_2, X_3) \text{ in } B'] \\ &= P[Y \text{ in } A]P[Z \text{ in } B]. \end{aligned}$$

A similar argument would establish the more general assertion.

Random *vectors* $\mathbf{X} = (X_1, \ldots, X_m)$ and $\mathbf{Y} = (Y_1, \ldots, Y_n)$ are said to be *independent* when the joint density of the combined vector $(\mathbf{X}, \mathbf{Y}) = (X_1, \ldots, X_m, Y_1, \ldots, Y_n)$ factors into the product of the densities of $\mathbf{X}$ and $\mathbf{Y}$. (Definitions for the discrete and general cases are in terms of the joint probability functions and joint distribution functions, respectively.) And again, if $\mathbf{X}$ and $\mathbf{Y}$ are independent, then $g(\mathbf{X})$ and $h(\mathbf{Y})$ are independent random variables—or independent random vectors, if g and h are vector functions.

Problems

2–59. A discrete distribution for (X, Y, Z) is defined by assigning equal probabilities to the following six points: (0, 0, 0), (0, 0, 1), (2, 0, 1), (2, 1, 0), (1, 2, 1), and (0, 2, 0).

(a) Determine the marginal distributions of X, Y, and Z.
(b) Determine the probability function for the joint marginal distribution of (X, Y).
(c) Determine the probability function for the joint conditional distribution of (X, Y) given $Z = 1$.

2–60. Let (X, Y, Z) have a uniform distribution in the unit sphere $x^2 + y^2 + z^2 \leq 1$. Derive the distribution of the random variable $R = (X^2 + Y^2 + Z^2)^{1/2}$.

2–61. Let (X, Y, Z) have a continuous joint distribution with constant density in the portion of the first octant ($x \geq 0$, $y \geq 0$, $z \geq 0$) that is bounded by the plane $x + y + z = 1$.

(a) Determine the constant value of the density.
(b) Determine the density of the marginal distribution of (X, Y).
(c) Determine the density of the marginal distribution of X.
(d) Determine the conditional density of (X, Y) given $Z = \frac{1}{2}$.

2–62. Write out the density or probability function for the joint distribution of $(X_1, \ldots, X_n)$, where X_i denotes the outcome of the ith in a sequence of independent trials of an experiment whose result

(a) Has the density function e^{-x}, for $x > 0$.
(b) Has the probability function $p^x(1 - p)^{1-x}$, for $x = 0, 1$, where p is a given number on (0, 1).

2–63. Show that the density function

$$f(x_1, \ldots, x_n) = 1, \qquad \text{for } 0 < x_i < 1, \quad i = 1, \ldots, n,$$

defines a distribution for the random vector $(X_1, \ldots, X_n)$ in which the marginal variables are independent.

2–64. Show that the density function

$$f(x_1, \ldots, x_n) = K, \qquad \text{for } x_i > 0, \quad x_1 + \cdots + x_n \leq 1,$$

defines a distribution in which the marginal variables are not independent.

2.2 Expectation

The terms *expectation, expected value, average value,* and *mean value* are all used for the same concept. In the case of a probability distribution in the value space of a random variable, they refer to what is the analog of the center of gravity of a mass distribution along a line. The term *expected value* will be used most frequently here, even though it is slightly misleading—the expected value of a given random variable may not even be a possible value, let alone an "expected" one.

Like the center of gravity of a mass distribution, the expected value is used as a measure of centering or location of a probability distribution. The term *average* can be motivated in the context of a long series of plays of a game of chance, as the reward per trial when the total winnings are distributed equally among the trials to determine an entry fee.

EXAMPLE 2–39. Suppose that a reward is offered, a number of dollars equal to the number of points showing when a die is tossed. If the game is to be fair, an entry fee should be charged for the privilege of playing the game. In a long series of plays, say N of them, the total amount of fees collected would be Nx dollars, if x is the fee for a single play. In this series of plays, the numbers 1, 2, 3, 4, 5, 6 would turn up approximately equally often as the number of points showing. If one assumes exactly $N/6$ of each, the total reward would be a number of dollars equal to

$$1 \cdot \frac{N}{6} + 2 \cdot \frac{N}{6} + \cdots + 6 \cdot \frac{N}{6}.$$

If the game is to be fair, this total should equal the total of the entry fees; setting it equal to Nx, one finds the equitable entry fee to be

$$x = 1 \cdot \tfrac{1}{6} + 2 \cdot \tfrac{1}{6} + \cdots + 6 \cdot \tfrac{1}{6} = \tfrac{7}{2},$$

or \$3.50. This is sometimes called the "mathematical expectation" of a single play of the game, but of course the actual amount won in a single play will never equal \$3.50—it will be either \$1, \$2, . . . , or \$6. The expectation or entry fee is in a sense an idealization, since in an actual sequence of N plays, the proportions will not be exactly equal to $N/6$, although close to it. The \$3.50 is then obtained only upon distributing the winnings in an *infinite* sequence of plays, an idealization that accounts for the qualifier "mathematical."

The number of dollars won in the above example is a random variable, and the same computation—multiplying values by proportions and summing—will be used in defining the expected value of more general random variables. In continuous cases it will be necessary to integrate instead of sum, and the appropriate definitions of expectation in general is given as an integral.

2.2.1 Simple Random Variables

A random variable that assumes one of a finite set of values is said to be *simple*. That is, there is a list of possible values $x_1, x_2, \ldots, x_k$ and corresponding probabilities $p_1, p_2, \ldots, p_k$ such that $p_1 + \cdots + p_k = 1$. The *expected value* of a simple random variable having these values and corresponding probabilities is defined to be

$$EX = x_1 p_1 + x_2 p_2 + \cdots + x_k p_k = \sum_{i=1}^{k} x_i p_i.$$

(The expected value will also be written as $E(X)$, especially when the name of the random variable is more complicated than just X.) The symbol μ or the more specific μ_X will also be used for this quantity, referring to the alternative terminology of *mean* value.

The weighted sum that defines expected value (the weights for the various values are the corresponding probabilities) is precisely the kind of weighted sum that defines center of gravity. If a mass m_1 is located on the x-axis at x_1, the mass m_2 at $x_2, \ldots$, and the mass m_k at x_k, the center of gravity of this system of masses is computed as follows:

$$\text{c.g.} = \frac{\sum x_i m_i}{\sum m_i} = \sum x_i \left(\frac{m_i}{m}\right),$$

where m is the total mass, and each sum extends from $i = 1$ to $i = k$. Notice that the relative mass—the proportion of the total mass at x_i—is the weight attached to x_i and, therefore, is the analog of the probability that $X = x_i$. Notice also that the sum of the relative masses equals 1, as does the sum of the probabilities.

The formula for EX can be rewritten in terms of probabilities in Ω if this sample space has a finite number of points. For, the probability p_i can be expressed as a sum of the probabilities of all points ω such that $X(\omega) = x_i$; therefore,

$$EX = \sum_i x_i p_i = \sum_{\omega \text{ in } \Omega} X(\omega) P(\omega).$$

EXAMPLE 2–40. The model for three independent tosses of a fair coin assigns equal probabilities (each $\frac{1}{8}$) to the eight distinct sequences making up the sample space. Let $X(\omega)$ denote the number of heads in the sequence ω [for example, $X(H, H, T) = 2$]. The mean value of the random variable X can be computed either as a sum over the sample space or as a sum over the value space. The sample points are: $HHH, HHT, HTH, THH, TTH, THT, HTT, TTT$, and the values of X are 3, 2, 2, 2, 1, 1, 1, 0, respectively. Thus, $p(3) = p(0) = \frac{1}{8}$ and $p(1) = p(2) = \frac{3}{8}$, so that

$$EX = 0 \cdot \tfrac{1}{8} + 1 \cdot \tfrac{3}{8} + 2 \cdot \tfrac{3}{8} + 3 \cdot \tfrac{1}{8} = \tfrac{3}{2}.$$

The calculation using the sample space is as follows:

$$EX = 0 \cdot \tfrac{1}{8} + 1 \cdot \tfrac{1}{8} + 1 \cdot \tfrac{1}{8} + 1 \cdot \tfrac{1}{8} + 2 \cdot \tfrac{1}{8} + 2 \cdot \tfrac{1}{8} + 2 \cdot \tfrac{1}{8} + 3 \cdot \tfrac{1}{8}.$$

The equivalence of these two computations of EX is at once apparent, the $\frac{3}{8}$ multiplying 1 corresponding to three terms in the second sum in which $\frac{1}{8}$ multiplies 1, and so forth.

As discussed in Section 2.1.6, a function $g(x)$ defines with $X(\omega)$ a new random variable $Y = g(X(\omega))$, written usually as $Y = g(X)$. Whether or not Ω is discrete, the space of X-values is a discrete probability space if X is a simple random variable, and Y can be thought of as a random variable on this probability space. But then, interpreting the X-space as Ω and the function $g(x)$ as the $X(\omega)$ in the formula preceding Example 2–40, one obtains:

$$EY = Eg(X) = \sum_j y_j P(Y = y_j) = \sum_i g(X_i) P(X = x_i).$$

If Ω is discrete, there is a third level at which the computation can be performed:

$$EY = E[g(X(\omega))] = \sum_{\omega \text{ in } \Omega} g(X(\omega))P(\omega).$$

The formula involving the probabilities $P(X = x_i)$ is ordinarily the simplest, since these probabilities are given, whereas the probabilities $P(Y = y_j)$ would have to be computed.

EXAMPLE 2–41. A cube has its six sides colored red, white, blue, green, yellow, and violet. It is assumed that these six sides are equally likely to show, when the cube is tossed. That is, one defines a probability space with elementary outcomes R, W, B, G, Y, V and probabilities $\frac{1}{6}$ for each of these outcomes; this space is Ω.

Consider now the random variable that assigns the number 1 to R and W, the number 2 to G and B, and the number 3 to Y and V:

$$X(R) = X(W) = 1;$$
$$X(G) = X(B) = 2;$$
$$X(Y) = X(V) = 3.$$

The space $\mathscr{X}$ of values of $X(\omega)$ can be thought of as consisting of the numbers 1, 2, and 3, each with probability $\frac{1}{3}$, the distribution induced by $X(\omega)$ from that on Ω.

Next, let $Y = (X - 2)^2$. The values $X = 1$ and $X = 3$ give rise to $Y = 1$; the value $X = 2$, to $Y = 0$. Thus, $\mathscr{Y}$ consists of the values 1 and 0 with probabilities $\frac{2}{3}$ and $\frac{1}{3}$, respectively. There are, then, three possible computations of

$$E(Y) = E[(X - 2)^2] = E[(X(\omega) - 2)^2].$$

One uses the distribution on Ω:

$$\tfrac{1}{6}[X(R) - 2]^2 + \cdots + \tfrac{1}{6}[X(V) - 2]^2 = \tfrac{2}{3};$$

another uses the distribution on $\mathscr{X}$:

$$\tfrac{1}{3}(1 - 2)^2 + \tfrac{1}{3}(2 - 2)^2 + \tfrac{1}{3}(3 - 2)^2 = \tfrac{2}{3};$$

and the third uses the distribution on $\mathscr{Y}$:

$$1 \cdot \tfrac{2}{3} + 0 \cdot \tfrac{1}{3} = \tfrac{2}{3}.$$

Following the mappings and transfers of probability makes it clear in this example that the computations are equivalent.

A special case of a simple random variable is that in which only one value is assumed: $P(X = k) = 1$. In such a case the expected value computation reduces to a single term and yields $EX = k$.

A special case of a function of a random variable is $Y = kX$. The mean value of Y can be computed from X-probabilities:

$$EY = \sum_i kx_i p_i = k \sum_i x_i p_i = kEX.$$

That is, the averaging operation is *homogeneous*—multiplying a random variable by a constant multiplies the mean value by that constant.

A frequently encountered type of random variable is one that is a sum of random variables. Given the random vector $(X(\omega), Y(\omega))$, then, consider the random variable $Z(\omega) = X(\omega) + Y(\omega)$. The mean value of Z can be computed in terms of the mean values of X and Y as follows:

$$\begin{aligned} EZ = E[X(\omega) + Y(\omega)] &= \sum [X(\omega) + Y(\omega)]P(\omega) \\ &= \sum X(\omega)P(\omega) + \sum Y(\omega)P(\omega) \\ &= EX + EY. \end{aligned}$$

This computation is readily extended by induction to the proposition that the average of any finite sum of random variables is the sum of the averages. Combining this *additivity* property of $E(\omega)$ with the homogeneity property just derived, one obtains the condition of *linearity* of the averaging operation:

$$E(aX + bY) = aEX + bEY.$$

[This derivation would seem to depend on the existence of an Ω on which both X and Y are defined. However, this sample space can be taken to be the space of points (x, y), with $X(x, y) = x$ and $Y(x, y) = y$. Moreover, the joint probabilities in this space can be taken to be anything at all with the given marginal probabilities for X and for Y, since the computation only involves those marginal probabilities.]

2.2.2 Expectation: Discrete and Continuous Variables

The general definition of expectation will be given in Section 2.2.5 (optional). For most purposes it will be sufficient to have working definitions for the important special cases of discrete and continuous random variables.

The expectation of a discrete random variable is defined using an obvious generalization of the definition for a simple random variable. Let X be discrete with possible values $x_1, x_2, \ldots$, and corresponding probabilities $p_1, p_2, \ldots$. The expected value of X is defined to be

$$EX = x_1 p_1 + x_2 p_2 + \cdots = \sum_i x_i p_i,$$

or in the notation of the probability function, $f(x_i) = p_i$,

$$EX = \sum_i x_i f(x_i).$$

This sum may be infinite and may not converge as an infinite sum. Indeed, if it does *not* converge *absolutely*, the mean EX is said not to exist, and one writes $E|X| = \infty$.

EXAMPLE 2–42. Consider the sample space Ω of all positive integers and the distribution defined by

$$P(\omega = k) = \frac{6}{\pi^2 k^2}, \qquad \text{for } k = 1, 2, \ldots.$$

(This is a legitimate distribution, since $\sum 1/k^2 = \pi^2/6$.) Then define the random variable

$$X(\omega) = \begin{cases} \omega, & \text{if } \omega \text{ is an odd integer,} \\ -\omega, & \text{if } \omega \text{ is even.} \end{cases}$$

The series for EX is convergent (as an alternating harmonic series):

$$\frac{6}{\pi^2}\left(1 - 2\cdot\frac{1}{2^2} + 3\cdot\frac{1}{3^2} - \cdots\right) = \frac{6}{\pi^2}\left(1 - \frac{1}{2} + \frac{1}{3} - \cdots\right),$$

but the series for $E|X|$ is divergent. (The series of positive terms and the series of negative terms in "EX" both diverge, although divergence of either one would make $E|X|$ divergent.) The mean EX does not exist.

Having seen the analogy between formulas for the discrete and continuous cases, one can easily conjecture the correct formula for expectation in the continuous case. The sum becomes an integral and the probability $f(x_i)$ becomes a probability element, $f(x)\,dx$:

$$EX = \int_{-\infty}^{\infty} xf(x)\,dx,$$

where $f(x)$ is the density function of X.

One is led to this same definition of EX by using a limiting argument involving discrete approximations to the continuous X: Assume first that $f(x)$ vanishes outside the interval (a, b), and let that interval be subdivided into n parts of width $h = (b - a)/n$; then let x_i denote the left endpoint of the ith subinterval. The discrete variable X^* with values $x_1, \ldots, x_n$ and probabilities

$$\begin{aligned} P(X^* = x_i) &= P(x_i < X < x_i + h) \\ &\doteq f(x_i) \cdot h \end{aligned}$$

is an approximation to X, with mean value

$$EX^* = \sum_i x_i P(X^* = x_i) \doteq \sum_i x_i f(x_i) h.$$

As n becomes infinite and h tends to 0, the approximation becomes ever better, and the mean of X^* tends to the definite integral of the function $xf(x)$ over (a, b) or, since $f(x) = 0$ outside that interval, over the whole axis as in the conjectured formula. If the distribution is not included in a finite interval, one defines the mean value of

$$X_{a,b} \equiv \begin{cases} X, & \text{if } a < X < b, \\ 0, & \text{otherwise,} \end{cases}$$

as above and passes to the limit as a becomes $-\infty$ and b becomes ∞. Again the result is

$$EX = \int_{-\infty}^{\infty} xf(x)\,dx.$$

EXAMPLE 2–43. If X denotes the stopping point of the spinning pointer, with $f(x) = 1$ on $0 < x < 1$, the mean value is the integral

$$EX = \int_0^1 x \cdot 1\,dx = \tfrac{1}{2}.$$

[The contribution to the integral over $(-\infty, \infty)$ outside $(0, 1)$ is 0 because $f(x)$ vanishes there.]

The analogy between probability distributions and mass distributions was discussed in Section 1.2.6. It was referred to again in Section 2.2.1 in connection with simple random variables, where the expectation was seen to be the analog of the center of gravity of a (point) mass distribution. In the continuous case, too, the notion of expectation is mathematically the same as that of center of gravity. Physics texts often gives the formula for the center of gravity of a distribution of mass along a line as

$$\text{c.g.} = \frac{\int x \, dM}{\int dM},$$

where $dM = m(x)\,dx$ is the element of mass and $m(x)$ is the density of mass at x. The denominator is the total mass, so that

$$f(x) = \frac{m(x)}{\text{total mass}}$$

is the relative mass density (with integral 1) and the center of gravity is the integral of $xf(x)$.

In Example 2–43, then, the distribution of probability with constant density on (0, 1) is analogous to a mass distribution with constant density—a uniform rod of unit length. The balance point of the rod is the midpoint, which is also the expected value of the probability distribution.

When a probability distribution is not confined to a finite interval, the integral defining expectation is an improper integral and may be divergent—even if the density itself has a convergent integral (as it must have to be a density). In such cases, the expectation is said not to exist. This occurs when *either* of the integrals

$$\int_0^\infty xf(x)\,dx \qquad \text{or} \qquad \int_{-\infty}^0 xf(x)\,dx$$

diverges. If both of these integrals converge, then so also does

$$\int_{-\infty}^\infty |x| f(x)\,dx = \int_0^\infty xf(x)\,dx - \int_{-\infty}^0 xf(x)\,dx.$$

That is, to say that EX exists is to say also that $E|X|$ exists. In the case of $E|X|$, divergence would mean that the limit defining the integral is infinite, and one writes $E|X| = \infty$.

EXAMPLE 2–44. The density function

$$f(x) = \frac{1/\pi}{1 + x^2}$$

defines a distribution for X. But the integral

$$\int_0^\infty \frac{x/\pi}{1 + x^2}\,dx = \lim_{B\to\infty} \left(\frac{1}{2\pi} \log(1 + x^2)\Big|_0^B\right)$$

is divergent. So the mean value does not exist, and $E|X| = \infty$. It might be noted that the distribution is symmetrical, and that the symmetric limit (called the *Cauchy principal value*) does exist:

$$\lim_{A\to\infty} \int_{-A}^A \frac{x/\pi}{1 + x^2}\,dx = 0.$$

However, it seems most expedient in the subsequent development to say that EX does not exist in this case. Indeed, it might be thought quite speculative to assert that the infinite moment of the positive part of the distribution would "balance" or cancel the infinite moment for the negative part.

The formulas for expectation in the discrete and continuous cases can be thought of as special cases of a single, more general formula. This point will be explained in the optional Sections 2.2.4 and 2.2.5, but the notation for the new kind of integral required is so suggestive it is given here:

$$EX = \int_{-\infty}^{\infty} x\, dF(x).$$

The "$dF(x)$" is exactly like the notation for the differential of $F(x)$, written in terms of the differential coefficient or derivative as

$$dF(x) = f(x)\, dx,$$

and indeed this notation gives the correct formula for EX when the derivative exists—when the distribution has a density. Moreover, the general integral reduces to the usual finite sum for expectations in discrete cases; and it can also handle cases that are neither discrete nor continuous.

The general notation will sometimes be used for simplicity of presentation. (Those who omit the material in Sections 2.2.4 and 2.2.5 need only mentally replace dF by $f\,dx$ or replace the integral by a sum, according to whether the random variable in question is continuous or discrete.)

Since a conditional distribution, say of X given $Y = y$, as obtained from a joint distribution of (X, Y), is again a distribution, the concept of expected value applies to conditional distributions. The mean value of a conditional distribution of a random variable is called the *conditional mean*, just to refer to the origin of the distribution as obtained by imposing a condition on one of two variables in a bivariate distribution; but the computation of a conditional mean is not different.

EXAMPLE 2–45. Let (X, Y) have the discrete distribution defined in the accompanying table (as in Problem 2–47).

Y \ X	1	2	3
2	$\frac{1}{12}$	$\frac{1}{6}$	$\frac{1}{12}$
3	$\frac{1}{6}$	0	$\frac{1}{6}$
4	0	$\frac{1}{3}$	0
	$\frac{1}{4}$	$\frac{1}{2}$	$\frac{1}{4}$

The conditional probabilities for $X = 1, 2, 3$ given $Y = 3$ are $\frac{1}{2}, 0, \frac{1}{2}$, respectively, so the conditional mean is

$$E(X \mid Y = 3) = 1 \cdot \tfrac{1}{2} + 2 \cdot 0 + 3 \cdot \tfrac{1}{2} = 2.$$

It can be seen, similarly, that the conditional mean, given $Y = 2$, and the conditional mean, given $Y = 4$, are also 2 and, moreover, that the "unconditional" mean, or the means of the marginal distribution of X, is also 2:

$$E(X) = 1 \cdot \tfrac{1}{4} + 2 \cdot \tfrac{1}{2} + 3 \cdot \tfrac{1}{4} = 2.$$

Observe, then, that the conditional mean *can* be independent of the condition and equal to the unconditional mean, even though the random variables are *not* independent.

EXAMPLE 2–46. Given $f(x, y) = 6(1 - x - y)$ for $0 < y < 1 - x$, $x > 0$, and $f(x, y) = 0$ elsewhere, the conditional density (see Problem 2–49) is $f(x \mid y) = 2(1 - x - y)/(1 - y)^2$, for $0 < x < 1 - y$. The conditional mean, given $Y = y$, is then the integral of x with respect to this conditional density:

$$E(X \mid Y = y) = \int_{-\infty}^{\infty} x f(x \mid y)\, dx$$

$$= \int_0^{1-y} \frac{2(1 - x - y)x}{(1 - y)^2}\, dx = \frac{1}{3}(1 - y).$$

The conditional mean $E(X \mid Y = y)$ is, of course, a function of the conditioning value y. This function is called the *regression function* of X on Y. It is used as a predictor for X when Y is given (see Chapter 10).

Problems

2–65. Determine the expected value of each of the following random variables, defined earlier in problems on p. 69:

(a) The number of defectives in a random selection of 4 articles from a lot of 10 in which 2 are defective (Problem 2–14). Given $f(0) = \frac{1}{3}, f(1) = \frac{8}{15}, f(2) = \frac{2}{15}$.
(b) The number of tests required to locate the 1 defective tube among 5 tubes (Problem 2–15). Given $f(x) = \frac{1}{5}$ for $x = 1, 2, 3$; $f(4) = \frac{2}{5}$.
(c) The number of correct identifications in the random assignment of 3 brand names to 3 cigarettes (Problem 2–16). Given $f(0) = \frac{1}{3}, f(1) = \frac{1}{2}, f(3) = \frac{1}{6}$.
(d) The number of points assigned in the Goren system to a card drawn at random from a standard deck. (See Example 2–28.) Given $f(x) = \frac{1}{13}$ for $x = 1, 2, 3, 4$; $f(0) = \frac{9}{13}$.

2–66. Three chips are drawn at random from five chips numbered 1 through 5, respectively (Problem 2–17). Let Y denote the smallest, and Z the largest number drawn. Let $R = Z - Y$ and verify that $ER = EZ - EY$ by computing each term.

2–67. Determine the expected total number of points that will show in the toss of two dice. (Use the formula for expected value of a sum.)

2–68. Determine the expected total number of points [see Problem 2–65(d)] in a hand of 13 cards dealt from a standard deck.

2–69. Compute the expected number of tails prior to the first heads in repeated tossings of a fair coin.

2–70. Compute the expected value of a random variable Y whose density function is $f(y) = \frac{1}{2}\exp(-|y|)$.

2–71. Compute the expected value of the random variable Z whose density is $f(z) = 1 - |z|$, for $|z| < 1$.

2–72. Determine the mean value of the random variable X with density $6x(1 - x)$ for $0 < x < 1$.

2–73. Determine the mean of the distribution with c.d.f. $(x + A)/(2A)$, for $|x| < A$.

2–74. Determine the mean value of the random variable θ whose distribution function is $F_\theta(u) = 1 - \cos u$ for $0 \leq u \leq \pi/2$. (See Example 2–21.)

2–75. Show that if X has a density function $f(x)$ that is symmetrical about $x = a$: $f(a - x) = f(a + x)$ for all x, then, if the mean value exists, it must be $EX = a$.

2–76. Given that (X, Y) has a joint density that is constant over the triangle with vertices (0, 0), (1, 0), (0, 1), and 0 outside this triangle, determine the regression function of Y on X and the regression function of X on Y. (See Example 2–31.)

2–77. Compute the conditional mean $E(X \mid Y = k)$ for the discrete bivariate distribution of Example 2–35:

$$f(x, y) = \frac{1}{6}, \qquad \text{for } x, y = 1, 2, 3, \quad \text{but } x \neq y.$$

2.2.3 Expectation of a Function of Random Variables

In discussing simple random variables in Section 2.2.1, it was found that the expected value of a random variable Y that is defined as a function of another random variable X can be computed on at least two levels—using the distribution of Y, or using the distribution of X:

$$E[g(X)] = \sum_j y_j P(Y = y_j) = \sum_i g(x_i)P(X = x_i).$$

The same situation obtains more generally; that is, $E[g(X)]$ can be computed in terms of the distribution of X or in terms of the distribution of Y.

In the general notation, covering all cases, the formula is as follows:

$$E[g(X)] = \int_{-\infty}^{\infty} g(x)\, dF(x).$$

In the discrete case this reduces to the obvious extension of the formula for simple random variables:

$$E[g(X)] = \sum_i g(x_i)f(x_i),$$

where the possible value x_i has probability $f(x_i)$. In the continuous case with density function $f(x)$, it reduces to

$$E[g(X)] = \int_{-\infty}^{\infty} g(x)f(x)\, dx.$$

The validity of the general formula for $E[g(X)]$ will be demonstrated in Section 2.2.5. Here it will be shown in the continuous case for functions $g(x)$ that have unique inverses. For such a function the random variable $Y = g(X)$ has been shown to have density

$$f_Y(y) = f_X(g^{-1}(y)) \frac{dg^{-1}(y)}{dy};$$

equivalently,

$$f_Y(y)\, dy = f_X(x)\, dx.$$

Hence, the change of variable $y = g(x)$ in the integral

$$E[g(X)] = \int_{-\infty}^{\infty} g(x) f_X(x)\, dx$$

yields

$$E[g(X)] = \int_{-\infty}^{\infty} y f_Y(y)\, dy = EY.$$

EXAMPLE 2–47. Suppose that X has a distribution with density function $K \exp(-\frac{1}{2}x^2)$, as in Example 2–38. Let $Y = |X|$. The mean value of Y is determined as follows:

$$\begin{aligned} EY &= \int_{-\infty}^{\infty} |x|\, dF_X(x) = K \int_{-\infty}^{\infty} |x| \exp(-\tfrac{1}{2}x^2)\, dx \\ &= K \int_0^{\infty} x \exp(-\tfrac{1}{2}x^2)\, dx - K \int_{-\infty}^{0} x \exp(-\tfrac{1}{2}x^2)\, dx \\ &= 2K \int_0^{\infty} x \exp(-\tfrac{1}{2}x^2)\, dx = 2K. \end{aligned}$$

Since K turns out to be $1/\sqrt{2\pi}$ (computed implicitly in Example 2–25), the result is $E|X| = \sqrt{2/\pi}$. Observe that the function $|x|$ does *not* have a single-valued inverse, so a simple change of variable does not change $\int |x| f(x)\, dx$ into $\int y f(y)\, dy$.

Occasionally, it is necessary to compute the mean of a function of a conditioned variable. This is simply the mean with respect to the weighting defined by the conditional density. Thus,

$$E[g(X) \mid A] = \int_{-\infty}^{\infty} g(x) f_{X|A}(x)\, dx.$$

Suppose now that $\{A_i\}$ is a partition of the sample space. The unconditional mean of $g(X)$ can be computed as an average of conditional means with respect to the probabilities of the partition sets:

$$E[g(X)] = \sum_i E[g(X) \mid A_i] P(A_i),$$

where only those A_i of positive probability need be included in the sum. This formula is an easy consequence of the result in Problem 2–51.

In Section 2.1.8, dealing with conditional distributions, it was found that one property of conditional probability functions is this:

$$P(A) = \sum_j P(A \mid Y = y_j) f_Y(y_j),$$

for the case of discrete Y. The analog for the continuous case was used as a criterion to be satisfied by a reasonable definition of conditional density, namely,

$$P(A) = \int_{-\infty}^{\infty} P(A \mid Y = y) f_Y(y)\, dy.$$

Upon comparison of these with the formula for the expected value of a function of a random variable, it is seen that these are special cases. That is,

$$P(A) = Eg(Y), \qquad \text{where } g(y) = P(A \mid Y = y).$$

Thus, the conditional probabilities are averaged over the possible values of the conditioning variable to obtain an unconditioned probability. The result is often written:

$$P(A) = E[P(A \mid Y)] \qquad \text{or} \qquad P(A) = E_Y[P(A \mid Y)].$$

This last result can be used, in turn, to show that averaging conditional expectations with respect to the conditioning variable yields the unconditional expectation. In the above formula, let A denote the event $\{X \le x\}$, and so obtain

$$F_X(x) = P(X \le x) = E[P(X \le x \mid Y)] = \int_{-\infty}^{\infty} F(x \mid y)\, dF_Y(y).$$

It then follows that

$$EX = \int_{-\infty}^{\infty} x\, dF_X(x) = \int_{-\infty}^{\infty} \left\{ \int_{-\infty}^{\infty} x\, dF(x \mid y) \right\} dF_Y(y)$$

$$= \int_{-\infty}^{\infty} E(X \mid y)\, dF_Y(y).$$

This is also written in the condensed form

$$EX = E_Y[E(X \mid Y)],$$

which means

$$EX = E[h(Y)], \qquad \text{where } h(y) = E(X \mid Y = y).$$

EXAMPLE 2–48. Let (X, Y) have the joint density

$$f(x, y) = \begin{cases} 2, & \text{for } x + y \le 1, \quad x \ge 0, \quad y \ge 0, \\ 0, & \text{elsewhere.} \end{cases}$$

The marginal density of Y (as in Example 2–31) is

$$f_Y(y) = \int_{-\infty}^{\infty} f(x, y)\, dx = \int_0^{1-y} 2\, dx = 2(1 - y), \qquad 0 < y < 1,$$

and the conditional density of X given $Y = y$ is

$$f(x \mid y) = \frac{f(x, y)}{f_Y(y)} = \frac{2}{2(1 - y)}, \quad \text{for } 0 < x < 1 - y.$$

The conditional mean of X given $Y = y$ is then

$$E(X \mid y) = \frac{1}{1 - y} \int_0^{1-y} x\, dx = \frac{1 - y}{2}.$$

Integrating this with respect to the distribution of Y yields the (unconditional) mean of X:

$$EX = E_Y[E(X \mid Y)] = \int_0^1 \frac{1 - y}{2} 2(1 - y)\, dy = \frac{1}{3}.$$

A function of two random variables also defines a random variable:

$$Z = g(X, Y),$$

and it is desirable to be able to compute EZ from the joint distribution of (X, Y) without going through the intermediate step of determining the distribution of Z. In the case of a discrete distribution with probability function

$$f(x, y) = P(X = x,\ Y = y),$$

the formula is

$$E[g(X, Y)] = \sum g(x, y) f(x, y),$$

where the sum extends over all possible pairs (x, y). This result is actually not so hard to derive directly; the more involved derivation is needed for the non-discrete cases. If (X, Y) has a continuous distribution in the plane with density function $f(x, y)$, the formula is

$$E[g(X, Y)] = \iint g(x, y) f(x, y)\, dx\, dy,$$

where the double integral extends over the whole xy-plane, and is evaluated in the usual way as an iterated integral (either with respect to the given coordinates, or in terms of some other convenient coordinates).

The sum of two random variables is an instance of a function of two random variables: $Z = X + Y$. It was already pointed out in the preceding section that the expected value of a sum is the sum of the expected values, and this can now be seen as follows, for the continuous case:

$$\begin{aligned} E(X + Y) &= \iint (x + y) f(x, y)\, dx\, dy \\ &= \iint x f(x, y)\, dx\, dy + \iint y f(x, y)\, dx\, dy \\ &= EX + EY, \end{aligned}$$

where all integrals extend from $-\infty$ to ∞. Thus, the additivity of the expected value is essentially the additivity of the double integral in terms of which $E(\cdot)$ is evaluated. One step in the above reasoning exploits the fact that X is also a special case of a function of (X, Y). For this function, the equivalence of the two ways of evaluating the mean is seen as follows:

$$EX = \iint xf(x, y)\,dy\,dx = \int x\left\{\int f(x, y)\,dy\right\}dx$$
$$= \int xf_X(x)\,dx = EX.$$

The product xy is another common instance of a function of two variables—and one might naïvely expect that the average product is the product of the averages; but this is not generally so, although it can happen. One class of bivariate distributions for which it is true is that in which the marginals are *independent.* In such a case, the average of the product of any function of one variable and any function of the other is the product of the averages. That is, if X and Y are *independent,*

$$E[g(X)h(Y)] = E[g(X)]\cdot E[h(Y)].$$

(In particular, this would mean that $E(XY) = (EX)(EY)$ whenever X and Y are independent.) This factorization of expectations follows easily from the factorization of the joint density into the product of the two marginals:

$$E[g(X)h(Y)] = \iint g(x)h(y)f(x, y)\,dx\,dy$$
$$= \iint g(x)h(y)f_X(x)f_Y(y)\,dx\,dy$$
$$= \int g(x)f_X(x)\left\{\int h(y)f_Y(y)\,dy\right\}dx$$
$$= E[g(X)]E[h(Y)].$$

EXAMPLE 2–49. The joint distribution of the colatitude and longitude variables (θ, ϕ) for the random orientation model in Example 2–23 was found there to have joint density

$$f_{\theta,\phi}(u, v) = \frac{1}{2\pi}\sin u, \qquad \text{for } 0 < v < 2\pi,\quad 0 < u < \frac{\pi}{2}.$$

According to Problem 2–55(b), the variables θ and ϕ are independent, with densities $\sin u$ and $1/(2\pi)$, respectively, on the appropriate intervals. If the model is used for the direction of a random force, the components would be of interest, one being proportional to $\sin\theta\cos\phi$. The expected value of this product is

$$E[\sin\theta\cos\phi] = E(\sin\theta)E(\cos\phi).$$

But

$$E(\cos\phi) = \int_0^{2\pi}\frac{\cos v\,dv}{2\pi} = 0,$$

so the expected product is 0.

Problems

2–78. Let X have a distribution on $-1 < x < 1$ whose density is constant there.

(a) Compute $E(X^2)$ in two ways.
(b) Compute $E|X|$ in your head by visualizing the distribution of $|X|$.

2–79. Let θ have density $f_\theta(u) = \sin u$ for $0 < u < \pi/2$ (as in Example 2–10).

(a) Compute $E(\cos \theta)$ in two ways.
(b) Compute $E(\cos^2 \theta)$.

2–80. Let Y have the probability function $f(k) = 2^{-k}$ for $k = 1, 2, \ldots$. Try to compute $E(2^Y)$.

2–81. Let (X, Y) have the joint density function

$$f(x, y) = (4xy)^{-1/2}, \qquad \text{for } 0 < y < x < 1$$

(and 0 elsewhere). Determine $E(Y \mid X = x)$, and then verify that $E[E(Y \mid X)] = EY$.

2–82. Compute $E(XY)$ when (X, Y) has the joint density

$$f(x, y) = 6(1 - x - y), \qquad \text{for } 0 < y < 1 - x < 1.$$

2–83. Let (X, Y) be distributed on the triangle $0 < x < y < 1$ with constant density there.

(a) Determine $E(X \mid Y = y)$ by looking at the cross section at y.
(b) From (a), show that $EX = EY/2$.
(c) Compute $E[(Y - X)^2]$.

2–84. Let (X, Y) be distributed on the unit circle $x^2 + y^2 \leq 1$ with constant density.

(a) Compute $E(XY)$.
(b) Show that $E(XY) = (EX)(EY)$, but that X and Y are not independent.

2–85. Show that $E(X \mid Y = y) = EX$ if X and Y are independent. Is the converse true? (Compare Example 2–45.)

2–86. Let X, Y, and Z be independent, each with the density function $f(\lambda) = e^{-\lambda}$ for $\lambda > 0$.

(a) Give the joint density of (X, Y, Z).
(b) Compute $E(X + Y \mid Z = z)$.
(c) Compute $E(Z \mid X + Y = k)$.
(d) Compute $E[(X + Y + Z)^2]$.
[*Hint*: Expand the square.]
(e) Compute $E(X \mid X + Y + Z = 1)$.
[*Hint*: Exploit the symmetry of the distribution and of the condition.]

★2.2.4 Riemann–Stieltjes Integrals

Consider a continuous function $g(x)$ defined on $a \leq x \leq b$, and another function $h(x)$, this one bounded and monotonically increasing (but not necessarily strictly so):

$$x < x' \qquad \text{implies} \qquad h(x) \leq h(x').$$

Let the interval $[a, b]$ be partitioned by points x_i:

$$a = x_0 < x_1 < \cdots < x_n = b,$$

and let M_i and m_i denote the least upper bound of $g(x)$ and the greatest lower bound of $g(x)$, respectively, on the ith subinterval of the partition, $x_{i-1} \leq x \leq x_i$.

The ordinary integral of elementary calculus is then defined for the function $g(x)$ as the common value of the following, when they are equal:

$$\underline{I} = \sup \sum m_i(x_i - x_{i-1})$$
$$\bar{I} = \inf \sum M_i(x_i - x_{i-1}),$$

where "sup" and "inf" mean, respectively, the least upper bound and greatest lower bound over all possible partitions of $[a, b]$. It can be shown, indeed, that $\underline{I} = \bar{I}$ whenever $g(x)$ is continuous, as assumed here. The integral defined in this way:

$$\int_a^b g(x)\, dx = \underline{I} = \bar{I}$$

is called a *Riemann integral.*

If in the defining expressions for $\underline{I}$ and $\bar{I}$ the interval length $x_i - x_{i-1}$ is replaced by the amount of change of the function $h(x)$ on that interval, namely, by $h(x_i) - h(x_{i-1})$, the quantities $\bar{I}$ and $\underline{I}$ are again equal, and define the *Riemann–Stieltjes integral* of $g(x)$ with respect to $h(x)$:

$$\int_a^b g(x)\, dh(x).$$

If the value $g(\xi_i)$ is used in place of the m_i, or M_i, where ξ_i is some point on the interval $[x_{i-1}, x_i]$, then the resulting sums and their limits over any sequence of successively finer partitions are squeezed between $\bar{I}$ and $\underline{I}$ and may be used to evaluate the integral when (as can be shown when h is monotonic and bounded as it is here) $\bar{I} = \underline{I}$. (This situation is analogous to that of the ordinary Riemann case.) For example, one could take the partition points to be equally spaced, of width $(b - a)/n$, and evaluate the function $g(x)$ at the right-hand endpoint of each subinterval to obtain:

$$\int_a^b g(x)\, dh(x) = \lim_{n \to \infty} \sum_{k=1}^{n} g(x_k)[h(x_k) - h(x_{k-1})].$$

If $h(x)$ is differentiable on $[a, b]$, the difference in its values at the endpoints of the interval $[x_{i-1}, x_i]$ can be expressed as the product of its derivative at some intermediate point ξ_i and the width of the interval (according to the mean value theorem):

$$h(x_i) - h(x_{i-1}) = h'(\xi_i)(x_i - x_{i-1}).$$

With this inserted into the above limit, the result looks much like an ordinary Riemann integral of $g(x)h'(x)$:

$$\lim_{n \to \infty} \sum_{k=1}^{n} g(x_k)h'(\xi_k)(x_k - x_{k-1}),$$

and so one would expect that

$$\int_a^b g(x)\,dh(x) = \int_a^b g(x)h'(x)\,dx.$$

That this is valid is a result of Duhamel's theorem.[5] Thus, the Riemann–Stieltjes integral not only contains the Riemann integral as a special case (that is, when $h(x) = x$), it also can be evaluated as a Riemann integral when the integrator function $h(x)$ is differentiable; the notation itself suggests the method: $dh(x)$ is replaced by $h'(x)\,dx$.

EXAMPLE 2–50. The function $g(x) = e^x$ is continuous, and the function $h(x) = x^2$ is differentiable on the interval [0, 1]. Hence,

$$\int_0^1 e^x\,d(x^2) = \int_0^1 2xe^x\,dx = 2.$$

If the integrator function $h(x)$ is constant on some subinterval of the interval of integration, the terms in the defining sum coming from that subinterval are all zero, since the differences in values of $h(x)$ would be zero. A special case of considerable importance is that in which $h(x)$ is a *step function*, that is, a function that increases only in jumps and is constant between jumps. Suppose that $h(x)$ jumps an amount j_k at the point η_k, for $k = 1, \ldots, m$, and is constant between these points. (The amount j_k of the "jump" in $h(x)$ is simply the difference between the limit from the left and the limit from the right. See Figure 2–19.) The only terms in the approximating sums that contribute something are those coming from partition subintervals containing the jump points—the other terms will all be zero because $h(x)$ does not change over those subintervals. Thus,

$$\int_a^b g(x)\,dh(x) = \lim_{n \to \infty} \sum_{k=1}^{m} g(x_{i_k})j_k = \sum_{k=1}^{m} g(\eta_k)j_k,$$

where x_{i_k} is the partition point just to the right of η_k, and approaches η_k as n becomes infinite. So the Riemann–Stieltjes integral reduces to an ordinary sum of values of $g(x)$ at the jump points, each weighted with the corresponding amount of the jump in $h(x)$.

[5] See D. V. Widder, *Advanced Calculus*, 2nd ed. (Englewood Cliffs, N.J.: Prentice–Hall, 1961), p. 173.

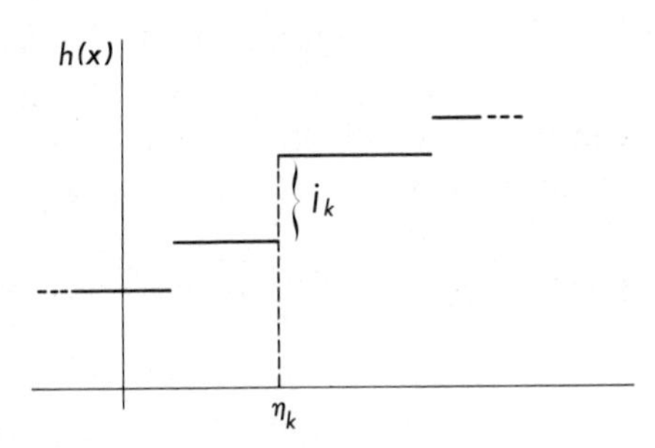

Figure 2–19

EXAMPLE 2–51. Let $F(x)$ denote the cumulative distribution function for the random variable that is the number of points showing in the toss of a die. This function jumps an amount $\frac{1}{6}$ at each of the values $x = 1, 2, 3, 4, 5, 6$, and is constant between these jump points. Hence,

$$\int_0^7 x\,dF(x) = 1(\tfrac{1}{6}) + 2(\tfrac{1}{6}) + \cdots + 6(\tfrac{1}{6}) = EX = \tfrac{7}{2}.$$

(The interval of integration could of course, have been any interval containing the points $1, \ldots, 6$, without altering the result.)

Suppose next that the integrator function $h(x)$ is continuous except for jumps j_i at x_i, $i = 1, \ldots, m$, and is differentiable between these jumps. Then $h(x)$ can be written as a sum of two functions, $h_1(x)$, a pure step function with the same jumps as $h(x)$, and $h_2(x)$, a continuous function that is differentiable except possibly at the finitely many points where $h(x)$ has jumps. It is not hard to show in terms of the definition of the integral that when the integrator function is a sum, the integral can be expressed as the sum of integrals in which the integrator functions are the summands of the original integrator. Thus,

$$\int_a^b g(x)\,dh(x) = \int_a^b g(x)\,dh_1(x) + \int_a^b g(x)\,dh_2(x)$$
$$= \sum_{i=1}^m g(x_i)j_i + \int_a^b g(x)h_2'(x)\,dx.$$

EXAMPLE 2–52. Let $g(x) = x$ and consider the following integrator function:

$$F(x) = \begin{cases} 0, & \text{for } x < 0, \\ 1 - .8e^{-x}, & \text{for } x \geq 0. \end{cases}$$

This is the distribution function for a model for the operating life of a certain type of equipment, introduced in Example 2–13 and sketched in Figure 2–10. This function $F(x)$ can be expressed as the sum of a step function with a single jump of height .2 at $x = 0$ and a function that is differentiable. Therefore, the integral over, say $(-1, 1)$, is

$$\int_{-1}^1 x\,dF(x) = 0 \times .2 + \int_0^1 .8xe^{-x}\,dx.$$

The case of an infinite interval of integration can be handled in the same fashion as it is for Riemann integrals, namely, by evaluating the integral over a truncated interval and passing to the limit as the right-hand and left-hand truncation points move independently to ∞ and $-\infty$, respectively:

$$\int_{-\infty}^{\infty} g(x)\, dh(x) = \lim_{\substack{A \to -\infty \\ B \to \infty}} \int_A^B g(x)\, dh(x).$$

EXAMPLE 2–53. Let X denote the number of tails preceding the first heads thrown in an infinite sequence of tosses of a fair coin. Then,

$$p_k = P(X = k) = \frac{1}{2^{k+1}}, \qquad k = 0, 1, 2, \ldots,$$

and the c.d.f. is a step function with jump p_k at the nonnegative integer k. Then,

$$\begin{aligned} \int_{-\infty}^{\infty} x\, dF(x) &= \lim_{\substack{A \to -\infty \\ B \to \infty}} \int_A^B x\, dF(x) \\ &= \lim_{N \to \infty} (0 \cdot p_0 + 1 \cdot p_1 + \cdots + N \cdot p_N) \\ &= \lim_{N \to \infty} \sum_{k=0}^{N} \frac{k}{2^{k+1}} = 1. \end{aligned}$$

★2.2.5 General Definition of Expectation

Let the random variable $X(\omega)$ be defined on the probability space $(\Omega, \mathcal{F}, P)$, and let $A_1, \ldots, A_m$ constitute a partition of Ω (i.e., these are mutually disjoint and exhaust Ω). Let M_i denote the least upper bound, and m_i the greatest lower bound of $X(\omega)$ on A_i, and define

$$\bar{E} = \inf \sum_i M_i P(A_i)$$

and

$$\underline{E} = \sup \sum_i m_i P(A_i),$$

where the inf and sup are taken over *all partitions* of Ω. The motivation for these definitions lies in the earlier definition for the case of a simple random variable. A general random variable rounded off to M_i in the set A_i would become a simple random variable with mean $\sum M_i P(A_i)$; and similarly, $\sum m_i P(A_i)$ is the mean of a simple random variable obtained by rounding off $X(\omega)$ to m_i on A_i. If the partition is a fine one, it would appear that these rounded-off variables could be thought of as approximations to $X(\omega)$, one on the high and one on the low side. Clearly, $\underline{E} \leq \bar{E}$; and if ever $\underline{E} = \bar{E}$, this common value is said to be the *expected value* (or mean, or average value) of X:

$$EX \equiv \underline{E} = \bar{E}, \qquad \text{(when these are equal)}.$$

One of the immediate consequences of this definition is that if $X(\omega)$ is nonnegative with probability 1, then its mean is nonnegative:

$$P(X(\omega) \geq 0) = 1 \quad \text{implies} \quad EX \geq 0,$$

For, the partition of Ω into the sets B and B', where $X \geq 0$ and $X < 0$, respectively, yields the nonnegative sum

$$m_1 P(B) + m_2 P(B') \geq 0,$$

since $P(B') = 0$ and the first term is nonnegative. This sum, on the other hand, does not exceed $\underline{E}$, so the desired result is established.

The evaluation of an expected value according to this definition is usually rather difficult. However, as asserted in Section 2.2.2, it can be accomplished in most cases of interest by means of ordinary sums and (Riemann) integrals. Indeed, the expected value can be expressed quite generally as a Riemann–Stieltjes integral.

To see this, consider first the case in which the distribution of $X(\omega)$ is contained on the interior of the finite interval $[a, b]$. Let this interval be partitioned by points x_i: $a = x_0 < x_1 < \cdots < x_n = b$, and define

$$A_i = \{\omega \mid x_{i-1} < X(\omega) \leq x_i\}, \qquad i = 1, 2, \ldots, n.$$

If A_0 denotes the set of all ω not in any of these A_i, the sets $A_0, A_1, \ldots, A_n$ constitute a partition of Ω, in which $P(A_0) = 0$. Let m_i and M_i denote the inf and sup of $X(\omega)$ on A_i, respectively; then, taking the sup over all partitions of Ω of this particular type, one has

$$\sup \sum m_i P(A_i) \leq \underline{E},$$

because $\underline{E}$ is the sup over *all* partitions of Ω. (It could never be smaller than the sup over a smaller family of partitions such as the $\{A_i\}$.) Now, because $m_i \geq x_{i-1}$ it follows that

$$\sum x_{i-1}[F(x_i) - F(x_{i-1})] \leq \sum m_i[F(x_i) - F(x_{i-1})] = \sum m_i P(A_i).$$

Taking the sup here over all partitions $a < x_1 < \cdots < b$ yields

$$\underline{I} = \sup \sum x_{i-1}[F(x_i) - F(x_{i-1})] \leq \sup \sum m_i P(A_i) \leq \underline{E}.$$

By similar reasoning involving M_i and infs, it follows that

$$\underline{I} \leq \underline{E} \leq \bar{E} \leq \bar{I}.$$

But $F(x)$ is monotonic and x is continuous, which means that $\underline{I} = \bar{I}$, and their common value is the Stieltjes integral of x with respect to $F(x)$. Since $\underline{E}$ and $\bar{E}$ are trapped between two numbers that are equal, it follows that they are equal to each other, defining the expectation EX, and to the common value of $\underline{I}$ and $\bar{I}$:

$$EX = \int_a^b x\, dF(x) = \int_{-\infty}^{\infty} x\, dF(x).$$

The case in which the distribution of X is not contained in a finite interval can be handled by introducing a truncated variable:

$$X_{A,B}(\omega) = \begin{cases} A, & \text{if } X(\omega) \le A, \\ X(\omega), & \text{if } A < X(\omega) < B, \\ B, & \text{if } X(\omega) \ge B, \end{cases}$$

and passing to the limit:[6]

$$EX = \lim_{\substack{A \to -\infty \\ B \to \infty}} E(X_{A,B}) = \lim_{\substack{A \to -\infty \\ B \to \infty}} \int_A^B x\, dF(x) = \int_{-\infty}^{\infty} x\, dF(x).$$

In summary, then, it is seen that if the Riemann–Stieltjes integral of x with respect to $F(x)$ exists, then so does the expected value of X, and they are equal.

It was seen earlier that if X and Y are simple random variables, or if they have a continuous joint distribution, then

$$E(aX + bY) = aEX + bEY.$$

This *linearity* of the averaging operation can be established in the general case by a limiting process, using the fact that an arbitrary random variable can be expressed as a limit of simple random variables. Special cases of linearity are these:

1. $E(-X) = -EX$.
2. $E(Y - X) = EY - EX$.

In view of the previously established fact that $X \ge 0$ with probability 1 implies $EX \ge 0$, there follows from (2):

3. $P(Y \ge X) = 1$ implies $EY \ge EX$.

Moreover, since $-X \le |X|$ and $X \le |X|$, it follows that

4. $|EX| \le E|X|$.

One virtue of the definition of expectation given here (as the common value of $\underline{E}$ and $\bar{E}$) is that it permits a general proof of the formula for the expected value of a function of a random variable. The proof is as follows.

Suppose that $Y = g(X)$ has a distribution contained on the interior of a finite interval $[c, d]$, and that X is similarly restricted to the interior of a finite interval $[a, b]$. The random variable Y is really $Y(\omega) = g(X(\omega))$, where ω is a generic point in an underlying sample space Ω (which, of course, could be the space of X-values, with $X(x) = x$). A partition

$$c = y_0 < y_1 < \cdots < y_m = d$$

[6] To do this, it would be necessary to know that the expected value of the limit is the same as the limit of the expected value, a point that will not be considered here.

of the interval $[c, d]$ induces a partition $A_1, \ldots, A_m$ of Ω:

$$A_j = \{\omega \mid y_{j-1} < Y(\omega) \leq y_j\}.$$

Taking the inf and sup (over all such partitions) of lower and upper approximating sums, as introduced in the preceding section, leads (as it did there) to

$$E(Y) = \int_{-\infty}^{\infty} y \, dF_Y(y).$$

However, suppose instead one considers a partition of $[a, b]$:

$$a = x_0 < x_1 < \cdots < x_n = b.$$

This also induces a partition of Ω, say, $B_1, \ldots, B_n$. For each partition set B_i let M_i denote the sup, and m_i the inf of the values of $g(X(\omega))$. These are also the sup and inf, respectively, of the values of $g(x)$ on (x_{i-1}, x_i). But then if one defines

$$\underline{J} = \sup \sum m_i P(B_i) = \sup \sum m_i [F(x_i) - F(x_{i-1})]$$

and

$$\bar{J} = \inf \sum M_i P(B_i) = \inf \sum M_i [F(x_i) - F(x_{i-1})],$$

taken over all partitions induced by partitions of the interval $[a, b]$, these define, when equal, the Stieltjes integral of $g(x)$ with respect to $F(x)$. But they also include between them the quantities called $\bar{E}$ and $\underline{E}$—the inf and sup over *all* partitions of Ω. That is, if $\underline{J} = \bar{J}$, then $\underline{J} = \underline{E} = \bar{E} = \bar{J}$, and so

$$E(Y) = E(g(X)) = \int_{-\infty}^{\infty} g(x) \, dF_X(x).$$

[According to the development given, this integral should extend over the interval from $x = a$ to $x = b$, although if the entire distribution is contained on that interval, there is no harm in writing the limits as $-\infty$ and ∞, since $dF(x)$ would vanish outside $[a, b]$. However, if the distribution is not confined to a finite interval, the truncation argument used earlier yields the above as a general formula.]

Problems

2–87. Evaluate:

(a) $\displaystyle\int_1^2 x^2 \, d(\log x)$. (b) $\displaystyle\int_0^{\infty} x \, d(1 - e^{-x})$.

2–88. Let F denote the c.d.f. of the number of points showing when a fair die is tossed, and evaluate the following:

(a) $\displaystyle\int_{-\infty}^{\infty} x \, dF(x)$. (b) $\displaystyle\int_{-\infty}^{\infty} x^2 \, dF(x)$.

2–89. Let $F(x) = 0$ for $x < 0$, and $1 - .8e^{-x}$ for $x \geq 0$. Evaluate:

(a) $\int_{-\infty}^{\infty} x \, dF(x)$. (b) $\int_{-\infty}^{\infty} e^{tx} \, dF(x)$, for $t < 1$.

2–90. Show that if $F(x)$ is nondecreasing, then

$$\int_a^b dF(x) = F(b) - F(a-).$$

2–91. Show that if $P(A) = 0$, then $E[I_A(X)g(X)] = 0$, where $I_A(x)$ is 1 or 0 accordingly as x is or is not in the set A.

[*Hint*: Evaluate $\sum m_i P(A_i)$ and $\sum M_i P(A_i)$ for the partition $A_1 = A$ and $A_2 = A^c$.]

2.3 Moments of a Distribution

When it exists, the quantity $E[(X - b)^k]$ is called the kth *moment* of the random variable X *about the point* $x = b$ (or the kth moment of the distribution of X). The *absolute* kth moment about $x = b$ is the quantity $E[|X - b|^k]$. Sometimes the term "moment" by itself is used to denote a moment about the particular point $x = 0$; such moments will be denoted

$$\mu'_k \equiv EX^k.$$

Moments about the mean value $\mu = \mu'_1$ will be denoted by

$$\mu_k \equiv E[(X - \mu)^k]$$

and are referred to as *central moments*. In any case, the exponent k will be called the *order* of the moment. It will be assumed here to be a nonnegative integer, although "fractional" moments are sometimes considered.

A moment may fail to exist because the expectation that would define it does not exist. (Recall that EY is said not to exist if $E|Y|$ is not finite.) However, if moments of order k exist, then moments of order j for $j \leq k$ also exist. This follows because of the inequality

$$|x^j| \leq |x|^k + 1 \qquad \text{if } j \leq k,$$

for then

$$E|X^j| = E|X|^j \leq E|X|^k + 1,$$

and so EX^j exists. [Note that $E[(X - b)^j]$ is expressible in terms of moments about zero of order j or smaller, and so exists if the jth moment about zero exists.]

EXAMPLE 2–54. In Example 2–43 it was seen that the random variable with density function

$$f(x) = \frac{1/\pi}{1 + x^2}$$

does not have a mean value, the integral being divergent that would define it. But then the distribution of X can have no higher-order moments either. This fact makes the distribution a useful counterexample in a number of theorems whose assumptions include the existence of one or more moments. It is referred to as the *Cauchy distribution.*

The terminology of "moments" stems from mechanics and from the previously mentioned analogy between probability distributions and mass distributions (Section 2.2.2). The *moment* of a point mass about another point (or an axis through the point) is the product of the mass and the distance to the point (or axis). If several point masses are involved, the total moment is the sum of the individual moments, and the average moment is that sum divided by the total mass. (If the distances are raised to the kth power, the result is called a kth moment.) Since an average moment involves relative masses, the quantity referred to as a moment of a probability distribution is precisely the average moment of the corresponding mass distribution.

The first moment of a distribution is its expectation, and the first central moment is always 0:

$$\mu_1 = E(X - \mu) = EX - \mu = 0.$$

In the analogy with mass distributions, the first moment corresponds to the center of gravity—the average first moment of the masses. And the vanishing of the first central moment corresponds to the fact that a mass distribution balances at the center of gravity.

The mean, then, measures what might be thought of as the center or middle of a distribution. It is a *parameter* of the distribution—a *location* parameter, which tells where a distribution is located along the axis by giving the location of its center. The second moment of a distribution, as will be seen in the next section, is related to the spread or dispersion of a distribution. The third moment has to do with skewness—a kind of asymmetry. Higher-order moments of a distribution measure or describe other, less intuitive aspects of a distribution.

Although the first and second moments do not completely describe or characterize a distribution (just as shoe size and weight do not alone characterize a person's body), they turn out to be its most important and useful parameters—for reasons that will emerge. Different distributions can, of course, have the same mean, and even the same first and second moments. It can even happen that different distributions have equal moments of all orders, although under rather general conditions the moments do uniquely define a distribution. The study of such questions is a problem of classical mathematics called, appropriately, the *moment problem*, on which there is an extensive literature.

2.3.1 The Variance

The second central moment of a distribution is commonly employed as a measure of its *dispersion* or *spread*; it is called the *variance* of the distribution, and is usually denoted by σ^2. In referring to a random variable X having the given distribution, the notation var X is used:

$$\text{var } X \equiv \sigma^2 \equiv E[(X - \mu)^2].$$

(As in the case of μ, a subscript may be applied to σ^2 if there is a possibility of confusion with other random variables: σ_X^2.) The term *variance* refers to the fact that observations on X are unpredictable and will vary from one trial to another, and the variance describes the extent of the variability in the values of X that can be observed.

Being an average *squared* deviation of X about its mean, the units of σ^2 are the square of the units of X. Thus, if X is a number of inches, var X is in square inches. The square root of var X, however, has the same units as X and so is a bit easier to appreciate intuitively. It is called the *standard deviation* of X (or of its distribution):

$$\sigma \equiv \sqrt{\text{var } X} = \sqrt{E[(X - \mu)^2]}.$$

It is a kind of average deviation—what is called in physics a *root-mean-square* (rms) average—and can be thought of as "typical" deviation from the mean. As will be discussed later, it can never exceed the maximum possible deviation about the mean.

(Other kinds of "typical deviation" about the mean might have been proposed to measure variability. Of course, the plain average, or expectation of the deviation $X - \mu$, would be zero; positive deviations exactly cancel the negative ones. Squaring the deviation $X - \mu$ is one way of preventing cancellation, but another would be just to ignore the sign and consider the absolute deviation, $|X - \mu|$. The *mean deviation* of a distribution is sometimes defined as the average absolute deviation, $E|X - \mu|$, and it too, measures variability. However, the fact that squaring is a smoother operation than taking absolute values makes the manipulation and analysis of variances much easier than the manipulation and analysis of mean deviations.)

EXAMPLE 2–55. Let X have the "triangular" density

$$f(x) = 1 - |x|, \qquad \text{for } |x| < 1.$$

Because of the symmetry about $x = 0$, the mean of the distribution is 0. The variance is then the second moment about $x = 0$:

$$\begin{aligned}\text{var } X = EX^2 &= \int_{-1}^{1} x^2(1 - |x|)\,dx \\ &= 2\int_0^1 x^2(1 - x)\,dx = \tfrac{1}{6}.\end{aligned}$$

The standard deviation is

$$\sqrt{\tfrac{1}{6}} \doteq .408,$$

which is between the smallest and largest possible deviations (0 and 1, respectively) about the mean. The mean deviation about the mean is

$$E|X - 0| = E|X| = 2\int_0^1 x(1 - x)\,dx = \tfrac{1}{3}.$$

The variance and the second moment about some point other than the mean are related by the following identity:

$$E[(X - a)^2] = \text{var}\, X + (\mu - a)^2,$$

called the *parallel axis theorem*, a name stemming from its application to mass distributions. The identity is established by adding and subtracting μ from $X - a$, regrouping and expanding the result as a binomial:

$$\begin{aligned}(X - a)^2 &= [(X - \mu) + (\mu - a)]^2 \\ &= (X - \mu)^2 + 2(X - \mu)(\mu - a) + (\mu - a)^2.\end{aligned}$$

The expected value of this is the sum of the expected values of the three terms on the right; the expected value of the first term is the variance, the expected value of $(\mu - a)^2$ is just that constant, and the expected value of the middle term is 0:

$$E[2(X - \mu)(\mu - a)] = 2(\mu - a)E(X - \mu) = 0.$$

The parallel axis theorem provides a characterization of the variance of a random variable as the *smallest* second moment, that is, as compared with second moments about other points than μ. For the mean square deviation about $x = a$ is equal to the variance *plus* the nonnegative quantity $(\mu - a)^2$, which assumes its smallest value (0) when $\mu = a$.

A useful special case of the parallel axis theorem is obtained upon setting a equal to 0:

$$\text{var}\, X = EX^2 - \mu^2.$$

Notice, then, that the expected square of a random variable is *not* the same as the square of the expected value, and that the difference of these two quantities defines the variance.

EXAMPLE 2–56. The probability distribution of X, the number of defectives in a random selection of 4 articles from 10, of which 2 are defective, obtained in Problem 2–65(a), is repeated in the following table. Also given are the products needed to compute the average and the average square of X.

k	$f(k)$	$kf(k)$	$k^2f(k)$
0	$\frac{5}{15}$	0	0
1	$\frac{8}{15}$	$\frac{8}{15}$	$\frac{8}{15}$
2	$\frac{2}{15}$	$\frac{4}{15}$	$\frac{8}{15}$
Total	1	$\frac{12}{15}$	$\frac{16}{15}$

From this one computes

$$EX = \tfrac{4}{5} \qquad \text{and} \qquad EX^2 = \tfrac{16}{15}.$$

The variance is then

$$\text{var } X = EX^2 - (EX)^2 = \tfrac{16}{15} - (\tfrac{4}{5})^2 = \tfrac{32}{75}.$$

The standard deviation is $\sqrt{\tfrac{32}{75}} \doteq .653$.

Given a joint distribution for (X, Y), the variance of the conditioned variable $X \mid Y = y$, written more simply as $X \mid y$, is called the *conditional variance* of X:

$$\begin{aligned} \text{var } (X \mid y) &= E[(X - \mu_{X|y})^2 \mid y] \\ &= E(X^2 \mid y) - \mu_{X|y}^2. \end{aligned}$$

This is a function of the value y given as the observed value of Y,

$$g(y) \equiv \text{var } (X \mid y),$$

and it might be expected (in view of a result for the mean) that averaging out the y:

$$E[g(Y)] = E[\text{var } (X \mid Y)]$$

would produce the unconditional quantity, var X. But the correct relation is more complicated. It can be derived from an application of the parallel axis theorem to the distribution of $X \mid y$:

$$E[(X - \mu_X)^2 \mid y] = \text{var } (X \mid y) + (\mu_{X|y} - \mu_X)^2.$$

Averaging in y with respect to the distribution of Y yields

$$E[(X - \mu_X)^2] = E[\text{var } (X \mid Y)] + E[(\mu_{X|Y} - \mu_X)^2].$$

The last term can be interpreted as a variance, the variance of $h(Y)$, where

$$h(y) = \mu_{X|y} \qquad \text{and} \qquad E[h(Y)] = E\mu_{X|Y} = \mu_X.$$

Thus,

$$\text{var } X = E[\text{var } (X \mid Y)] + \text{var } (\mu_{X|Y}),$$

or in words, the variance of X is the mean of its conditional variance plus the variance of its conditional mean.

EXAMPLE 2–57. In Examples 2–31 and 2–48, the bivariate distribution defined by the density

$$f(x, y) = 2, \qquad \text{for } 0 < x + y < 1, \quad x > 0, \quad y > 0,$$

was found to have the conditional density

$$f(x \mid y) = \frac{1}{1 - y}, \qquad \text{for } 0 < x < 1 - y,$$

and conditional mean

$$E(X \mid y) = \frac{1 - y}{2}.$$

The conditional variance of X given $Y = y$ is

$$E(X^2 \mid y) - \mu_{X|y}^2 = \int_0^{1-y} \frac{x^2\,dx}{1 - y} - \frac{(1 - y)^2}{4} = \frac{(1 - y)^2}{12},$$

and the expected value of this variance with respect to the Y-density is

$$\int_0^1 \frac{(1 - y)^2}{12} 2(1 - y)\,dy = \frac{1}{24}.$$

The variance of X is

$$\int_0^1 x^2 2(1 - x)\,dx - \left(\frac{1}{3}\right)^2 = \frac{1}{18},$$

and the variance of $E(X \mid Y)$ is

$$\text{var}\left(\frac{1 - Y}{2}\right) = \frac{1}{4}\,\text{var}\,Y = \frac{1}{72}.$$

Thus, the variance of X is the mean of its conditional variance plus the variance of its conditional mean:

$$\frac{1}{18} = \frac{1}{24} + \frac{1}{72}.$$

2.3.2 Chebyshev and Related Inequalities

The Chebyshev inequality is a useful theoretical tool as well as a relation that connects the variance of a distribution with the intuitive notion of dispersion in a distribution. It and various other inequalities follow from a somewhat more general inequality.

Basic Inequality. *Let b denote a positive constant and $h(x)$ a nonnegative function.*[7] *Then*

$$P[h(X) \geq b] \leq \frac{1}{b} E[h(X)],$$

provided that the expectation exists.

PROOF: Let A denote the set

$$A = \{x \mid h(x) \geq b\}.$$

Then if $0 < P(A) < 1$,

$$\begin{aligned} E[h(X)] &= E[h(X) \mid A]P(A) + E[h(X) \mid A^c]P(A^c) \\ &\geq E[h(X) \mid A]P(A) \geq bP(A). \end{aligned}$$

[7] The "x" here could also be a random vector, $(x_1, \ldots, x_n)$, with h a function of n variables.

The first inequality follows because $h(x) \geq 0$, and the second because $h(x) \geq b$ for x on A. If $P(A) = 0$, the basic inequality is trivially true; and if $P(A) = 1$, the second term in the expression for $E[h(X)]$ is missing, so that again the desired result follows.

With $h(x) = (x - \mu)^2$ and $c = \sqrt{b}$, the basic inequality reduces to the Chebyshev inequality.

Chebyshev's Inequality. *For any constant $c > 0$,*

$$P(|X - \mu| \geq c) \leq \frac{\sigma_X^2}{c^2}.$$

Essentially the same inequality can be expressed in other useful forms:

(i) $P(|X - \mu| < c) \geq 1 - \dfrac{\sigma^2}{c^2}$.

(ii) $P(|X - \mu| \geq k\sigma) \leq \dfrac{1}{k^2}$.

EXAMPLE 2–58. If Z has the distribution function given in Table 1 (see Appendix), in which the mean is 0 and the variance is 1, then

$$P(|Z - EZ| > 3\sigma_Z) = P(|Z| > 3) = .0026,$$

which is, indeed, less than $(\frac{1}{3})^2 = \frac{1}{9}$.

This example shows that the Chebyshev bound is sometimes very crude—depending on how much is thrown away in the proof of the inequality. On the other hand, for *all* distributions (with second moments) the probability outside ± 3 standard deviations on either side of the mean does not exceed $\frac{1}{9}$. For some distributions this probability may be 0, and for the one in the preceding example it is .0026. In the next example there is given a distribution in which the Chebyshev bound is actually attained for a certain value of k, showing that in the general inequality the bound could not be improved without further assumptions on the distribution.

EXAMPLE 2–59. Suppose that X assumes the values 1 and -1, each with probability .5. Then $EX = 0$ and $\text{var } X = 1$, and

$$P(|X - EX| \geq \sigma_X) = 1 - P(-1 < X < 1) = 1.$$

But 1 is precisely the Chebyshev bound for this probability, and so the inequality becomes an equality.

In that at most $\frac{1}{4}$ of a distribution can be placed outside the range $\mu \pm 2\sigma$, at most $\frac{1}{9}$ outside the range $\mu \pm 3\sigma$, at most $\frac{1}{100}$ outside the range $\mu \pm 10\sigma$, etc., the Chebyshev inequality does indeed relate the standard deviation to the dispersion of probability about the center of the distribution. Another aspect of this relationship is the following fact:

$$\sigma = 0 \quad \text{implies} \quad P(X = \mu) = 1.$$

This is seen by setting $\sigma = 0$ in the alternate form (i) of Chebyshev's inequality:

$$P(|X - \mu| < c) = 1, \qquad \text{for } c > 0,$$

and then, with $c = 1/n$, passing to the limit as $n \to \infty$:

$$P(X = \mu) = \lim_{n \to \infty} P\left(|X - \mu| < \frac{1}{n}\right) = 1.$$

Yet another aspect of the relationship between variance and the dispersion of probability is the fact that a standard deviation is a deviation about the mean that cannot exceed the largest deviations that will be encountered. That is,

$$P(|X - \mu| > K) = 0 \quad \text{implies} \quad \sigma \le K.$$

To see this, let

$$B = \{x \mid (x - \mu)^2 \le K^2\}$$

and decompose σ^2 along the lines of the argument used in establishing the basic inequality:

$$\sigma^2 = E[(X - \mu)^2 \mid B]P(B) + E[(X - \mu)^2 \mid B^c]P(B^c) \le K^2.$$

Problems

2–92. Compute the variance of the number of tests required to locate the one defective article among five articles [cf. Problem 2–65(b)], given $f(x) = \frac{1}{5}$ for $x = 1, 2, 3$, and $f(4) = \frac{2}{5}$.

2–93. Compute the variance for the density function of Example 2–12: $f(x) = e^{-x}$, for $x > 0$.

2–94. Compute the standard deviation of the random variable whose c.d.f. is $F(x) = x^2$ for $0 < x < 1$ (and 0 for $x < 0$ and 1 for $x > 1$).

★2–95. Compute the variance of the distribution in Problem 2–89: $F(x) = 0$ for $x < 0$, and $1 - .8e^{-x}$ for $x \ge 0$.

2–96. Determine the mean and variance of the random variable $Y = (X - \mu)/\sigma$, where μ and σ^2 are the mean and variance of X, respectively. (This linear transformation is called a *standardizing* transformation.)

2–97. Obtain an expression for the third central moment μ_3 in terms of moments about zero. Show also that

$$\mu_3 = \mu_3' - 3\mu\sigma^2 - \mu^3.$$

2–98. Show that if X is a continuous random variable, the smallest absolute first moment is the one about the median.

[*Hint*: Write $E|X - a|$ as the sum of an integral over $x \leq a$ and one over $x > a$, and then minimize with respect to a.]

2–99. Consider the discrete distribution for X with $f(-a) = f(a) = \frac{1}{8}$ and $f(0) = \frac{3}{4}$. Compute $P(|X| \geq 2\sigma)$ and compare with the bound given by the Chebyshev inequality for this probability.

2–100. Refer to Table II of the Appendix. The row marked "4 degrees of freedom" gives percentiles of a certain distribution with mean 4 and variance 8. Compute $P(|X - 4| > 8)$ approximately, using table entries, and determine also the Chebyshev bound for this probability.

2–101. Show that $P(|X| > b)$ does not exceed either EX^2/b^2 or $E|X|/b$. With $f(x) = e^{-x}$ for $x > 0$, show that one bound is better when $b = 3$ and the other when $b = \sqrt{2}$.

2–102. For any sequence of random variables, $X_1, \ldots, X_n$, show

$$P(\max |X_i| > K) \leq \frac{1}{K^2} \sum E(X_i^2).$$

[*Hint*: Show that if $Y_i \geq 0$, $\{\max Y_i > 1\} \subset \{\sum Y_i > 1\}$, and then let $Y_i = |X_i|/K$.]

2–103. Show that if $EX^2 = 0$, then $P(X = 0) = 1$.

2–104. Show that $\sigma > K > 0$ implies $P(|X| > K) > 0$.

[*Hint*: Observe that $\sigma^2 \leq EX^2$ and decompose EX^2 into integrals over $|x| > K$ and $|x| < K$.]

2–105. Show that $E[\text{var}(Y \mid X)] = \text{var } Y$, if and only if $E(Y \mid x)$ is essentially independent of x (i.e., if $E(Y \mid X) = EY$ with probability 1).

2–106. Let (X, Y) have the discrete distribution given in the following table of probabilities:

Y \ X	0	2	4
1	$\frac{1}{4}$	0	$\frac{1}{4}$
3	$\frac{1}{12}$	$\frac{1}{3}$	$\frac{1}{12}$

Verify that $E[\text{var}(X \mid Y)] = \text{var } X$.

2.3.3 Covariance and Correlation

An expectation of the form $E[(X - a)^r(Y - b)^s]$ is called a *product moment* or a *mixed moment* if the integers r and s are greater than 0. In a bivariate distribution, the second-order mixed moment is of particular utility and importance; when the deviations are taken about the means, it is called the *covariance* of the distribution:

$$\text{cov}(X, Y) = E[(X - \mu_X)(Y - \mu_Y)] = E(XY) - (EX)(EY).$$

Notice that cov (X, X) = var X, suggesting the notation $\sigma_{X,Y}$ for the covariance.

The term *covariance* itself comes from the notion that this quantity purports to measure "covariation"—to be indicative of the degree to which the variables are concordant or coherent. If X tends to be large when Y is large, and small when Y is small (in the algebraic sense), then the covariance will be positive; if large values of X tend to correspond to small values of Y, the covariance will be negative; if there is no such tendency—if knowing that X is large does not give much information as to the tendency of Y, the covariance will be close to zero.

EXAMPLE 2–60. Consider the discrete distribution of (X, Y) defined in the following table of probabilities:

Y \ X	6	8	10
1	.2	0	.2
2	0	.2	0
3	.2	0	.2

The expected value of X is 8, and that of Y is 2. (These are evident from the symmetry of the marginal distributions.) The expected product is obtained by calculating the sum of the various possible products, each weighted with the probability for that particular pair of values:

$$E(XY) = 6 \times .2 + 10 \times .2 + 16 \times .2 + 18 \times .2 + 30 \times .2 = 16.$$

The covariance is then

$$\text{cov}\,(X, Y) = E(XY) - (EX)(EY) = 16 - 16 = 0.$$

This covariance vanishes because of the symmetry in the distribution; there is a positive product $(X - \mu_X)(Y - \mu_Y)$ for each negative one, and they average out to 0 because they are weighted equally.

The covariance is really a poor measure of coherence, in that it is sensitive to the scale of measurement adopted—being multiplied by any scale factor introduced in one variable or the other. For example, if X and Y are measured in numbers of inches, the covariance will be 144 times what it would be if they were measured in numbers of feet. To obtain a measure of coherence that does not have this defect, the *correlation coefficient* is used, denoted and defined as follows:

$$\rho_{X,Y} \equiv \frac{\sigma_{X,Y}}{\sigma_X \sigma_Y}.$$

Here a scale change in X would introduce a factor in σ_X as well as in $\sigma_{X,Y}$, and these would cancel.

Since the covariance is the average product minus the product of the averages, it follows that when these are equal the covariance vanishes. But they *are* equal when X and Y are independent:

$$\sigma_{X,Y} = E(XY) - (EX)(EY) = 0, \qquad (X \text{ and } Y \text{ independent}).$$

Moreover, when the covariance vanishes, the correlation is 0; so independence implies that the variables are uncorrelated (i.e., have zero correlation). Example 2–60 however, shows that a correlation of 0 does not imply independence; for in that example, the variables cannot be independent with zero entries in the table of joint probabilities. Thus, although in everyday speech the concepts of "dependence" and "relationships" are not clearly distinguished, here their technical meanings are quite distinct.

For X and Y to be correlated is for them to have a particular kind of dependence. But there are other kinds of dependence. In Example 2–60, for instance, knowing that $Y = 1$ limits the range of possible values of X; but it does not tip the scales, in betting, to either $X = 6$ or $X = 10$. On the other hand, when there is a positive correlation, then the information that Y has a value greater than its mean increases the odds on values of X that are greater than its mean.

Now, if $\rho_{X,Y}$ is to measure coherence of X and Y, it is natural to want to know what it means for it to have the value .6, and indeed, to know just how big a correlation can get.

That the correlation coefficient is actually limited in magnitude to 1 can be seen from the following inequality:

$$|\sigma_{X,Y}| \leq \sigma_X \sigma_Y,$$

which yields the asserted bound for $\rho_{X,Y}$ upon division by $\sigma_X \sigma_Y$:

$$|\rho_{X,Y}| \leq 1.$$

To prove this, consider the nonnegative random variable $(U - kV)^2$, where U and V are random variables and k is a real constant. The expectation of this variable is surely nonnegative:

$$0 \leq E[(U - kV)^2] = k^2 E(V^2) - 2kE(UV) + E(U^2).$$

In order for a quadratic function of k, such as this, to be nonnegative for all real k, it is necessary and sufficient for its discriminant to be nonpositive:

$$[E(UV)]^2 - E(U^2)E(V^2) \leq 0.$$

This is a form of the *Schwarz inequality*. Setting $U = X - EX$ and $V = Y - EY$ yields the desired result.

The significance of the extreme cases, $\rho = +1$ or -1, is of interest. In either case the discriminant above is actually 0, which would mean that the parabola representing the quadratic in k just touches the k-axis at some point. That is, there is some value of k such that the quadratic is equal to 0:

$$E[(U - kV)^2] = 0, \qquad \text{for some } k.$$

But this in turn implies that

$$P(U = kV) = 1, \qquad \text{for some } k.$$

That is, a correlation of $+1$ or -1 implies that $X - EX$ is, with probability 1, just a multiple of $Y - EY$; in other words, X and Y are linearly related with probability 1. The joint distribution of (X, Y) is then concentrated along the straight line representing that linear relationship. And then, of course, if the value of one variable is given, the value of the other variable can be deduced. (The joint distribution is bivariate only in the singular sense, and one variable is unessential.) For these reasons, the correlation $\rho_{X,Y}$ is often called a coefficient of *linear* correlation.

EXAMPLE 2–61. Consider the density function

$$f(x, y) = \begin{cases} 8xy, & \text{if } 0 \leq x \leq y \leq 1, \\ 0, & \text{otherwise.} \end{cases}$$

The marginal density functions are $4x(1 - x^2)$ and $4y^3$, each defined on $[0, 1]$. From these can be computed the means and variances: $E(X) = \frac{8}{15}$, $E(Y) = \frac{4}{5}$, $\operatorname{var} X = \frac{11}{225}$, and $\operatorname{var} Y = \frac{2}{75}$. The expected product is

$$E(XY) = \int_0^1 \int_0^y xy\, 8xy\, dx\, dy = \frac{4}{9},$$

and then

$$\rho_{X,Y} = \frac{E(XY) - E(X)E(Y)}{[(\operatorname{var} X)(\operatorname{var} Y)]^{1/2}} = \frac{4/9 - 32/75}{[(2/75)(11/225)]^{1/2}} = \frac{4}{\sqrt{66}}.$$

2.3.4 Variance of a Sum

Consider first a sum of two random variables, $Z = X + Y$. Since the expected value of the sum is the sum of the expectations,

$$Z - EZ = (X - EX) + (Y - EY).$$

The variance of the sum is the average square of this deviation:

$$\begin{aligned} \operatorname{var}(X + Y) &= E[(Z - EZ)^2] \\ &= E[(X - EX)^2 + (Y - EY)^2 + 2(X - EX)(Y - EY)] \\ &= \operatorname{var} X + \operatorname{var} Y + 2 \operatorname{cov}(X, Y). \end{aligned}$$

So the variance is *not* additive, unless, perchance, the variables are *uncorrelated*, in which case

$$\operatorname{var}(X + Y) = \operatorname{var} X + \operatorname{var} Y, \qquad (X \text{ and } Y \text{ uncorrelated}).$$

And surely, if X and Y are independent, this relation holds, since they are then also uncorrelated.

If $Z = aX + bY$, the above formula for the variance of a sum is applied to the variables aX and bY to obtain

$$\text{var}\,(aX + bY) = a^2 \text{ var } X + b^2 \text{ var } Y + 2ab \text{ cov }(X, Y).$$

In particular, notice that if X and Y are uncorrelated, then

$$\text{var}\,(X - Y) = \text{var}\,[X + (-Y)] = \text{var } X + \text{var } Y,$$

and not (as might be naïvely expected) the difference of the variances.

The preceding discussion provides the possibility of an instructive and helpful geometric representation for random variables that have finite variances. Consider two vectors in the plane whose lengths are proportional to σ_X and σ_Y, respectively, and separated by an angle θ whose cosine is $\rho_{X,Y}$. The length of the diagonal of the parallelogram determined by them can be computed using the law of cosines (see Figure 2–20):

$$\begin{aligned} d^2 &= \sigma_X^2 + \sigma_Y^2 - 2\sigma_X\sigma_Y \cos(\pi - \theta) = \sigma_X^2 + \sigma_Y^2 + 2\sigma_X\sigma_Y\rho_{X,Y} \\ &= \sigma_X^2 + \sigma_Y^2 + 2\sigma_{X,Y} = \text{var}\,(X + Y). \end{aligned}$$

Thus, the length of the diagonal d is the standard deviation of $X + Y$, and so the addition of the random variables X and Y corresponds to vector addition (according to the parallelogram law) of the vectors representing them. In terms of this representation, a correlation of 0 corresponds to perpendicularity of the vectors X and Y, in which case $\text{var}\,(X + Y) = \text{var } X + \text{var } Y$. A correlation of 1 corresponds to the pointing of the vectors in the same direction—one is a multiple of the other, and a correlation of -1 corresponds to vectors pointing in opposite directions. All of this may be stated more precisely by saying that random variables with finite variance constitute a linear vector space, with an "inner product" of two random variables defined as their covariance.

Consider next the case of n random variables, $X_1, X_2, \ldots, X_n$. Reasoning exactly as in the case of two random variables, one obtains

$$\begin{aligned} \text{var}\,(X_1 + \cdots + X_n) &= E\left\{\left[\sum_i (X_i - EX_i)\right]^2\right\} \\ &= E\left\{\sum_j \sum_i (X_i - EX_i)(X_j - EX_j)\right\} \\ &= \sum_j \sum_i E[(X_i - EX_i)(X_j - EX_j)] \\ &= \sum_j \sum_i \text{cov}\,(X_i, X_j), \end{aligned}$$

where, of course, $\text{cov}\,(X_i, X_i) = \text{var } X_i$. The terms in this sum can be thought of as laid out in a square array, corresponding to $i = 1, \ldots, n$ horizontally and to $j = 1, \ldots, n$ vertically, with $i = j$ along the main diagonal. Of the n^2 terms, n are

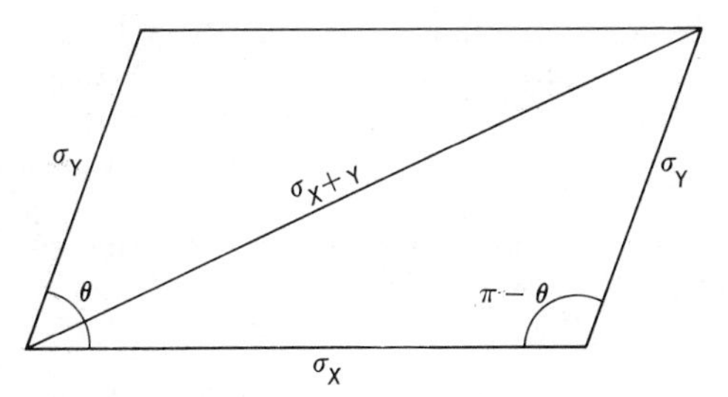

Figure 2–20. Vector representation of random variables.

variances and the remaining $n^2 - n$ are covariances; the covariances above the diagonal have matching covariances below the diagonal, since $\text{cov}(X_i, X_j) = \text{cov}(X_j, X_i)$. Finally, then, the variance of the sum can be written as follows:

$$\begin{aligned}\text{var}\left(\sum_i X_i\right) &= \sum_i \text{var}\, X_i + \sum_{i \neq j} \text{cov}(X_i, X_j) \\ &= \sum_i \text{var}\, X_i + 2 \sum_{i > j} \text{cov}(X_i, X_j).\end{aligned}$$

In the important special cases in which the X's are mutually uncorrelated, each covariance is 0 and the variance becomes additive:

$$\text{var}\left(\sum_i X_i\right) = \sum_i \text{var}\, X_i, \qquad (X_i\text{'s pairwise uncorrelated}).$$

One of the important applications of this last result is in the realm of error analysis, in which it is often possible to express a system error as a linear combination of independent, component errors (although sometimes only to a first approximation). The following examples illustrate this.

EXAMPLE 2–62. When resistors are placed in "series," the resistance of the combination is the sum of the individual resistances. Suppose that two 400-ohm and one 200-ohm resistors are placed in series and that they are "5 per cent" resistors. Let us assume that a "5 per cent" resistor is one of a population of resistors whose resistances have a probability distribution about the nominal resistance with a standard deviation that is 5 per cent of the nominal resistance. Then the standard deviations of the given resistances are 20, 20, and 10 ohms, respectively. The standard deviation of the resistance of the series combination is

$$(20^2 + 20^2 + 10^2)^{1/2} = 30 \text{ ohms},$$

which is 3 per cent of the nominal series resistance of 1000 ohms. The series combination (considered as one of possible series combinations made up at random from the populations of 400-ohm and 200-ohm resistors) has a precision that is better than that of the components.

[It must be admitted that the "5 per cent" label on a resistor is probably not to be interpreted as one standard deviation. This might even vary with the individual manufacturer; but if the measure of variability quoted is *proportional* to the standard deviation, the preceding calculation of percentage would remain valid. At any rate, it is usually not appropriate to add the absolute values of the tolerance figures given (to obtain 50 ohms or 5 per cent in the example cited), since the probability is small that the errors would combine without some cancellation of positive and negative errors.]

EXAMPLE 2–63. Suppose that a quantity is to be computed as the quotient of two measured quantities,

$$A = \frac{B}{C}.$$

If B and C contain errors,

$$B = B_0 + \beta, \qquad C = C_0 + \gamma$$

the computed value of A will be correspondingly in error:

$$A = A_0 + \alpha = \frac{B_0}{C_0} + \alpha,$$

where then

$$\alpha = A - A_0 = \frac{B_0 + \beta}{C_0 + \gamma} - \frac{B_0}{C_0} \doteq \frac{1}{C_0^2}(C_0\beta - B_0\gamma).$$

This last approximation to the increment in A is just the differential. The variance of the error in A is approximately the variance of this linear combination of β and γ,

$$\text{var}\ \alpha \doteq \frac{1}{C_0^2}(\text{var}\ \beta) + \frac{B_0^2}{C_0^4}(\text{var}\ \gamma).$$

Problems

2–107. Given the discrete distribution of Problem 2–106:

Y \ X	0	2	4
1	$\frac{1}{4}$	0	$\frac{1}{4}$
3	$\frac{1}{12}$	$\frac{1}{3}$	$\frac{1}{12}$

(a) Compute the correlation coefficient.
(b) Compute the covariance of X and Y given $X \neq 4$.

2–108. Two beads are selected at random (no replacement) from a bowl containing four white, one red, and two black beads. Let X denote the number of red and Y the number of black beads in the selection. Determine cov (X, Y).

2–109. Show that X and Y are uncorrelated, given their joint distribution according to the following table of probabilities. Are they independent?

Y \ X	−1	0	1
0	.1	.1	.1
2	.1	.2	.1
4	.1	.1	.1

2–110. Compute the coefficient of correlation in the bivariate distribution with density $f(x, y) = 2$ for $x + y \leq 1$, $x \geq 0$, $y \geq 0$.

2–111. Consider the random orientation (θ, ϕ), as considered in Example 2–23 with joint density function $f_{\theta,\phi}(u, v) = (\sin u)/2\pi$ for $0 < \theta < \pi/2$, $0 < \phi < 2\pi$. Show that the variables $X = \cos\phi \sin\theta$ and $Y = \sin\phi \sin\theta$ are uncorrelated, even though [as shown in Problem 2–55(b)] they are not independent.

2–112. Use the Schwarz inequality to show that if $E(X^2)$ is finite, so is $E|X|$.

2–113. Show that $\text{cov}(X, X + Y) = \text{var } X + \text{cov}(X, Y)$. More generally, show that

$$\text{cov}\left(\sum_i a_i X_i, \sum_j b_j Y_j\right) = \sum_i \sum_j a_i b_j \,\text{cov}(X_i, Y_j).$$

2–114. Show that $X + Y$ and $X - Y$ are uncorrelated if and only if $\text{var } X = \text{var } Y$.

2–115. Show that if $\rho^2 = 1$, the variance of $X + Y$ is $(\sigma_X \pm \sigma_Y)^2$, the + or − depending on whether $\rho = 1$ or $\rho = -1$.

2–116. Show that if $\text{var } X = \text{var } Y$, then $\rho_{X,X+Y} = [\frac{1}{2}(1 + \rho_{X,Y})]^{1/2}$. (Interpret this in light of the geometric representation of a random variable as a vector, discussed in Section 2.3.3.)

2–117. Thirteen cards are drawn at random (without replacement, as in dealing) from a standard bridge deck. Let $X_1, \ldots, X_{13}$ denote the numbers of points assigned to the cards drawn [as in Problem 2–65(d)]: (4 for Ace, 3 for King, 2 for Queen, 1 for Jack, and 0 for anything else).

(a) Compute the expected total number of points.

(b) Determine the variance of the total number of points.

[*Hint*: First compute $\text{var } X_i$ to be 290/169 and $\text{cov}(X_i, X_j)$ to be $-290/(169 \cdot 51)$, and then use these in the general formula for the variance of a sum.]

2–118. Gears A, B, and C have basic widths 0.5, 0.3, and 0.7 in., with standard deviations of 0.001, 0.004, and 0.002 in., respectively. If these are assembled side by side on a single shaft, what is the basic width of the assembly and the standard deviation of this total width?

2–119. The position error (in nautical miles) in a certain guidance system after a given time of operation is given by

$$\delta = 5\delta_a + 20\delta_g,$$

where δ_a and δ_g are independent random errors in an accelerometer and in a gyro (measured in ft/sec^2 and deg/hr, respectively.) Determine the standard deviation of position error corresponding to standard deviations of 0. 1 ft/sec^2 and 0.05 deg/hr in δ_a and δ_g, respectively.

2–120. Obtain an approximate expression for the variance of the error in computing a product AB, using measured values of A and B that involve independent errors α and β about the actual values A_0 and B_0.

2.4 Generating Functions

A variety of functions that "generate" certain aspects of a probability distribution can be defined for a given distribution. They are often thought of as "transforms" of the density function or probability function defining the distribution, and they have a particular usefulness in connection with sums of independent random variables—a usefulness that usually stems from the simple algebraic law: $a^{x+y} = a^x a^y$. Of those that will be considered, the probability generating function is particularly useful in certain kinds of combinatorial problems. The factorial moment generating function and the moment generating function can be used to generate the moments of a distribution—a single integration (followed by a series expansion) replacing all of the integrations necessary to obtain the various moments. The characteristic function is most useful as a theoretical tool—proving theorems about sums of independent random variables that will be crucial in obtaining probability distributions of statistics commonly used in inference problems.

2.4.1 Moment Generating Functions

The *moment generating function* of the distribution of a random variable X is defined formally as follows:

$$\psi(t) = E(e^{tX}) = \int_{-\infty}^{\infty} e^{tx}\, dF(x).$$

[If necessary to avoid confusion, a subscript will link this function to the random variable: $\psi_X(t)$.] When it exists, this expectation depends on the choice of t, and so defines a function of t, as the notation suggests. For $t = 0$ it always exists: $\psi(0) = 1$, but for other values of t it may or may not exist, depending on the distribution.

If the exponential function in the integrand of $\psi(t)$ is replaced by its power series expansion, there follows

$$\begin{aligned}\psi(t) &= \int_{-\infty}^{\infty} \sum_0^{\infty} \frac{(tx)^k}{k!}\, dF(x) = \sum_0^{\infty} \frac{t^k}{k!} \left(\int_{-\infty}^{\infty} x^k\, dF(x)\right)\\ &= \sum_0^{\infty} E(X^k) \frac{t^k}{k!},\end{aligned}$$

provided that the interchange of summation and integral is permitted. If it is, then, the kth moment of a distribution is simply the coefficient of $t^k/k!$ in the power series expansion of the moment generating function. The idea is that in integrating e^{tx} one is simultaneously integrating all powers of x, with the hope of recovering the individual integrals as coefficients in the expansion of the result. The coefficient of $t^k/k!$ in a Maclaurin power series expansion (as derived in differential calculus) is the kth derivative at 0:

$$E(X^k) = \psi^{(k)}(0).$$

Thus, if the function is one whose power series is not well known, the coefficients can be found by differentiation. This result may also be seen by differentiating $\psi(t)$ k times:

$$\frac{d^k}{dt^k} E(e^{tX}) = E(X^k e^{tX}),$$

provided that the operations of differentiation and expectation (which is an integration) can be interchanged. Justification of these various formal manipulations will not be carried out here; but it can be shown that if $\psi_X(t)$ exists in a proper interval about $t = 0$, then X has moments of all orders, and they can be computed as described.

EXAMPLE 2–64. Consider a simple random variable X having just two possible values, 1 with probability p and 0 with probability $1 - p$. The moment generating function is

$$\begin{aligned}\psi(t) &= E(e^{tX}) = e^t \cdot p + e^0 \cdot (1 - p) = pe^t + (1 - p)\\ &= p\left(1 + \frac{t}{1!} + \frac{t^2}{2!} + \cdots\right) + (1 - p)\\ &= 1 + p\left(\frac{t}{1!}\right) + p\left(\frac{t^2}{2!}\right) + \cdots.\end{aligned}$$

The coefficients of $t^k/k!$ are clearly all equal to p (for $k = 1, 2, \ldots$), and so $E(X^k) = p$, except for $k = 0$. But, of course, this result can be obtained directly with very little effort, for $X^k = X$ when $X = 0$ or $X = 1$ as is the case here, and hence $EX^k = EX$.

EXAMPLE 2–65. Successive integrations by parts show that for any nonnegative integer k,

$$\int_0^\infty x^k e^{-x}\, dx = k!$$

Therefore, the function $x^k e^{-x}/k!$ can be used as a density on $x > 0$. If X has such density, its moment generating function is

$$\begin{aligned}\psi(t) = E(e^{tX}) &= \int_0^\infty e^{tx} x^k e^{-x} \frac{dx}{k!}\\ &= \frac{1}{k!}\int_0^\infty x^k e^{-(1-t)x}\, dx.\end{aligned}$$

The change of variable $y = (1 - t)x$ yields

$$\psi(t) = \frac{1}{k!\,(1 - t)^{k+1}} \int_0^\infty y^k e^{-y}\, dy = \frac{1}{(1 - t)^{k+1}},$$

the computation being valid for $t < 1$. This function is easily differentiated:

$$\psi'(t) = \frac{d}{dt}(1 - t)^{-(k+1)} = (k + 1)(1 - t)^{-k-2},$$

$$\psi''(t) = -(k + 1)(-k - 2)(1 - t)^{-k-3}$$

and so on. Evaluation at $t = 0$ yields

$$EX = \psi'(0) = k + 1,$$

and

$$EX^2 = \psi''(0) = (k + 1)(k + 2).$$

The same results could have been obtained from the binomial expansion of $\psi(t)$:

$$(1 - t)^{-(k+1)} = 1^{-(k+1)} + (-k - 1)1^{-k-2}(-t) + \frac{(-k - 1)(-k - 2)}{2!} 1^{-k-3}(-t)^2 + \cdots.$$

The moment EX^i is the coefficient of $t^i/i!$. (But, of course, knowing the expansion is equivalent to knowing the derivatives.)

Again, in this case, the direct calculation of the moments as integrals would have been just as easy:

$$EX^r = \int_0^\infty x^r x^k \frac{e^{-x}}{k!} dx = \frac{(k + r)!}{k!} = (k + r)_r.$$

Thus, $EX = k + 1$, $EX^2 = (k + 2)(k + 1)$, as before. Nevertheless, there are instances (cf. Problem 2–127) in which the averaging of e^{tX} is actually simpler than the averaging of the powers of X.

If X and Y are *independent* random variables, the moment generating function of the sum $X + Y$ is a particularly simple combination of the moment generating functions of the summands, namely, their product:

$$\psi_{X+Y}(t) = E(e^{t(X+Y)}) = E(e^{tX}e^{tY})$$
$$= E(e^{tX})E(e^{tY}) = \psi_X(t)\psi_Y(t).$$

Thus, the moment generating function of a sum of two independent random variables is the product of their moment generating functions. Finite induction extends this result to the sum of any finite number of independent random variables: if $X_1, \ldots, X_n$ are *independent*, then

$$\psi_{\Sigma X_i}(t) = \prod_{i=1}^{n} \psi_{X_i}(t).$$

If, moreover, the summands have identical distributions, say, with common moment generating function $\psi_X(t)$, then

$$\psi_{\Sigma X_i}(t) = [\psi_X(t)]^n.$$

EXAMPLE 2–66. Let $X_1, \ldots, X_n$ be independent random variables, each with the distribution: $P(X = 1) = p$, $P(X = 0) = 1 - p$. The moment generating function of this common distribution is

$$\psi_X(t) = E(e^{tX}) = e^t \cdot p + e^0 \cdot q,$$

where $q = 1 - p$, and so the moment generating function of the sum is

$$\psi_{\Sigma X_i}(t) = [\psi_X(t)]^n = (pe^t + q)^n.$$

From this, one can calculate the mean of the sum as the value of the derivative at $t = 0$: $E(\sum X_i) = np$. (This could also be easily calculated directly as the sum of the means.)

If one obtains the moment generating function of a distribution indirectly (as in the preceding example, where it was found by means other than direct evaluation of its integral formulation), one can then calculate the moments of that distribution; but the question would remain as to the density function or probability function of the distribution. There is a "uniqueness theorem" for moment generating functions,[8] which says that there can only be one distribution leading to a given moment generating function (under certain conditions). Thus, if the moment generating function of a random variable Y is obtained indirectly but is recognized as the moment generating function of a known distribution, then that distribution is the distribution of Y.

It is not hard to find examples of distributions for which moment generating functions do not exist. Indeed, a distribution that does not have moments of all orders would not have a moment generating function.

2.4.2 The Factorial Moment Generating Function

Closely related to the moment generating function is a function that generates *factorial* moments. This function is defined as follows:

$$\eta_X(t) = E(t^X) = E[e^{X(\log t)}] = \psi_X(\log t).$$

Since $\log 1 = 0$, it is the point $t = 1$ that might be expected to be of interest, and that produces the factorial moments from the derivatives. Formally, one has

$$\eta'(t) = E(Xt^{X-1}), \qquad \eta''(t) = E[X(X-1)t^{X-2}], \qquad \ldots,$$

and hence

$$\eta'(1) = E(X), \qquad \eta''(1) = E[X(X-1)], \qquad \eta'''(1) = E[X(X-1)(X-2)],$$

and so on. In general, then (assuming that the indicated moments exist, and that the steps of differentiation and averaging can be interchanged),

$$\eta^{(k)}(1) = E[X(X-1)\cdots(X-k+1)] = E[(X)_k].$$

This is what is called the kth *factorial moment.*

The first factorial moment is the same as the first moment. The second factorial moment is a combination of the first two moments:

$$E[X(X-1)] = E(X^2) - EX,$$

from which the variance can be computed by adding EX to obtain $E(X^2)$ and then subtracting $(EX)^2$:

$$\text{var } X = \eta''(1) + EX - (EX)^2.$$

[8] Because the characteristic function is more convenient as a theoretical tool, it is the uniqueness theorem for characteristic functions that is usually given—as it is in Section 2.4.4. When it exists, the moment generating function can be expressed in terms of the characteristic function, and its uniqueness then follows that of the characteristic function.

EXAMPLE 2–67. For the distribution of the sum $X_1 + \cdots + X_n$ in Example 2–66, the moment generating function was found to be $(pe^t + q)^n$. The factorial moment generating function is then

$$\eta(t) = \psi(\log t) = (pt + q)^n.$$

The second derivative of this is

$$\eta''(t) = n(n - 1)(pt + q)^{n-2}p^2,$$

whence

$$E[X(X - 1)] = \eta''(1) = n(n - 1)p^2,$$

since $p + q = 1$. And then (since $EX = np$, as seen in Example 2–66),

$$\text{var } X = n(n - 1)p^2 + np - (np)^2 = np(1 - p).$$

Like the moment generating function, the factorial moment generating function has the property that for a sum of independent random variables, it is just the product of the factorial moment generating functions of the summands:

$$\eta_{\Sigma X_i}(t) = E(t^{\Sigma X_i}) = E\left(\prod t^{X_i}\right) = \prod \eta_{X_i}(t).$$

Given a probability distribution, one may well wonder whether to use the moment generating function or the factorial moment generating function—or to calculate desired moments directly. The form of the density or probability function is the clue; sometimes it combines most neatly with e^{tx} and sometimes with t^x—and sometimes it is messy with either one.

2.4.3 Probability Generating Functions

Consider a discrete random variable X whose possible values are nonnegative integers: 0, 1, 2,.... The factorial moment generating function of the distribution of X is then

$$\eta_X(t) = E(t^X) = \sum_{k=0}^{\infty} t^k P(X = k).$$

This means that in the Maclaurin power series expansion of $\eta_X(t)$, the coefficient of t^k is the probability that the value of X is k. That is, whereas the expansion of $\eta(t)$ about the point $t = 1$ produces the factorial moments, the expansion about $t = 0$ produces the probabilities. A combination of this property with the one relating to sums of independent random variables provides a technique for calculating probabilities, as shown in the following example.

EXAMPLE 2–68. Let X denote the number of points on a die. The distribution of X, for a fair die, has the generating function

$$E(t^X) = \frac{t + t^2 + t^3 + t^4 + t^5 + t^6}{6} = \frac{t(1 - t^6)}{6(1 - t)}.$$

If this die is tossed three times, with results X_1, X_2, X_3, and if these are independent random variables, then the total number of points thrown is a random variable Y whose factorial moment generating function is the cube of that for the outcome of a single toss:

$$E(t^Y) = E(t^{X_1})E(t^{X_2})E(t^{X_3}) = \left(\frac{t(1-t^6)}{6(1-t)}\right)^3$$

$$= \frac{t^3}{216}\sum_{k=0}^{3}\binom{3}{k}(-t^6)^k \sum_{j=0}^{\infty}\binom{-3}{j}(-t)^j,$$

the last sum being the extension of the binomial expansion to the case of negative integer exponents, where

$$\binom{-n}{k} = \frac{(-n)(-n-1)\cdots(-n-k+1)}{k!}.$$

This expansion can be shown to converge to the right value when $|t| < 1$. Writing the expression for $E(t^Y)$ as a double sum, one has

$$E(t^Y) = \frac{1}{216}\sum_{k=0}^{3}\sum_{j=0}^{\infty}(-1)^{j+k}\binom{3}{k}\binom{-3}{j}t^{6k+j+3}.$$

From this, one can read, for instance, the probability that the total number of points is 7, as the coefficient of t^7. This power only occurs for $k = 0$ and $j = 4$:

$$P(Y = 7) = \frac{1}{216}\binom{3}{0}\binom{-3}{4} = \frac{(-3)(-4)(-5)(-6)}{216 \cdot 24} = \frac{15}{216}.$$

Problems

2–121. A random variable X assumes the value b with probability 1.

(a) Determine the moment generating function of X and from it the moments of X.
(b) Determine the central moments of the distribution.

2–122. Write the integral that would define the moment generating function for the density,

$$f(x) = \frac{1/\pi}{1 + x^2},$$

and examine it for convergence. (Would you expect it to converge in view of Example 2–44?)

2–123. Obtain a formula for the moment generating function of $Y = aX + b$ in terms of $\psi_X(t)$.

2–124. Given $\psi_X(t) = (1 - t^2)^{-1}$, compute var X.

2–125. Compute the factorial moment generating function of the discrete distribution defined by the probability function $f(k) = 1/2^{k+1}$ for $k = 0, 1, \ldots$. Determine the mean and variance of the distribution.

2–126. Determine the moment generating function of the distribution defined by the density $f(x) = 1$ for $0 < x < 1$. Determine from it the mean and variance.

2–127. Obtain the moment generating function of the distribution with density $f(x) = K \exp(-\frac{1}{2}x^2)$.

[*Hint*: It is not necessary to know K. Complete the square in the exponent and exploit the fact that the integral over $(-\infty, \infty)$ of $\exp(-\frac{1}{2}x^2)$ is the same as the integral of $\exp[-\frac{1}{2}(x - a)^2]$.]

2–128. Let $X_1, \ldots, X_n$ be independent, each having the distribution of X in Problem 2–125. Determine the factorial moment generating function of the sum $\sum X_i$. Obtain its power series expansion and read from it the probability that $\sum X_i = m$.

2–129. Derive the moment generating function of the sum $Y = X_1 + \cdots + X_n$, where the X's are independent random variables each with density e^{-x} for $x > 0$. By recognizing the result as the moment generating function of a previously encountered distribution, deduce the distribution of Y.

2–130. Obtain the moment generating function of the sum of n independent variables each of which has the density of Problem 2–127. Compute from it the variance of the sum.

2–131. Determine the probability that the total number of points showing in a toss of three dice is 9. (Compare Example 2–68.)

2–132. Three numbers are drawn at random from the integers 0, 1, . . . , 9. Assuming independent distributions, determine the probability that the sum of the numbers selected is 10.

2.4.4 The Characteristic Function

It has been pointed out that the moment generating function of a distribution may or may not exist. A transform that generates moments in similar fashion but which always exists is the *characteristic function*:

$$\phi(t) = E(\cos tX) + iE(\sin tX),$$

where i is the imaginary unit, $\sqrt{-1}$. Because of the deMoivre formula: $e^{i\theta} = \cos\theta + i\sin\theta$, the definition is written formally as

$$\phi(t) = E(e^{itX}).$$

This will be defined for all real t because the sine and cosine functions are bounded in magnitude by 1. For example,

$$E|\cos tX| = \int_{-\infty}^{\infty} |\cos tx|\, dF(x) \leq \int_{-\infty}^{\infty} dF(x) = 1.$$

EXAMPLE 2–69. Consider the distribution defined by the density e^{-x} for $x > 0$. The characteristic function is computed as follows:

$$\begin{aligned}\phi(t) &= \int_0^\infty e^{-x}\cos tx\, dx + i\int_0^\infty e^{-x}\sin tx\, dx \\ &= \frac{1}{1+t^2} + i\frac{t}{1+t^2} = \frac{1}{1-it}.\end{aligned}$$

Observe that the same result would follow if one treated e^{itx} formally, just as he would e^{tx}:

$$\int_0^\infty e^{itx}e^{-x}\,dx = \int_0^\infty e^{-x(1-it)}\,dx = \frac{1}{1-it}.$$

The phenomenon in the above example—that the formal computation with e^{itx} as though it were not complex gives the correct result—is not an accident. Indeed, the very symbolism $E(e^{itX})$ suggests that the moment generating function and the characteristic functions are related:

$$\phi(t) = \psi(it).$$

That this is true when $\psi(t)$ exists can be demonstrated using complex variable theory. Moreover, it can be shown that differentiating under the "E" is permitted k times when X has k finite moments, so that

$$E(X^k) = i^{-k}\phi^{(k)}(0).$$

If moments up to a certain order, say r, exist, then it is possible to express $\phi(t)$ as a Maclaurin series with a remainder—even though the complete series expansion would not exist. Theorem 4 asserts that for a characteristic function, the remainder term is better behaved than it is in such expansions generally.

Although the characteristic function does generate moments, its principal use is as a tool in deriving distributions. For this purpose it is necessary to know several facts about characteristic functions. It is beyond the scope of this treatment to present proofs of these facts, but their statements are not hard to comprehend.

Theorem 1. Inversion Formula. *If $x - h$ and $x + h$ are any two points of continuity of $F(x)$, the increment over the interval between them is given by the formula:*

$$F(x+h) - F(x-h) = \lim_{T\to\infty} \frac{1}{\pi}\int_{-T}^{T} \frac{\sin ht}{t}\, e^{-itx}\phi(t)\,dt.$$

Theorem 2. Uniqueness Theorem. *To each characteristic function there corresponds a unique distribution function having that characteristic function.*

Theorem 3. Continuity Theorem. *If distribution functions $\{F_n(x)\}$ converge to a distribution function $F(x)$, the corresponding characteristic functions $\{\phi_n(t)\}$ converge to the characteristic function of $F(x)$. Conversely, if a sequence of characteristic functions $\{\phi_n(t)\}$ converge to $\phi(t)$ which is continuous at $t = 0$, then $\phi(t)$ is a characteristic function, and the corresponding distribution functions $\{F_n(x)\}$ converge to the distribution function determined by $\phi(t)$.*

Theorem 4. Expansion with Remainder. *If $E(|X|^k)$ exists, so does $E(|X|^j)$ for $j = 0, 1, \ldots, k-1$, and*

$$\phi(t) = 1 + E(X)(it) + E(X^2)\frac{(it)^2}{2!} + \cdots + E(X^k)\frac{(it)^k}{k!} + o(t^k),$$

where $o(u)$ denotes a function such that $o(u)/u \to 0$ as $u \to 0$.

The fact that there is an inversion formula implies Theorem 2, that there is a uniqueness property; for, the recovery of $F(x)$ by means of the inversion formula is a uniquely defined operation, except at points of discontinuity of $F(x)$, where uniqueness is provided by the requirement of right-continuity of $F(x)$. Theorem 3 will permit obtaining limiting distributions by obtaining their characteristic functions as limits of characteristic functions.Theorem 4 relates the existence of moments to the smoothness (existence of derivatives) of the characteristic function.

EXAMPLE 2–70. Consider again the density function of

$$f(x) = [\pi(1 + x^2)]^{-1}.$$

The characteristic function of the distribution defined by this density is

$$\phi(t) = \frac{1}{\pi}\int_{-\infty}^{\infty} \frac{\cos tx}{1 + x^2}\,dx + \frac{i}{\pi}\int_{-\infty}^{\infty} \frac{\sin tx}{1 + x^2}\,dx = e^{-|t|}.$$

(The evaluation of these integrals is not elementary, but their values can be found in tables of definite integrals.) This characteristic function is not differentiable at $t = 0$, corresponding to the fact that not even the first moment of the distribution exists.

The inversion formula, given in the general case in Theorem 1, provides not only the uniqueness claimed in Theorem 2, but also a means of obtaining the c.d.f. of a distribution when that distribution is derived in terms of its characteristic function. In the case of a continuous distribution, the density function is given by

$$\begin{aligned} f(x) &= \lim_{h\to 0} \frac{F(x + h) - F(x - h)}{2h} \\ &= \lim_{h\to 0}\lim_{T\to\infty} \frac{1}{\pi}\int_{-T}^{T} \frac{\sin ht}{2ht}\, e^{-itx}\phi(t)\,dt. \end{aligned}$$

If limits can be moved in and out at will, it would follow that

$$f(x) = \lim_{T\to\infty} \frac{1}{2\pi}\int_{-T}^{T} e^{-itx}\phi(t)\,dt.$$

This can be established and expresses the density $f(x)$ as the same kind of transform of $\phi(t)$ as $\phi(t)$ is of the density (the only essential difference is in the sign of the exponent). The functions $f(x)$ and $\phi(t)$ are said to constitute a *Fourier transform* pair in which the first one is the (direct) Fourier transform of the second one and the second one is the inverse Fourier transform of the first one. (Which one is direct and which one is inverse is arbitrary.) The formulas for f and ϕ in terms of each other are analogous to the formulas for the Fourier series of a periodic function and for the coefficients in that series.

EXAMPLE 2–71. If $\phi(t) = e^{-|t|}$, as in Example 2–70, then

$$f(x) = \lim_{T\to\infty} \frac{1}{2\pi} \int_{-T}^{T} e^{-|t|-itx}\, dt$$

$$= \lim_{T\to\infty} \frac{1}{\pi} \int_{0}^{T} e^{-t} \cos tx\, dt = \frac{1/\pi}{1+x^2},$$

which is the density function assumed in Example 2–70—thereby completing the circle.

2.4.5 Multivariate Generating Functions

Multivariate analogs of univariate moment generating and characteristic functions can be defined so as to serve similar purposes. The *bivariate moment generating function* of the distribution of (X, Y) is defined to be

$$\psi(s, t) = E(e^{sX+tY}).$$

The derivatives of this function evaluated at (0, 0) produce the various moments of the distribution—marginal and mixed:

$$\frac{\partial^{h+k}}{\partial s^h \partial t^k} \psi(s, t)\bigg|_{s=t=0} = E(X^h Y^k),$$

whenever the implied differentiation under the averaging operation is legitimate. Again these moments are coefficients in the expansion of the moment generating function. The moment generating function for the random vector $(X_1, \ldots, X_n)$ is defined to be

$$\psi(t_1, \ldots, t_n) \equiv E[\exp(t_1X_1 + \cdots + t_nX_n)],$$

and the characteristic function, similarly, is

$$\phi(t_1, \ldots, t_n) \equiv E[\exp(it_1X_1 + \cdots + it_nX_n)].$$

The marginal distributions for subsets of the n components have moment generating or characteristic functions obtained by setting equal to 0 those t's that correspond to the unused variables. For example, the marginal distribution of (X_1, X_3) has the characteristic function

$$\begin{aligned}\phi_{X_1,X_3}(t_1, t_3) &= E[\exp(it_1X_1 + it_3X_3)]\\ &= \phi(t_1, 0, t_3, 0, \ldots, 0).\end{aligned}$$

When the components of a random vector are *independent*, the various generating functions factor into the product of the univariate generating functions of the components. Thus, if X_1, X_2, X_3 are independent, then

$$\psi(t_1, t_2, t_3) = \psi_{X_1}(t_1)\psi_{X_2}(t_2)\psi_{X_3}(t_3).$$

That factorization, conversely, implies independence follows from the fact that there is also a uniqueness theorem in the case of multivariate distributions. That is, if (in the case of two variables)

$$\phi_X(s)\phi_Y(t) = \phi_{X,Y}(s, t),$$

then X and Y are independent. For then

$$\begin{aligned}\phi_{X,Y}(s, t) &= E(e^{isX})E(e^{itY}) \\ &= \int_{-\infty}^{\infty} e^{isx}f_X(x)\,dx \int_{-\infty}^{\infty} e^{ity}f_Y(y)\,dy \\ &= \int_{-\infty}^{\infty}\int_{-\infty}^{\infty} e^{i(sx+ty)}f_X(x)f_Y(y)\,dx\,dy,\end{aligned}$$

which is the characteristic function of a bivariate distribution with density $f_X(x)f_Y(y)$, From the uniqueness theorem, then

$$f_{X,Y}(x, y) = f_X(x)f_Y(y),$$

and this implies independence of X and Y. (In this last argument, the X and Y could themselves be multivariate, with appropriate multiple integrals replacing the single integrals.) The same reasoning would be used in showing that if $X_1, \ldots, X_n$ are independent random variables (or random vectors) whose joint characteristic function is the product of the characteristic functions of the individual X's, then these X's are independent, and conversely.

Problems

2–133. Determine the characteristic function of the distribution defined by the density $f(x) = \lambda e^{-\lambda x}$, for $x > 0$, in terms of the given positive constant λ.

2–134. Determine the characteristic function of the distribution defined by the density $K \exp(-\frac{1}{2}x^2)$. (Compare Problem 2–127.)

2–135. Determine the characteristic function of the distribution defined by the density function

$$f(x) = \frac{a/\pi}{(x - m)^2 + a^2}$$

for given constants a and m (with $a > 0$), by making a change of variable in the integral defining it and using the result of Example 2–70.

2–136. Given that X has a finite mean μ and finite variance σ^2.

(a) Write out the expansion with remainder given by Theorem 4 for characteristic functions, carrying out as far as the assumptions on X permit.
(b) Write the expansion of $\phi_Y(t)$, where $Y = X - \mu$.

2–137. Express the characteristic function of $aX + b$ in terms of the characteristic function of X, and apply it in the particular case $Y = X - \mu$. Write the factor $e^{-i\mu t}$ as a finite expansion with remainder $o(t^2)$, using Theorem 4, after recognising that it is the characteristic function of the distribution of the constant random variable $-\mu$. Multiply this expansion by the expansion for $\phi_X(t)$, assuming (as in Problem 2–136) the existence of the first two moments, to obtain again the expansion for Y in Problem 2–136.

2–138. Let $X_1, \ldots, X_n$ be independent, identically distributed random variables, and let $\phi(t)$ denote the characteristic function of their common distribution. Obtain a general expression for the characteristic function of their average:

$$\bar{X} \equiv \frac{1}{n}(X_1 + \cdots + X_n)$$

in terms of $\phi(t)$. Also write out the expression explicitly in the special cases in which the common distribution has density

(a) $\lambda e^{-\lambda x}$, $x > 0$, as in Problem 2–133.
(b) $Ke^{-x^2/2}$, as in Problem 2–134.
(c) $[\pi(1 + x^2)]^{-1}$, as in Example 2–70.

2–139. Let $X_1, \ldots, X_n$ be independent, identically distributed variables and $\bar{X}$ their average, as in Problem 2–138. These X's can be thought of as "observations," the results of n trials of the experiment defined by their common distribution, and it will be seen in later chapters that the deviations $X_1 - \bar{X}, \ldots, X_n - \bar{X}$ are important in inference. Determine the multivariate moment generating function of those deviations, both in general (in terms of the m.g.f. of the common distribution) and in the special case in which that common distribution has density $Ke^{-x^2/2}$ (Problem 2–127).

2.5 Limit Theorems

The theory of "large" samples in statistical inference makes use of certain results concerning the limiting behavior of sequences of random variables and probability distributions. The rigorous treatment of such "limit theorems" is not elementary, and the mathematical background assumed here permits only a sketchy presentation of useful results. (More complete discussions may be found in Cramér [5] and Loève [8].)

Consider an infinite sequence of random variables defined on a probability space: $Y_1, Y_2, \ldots$. (This space may be taken to be the space of infinite sequence of real numbers, in which case the Y's are coordinate functions.) It is desirable to define what might be meant by the *limit* of the sequence; but random variables are functions, and there are several ways in which a given sequence of functions might be considered to approach a limit function.

1. *Convergence in distribution*: The sequence $\{Y_n\}$ is said to converge to Y *in distribution*, if and only if at each point λ where $F_Y(\lambda)$ is continuous,

$$\lim_{n\to\infty} F_{Y_n}(\lambda) = F_Y(\lambda).$$

[An important corollary of convergence in distribution is that for large but finite n, the probability $P(Y_n \le \lambda)$ can be approximated by the probability $F_Y(\lambda)$, which may be considerably simpler.]

2. *Convergence in quadratic mean*: The sequence $\{Y_n\}$ is said to converge to Y *in the mean*, or *in quadratic mean*, or *in mean square*, if and only if the average squared difference tends to 0:

$$\lim_{n\to\infty} E[(Y_n - Y)^2] = 0.$$

(In mathematics, this is called L_2-*convergence*; the "E" is an integral and the random variables are functions on Ω.)

3. *Convergence in probability* (weak convergence): The sequence $\{Y_n\}$ is said to converge to Y *in probability*, if and only if for any $\epsilon > 0$,

$$\lim_{n\to\infty} P(|Y_n - Y| \ge \epsilon) = 0.$$

(In mathematics, this is referred to as *convergence in measure*.)

4. *Almost sure convergence* (strong convergence): The sequence $\{Y_n\}$ converges to Y *almost surely*, or with probability 1, if and only if

$$P\left(\lim_{n\to\infty} Y_n = Y\right) = 1.$$

These various types of convergence are not all completely equivalent, but there are some important interrelations:

Theorem A. *If $Y_n \to Y$ in quadratic mean, then $Y_n \to Y$ in probability.*

This is established by means of the basic inequality in Section 2.3.2. In the notation used there, set $h(x) = x^2$, $b = \epsilon^2$ (for any given $\epsilon > 0$), and $X = |Y_n - Y|$. Then

$$P(|Y_n - Y| > \epsilon) \le \frac{1}{\epsilon^2} E[(Y_n - Y)^2].$$

And, clearly, if the right side tends to 0, so does the left.

Theorem B. *If k is a constant, then convergence of Y_n to k in distribution is equivalent to convergence in probability.*

There are two implications to be established here, and they follow from the following inequalities, given $\epsilon > 0$:

$$P(|Y_n - k| \ge \epsilon) = [1 - F_n(k + \epsilon)] + F_n(k - \epsilon) + P(Y_n = k + \epsilon)$$

and

$$P(|Y_n - k| \geq 2\epsilon) \leq [1 - F_n(k + \epsilon)] + F_n(k - \epsilon) - P(Y_n = k - \epsilon).$$

The first of these relations shows that if Y_n tends to k in probability, then $F_n(k + \epsilon) \to 1$ and $F_n(k - \epsilon) \to 0$, which is convergence to k in distribution. The converse implication follows from the second relation.

Theorem C. *Almost sure (strong) convergence implies convergence in probability (weak convergence).*

This will not be proved here. (See Loève [18] for this and other relationships.)

2.5.1 Laws of Large Numbers

A basic notion in the formulation of a probability model is that the probability of an event is intended to embody the observed phenomenon of long-run stability in the relative frequency of occurrence of the event in a sequence of trials of the experiment. It is then certainly of interest to determine whether stability is a mathematical consequence of the axioms in the model that has been developed. That this is so is a result referred to as a *law of large numbers.*

Thus, in the particular case of trials in which an event A occurs with probability p and does not occur with probability $1 - p$, the law of large numbers would assert that

$$\lim_{n\to\infty} \frac{Y}{n} = p,$$

in some sense, where Y denotes the number of times in n trials that A does occur—the frequency of the event A. In fact, if the trials are independent experiments, the relative frequency of A does tend toward the probability of A in the mean, in probability, and almost surely. Only convergence in probability will be shown here, but this will be done for a more general law.

Consider, then, a sequence $X_1, X_2, \ldots$ of random variables, identically distributed, and any finite set of which are independent. The *weak law of large numbers* is the following theorem.

Khintchine's Theorem. *If the common distribution of the independent, identically distributed variables $X_1, X_2, \ldots$ has a finite first moment μ, then the sequence of averages,*

$$Y_n = \frac{1}{n}(X_1 + \cdots + X_n)$$

converges to μ in probability.

The proof is a simple application of the Chebyshev inequality if it can be assumed that second moments exist, as they usually do. More generally however, the asserted convergence holds even when the second moment is not finite, and this will be demonstrated using the expansion theorem for characteristic functions and the following fact from calculus:

Lemma. *Let $o(x)$ denote any function such that $o(x)/x$ has the limit 0 as x tends to 0. Then*

$$\lim_{x \to 0} [1 + ax + o(x)]^{1/x} = e^a.$$

To establish this, let $y = ax + o(x) = x[a + o(1)]$, where $o(1)$ is a function that tends to 0 with x (as does y). Then

$$\begin{aligned}[1 + ax + o(x)]^{1/x} &= [(1 + y)^{1/y}]^{a+o(1)} \\ &= [(1 + y)^{1/y}]^a[(1 + y)^{1/y}]^{o(1)}.\end{aligned}$$

As x tends to 0, the first factor tends to e^a and the second to 1 (its logarithm tends to 0), which gives the desired result.

Returning to the proof of Khintchine's theorem, one first obtains the characteristic function of Y_n (as in Problem 2–138):

$$\phi_{Y_n}(t) = \left[\phi\left(\frac{t}{n}\right)\right]^n,$$

where $\phi(t)$ is the characteristic function of the common distribution of the X's. From Theorem 4 (expansion with remainder, in the case where EX exists):

$$\phi(t) = 1 + i\mu t + o(t),$$

so that

$$\phi_{Y_n}(t) = \left[1 + \frac{i\mu t}{n} + o\left(\frac{t}{n}\right)\right]^n.$$

This converges, by the lemma (with $x = 1/n$), to $e^{i\mu t}$, the characteristic function of the distribution of the constant random variable μ. But then, by the uniqueness theorem, it follows that Y_n converges to μ in distribution and therefore (by Theorem B) also in probability, as was to be shown.

[The above argument required use of the lemma in the case where a is complex. This can be justified by means of complex variable theory, a sophistication that seems hard to avoid in establishing general results; for, working with distributions that do not have moments of all orders leads one to the complex-valued $\phi(t)$. An alternative derivation of the weak law of large numbers, which also assumes that certain results for real functions hold also for complex valued functions, will be given in Problem 2–145. A derivation based on the Chebyshev inequality, valid when second moments exist, will be called for in Problem 2–140.]

The special case of the law of large numbers referred to at the outset is obtained from the general case when the common distribution of the X's is that defined by the probability function

$$f(1) = P(A), \qquad f(0) = P(A^c).$$

That is, X_i is 1 or 0 according to whether A does or does not occur. Here, the sum $X_1 + \cdots + X_n$ is just the frequency of A's in the n trials, and Y_n is the relative frequency of A. The expected value of X_i is easily seen to be $P(A)$.

Even though (according to the laws of large numbers) the average, $Y_n = (X_1 + \cdots + X_n)/n$, of independent and identically distributed random variables tends toward the common expected value, it does not follow that the *sum* $S_n = X_1 + \cdots + X_n$ is necessarily close to $nE(X)$. This is because the sum has a large variance:

$$\text{var } S_n = n \text{ var } X,$$

which (as $n \to \infty$) indicates a large dispersion even when var X is finite.

2.5.2 The Central Limit Theorem

It is a remarkable fact that the random variable Y_n, defined as the arithmetic mean of a sequence of n-independent replicas (in distribution) of a random variable X, has a distribution whose shape tends to a limiting shape that is *independent of the distribution of* X, so long as X has a finite variance, as n becomes infinite. However, to study this limiting shape, it is necessary to modify Y_n so that the limiting distribution is not singular—the limiting distribution function of Y_n itself is a step function with a single step at EX. On the other hand, the variable

$$nY_n = S_n = X_1 + X_2 + \cdots + X_n$$

has both a mean ($n\mu$) and a variance ($n\sigma^2$, where $\sigma^2 = \text{var } X$) that become infinite with n; thus one loses track of the shape of the distribution because it flattens out and moves off to infinity.

A device that keeps the distribution from shrinking or expanding excessively is that of standardization—forming a linear function that has mean zero and variance 1:

$$Z_n \equiv \frac{Y_n - \mu}{\sigma/\sqrt{n}} = \frac{S_n - n\mu}{\sqrt{n}\,\sigma}.$$

It is readily verified that $EZ_n = 0$ and var $Z_n = 1$, since var $Y_n = \sigma^2/n$. Having fixed mean and variance, the random variable Z_n has a distribution whose shape can be examined as n becomes infinite.

The standardized Z_n can be written

$$Z_n = \sum_1^n U_i, \qquad \text{where} \qquad U_i = \frac{X_i - \mu}{\sqrt{n}\,\sigma}.$$

Since $E(X_i - \mu) = 0$ and $\text{var}\,(X_i - \mu) = \sigma^2$, it follows that

$$\phi_{X_i-\mu}(t) = 1 - \frac{\sigma^2 t^2}{2} + o(t^2),$$

whence

$$\phi_{U_i}(t) = \phi_{X_i-\mu}\left(\frac{t}{\sqrt{n\sigma^2}}\right) = 1 - \frac{t^2}{2n} + o\left(\frac{t^2}{n}\right).$$

The characteristic function of Z_n is then the nth power:

$$\phi_{Z_n}(t) = [\phi_{U_i}(t)]^n = \left[1 - \frac{t^2}{2n} + o\left(\frac{t^2}{n}\right)\right]^n.$$

But according to the lemma of the preceding section (again with $x = 1/n$) the limit as n becomes infinite is $e^{-t^2/2}$, the characteristic function of the distribution with density $Ke^{-z^2/2}$, as found in Problem 2–134. By the Continuity Theorem and Uniqueness Theorem for characteristic functions, then, the sequence Z_n converges in distribution to the distribution defined by this density—called the *standard normal* distribution, to be taken up in more detail in the next chapter. It is shown there, too, that $K = 1/\sqrt{2\pi}$, a fact that is already implicit in Example 2–25. At this point, it suffices to refer to Table I in the Appendix for values of the c.d.f. of the limiting distribution.

The above discussion can be summarized in the following theorem.

Central Limit Theorem. *Let $X_1, X_2, \ldots$ be a sequence of identically distributed random variables with mean μ and variance σ^2 (both finite), any finite number of which are independent. Let $S_n = X_1 + \cdots + X_n$. Then for each z,*

$$\lim_{n\to\infty} P\left(\frac{S_n - n\mu}{\sqrt{n}\,\sigma} \le z\right) = \Phi(z) = \frac{1}{\sqrt{2\pi}}\int_{-\infty}^{z} \exp\left(-\frac{1}{2}u^2\right) du.$$

Since, according to the definition of a limit, the limitand can be made arbitrarily close to the limit by taking n large enough, it follows that the distribution function of the standardized sum can be approximated with the aid of the standard normal table. In particular, setting $z = (y - n\mu)/\sqrt{n\sigma^2}$, one has

$$P(X_1 + \cdots + X_n \le y) \doteq \Phi\left(\frac{y - n\mu}{\sqrt{n}\,\sigma}\right),$$

which is the useful formulation of the result.

EXAMPLE 2–72. People using a certain elevator are considered to be drawn randomly from a large population of people with mean 175 lb and standard deviation 20 lb. What is the probability that 16 persons would have a combined weight exceeding the load limit of 3000 lb?

Assuming the total weight W of 16 persons to be approximately normally distributed with mean $16 \times 175 = 2800$ lb and variance $16 \times 400 = 80^2$ lb^2, one finds

$$P(W > 3000 \text{ lb}) = 1 - P(W < 3000 \text{ lb}) = 1 - F_W(3000)$$
$$\doteq 1 - \Phi\left(\frac{3000 - 2800}{80}\right) = 0.0062,$$

from Table I, with $z = \frac{200}{80}$.

The Central Limit Theorem is called "central" because it is central to the distribution theory necessary for statistical inference. It is a remarkable result, in that the limiting distribution is the same no matter what the common distribution of the summands, so long as the variance is finite. This common distribution can be discrete, or not symmetric, or not unimodal (that is, with more than one maximum in its density). However, the more nearly normal the summands, the fewer the number of summands necessary to achieve approximate normality. In particular, it will be shown in the next chapter that if the summands are normal themselves, then the sum is exactly normal for any finite number of terms. Because of this dependence of the rate of approach to normality on the underlying distribution of the summands, it is not easy to specify the size of n for a good approximation. However, in most cases, an n of 25 or more is adequate for approximations to two decimal places. Experience with problems and examples will help to give some idea of how well the approximation works.

The form of central limit theorem given above is not the most general in that the same conclusion of asymptotic normality of a sum is valid under weaker conditions. It is not necessary that the terms in the sum be identically distributed, provided their third moments satisfy a certain mild condition, for instance.[9] There are also circumstances under which the assumption of independence of the terms can be relaxed.

Problems

2–140. Deduce the weak law of large numbers for the case in which the variance is finite by showing that the arithmetic mean approaches the expected value in quadratic mean.

2–141. Show that if $X_1, X_2, \ldots$ are independently and identically distributed with finite kth moment, then $(X_1^k + \cdots + X_n^k)/n$ converges in probability to $E(X^k)$, where X refers to the common distribution of the X_i's.

2–142. Booklets are packaged in bundles of 100 by weighing them. Suppose that the weight of each booklet is a random variable with mean 1 oz and standard deviation .05 oz. What is the probability that a bundle of 100 booklets weighs more than 100.5 oz (and so would be considered to have more than 100 booklets)? Assume independence of the weights.

[9] See H. Cramér [5], pp. 213 ff.

• **2–143.** In adding n real numbers, each is rounded off to the nearest integer. Assuming that the round-off error is a continuous random variable with constant density on the interval $(-.5, .5)$, determine the probability that the error in the sum is no greater than $\sqrt{n}/2$ in magnitude, for large n. (Assume independence of the errors.)

2–144. A sign in an elevator reads: "Capacity, 3000 lb or 20 persons." Assume a standard deviation of 20 lb for the weight of a person drawn at random from all the people who might ride this elevator, and calculate approximately this person's expected weight, given that the probability that a full load of 20 persons will weigh more than 3000 lb is .20.

2–145. Assuming that the logarithm function is defined for complex arguments, and that the usual rules for derivatives of logarithms and for evaluating indeterminate forms carry over to complex functions, use l'Hospital's rule to show:

(a) If $\mu = -i\phi'(0)$ exists, then $[\phi(t/n)]^n \to e^{i\mu t}$ as $n \to \infty$. (See Khintchine's theorem.)
(b) If $\mu = 0$ and σ^2 is finite, so that $\phi''(0) = -\sigma^2$, then

$$\left[\phi\left(\frac{t}{\sqrt{n}\,\sigma}\right)\right]^n \to e^{-t^2/2}$$

as $n \to \infty$. (See the Central Limit Theorem.)

[*Hint*: In (a) take the limit of the logarithm divided by t. In (b) take the limit of the logarithm divided by t^2.]

3

SOME PARAMETRIC FAMILIES OF DISTRIBUTIONS

A number of probability distributions have been encountered in the problems and examples of the preceding chapter that are sufficiently important to warrant an identification by name and a more systematic study of certain of their properties and applications. Their significance lies in their utility as workable models for many important practical situations. Although useful in application, they represent idealizations, as do most mathematical models, implying conditions that cannot really be met in practice. Or perhaps a safer generality would be that one simply cannot know whether the conditions are or are not met in practice. Indeed, one of the purposes of statistical inference is to exploit experimental data to decide whether a certain ideal model can be assumed to be the one representing an actual phenomenon.

Specific names for probability models are usually assigned not to particular distributions but to "classes" of distributions. The distributions in a class would have certain aspects in common and be related, ordinarily, through their representation by a density or probability function involving a "parameter"—an extra variable in the formula whose particular values define the particular distributions in the class. The parameter values then serve to index the various members of the class or family, and so it is called a parametric class or parametric family.

Although many important classes of distributions arise in the study of the distributions of statistics used in inference, the classes taken up in this chapter are more directly related to the basic experiment at hand.

3.1 Distributions for Bernoulli Trials

A *Bernoulli experiment* is one in which there are just two outcomes of interest—some event A either happens or does not happen. Some examples of such experiments include tossing a coin (resulting in heads or tails), determining the sex of a randomly drawn person or animal (male or female), testing a product (good or defective), playing a game (win or lose), comparing one variable with another (larger or smaller), giving an inoculation (takes or does not take), carrying out a mission (success or failure), and so on.

The indicator function of the event A is called a *Bernoulli random variable*:

$$X(\omega) = \begin{cases} 1, & \text{if } A \text{ occurs,} \\ 0, & \text{if } A^c \text{ occurs.} \end{cases}$$

Because the variable has just two possible values, it is discrete whether or not Ω is discrete. Its distribution is completely defined and characterized by a single number, the probability $p = P(A)$. Sometimes the occurrence of A is referred to as a *success* (i.e., one "succeeds" in getting an A), and p is the probability of success. The probability of failure of A^c is, of course, $1 - p$, usually called q.

EXAMPLE 3–1. An urn contains four white beads and six black beads. When one bead is drawn at random from the urn, it can be classified as either white or black, and the function

$$X(\omega) = \begin{cases} 1, & \text{if } \omega \text{ is black,} \\ 0, & \text{if } \omega \text{ is white,} \end{cases}$$

is a Bernoulli random variable. The probability of success (drawing a black bead) is the proportion of black beads in the urn: $p = \frac{6}{10}$ and $q = \frac{4}{10}$.

The probability function of a Bernoulli variable can be represented simply in a table,

k	$f(k)$
0	q
1	p

or it can be given by a formula,

$$f(k) = P(X = k) = p^k q^{1-k}, \qquad \text{for } k = 0 \text{ or } 1.$$

It may seem like an unnecessary sophistication to give a formula for something so simple, but the formula does come in handy in studying methods of inference about Bernoulli experiments.

The moments are easy to compute, because when X can be only 0 or 1, then $X^k = X$ (if $k > 0$):

$$EX^k = EX = 1 \cdot p + 0 \cdot q = p.$$

The variance is (as always) the mean square minus the square of the mean:

$$\text{var } X = EX^2 - (EX)^2 = p - p^2 = pq.$$

Using the moments $EX^k = p$ as coefficients of $t^k/k!$ in a series, to reverse the usual computation, yields the moment generating function:

$$1 + pt + \frac{pt^2}{2} + \frac{pt^3}{3!} + \cdots = q + p\left(1 + t + \frac{t^2}{2} + \frac{t^3}{3!} + \cdots\right)$$
$$= q + pe^t = E(e^{tX}) = \psi(t).$$

When a Bernoulli experiment is performed over and over again, this is said to be a sequence of *Bernoulli trials*. If it is really the same experiment each time, the value of p is constant over the sequence of trials, and one speaks of them as *identical* Bernoulli trials. (This does *not* mean that the results are the same, but simply that the experiments are the same.) In a finite sequence of Bernoulli trials, the indicator functions for the trials keep count of the event A of interest, and their *sum* over n trials is precisely the number of "successes" among those trials.

3.1.1 The Binomial Distribution

Suppose now that the random variables $X_1, \ldots, X_n$, denoting the results of n Bernoulli trials, are *independent* random variables; that is, the trials are independent experiments. With a constant p the X's are identically distributed, each with factorial moment generating function

$$\eta(t) = \psi(\log t) = pt + q.$$

The number of successes among the n trials, which is the sum of the 0's and 1's resulting from the individual trials,

$$S_n = X_1 + X_2 + \cdots + X_n,$$

has a moment generating function that is the product of the moment generating functions of the summands:

$$\eta_{S_n}(t) = [\eta(t)]^n = (pt + q)^n$$
$$= \sum_{k=0}^{n} \binom{n}{k} p^k q^{n-k} t^k.$$

The probabilities for the various possible values of S_n (namely, $0, 1, 2, \ldots, n$) are then the coefficients of corresponding powers of t (see Section 2.4.3):

$$P(S_n = k) = P(k \text{ successes in } n \text{ trials}) = \binom{n}{k} p^k q^{n-k}.$$

This result is known as the *binomial formula*. A random variable with this distribution is said to be *binomially distributed.*

The binomial formula can be deduced more directly from the independence of the trials. The probability of a particular pattern of k successes among the n trials is just the product of n factors, a p for each success and a q for each failure:

$p^k q^{n-k}$. The number of distinct patterns in which there are exactly k successes is $\binom{n}{k}$, and each of these has probability $p^k q^{n-k}$. Hence, the probability of k successes in n independent trials is $\binom{n}{k} p^k q^{n-k}$, as obtained previously.

The moments of S_n can be obtained from its factorial moment generating function; in particular,

$$ES_n = \eta'(1) = np$$

and

$$\text{var } S_n = \eta''(1) + ES_n - (ES_n)^2 = n(n-1)p^2 + np - n^2p^2 = npq.$$

The mean and variance can also be computed from the moments of the common distribution of the results of the individual trials:

$$ES_n = EX_1 + \cdots + EX_n = p + \cdots + p = np$$

and

$$\text{var } S_n = \text{var } X_1 + \cdots + \text{var } X_n = pq + \cdots + pq = npq,$$

the latter exploiting the assumed independence of the trials.

EXAMPLE 3–2. Let X denote the number of sixes thrown in 12 tosses of a die. Assuming independence, the "binomial formula" derived above gives

$$P(X = k) = \binom{12}{k}\left(\frac{1}{6}\right)^k \left(\frac{5}{6}\right)^{12-k},$$

since p, the probability of a six in a single toss, is $\frac{1}{6}$. The expected number of sixes is

$$E(X) = np = 12 \cdot \frac{1}{6} = 2.$$

The variance of the distribution is $npq = \frac{5}{3}$. The probability of any event described in terms of X can be computed as a suitable sum of binomial probabilities. For instance,

$$\begin{aligned} P(\text{at least 3 sixes}) &= 1 - P(\text{2 or fewer sixes}) \\ &= 1 - \left[\binom{12}{2}\left(\frac{1}{6}\right)^2\left(\frac{5}{6}\right)^{10} + \binom{12}{1}\left(\frac{1}{6}\right)^1\left(\frac{5}{6}\right)^{11} + \binom{12}{0}\left(\frac{1}{6}\right)^0\left(\frac{5}{6}\right)^{12}\right] \\ &\doteq .32. \end{aligned}$$

3.1.2 The Negative Binomial Distribution

Suppose, instead of performing a fixed or given number of trials, one performs independent Bernoulli trials repeatedly, until a given number of successes are observed, and then stops. In this setting, the total number of trials required is *random*, equal to the (random) number of failures encountered before the given number of successes plus that number of successes.

To attack a simpler problem first, consider the number of failures encountered prior to the *first* success, and call this X. The probability that k failures occur in a row followed by a success is the probability that X takes on the value k:

$$P(X = k) = q \cdots q \cdot p = q^k p, \qquad \text{for } k = 0, 1, \ldots.$$

The distribution defined by these probabilities is said to be *geometric*, since the probabilities are terms in a geometric series:

$$p + pq + pq^2 + \cdots = p\frac{1}{1-q} = 1.$$

The factorial moment generating function is

$$\eta(t) = E(t^X) = \sum_0^\infty t^k q^k p = \frac{p}{1 - tq}$$

$$= \frac{1}{1 - (t-1)q/p} = \sum_0^\infty \left(\frac{q}{p}\right)^k (t-1)^k.$$

The factorial moments, $\eta^{(k)}(1)$, are the coefficients of $(t-1)^k/k!$ in this Taylor expansion about $t = 1$:

$$E[(X)_k] = E[X(X-1)\cdots(X-k+1)] = k!\left(\frac{q}{p}\right)^k.$$

Thus, the mean is q/p and the variance q/p^2.

The random variable $X + 1$, the number of trials needed to get the first success, has mean $q/p + 1 = 1/p$, which is an intuitively appealing result. For example, if the probability of success at each trial is $\frac{1}{6}$, then the mean number of trials to get the first success is the reciprocal, 6.

Consider next S_r, the number of failures encountered prior to the rth success. This is the number X_1 encountered prior to the first success plus the number X_2 encountered after the first but prior to the second success, . . . , plus the number X_r encountered after the $(r-1)$st, but prior to the rth success:

$$S_r = X_1 + X_2 + \cdots + X_r.$$

This is a sum of r independent variables, each having the geometric distribution. The generating function is therefore the rth power,

$$\eta_{S_r}(t) = [\eta(t)]^r = p^r(1 - qt)^{-r} = p^r \sum_{k=0}^\infty \binom{-r}{k}(-qt)^k,$$

where the negative binomial coefficient is defined (as might be expected) to be

$$\binom{-r}{k} = \frac{(-r)_k}{(k)_k} = \frac{(-r)(-r-1)\cdots(-r-k+1)}{k!}$$

$$= (-1)^k \binom{r+k-1}{k}.$$

(The formal binomial series for a negative power of a binomial $1 + x$ can be shown to converge to that function for $|x| < 1$.) The coefficient of t^k in the factorial moment generating function is the probability of k failures prior to the rth success:

$$P(S_r = k) = \binom{r + k - 1}{k} p^r q^k.$$

[This could have been obtained more directly by noting that there are exactly k failures prior to the rth success if and only if the first $r + k - 1$ trials are made up of k failures and $r - 1$ successes in some order, and the $(r + k)$th trial is a success.] The variable S_r is said to have a *negative binomial* distribution because the probabilities of its various possible values are terms in the expansion of a negative power of a binomial.

The mean and variance of S_r can be obtained (as in the case of the binomial variable) either from the mean and variance of the X's that make it up,

$$ES_r = rEX = \frac{rq}{p}, \qquad \text{var } S_r = r \text{ var } X = \frac{rq}{p^2},$$

or from the coefficients in the expansion of the factorial moment generating function about $t = 1$. (See Problem 3–13.)

EXAMPLE 3–3. Let T denote the number of trials necessary to obtain a number of 2's and 3's totaling 12, in independent tosses of an ordinary die. Here "success" is the throwing of a 2 or a 3, with probability $\frac{1}{3}$ at each trial. The variable T can be expressed as $S + 12$, where S is the number of "failures" prior to 12 successes. Thus, the values of T are 12, 13, . . ., with probabilities

$$P(T = k + 12) = \binom{11 + k}{k} \left(\frac{1}{3}\right)^{12} \left(\frac{2}{3}\right)^k, \qquad \text{for } k = 0, 1, \ldots.$$

The expected value of T is $ES + 12$, or

$$ET = 12 \cdot \frac{2/3}{1/3} + 12 = 36.$$

Problems

3–1. One hundred owners of car C are given blindfolded rides in each of cars C and L to test the "ride." If the two cars really ride the same, the probability that an owner would say car L rides better, given that he notes a difference, is $\frac{1}{2}$. Assuming that the 100 owners do all state a preference, what is the distribution of the number of them who choose car L? Determine the mean and standard deviation of this number. Write a formula for the probability that exactly 50 out of the 100 choose L, and calculate it with the aid of Tables XIII and XV in the Appendix.

3–2. A person attempts to predict the fall of a coin in each of several successive trials. If he really has no clairvoyant powers and is only guessing, what is the probability that he will predict correctly in 4 out of 4 successive trials? In at most 2 out of 10 trials? In 8 or more out of 10 successive trials?

3–3. Let X have the density $f(x) = [\pi(1 + x^2)]^{-1}$, and let Y denote the indicator function of the event $\{X \geq 1\}$. What is the distribution of Y?

3–4. The reliability (probability of successful functioning) of a certain automatic cutoff device is assumed to be $\frac{11}{12}$. What is the reliability of a system consisting of four such devices so arranged that any one would provide the cutoff if functioning properly?

3–5. Determine the reliability of a certain piece of equipment that consists of four components, each having reliability 0.9 so arranged that each component must work if the equipment is to work.

3–6. A random variable X is binomially distributed, with mean 3 and variance 2. Compute $P(X = 7)$.

3–7. Determine the probability that all of five articles taken from a production line are good, given that the probability of a defective article's being produced is 0.1. What is the probability that at most one of five has a defect? What is the average number of defectives in a lot of 50 articles?

3–8. Determine the rth factorial moment of a binomial distribution.

3–9. Evaluate:

$$\sum_0^{20} k^2 \binom{20}{k} (.3)^k (.7)^{20-k}.$$

3–10. (a) A die is cast repeatedly until a 6 shows. What is the expected number of throws necessary?

(b) Two dice are cast repeatedly. What is the expected number of casts required to obtain a 7? To obtain a 7 or an 11?

3–11. Make a table of values and probabilities for the negative binomial variable defined as the number of failures encountered prior to the fifth success in a series of independent Bernoulli trials with $p = \frac{1}{2}$. What are the most likely values?

3–12. Show:

$$\sum_0^{\infty} \frac{\binom{4+k}{k}}{2^k} = 32.$$

3–13. Obtain the mean and variance of the negative binomial distribution from the coefficients in the expansion of the factorial m.g.f. about $t = 1$.

[*Hint:* Use the form $(1 - [t - 1]q/p)^{-r}$, and write the first three terms in a negative binomial expansion.]

3–14. Use generating functions to show that if X and Y are *independent*, then

(a) $X + Y$ is binomial if X and Y are each binomial with the same p.

(b) $X + Y$ is negative binomial if X and Y are each negative binomial with the same p.

Interpret each result in terms of repeated Bernoulli trials.

3–15. Let $X_1, X_2, \ldots, X_n$ be independent and identically distributed with common c.d.f. $F(x)$. Let S_n denote the number of these observations that do not exceed x, a given, fixed value. What is the distribution of S_n? Show that S_n/n approaches $F(x)$ in quadratic mean and, therefore, in probability.

[*Hint:* Define Y_i to be the indicator function of the event $X_i \leq x$ and examine the sum $\sum Y_i$.]

3.1.3 Sampling Without Replacement

One situation in which Bernoulli trials are encountered is that in which an object is drawn at random from a collection of objects of two types—say, black and white beads in an urn, as in Example 3–1. In order to repeat this experiment so that the results are independent and identically distributed, it is necessary to replace each bead drawn and to mix the beads before the next one is drawn. This process is referred to as *sampling with replacement*. If the sampling is done *without* replacement of the beads drawn, the resulting "trials" are still of the Bernoulli type but no longer independent.

EXAMPLE 3–4. Four balls are drawn, one at a time, at random and without replacement from 10 balls in a container, three black and seven white. The probability that the third ball drawn is black can be computed as follows:

$$\begin{aligned} P(\text{3rd ball black}) &= P(WWB) + P(WBB) + P(BWB) + P(BBB) \\ &= \frac{7}{10}\cdot\frac{6}{9}\cdot\frac{3}{8} + \frac{7}{10}\cdot\frac{3}{9}\cdot\frac{2}{8} + \frac{3}{10}\cdot\frac{7}{9}\cdot\frac{2}{8} + \frac{3}{10}\cdot\frac{2}{9}\cdot\frac{1}{8} \\ &= \frac{3}{10}, \end{aligned}$$

which is the same as the probability that the first ball drawn is black. (The symbol "WWB," for example, means the event that the first ball drawn is white, the second is white, and the third is black.) It should not be surprising that this probability for black is the same on the third draw as on the first, when it is realized that the probability referred to is the marginal probability or absolute probability, and *not* a conditional probability given the results of the first two drawings. When each drawing is considered by itself, it may as well be thought of as made from the whole collection initially available—when no information is at hand concerning the results of previous trials.

In the general case, n objects are to be drawn at random, one at a time, from a collection of N objects, M of one kind and $N - M$ of another kind. The one kind of object will be thought of as "success," and coded 1; the other kind is coded 0. Let $X_1, \ldots, X_n$ denote the sequence of coded outcomes; that is, X_i is 1 or 0 according to whether the ith draw results in success or failure. The total number S_n of successes in n trials is just the sum of the X's:

$$S_n = X_1 + \cdots + X_n,$$

as it was in the case of independent Bernoulli trials.

As suggested in Example 3–4, the X's here are also identically distributed. That is, the probability "p" of a 1 on the ith trial is the same at each trial:

$$p = P(X_i = 1) = \frac{M}{N}.$$

To see this, observe first that the probability of a given sequence of objects (thinking of the N objects as distinct) is

$$\frac{1}{N} \cdot \frac{1}{N-1} \cdots \frac{1}{N-n+1} = \frac{1}{(N)_n}$$

so that the $(N)_n$ sequences are equally likely. The probability that an object of type "1" occurs in the ith position in the sequence is the fraction of sequences having that property:

$$P(X_i = 1) = \frac{M(N-1)_{n-1}}{(N)_n} = \frac{M}{N}.$$

[In the numerator, M is the number of ways of filling the ith position with an object coded 1, and $(N-1)_{n-1}$ is the number of ways of filling the remaining $n-1$ places in the sequence from the $N-1$ remaining objects.]

It was seen in Chapter 1 (Example 1–44 and Problem 1–54) that with regard to the number of successes or 1's among the n objects drawn, it does not matter whether the model assumed is that of drawing one at a time at random, or that of simultaneously drawing n at random. And from the standpoint of the latter model, the probability function of S_n is readily seen (by methods used in Chapter 1) to be

$$P(S_n = k) = f(k; N, M, n) = \frac{\binom{M}{k}\binom{N-M}{n-k}}{\binom{N}{n}}.$$

The possible values of S_n are included in the list $0, 1, \ldots, n$, and the given formula works for these values if it is understood that

$$\binom{a}{b} = 0, \qquad \text{when } b > a > 0.$$

A random variable with the probability function $f(k; N, M, n)$ is said to have a *hypergeometric* distribution.

The mean of the hypergeometric distribution is easily obtained from the representation of a hypergeometric variable as a sum of the Bernoulli trials that make it up:

$$E(S_n) = E(X_1 + \cdots + X_n) = p + p + \cdots + p = np = n \cdot \frac{M}{N}.$$

That is, whether the objects drawn are replaced or not, the expected number of type "1" among the n objects drawn is the same—just the number drawn times the probability of success at each trial. However, the variance of S_n is not the sum of the variances of the X's, since the latter are not independent. One method of calculation of var S_n is suggested in Problem 3–29, but it can also be obtained from a calculation of the general factorial moment.

In the following derivation, it will simplify the manipulations if k is permitted to range over the integers from 0 to N. This requires the understanding that $\binom{a}{b} = 0$ when $0 < a < b$, as before, and also when $b < 0$.

The factorial moments of the hypergeometric distribution can be computed as follows:

$$E[(S_n)_r] = \sum_{k=r}^{N} (k)_r f(k; N, M, n) = \sum_{k=r}^{N} (k)_r \frac{\binom{N-M}{n-k}\binom{M}{k}}{\binom{N}{n}}.$$

By using the identity

$$(k)_r \binom{M}{k} = (M)_r \binom{M-r}{k-r},$$

one obtains

$$\begin{aligned}
\sum_{k=r}^{N} (k)_r \binom{N-M}{n-k}\binom{M}{k} &= \sum_{k=r}^{N} (M)_r \binom{M-r}{k-r}\binom{N-M}{n-k} \\
&= (M)_r \sum_{j=0}^{N} \binom{M-r}{j}\binom{(N-r)-(M-r)}{(n-r)-j} \\
&= (M)_r \binom{N-r}{n-r} \sum_{j=0}^{N} f(j; N-r, M-r, n-r) \\
&= (M)_r \binom{N-r}{n-r} = \frac{(M)_r (n)_r}{(N)_r} \binom{N}{n}.
\end{aligned}$$

Thus, if $N \geq r$,

$$E[(S_n)_r] = \frac{(M)_r (n)_r}{(N)_r}.$$

In particular,

$$ES_n = \frac{nM}{N} \quad \text{and} \quad E[S_n(S_n - 1)] = \frac{M(M-1)n(n-1)}{N(N-1)}.$$

From these one can compute (for $N > 1$)

$$\begin{aligned}
\text{var } X S_n = E[S_n(S_n - 1)] + ES_n - (ES_n)^2 &= n \cdot \frac{M}{N} \cdot \frac{N-M}{N} \cdot \frac{N-n}{N-1} \\
&= npq \frac{N-n}{N-1},
\end{aligned}$$

where $p = M/N$, the probability at each trial that the object drawn is of the type of which there are initially M.

This last formula differs from the formula for the variance of S_n when drawing is done with replacement and mixing (the binomial case) by the extra factor $(N - n)/(N - 1)$. This factor is 1 when $n = 1$ (in which case there is no distinction between the two schemes), and is 0 for $n = N$. In the latter case, the entire group of available objects is drawn; and the random variable S_n is no longer variable but constantly equal to M, so that its variance is 0.

EXAMPLE 3–5. A lot of 10 contains 3 defective and 7 good articles. Suppose that 4 articles are drawn from the lot without replacement. The probability function for the number of defective articles in the sample of 4 is

$$f(k; 10, 3, 4) = \frac{\binom{3}{k}\binom{7}{4-k}}{\binom{10}{4}}.$$

In the following table are given the values of this function (multiplied through by 30 for simplicity) for the possible values $k = 0, 1, 2, 3$, together with products needed for computing the mean and variance directly:

k	$30f(k)$	$30kf(k)$	$30k^2f(k)$
0	5	0	0
1	15	15	15
2	9	18	36
3	1	3	9
	30	36	60

The expected value is then

$$E(S_4) = \frac{36}{30} = 4 \cdot \frac{3}{10},$$

and the variance,

$$\text{var } S_4 = \frac{60}{30} - \left(\frac{36}{30}\right)^2 = 4 \cdot \frac{3}{10} \cdot \frac{7}{10} \cdot \frac{10 - 4}{10 - 1}.$$

These results are seen to be the same as what would be obtained using the formulas for mean and variance derived above.

Another interesting feature of the factor that distinguishes the variance in the no-replacement case from the variance in the replacement case is that for fixed p (i.e., a fixed proportion M/N of the objects of type "1") and fixed n, the factor tends to 1 as N becomes infinite. This is not surprising when it is realized that when the population of objects is very large, the removal of a small number of them does not appreciably alter the proportions in what is left, and these proportions are then almost independent of what was drawn earlier.

Indeed, it will now be demonstrated that the hypergeometric probability function tends toward the binomial probability function:

$$f(k; N, M, n) \to \binom{n}{k} p^k q^{n-k},$$

for fixed k and n, and fixed $p = M/N$. (Both M and N become infinite, in a fixed ratio.) It is simply a matter of expressing the hypergeometric probability in terms of factorials, judiciously grouping factors, and observing what happens:

$$\frac{\binom{M}{k}\binom{N-M}{n-k}}{\binom{N}{n}} = \frac{(M)_k(N-M)_{n-k}n!}{k!\,(n-k)!\,(N)_n}$$

$$= \frac{M}{N}\frac{M-1}{N-1}\cdots\frac{M-k+1}{N-k+1}\frac{N-M}{N-k}\cdots$$

$$\frac{N-M-n+k+1}{N-n+1}\binom{n}{k}$$

$$\to p \cdot p \cdots p \cdot q \cdots q\binom{n}{k} = p^k q^{n-k}\binom{n}{k}.$$

The usefulness of the approach of the hypergeometric to the binomial distribution lies in the possibility of estimating a hypergeometric probability with an easier to compute binomial probability, when the number of objects drawn is considerably less than the number available. From a practical point of view, it is important to know "how fast" the hypergeometric distribution tends to the binomial. To give some hint of this, the calculations in the next example are presented.

EXAMPLE 3–6. Presented in the following table are hypergeometric probabilities of 0, 1, 2, 3, 4, and 5 defective articles in a selection of 10 articles from populations of sizes $N = 50$, 100, 200, and ∞, half of which are defective. The last, $N = \infty$, is really the binomial case for $n = 10$ and $p = 0.5$.

k	$N = 50$	$N = 100$	$N = 200$	$N = \infty$ (binomial)
0	.0003	.0006	.0008	.0010
1	.0050	.0072	.0085	.0098
2	.0316	.0380	.0410	.0439
3	.1076	.1131	.1153	.1172
4	.2181	.2114	.2082	.2051
5	.2748	.2539	.2525	.2461

3.1.4 Approximate Binomial Probabilities (Moderate p)

The Central Limit Theorem is useful in providing a means of approximate computation of binomial probabilities when a direct computation is tedious. This application hinges upon the fact that a binomially distributed random variable has the distribution of a sum of independent, identical Bernoulli variables. That is, if $X_i = 0$ with probability $1 - p$ and $X_i = 1$ with probability p, and if $X_1, \ldots, X_n$ are independent, then their sum has the binomial distribution (n, p), with mean np and variance $np(1 - p)$. According to the Central Limit Theorem, then, the distribution of the sum S_n, and hence the binomial distribution, is asymptotically normal for large n. More precisely, for fixed p,

$$\lim_{n\to\infty} P\left(\frac{S_n - np}{\sqrt{np(1-p)}} \le z\right) = \Phi(z).$$

Hence,

$$P(S_n \le x) = P\left(\frac{S_n - np}{\sqrt{np(1-p)}} \le \frac{x - np}{\sqrt{np(1-p)}}\right) \doteq \Phi\left(\frac{x - np}{\sqrt{np(1-p)}}\right).$$

EXAMPLE 3–7. Consider a binomial distribution with $n = 8$ and $p = \frac{1}{2}$. The actual distribution function is

$$P(X \le x) = \sum_{k\le x} \binom{8}{k}\left(\frac{1}{2}\right)^k\left(\frac{1}{2}\right)^{n-k},$$

and the normal approximation is

$$P(X \le x) \doteq \Phi\left(\frac{x-4}{\sqrt{2}}\right).$$

The graphs of these two functions are shown in Figure 3–1.

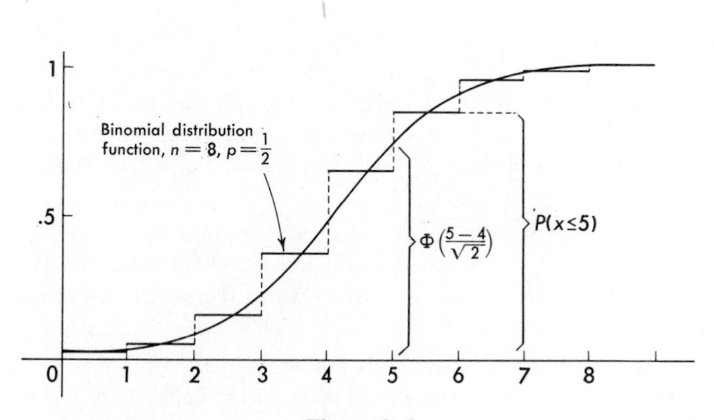

Figure 3–1

It is seen in Figure 3–1 that the normal approximation to the binomial is good when x is about halfway between integers; but it is *at* the integers that the value of $F(x)$ is needed. A better approximation to the value $F(k)$ is obtained by taking the ordinate on the continuous curve one-half unit to the right of k:

$$F(k) = P(X \leq k) \doteq \Phi\left(\frac{k + \frac{1}{2} - np}{\sqrt{np(1-p)}}\right).$$

With this "continuity correction," the normal approximation can be quite good even for rather small values of n, as in the next example.

EXAMPLE 3–8. Consider a binomial distribution with $n = 4$ and $p = \frac{1}{2}$:

$$P(X \leq k) \doteq \Phi\left[\frac{k + \frac{1}{2} - 2}{\sqrt{1}}\right].$$

These approximate values and the values computed from the actual distribution function are given in the following table.

		Normal Approximation	
k	$F_X(k)$	With Continuity Correction	Without Continuity Correction
0	.0625	.0668	.0228
1	.3125	.3085	.1587
2	.6875	.6915	.5000
3	.9375	.9332	.8413
4	1.0000	.9938	.9772

When p is near .5, the binomial distribution is rather symmetric and the normal approximation is useful even for quite small n. However, values of p near 0 or 1 result in a somewhat skewed distribution and a larger value of n is required for an acceptable approximation.

Problems

3–16. Determine the probability function for the number of white beads among five beads drawn at random from a bowl containing four white and seven black beads. Use this to compute the mean and variance and check the results using the formulas.

3–17. Determine the probability that 3 out of 5 articles drawn at random from 100 articles are defective, given that 10 in the lot of 100 are defective. Compare this with the corresponding answer for the case in which the articles are drawn one at a time with replacement and mixing.

3–18. In a bridge hand of 13 cards, what is the expected number of Aces? The expected number of face cards? (Of the 52 cards in the deck, 4 are Aces and 12 are face cards.)

3–19. In a population of 10,000 voters, 45 per cent favor a certain proposal. What is the probability that among 10 voters chosen at random without replacement, 6 or more favor the proposal?

3–20. Show that for fixed n and fixed $p = M/N$ the rth factorial moment of the hypergeometric distribution tends to that of a binomial distribution as N becomes infinite.

3–21. Of 20 cups of coffee, 15 are brewed in the usual way and 5 are made from instant coffee. After tasting all 20 cups, a taster selects 5 that he thinks are the ones made from instant coffee. What is the probability that if his selection is random (made by "pure chance"), exactly k of the 5 he selects are made from instant coffee?

3–22. Three chips are selected at random from ten chips, of which five are red, three white, and two black. Let X denote the number of red, and Y the number of white, chips among the three selected. Determine the following:

(a) $P(X = 1, Y = 1)$.
(b) The marginal distribution of X.
(c) $E(X + Y)$.

3–23. Evaluate:

$$\sum_0^4 k \binom{4}{k} \binom{8}{6 - k}.$$

3–24. A coin is to be tossed 100 times. Determine approximately probabilities that

(a) Less than 50 heads turn up.
(b) Exactly 50 heads turn up.
(c) More than 40 but less than 60 heads turn up.

3–25. Determine approximately the probability that in at least 28 of 72 tosses of a die the outcome is a 1 or a 2.

3–26. Given that 55 per cent of the votes in a city of 50,000 voters will vote for candidate A, what is the probability that a random selection of 100 voters would not show a majority in favor of A?

3–27. A bowl contains three white and seven black beads, from which one at a time is drawn. Determine the distribution of the number of black beads drawn before a white one is drawn (a) if the beads drawn are replaced and mixed with the rest after each drawing, and (b) if the beads drawn are not replaced.

3–28. In random sampling without replacement from M black and $N - M$ white beads, express the probability of a black bead on the jth draw as an average of conditional probabilities given k black beads among the first $j - 1$ beads drawn. Evaluate this average (by observing that k is hypergeometric and using the appropriate formula for its mean) and again obtain the result that at each trial, $p = M/N$.

3–29. Calculate the variance of the hypergeometric S_n using the formula for the variance of a sum of dependent variables, and evaluating cov (X_i, X_j) as cov $(X_1\ X_2)$, since it is the same for all $i \neq j$.

3.2 The Poisson Process

An experiment of chance that continues in time (or in space) and is observed as it unfolds is sometimes called a *stochastic process*, or a *random process* (or, simply, a *process*). A snapshot or observation at each instant of time (or at each point is space) is an experiment of chance. If one is observing the time variation of some numerical variable, the instantaneous values make up a family of random variables indexed by time as a parameter. An observation on the complete process is a *function* of time.

One important process is the *Poisson process*, used to describe a wide variety of phenomena that share certain characteristics, phenomena in which some kind of "happening" takes place sporadically over a period of time in a manner that is commonly thought of as "at random." Examples of happenings for which the model is found useful include arrivals at a service counter, flaws in a long manufactured tape or wire, clicks or counts recorded by a Geiger counter near a radioactive substance, and breakdowns or failures of a piece of equipment or component (when failures are corrected as they occur and the equipment is put back into operation). Just to have a convenient reference term in the general discussion, the happening will be referred to as an "event."[1] And, even though in some cases the observed functions are functions of position in space (or on a line or plane), the general discussion will be phrased in terms of time variation.

The particular model for such processes that goes by the name *Poisson* is that in which the following postulates can be assumed to hold. In stating them the notation $P_n(h)$ will be used for the probability that n events occur in an interval of width h.

Poisson Postulates

1. *Events defined according to the numbers of events in nonoverlapping intervals of time are* independent.
2. *The probability structure of the process is time invariant.*
3. *The probability of exactly one event in a small interval of time is approximately proportional to the size of the interval:*

$$P_1(h) = \lambda h + o(h), \qquad \text{as } h \to 0.$$

4. *The probability of more than one event in a small interval is negligible in comparison with the probability of one event in that interval:*

$$\sum_{n>1} P_n(h) = o(h), \qquad \text{as } h \to 0.$$

[1] This use of the word *event* is not the same as that introduced in Chapter 1 (as a set of outcomes).

The assumption of time invariance means that the probability distribution of any set of observations on the process is dependent only on the time spacing of the observations and not on the location of the reference origin on the time-axis. (This justifies using only the *width* of the time interval in the notation for the probability of n events in an interval of given width.) The notation $o(h)$, read "little oh of h," was introduced in Section 2.5.1, and stands for any function of h that "goes to 0 faster than h" in this sense:

$$\lim_{h \to 0} \frac{o(h)}{h} = 0.$$

[In particular, h^2, $\sin^2 h$, and $1 - \exp(-h^2)$ all have this property, and might be symbolized by $o(h)$. Notice that $a[o(h)] + b[o(h)] = o(h)$.] The third and fourth postulates could then be written in the following form:

3. $[P_1(h) - \lambda h]/h \to 0$ as $h \to 0$, for some $\lambda > 0$.
4. $\sum_{n>1} P_n(h)/h \to 0$ as $h \to 0$.

Postulates 1 through 4 will be shown to define $P_n(t)$.

3.2.1 The Poisson Random Variable

Of particular interest in connection with a Poisson process is the random variable

$$X = \text{number of events in an interval of width } t,$$

for a given t. The probability function for this discrete random variable is, in the Poisson notation,

$$f(n) = P(n \text{ events in time } t) = P_n(t).$$

The derivation of $P_n(t)$ from the Poisson postulates uses a differential approach, since the postulates deal with small increments of time. Consider then the interval $(0, t + h)$, decomposed into the intervals $(0, t)$ and $(t, t + h)$. The condition that n events occur in $(0, t + h)$ can be expressed as a union of $n + 1$ events, according to how many fall in $(0, t)$ and how many in $(t, t + h)$. Thus,

$$\begin{aligned} P_n(t + h) &= P[n \text{ events in } (0, t + h)] \\ &= P[n \text{ events in } (0, t) \text{ and none in } (t, t + h)] \\ &\quad + P[n - 1 \text{ events in } (0, t) \text{ and 1 in } (t, t + h)] \\ &\quad + \cdots + P[0 \text{ events in } (0, t) \text{ and } n \text{ in } (t, t + h)]. \end{aligned}$$

By Postulate 1, each probability on the right can be factored and written as the probability of a certain number of events in $(0, t)$ times the probability of a certain number in $(t, t + h)$, since these intervals are nonoverlapping. Hence, for $n > 0$,

$$P_n(t + h) = P_n(t)[1 - \lambda h + o(h)] + P_{n-1}(t)[\lambda h + o(h)] + P_{n-2}(t)o(h) + \cdots + P_0(t)o(h).$$

and for $n = 0$,

$$P_0(t + h) = P_0(t)[1 - \lambda h + o(h)].$$

Transposing the $P_n(t)$, dividing by h, and passing to the limit as $h \to 0$, one obtains the derivative of $P_n(t)$:

$$\begin{cases} P_n'(t) = -\lambda P_n(t) + \lambda P_{n-1}(t), & n = 1, 2, 3, \ldots \\ P_0'(t) = -\lambda P_0(t). \end{cases}$$

The appropriate initial conditions are $P_0(0) = 1$ and $P_n(0) = 0$ for $n > 0$.

Since $P_0'/P_0 = -\lambda$, it follows (in view of the initial condition) that

$$P_0(t) = e^{-\lambda t}.$$

Substitution of this in the equation for $P_1'(t)$ yields

$$P_1'(t) + \lambda P_1(t) = \lambda e^{-\lambda t}.$$

This becomes integrable upon multiplication by $e^{\lambda t}$, with the result

$$e^{\lambda t}P_1(t) = \lambda t + \text{const.}$$

The initial condition $P_1(0) = 0$ then implies

$$P_1(t) = \lambda t e^{-\lambda t}.$$

Substitution of this into the equation for $P_2'(t)$ results in still another first order linear differential equation, for P_2, and so on. The results can be summarized in the following expression for $P_n(t)$, which can be derived by induction as outlined, or shown to satisfy the differential relations for $P_n(t)$ by direct substitution:

$$P_n(t) = \frac{e^{-\lambda t}(\lambda t)^n}{n!}.$$

For a given time interval t, the quantities λ and t always occur in the combination λt, so there is essentially just this single parameter, $m \equiv \lambda t$. The probabilities for $X = 0, 1, 2, \ldots$ add up to 1, as they must:

$$e^{-m} + e^{-m}m + e^{-m}\frac{m^2}{2!} + \cdots = e^{-m}\sum_{k=0}^{\infty}\frac{m^k}{k!} = e^{-m}e^{m} = 1.$$

That is, the Poisson probabilities are proportional to the terms in the series expansion of the function e^m. Poisson probabilities are given in Table XIII for a number of values of m. More precisely, it gives *cumulative* probabilities:

$$F(c) = \sum_{k=0}^{c} e^{-m} \frac{m^k}{k!}.$$

Probabilities of individual values can be obtained by taking differences in successive tabulated probabilities:

$$P(X = c) = F(c) - F(c - 1) = P(X \leq c) - P(X \leq c - 1).$$

The moments of the distribution are readily calculated from the factorial moment generating function:

$$\eta(t) = E(t^X) = \sum_0^\infty t^k f(k) = \sum_0^\infty e^{-m} \frac{(tm)^k}{k!} = e^{m(t-1)}.$$

From the rth derivative,

$$\eta^{(r)}(t) = m^r e^{m(t-1)},$$

the rth factorial moment is obtained by substituting $t = 1$:

$$E[(X)_r] = \eta^{(r)}(1) = m^r,$$

so that

$$EX = m, \qquad E(X^2 - X) = m^2, \qquad \text{var } X = m^2 + m - m^2 = m.$$

The mean and variance are seen to be equal, each having the value m, the parameter that indexes the family of Poisson distributions.

The parameter λ in the Poisson process can now be interpreted in terms of the number of events in a unit interval; for, since

$$E(\text{number of events in } t) = EX = \lambda t,$$

it follows with $t = 1$ that

$$E(\text{number of events in unit time}) = \lambda.$$

Thus, although the probability of an event in an interval is proportional to the width of the interval only for very tiny intervals, the *expected* number of events in an interval is proportional to the width of the interval, for intervals of any size. Cutting the interval in half, for instance, halves the expected number of events.

EXAMPLE 3–9. Customers enter a waiting line "at random" at a rate of 4 per min. Assuming that the number entering the line in any given time interval has a Poisson distribution, one can determine, say, the probability that at least one customer enters the line in a given half-minute interval. Taking a minute as the unit of time, one has $\lambda = 4$, and hence the average number of arrivals per half minute is $\lambda/2 = 2$. Therefore,

$$P(\text{at least one arrival in a half-minute interval})$$
$$= 1 - P(\text{none arrives in a half-minute interval})$$
$$= 1 - e^{-\lambda t}\frac{(\lambda t)^0}{0!} = 1 - e^{-2} \doteq 0.865.$$

Problems

3–30. Weak spots occur in a certain manufactured tape on the average of 1 per 1000 ft. Assuming a Poisson distribution of the number of weak spots in a given length of tape, what is the probability that

(a) A 2400-ft roll will have at most two defects?
(b) A 1200-ft roll will have no defects?
(c) In a box of five 1200-ft rolls, two have just one defect and the other three have none?

3–31. Given that X has a Poisson distribution with variance 1, calculate $P(X = 2)$.

3–32. A Geiger counter records on the average of 40 counts per min when in the neighborhood of a certain weakly radioactive substance. Determine the probability that

(a) There will be two counts in a 6-sec period.
(b) There will be k counts in a T-sec period.

3–33. Telephone calls are being placed through a certain exchange at random times on the average of 4 per min. Assuming a Poisson law, determine the probability that in a 15-sec interval there are 3 or more calls.

3–34. Flaws in the plating of large sheets of metal occur at random on the average of one in a section of area 10 sq ft. What is the probability that a sheet 5 ft by 8 ft will have no flaws? At most one flaw?

3–35. Show directly that the Poisson formula obtained for $P_n(t)$ does satisfy Postulates 3 and 4, which were used in deriving it.

3–36. Evaluate:

$$\sum_0^\infty \frac{(k^2 - k)3^k}{k!}.$$

3–37. Show that the sum of independent Poisson variables has a Poisson distribution.

3–38. Men arrive at a service counter according to a Poisson process at an average of 6 per hr, women according to a Poisson process at an average of 12 per hr, and children according to a Poisson process at an average of 12 per hr. Determine the probability that at least two customers (without regard to sex or age) arrive in a 5-min period.

3.2.2 Exponential and Gamma Distributions

Consider a Poisson process with parameter λ, and, changing the language to that of a more specific setting, let the "events" being counted be failures of a piece of equipment. (Assume that the equipment, when it fails, is immediately repaired and put back into operation.) Let L denote the time to the next failure as measured from any given instant of time, which may as well be called $t = 0$. This is the "future life" of the equipment "now" in operation. The distribution of L is readily obtained by expressing the event $L > t$ in terms of a Poisson random variable:

$$\begin{aligned} F_L(t) &= P(L \leq t) = 1 - P(L > t) \\ &= 1 - P(\text{no failure in } (0, t)) = 1 - e^{-\lambda t}, \qquad \text{for } t > 0. \end{aligned}$$

The distribution defined by this c.d.f. is called an *exponential*, or a *negative exponential*, or a *Laplace* distribution. It is a continuous distribution, with density

$$f_L(t) = F_L'(t) = \lambda e^{-\lambda t}, \qquad \text{for } t > 0.$$

Its moment generating function, defined for $s < \lambda$, is

$$\begin{aligned} \psi_L(s) = E(e^{sL}) &= \int_0^\infty e^{st}(\lambda e^{-\lambda t})\, dt \\ &= \frac{1}{1 - s/\lambda} = \sum_0^\infty \frac{s^k}{\lambda^k}, \end{aligned}$$

and the kth moment is, therefore,

$$E(L^k) = \frac{k!}{\lambda^k}.$$

The mean and variance are then

$$EL = \frac{1}{\lambda} \qquad \text{and} \qquad \text{var } L = \frac{1}{\lambda^2}.$$

Notice that the mean time to failure is the reciprocal of the mean number of failures per unit time! If, for instance, the mean number of failures per hour is 6, then the mean time to failure is $\frac{1}{6}$ hr, or 10 min.

Although possibly unnoticed, it is significant that in the above derivation, the time L was measured from any instant at which a watch was placed on the equipment, whether just after the equipment was put in operation following a previous failure, or after it had been operating for a long time. That is, the future life of equipment whose breakdown pattern follows the Poisson law is a random variable whose distribution is independent of how long the equipment has been operating. In such a case, preventive maintenance involving replacement of the equipment or component while it is still working would be of no advantage. This can come about only because failures in such cases are caused by external sources and not by a wearing out of the equipment. If wear-out is a cause of failure, the Poisson model is not appropriate.

The time to the rth failure after a given instant of time (say, $t = 0$) is a random variable that is the sum of the exponential variables measuring the time to the first failure, the time to the second failure measured from the first, and so forth. Denoting this total time by S_r, one can express it as

$$S_r = L_1 + L_2 + \cdots + L_r,$$

where L_i is the time from the $(i - 1)$st to the ith failure and has the exponential distribution. Because the inequalities in the event $\{L_1 > t_1, \ldots, L_r > t_r\}$ are equivalent to events relating to nonoverlapping intervals of time, the random variables $L_1, \ldots, L_r$ are independent, and the moment generating function of their sum S_r is, therefore,

$$\psi_{S_r}(s) = [\psi_L(s)]^r = \left(1 - \frac{s}{\lambda}\right)^{-r}.$$

Upon comparing this with the moment generating function of X in Example 2–65, it is apparent that $X = \lambda S_r$ (with $r = k + 1$) so that

$$f_{S_r}(t) = \lambda f_X(\lambda t) = \frac{\lambda^r t^{r-1} e^{-\lambda t}}{(r-1)!}, \qquad (t > 0).$$

A distribution with such a density is called a *gamma distribution*; it will be encountered again in Chapter 7. The mean and variance of S_r are easily computed from its structure as a sum of independent, exponential L's:

$$ES_r = r\,EL = \frac{r}{\lambda}, \qquad \text{var } S_r = r \text{ var } L = \frac{r}{\lambda^2}.$$

EXAMPLE 3–10. Consider the Poisson process describing the arrivals of customers at a service counter, as in Example 3–9, with an average of 4 arrivals per min. The average time to an arrival is then $\frac{1}{4}$ min, and the density of the time to an arrival is $4e^{-4t}$, for $t > 0$. The time to the tenth arrival has the density

$$f_{S_{10}}(t) = \frac{4^{10} t^9 e^{-4t}}{9!}, \qquad \text{for } t > 0,$$

with mean 2.5 min (i.e., 10 times $\frac{1}{4}$ min).

3.2.3 Approximating Binomial Probabilities (Small p)

Besides being a useful probability model in its own right, the Poisson distribution can be used in approximating binomial probabilities in those skewed cases in which an unusually large n would be required for a successful use of the Central

Limit Theorem. Perhaps the most natural way to see this is through an alternative derivation of the Poisson formula from the Poisson postulates, a bit more heuristic than the one given earlier.

Consider a Poisson process characterized by the constant λ as the mean number of "events" (whatever happenings are being observed) per unit time. Let Y denote the number of arrivals in an interval $(0, t)$, and let this interval be subdivided into n equal parts, of width $h = t/n$. Consider now the n "Bernoulli trials" corresponding to these n subdivision intervals—"success" corresponding to an event in a given subinterval and "failure" corresponding to no event. The trials are not exactly Bernoulli, since there could be *more* than one event in a subinterval. But n will be allowed to increase without limit, so that the subinterval length h will shrink to 0; and for small h it is approximately true that either one event or no events occur in an interval of width h, with probabilities:

$$p = P(1 \text{ event in } h) \doteq \lambda h,$$
$$1 - p = P(\text{no events in } h) \doteq 1 - \lambda h.$$

There will be k arrivals in $(0, t)$, then, if there is one event in each of k subintervals. The probability of this is approximately binomial:

$$P(k \text{ "successes" in } n \text{ trials}) = \binom{n}{k} p^k (1 - p)^{n-k}.$$

As n becomes infinite, the approximations become better and better. Putting $p = \lambda h = \lambda t/n$, one obtains

$$P[k \text{ events in } (0, t)] \doteq \binom{n}{k}\left(\frac{\lambda t}{n}\right)^k \left(1 - \frac{\lambda t}{n}\right)^{n-k}$$
$$= \frac{n}{n} \cdot \frac{n-1}{n} \cdots \frac{n-k+1}{n} \cdot \frac{(\lambda t)^k}{k!} \left(1 - \frac{\lambda t}{n}\right)^{-k} \left(1 - \frac{\lambda t}{n}\right)^n.$$

As n becomes infinite, the first k factors on the right tend to 1, the next is fixed, the next tends to 1, and last factor can be written

$$\left[\left(1 - \frac{\lambda t}{n}\right)^{-n/\lambda t}\right]^{-\lambda t},$$

which tends to $\exp(-\lambda t)$, since the quantity in brackets converges to e. Hence,

$$P(Y = k) = P[k \text{ events in } (0, t)] = \frac{(\lambda t)^k}{k!} e^{-\lambda t},$$

as was derived earlier.

What has been shown, in the course of the above derivation of the Poisson formula, is the following:

$$\lim_{\substack{n \to \infty \\ np \text{ fixed}}} \binom{n}{k} p^k (1 - p)^{n-k} = e^{-np} \frac{(np)^k}{k!}.$$

Thus, for large but finite n and small p, one can approximate a binomial probability (n, p) with a Poisson probability in which $m = np$ is the mean of both distributions. The following example gives some indication as to the degree of success of the approximation.

EXAMPLE 3–11. In the following table are values of the probability function for each of two binomial distributions, one in which $n = 10$ and $p = .1$, and the other in which $n = 20$ and $p = .05$. In each case $np = 1$, and the Poisson approximations using $\lambda t = np = 1$ are also given. (The table is clearly not quite complete.)

k	Poisson $m = 1$	Binomial (10, .1)	Binomial (20, .05)
0	.368	.349	.358
1	.368	.387	.377
2	.184	.194	.187
3	.061	.057	.060
4	.015	.011	.013
5	.0031	.0015	.0022

Although the approximation scheme here is given for the case of small p, it applies equally well for values of p near 1—just by interchanging the basic Bernoulli coding so that p becomes q, and vice versa. When p is near 1, then q is small.

EXAMPLE 3–12. Consider the binomial distribution for $n = 20$ and $p = .8$. If X denotes the number of successes in 20 independent Bernoulli trials, and the probability of success is .8 in each trial, then

$$\begin{aligned} P(X = 16) &= P(16 \text{ successes in 20 trials}) \\ &= P(4 \text{ failures in 20 trials}) \\ &= \binom{20}{4}(.2)^4(.8)^{16} \\ &= e^{-4}\frac{4^4}{4!} = .629 - .433 = .196. \end{aligned}$$

The Poisson parameter here is $20 \times .2 = 4$, and Table XIII in the Appendix was used to obtain the cumulative probabilities at 4 and at 3 (whose difference gives the probability for 4).

It is, of course, true that when n is "large," the normal distribution can be used to approximate binomial probabilities. But when p is small or close to 1, the binomial distribution is quite skewed (one way or the other), and it takes a larger n to get a reasonable degree of approximation than when p is moderate. The next example gives both the normal and Poisson approximations in one case, for comparison.

EXAMPLE 3–13. Cumulative probabilities for a binomial distribution with $n = 8$ and $p = \frac{1}{8}$ are listed in the following table. In addition to the actual probabilities (exact to four decimal places), the Poisson approximations with $m = np = 1$ and the normal approximations with $\mu = np = 1$ and $\sigma^2 = npq = \frac{7}{8}$ are given.

k	Binomial	Normal	Poisson
0	.3436	.2965	.3679
1	.7363	.7035	.7358
2	.9327	.9456	.9197
3	.9888	.9962	.9810
4	.9988	.9999	.9963
5	.9999	1.0000	.9994

Even in this instance in which p is not really very small, the Poisson approximation is much better for small k and almost as good for large k. The smaller the p, the more preferable is the Poisson approximation.

The Poisson process has been seen to be a continuous-time analog of a sequence of Bernoulli trials with small p—in the one case arrivals or breakdowns or flaws occur at random, unpredicted instants of time, and in the other heads or "success" occurs occasionally as the trials proceed, in unpredictable fashion. The analogy can be carried further. The geometric distribution describes the "waiting time" to heads in a Bernoulli sequence, and the negative exponential distribution describes the waiting time to an arrival or breakdown in a Poisson process. The negative binomial distribution describes the "waiting time" to the rth heads in a Bernoulli sequence, and the gamma distribution describes the waiting time to the rth arrival in a Poisson process. Thus, in a sense:

$$\text{geometric:binomial:negative binomial} = \text{exponential:Poisson:gamma}.$$

Problems

3–39. Determine the median of the exponential distribution.

3–40. Customers join a waiting line according to the Poisson law, with an average time between arrivals of 2 min.

(a) What is the probability that 6 min will elapse with no customers arriving?
(b) What is the probability that in a 6-min interval at most two customers arrive?

3–41. In a certain electronic device there are ten tubes each having a life distribution that is exponential with mean 50 hr. The device fails if any one of the tubes fails, and when it fails, the tube that has caused the failure is replaced and the device turned on again. Determine the distribution of the time from one failure of the device to the next, assuming that tubes fail independently of one another.

3–42. Let the times between failures of a certain device have a common distribution function $F(x)$; the probability of successful operation (no failures) for a period t is called the *reliability* of the device:

$$R(t) = 1 - F(t).$$

The *hazard* is defined in terms of reliability:

$$H(t) = -\frac{d}{dt}[\log R(t)].$$

(This can be interpreted as the rate of "dying" among many such systems relative to the number still operating.) Show that if the hazard is constant, the distribution of system life (or time between failures) is the exponential, and conversely.

3–43. Show that if the times between successive "events" have the exponential, distribution with parameter λ, then

$$P(\text{one or more events in } h) = \lambda h + o(h).$$

3–44. A certain machine manufactures bolts and turns out defective bolts on the average of 1 per 200 bolts. They are packaged in boxes of 50. Determine the probability that a box has at most 1 defective bolt. Determine also the probability that of the 100 boxes in a carton, no box has more than 1 defective bolt.

3–45. Given that 1500 out of 50,000 people in a certain city are watching a certain television program, what is the probability that of 200 people called at random fewer than 4 are watching the program?

3–46. A man holds 5 tickets in a lottery in which 1000 tickets are sold. Ten tickets are to be drawn for prizes. What is the probability that the man wins at least one prize?

3–47. Use appropriate approximations and tables to evaluate each of the following:

(a) $P(X = 52)$, where X is binomial, $n = 100$, $p = \frac{1}{2}$.
(b) $P(X \leq 3)$, where X is binomial, $n = 100$, $p = .05$.
(c) $P(X \leq 50)$, where X is Poisson, $m = 64$.

[*Hint:* A Poisson variable with $m = 64$ is approximately normal.]

3–48. Each of two units has an exponential distribution of time to failure with mean 1 hr and is placed in a system so that the system fails if and only if *both* units fail. What is the expected time to failure of the system?

3–49. Let L denote the life (or time to failure) of a certain piece of equipment, and suppose that it follows the exponential law with mean $1/\lambda$. Determine the conditional distribution of future life, $L - L_0$, given that the equipment is still operating after L_0 units of time have elapsed. That is, compute

$$P\{L - L_0 \leq x \mid L > L_0\}.$$

3–50. Show that the distribution function of S_r, the time to the rth "failure" in a Poisson process, is the following:

$$F_{S_r}(x) = 1 - \sum_{k=0}^{r-1} e^{-\lambda x}\frac{(\lambda x)^k}{k!}, \qquad x > 0,$$

in two ways:

(a) Differentiate $F_{S_r}(x)$ to obtain the previously given density of S_r and check that the additive constant is correct.

(b) Calculate $P(S_r \le x)$ by expressing the event $\{S_r \le x\}$ in terms of a Poisson random variable.

3–51. Given that the times between successive "failures" are independent, identical, exponential random variables, show that the probability of more than one failure in time h is $o(h)$ as h tends to 0. (That is, that Poisson Postulate 4 is satisfied.)

[*Hint:* Let X_1 be the time to the first, and X_2 the subsequent time to the second failure after the beginning of the interval of width h and evaluate $P(X_1 + X_2 < h)$.]

3.3 The Normal and Related Distributions

The "standard normal" distribution was encountered in the Central Limit Theorem as the limiting distribution of a standardized sum of independent, identically distributed summands, as the number of terms increases without limit. The adjective "normal" will also be applied to distributions related to the standard normal distribution by a change of scale and/or a translation. The form of the density of these distributions is such as to permit an extensive analysis of distributions occurring in inference about normal populations. This in itself would not be sufficient justification for the prominent role of normal inference in statistical theory, but it happens that the normal model is often a workable representation for actual phenomena. It has been said that the Central Limit Theorem tends to account for the occurrence of approximately normal distributions in practice, since random phenomena are often additive combinations of several contributory variables.

To review, the *standard normal density function* is the following:

$$f(z) = \frac{1}{\sqrt{2\pi}} \exp\left(-\frac{1}{2} z^2\right).$$

The constant multiplier, which makes the area under the graph of the density function equal to 1, is most easily determined as what is needed to make the volume under the joint density surface of a pair of independent, standard normal variables equal to 1. This joint density is just the product of marginal densities:

$$f(x, y) = \frac{1}{2\pi} \exp\left(-\frac{1}{2}(x^2 + y^2)\right),$$

and the volume under the surface that represents it is

$$\int_{-\infty}^{\infty}\int_{-\infty}^{\infty} f(x, y)\, dx\, dy = \frac{1}{2\pi}\int_{-\infty}^{\infty}\int_{-\infty}^{\infty} \exp\left[-\frac{1}{2}(x^2 + y^2)\, dx\, dy\right]$$

$$= \frac{1}{2\pi}\int_{0}^{2\pi}\int_{0}^{\infty} \exp\left(-\frac{1}{2} r^2\right) r\, dr\, d\theta = 1.$$

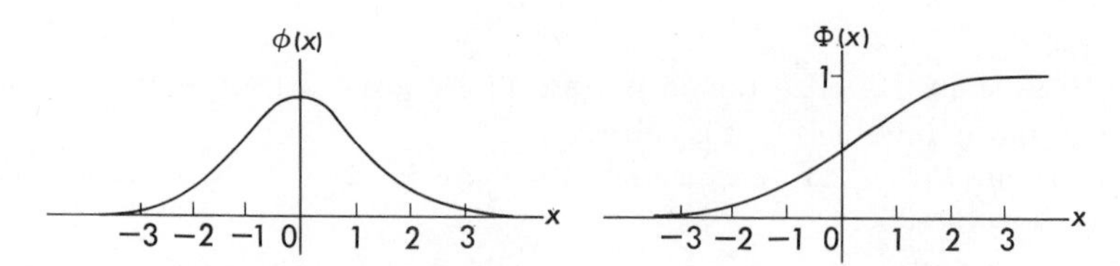

Figure 3–2. Standard normal density and c.d.f.

The distribution function of a standard normal distribution will be denoted by $\Phi(z)$,

$$\Phi(z) = \int_{-\infty}^{z} \frac{\exp(-\frac{1}{2}u^2)}{\sqrt{2\pi}}\,du,$$

and is tabulated for $-4 \leq z \leq 4$ in Table I of the Appendix. Graphs of the standard normal density $\phi(z)$ and the standard normal c.d.f. $\Phi(z)$ are shown in Figure 3–2. Certain percentiles are given in Table Ia, and certain two-tail probabilities are given in Table Ib. The corresponding moment generating function was found in Problem 2–127 to be

$$\psi_Z(t) = e^{t^2/2} = \sum_0^\infty \frac{(t^2/2)^k}{k!} = \sum_0^\infty \frac{(2k)!}{2^k k!}\frac{t^{2k}}{(2k)!},$$

from which it is seen that the odd order moments are 0 and that the even order moments are given by the formula

$$E(Z^{2k}) = (2k-1)(2k-3)\cdots 5\cdot 3\cdot 1, \qquad \text{for } k = 1, 2, \ldots.$$

3.3.1 The General Normal Distribution

A random variable is said to be *normally distributed* if it can be expressed as a linear transformation of a standard normal variable. Thus, if Z is standard normal, any variable of the form

$$X = aZ + b,$$

for $a \neq 0$, is normally distributed. Clearly,

$$EX = b \qquad \text{and} \qquad \text{var } X = a^2.$$

Moreover, a can be taken as positive without loss of generality, since $-Z$ is standard normal if Z is standard normal (see Problem 3–55). The expression for X in terms of Z can then be written in the form

$$X = \sigma Z + \mu \qquad \text{or} \qquad Z = \frac{X-\mu}{\sigma},$$

where $\mu = EX$ and σ is the standard deviation of X. Central moments of X are as follows:

$$E[(X-\mu)^n] = \sigma^n E(Z^n) = \begin{cases} 0, & \text{if } n \text{ is odd}, \\ (2k-1)\cdots 5\cdot 3\cdot 1\sigma^{2k}, & \text{if } n = 2k, \end{cases}$$

for $k = 1, 2, \ldots$. A particular moment that will be encountered frequently in studying inference about normal models is the fourth central moment,

$$E[(X-\mu)^4] = 3\sigma^4.$$

The general normal c.d.f. is expressible in terms of the standard normal c.d.f.,

$$\begin{aligned} P(X \le x) &= P(\sigma Z + \mu \le x) \\ &= P\left(Z \le \frac{x-\mu}{\sigma}\right) = \Phi\left(\frac{x-\mu}{\sigma}\right). \end{aligned}$$

From this it is evident that a single table, the one for the *standard* normal distribution, suffices for the calculation of general normal probabilities. To find $P(X \le x)$, simply enter the standard normal table (Table I in the Appendix) at the "standardized" argument $(x - \mu)/\sigma$, which measures x as being so many standard deviations away from the mean μ.

EXAMPLE 3–14. Consider the computation of $P(|X - 9| > 1)$, where X is normally distributed with mean 10 and variance 4. [This is sometimes written as follows: $X \overset{d}{=} \mathcal{N}(10, 4)$.] Table I, Appendix, can be used if one first expresses the event in terms of events of the type $X < c$ and then evaluates their probabilities by expressing them in terms of the standard normal c.d.f.:

$$\begin{aligned} P(|X-9| > 1) &= 1 - P(|X-9| < 1) = 1 - P(8 < X < 10) \\ &= 1 - [F_X(10) - F_X(8)] = 1 - \Phi\left(\frac{10-10}{2}\right) + \Phi\left(\frac{8-10}{2}\right) \\ &= 1 - .5 + .1587 = .6587. \end{aligned}$$

(Here, as in other instances in which tables are used, the equality is only approximate —up to the accuracy of the table.)

The general normal density function, the density of the variable $X = \sigma Z + \mu$, is calculated by expressing it in terms of the density of the standard normal Z:

$$f_{\sigma Z+\mu}(x) = \frac{1}{\sigma} f_Z\left(\frac{x-\mu}{\sigma}\right) = \frac{1}{\sqrt{2\pi}\,\sigma} \exp\left[-\frac{1}{2\sigma^2}(x-\mu)^2\right].$$

This is essentially an exponential function of a quadratic exponent in which the coefficient of the square term is negative. Any such function, with suitable normalization by a multiplicative constant, can serve as a normal density, as will be evident from the following example.

EXAMPLE 3–15. Consider the function $\exp(-3x^2 + 6x)$. This can be used as the basis of a density. The exponent can be rewritten by completing the square:

$$-3x^2 + 6x = -3(x-1)^2 + 3 = -\frac{(x-1)^2}{1/3} + 3.$$

Upon comparison with the exponent of the general normal density, it is apparent that this is a special case in which $\mu = 1$ and $2\sigma^2 = \frac{1}{3}$. Thus, the following function, which is a constant times the given one, is a normal density:

$$\frac{1}{\sqrt{2\pi}\,\sqrt{1/6}} \exp\left(-\frac{(x-1)^2}{2(1/6)}\right).$$

The moment generating function of a general normal distribution is readily obtained from that of the standard normal distribution,

$$\psi_X(t) = E[\exp(tX)] = E[\exp t(\sigma Z + \mu)] = \exp(t\mu + \tfrac{1}{2}\sigma^2 t^2),$$

and the characteristic function of the distribution is then

$$\phi_X(t) = \psi_X(it) = \exp(it\mu - \tfrac{1}{2}\sigma^2 t^2).$$

With this one can determine the distribution of a sum of independent, normal distributed variables. For, if $X_1, X_2, \ldots, X_n$ are independent, identically distributed variables with a common normal distribution $\mathcal{N}(\mu, \sigma^2)$, and S_n denotes their sum, then

$$\phi_{S_n}(t) = [\phi_X(t)]^n = \{\exp(it\mu - \tfrac{1}{2}\sigma^2 t^2)\}^n = \exp[i(n\mu)t - \tfrac{1}{2}(n\sigma^2)t^2],$$

which is the characteristic function of a normal distribution with mean $n\mu$ and variance $n\sigma^2$. This, then, is the distribution of S_n. (See also Problem 3–65.)

3.3.2 The Lognormal Distribution

The random variable X is said to have a *lognormal* distribution if $\log X$ is normally distributed, that is, if X is of the form e^Y where Y is normal. The pertinent properties of a lognormal distribution can then be derived from properties of the normal distribution. In particular, the density function is computed as follows for $x > 0$:

$$f_X(x) = \frac{d}{dx} P(e^Y < x) = \frac{d}{dx} F_Y(\log x)$$

$$= \frac{1}{x} f_Y(\log x) = \frac{1}{\sqrt{2\pi}\,\sigma x} \exp\left(\frac{-(\log x - \mu)^2}{2\sigma^2}\right).$$

The μ and σ^2 here are the mean and variance of the normally distributed Y, that is, of $\log X$. The moments of the lognormal X itself can be given in terms of μ and σ^2, which are then parameters of the distribution of X (but which do not have simple, intuitive interpretations).

The kth moment of X about 0 is expressible in terms of the moment generating function of $\log X$:

$$E(X^k) = E(e^{kY}) = \psi_Y(k) = \exp(k\mu + \tfrac{1}{2}\sigma^2 k^2).$$

In particular,

$$EX = \exp(\mu + \tfrac{1}{2}\sigma^2),$$

and

$$\operatorname{var} X = E(X^2) - (EX)^2 = \{\exp(\sigma^2) - 1\} \exp(2\mu + \sigma^2).$$

The lognormal distribution finds application in a wide variety of fields: physics, engineering, economics, biology, astronomy, sociology, and even philology.[2] It is used as the distribution of incomes, of household size, of particle size, body weight, results of endurance tests, and so on.

3.3.3 Rayleigh and Maxwell Distributions

A number of other distributions arise in connection with normal distributions. Those encountered in inference problems are taken up in a later chapter. Here mention will be made of two distributions encountered in model building with normal components.

The bivariate distribution defined by two independent, normal variates with mean zero and common variance is sometimes called *circular normal*, with density

$$f(x, y) = \frac{1}{2\pi\sigma^2} \exp\left(-\frac{x^2 + y^2}{2\sigma^2}\right).$$

The radial distance out to the point (X, Y):

$$Z = (X^2 + Y^2)^{1/2}$$

has a distribution whose density is found by differentiating the c.d.f.:

$$f_Z(z) = \frac{d}{dz} P(X^2 + Y^2 < z^2) = \frac{1}{2\pi\sigma^2} \frac{d}{dz} \int_A\!\!\int \exp\left[-\frac{1}{2\sigma^2}(x^2 + y^2)\right] dx\, dy$$

$$= \frac{1}{\sigma^2} \frac{d}{dz} \int_0^z r \exp\left[-\frac{r^2}{2\sigma^2}\right] dr = \frac{z}{\sigma^2} \exp\left[-\frac{z^2}{2\sigma^2}\right], \qquad \text{for } z > 0.$$

(The region denoted by A is the interior of a circle of radius z and center at the origin. The integral over this region was expressed in terms of polar coordinates and the θ-integration performed before differentiating with respect to z.) The distribution with this density is called a *Rayleigh distribution.*

A trivariate distribution for (X, Y, Z) in which these components are independent normal variates with mean 0 and common variance σ^2 might be called a spherical normal distribution. The distance from the origin out to the point (X, Y, Z):

$$R = (X^2 + Y^2 + Z^2)^{1/2}$$

[2] See J. Aitchison and J. A. C. Brown, *The Lognormal Distribution* (New York: Cambridge University Press, 1957).

is a random variable whose distribution can be derived in just about the same way as the Rayleigh distribution was derived above. The result is the density

$$f_R(u) = \sqrt{\frac{2}{\pi}} \left(\frac{u^2}{\sigma^3}\right) \exp\left(-\frac{u^2}{2\sigma^2}\right), \qquad \text{for } u > 0,$$

defining what is sometimes called the *Maxwell distribution.* The derivation is called for in Problem 3–62.

Problems

3–52. Given that X is normal with mean 10 and variance 4, compute

$$P(|X - 10| > 3).$$

3–53. Given that X is normally distributed with mean 10 and that $P(X > 12)$ is .1587, determine $P(9 < X < 11)$.

3–54. Given that X is normal with mean 30 and variance 24, compute

(a) $P(X > 27)$.
(b) $P(|X - 28| < 2)$.

3–55. Show that if Z has a normal distribution with zero mean, so does $-Z$.

3–56. Determine the 1st and 3rd quartiles (i.e., the 25th and 75th percentiles) of a standard normal distribution. From these determine the 1st and 3rd quartiles of a normal distribution with mean 10 and variance 4.

3–57. Suppose that in tabulating observations on a normal random variable with mean 10 and variance 4, the range of possible values is divided into 10 parts by the points 6, 7, . . ., 14. (The first "class interval" goes from $-\infty$ to 6, the next from 6 to 7, etc.) Determine the probabilities of each interval.

3–58. Using characteristic functions, show that if X is normal, so is $Y = aX + b$ for any constants a and b, with $a \neq 0$.

3–59. Determine the density function and expectation of the random variable $|X - \mu|$, where X is normal with mean μ and variance σ^2.

3–60. (a) Show that if X is a lognormal, so is X^α, for any real α.
(b) Determine the median of a lognormal variable X in terms of the parameters μ and σ^2 (mean and variance of log X).

3–61. It is found that in a certain rock-crushing process, the "diameters" d of the crushed rocks have approximately a lognormal distribution, with mean diameter 1.5 in and standard deviation .3 in.

(a) What percentage of rocks would have diameters exceeding 2 in? Less than 1 in?
(b) Assuming rock weights to be proportional to d^3, what is the expected weight of a rock (in terms of the constant of proportionality)?

3–62. Derive the density of Maxwell distribution.

3–63. Obtain formulas for the mean and variance of the Rayleigh distribution.

3–64. Discuss the distribution of Z^2, where Z is standard normal. (Obtain the density, m.g.f., mean, and variance.)

3–65. Using characteristic functions, show that any linear combination of independent, normal variables is again normal.

3.3.4 The Chi-Square Distribution

The chi-square distribution is defined as the distribution of a sum of squares of independent standard normal variables. Beginning with the special case of the square of a *single* standard normal variable, let X be normally distributed with zero mean and unit variance. The characteristic function of X^2 is then

$$E[\exp(itX^2)] = \frac{1}{\sqrt{2\pi}} \int_{-\infty}^{\infty} \exp\left(itx^2 - \frac{1}{2}x^2\right) dx = (1 - 2it)^{-1/2}.$$

From this can be obtained the distribution of a sum of squares of k independent standard normal variables $X_1, \ldots, X_k$:

$$\chi^2 = X_1^2 + X_2^2 + \cdots + X_k^2.$$

The characteristic function of the distribution of χ^2 is the kth power of the characteristic function of a single term:

$$\phi_{\chi^2}(t) = (1 - 2it)^{-k/2}.$$

The density of this distribution will be derived in Chapter 7. It is of the form

$$f_{\chi^2}(u) = (\text{const})u^{k/2-1}e^{-u/2}.$$

(Observe that when $k/2$ is an integer, the distribution is what was called a gamma distribution in Section 3.2.2.) Its cumulative distribution function is given in Table II, which is a table of percentiles for various values of k, called the number of *degrees of freedom*.

The mean and variance of a chi-square distribution can be read from the series expansion for the characteristic function:

$$\phi_{\chi^2}(t) = (1 - 2it)^{-k/2} = 1 + k(it) + (k^2 + 2k)\frac{(it)^2}{2!} + \cdots.$$

They are as follows:

$$E(\chi^2) = k \quad \text{and} \quad \text{var}\,\chi^2 = 2k.$$

In Figure 3–3 are sketched the graphs of the chi-square density functions for several values of the parameter k. Observe that for $k = 1$, the density is infinite at $x = 0$, and that for $k = 2$, the density is that of the exponential distribution. As k increases, the mean shifts to the right and the variance increases. As $k \to \infty$, the shape approaches that of the normal density, for according to the

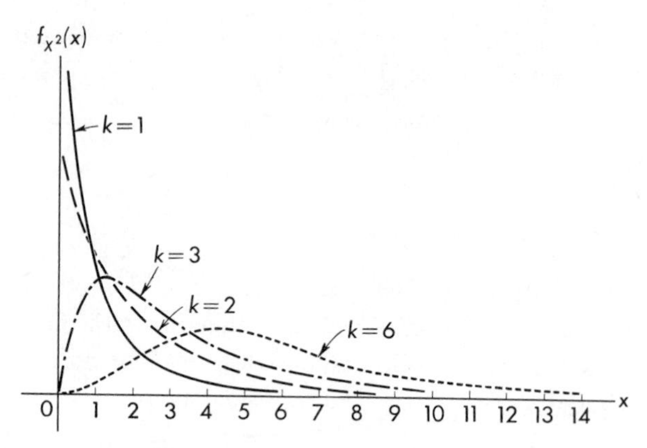

Figure 3–3. Chi-square densities.

Central Limit Theorem, the distribution of the sum of k independent, identically distributed variables is asymptotically normal. The mean and variance of this asymptotic distribution are, respectively, k and $2k$.

Tables of chi-square percentiles (like Table II, Appendix) take advantage of the fact that the distribution is asymptotically normal. Chi-square percentiles can be approximated, using normal percentiles as follows: For any given probability p,

$$p = P(\chi^2 < \chi_p^2) = P\left(\frac{\chi^2 - k}{\sqrt{2k}} < \frac{\chi_p^2 - k}{\sqrt{2k}}\right)$$

$$\doteq P\left(Z < \frac{\chi_p^2 - k}{\sqrt{2k}}\right),$$

where Z is standard normal. Thus, if z_p is the 100 pth percentile of the standard normal distribution,

$$z_p \doteq \frac{\chi_p^2 - k}{\sqrt{2k}} \quad \text{or} \quad \chi_p^2 \doteq \sqrt{2k}z_p + k.$$

A somewhat better approximation results from a modification using the variable

$$Y = \sqrt{2\chi^2} - \sqrt{2k - 1}.$$

This has the distribution function

$$P(Y < x) = P(\sqrt{2\chi^2} < \sqrt{2k - 1} + x)$$

$$= P\left(\chi^2 < \frac{2k - 1}{2} + x\sqrt{2k - 1} + \frac{x^2}{2}\right)$$

$$\doteq P(\chi^2 < k + \sqrt{2k}x) = P\left(\frac{\chi^2 - k}{\sqrt{2k}} < x\right).$$

The approximation holds for large k, and Y has asymptotically the standard normal distribution. So, as before, for any p,

$$\sqrt{2\chi_p^2} - \sqrt{2k - 1} \doteq z_p,$$

or, finally,

$$\chi_p^2 \doteq \tfrac{1}{2}(z_p + \sqrt{2k - 1})^2.$$

EXAMPLE 3–16. In computing the 95th percentile of the chi-square distribution with 50 degrees of freedom, the first approximate formula given above yields

$$\chi_{.95}^2 = \sqrt{100} \times 1.645 + 50 = 66.45.$$

The modified formula yields

$$\chi_{.95}^2 = \tfrac{1}{2}(1.645 + \sqrt{100 - 1})^2 = 67.2,$$

which is more nearly correct, the actual value being closer to 67.5.

The chi-square distribution has a number of uses in problems of statistical inference concerning normal distributions and also in problems of goodness of fit and in contingency tables.

Problems

3–66. Using a normal approximation, estimate:

(a) The 80th percentile of a chi-square distribution with 60 degrees of freedom.
(b) $P(\chi^2 > 60)$, where χ^2 has the chi-square distribution with 50 degrees of freedom.

3–67. Suppose that Y/σ^2 has a chi-square distribution with 10 degrees of freedom. Determine the density function, mean, and variance of Y.

3–68. Show that a sum of independent chi-square variables has again the chi-square distribution, the number of degrees of freedom of the sum being the sum of the numbers of degrees of freedom of the summands.

3–69. Show that if U is uniform on [0, 1], the variable $-2 \log U$ has a chi-square distribution with 2 degrees of freedom.

3–70. Show that if Z has the Maxwell distribution with parameter σ^2, then Z^2/σ^2 has the chi-square distribution with 3 degrees of freedom.

3.4 The Multinomial Distribution

The multinomial distribution is a generalization of the binomial distribution to instances in which the independent, identical "trials" involved are experiments with more than two outcomes. The binomial distribution arose in connection with a sequence of independent Bernoulli trials, each of which has two possible outcomes, as the number of times one particular outcome occurs in a given number of trials. Consider now a sequence of independent trials of an experiment that has k outcomes, $A_1, A_2, A_3, \ldots, A_k$ with corresponding probabilities $p_1, p_2, \ldots, p_k$. These probabilities are, of course, related:

$$p_1 + p_2 + \cdots + p_k = 1$$

so that if $k - 1$ of them are specified, the remaining one is automatically determined.

Consider the random variables $X_1, \ldots, X_k$, where

$$X_i = \text{frequency of } A_i \text{ among } n \text{ trials,}$$

which is, of course, binomially distributed (n, p_i). Since

$$X_1 + \cdots + X_k = n,$$

it follows that the joint distribution of $(X_1, \ldots, X_k)$ would be concentrated on this hyperplane. That is, there are really essentially only $k - 1$ variables, the remaining one being determined from the fact that they all must add up to n. Either the joint distribution of $(X_1, \ldots, X_k)$ or the joint distribution of $k - 1$ of these, say, $(X_1, \ldots, X_{k-1})$, is called *k-nomial*, the term *multinomial* then referring to any such distribution. In the case $k = 2$, with $A_1 =$ "success" and $A_2 =$ "failure," the distribution of the number of successes has already been called *binomial*, which fits into the general terminology.

The probability function of the k-nomial distribution can be derived with the same kind of reasoning as used in the binomial case. For a particular sequence of n trials, of which f_1 result in A_1, f_2 result in $A_2, \ldots,$ and f_k result in A_k, the probability is just the product of corresponding p's:

$$p_1^{f_1} p_2^{f_2} \cdots p_k^{f_k}.$$

But such a sequence of results can come in many patterns—the number of which is the number of ways of arranging n objects, f_1 of one kind, $\ldots$, and f_k of the kth kind, namely, $n!$ divided by a factorial for each group of like objects. The total probability for all sequences with the given frequencies is then

$$P(X_1 = f_1, \ldots, \text{and } X_k = f_k) = \frac{n!}{f_1! \cdots f_k!} p_1^{f_1} p_2^{f_2} \cdots p_k^{f_k},$$

provided, of course, $f_1 + \cdots + f_k = n$. This is the joint probability function of the distribution.

The term *multinomial* is appropriate, since (as in the particular case $k = 2$) these probabilities just derived are terms in a multinomial expansion:

$$(p_1 + \cdots + p_k)^n = \sum \frac{n!}{f_1! \cdots f_k!} p_1^{f_1} \cdots p_k^{f_k} = 1,$$

the sum extending over all sets of nonnegative integers that sum to n.

The moment generating function of the distribution of $(X_1, \ldots, X_k)$ is

$$\psi(t_1, \ldots, t_k) = E[\exp(t_1 X_1 + \cdots + t_k X_k)]$$
$$= (p_1 e^{t_1} + \cdots + p_k e^{t_k})^n.$$

The marginal m.g.f. of $(X_1, \ldots, X_{k-1})$ is obtained by setting $t_k = 0$:

$$\psi(t_1, \ldots, t_{k-1}, 0) = (p_1 e^{t_1} + \cdots + p_{k-1} e^{t_{k-1}} + p_k)^n.$$

This distribution has marginal distributions of the same type. In particular, the univariate marginals are binomial:

$$E(e^{t_i X_i}) = (p_i e^{t_i} + 1 - p_i)^n,$$

as observed at the outset (from the definition of X_i).

EXAMPLE 3–17. A die is tossed 12 times. Let X_i denote the number of tosses in which i dots turn up, for $i = 1, \ldots, 6$. Then $p_i = \frac{1}{6}$, and $E(X_i) = \frac{12}{6} = 2$, for each i. A sample probability computation is

$$P(X_1 = 2, X_2 = 2, \ldots, X_6 = 2) = \frac{12!}{2!\,2! \cdots 2!}\left(\frac{1}{6}\right)^2\left(\frac{1}{6}\right)^2 \cdots \left(\frac{1}{6}\right)^2$$
$$= \frac{1925}{549{,}872}$$

(which shows, incidentally, that the "expected" outcome is not very likely). The marginal distribution, say, of X_1 and X_2, is the trinomial distribution defined by

$$P(X_1 = f_1 \text{ and } X_2 = f_2) = \frac{12!}{f_1!\,f_2!\,(12 - f_1 - f_2)!} p_1^{f_1} p_2^{f_2} p^f,$$

where $p = 1 - p_1 - p_2 = \frac{4}{6}$ and $f = 12 - f_1 - f_2$.

EXAMPLE 3–18. In recording the value of a continuous random variable X, it is necessary to "round off," a process that in effect divides the value space of X into mutually disjoint intervals $I_1, I_2, \ldots, I_k$. These constitute a partition of the value space, and as a succession of values of X are obtained, in repeated trials of the underlying experiment, one notes only the interval in which each value falls. The number of observations in a given interval I_j, called the frequency f_j of that interval, is a binomial variable if the observations are independent. The *joint* distribution of the frequencies $(f_1, \ldots, f_k)$ is multinomial, with

$$p_j = P(X \text{ falls in } I_j) = \int_{I_j} f(x)\,dx,$$

where $f(x)$ is the density of the distribution of X.

3.5 The Exponential Family

A one-parameter family of distributions that can be written (by suitable choice of functions) in the form

$$f(x; \theta) = B(\theta)h(x) \exp[Q(\theta)R(x)]$$

is said to belong to the *exponential family* of distributions. Most of the distributions encountered so far belong to this family. The table on p. 199 indicates the choice of functions corresponding to each of these various distributions. [Although the name θ for the indexing parameter is used at the head of the columns, the names (p, λ, m) used in the earlier discussions of particular cases appear in the table.]

Beside showing why it is that these models have so much in common when it comes to inference about them, recognizing them to be just special cases of a more general model makes it possible to derive once and for all many results that would otherwise have to be obtained in each particular case.

There is also an "exponential family" of distributions indexed by a multi-dimensional parameter $\boldsymbol{\theta} = (\theta_1, \ldots, \theta_k)$, namely, that defined by densities of the form

$$f(x; \boldsymbol{\theta}) = B(\boldsymbol{\theta})h(x) \exp [Q_1(\boldsymbol{\theta})R_1(x) + \cdots + Q_k(\boldsymbol{\theta})R_k(x)].$$

The beta distribution (see Problem 3–77) and the normal distribution (see the following example) belong to this family.

EXAMPLE 3–19. The general normal distribution is defined by the following density function, depending on the vector parameter $\boldsymbol{\theta} = (\mu, \sigma^2)$:

$$\begin{aligned} f(x; \mu, \sigma^2) &= \frac{1}{\sqrt{2\pi\sigma^2}} \exp\left(-\frac{(x-\mu)^2}{2\sigma^2}\right) \\ &= \frac{1}{\sqrt{2\pi\sigma^2}} \exp\left(-\frac{\mu^2}{2\sigma^2}\right) \exp\left[-\frac{x^2}{2\sigma^2} + x\left(\frac{\mu}{\sigma^2}\right)\right]. \end{aligned}$$

This is seen to belong to the exponential family, with the following identifications:

$$B(\boldsymbol{\theta}) = \frac{1}{\sqrt{2\pi\sigma^2}} \exp\left[-\frac{\mu^2}{2\sigma^2}\right], \qquad h(x) = 1,$$

$$Q_1(\boldsymbol{\theta}) = -\frac{1}{2\sigma^2}, \qquad R_1(x) = x^2,$$

$$Q_2(\boldsymbol{\theta}) = \frac{\mu}{\sigma^2}, \qquad R_2(x) = x.$$

A multivariate density for the random vector $\mathbf{X} = (X_1, \ldots, X_n)$ is said to belong to the exponential family of distributions if it is of the following form, in which $\boldsymbol{\theta} = (\theta_1, \ldots, \theta_k)$ and $\mathbf{x} = (x_1, \ldots, x_n)$:

$$f(\mathbf{x}; \boldsymbol{\theta}) = B(\boldsymbol{\theta})h(\mathbf{x}) \exp [Q_1(\boldsymbol{\theta})R_1(\mathbf{x}) + \cdots + Q_k(\boldsymbol{\theta})R_k(\mathbf{x})].$$

The joint density of n independent observations on a random variable X whose distribution belongs to the univariate exponential family is of the above form. For, if the density of X (in the case of a single parameter) is

$$f_X(x) = C(\theta)g(x) \exp [Q(\theta)S(x)],$$

Name	$f(x; \theta)$	$B(\theta)$	$Q(\theta)$	$R(x)$	$h(x)$
Bernoulli	$p^x(1 - p)^{1-x}$	$1 - p$	$\log \frac{p}{1 - p}$	x	1
Binomial	$\binom{n}{x} p^x(1 - p)^{n-x}$	$(1 - p)^n$	$\log \frac{p}{1 - p}$	x	$\binom{n}{x}$
Geometric	$p(1 - p)^x$	p	$\log (1 - p)$	x	1
Negative binomial	$\binom{r + x - 1}{x} p^r(1 - p)^x$	p^r	$\log (1 - p)$	x	$\binom{r + x - 1}{x}$
Poisson	$e^m \frac{m^x}{x!}$	e^{-m}	$\log m$	x	$\frac{1}{x!}$
Exponential	$\lambda e^{-\lambda x}$	λ	$-\lambda$	x	1
Normal $(0, \theta)$	$(2\pi\theta)^{-1/2} \exp\left[-\frac{x^2}{2\theta}\right]$	$(2\pi\theta)^{-1/2}$	$-(2\theta)^{-1}$	x^2	1
Normal $(\theta, 1)$	$(2\pi)^{-1/2} \exp[-\frac{1}{2}(x - \theta)^2]$	$(2\pi)^{-1/2} \exp(-\frac{1}{2}\theta^2)$	θ	x	$\exp(-\frac{1}{2}x^2)$
Gamma	$\lambda^n x^{n-1} \cdot \frac{e^{-\lambda x}}{(n - 1)!}$	$\frac{\lambda^n}{(n - 1)!}$	$-\lambda$	x	x^{n-1}
Rayleigh	$\frac{x}{\theta^2} \exp\left[-\frac{x^2}{2\theta^2}\right]$	$\frac{1}{\theta^2}$	$-\frac{1}{2\theta^2}$	x^2	x

then the joint density of n independent replicas of X is

$$f(\mathbf{x}; \theta) = \prod f(x_i; \theta) = [C(\theta)]^n \exp\left[Q(\theta) \sum S(x_i)\right] \prod g(x_i),$$

in which the sum and product extend from $i = 1$ to $i = n$. Clearly, with $B(\theta) = [C(\theta)]^n$, $R(\mathbf{x}) = \sum S(x_i)$, and $h(\mathbf{x}) = \prod g(x_i)$, this multivariate density belongs to the exponential family.

To avoid the conclusion that all distributions belong to the exponential family, mention should be made of some that do not: the Cauchy family of distributions (Problems 2–135) and the family of distributions uniform on $[0, \theta]$.

Problems

3–71. One black, three white, and two red balls are placed in a container. One ball is selected at random and replaced, and then a second ball is selected at random. Let X and Y denote, respectively, the number of red and white balls that turn up. Construct a probability table for (X, Y) and determine from it the marginal distributions of X and Y. (Notice that these are binomial distributions.)

3–72. Four independent observations are to be made on a random variable with density $f(x) = 1 - |x|$, $-1 < x < 1$. Suppose that the interval from -1 to 1 is divided into four class intervals of equal length. What is the probability that one observation will fall in the leftmost class interval, one in the next, two in the next, and none in the rightmost class interval?

3–73. A group of students consists of 5 from each of the four classes (freshman sophomore, junior, senior). Determine the probability that a committee of 8 chosen at random from these 20 students represents the classes equally.

3–74. Determine an approximate answer to the preceding problem with the change that the selection is made from the whole school, which has 5000 students in each of the four classes.

3–75. Show that the Maxwell distribution belongs to the exponential family (see Section 3.3.3).

3–76. Given that $(X_1, \ldots, X_k)$ has a multinomial distribution with parameters $(p_1, \ldots, p_k)$, where $\sum p_i = 1$, use the moment generating function to derive the covariance of X_1 and X_2 to be $-np_1p_2$.

3–77. The *beta distribution* with parameters (r, s) is defined by the density

$$f(x; r, s) = (\text{const})x^{r-1}(1 - x)^{s-1}, \qquad 0 < x < 1.$$

The constant multiplier is a function of r and s, call it $1/B(r, s)$ (see Section 7.1.2). Show that this distribution belongs to the exponential family by identifying the various functions in the general form of density.

3–78. Show that if X_1 and X_2 are independent, discrete, and identically distributed with density $B(\theta)h(x) \exp[Q(\theta)R(x)]$, then $R(X_1) + R(X_2)$ has a distribution in the exponential family.

4

REDUCTION OF DATA

Statistical inference is the drawing of conclusions from statistical evidence, from data. These conclusions are concerned with the way things are—with the state of affairs, called the *state of nature*, which governs the generation of data and gives rise to the particular data at hand that constitute the evidence. Inference is usually less than definitive, because it amounts to reasoning from the particular to the general, a process that involves some degree of risk. But we are continually confronted with data, with experiences and experimental results, which are the only basis we have for making necessary plans and decisions. We have no choice but to form our notion of governing laws on the basis of data and to act accordingly.

Because experimental results derive from certain governing laws and because those laws then determine the nature of the data, it is not unreasonable to assume that it is possible to learn something about the laws—about "nature"—from the data. That is, the data should contain some information about the state of nature. This will indeed be the case if the data are gathered intelligently so that the experiments performed to obtain the data are related in a known way to the state of nature.

The state of nature—which refers, not to all of "nature" but to that aspect of it that governs a particular phenomenon of interest—is often usefully defined by a probability model, a model that defines or induces the probability distribution of the data to be used in inference. The set of all possible states of nature will be called Θ, and an individual state or particular probability model will be referred to as θ. In many instances θ will be a real-valued parameter of a probability distribution, but more generally it is simply a label for a particular model.

4.1 Sampling

The process of gathering data or of obtaining results from several performances of an experiment of chance is called *sampling*. The results themselves are called *observations*, and the collection of observations is called a *sample*.

Although the result of an experiment of chance may be more general, attention will be focused on the case in which the result of each performance of the experiment is a single random variable.

4.1.1 Random Sampling

Some of the terminology of sampling stems from the following situations that are frequently encountered in statistics:

1. Objects are drawn one at a time from an actual, finite collection of objects called a *population*, and a particular characteristic of interest is determined for each object drawn. After each observation, and before the next drawing, the object just drawn is replaced and the population of objects is thoroughly mixed.
2. Objects are drawn from an actual, finite population as in (1) except that the objects are *not* replaced.

The population of objects is frequently a collection of people, and the observed characteristic may be such a thing as weight, eye color, political preference, and so forth, or a combination of these. The basic probability space—the experiment of chance—is this collection of people or objects, although probability would naturally be transferred to the space of "values" of the characteristic of interest; this value space itself can be conceived of as the basic probability space.

When objects are drawn in such a way that at each drawing all remaining objects are equally likely to be chosen, the sampling is called *random*, a usage that conforms to the layman's notion of "selecting at random." It is to make each drawing random that the population should be mixed when objects drawn are replaced. With this understanding that each selection is random, the sampling in (1) is called *random sampling with replacement*; and in (2), *random sampling without replacement*, or sometimes *simple random sampling*.

In a sense, random sampling without replacement is better than random sampling with replacement, since in the former case objects that have been drawn are not put back into the pool of available objects to confound things. To take an extreme case, suppose that there are only two objects in the population; when one is drawn, selection of a second object would furnish complete information about the original population *if* the first were not replaced. Drawing without replacement is also sometimes more convenient in that the mixing required with the replacement of objects is not always easy to achieve. On the other hand, as will be seen in a moment, the mathematically simpler process is sampling with replacement. Of course, if a population is enormous with respect to the size of the sample to be drawn, it is practically immaterial whether the objects drawn are or are not replaced; sampling without replacement merges into sampling with replacement as the population size becomes infinite. The theory of one could then be used with the practice of the other.

Suppose that when an object is drawn, the characteristic measured or otherwise ascertained is X. This is random quantity whose distribution is determined by the proportions of the various values of X among the objects in the population and the agreement that the objects are equally likely. It is this distribution of X that will be called the *population distribution.*

EXAMPLE 4–1. In a group of 100 freshmen, 73 are eighteen, 22 are seventeen, 4 are nineteen, and 1 is sixteen years old. Selecting one freshman from the group at random is represented by assigning probability .01 to each freshman. The population random variable X that is of interest is the age of a freshman drawn, and this variable has the following probability table:

Age	16	17	18	19
Probability	.01	.22	.73	.04

The following notation will be used consistently: X_1 denotes the characteristic X of the first object drawn; X_2 denotes the characteristic X of the second object drawn; and so on. A random sample is then written in the form $(X_1, X_2, \ldots, X_n)$, listing the values of the observed characteristic as they are obtained, where n is the size of the sample. This vector of the n observations in a sample will also be written as $\mathbf{X}$.

It is clear that in either (1) or (2), the quantity X_1 is a random variable with the population distribution—the distribution of X. It is also clear that in (1), *all* observations $X_1, \ldots, X_n$ have this population distribution as a common distribution because prior to each drawing the population of objects is restored to its original condition. It is not quite so clear, but true, that also in (2) the observations share a common distribution. This statement refers to the marginal distributions of the observations—to the distribution of X_7, for instance, unconditioned by information as to the values of any previous or subsequent X's.

EXAMPLE 4–2. A bowl contains four beads numbered from 1 to 4. Two are drawn at random, one at a time. Let X_1 denote the number on the first bead drawn and X_2 the number on the second bead drawn. There are 12 possible samples:

$$(1, 2),\quad (1, 3),\quad (1, 4),\quad (2, 3),\quad (2, 4),\quad (3, 4),$$
$$(2, 1),\quad (3, 1),\quad (4, 1),\quad (3, 2),\quad (4, 2),\quad (4, 3).$$

As discussed in Example 1–48, each of these 12 outcomes has probability $\frac{1}{12}$. From this, one can compute the distributions of X_1 and X_2. For instance,

$$P(X_1 = 1) = P[(1, 2), (1, 3), \text{ or } (1, 4)] = \tfrac{1}{4}.$$

Similarly,

$$P(X_2 = 1) = P[(2, 1), (3, 1), \text{ or } (4, 1)] = \tfrac{1}{4}.$$

In like fashion it is found that for X_1, each of the possible values 1, 2, 3, and 4 has probability $\frac{1}{4}$, and that X_2 has exactly the same distribution—the population distribution.

Thus the basic difference between sampling of types (1) and (2) is not in the marginal distributions of the individual observations, for in both cases these observations are identically distributed. However, in case (1) the result of any observation is not affected by the results of any other observations; the observations are *independent* random phenomena. In case (2) the observations are *not* independent.

There is another kind of commonly occurring situation, mechanically different from (1) and (2), in which the results are mathematically of the same type as (1), random sampling with replacement:

3. Observations are obtained as the result of repeated, independent performances of an experiment, under conditions that are identical with respect to those factors that can be controlled.

This description includes (1) as a special case, but now does not necessarily refer to a tangible "population" from which an object is to be selected. However, one may imagine an infinite population of possible results. Performing the experiment selects one of these results, and performing the experiment again selects a result from the same collection of possible results as was available in the first trial. That is, repeating the experiment under "identical" conditions means that the first result is "replaced" and is again one of the candidates to be "drawn" the next time. In both (1) and (3), then, the observations are identically distributed and independent. The term *random sampling* without further qualification will denote such a process:

Definition. *A* random sample *from a population random variable X is a set of independent, identically distributed, random variables $X_1, \ldots, X_n$, each with the distribution of X.*

A random sample (with replacement) is simpler to treat mathematically than is a sample obtained without replacement from a finite population. This is a result of the independence in a random sample, which implies that the joint density function of the sample observations is a product of their marginal distribution densities:

$$f_{\mathbf{X}}(u_1, \ldots, u_n) = f_{X_1}(u_1) \cdots f_{X_n}(u_n) = \prod_{i=1}^{n} f_X(u_i),$$

where X denotes the population random variable, whose distribution is shared by each of the observations.

EXAMPLE 4–3. The life of a certain type of electron tube in a given application has an exponential distribution with mean life of 10 hr. The life of each tube operated in this way then has the density

$$f_X(u) = .1e^{-.1u}, \qquad u > 0.$$

The joint density function of the lives $X_1, \ldots, X_n$ of n such tubes operated independently is

$$f_{\mathbf{X}}(u_1, \ldots, u_n) = (.1)^n \prod_{i=1}^{n} e^{-.1u_i} = (.1)^n \exp\left[-.1 \sum_{i=1}^{n} u_i\right].$$

Whether an actual sampling process is faithfully described by a mathematical model for sampling is usually a moot question and one that will not be considered here. In the last analysis it is really unanswerable, and about all that can be done is to make every effort to see that the process used in collecting data conforms as nearly as can be determined to a definite mathematical model; otherwise, the grounds for inference are wobbly indeed. The matter of how to make such efforts will not be considered in this book; in the case of scientific experiments, their consideration would require knowledge of the field of application, and in the case of sampling from actual populations, the subject is so extensive as to be beyond our scope.

4.1.2 Frequency Tabulations

In obtaining observations on a discrete random variable X with possible values $x_1, x_2, \ldots$, one naturally finds that only these values occur in the sample. Indeed, even if the list of possible values is considered to be infinite, only a finite number occur in a finite sample, so that the list of actually observed distinct values can be taken to be $x_1, \ldots, x_k$ for some k.

As sampling proceeds, each result can be recorded by making a mark opposite the value observed. The number of marks for a given x_i, when the sampling is finished, is called the *frequency* of x_i and is denoted by f_i. Thus, the frequency f_i is the number of times the value x_i occurs ($i = 1, \ldots, k$), and clearly

$$f_1 + \cdots + f_k = n,$$

where n is the sample size. A table of frequencies is often used to present the data, summarized in this way:

Possible Value	Frequency
x_1	f_1
$\vdots$	$\vdots$
x_k	f_k

In the case of a continuous variable one cannot, ideally, anticipate the values that might be encountered. However, in the reading of scales on measuring devices and in writing down numbers, observations are necessarily rounded off. The result is that recorded observations are actually observations from a *discrete* population that only approximates the true, continuous population. There will usually be repeated values, and the presentation of sample results in a frequency table is again appropriate. Indeed, it is often useful, by regrouping observations, to use what is in effect a rounding off that is cruder than that inherent in the measurement process. To regroup, a range of values that includes all the data is partitioned into *class intervals*, usually taken to be of equal size for convenience. The observations in each class interval are represented by, say, the midpoint of that interval, and a frequency is assigned that value according to the number of the original observations that fall in the interval. The regrouped data are patently different from the original data but may serve to give a more digestible delineation of the sample's significant features if the regrouping is not carried to the extreme (with all observations in one class interval).

The content of a frequency tabulation is often presented in graphical form as a *histogram*. This is a plot in which each frequency is represented by a bar whose height is proportional to the frequency and whose base is the corresponding class interval. The bar, which spreads over the class interval, is used to point up the fact that the observations involved in a given frequency could have come from anywhere in the corresponding interval. (By the same token, if the frequency tabulation is for results of a discrete variable, the bar is not really appropriate. A rod with height proportional to frequency at the corresponding value would be used instead.)

EXAMPLE 4–4. The following are hemoglobin levels of 50 cancer patients at a veterans' hospital (in gm/100 ml):

13.6	14.8	13.7	14.2	11.5
11.9	13.8	14.6	14.2	12.7
13.4	11.5	11.9	14.8	12.7
12.4	15.3	15.2	13.5	15.0
12.4	12.0	13.8	11.7	10.0
13.2	15.5	14.0	13.5	15.0
12.7	12.9	13.7	15.1	13.5
15.7	12.7	15.7	10.9	14.0
14.8	14.0	13.8	12.7	11.9
12.0	11.4	11.1	13.7	13.2

These numbers extend from 10.0 to 15.7, and a class interval width of 0.9 would require about seven class intervals to include all the data. Taking the class interval boundaries as 9.95, 10.85, and so forth, avoids the embarrassment of having observations fall on the boundaries. One possible frequency tabulation can then be shown as follows:

Class Interval	Midpoint	Frequency
9.95–10.85	10.4	1
10.85–11.75	11.3	6
11.75–12.65	12.2	7
12.65–13.55	13.1	12
13.55–14.45	14.0	12
14.45–15.35	14.9	9
15.35–16.25	15.8	3
	Total	50

The histogram for this frequency distribution is plotted in Figure 4–1.

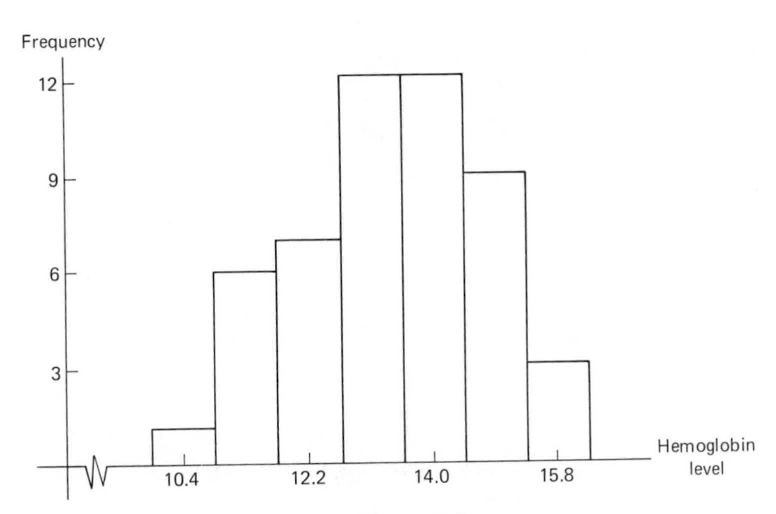

Figure 4–1

4.1.3 The Sample Distribution Function

Imagine masses of equal weight placed at the various observed values in a given sample $(x_1, \ldots, x_n)$ of observations on a random variable X. This placement defines a probability-type distribution with weight $1/n$ at each observed value, and the corresponding c.d.f. is called the *sample distribution function.* This function will be denoted, rather imprecisely, by $F_n(x)$ and defined formally as

$$F_n(x) = \frac{1}{n} \times \text{(number of observations not exceeding } x\text{)}.$$

As the c.d.f. of a probability distribution (although that distribution is just a mathematical entity and not the model for any aspect of the real experiment being studied), this sample distribution function has all the properties of a distribution function. And because the sample distribution is discrete, the sample distribution function is a *step function*, rising in steps of $1/n$ at the observed values. (If there are repeated values—as there would tend to be if the population is discrete or if the data from a continuous population are given in a frequency table—the step height at x_i would be the relative frequency, f_i/n.)

The sample distribution function is not only an analog of the population distribution function, but also converges thereto, in a well-defined sense. Since the sample c.d.f. is a random function (it is a function, but depends on the random observations in the sample), the convergence can only be asserted in terms of probability. It can be shown that the sample distribution function converges to the population c.d.f. *uniformly in x*, with probability 1.[1] Here it will be shown only that the sample distribution function converges in quadratic mean to the population c.d.f. *at each x*. This follows from the fact that $nF_n(x)$ is just the number of observations not exceeding x and is, therefore, a binomial random variable with

$$p = P(X \leq x) = F(x).$$

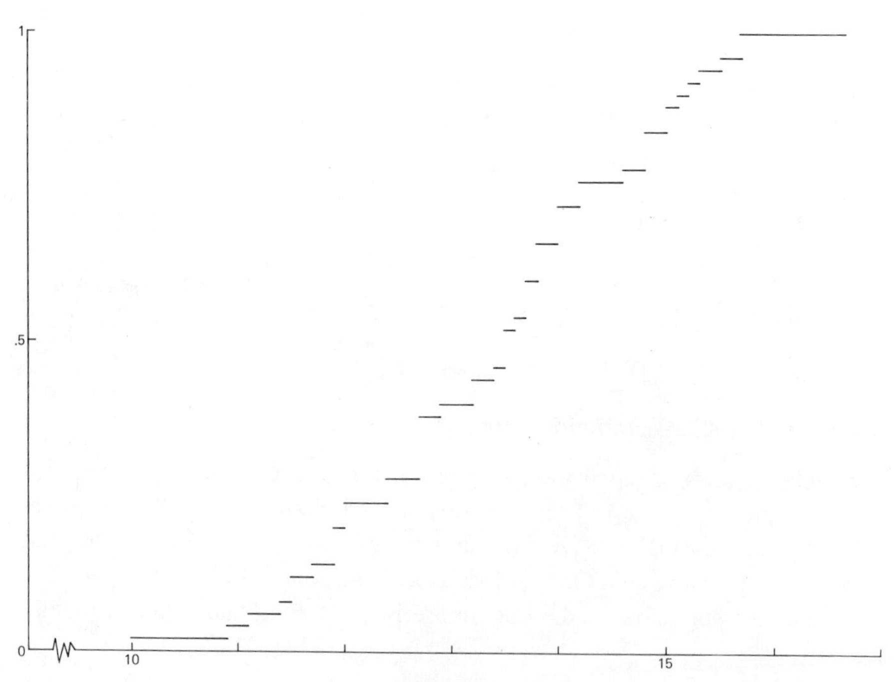

Figure 4–2

[1] M. Loève, [18], p. 20.

For, then

$$E[nF_n(x)] = nF(x) \qquad \text{and} \qquad \text{var}\,[nF_n(x)] = nF(x)[1 - F(x)]$$

so that

$$E[F_n(x)] = F(x) \qquad \text{and} \qquad \text{var}\,F_n(x) = \frac{F(x)[1 - F(x)]}{n}.$$

Since the expectation is $F(x)$ and the variance tends to zero as n becomes infinite, the assertion is established.

EXAMPLE 4–5. The hemoglobin data given in Example 4–4 are shown in Figure 4–2 by means of the sample distribution function. Notice that although most step heights are $\frac{1}{50} = .02$, some are multiples of this, indicating repeated values in the data.

4.2 Statistics

In assimilating the information contained in a sample, in describing and comparing samples, and in making decisions on the basis of the results of a sample, it is convenient to be able to use, rather than the complete list of observations as they were obtained, some more readily comprehended measures computed from the sample. It will be seen subsequently that there are principles that lead naturally to certain measures as being appropriate to certain problems. Here are presented some of the sample measures of characteristics that actually do find many useful applications. This will provide a "vocabulary" of some of the simpler, common sample characteristics for purposes of illustration. For the present these can be thought of as being constructed by using intuition, and historically this is usually the way they were introduced. Actually, more than just furnishing illustrations, intuitively derived sample measures and procedures based on them frequently turn out to work pretty well. But they can and should be compared with others developed more systematically.

The term *statistic* denotes a descriptive measure computed from the observations in a sample. The value of a given statistic then depends on the values of the observations; hence the following:

Definition. *A* statistic *is a function of the observations in a sample.*

In particular, the computation of a statistic does not require knowledge of any unknown population characteristics.

The term *statistic* applies to the *relationship* between independent and dependent variables—or to the random variable defined by the functional relation. If $(x_1, \ldots, x_n)$ is a possible sample point (a possible list of observations in a sample), the functional relationship

$$y = t(x_1, \ldots, x_n)$$

provides a transformation or mapping from the space of all sample points to the space of values of the function. This mapping induces a probability distribution in the latter space and thereby defines a random variable:

$$Y = t(X_1, \ldots, X_n) = t(\mathbf{X}).$$

Computation of a statistic from a set of observations constitutes a reduction of the data to a single number, or to a vector of numbers, if the function defining the statistic is vector-valued. In the process of such reduction, certain information about the population may be lost; but ideally the measure computed would be chosen so that the information lost is not pertinent to the particular problem at hand and the measure is still sufficient to handle it. A notion of "sufficiency" that makes this notion more precise will be taken up in Section 4.4.

The population "variable," and hence the observations in a sample, need not be numerical (e.g., a coin may fall heads or tails). But the term *statistic* usually refers to a numerical function of the observations. The statistics to be introduced at this point happen to be functions defined for numerical data, but numerical statistics for categorical data will be encountered in later chapters.

4.2.1 Statistics Based on Order

Frequently, in an attempt to create order out of the chaos of a mass of data, the observations are put in *numerical* order. (This would assume, of course, that the sample space is numerical.) The result is a permutation of the original observations and will be denoted by $[X_{(1)}, \ldots, X_{(n)}]$. That is, $X_{(1)}$ is the smallest observation, $X_{(2)}$ is the next smallest, and so on. The vector of these ordered observations is sometimes referred to as *the order statistic*, although various other quantities based on order are also thought of as order statistics. There may be some ambiguity in the order statistic if there are ties among the observations. This should not happen in the case of observations from a continuous population but may happen because of round-off in reading and recording measurements. (The order statistic is not usually used in the case of a discrete population.)

The sample *median* and *quartiles* are defined to parallel the previously defined, corresponding population parameters. Thus, the median is the middle value or a value such that half the sample observations are to the left and half to the right of it. More precisely, if the number of observations is odd, the median is the middle observation when the observations are in numerical order; if it is even, the median is (arbitrarily) defined as the average of the middle two of the ordered observations:

$$\text{median} = \begin{cases} X_{(k)}, & \text{if } n = 2k - 1 \quad \text{(odd)}, \\ \frac{1}{2}[X_{(k)} + X_{(k+1)}], & \text{if } n = 2k \quad \text{(even)}. \end{cases}$$

The quartiles will be the three values dividing the ordered observations into four groups of equal size. Deciles and percentiles would be defined similarly, but usually only if there are a great many observations in the sample.

The sample median, like the population median, is a measure of the middle or center or location of the observations. Another measure of location sometimes used, based on order statistics, is the *midrange*:

$$\text{midrange} = \tfrac{1}{2}[X_{(1)} + X_{(n)}],$$

which is simply the value (not necessarily one that is observed) halfway along the scale from the smallest observed value to the largest. (The median is also halfway, but in terms of numbers of observations rather than in terms of the scale of values.) Still another measure of location is the average of the 1st and 3rd quartiles: $(Q_1 + Q_3)/2$.

The *range* of a sample is defined to be the difference between the smallest and largest observations:

$$R \equiv X_{(n)} - X_{(1)}.$$

It measures spread or dispersion in a sample, as does the *interquartile range*: $Q_3 - Q_1$.

EXAMPLE 4–6. Consider the following observations:

$$31, 28, 27, 32, 36, 33, 29, 35, 24, 33.$$

The smallest is $x_{(1)} = 24$, the largest is $x_{(10)} = 36$, and the order statistic is

$$(24, 27, 28, 29, 31, 32, 33, 33, 35, 36),$$

or, to indicate the permutation of the original order of observation:

$$(x_9, x_3, x_2, x_7, x_1, x_4, x_6, x_{10}, x_8, x_5).$$

The range is the difference $36 - 24 = 12$, and the median is the average of the fifth smallest and sixth smallest observations:

$$\tfrac{1}{2}(31 + 32) = 31.5.$$

(One would usually not bother to define quartiles in a sample this small.) The midrange is 30, halfway from 24 to 36.

4.2.2 Sample Moments

The first two moments of the sample distribution (as characterized by the sample distribution function) are perhaps the most frequently used of all statistics. The *sample mean* is the first moment:

$$\bar{X} \equiv \frac{1}{n}\sum_{i=1}^{n} X_i,$$

with the usual property that the first moment about this value is zero:

$$\frac{1}{n}\sum_{i=1}^{n}(X_i - \bar{X}) = 0.$$

The second moment of the sample distribution function about the sample mean is called the *sample variance*:

$$S^2 = \frac{1}{n}\sum_{i=1}^{n}(X_i - \bar{X})^2 = \frac{1}{n}\sum_{i=1}^{n}X_i^2 - \bar{X}^2.$$

The positive square root of the sample variance is called the *sample standard deviation*:

$$S = \sqrt{S^2}.$$

The notation S_x will be used for standard deviation (and S_x^2 for variance) when a subscript is needed for more explicit identification of S as relating to observations $X_1, \ldots, X_n$.

It is useful to observe and easy to show that under the important particular kind of transformation of data called *linear*, the sample mean undergoes exactly the same transformation, and the sample variance is multiplied by the square of the scale factor. That is, if $Y_i = aX_i + b$, then $\bar{Y} = a\bar{X} + b$ and $S_y^2 = a^2S_x^2$. These relations can be seen directly, but they follow from the fact that a sample distribution can be considered mathematically as a probability distribution. (They have already been seen to hold for probability distributions.) This same reasoning shows that the "parallel axis theorem,"

$$S^2 = \frac{1}{n}\sum(X_i - a)^2 - (\bar{X} - a)^2,$$

holds as a special case of the parallel axis theorem for probability distributions.

When data are given in a frequency table, the summation of like values is accomplished by multiplying the common value by its frequency. Thus, for the mean,

$$\bar{X} = \frac{1}{n}(\text{sum of the observations}) = \frac{1}{n}\sum_{i=1}^{k}f_i x_i,$$

and for the variance,

$$S^2 = \frac{1}{n}\sum_{i=1}^{k}f_i x_i^2 - (\bar{X})^2.$$

(If the frequencies in these computations are those resulting from regrouping the original data into wider class intervals, the computed mean and variance are not quite the mean and variance of the original data. Sometimes "Sheppard's corrections" are applied, which for the above moments amount to leaving the mean as it is and subtracting from the computed variance $h^2/12$, where h is the class interval width.[2])

Computation is often simpler if the data are transformed so that the new origin is near the mean and the new unit is the class interval width, as in the following example.

EXAMPLE 4–7. The frequency table for the hemoglobin data of Example 4–4 is repeated here, along with a column of values transformed to simplify the computations, and columns used in computing the mean and variance. The transformation used puts the origin at 13.1 and makes the class interval width a unit:

$$y_i = \frac{x_i - 13.1}{.9}, \qquad \text{or} \qquad x_i = .9y_i + 13.1.$$

x_i	f_i	y_i	$f_i y_i$	$f_i y_i^2$
10.4	1	−3	−3	9
11.3	6	−2	−12	24
12.2	7	−1	−7	7
13.1	12	0	0	0
14.0	12	1	12	12
14.9	9	2	18	36
15.8	3	3	9	27
Total	50		17	115

Then

$$\bar{Y} = \tfrac{17}{50} \qquad \text{and} \qquad \bar{X} = .9\bar{Y} + 13.1 = 13.406,$$

and

$$\overline{Y^2} = \tfrac{115}{50}, \qquad S_y^2 = 2.3 - (.34)^2 = (1.48)^2$$

so that

$$S_X = .9S_y = 1.33.$$

(The bar over the Y^2 was used to denote the average square.) Observe that there is no total in the x_i or y_i column, since the sum of the *distinct* values has no significance. Notice also that in computing the average square, only the y is squared (and not the frequency).

Problems

4–1. Compute the median, mean, range, and standard deviation of the following observations: −3, 2, 5, 0, −4.

4–2. Given that $X_1^2 + \cdots + X_{10}^2 = 160$ and that $S^2 = 12$, determine two possible values of $\bar{X}$, with $n = 10$.

[2] See H. Cramér, [5], pp. 359 ff.

4–3. Given the following 30 test scores: 51, 52, 52, 58, 59, 59, 61, 62, 63, 69, 72, 74, 76, 80, 80, 80, 81, 81, 82, 83, 83, 84, 86, 87, 87, 88, 88, 89, 90, 94.

(a) Determine the mean and standard deviation and plot the sample distribution function.

(b) Round off the above scores to the nearest multiple of 5 and compute the mean and standard deviation of the rounded-off data.

(c) Determine the median and the midrange of the given scores.

4–4. The mean and standard deviation of the numbers $x_1, \ldots, x_n$ are 5 and 2, respectively. Determine the average of their squares and the mean and standard deviation of $y_1, \ldots, y_n$, where $y_i = 3x_i + 4$.

4–5. Write out the joint density or probability function of the observations $X_1, \ldots, X_n$ in a random sample from

(a) A uniform distribution on $[a, b]$.

(b) A normal distribution (μ, σ^2).

(c) A Bernoulli distribution (p).

(d) A Poisson distribution with expectation m.

(e) A geometric distribution with parameter p.

4–6. Write the joint probability function of $X_1, \ldots, X_n$, the results of random sampling *without* replacement from a population of size N including M objects coded 1 and $N - M$ coded 0.

4–7. Show the following:

(a) $\displaystyle \frac{1}{n}\sum_{i=1}^{n} X_i^2 - \bar{X}^2 = \frac{1}{n}\sum_{i=1}^{n} (X_i - \bar{X})^2.$

(b) If $Y_i = aX_i + b$, then $\bar{Y} = a\bar{X} + b$ and $S_y = |a|S_x$.

(c) The second moment of a set of observations is smallest when taken about their mean.

4–8. Using class intervals of width 1.1 and starting at 9.85, construct a frequency tabulation and histogram for the hemoglobin data in Example 4–4. Compute the mean and standard deviation from your tabulation. (Observe the differences in histograms and in the computed statistics resulting from the different systems of class intervals.)

4.3 Sampling Distributions

A statistic $T = t(X_1, \ldots, X_n)$, as a function of sample observations that are random variables, is also a random variable.[3] Its value will vary from sample to sample. Its distribution is the model for the pattern of variation in its values and is called the *sampling distribution* of T.

[3] The function t would have to be such that T is a measurable function on the sample space, and this will be implicitly assumed henceforth.

In order to appreciate the significance of a value of T as calculated from a particular sample, it is helpful to know something about the nature of T as a random variable, that is, about what might have happened and would happen in other samples. Moreover, what is really relevant is how the sampling distribution of T relates to the state of nature that is the object of a given statistical study.

Depending on the nature of the function $t(x_1, \ldots, x_n)$ that defines T, the distribution of T may be easy or difficult—or perhaps impossible—to derive theoretically. Sometimes all that can be done is to obtain an empirical distribution for T by artificial sampling experiments.

4.3.1 Distribution of Sample Moments

The sample mean is one of the easiest statistics to study, a simple function of the sample observations:

$$\bar{X} = \frac{1}{n}(X_1 + \cdots + X_n).$$

The mean and variance of the distribution of $\bar{X}$ are readily expressed in terms of population moments:

$$E(\bar{X}) = \frac{1}{n}\sum EX_i = \mu,$$

which is valid so long as the X's are identically distributed (with mean μ); and

$$\text{var } \bar{X} = \frac{1}{n^2}\sum \text{var } X_i = \frac{\sigma^2}{n}$$

(where σ^2 is the population variance), a formula that holds when the X's are pairwise uncorrelated—which they are when the sample is a random sample. In the case of a simple random sample—sampling without replacement from a finite population of size N—the variance of $\bar{X}$ is[4]

$$\text{var } \bar{X} = \frac{\sigma^2}{n}\frac{N - n}{N - 1}.$$

Not only the first two moments, but also the complete distribution of $\bar{X}$, as defined by its characteristic function, can be expressed in terms of the population distribution:

$$\phi_{\bar{X}}(t) = \left[\phi_X\left(\frac{t}{n}\right)\right]^n,$$

where $\phi_X(t)$ is the population characteristic function, and where again it is assumed that the observations comprise a random sample. The characteristic

[4] See S. S. Wilks, [22], p. 219.

function may or may not be easy to invert to obtain the density or the c.d.f. of $\bar{X}$. In the important normal case, however, there is no problem—the distribution of $\bar{X}$ is again normal (cf. Section 3.3.1 and Problem 4–13).

The Central Limit Theorem shows that the asymptotic or large-sample distribution of the mean of a random sample from *any* population with finite variance is normal. That is, the limiting distribution of

$$Z = \frac{\bar{X} - \mu}{\sigma/\sqrt{n}}$$

is standard normal so that approximate probabilities of events expressed in terms of $\bar{X}$ can be obtained from Table I of the Appendix.

EXAMPLE 4–8. In a population divided equally between blacks, coded 1, and whites, coded 0, the mean is $p = \frac{1}{2}$ and the variance, $pq = \frac{1}{4}$. The distribution of the proportion of blacks in a sample of 100 is approximately normal with mean $p = \frac{1}{2}$ and variance $\sigma^2/n = \frac{1}{400}$. The probability that the sample proportion will exceed, say, .55 would be

$$P(\bar{X} > .55) \doteq 1 - \Phi\left(\frac{.55 - .50}{1/20}\right) = \Phi(-1) \doteq 0.16.$$

The distribution of the sample variance, as might be expected, is not quite so easy, although its mean is readily computed:

$$E(S^2) = E\left(\frac{1}{n}\sum (X_i - \mu)^2 - (\bar{X} - \mu)^2\right)$$

$$= \sigma^2 - \text{var}\,\bar{X} = \sigma^2\left(1 - \frac{1}{n}\right).$$

The variance of S^2 can be shown to be

$$\text{var}\, S^2 = \frac{\mu_4 - \mu_2^2}{n} - \frac{2(\mu_4 - 2\mu_2^2)}{n^2} + \frac{\mu_4 - 3\mu_2^2}{n^3},$$

where μ_k denotes the kth population moment about the population mean. It will be convenient to have a reduction of this expression for the particular case of a normal population, in which case $\mu_4 = 3\mu_2^2$:

$$\text{var}\, S^2 = \frac{2\sigma^4(n-1)}{n^2}, \qquad \text{(random sample from normal population)}.$$

Higher-order sample central moments can be treated with ever-increasing complexity of detail.[5]

In Problem 4–12, it will be shown that sample moments about zero are asymptotically normal and tend in probability to corresponding population moments. With this information and the fact that sample central moments are expressible

[5] See H. Cramér, [5], p. 348.

as polynomial functions of sample moments about zero, one can show that the central moments are also asymptotically normal and tend in probability to the corresponding population central moments.[6]

Problems

4–9. A population consists of four chips numbered 1, 2, 3, and 4. Let (X_1, X_2) denote a simple random sample (random sampling without replacement), and let $(X_{(1)}, X_{(2)})$ denote the corresponding order statistic.

(a) Determine the probability table for the joint distribution of $X_{(1)}$ and $X_{(2)}$.
(b) Determine the distribution of the range.
(c) Determine the distribution of the sample mean.

4–10. Determine the probability that the average score on a certain aptitude test given to a group of 100 incoming freshmen exceeds 70, given (from long experience with the test) that the population of test scores has mean 68 with standard deviation 10.

4–11. Two independent random samples are taken from a normal population with mean 150 and variance 28.6. The sample sizes are 10 and 25, and the corresponding sample means are $\bar{X}_1$ and $\bar{X}_2$. Determine the following:

(a) $E(\bar{X}_1 - \bar{X}_2)$. (b) var $(\bar{X}_1 - \bar{X}_2)$. (c) $P(|\bar{X}_1 - \bar{X}_2| > 4)$.

4–12. The kth sample moment about zero, $\sum X_i^k/n$, is a sum of independent variables. Express its mean and variance in terms of population moments, and determine the asymptotic distribution. Show also that it converges in probability to the corresponding population moment.

4–13. Use characteristic functions to show that

(a) The mean of a random sample from a normal population is normally distributed.
(b) The mean of a random sample from a Cauchy population:

$$f(x;\theta) = [\pi(1 + (x - \theta)^2)]^{-1}$$

has the same distribution as the population. (See Problem 2–138.)

4–14. Consider a random sample of size n without replacement from a population of size N. Given the formula for var $\bar{X}$ in the text, show that

$$E(S^2) = \frac{n-1}{n}\frac{N}{N-1}\sigma^2.$$

Verify both formulas for the situation of Problem 4–9. (Since the distribution of R is already obtained there, you may find it useful that for $n = 2$, $S^2 = R^2/4$.)

4.3.2 Components of the Order Statistic

The c.d.f. of the largest observation $X_{(n)}$ in a random sample of size n from a population with c.d.f. $F(x)$ can be derived by rewriting the event that defines it:

$$\{X_{(n)} \le y\} = \{X_1 \le y, \ldots, \text{ and } X_n \le y\}.$$

[6] Ibid., p. 365.

(Each condition implies the other.) The independence of the observations then permits factorization of the probability of the event on the right:

$$P(X_{(n)} \le y) = P(X_1 \le y, \ldots, \text{ and } X_n \le y)$$
$$= P(X_1 \le y) \cdots P(X_n \le y) = [F(y)]^n.$$

In much the same fashion one can obtain the c.d.f. of the kth smallest observation in terms of the population c.d.f.:

$$P[X_{(k)} \le y] = P(k \text{ or more of the } n \text{ observations are } \le y)$$
$$= \sum_{j=k}^{n} \binom{n}{j} [F(y)]^j [1 - F(y)]^{n-j},$$

the individual terms in this sum being probabilities that in n independent trials precisely j result in an observation that does not exceed y. [The individual trials are of the Bernoulli type with $p \doteq F(y)$.]

The density function of $X_{(k)}$ can be obtained from the above distribution function by differentiating with respect to y:

$$f_{X_{(k)}}(y) = \sum_{j=k}^{n} \binom{n}{j} j[F(y)]^{j-1} f(y)[1 - F(y)]^{n-j}$$
$$+ \sum_{j=k}^{n} \binom{n}{j} (n - j)[F(y)]^j [1 - F(y)]^{n-j-1}[-f(y)]$$
$$= nf(y)\left\{\sum_{j=k}^{n} \binom{n-1}{j-1} [F(y)]^{j-1}[1 - F(y)]^{n-j}\right.$$
$$\left. - \sum_{j=k}^{n-1} \binom{n-1}{j} [F(y)]^j [1 - F(y)]^{n-j-1}\right\}.$$

Letting $j = m - 1$ in the second sum results in terms identical with those in the first sum, but from $m = k + 1$ to n. These then cancel except for the term $j = k$ in the first sum:

$$f_{X_{(k)}}(y) = nf(y)\binom{n-1}{k-1}[F(y)]^{k-1}[1 - F(y)]^{n-k}.$$

Putting $k = n$, one obtains the distribution and density functions for the largest observation $X_{(n)}$:

$$F_{X_{(n)}}(y) = [F(y)]^n, \qquad f_{X_{(n)}}(y) = n[F(y)]^{n-1} f(y),$$

and putting $k = 1$, the distribution and density functions of $X_{(1)}$:

$$F_{X_{(1)}}(y) = 1 - [1 - F(y)]^n, \qquad f_{X_{(1)}}(y) = n[1 - F(y)]^{n-1} f(y).$$

With a little more work one can obtain the joint distribution of any *pair* of the ordered observations. The extra work is perhaps a minimum in the case of the smallest and largest observations, whose joint distribution function is

$$F_{X_{(1)}, X_{(n)}}(u, v) = P[X_{(1)} \le u \text{ and } X_{(n)} \le v].$$

To evaluate this, first observe that

$$\{X_{(n)} \le v\} = \{X_{(1)} \le u \text{ and } X_{(n)} \le v\} + \{X_{(1)} > u \text{ and } X_{(n)} \le v\}.$$

The events on the right are disjoint, so the probability of the sum is the sum of the probabilities; transposing the second term, then, one obtains

$$\begin{aligned} F_{X_{(1)},X_{(n)}}(u, v) &= P[X_{(n)} \le v] - P[X_{(1)} > u \text{ and } X_{(n)} \le v] \\ &= [F(v)]^n - P(\text{all } X_i \text{ lie between } u \text{ and } v) \\ &= \begin{cases} [F(v)]^n - [F(v) - F(u)]^n, & \text{if } u \le v, \\ [F(v)]^n, & \text{if } u > v. \end{cases} \end{aligned}$$

The joint density is found by differentiation:

$$f_{X_{(1)},X_{(n)}}(u, v) = \begin{cases} n(n-1)[F(v) - F(u)]^{n-2} f(u) f(v), & \text{if } u \le v, \\ 0, & \text{if } u > v. \end{cases}$$

4.3.3 The Midrange and Range

From the joint distribution of the smallest and largest observations in a random sample one can derive the distribution of functions of those statistics. The midrange,

$$A = \tfrac{1}{2}[X_{(1)} + X_{(n)}],$$

is sometimes useful in location problems. Its distribution function is

$$F_A(x) = P(A \le x) = P[X_{(1)} + X_{(n)} \le 2x] = \iint_R f(u, v)\, du\, dv,$$

where R denotes the region (Figure 4–3) in the uv plane defined by the inequalities $u + v \le 2x$ and $u \le v$, and $f(u, v)$ denotes the joint density of $X_{(1)}$ and $X_{(n)}$. Making the change of variable $u = s$ and $v = t - s$, one obtains

$$F_A(x) = \int_{-\infty}^{2x} \int_{-\infty}^{t/2} f(s, t - s)\, ds\, dt.$$

The derivative with respect to x, which appears only in the upper limit of the outer integral, is the inner integral evaluated at $t = 2x$ times the derivative of $2x$ with respect to x:

$$\begin{aligned} f_A(x) &= 2 \int_{-\infty}^{x} f(s, 2x - s)\, ds, \\ &= 2n(n-1) \int_{-\infty}^{x} [F(2x - s) - F(s)]^{n-2} f(s) f(2x - s)\, ds. \end{aligned}$$

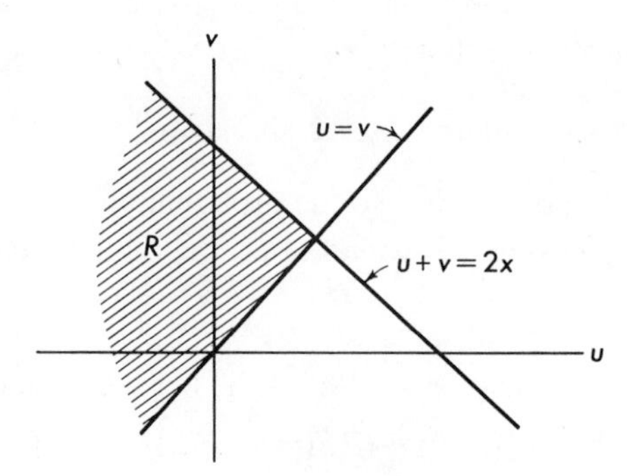

Figure 4–3

EXAMPLE 4–9. Let X be uniform on [0, 1]. For $x < 0$ the $f(s)$ in the above integral formula for $f_A(x)$ vanishes, since $s < x < 0$. For $x > 1$, $2x - s > x > 1$ (for $s < x$) and, therefore, the factor $f(2x - s)$ vanishes. For $0 < x < 1$, it is quickly discovered that different calculations are required in the left and right halves of the interval.

For $0 < x < \frac{1}{2}$, one has $0 < s < x < \frac{1}{2}$ and also $0 < x < 2x - s < 1 - s < 1$. Thus, both s and $2x - s$ are within [0, 1], where the density of X is 1, and the c.d.f. is x. Consequently, for $0 < x < \frac{1}{2}$,

$$f_A(x) = 2n(n - 1) \int_0^x [(2x - s) - s]^{n-2}\, ds = n2^{n-1}x^{n-1}.$$

On the interval $\frac{1}{2} < x < 1$, though, one finds $2x - 1 > 0$; but then since $f(2x - s)$ is zero for $2x - s > 1$ or $2x - 1 > s$, it follows that the s-integration really starts at $2x - 1$ instead of at 0. Further, on this reduced interval $(2x - 1, x)$ over which the integral is taken, one has $1 > 2x - s > x > \frac{1}{2}$ and $0 < 2x - 1 < s < x < 1$ so that both s and $2x - s$ are on [0, 1), where the density is 1 and the c.d.f. is x:

$$f_A(x) = 2n(n - 1) \int_{2x-1}^x [(2x - s) - s]^{n-2} ds = n2^{n-1}(1 - x)^{n-1}.$$

This is the reflection about $x = \frac{1}{2}$ of the earlier formula for the left half of the interval. That is, $f_A(x)$ is symmetric about $x = \frac{1}{2}$, which should certainly not offend one's intuition.

The sample range, $R = X_{(n)} - X_{(1)}$, can also be treated using the joint distribution of the smallest and largest observations. Its distribution function is

$$F_R(r) = P[X_{(n)} - X_{(1)} \le r] = \iint_S f(u, v)\, du\, dv,$$

where $f(u, v)$ is the joint density as before, and S is the region in the uv plane defined by $0 < v - u < r$. Then

$$F_R(r) = \int_{-\infty}^{\infty} \int_u^{u+r} f(u, v)\, dv\, du,$$

and the density function is the derivative of this with respect to r:

$$f_R(r) = \int_{-\infty}^{\infty} f(u, u + r)\, du$$

$$= n(n-1) \int_{-\infty}^{\infty} [F(u + r) - F(u)]^{n-2} f(u) f(u + r)\, du,$$

which is valid for $r > 0$. If $r < 0$, both $F_R(r)$ and $f_R(r)$ are 0.

EXAMPLE 4–10. Consider the uniform distribution on $[a, b]$. Since the density function vanishes outside $[a, b]$, the product $f(u)f(u + r)$ vanishes outside $[a, b - r]$. On this latter interval both $f(u)$ and $f(u + r)$ are equal to $1/(b - a)$, and

$$F(u + r) - F(u) = \int_{u}^{u+r} \frac{1}{b - a}\, du = \frac{r}{b - a}.$$

Hence,

$$f_R(r) = \frac{n(n-1)}{(b - a)^n} r^{n-2}(b - r - a), \qquad \text{for } 0 < r < b - a.$$

4.3.4 Sample Percentiles

Unless np is an integer, there is a unique 100pth percentile of the sample distribution function of a sample of size n. It is, to be specific, the kth smallest observation $X_{(k)}$, where $k = [np] + 1$, the quantity $[np]$ denoting the greatest integer not exceeding np. If np is an integer, the 100pth percentile is not uniquely defined and can be taken to be any value between $X_{(np)}$ and $X_{(np+1)}$. When the percentile is uniquely defined, it is one of the ordered observations whose density function has been given in Section 4.3.2. in terms of the population density. It can be shown that the *asymptotic* distribution is normal with mean x_p, the corresponding population percentile, and variance

$$\frac{1}{[f(x_p)]^2} \frac{p(1 - p)}{n},$$

where $f(x)$ is the population density function,[7] assumed to have a continuous derivative in the neighborhood of x_p.

EXAMPLE 4–11. The median of a sample is the sample 50th percentile. The asymptotic normal distribution of the sample median then has $x_{.5}$, the population median, as its expected value, and the variance is $[4nf^2(x_{.5})]^{-1}$. If the population is normal (μ, σ^2), the sample median is asymptotically normal $(\mu, \pi\sigma^2/2n)$, since then

$$f(x_{.5}) = f(\mu) = \frac{1}{\sqrt{2\pi}\,\sigma} \exp\left[-\frac{1}{2\sigma^2}(\mu - \mu^2)\right] = \frac{1}{\sqrt{2\pi}\,\sigma}.$$

[7] See H. Cramér, [5], pp. 367 ff.

4.3.5 The Monte Carlo Method

One of the challenges in developing any statistical procedure is that of determining the distribution of such statistics as might be proposed for use in inference, under the various admissible assumptions about the state of nature. The following example illustrates the empirical approach (sometimes called a "Monte Carlo" method) to a sampling distribution, an approach that is often resorted to when the ideal or mathematical sampling distribution seems intractible.

A computer can be programmed to yield a sequence of numbers that, for practical purposes, are independent observations on a random variable whose distribution is uniform on (0, 1). This was done to obtain 400 samples of size 5 from this uniform population. For each sample the mean, standard deviation, and range were computed, resulting in a "sample" of 400 sample means, one of 400 sample standard deviations, and one of 400 sample ranges. These were recorded in frequency distributions and plotted as histograms in Figures 4–4, 4–5, and 4–6. The actual distributions have smooth density functions, and similar plots for ever-larger numbers of samples would tend toward the ideal densities. (Smooth curves suggested by the histograms have been superimposed on them to show what the ideal is apt to be like.) The true or ideal sampling distribution of the range was derived in Section 4.3.3; the sampling distribution of the mean can be derived using characteristic functions; and that of the standard deviation would seem to defy any but a numerical approach.

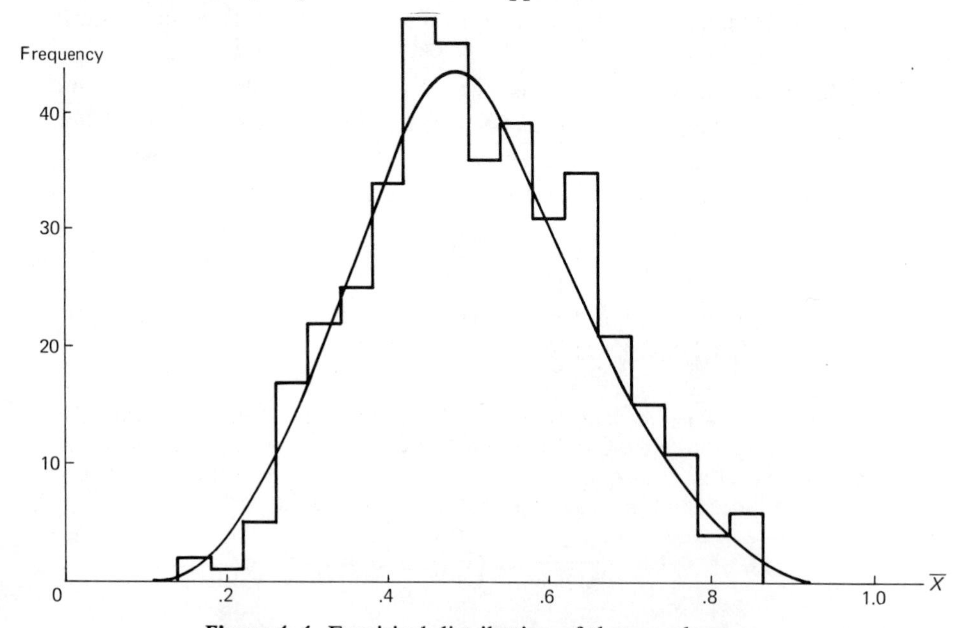

Figure 4–4. Empirical distribution of the sample mean.

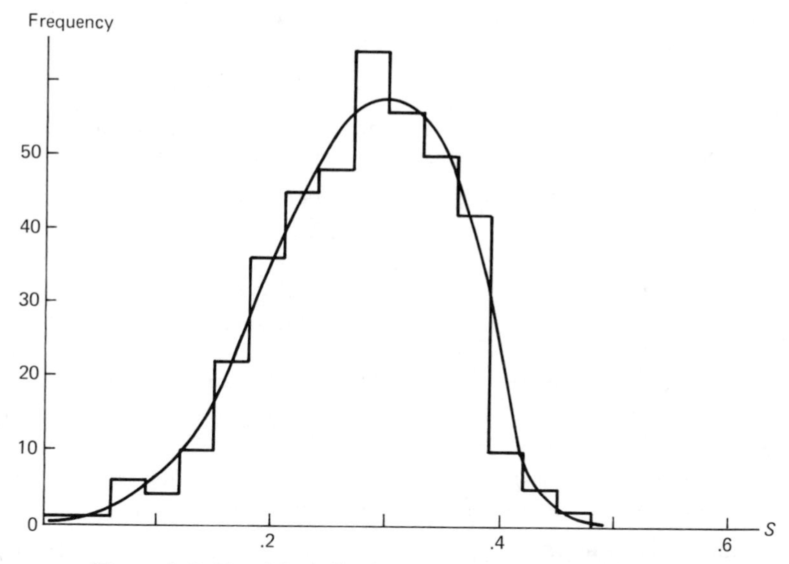

Figure 4–5. Empirical distribution of the standard deviation.

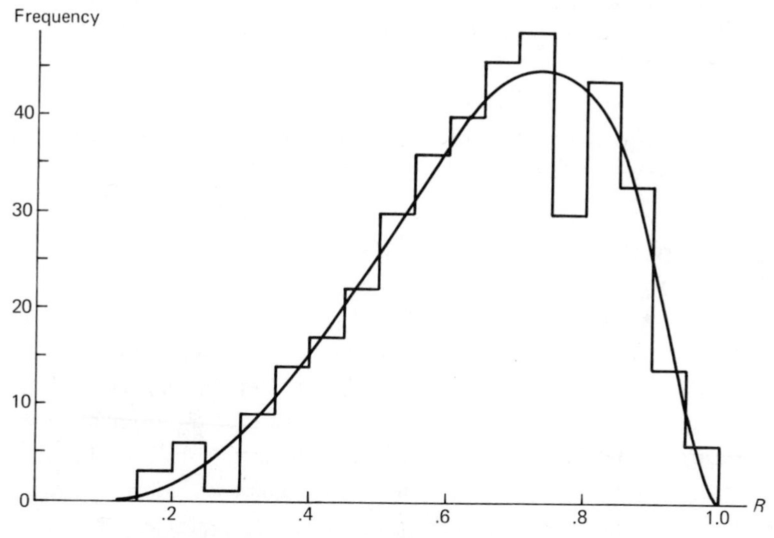

Figure 4–6. Empirical distribution of the range.

Problems

4–15. Compute the mean and variance of the smallest and of the largest observations of a random sample from a uniform distribution on [0, 1].

[*Hint:* The moments for one can be obtained from those of the other by exploiting the symmetry.]

4–16. Determine the mean and variance of the largest observation from a uniform distribution on $[a, b]$.

[*Hint:* Make a linear transformation and use the results from Problem 4–15.]

4–17. Determine the mean and variance of the midrange of a random sample from a uniform distribution on $\theta - \frac{1}{2} < x < \theta + \frac{1}{2}$, in terms of the parameter θ.

[*Hint:* The population distribution is that of Example 4–9 except for a shift in location.]

4–18. Identical units, each with time to failure having the c.d.f. $F(x)$ and reliability function $R(x) = 1 - F(x)$ (see Example 2–13) are operated in a system so that the system fails if and only if all units fail. (The units are said to be in *parallel*.) Show that the time to failure of the system is the maximum of the times to failure of the individual units, and so obtain a formula for system reliability in terms of $R(x)$. Obtain also a corresponding formula for the case in which the individual units have possibly different life characteristics—that is, the reliability of the ith unit is $R_i(x)$.

4–19. Paralleling Problem 4–18, study the case of units in *series*—that is, units operated in a system so that the system fails as soon as one unit fails.

4–20. Compute the mean and variance of the range of a random sample from a uniform population on (a, b).

4–21. Derive the joint density function of $X_{(1)}$ and $X_{(2)}$ in the case of a random sample from a population with c.d.f. $F(x)$, along the lines of the derivation for $(X_{(1)}, X_{(n)})$ given in the text.

[*Hint:* Decompose the event $X_{(1)} \leq u$.]

4–22. Determine the asymptotic distribution of the median of a sample random sample from a Cauchy distribution with location parameter θ.

4.4 Sufficient Statistics

The notion of sufficiency was introduced, rather imprecisely, in an earlier discussion of the reduction of data. It was pointed out that reducing data may involve a loss of information pertinent to the problem at hand, and a statistic to be used in inference should not entail a reduction of the original sample of such a nature or degree as to be no longer sufficient to handle the problem. Although the notion will now be made more precise, a rigorous treatment of sufficiency in the general case requires a rigorous treatment of conditional probability as a basis. Not having this, one can only reason intuitively and simply state results (except in the discrete case, in which it is possible to be quite precise and rigorous with the background assumed).

4.4.1 Statistics and Partitions

In Section 1.1.5 the notion of a *partition* as induced by a function on a sample space was introduced. A statistic, being a function on the space of values of the data Z (whether this be a single random variable or a vector of observations as in a random sample), defines a partition of that space—into mutually disjoint "partition sets" whose union is the whole space. A *partition set* is determined and identified by a particular value of the function, or in the present case, by a value of the statistic, as the set of points on which the function assumes that value. Distinct values define distinct partition sets.

Two statistics $T_1 = g(Z)$ and $T_2 = h(Z)$ that define the same partition are in 1–1 correspondence—there is a function $T_1 = s(T_2)$ with a unique inverse, $T_2 = s^{-1}(T_1)$. This function may not be easy to write down, but it does exist; for, given any value of T_2, one can go to the corresponding partition set and obtain as the value of $s(T_2)$ the T_1-value that defines that partition set.

Given a partition, any assignment of a number to each set such that no two sets have the same number assigned defines a statistic. For, that number is simply assigned to each point of the set it identifies.

EXAMPLE 4–12. Consider a random variable Z with four possible values, z_1, z_2, z_3, and z_4. The statistic T_1 defined by

$$T_1 = \begin{cases} 0, & \text{if } Z = z_1, \text{ or } z_2, \\ 1, & \text{if } Z = z_3, \\ 2, & \text{if } Z = z_4, \end{cases}$$

defines the partition of the Z-space into the sets $\{z_1, z_2\}$, $\{z_3\}$, and $\{z_4\}$. The same partition would be defined by

$$T_2 = \begin{cases} 3, & \text{if } Z = z_1 \text{ or } z_2, \\ 17, & \text{if } Z = z_3, \\ 7, & \text{if } Z = z_4, \end{cases}$$

or, indeed, by any statistic that has distinct values on the sets $\{z_1, z_2\}$, $\{z_3\}$, and $\{z_4\}$. The statistic T_1 is a function of T_2, and vice versa, with the following correspondences between values of one and values of the other: $0 \leftrightarrow 3$, $1 \leftrightarrow 17$, and $2 \leftrightarrow 7$.

A partition Π_1 is said to be a *reduction* of a partition Π_2 if each partition set of Π_1 is precisely the union of sets of Π_2. In such a case, a statistic T_1 that defines Π_1 must be a function of any statistic T_2 that defines Π_2, but not a function with a unique inverse unless the two partitions are exactly the same. That is, each value of T_2 leads to and so defines a corresponding value of T_1—a value of T_2 picks out one of the partition sets of Π_2, which in turn is contained entirely in some partition set of Π_1, and is then assigned the value that T_1 has on that set. If Π_1 is a reduction of Π_2, one says that the statistic T_1 is a reduction of T_2—meaning, then, that it is a function of T_2.

If T_1 is a reduction of T_2 and T_2 is a reduction of T_1, then these statistics must define the same partition and so be related by a function that is a 1–1 correspondence. Given any two statistics (or corresponding partitions), it is not always the case that one is a reduction of the other. Thus, partitions of a given space are only partially ordered.

EXAMPLE 4–13. Continuing Example 4–12, let T_3 be defined by

$$T_3 = \begin{cases} 4, & \text{if } Z = z_1 \text{ or } z_2, \\ 5, & \text{if } Z = z_3 \text{ or } z_4. \end{cases}$$

The partition determined by T_3 consists of the sets $\{z_1, z_2\}$ and $\{z_3, z_4\}$, which is clearly a reduction of the partition defined by T_2 (or by T_1) in the preceding example, since $\{z_3, z_4\}$ is the union of $\{z_3\}$ and $\{z_4\}$. On the other hand, the statistic

$$T_4 = \begin{cases} 6, & \text{if } Z = z_1 \text{ or } z_3, \\ 8, & \text{if } Z = z_2 \text{ or } z_4, \end{cases}$$

is not a reduction of T_2—nor is T_2 a reduction of T_4.

EXAMPLE 4–14. Consider a sample (X_1, X_2, X_3) from a continuous population. The statistic $S = X_1 + X_2 + X_3$ defines a partition consisting of planes of the form $x_1 + x_2 + x_3 = \text{const.}$ Precisely the same partition is defined by the sample mean, $\bar{X} = S/3$; the $\bar{X}$-label for any set is just $\frac{1}{3}$ of the S-label for that set, but sets with distinct $\bar{X}$-labels have distinct S-labels. The statistic $T = X_1^2 + X_2^2 + X_3^2$ defines the partition whose sets are spheres centered at the origin, each sphere being labeled by the square of its radius. The statistic $(\bar{X}, T)$ defines partition sets that are loci of the simultaneous system

$$\begin{cases} x_1 + x_2 + x_3 = a, \\ x_1^2 + x_2^2 + x_3^2 = b, \end{cases}$$

for various choices of the two constants. When the two surfaces defined by these two equations (namely, a plane and a sphere) intersect, the intersection is a circle; the union of all the circles in this two-parameter (a, b) family of circles is the whole sample space.

The conditional probability of an event, given $T = t$, where T is some statistic, or function of the observations $\mathbf{X} = (X_1, \ldots, X_n)$, is the same as the conditional probability given $U = u$, where U is any statistic defining the same partition as T [so that U is some function of T, say $h(T)$, and then $u = h(t)$]. That is to say, the conditional probability is really *defined by the partition set* that is known to have "occurred," and is not dependent on the particular statistic that is used to define that partition set. The quantity $P(A \mid T = t)$ can thus be interpreted as the probability that condition A is satisfied by the sample point $\mathbf{X}$, given that $\mathbf{X}$ falls in the partition set whose T-label happens to be t.

4.4.2 Definition of Sufficiency

Before taking up the general definition, consider the following example, a discrete case that illustrates the main idea without the complications of general notation.

EXAMPLE 4–15. Consider the family of Bernoulli distributions indexed as usual by the probability p of "success" in a single trial:

$$P(X = x) = p^x(1 - p)^{1-x}, \qquad x = 0 \text{ (failure)}, 1 \text{ (success)}.$$

There are eight distinct samples of size three, and the probability function of (X_1, X_2, X_3) is

$$P(\mathbf{X} = \mathbf{x}) = P(X_1 = x_1, X_2 = x_2, X_3 = x_3) = p^k(1 - p)^{3-k},$$

where $x_1 + x_2 + x_3 = k$, the number of 1's among the values (x_1, x_2, x_3). Let the statistic T denote the number of successes in three trials: $T = X_1 + X_2 + X_3$, with probability function

$$P(T = k) = P(k \text{ successes in 3 trials}) = \binom{3}{k} p^k(1 - p)^{3-k}.$$

The *conditional* probability of obtaining (x_1, x_2, x_3), given that $T = k$, is the quotient

$$p(\mathbf{x} \mid T = k) = P(\mathbf{X} = \mathbf{x} \mid T = k) = \frac{P(\mathbf{X} = \mathbf{x} \text{ and } T = k)}{P(T = k)}.$$

The numerator is 0 unless $x_1 + x_2 + x_3 = k$, in which case the condition $T = k$ is redundant; hence,

$$p(\mathbf{x} \mid T = k) = \begin{cases} 0, & \text{if } x_1 + x_2 + x_3 \neq k, \\ \dfrac{p^k(1 - p)^{3-k}}{\binom{3}{k} p^k(1 - p)^{3-k}} = \dfrac{1}{\binom{3}{k}}, & \text{if } x_1 + x_2 + x_3 = k. \end{cases}$$

These probabilities for the various samples and conditions are itemized in the following tabulation:

Sample	T	Probability	Probability Given $T = 0$	Probability Given $T = 1$	Probability Given $T = 2$	Probability Given $T = 3$
(0, 0, 0)	0	$(1 - p)^3$	1	0	0	0
(0, 0, 1)	1	$p(1 - p)^2$	0	$\frac{1}{3}$	0	0
(0, 1, 0)	1	$p(1 - p)^2$	0	$\frac{1}{3}$	0	0
(1, 0, 0)	1	$p(1 - p)^2$	0	$\frac{1}{3}$	0	0
(0, 1, 1)	2	$p^2(1 - p)$	0	0	$\frac{1}{3}$	0
(1, 0, 1)	2	$p^2(1 - p)$	0	0	$\frac{1}{3}$	0
(1, 1, 0)	2	$p^2(1 - p)$	0	0	$\frac{1}{3}$	0
(1, 1, 1)	3	p^3	0	0	0	1

Note that the distribution of (X_1, X_2, X_3), given any particular value of T, *does not involve the parameter p.*

One can imagine that a sample point, say (1, 1, 0), is obtained in this way: First, a game is played that gives the value of T, namely, $T = 2$. Then, among the three samples for which the sample sum is 2, the particular result (1, 1, 0) comes from playing a second game—tossing a three-sided object with equal probabilities. This second game does not depend on p, and it would seem quite pointless to play the second game whose outcome depends in no way on the parameter of interest. That is, the information relevant to p in a particular sample is contained in the sample sum, and it does no good to know which particular sample with that sum is the one that was actually observed.

The statistic T is said to be *sufficient* for the family of Bernoulli distributions because knowing T is just as good as knowing the actual sample insofar as knowledge of p is concerned. But notice that the same would be true of $T/3$, or of T^2, or of any other function of T that defines the same partition. Sufficiency is basically a property of the partitioning of the space of possible samples into these four sets:

$$\{(0, 0, 0)\}$$
$$\{(0, 0, 1), (0, 1, 0), (1, 0, 0)\}$$
$$\{(0, 1, 1), (1, 0, 1), (1, 1, 0)\}$$
$$\{(1, 1, 1)\}.$$

On any one of these partition sets, T is constant—a different constant for each set; and so T serves to identify the sets. Similarly, $T/3$ would also serve to identify them. But no matter how these partitions sets are coded, the conditional distribution of X, given that the result is in a particular partition set, does not depend on p, and one is no better off knowing exactly which sample point occurred than he is knowing which partition set occurred.

Definition. *A statistic* $T = t(\mathbf{X})$ *is said to be* sufficient *for a family of distributions if and only if the conditional distribution of* $\mathbf{X}$ *given the value of* T *is the same for all members of the family.*

If the family is indexed by a (real or vector) parameter θ, the conditional distribution of $\mathbf{X}$, given the value of T, is to be independent of θ.

The statement that $T = t(\mathbf{X})$ has a certain value, say, $T = k$, is interpreted geometrically to mean that the sample point $\mathbf{X}$ lies on the partition set $t(x_1, \ldots, x_n) = k$, which is the set of *all* points for which $t(\mathbf{x}) = k$. Sufficiency of T for a family of distribution means that the conditional distribution of probability on such partition sets induced by one member of the family is the same as that induced by any other. Knowledge of $\mathbf{X}$ in more detail than just the value of $t(\mathbf{X})$ does not help in chasing down the particular member of the family that is the true state of nature. It is actually the *partition* that is sufficient, and any function $g(x)$ that has distinct values on distinct partition sets is a sufficient statistic. The most that a sample can say regarding the underlying state of nature is contained in the statement that the sample point is on some one of the partition sets, and no pertinent information is discarded in a reduction $T = t(\mathbf{x})$ that defines the partition.

The significance of a sufficient statistic T in statistical problems is that by restricting attention to procedures based on T, no information is overlooked. For, as was argued in the illustrative example, the original experiment of chance which determines a sample point $\mathbf{x}$ can be thought of as first an experiment that determines the value of T, followed by one that does not depend on which member of the family of distributions is the one that governs. That is, to determine the sample and hence any other statistic, one can first play the game to find T and then play the second game, which is the same for all states of nature considered. Thus, playing the second game, which is unrelated to the state of nature, yields no information that would permit a better inference about nature than one based only on the value of the sufficient statistic T.

EXAMPLE 4–16. Let X_1 and X_2 be independent Poisson variables, each with mean m. Their joint probability function, for nonnegative integers a and b, is

$$f(a, b) = P(X_1 = a, X_2 = b) = e^{-2m} \frac{m^{a+b}}{a!\, b!}.$$

The statistic $X_1 - X_2$ is *not* sufficient, as can be seen by calculating the conditional probability function given the value of $X_1 - X_2$. Indeed, it will suffice to show the calculation for the given value $X_1 - X_2 = 0$:

$$\begin{aligned} P(X_1 = X_2) &= f(0, 0) + f(1, 1) + f(2, 2) + \cdots \\ &= e^{-2m}\left[1 + m^2 + \frac{m^4}{(2!)^2} + \frac{m^6}{(3!)^2} + \cdots\right], \end{aligned}$$

and

$$P(X_1 = a, X_2 = b \mid X_1 = X_2) = \begin{cases} \dfrac{f(a, a)}{P(X_1 = X_2)}, & \text{if } a = b, \\ 0, & \text{if } a \neq b. \end{cases}$$

And because

$$\frac{f(a, a)}{P(X_1 = X_2)} = (\text{const})\, m^{2a} \Big/ \sum_0^{\infty} (m^{2k}/(k!)^2),$$

which surely does depend on m, the condition for sufficiency is not satisfied.

The statistic T in the definition of sufficiency can be a vector function of the observations. For instance, it might be the pair $(\bar{X}, S^2)$, or it might be the order statistic, $(X_{(1)}, \ldots, X_{(n)})$, or it could be even the original sample point itself: $(X_1, \ldots, X_n)$. With this understanding, it is clear that the sample itself is always (trivially) a sufficient statistic for any family of distributions, corresponding to the partition of the sample space into its individual points.

EXAMPLE 4–17. Consider a random sample $(X_1, \ldots, X_n)$ from a discrete population with probability function $f(x; \theta)$. The statistic

$$T = (X_1, \ldots, X_{n-1})$$

is *not* sufficient. For

$$P[\mathbf{X} = \mathbf{x} \mid T = (k_1, \ldots, k_{n-1})] = \frac{P(\mathbf{X} = \mathbf{x})}{P(T = (k_1, \ldots, k_{n-1}))} = P(X_n = x_n)$$

because the observations are independent. (This formula holds if $x_1 = k_1, \ldots, x_{n-1} = k_{n-1}$; otherwise, the probability is 0.) But this conditional probability given the value of T is just the probability function of the nth observation, $f(x_n, \theta)$, which *does* depend on θ.

Suppose that T is sufficient (for a given family of distributions), and that $T = g(U)$. Then, intuitively at least, U should be sufficient. For $g(U)$ is a reduction of U and could not add "information" about the population to what is contained in U; indeed, it would usually reduce the information available. So if U were not sufficient, a reduction $T = g(U)$ ought not to be so. And ordinarily this is the case; an argument using the factorization criterion of the next section will show it to be so in situations where the criterion is valid.

Because using a statistic that is a reduction of a given sample usually simplifies both the methodology and the theory, it is natural to wonder how much the data can be reduced without sacrificing sufficiency. A statistic is said to be *minimal sufficient* if it is sufficient and if no reduction of it is sufficient. A technique for producing minimal sufficient statistics will be given in Section 4.4.4.

Problems

4–23. Consider a statistical problem with just three states of nature, and with data Z having the following probability table:

z_i	$f(z_i \mid \theta_1)$	$f(z_i \mid \theta_2)$	$f(z_i \mid \theta_3)$
z_1	.4	.6	.2
z_2	.2	.3	.1
z_3	.4	.1	.7

Consider the partition whose partition sets are $A = \{z_3\}$ and $B = \{z_1, z_2\}$.

(a) Determine the conditional distributions in the sample space defined by $f(z_i \mid A)$ and $f(z_i \mid B)$.

(b) Construct a statistic that defines the partition A, B; is it sufficient?

(c) Consider the statistic:

$$T(Z) = \begin{cases} 0, & \text{if } Z = z_1, \\ 1, & \text{if } Z = z_2 \text{ or } z_3. \end{cases}$$

Is this sufficient?

4–24. Determine the partition of the plane defined by the order statistic for a sample (X_1, X_2), that is, by $T = (X_{(1)}, X_{(2)})$. Determine also the partition sets defined by $U = X_1 + X_2$. Is T a reduction of U, or conversely?

4–25. Describe the partitions defined by $\bar{X}$ and $\bar{X}^2$, where $\bar{X}$ is the mean of a sample of size three, and observe how one is a reduction of the other.

4–26. An urn contains five beads, marked 1, 2, 3, 4, and 5, respectively. Three are drawn, one at a time, without replacement. Let X_i be the number on the ith bead drawn.

(a) List the points in the sample space (X_1, X_2, X_3) that lead to the particular order statistic (1, 2, 5).

(b) List the 10 possible order statistic vectors and their corresponding probabilities of occurrence.

(c) Determine the partitions of the sample space for the order statistic [in (b)] corresponding to each of these statistics:

(i) $(X_{(1)}, X_{(3)})$. (ii) $X_{(1)}$. (iii) $X_{(3)}$. (iv) $X_{(3)} - X_{(1)}$.

4–27. In sampling without replacement from N objects, of which M are black and $N - M$ are white, the probability of a particular sequence of n observations was found in Problem 1–54 to be given by the formula

$$\frac{\binom{M}{k}\binom{N-M}{n-k}}{\binom{n}{k}\binom{N}{n}}, \qquad k = 0, 1, \ldots, n,$$

where k is the number of black objects in the sequence. Determine the conditional distribution in the sample space (i.e., the conditional probability of any given sample sequence) given the value of Y, the number of black objects in the sequence of n observations. Is Y sufficient? (The population parameter here is M, the numbers N and n being given constants.)

4–28. A sample (X_1, X_2) has the distribution given in the accompanying table. The parameter M can take on any of the values 0, 1, 2, 3, 4.

Sample	Probability
(0, 0)	$(12 - 7M + M^2)/12$
(0, 1)	$(4M - M^2)/12$
(1, 0)	$(4M - M^2)/12$
(1, 1)	$(M^2 - M)/12$

(a) Is (X_1, X_2) a *random* sample?
(b) Is $X_1 + X_2$ sufficient?
(c) Is X_1X_2 sufficient?

4–29. Given a random sample $(X_1, \ldots, X_n)$ from a Poisson population with mean m, show that $X_1 + \cdots + X_n$ is sufficient, but that X_1 is not.

4–30. Let (X_1, X_2) be a random sample from a discrete distribution that is uniform on the integers $1, 2, \ldots, N$ (where N is a parameter). Calculate the conditional probabilities in the sample space given $X_1 + X_2 = 4$ (say), and conclude from your result that the sum is not sufficient for N.

4.4.3 The Factorization Criterion

To show that a statistic is sufficient, it is often easier, rather than to determine the conditional distribution in the sample space given the value of the statistic, to use a necessary and sufficient condition concerning the form of the density or probability function of the observations.

The joint density functions of the observations $(X_1, \ldots, X_n)$ in a sample, or their joint probability function in the discrete case, will be said to satisfy the *factorization criterion* in terms of a statistic $T = t(\mathbf{X})$ if and only if it can be expressed in the form

$$f(\mathbf{x}; \theta) = g(t(\mathbf{x}), \theta)h(\mathbf{x}),$$

the product of a factor that depends on $\mathbf{x}$ only through the value of $t(\mathbf{x})$ and a factor independent of the parameter θ. It is fairly straightforward to show in the discrete case that the factorization criterion is necessary and sufficient for $T = t(\mathbf{X})$ to be sufficient for θ. This will now be done.

Suppose, then, that $f(\mathbf{x}; \theta) = P_\theta(\mathbf{X} = \mathbf{x})$, and assume first that the factorization criterion is satisfied:

$$P_\theta(\mathbf{X} = \mathbf{x}) = g[t(\mathbf{x}), \theta]h(\mathbf{x}).$$

The probability function of $T = t(\mathbf{x})$ can be computed by summing over those $\mathbf{x}$ for which $t(\mathbf{x}) = t_0$:

$$\begin{aligned} P_\theta[T = t_0] = \sum_{t(\mathbf{x})=t_0} P_\theta(\mathbf{X} = \mathbf{x}) &= \sum_{t(\mathbf{x})=t_0} g[t(\mathbf{x}), \theta]h(\mathbf{x}) \\ &= g[t_0, \theta] \sum_{t(\mathbf{x})=t_0} h(\mathbf{x}). \end{aligned}$$

The conditional distribution of $\mathbf{X}$, given $T = t_0$, is then

$$P_\theta(\mathbf{X} = \mathbf{x} \mid T = t_0) = \frac{P_\theta(\mathbf{X} = \mathbf{x} \text{ and } T = t_0)}{P_\theta(T = t_0)} = \begin{cases} 0, & \text{if } t(\mathbf{x}) \neq t_0, \\ \dfrac{P_\theta(\mathbf{X} = \mathbf{x})}{P_\theta(T = t_0)}, & \text{if } t(\mathbf{x}) = t_0, \end{cases}$$

where, if $t(\mathbf{x}) = t_0$,

$$\frac{P_\theta(\mathbf{X} = \mathbf{x})}{P_\theta(T = t)} = \frac{g[t_0, \theta]h(\mathbf{x})}{g[t_0, \theta] \sum_{t(\mathbf{x})=t_0} h(\mathbf{x})},$$

which is independent of θ. Hence, T is sufficient.

Conversely, suppose the conditional probability function, given $T = t_0$, to be independent of θ; let it be denoted by $c(\mathbf{x}, t_0)$:

$$P(\mathbf{X} = \mathbf{x} \mid T = t_0) \equiv c(\mathbf{x}, t_0).$$

[It can be assumed that $T = t_0$ has positive probability; for at any point $\mathbf{x}$ such that $P[t(\mathbf{X}) = t(\mathbf{x})] = 0$ one would have

$$P(\mathbf{X} = \mathbf{x}) \leq P[t(\mathbf{X}) = t(\mathbf{x})] = 0,$$

and the factorization is trivial: $0 = 0$.] At any given sample point $\mathbf{x}$, let $t(\mathbf{x}) = t_0$; then adding the consistent restriction $t(\mathbf{X}) = t_0$ does not alter the event $\mathbf{X} = \mathbf{x}$:

$$\begin{aligned} P_\theta(\mathbf{X} = \mathbf{x}) &= P_\theta[\mathbf{X} = \mathbf{x}, t(\mathbf{X}) = t_0] \\ &= P[\mathbf{X} = \mathbf{x} \mid t(\mathbf{X}) = t_0]P_\theta[t(\mathbf{X}) = t_0] \\ &= c(\mathbf{x}, t_0)P_\theta[t(\mathbf{X}) = t(\mathbf{x})]. \end{aligned}$$

This is the desired factorization, since $P_\theta[t(\mathbf{X}) = t(\mathbf{x})]$ depends on $\mathbf{x}$ only through the value of $t(\mathbf{x})$.

The factorization criterion is often used in the following way: A joint density function $f(\mathbf{x}; \theta)$ is inspected to determine whether there is any factorization of the type required, in terms of some function $t(\mathbf{x})$. If there is, then $T = t(\mathbf{X})$ is a sufficient statistic.

EXAMPLE 4–18. Consider a random sample $(X_1, \ldots, X_n)$, where each observation has the geometric distribution, defined (for $0 < p < 1$) by

$$f(x; p) = p^x(1 - p), \qquad x = 0, 1, \ldots.$$

The joint probability function of the observations is

$$f(\mathbf{x}; p) = \prod_1^n p^{x_i}(1 - p) = (1 - p)^n p^{\Sigma x_i} = g\left(\sum x_i, p\right)h(x),$$

where $h(\mathbf{x}) = 1$ and

$$g(t, p) = (1 - p)^n p^t.$$

Because $f(\mathbf{x}; p)$ factors in this way, the statistic $\sum X_i$ is sufficient.

The proof that the factorization criterion can be used to establish sufficiency in the general case is more involved, and conditions of regularity are needed that will not be given here. Various theorems have been given (by Fisher, Neyman, Halmos and Savage, and Bahadur—see Lehmann [16], p. 19 or Zacks [23], pp. 44 ff.), and these cover all the cases that will be encountered here. The essential idea of these proofs is similar to that of the discrete case, except that the conditional distribution is not so easy to compute.

EXAMPLE 4–19. Consider a problem with two states of nature, θ_0 and θ_1, and with data $\mathbf{X}$. Let $f_0(\mathbf{x})$ and $f_1(\mathbf{x})$ be the density function of $\mathbf{X}$ under θ_0 and θ_1, respectively. (In the discrete case the f's would be probability functions.) Then, setting

$$g(\lambda, \theta) = \begin{cases} \sqrt{\lambda}, & \text{if } \theta = \theta_0, \\ \dfrac{1}{\sqrt{\lambda}}, & \text{if } \theta = \theta_1, \end{cases}$$

and

$$h(\mathbf{x}) = [f_0(\mathbf{x})f_1(\mathbf{x})]^{1/2},$$

one finds that

$$T \equiv t(\mathbf{X}) = \frac{f_0(\mathbf{X})}{f_1(\mathbf{X})}$$

is sufficient, by the factorization criterion. For,

$$g(t(\mathbf{x}), \theta)h(\mathbf{x}) = \begin{cases} [f_0(\mathbf{x})f_1(\mathbf{x})]^{1/2}\left(\dfrac{f_0(\mathbf{x})}{f_1(\mathbf{x})}\right)^{1/2} = f_0(\mathbf{x}), & \text{if } \theta = \theta_0, \\ [f_0(\mathbf{x})f_1(\mathbf{x})]^{1/2}\left(\dfrac{f_1(\mathbf{x})}{f_0(\mathbf{x})}\right)^{1/2} = f_1(\mathbf{x}), & \text{if } \theta = \theta_1, \end{cases}$$

which is precisely the density of the data, for each state. The statistic T whose sufficiency is herewith established is called the *likelihood ratio* statistic.

EXAMPLE 4–20. Consider a uniform population on $(0, \theta)$. A random sample of size n has density

$$f(\mathbf{x}; \theta) = \begin{cases} \dfrac{1}{\theta^n}, & \text{if } 0 < x_i < \theta \quad \text{for} \quad i = 1, \ldots, n, \\ 0, & \text{if } \theta \le x_i \quad \text{or} \quad x_i < 0 \quad \text{for any } i. \end{cases}$$

This can be written in the form

$$f(\mathbf{x}; \theta) = g(x_{(n)}, \theta)h(\mathbf{x}),$$

where

$$g(u, \theta) = \begin{cases} \dfrac{1}{\theta^n}, & \text{if } \theta > u, \\ 0, & \text{if } \theta \le u, \end{cases}$$

and

$$h(\mathbf{x}) = \begin{cases} 1, & \text{if } x_i > 0 \text{ for all } i, \\ 0, & \text{if any } x_i \le 0. \end{cases}$$

Thus, the factorization criterion is satisfied in terms of the statistic $T = X_{(n)}$, the largest of the n observations. Therefore, this statistic is sufficient for θ. [This is a case in which the interval of positive density depends on the parameter θ. It is possible to formulate a factorization theorem to include it, although it is not covered by the simpler versions sometimes given. On the other hand, it can be shown that the conditional distribution given the value of $X_{(n)}$ is independent of θ; the calculation will be suggested in Problem 11–4.]

It is usually not easy to use the factorization criterion to show that a statistic T is *not* sufficient, for this would mean showing that $f(\mathbf{x}; \theta)$ cannot be factored in the right way. (As was done in the previous examples, the insufficiency of T can be shown by noting that the conditional distribution of $\mathbf{X}$ given T depends on θ, when this conditional distribution can be computed.)

The factorization criterion shows easily that if $T = c(U)$ and T is sufficient, then U is sufficient. For, sufficiency of T implies

$$f(\mathbf{X}; \theta) = g(T, \theta)h(\mathbf{X}) = g(c(U), \theta)h(\mathbf{X}) = \tilde{g}(U, \theta)h(\mathbf{X}),$$

which shows the sufficiency of U.

In Example 4–17, it was seen that given the n observations in a random sample, the vector of $n - 1$ of them could not be sufficient for any nontrivial family of distributions. It is now apparent that no function of those $n - 1$ observations could be sufficient, since it would be a further reduction of an already insufficient statistic.

4.4.4 Constructing Minimal Sufficient Statistics

A statistic is said to be *minimal sufficient* if it is sufficient and if any reduction of the partition of the sample space defined by it is not sufficient. A technique for obtaining a minimal sufficient partition has been devised by Lehmann and Scheffé[8] and will be presented next. Once the partition is obtained, of course, a minimal sufficient statistic can be defined by assigning distinct numbers to distinct partition sets.

In constructing sets of a partition that is to be sufficient for the family of sample densities (or probability functions) $f(\mathbf{x}; \theta)$, for θ in some parameter space Θ, there is associated with each point $\mathbf{x}$ in the sample space a *set* $D(\mathbf{x})$:

$$D(\mathbf{x}) = \{\mathbf{y} \mid f(\mathbf{y}; \theta) = k(\mathbf{y}, \mathbf{x}) f(\mathbf{x}; \theta)\},$$

where $k(\mathbf{y}, \mathbf{x})$ is not 0 and is not dependent on θ. That is, $D(\mathbf{x})$ is the set of all sample points $\mathbf{y}$ "equivalent" to $\mathbf{x}$ in the sense that the ratio of the densities at $\mathbf{y}$ and at $\mathbf{x}$ does not involve the state of nature. The reason for writing the definition in terms of a product rather than a ratio is to take into account the points for which $f(\mathbf{x}; \theta)$ is 0. Indeed, all points such that $f(\mathbf{x}; \theta) = 0$ for all θ will be equivalent, and so lie in the same "D"; call it D_0. Every $\mathbf{x}$ will lie in some D, namely, in $D(\mathbf{x})$, and there is no overlapping of the D's—so that they do indeed constitute a partition of the sample space. For if two D's, say $D(\mathbf{x})$ and $D(\mathbf{y})$, have a point $\mathbf{z}$ in common, then $\mathbf{z}$ is equivalent to both $\mathbf{x}$ and $\mathbf{y}$, which are then equivalent to each other and define the same D. (The "equivalence" of sample points being used here is an equivalence relation in the mathematical sense, being reflexive, symmetric, and transitive. The partition sets are then *equivalence classes* under this relation.)

Defining the D's thus defines a partition of the sample space, and this partition is asserted to be the minimal sufficient partition. A rigorous proof of this assertion requires measure theory and will not be attempted. However, the basic idea of the proof is not difficult. It involves showing (a) that the partition is sufficient, and (b) that any reduction of it is not sufficient.

To show sufficiency, choose for each set D of the partition a representative point $\mathbf{x}_D$. Let $G(\mathbf{x})$ denote the mappings from a given point $\mathbf{x}$ to the set $D(\mathbf{x})$ in which it lies and then to the representative point $\mathbf{x}_D$ of that set. That is, $G(\mathbf{X})$ is a statistic and defines the partition. Now, for any partition set D except D_0, and for any $\mathbf{x}$ in D,

$$f(\mathbf{x}; \theta) = k(\mathbf{x}, \mathbf{x}_D) f(\mathbf{x}_D; \theta) = k[\mathbf{x}, G(\mathbf{x})] f[G(\mathbf{x}); \theta].$$

But then for all $\mathbf{x}$,

$$f(\mathbf{x}; \theta) = h(\mathbf{x}) g[G(\mathbf{x}), \theta],$$

where

$$h(\mathbf{x}) = \begin{cases} 0, & \text{if } \mathbf{x} \text{ is in } D_0, \\ k[\mathbf{x}, G(\mathbf{x})], & \text{otherwise,} \end{cases}$$

[8] E. Lehmann and H. Scheffé, "Completeness, Similar Regions, and Unbiased Estimation," in *Sankhya* **10**, 327 ff. (1950).

and

$$g[G(\mathbf{x}), \theta] = f[G(\mathbf{x}); \theta].$$

The function $h(\mathbf{x})$ does not involve θ, and the function $g[G(\mathbf{x}), \theta]$ depends on $\mathbf{x}$ only through the values of the function $G(\mathbf{x})$. Thus $G(\mathbf{X})$ is sufficient, as is the given partition, which it defines.

To see the minimality of the sufficient statistic $G(\mathbf{X})$ and the corresponding partition, consider any other sufficient statistic $t(\mathbf{X})$ with corresponding partition sets E. The minimality follows when it is shown that each set E is contained in some D of the constructed partition (except possibly for points in a set of probability 0). Let $\mathbf{x}$ and $\mathbf{y}$ be points in E so that $t(\mathbf{x}) = t(\mathbf{y})$. Since $t(\mathbf{X})$ is sufficient, the joint density at $\mathbf{x}$ can be factored in the form

$$f(\mathbf{x}; \theta) = r(\mathbf{x})s[t(\mathbf{x}), \theta] = r(\mathbf{x})s[t(\mathbf{y}), \theta],$$

and, likewise, at $\mathbf{y}$:

$$f(\mathbf{y}; \theta) = r(\mathbf{y})s[t(\mathbf{y}), \theta].$$

If $r(\mathbf{x}) \neq 0$,

$$f(\mathbf{y}; \theta) = r(\mathbf{y})\frac{f(\mathbf{x}; \theta)}{r(\mathbf{x})} = k(\mathbf{y}, \mathbf{x})f(\mathbf{x}; \theta),$$

where

$$k(\mathbf{y}, \mathbf{x}) = \frac{r(\mathbf{y})}{r(\mathbf{x})}$$

is not zero if $r(\mathbf{y})$ is not zero. Hence, if $r(\mathbf{y}) \neq 0$, $\mathbf{x}$ and $\mathbf{y}$ belong to the same D. Thus, all of E is contained in D, except possibly for those points $\mathbf{x}$ such that $r(\mathbf{x}) = 0$; but for such points, $f(\mathbf{x}; \theta) = 0$ for all θ, and the totality of all such points has probability 0.

EXAMPLE 4–21. Consider the Bernoulli family with probability function

$$f(x; p) = p^x(1 - p)^{1-x}, \qquad x = 0, 1.$$

For a sample $\mathbf{X}$ of independent observations, the joint probability function at $\mathbf{x}$ is

$$f(\mathbf{x}; p) = p^{\Sigma x_i}(1 - p)^{n - \Sigma x_i},$$

and at the point $\mathbf{y}$ it is

$$f(\mathbf{y}; p) = p^{\Sigma y_i}(1 - p)^{n - \Sigma y_i}.$$

The ratio is

$$\frac{f(\mathbf{x}; p)}{f(\mathbf{y}; p)} = \left(\frac{p}{1 - p}\right)^{\Sigma x_i - \Sigma y_i},$$

which is independent of p if and only if $\sum x_i = \sum y_i$. Thus, points whose coordinates have the same sum lie in the same set of the minimal sufficient partition. The sum $X_1 + \cdots + X_n$ is, therefore, a minimal sufficient statistic.

EXAMPLE 4–22. Consider the random sample $\mathbf{X}$, where each observation is normal (μ, σ^2). The ratio of the joint density function at $\mathbf{x}$ to its value at a point $\mathbf{y}$ is

$$\frac{f(\mathbf{x}; \mu, \sigma^2)}{f(\mathbf{y}; \mu, \sigma^2)} = \exp\left\{-\frac{1}{2\sigma^2}\left[\sum x_i^2 - \sum y_i^2 - 2\mu\left(\sum x_i - \sum y_i\right)\right]\right\}.$$

This ratio is independent of the parameters (μ, σ^2) if and only if *both* $\sum x_i^2 = \sum y_i^2$ and $\sum x_i = \sum y_i$. Therefore, $(\sum X_i^2, \sum X_i)$ is a minimal sufficient statistic for the normal family. Since these uniquely determine $\bar{X}$ and S^2 and conversely, the pair $(\bar{X}, S^2)$ is also minimal sufficient.

In each of the above examples, there is a minimal sufficient statistic of the same dimension as the parameter indexing the family of states. This is not always the case; in the case of a Cauchy family, for instance, the minimal sufficient statistic is the order statistic, and any further reduction sacrifices sufficiency, as will be shown next.

EXAMPLE 4–23. Let $\mathbf{X}$ be a random sample from a population defined by the Cauchy density with location parameter θ:

$$f(x; \theta) = \frac{1/\pi}{1 + (x - \theta)^2}.$$

Two sample points $\mathbf{x}$ and $\mathbf{y}$ will lie on the same partition set of the minimal sufficient partition if and only if the ratio

$$\frac{f(\mathbf{x}; \theta)}{f(\mathbf{y}; \theta)} = \prod_{j=1}^{n} \frac{1 + (y_j - \theta)^2}{1 + (x_j - \theta)^2}$$

is independent of θ. The numerator is a polynomial of degree $2n$ in θ, as is the denominator; and to say that the ratio does not depend on θ is to say that the polynomials are identical (the leading coefficients being equal). This in turn means that the set of zeros of the numerator polynomial, $y_j \pm i$ (for $j = 1, \ldots, n$), is the same as the set of zeros of the denominator polynomial, $x_j \pm i$ (again for $j = 1, \ldots, n$). This is true if and only if the real numbers $(x_1, \ldots, x_n)$ are a permutation of the numbers $(y_1, \ldots, y_n)$. A partition set of the minimal sufficient partition, therefore, consists of the $n!$ permutations of n real numbers; this partition is defined by the order statistic, $(X_{(1)}, \ldots, X_{(n)})$.

4.4.5 The Exponential Family

The *exponential family* of distributions (introduced in Section 3.5) consists of those distributions with densities or probability functions expressible in the following form:

$$f(x; \theta) = B(\theta)h(x) \exp [Q(\theta)R(x)].$$

The Bernoulli, binomial, Poisson, geometric, and gamma distributions, for instance, were seen to be included in this family.

A minimal sufficient statistic for the family is found by using the technique given in Section 4.4.4. The joint density function (or probability function) for a random sample **X** is

$$f(\mathbf{x}; \theta) = B^n(\theta) \exp\left[Q(\theta) \sum R(x_i)\right] \prod h(x_i).$$

The ratio of this density at **x** to its value at **y** is

$$\frac{f(\mathbf{x}; \theta)}{f(\mathbf{y}; \theta)} = \exp\left[Q(\theta)\left\{\sum R(x_i) - \sum R(y_i)\right\}\right] \prod \frac{h(x_i)}{h(y_i)}.$$

This is independent of θ if and only if $\sum R(x_i) = \sum R(y_i)$ and, therefore, $t(\mathbf{X}) = \sum R(X_i)$ is minimal sufficient.

The statistic $t(\mathbf{X})$ has itself a distribution that belongs to the exponential family. To show this in the continuous case would require transformation of a multiple integral; the verification in the discrete case is simpler and proceeds as follows. The probability function for $t(\mathbf{X})$ is

$$\begin{aligned} p(t; \theta) &= P\left(\sum R(X_i) = t\right) \\ &= \sum_{\Sigma R(x_i) = t} B^n(\theta) \exp\left[Q(\theta) \sum R(x_i)\right] \prod h(x_i) \\ &= b(\theta) H(t) \exp\left[t Q(\theta)\right], \end{aligned}$$

where $b(\theta) = B^n(\theta)$ and

$$H(t) = \sum_{\Sigma R(x_i) = t} \prod h(x_i).$$

Clearly, then, $p(t; \theta)$ belongs to the exponential family.

EXAMPLE 4–24. The number of trials preceding the first "heads" in a sequence of independent Bernoulli trials has the probability function

$$f(x; p) = p(1 - p)^x = pe^{x \log(1-p)}.$$

The statistic $X_1 + \cdots + X_n$, the sum of n independent observations, is then minimal sufficient and has a probability function that is again of the exponential family. This sum is the number of "tails" occurring before the nth "heads" and has the negative binomial distribution—which has been shown to be in the exponential family in Section 3.5.

For the case of a multidimensional parameter $\boldsymbol{\theta} = (\theta_1, \ldots, \theta_k)$, the exponential family of distributions has a density (or probability function) of the form

$$f(x; \boldsymbol{\theta}) = B(\boldsymbol{\theta}) h(x) \exp\left[Q_1(\boldsymbol{\theta}) R_1(x) + \cdots + Q_k(\boldsymbol{\theta}) R_k(x)\right].$$

Given a random sample **X** the k-dimensional statistic

$$\left[\sum R_1(X_i), \ldots, \sum R_k(X_i)\right]$$

(sums extending from $i = 1$ to $i = n$, the sample size) is a minimal sufficient statistic for the family.

It is interesting that under certain regularity assumptions, and if the set in which $f(x; \boldsymbol{\theta})$ is positive does not depend on $\boldsymbol{\theta}$, then if there exists a k-dimensional sufficient statistic on a random sample of size n from $f(x; \boldsymbol{\theta})$ with $n > k$, the distribution $f(x; \boldsymbol{\theta})$ must belong to the exponential family.[9] In this sense, the exponential family consists of those distributions for which there is a sufficient statistic of the same dimension as that of the parameter.

Problems

4–31. Use the factorization criterion to determine, in each case, a sufficient one-dimensional statistic (i.e., a single real-valued function of the observations) based on a random sample of size n.

(a) From a Bernoulli population, $f(x; p) = p^x(1 - p)^{1-x}$, $x = 0, 1$.
(b) From a geometric population, $f(x; p) = p(1 - p)^x$, $x = 0, 1, \ldots$.
(c) From an exponential population, $f(x; \lambda) = \lambda e^{-\lambda x}$, $x > 0$.

4–32. Determine a sufficient statistic for a sample obtained by sampling without replacement from a finite Bernoulli population, using the factorization criterion (see Problem 4–27).

4–33. Determine, by use of the factorization criterion, a sufficient statistic based on a random sample of size n for the Rayleigh family (Section 3.3.3), with density

$$f(z; \sigma^2) = \frac{z}{\sigma^2} \exp\left[-\frac{z^2}{2\sigma^2}\right], \qquad z > 0.$$

4–34. Use the factorization criterion to determine a two-dimensional statistic (i.e., two real-valued functions of the observations), based on a random sample of size n, that is sufficient for the normal family of distributions with parameters (μ, σ^2).

4–35. Use the factorization criterion to determine a sufficient statistic based on a random sample from the family of distributions for a random variable with a finite list of possible values: $x_1, \ldots, x_k$, with corresponding probabilities: $p_1, \ldots, p_k$. (See Section 3.4.)

4–36. Determine a minimal sufficient statistic based on a random sample from each of the following populations:

(a) Bernoulli, with parameter p.
(b) Normal with known variance.
(c) Normal with known mean.
(d) Exponential with mean $1/\lambda$.
(e) Rayleigh: $f(z; v) = \dfrac{z}{v} \exp\left\{-\dfrac{z^2}{2v}\right\}$, $z > 0$.
(f) Maxwell: $f(x; \sigma) = \dfrac{2x^2}{\pi\sigma^3} \exp\left\{-\dfrac{x^2}{2\sigma^2}\right\}$, $x > 0$.

[9] See E. Lehmann, [16], p. 51.

4–37. Obtain the minimal sufficient partition for the family of three distributions in Problem 4–23: (.4, .2, .4), (.6, .3, .1), and (.2, .1, .7) for (z_1, z_2, z_3).

4–38. Determine a minimal sufficient statistic for sampling without replacement from a finite Bernoulli population.

4–39. Show that the sufficient statistic obtained in Problem 4–35 is actually minimal sufficient.

4–40. Derive the likelihood ratio (cf. Example 4–19) as a *minimal* sufficient statistic for a problem with just two states of nature.

4–41. Let X be uniform on $\theta - \frac{1}{2} < x < \theta + \frac{1}{2}$, and consider a random sample $\mathbf{X}$ of size n. Show that $f(\mathbf{X}; \theta)$ depends on the observations only through the value of $(X_{(1)}, X_{(n)})$. Assuming that the factorization theorem applies, what conclusion can be drawn?

4.5 Likelihood

Suppose that the generation of data Z (whether Z be a single observation or a vector of observations in a sample) is described by a model given by a probability function $f(z; \theta)$ or by a density function $f(z; \theta)$. The "θ" may be a particular value of some parameter or just the name for one of the models considered admissible as a description of the experiment that yields Z. For a given model, $f(z; \theta)$ gives the probabilities of encountering the various z-values; it is a function of z, and the term *probability function* or *density function* refers to that dependence.

For a *given* data point z, the probability or density $f(z; \theta)$ is a function of θ. It gives the probability of that value z (or the probability density at z in the continuous case) under the various models as indexed by θ.

Definition. *For a given z, the* likelihood function $L(\theta)$ *is any function of θ proportional to $f(z; \theta)$.*

If it is slightly inaccurate to speak of "the" likelihood function, when it is only uniquely determined up to a constant factor (that may depend on z), let it be understood that the term refers to the equivalence class of functions proportional to $f(z; \theta)$.

EXAMPLE 4–25. Suppose there are just three competing models for a certain discrete experiment with three outcomes, as defined in the following probability table (from Problem 4–37):

	θ_1	θ_2	θ_3
z_1	.4	.6	.2
z_2	.2	.3	.1
z_3	.4	.1	.7

The columns define the three probability functions, and the rows define the three likelihood functions. Thus, for the observation $Z = z_1$ the likelihood function has just the three values .4, .6, and .2. Notice that these values do not add up to 1 (as do the probabilities in the columns); for although they are probabilities, they are probabilities of the *same* outcome, under different models. Moreover, what is significant is not the values, but that they are in the proportion 2:3:1. (Incidentally, the likelihood function for z_2 is the same as for z_1.)

EXAMPLE 4–26. The joint density function of n independent observations from a normal population with unit variance is

$$f(\mathbf{x};\mu) = (2\pi)^{-n/2}\exp\left[-\tfrac{1}{2}\sum(x_i-\mu)^2\right].$$

For given $\mathbf{x}$, this is the likelihood function:

$$\begin{aligned} L(\mu) = f(\mathbf{x};\mu) &= C\exp\left[-\tfrac{1}{2}\sum x_i^2 + n\mu\bar{x} - \tfrac{1}{2}n\mu^2\right] \\ &= C'\exp\left[-\tfrac{1}{2}n(\mu-\bar{x})^2\right]. \end{aligned}$$

As a function of μ this is similar to a normal density with mean $\bar{x}$, but it is not a density.

4.5.1 The Likelihood Principle

The likelihood function at the point z is (generally) a different function from what it is at the point z'. The notation $L(\theta)$ obscures this, and perhaps a better notation would be

$$L(\theta; z) = f(z; \theta).$$

(The argument preceding the semicolon, in each case, is the primary one, as far as the term *function* is concerned.) Thinking of the likelihood function as defined at each value z of the random variable Z, one has essentially a *random function*, $L(\theta; Z)$, that is, a function of θ determined by the outcome of the experiment Z.

In a generalized sense, then, for a sample $\mathbf{X}$ the likelihood function is a "statistic";[10] and as such, it is minimal sufficient. Indeed, the essence of the construction of a minimal sufficient partition is that two points $\mathbf{x}$ and $\mathbf{x}'$ are in the same partition set if the likelihood functions at $\mathbf{x}$ and $\mathbf{x}'$ are proportional. In view of this, it is not unreasonable to require that two observed results $\mathbf{x}$ and $\mathbf{x}'$ should impel the same statistical inference if the likelihood functions at these points are the same.

A somewhat stronger version of such a requirement is espoused by some statisticians:

[10] This may seem to violate the earlier warning that a statistic should not depend on a population parameter; but it is an entire *function* that is assigned, by $L(\mathbf{X}; \theta)$, to the outcome $\mathbf{X} = \mathbf{x}$, and that function is computable from the sample alone.

Likelihood Principle. *Suppose that a sampling process A with outcome* $\mathbf{X}$ *and a sampling process B with outcome* $\mathbf{Y}$ *are governed by the same basic probability model, one of a family indexed by* θ. *If the experimental results* $\mathbf{X} = \mathbf{x}$ *and* $\mathbf{Y} = \mathbf{y}$ *are such that the likelihood functions are the same,*

$$L_A(\theta; \mathbf{x}) = L_B(\theta; \mathbf{y}),$$

then the inference based on the one experimental result should be the same as that based on the other.

This principle asserts somewhat more than what might be termed the *principle of sufficiency*: that inferences may and should be based on a sufficient statistic. It says that even if different methods are used to sample a population, one should draw the same inferences if the likelihood functions turn out to be the same. (See Cox and Hinkley, [4], p. 39.)

EXAMPLE 4–27. If a Bernoulli experiment is repeated three times in independent trials, the probability of a given $\mathbf{x}$ is

$$f(\mathbf{x}; p) = p^{x_1+x_2+x_3}(1 - p)^{3-x_1-x_2-x_3}.$$

With $\mathbf{x}$ fixed, this function of p is the likelihood function;

$$L(p) = Cp^{\Sigma x_i}(1 - p)^{3-\Sigma x_i}.$$

If a binomial experiment with $n = 3$ and this same p is performed once, the probability of the outcome k is

$$f(k; p) = \binom{3}{k} p^k(1 - p)^{3-k}.$$

If $k = x_1 + x_2 + x_3$, this defines the same likelihood function as the result (x_1, x_2, x_3) of three Bernoulli trials.

Suppose, however, that the basic Bernoulli experiment is performed in independent trials until k successes are obtained. The probability that it takes three trials is

$$f(n; p) = \binom{n-1}{k-1} p^k(1 - p)^{n-k}$$

so that for this outcome ($n = 3$), the likelihood function is again

$$L(p) = Cp^k(1 - p)^{3-k}.$$

The likelihood principle would assert that even though the sample size was determined by a rule involving the observations (which are random), the inferences should be the same so long as one observed k successes in three trials.

4.5.2 Maximum Likelihood

It is an old and recurrent notion that for given data x, the likelihood at θ is a measure of the degree to which the data support θ as the "explanation" of the data—as the model for the process that generated the data. Thus, if $L(\theta) = 2L(\theta')$, the feeling is that θ is better supported by the data than θ' because the observed outcome is twice as likely under θ as under θ'. However appealing this may sound, it should not be confused with deductive logic. If the notion is used (as it will be) to suggest methods of inference, those methods will have to be judged on the basis of their performance.

EXAMPLE 4–28. A thumbtack, when tossed in the air and allowed to fall on a flat surface, will assume one of two final positions, point up (U) or point down (D). This sample space, $\{U, D\}$, is evident upon inspection of the tack and a study of its equilibrium positions. But it is also evident that one cannot predict the outcome of a toss, and it is natural to assume a probability model of Bernoulli type—probability p for point up and probability $1 - p$ for point down. Suppose, to learn about p, the tack is tossed twice and falls both times with point down. Given the outcome (D, D), the likelihood function is

$$L(p) = (1 - p)^2.$$

Thus, the likelihood of $p = 0$ is 1, and the likelihood of $p = \frac{1}{2}$ is $\frac{1}{4}$. On the likelihood scale, then, the model $p = 0$ has (in the data) four times the support of the model $p = \frac{1}{2}$.

The notion that likelihood is a valid basis for comparing two states of nature as possible explanations of observed phenomena does turn out to be fruitful; its acceptance leads to statistical procedures that are good from various points of view. The natural extension of this notion would seem, in considering more than two values of θ, to be that the state or θ-value with *maximum* likelihood should be regarded as especially significant. This gives rise to the following.

Maximum Likelihood Principle. *A statistical inference or procedure should be consistent with the assumption that the best explanation of a set of data is provided by $\hat{\theta}$, a value of θ that maximizes the likelihood function.*

In maximizing the likelihood function, it may be that there is not a unique maximum point. But any value $\hat{\theta}$ for which $L(\hat{\theta})$ is unsurpassed in value is called a *maximum likelihood state*.

If θ is a single real parameter, it would be typical that $L(\theta)$ is differentiable and finite, so that the derivative $L'(\theta)$ vanishes at $\hat{\theta}$, or (since $\log L$ has its maxima at the same points as L) that

$$\frac{\partial}{\partial \theta} \log L(\theta) = 0.$$

This conditional equation (or the equivalent $L'(\theta) = 0$) is called the *likelihood equation*. (One reason for writing it in terms of $\log L$ is that this is often easier to differentiate.) The usual procedure for determining $\hat{\theta}$ is to solve the likelihood equation and examine its roots as possible maximizing values of θ. However, it can easily happen that the maximizing value of θ occurs as a point where $L'(\theta)$ does not exist or is not 0, and one should be on the alert for such possibilities.

EXAMPLE 4–29. In the preceding example, the likelihood of p corresponding to two successive failures in two Bernoulli trials was given as

$$L(p) = (1 - p)^2, \qquad 0 \le p \le 1.$$

If one blindly differentiates this and sets the derivative equal to 0:

$$L'(p) = 2(1 - p) = 0,$$

he obtains the result $p = 1$. This is clearly a poor explanation of two failures, and it is in fact a *minimizing* value of p. The maximum of $L(p)$ occurs, over the restricted region $0 \le p \le 1$, at $p = 0$, where the derivative is not 0. And $p = 0$ is then the maximum likelihood state.

It was pointed out that the likelihood function is a statistic, depending on the observed sample $\mathbf{X}$. Thus, a value of θ that maximizes the likelihood will depend on $\mathbf{X}$; therefore, $\hat{\theta}$ is a statistic—a function of the observations in a sample. Moreover, it is a statistic that is ordinarily a function of any sufficient statistic.

To see this, suppose that $T = t(\mathbf{X})$ is sufficient so that

$$L(\theta) = f(\mathbf{X}; \theta) = g[t(\mathbf{X}), \theta]h(\mathbf{X}).$$

Then

$$\log L(\theta) = \log g[t(\mathbf{X}), \theta] + \log h(\mathbf{X}),$$

and maximizing $\log L$ is equivalent to maximizing $\log g[t(\mathbf{X}), \theta]$. A statistic $\hat{\theta}$ that accomplishes this will, of course, depend on $\mathbf{X}$ only through the value of the sufficient statistic T, provided that there is a unique maximum. A pathological exception may occur if $\hat{\theta}$ is not unique, as in the next example.

EXAMPLE 4–30. Let $\mathbf{X}$ be a random sample from a population that is uniform on $(\theta - \frac{1}{2}, \theta + \frac{1}{2})$, as in Problem 4–41. It was found there that $(X_{(1)}, X_{(n)})$ is sufficient. The likelihood function is

$$L(\theta) = g[X_{(1)}, X_{(n)}; \theta] = \begin{cases} 1, & \text{if } X_{(n)} - \frac{1}{2} < \theta < X_{(1)} + \frac{1}{2} \\ 0, & \text{elsewhere.} \end{cases}$$

Its maximum is assumed for any value of θ between $X_{(n)} - \frac{1}{2}$ and $X_{(1)} + \frac{1}{2}$. Also, the statistic

$$(\cos^2 X_1)(X_{(n)} - \tfrac{1}{2}) + (\sin^2 X_1)(X_{(1)} + \tfrac{1}{2})$$

maximizes the likelihood function but depends on X_1 as well as on the sufficient statistic.

Suppose that a new parameter ξ is introduced to index the same family of states of nature as does θ. (For instance, in the exponential distribution, sometimes it is convenient to use the mean θ as the parameter, and sometimes $\lambda = 1/\theta$.) Then there is a one-to-one correspondence between labels—a function g with unique inverse:

$$\xi = g(\theta), \qquad \theta = g^{-1}(\xi).$$

The likelihood function of ξ will be

$$\tilde{L}(\xi) = f(x; g^{-1}(\xi)) = L(g^{-1}(\xi)).$$

And if the maximum of $L(\theta)$ is attained when $\theta = \hat{\theta}$, then the maximum of $\tilde{L}(\xi)$ is attained when $\xi = \hat{\xi} \equiv g(\hat{\theta})$. For given any ξ, with $\xi = g(\theta)$,

$$\tilde{L}(\xi) = L(g^{-1}(\xi)) = L(\theta) \leq L(\hat{\theta}) = L(g^{-1}(\hat{\xi})) = \tilde{L}(\hat{\xi}),$$

so $\tilde{L}(\xi)$ is maximized by $\hat{\xi}$. Thus, the model that maximizes the likelihood is the same no matter which of the equivalent parametrizations is used.

It is generally recognized that whereas the likelihood may be appropriate for describing the evidential meaning in a set of data with regard to the state θ, an investigator will usually have more than data on which to base his inference. He has some knowledge of the mechanics of the experiment, perhaps some past experience with it or with similar experiments, some common sense, and possibly even educated hunches that should not be ignored. One way of incorporating such "prior" information or convictions will be taken up in Chapter 8.

Problems

4–42. Write out likelihood functions for a random sample of size n from each of the following populations:

(a) Normal (μ, σ^2). (b) Poisson (m). (c) Bernoulli (p).
(d) Exponential (λ). (e) Cauchy (θ).

4–43. Let Y denote the number of defectives in a sample (without replacement) of two articles from a lot of five, an unknown number (M) of which are defective. Give the likelihood of M corresponding to each possible value of Y.

4–44. Determine the likelihood function, given a random sample of size n, if the population is uniform on $[0, \theta]$.

4–45. Suppose that experimenter A observes a Poisson process and records t_0 of elapsed time before the kth event occurs (he has set k in advance). And suppose that experimenter B observes the process, counting the number of events in a time t_0 (which he has set in advance), and records k events in that time period. Show that their likelihood functions are the same (if their t_0's and k's happen to agree).

4–46. Determine the maximum likelihood state of nature for random samples of size n from each population:

(a) Normal (μ, σ^2). (b) Poisson (θ). (c) Uniform on $[0, \theta]$.

4–47. Determine the maximum likelihood state of nature for an exponential population that is

(a) Indexed by the mean:

$$f(x; \theta) = \frac{1}{\theta} e^{-x/\theta}.$$

(b) Indexed by the reciprocal of the mean:

$$f(x; \lambda) = \lambda e^{-\lambda x}.$$

4–48. Show that if $f(x; \theta) = B(\theta)h(x)e^{Q(\theta)R(x)}$, then any solution of the likelihood equation depends on $\mathbf{X}$ (a random sample of size n) only through the value of $\sum R(X_i)$.

4.5.3 Monotone Likelihood Ratio

If likelihood is accepted as measuring the support in a sample for the state θ, then the *likelihood ratio*:

$$\Lambda \equiv \lambda(\mathbf{X}; \theta, \theta') \equiv \frac{L(\theta; \mathbf{X})}{L(\theta'; \mathbf{X})}$$

is pertinent in comparing the relative merits of two states θ and θ' in explaining the data $\mathbf{X}$. That is, large values of Λ would point to θ and small values to θ'.

Suppose now that θ is a single real parameter. Sometimes there is a statistic $T = t(\mathbf{X})$ such that for *every* pair of values θ and θ', with $\theta > \theta'$, the likelihood ratio Λ is a monotone function of T. If it is monotone increasing, then the larger the T, the larger the Λ, and so a large value of T would tend to support the larger of any two parameter values.

Definition. *A family of distributions indexed by the real parameter θ is said to have a* monotone likelihood ratio *if there is a statistic T such that for each pair of values θ and θ', with $\theta > \theta'$, the likelihood ratio $L(\theta)/L(\theta')$ is a nondecreasing function of T.*

(The "nondecreasing" could have been replaced by "monotone," but it is convenient to specify the direction of monotoneity. Since a nonincreasing function of T would be a nondecreasing function of $-T$, there is no loss of generality.)

It is often the case that the distribution of a statistic T is located by θ so that large θ-values make large T-values more probable. But in inference one must reason from an observed T-value to θ and would like to be able to infer large θ-values (say) from an observed value of T that is large; the above argument would suggest that this inference is a proper one in the case of a monotone likelihood ratio. It will be seen that the desired linking of θ- and T-values turns out to have some justification, in the monotone likelihood ratio case, when the statistical problem has a certain type of structure. This matter will be discussed further in Chapters 6 and 8.

EXAMPLE 4–31. The likelihood ratio for a random sample $\mathbf{X}$ from a normal population with unit variance is

$$\begin{aligned}\Lambda &= \frac{\exp\{-\frac{1}{2}\sum (X_i - \mu)^2\}}{\exp\{-\frac{1}{2}\sum (X_i - \mu')^2\}} \\ &= \exp\left\{-\tfrac{1}{2}\left[\sum X_i^2 - 2\mu \sum X_i + n\mu^2 - \sum X_i^2 + 2\mu' \sum X_i - n\mu'^2\right]\right\} \\ &= \exp\left[-\tfrac{n}{2}(\mu^2 - \mu'^2) + (\mu - \mu')\sum X_i\right].\end{aligned}$$

If $\mu > \mu'$, this is clearly a monotone increasing function of $\bar{X}$. It might be expected, then, that a large sample mean would point to a large population mean. (Of course, a large μ does tend to produce a large $\bar{X}$, but the inference from $\bar{X}$ to μ is not so obvious.)

EXAMPLE 4–32. If X has a distribution in the one-parameter exponential family,

$$f(x;\theta) = B(\theta)h(x)e^{Q(\theta)R(x)},$$

the likelihood ratio (for a random sample X) is

$$\Lambda = \frac{B^n(\theta)}{B^n(\theta')}\exp\left\{[Q(\theta) - Q(\theta')]\sum R(X_i)\right\}.$$

And if Q is an increasing function of θ, then Λ is an increasing function of $\sum R(X_i)$, a minimal sufficient statistic, and the family has a monotone likelihood ratio.

EXAMPLE 4–33. Consider a single observation from a Cauchy population, with

$$f(x;\theta) = \frac{1/\pi}{1 + (x - \theta)^2}.$$

The likelihood ratio is

$$\Lambda = \frac{1 + (X - \theta')^2}{1 + (X - \theta)^2},$$

which is not a monotone function of X. This is evident upon inspection of Figure 4–7. The numerator has the value d'^2 and the denominator d^2, where d' and d are the hypotenuses of right triangles with legs 1 and $(x - \theta')$ or $(x - \theta)$. For the value of x shown in the figure, $d' > d$; as x varies from $-\infty$ to ∞, the ratio starts at 1, is less than 1 for $x < (\theta + \theta')/2$, is greater than 1 for $(\theta + \theta')/2 < x$, and tends to 1 again as x becomes infinite. So, even though large θ would tend to produce a large X-value, one may run into trouble in trying to infer a large θ from an observation X that is large.

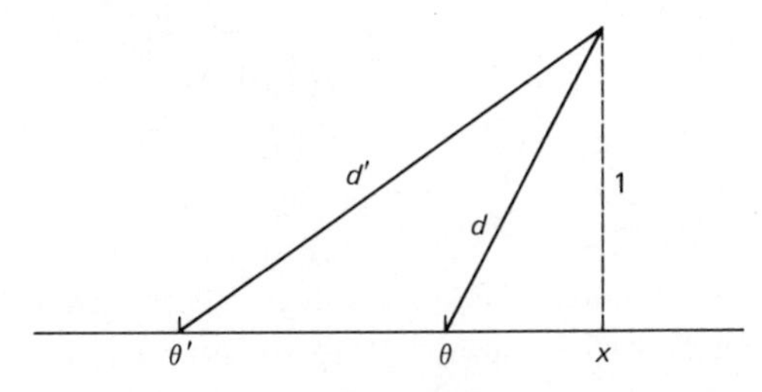

Figure 4–7

4.5.4 Information in a Sample

Again, let it be assumed that θ is a real parameter. (The notion of "information" can also be defined in the case of a vector parameter, but this will not be done here. See Cramer [5], p. 490, or Zacks [23], p. 194.)

The logarithm of the likelihood function is an additive function over independent data. That is, if $\mathbf{X}$ and $\mathbf{Y}$ are independent,

$$L(\theta; \mathbf{X}, \mathbf{Y}) = L_1(\theta; \mathbf{X})L_2(\theta; \mathbf{Y}),$$

and so

$$\log L(\theta; \mathbf{X}, \mathbf{Y}) = \log L_1(\theta; \mathbf{X}) + \log L_2(\theta; \mathbf{Y}).$$

The log-likelihood function is called by some the *support* of θ.

The unspecified constant multiplier in a likelihood function becomes an additive constant in the support. So it is not surprising that the derivative (with respect to θ) of the log-likelihood, which ignores additive constants, is a relevant quantity. It defines a (generalized) statistic called the *score*:

$$V = v(\mathbf{X}; \theta) \equiv \frac{\partial}{\partial\theta} \log L(\theta; \mathbf{X}) = \frac{L'(\theta)}{L(\theta)},$$

where the prime denotes differentiation with respect to θ. The statistic V is sufficient—knowing its value is equivalent to knowing the likelihood. [It should be mentioned that $v(\mathbf{x}; \theta)$ is defined only for $\mathbf{x}$ such that $f(\mathbf{x}; \theta) \neq 0$. Let it be assumed that V is defined to be 0 at points for which $f(\mathbf{x}; \theta) = 0$; the set of such points has probability 0.]

Some facts about V will now be quoted and demonstrated informally. They will be given as "theorems," for convenience of reference, but their statements are incomplete—omitting regularity conditions that make the assertions valid in general (cf. Zacks [23], pp. 182 ff.).

Theorem 1. *The mean score is* 0*:*

$$E(V) = E[v(\mathbf{X}; \theta)] = 0.$$

In the continuous case, this result follows, provided that one can interchange integration and differentiation (which suggests the need for some regularity conditions):

$$E(V) = \int \frac{f'(\mathbf{x}; \theta)}{f(\mathbf{x}; \theta)} f(\mathbf{x}; \theta)\, d\mathbf{x}$$

$$= \frac{d}{d\theta} \int f(\mathbf{x}; \theta)\, d\mathbf{x} = 0,$$

where again the prime denotes $d/d\theta$. (The integrals here are of the dimension of $\mathbf{X}$; the boldface $\mathbf{x}$ in $d\mathbf{x}$ calls attention to this. The range of integration is the whole $\mathbf{x}$-space except for the set of points where $f = 0$.)

EXAMPLE 4–34. Consider a population that is uniform on $0 < x < \theta$, with density

$$f(x; \theta) = \frac{1}{\theta}, \qquad \text{for } 0 < x < \theta.$$

Then

$$\frac{\partial}{\partial\theta} \log f(x; \theta) = -\frac{1}{\theta}, \qquad \text{for } 0 < x < \theta,$$

and so $E(V) \neq 0$. The given density does not satisfy the conditions permitting interchange of differentiation and integration. (The dependence of the set of x for which the density is not 0 on the parameter is a situation that usually calls for special attention.)

The variance of the score function is sometimes called the *information* (or *Fisher information*) in the experiment that yields the sample $\mathbf{X}$:

$$I_{\mathbf{X}}(\theta) = \text{var } V = E\left\{\left[\frac{\partial}{\partial\theta} \log f(\mathbf{X}; \theta)\right]^2\right\}.$$

The notation might be misleading; for a given experiment with outcome $\mathbf{X}$, the information does *not* depend on the particular observation $\mathbf{X} = \mathbf{x}$, since this variable has been averaged out. But the information in one experiment with outcome $\mathbf{X}$ will usually be different from that in another experiment with outcome $\mathbf{Y}$.

Theorem 2. *Information is additive over independent experiments. Thus, if* $\mathbf{X}$ *and* $\mathbf{Y}$ *are independent, then*

$$I_{\mathbf{X}}(\theta) + I_{\mathbf{Y}}(\theta) = I_{\mathbf{X},\mathbf{Y}}(\theta).$$

This follows because the score is additive, and because the variance of the sum of independent variables is the sum of their variances.

Corollary. *The information in a random sample of size n is just n times the information in a single observation:*

$$I_{\mathbf{X}}(\theta) = nI_{\mathbf{X}}(\theta).$$

Theorem 3. *The information provided by a* sufficient *statistic* $T = t(\mathbf{X})$ *is the same as that in the sample* $\mathbf{X}$*:*

$$I_T(\theta) = I_{\mathbf{X}}(\theta).$$

This is a consequence of the factorization theorem; the sufficiency of T implies

$$f(\mathbf{X}; \theta) = g(t(\mathbf{X}), \theta)h(\mathbf{X}),$$

where $g(t, \theta)$ is the density of T. Then, since $h' = 0$ (h is independent of θ), it follows that

$$\frac{\partial}{\partial\theta} \log f(\mathbf{X}; \theta) = \frac{\partial}{\partial\theta} \log g(T; \theta),$$

which implies the asserted equality of informations.

Theorem 4. *The information can be computed by the alternative formula:*

$$I_{\mathbf{X}}(\theta) = -E\left[\frac{\partial^2}{\partial\theta^2} \log f(\mathbf{X}; \theta)\right] = -E\left(\frac{\partial V}{\partial\theta}\right).$$

To see this, first compute (with primes meaning $\partial/\partial\theta$)

$$\begin{aligned}\frac{\partial^2}{\partial\theta^2} \log f = \frac{\partial}{\partial\theta}\left(\frac{f'}{f}\right) &= \frac{ff'' - (f')^2}{f^2} \\ &= \frac{f''}{f} - \left(\frac{f'}{f}\right)^2.\end{aligned}$$

And then, because

$$E\left(\frac{f''}{f}\right) = \int f''\, d\mathbf{x} = \frac{d^2}{d\theta^2}\int f\, d\mathbf{x} = 0,$$

the asserted equality follows upon taking expectations.

EXAMPLE 4–35. Let X have a normal distribution with mean zero and variance $\sigma^2 = \theta$:

$$f(x; \theta) = (2\pi\theta)^{-1/2} e^{-x^2/(2\theta)}.$$

Then

$$\log f = -\frac{1}{2}\log(2\pi) - \frac{1}{2}\log\theta - \frac{x^2}{2\theta}$$

and

$$V = -\frac{1}{2\theta} + \frac{X^2}{2\theta^2}.$$

The information in a single observation X is

$$I_X(\theta) = E(V^2) = \frac{E(X^2 - \theta)^2}{4\theta^4} = \frac{\operatorname{var}(X^2)}{4\theta^4} = \frac{1}{2\theta^2}.$$

Alternatively,

$$I_X(\theta) = -E\left(\frac{\partial V}{\partial\theta}\right) = E\left(\frac{X^2}{\theta^3} - \frac{1}{2\theta^2}\right) = \frac{1}{2\theta^2}.$$

The information in a random sample of size n is

$$I_{\mathbf{X}}(\theta) = nI_X(\theta) = \frac{n}{2\theta^2}.$$

The statistic $T = \sum X_i^2$ is sufficient, and T/θ has a chi-square distribution with n degrees of freedom (being the sum of squares of n independent standard normal variates). Thus,

$$f_T(t;\theta) = (\text{const})\,\frac{1}{\theta}\left(\frac{t}{\theta}\right)^{n/2-1} e^{-t/(2\theta)}.$$

Taking the logarithm and differentiating twice with respect to θ, one obtains

$$\frac{\partial^2(\log f_T)}{\partial\theta^2} = -\frac{t}{\theta^3} + \frac{n}{2\theta^2}$$

so that

$$I_T(\theta) = -E\left\{\frac{\partial^2 \log f_T}{\partial\theta^2}\right\} = \frac{n}{2\theta^2},$$

which is the same as the information in $\mathbf{X}$ (as predicted by Theorem 3).

The following theorem, expanding on Theorem 3, makes precise the notion that the reduction of a sample to a statistic may lose information relative to the state θ, but that there is no loss of information if and only if sufficiency is maintained in the process of data reduction.

Theorem 5. *If $T = t(\mathbf{X})$, then $I_T(\theta) \le I_{\mathbf{X}}(\theta)$, with equality holding if and only if T is sufficient.*

In order to show this, it is convenient to have the following lemma. As before, let $g(t;\theta)$ denote the density or probability function of the statistic T.

LEMMA

$$E\left\{\frac{\partial}{\partial\theta}\log f(\mathbf{X};\theta) \mid T = t\right\} = \frac{\partial}{\partial\theta}\log g(t;\theta).$$

The discrete case is all that will be attempted here. In that case,

$$g(t;\theta) = P_\theta(t(\mathbf{X}) = t) = \sum_{t(\mathbf{x})=t} f(\mathbf{x};\theta)$$

and

$$g' = \sum f',$$

the sums again being taken over points $\mathbf{x}$ such that $t(\mathbf{x}) = t$. The conditional probability of $\mathbf{x}$ given $T = t$ is

$$f_{\mathbf{X}|T=t}(\mathbf{x}) = \frac{f(\mathbf{x};\theta)}{g(t;\theta)}, \qquad \text{for } t(\mathbf{x}) = t,$$

and so

$$\begin{aligned} E\left\{\frac{\partial}{\partial\theta}\log f \mid T = t\right\} &= \sum_{t(\mathbf{x})=t}\left(\frac{\partial}{\partial\theta}\log f\right)\frac{f}{g} \\ &= \frac{\sum f'}{g} = \frac{g'}{g} = \frac{\partial}{\partial\theta}\log g(t;\theta), \end{aligned}$$

as asserted by the lemma.

The proof of Theorem 5, given the lemma, then proceeds as follows:

$$0 \le E\left\{\left[\frac{\partial}{\partial\theta}\log f(\mathbf{X};\theta) - \frac{\partial}{\partial\theta}\log g(t(\mathbf{X});\theta)\right]^2\right\} = I_{\mathbf{X}}(\theta) + I_T(\theta) - U,$$

where

$$U = 2E\left\{\frac{\partial}{\partial\theta}\log f(\mathbf{X};\theta)\frac{\partial}{\partial\theta}\log g(t(\mathbf{X});\theta)\right\}.$$

This expectation can be written as the expected value of the conditional expectation given $t(\mathbf{X}) = t$, under which condition the second factor in the braces is a constant and factors outside the conditional expectation:

$$\begin{aligned} U &= 2E\left\{\frac{\partial}{\partial\theta}\log g(t;\theta)E\left[\frac{\partial}{\partial\theta}\log f \mid T = t\right]\right\} \\ &= 2E\left\{\left[\frac{\partial}{\partial\theta}\log g(t;\theta)\right]^2\right\} = 2I_T(t), \end{aligned}$$

by the lemma. The desired inequality then follows at once. Proof of the assertion about the case of equality will be left to the reader.

Problems

4–49. Verify that the "$Q(\theta)$" is monotonic in each of the cases given in the list of special cases of the exponential family in Section 3.5. (Conclusion?)

4–50. Show that the hypergeometric distribution has a monotone likelihood ratio. [*Hint:* Show that for any M, the ratio $f(k; N, M+1, n)/f(k; N, M, n)$, in the notation of Section 3.1.3, is a monotonic function of k; and then use the fact that the product of increasing functions is an increasing function.]

4–51. Compute the information in a random sample of size n from each population:

(a) Normal $(\mu, 1)$. (b) Bernoulli (p). (c) Exponential (λ).

4–52. Compute $I_X(\theta)$ for a Cauchy population with location parameter θ. Compute also $I_{\bar{X}}(\theta)$, the information in the mean of a random sample of size n (cf. Problem 4–13). (Conclusion?)

4–53. Let X have a distribution in the exponential family:

$$\log f(x;\theta) = \log B(\theta) + \log h(x) + Q(\theta)R(x).$$

(a) Show that $E[R(X)] = -\dfrac{B'}{BQ'}$.

(b) Show that $I_X(\theta) = (B'/B)^2 - B''/B + (Q''/Q')(B'/B)$. [Could you compute var $R(X)$ from this result?]

5

ESTIMATION

The estimation problems to be considered are those called *parametric*. The data will consist of observations $X_1, \ldots, X_n$ whose distribution has the density or probability function $f(x; \theta_1, \ldots, \theta_k)$, a function of known form depending on the unknown real parameters $\theta_1, \ldots, \theta_k$. The values of the vector $\boldsymbol{\theta}$ are often restricted by the nature of the problem; for instance, a population variance is always nonnegative and a Bernoulli parameter is a number on the interval $0 \leq p \leq 1$. The set of admissible parameter values is termed the *parameter space*. But it may be that only some, not all, of the parameters are to be estimated, and so the value or vector that is announced as an estimate may be further restricted to a subspace of the parameter space.

The problem of estimation is to devise means of using sample observations to construct good estimates of one or more of the parameters. It is to be expected that "information" in the sample concerning those parameters will make an estimate based on the sample generally better than a sheer guess; however, the information in a finite sample is not complete, and there will usually be some error in the estimation.

The value announced as an estimate of a parameter is to be based on the available data, and a scheme that operates on the observations $(X_1, \ldots, X_n)$ to produce an estimated value of $\boldsymbol{\theta}$ is a *function*—a mapping from the data space to the space of admissible parameter values.

Definition. *An* estimator *is a function on the space of observations* $(X_1, \ldots, X_n)$ *with values in the parameter space.*

If the X's are real numbers (and they usually are), the space of observations can be taken to be R_n (Euclidean n-space). If there are k parameters to be estimated, the values of the estimator function are k-vectors. In any case, an estimator is what has already been termed a *statistic*; calling it an "estimator" simply emphasizes the use of the statistic.

5.1 Criteria for Estimators

In using any given estimator, it is natural to want to know how successful it will be in its task of estimation—to know, if possible, how close the estimated value comes to the parameter being estimated. However, it may seem impossible to assess the quality of an estimate, because one cannot tell how close his estimate from a particular sample is to the true value of θ, which is unknown. If one is willing to settle for an evaluation of the performance of an estimator in terms of probability—or equivalently, on how well it performs over a long sequence of trials in which it is used, then something can be said about the degree of success.

As a function of the observations in a sample (a random variable), an estimator will sometimes give a value that is close to the true value of θ and sometimes one that is far from it. It depends on the sample. The "distance" from a function (the estimator) to a constant (the true value of the parameter) can be defined in many ways; the more common measures are some functional of their difference. In estimating the parameter θ by the statistic T, the difference $T - \theta$ is called the *error*. A simple average of the error is not useful because positive and negative components of error could be large and yet cancel each other out. A measure that would not suffer this defect would be the average distance, $E|T - \theta|$, or more generally, the average of the kth power of the distance $E|T - \theta|^k$. When T and θ are one-dimensional, which will here be the usual case, the distance between T and θ is just the absolute value of their difference, and the measure used is

$$E|T - \theta|^k = \int_{-\infty}^{\infty} |t - \theta|^k f_T(t)\,dt$$
$$= \int_{-\infty}^{\infty} \cdots \int_{-\infty}^{\infty} |t(\mathbf{x}) - \theta|^k f_{\mathbf{X}}(\mathbf{x})\,d\mathbf{x},$$

where $T = t(\mathbf{x})$. The simplest such measure to analyze is that given by $k = 2$, the average squared error, which will be considered in detail next.

5.1.1 Mean Squared Error

The most commonly used measure of performance of an estimator is the average squared error, or *mean squared error*:

$$\text{m.s.e.} = E[(T - \theta)^2].$$

In the case of estimating several parameters $(\theta_1, \ldots, \theta_k)$ by a statistic $(T_1, \ldots, T_k)$, one might use a generalized mean squared error of the form

$$\sum a_i E[(T_i - \theta_i)^2].$$

For $k = 1$ the mean squared error, which is a second moment of a random variable, can be decomposed into two parts by the parallel axis theorem, the variance of the estimator and another nonnegative term:

$$E[(T - \theta)^2] = \text{var}\,T + (ET - \theta)^2.$$

The quantity $ET - \theta$ measures the amount by which the center of the distribution of T differs from θ and is called the *bias* in T (as an estimator of θ),

$$b_T(\theta) = ET - \theta.$$

The mean squared error cannot ordinarily be made 0, but it is often possible, by judicious choice of estimator, to make either var $T = 0$ or $b_T^2(\theta) = 0$. However, there is no particular virtue in making one of these terms vanish; it is their *sum* that counts. That is, the mean squared error will be small if and only if *both* the variance of T and the bias in T are small.

An estimator with zero bias is said to be *unbiased*, and the everyday connotations of the words "bias" and "unbiased" have led to a common belief—even conviction—that an estimator should be unbiased to be any good. There is clearly no basis for this belief as a general rule.

EXAMPLE 5–1. It was seen in Chapter 4 that the expected value of the sample mean is the population mean, when the latter exists. Thus

$$b_{\bar{X}}(\mu) = E\bar{X} - \mu = 0,$$

and so $\bar{X}$ is always unbiased in estimating μ. Its mean squared error is then just the variance,

$$\text{m.s.e.} = E[(\bar{X} - \mu)^2] = \text{var } \bar{X} = \frac{\sigma^2}{n}.$$

Now the statistic $T \equiv X_1$ (the first observation in the sample) is also unbiased,

$$EX_1 = \mu = 0,$$

but its mean squared error is

$$\text{m.s.e.} = E(X_1 - \mu)^2 = \sigma^2.$$

So if $n > 1$, X_1 is not much good by comparison with $\bar{X}$, even though it is unbiased. On the other hand, the statistic $T' = 0$ has zero variance, but its mean squared error, which is the square of the bias, will usually be large.

In searching for estimators, it sometimes happens that a statistic T is discovered whose mean is proportional to θ: $ET = c\theta$. In this case, it is common practice to use T/c as an estimator, with the excuse that it is at least unbiased. This is often a useful device. However, it can happen that if T is already close to being unbiased, multiplying it by a factor to make it exactly unbiased can deteriorate the mean squared error. The next example illustrates this point.

EXAMPLE 5–2. The mean and variance of the sample variance were given in Section 4.3.1. The expectation is

$$E(S^2) = E\left(\frac{1}{n}\sum(X_i - \bar{X})^2\right) = \sigma^2\left(1 - \frac{1}{n}\right),$$

and so S^2 has a slight negative bias, $-\sigma^2/n$. Its mean squared error is

$$\text{m.s.e.} = \text{var } S^2 + [ES^2 - \sigma^2]^2,$$

which, in the case of a *normal* population, reduces to

$$\text{m.s.e.} = \frac{2\sigma^4(n-1)}{n^2} + \left(-\frac{\sigma^2}{n}\right)^2$$
$$= \frac{2n-1}{n^2}\sigma^4.$$

The estimator

$$\tilde{S}^2 = \frac{n}{n-1}S^2 = \frac{1}{n-1}\sum(X_i - \bar{X})^2$$

is often used to estimate σ^2 and is called the "sample variance," with the excuse that it is unbiased—which it is:

$$E(\tilde{S}^2) = \frac{n}{n-1}E(S^2) = \sigma^2.$$

However, the mean squared error is now larger:

$$\text{m.s.e.} = \text{var}\ \tilde{S}^2 = \frac{n^2}{(n-1)^2}\frac{2(n-1)}{n^2}\sigma^4$$
$$= \frac{2\sigma^4}{n-1} > \frac{2n-1}{n^2}\sigma^4$$

(for any positive integer n). So if the population is normal, $\tilde{S}^2$ is not as good as S^2 in the sense of mean squared error. (See also Problem 5–7.)

The mean squared error gives a bound for the probability that the absolute error of estimation exceeds a given amount, by (a form of) the Chebyshev inequality:

$$P(|T - \theta| > \epsilon) \leq \frac{1}{\epsilon^2}E[(T-\theta)^2].$$

The smaller the mean squared error, the less likely it is that the estimator T will stray far from θ.

EXAMPLE 5–3. Let Y denote the sample sum in a sequence of n independent Bernoulli trials. The statistic $T = Y/n = \bar{X}$ is unbiased in estimating the Bernoulli parameter p, and so the m.s.e. is the variance:

$$\text{m.s.e.} = E\left(\frac{Y}{n} - p\right)^2 = \text{var}\left(\frac{Y}{n}\right) = \frac{pq}{n}.$$

If $n = 100$, for instance,

$$P\left(\left|\frac{Y}{n} - p\right| > .1\right) \leq \frac{pq}{.01n} \leq .25,$$

or less than 1 chance in 4 that the sample proportion misses the population proportion by more than .1. In this case a more precise bound can be calculated, since Y/n is approximately normal:

$$P\left(\left|\frac{Y}{n} - p\right| > .1\right) \leq P(|Z| > 2) = .0456,$$

where Z is standard normal (and $\sigma_{Y/n} = \sqrt{pq/100} \leq .05$).

The mean squared error will usually depend on one or more unknown parameters and will not, therefore, linearly order all candidate estimators. Thus, the problem of determining an estimator with uniformly smallest mean squared error as "optimal" cannot be fairly posed. Sometimes one can find an estimator with uniformly smallest variance in the class of *unbiased* estimators (see Sections 5.1.3 and 5.1.4) or in an even smaller class (Section 12.2.2); but apart from the fact that the optimality problem then has a solution, the restriction to unbiased estimators appears to be artificial—and may rule out some very good estimators, as evident in Example 5–2.

Despite its dependence on unknown parameters, the m.s.e. is often useful in assessing the performance of an estimator. This dependence is not an embarrassment in the case of large samples, for in such a case, any unknown parameters can be replaced by reasonable sample estimates without introducing a significant error in the m.s.e. With such replacement, the standard deviation of an estimator becomes what is called its *standard error*. The standard error is relevant for estimators of zero or small bias that are (as is often the case) approximately normally distributed, for then its meaning as a measure of reliability is readily comprehended.

EXAMPLE 5–4. Consider a large sample from a population with unknown mean μ and unknown variance σ^2. Because the sample mean converges in probability to the population mean, it is reasonable to approximate the latter by the former for a large sample. The estimator $\bar{X}$ is unbiased, having mean μ, so the risk is the variance of $\bar{X}$, which involves σ^2:

$$\text{var } \bar{X} = \frac{\sigma^2}{n} \doteq \frac{S^2}{n}.$$

The standard error of estimate using $\bar{X}$, called also the *standard error of the mean*, is the square root: $S/\sqrt{n}$.

5.1.2 Consistency

An estimator is often automatically defined for any sample size. For example, the sample moments are defined for any n, so a sample moment is in effect a *sequence* of estimates—if one assumes an infinite stream of observations X_1, $X_2, \ldots$. In particular, $\bar{X}$ defines such a sequence:

$$T_n \equiv \frac{1}{n}\sum X_i.$$

It is at least intuitively reasonable to expect that the larger the sample, the better the inference one could expect to make; and in line with this, it seems that a good estimator T_n should have the property that its mean squared error decreases to 0 as more and more observations are incorporated into its computation:

$$\lim_{n\to\infty} E[(T_n - \theta)^2] = 0.$$

If this condition holds the sequence $\{T_n\}$ is said to be *consistent in quadratic mean* (q.m.). It will hold, of course, if and only if *both* the variance of T_n *and* its bias tend to 0 as n becomes infinite.

It was seen in Section 2.5 that convergence in quadratic mean to a constant implies convergence in probability to that constant. Thus, if $\{T_n\}$ is consistent in the above sense, it is also true that T_n tends to θ in probability:

$$\lim P(|T_n - \theta| \geq \epsilon) = 0, \qquad \text{for any } \epsilon > 0.$$

This convergence in probability of T_n to θ is the condition that traditionally defines consistency of $\{T_n\}$. Since existence of second moments is not required, it is more widely applicable. "Consistent" without a modifier will mean consistent in this latter sense.

EXAMPLE 5–5. Sample moments tend in probability to corresponding population moments, a fact that was discussed in Section 4.3.1. Sample moments, therefore, are consistent estimates of the corresponding population moments. In particular, $\bar{X}$ is a consistent estimate of μ, and S^2 is a consistent estimate of σ^2. (The unbiased version $\tilde{S}^2$ given in Example 5–3 is also consistent, differing from S^2 by a factor that tends to 1 as n becomes infinite.)

EXAMPLE 5–6. Consider the one-parameter family of Cauchy distributions defined by the density

$$f(x; \theta) = \frac{1/\pi}{1 + (x - \theta)^2}.$$

This is symmetrical about θ, but $\bar{X}$ is not a consistent estimate of θ. For, the characteristic function of the sample mean is

$$\exp\left[-n\left|\frac{t}{n}\right| + n\left(\frac{it\theta}{n}\right)\right] = \exp\left[-|t| + it\theta\right],$$

which is the characteristic function of the population itself (according to Problem 2–138). That is, the distribution of the sample mean is the same as that of the population, no matter how large the sample. The sample mean cannot converge in probability to any constant, let alone θ.

If a statistic T_n has a variance that tends to 0, while its expectation converges to k, a number different from θ, then T_n converges in probability—but to the wrong value, k. Consequently, even though the significance of bias has been minimized, an estimator that has a bias which does not disappear as n becomes infinite would have to be modified to remove the bias in order to obtain an estimator which is consistent.

Problems

5–1. Show that the average of the first two observations in a random sample is unbiased but not consistent.

5–2. Determine a condition on coefficients a_i so that the linear combination $\sum a_i X_i$ is an unbiased estimate of EX.

5–3. Show that if T is an unbiased estimate of θ, then $aT + b$ is an unbiased estimate of $a\theta + b$. Is T^2 an unbiased estimate of θ^2?

5–4. The distribution of $X_{(n)}$, the largest of the n observations in a random sample from a population that is uniform on $[0, \theta]$ was seen in Problem 4–15 to have density

$$f_{X(n)}(u) = \frac{nu^{n-1}}{\theta^n}, \qquad \text{for } 0 < u < \theta.$$

(a) Show that $X_{(n)}$ is a consistent estimate of θ.
(b) Determine a multiple of $\bar{X}$ that is unbiased and obtain its mean squared error.
(c) Determine a multiple of $X_{(n)}$ that is unbiased and compute its mean squared error. (Conclusions?)

5–5. Determine the standard error of estimate in the relative frequency of success in n independent Bernoulli trials, as an estimate of the probability p of success in a single trial.

5–6. Determine the standard error of the mean as an estimate of the mean of an exponential population.

5–7. Show that $A \sum (X_i - \bar{X})^2$ has smallest mean squared error in estimating the variance of a normal population when the constant A has the value $1/(n + 1)$.

5–8. Let Y denote the number of success in n Bernoulli trials with parameter p. Determine the mean squared error in estimating p by each of the estimators $T_1 = Y/n$ and $T_2 = (Y + 1)/(n + 2)$. Is one estimator better than the other? (Sketch the m.s.e. of each, as a function of p, for $n = 4$ and for $n = 8$.)

5–9. Consider the population with density $\lambda e^{-\lambda x}$, for $x > 0$. Because $EX = 1/\lambda$, the reciprocal of $\bar{X}$ is a natural estimator for λ.

(a) Compute $E(1/\bar{X})$ and then construct an unbiased estimate of λ.
(b) Show that the multiple of $1/\bar{X}$ with smallest m.s.e. is $(n - 2)/\sum X_i$.

5.1.3 Efficiency

If an estimator T has a mean squared error that is smaller than the mean squared error of T' in estimating θ from a given sample, the estimate T is thought of as making more "efficient" use of the observations. Thus, it is said that an estimator T of θ is more efficient than T' if

$$E[(T - \theta)^2] \leq E[(T' - \theta)^2],$$

with strict inequality for some θ. The relative efficiency of T' with respect to T is the ratio

$$e(T', T) = \frac{E[(T - \theta)^2]}{E[(T' - \theta)^2]}.$$

This would generally depend on θ, but it turns out frequently to be independent of θ. In the case of unbiased estimators, it is just the ratio of their variances; and one with minimum variance would be most efficient.

EXAMPLE 5–7. The linear combination of observations $\sum a_i X_i$ is an unbiased estimate of $E(X)$ if $\sum a_i = 1$. The particular combination that is most efficient is the one which minimizes

$$\operatorname{var}\left(\sum a_i X_i\right) = \sum a_i^2 \operatorname{var} X_i = (\operatorname{var} X) \sum a_i^2$$

or the one that minimizes $\sum a_i^2$, subject to $\sum a_i = 1$.

For such restricted minimization problems, it will be convenient to have the tool of the method of Lagrange's multipliers. It is shown in advanced calculus that the minimum of $g(\mathbf{y})$ subject to $h(\mathbf{y}) = K$ is found by locating the minimum of the function $g(\mathbf{y}) - \lambda h(\mathbf{y})$. This will not be proved here; but it is easily seen that if $\mathbf{y}$ satisfies $h(\mathbf{y}) = K$ and minimizes $g(\mathbf{y}) - \lambda h(\mathbf{y})$ for some λ, then for any other $\mathbf{y}'$ such that $h(\mathbf{y}') = K$,

$$g(\mathbf{y}) - \lambda h(\mathbf{y}) \leq g(\mathbf{y}') - \lambda h(\mathbf{y}'),$$

or, since $h(\mathbf{y}) = h(\mathbf{y}')$,

$$g(\mathbf{y}) \leq g(\mathbf{y}').$$

Thus $\mathbf{y}$ is the desired minimizing quantity.

Applying this method in the case at hand, one minimizes $\sum a_i^2 - \lambda \sum a_i$. The derivative of this with respect to a_j must vanish:

$$2a_j - \lambda = 0, \qquad j = 1, \ldots, n.$$

The minimizing a's are, therefore, all equal, and equal to $1/n$. The sample mean is thus the most efficient unbiased linear combination of the observations in a random sample.

An absolute measure of efficiency of an estimate would involve a comparison of its mean square deviation from the parameter being estimated with a lower bound or absolute minimum of such mean square deviations if one that is not zero exists. The "information inequality" is aimed at providing such a lower bound.

The level of presentation here precludes a rigorous derivation of the information inequality, but the following manipulations for the continuous case indicate the line of reasoning used to establish the inequality.[1]

The statistic $T = t(\mathbf{X})$ based on a sample $\mathbf{X}$ from $f(x; \theta)$ is considered as an estimator for the parameter θ, assumed now to be one-dimensional. Let the joint density function of the sample observations be

$$f(x_1, \ldots, x_n; \theta) = f(\mathbf{x}; \theta).$$

Let V denote, as in Chapter 4, the random variable called the *score*:

$$V = \frac{\partial}{\partial \theta} \log f(\mathbf{X}; \theta).$$

[1] See H. Cramér [5], Section 32.3.

The mean value of this variable is $EV = 0$ (Section 4.5.4). The variance of V is, therefore, its expected square, and the covariance of the random variables V and T is their expected product:

$$\begin{aligned}\operatorname{cov}(V, T) = E(VT) &= E\left[T \frac{\partial}{\partial \theta} \log f(\mathbf{X}; \theta)\right] \\ &= \int t(\mathbf{x}) \frac{1}{f(\mathbf{x}; \theta)} \left(\frac{\partial}{\partial \theta} f(\mathbf{x}; \theta)\right) f(\mathbf{x}; \theta)\, d\mathbf{x} \\ &= \frac{d}{d\theta} E(T) = \frac{d}{d\theta}[\theta + b_T(\theta)] = 1 + b_T'(\theta),\end{aligned}$$

where $b_T(\theta)$ is the bias in T.

The information inequality now results from the fact that a correlation is numerically bounded by 1—that is, from Schwarz' inequality (see Section 2.3.3):

$$\operatorname{var} T \geq \frac{[\operatorname{cov}(V, T)]^2}{\operatorname{var} V} = \frac{[1 + b_T'(\theta)]^2}{I(\theta)},$$

where

$$I(\theta) = \operatorname{var} V = E\left\{\left[\frac{\partial}{\partial \theta} \log f(\mathbf{X}; \theta)\right]^2\right\},$$

called in Section 4.5.4 the *information* in the sample. (The information inequality is also known as the Cramér–Rao inequality, or the Frechét inequality.)

The validity of the above derivation depends on fulfillment of conditions that permit interchange of integration and differentiation operations, on the existence and integrability of the various partial derivatives, on the differentiability of $b_T(\theta)$, and on the nonvanishing of $I(\theta)$. In the case of a discrete random variable with finitely many values, the $f(x; \theta)$ is a probability, the expectations are finite sums, and the interchange of differentiation and summation is permitted.

The information inequality can also be expressed in terms of the mean squared error:

$$\text{m.s.e.}(T) = \operatorname{var} T + b_T^2(\theta) \geq \frac{[1 + b_T'(\theta)]^2}{I(\theta)} + b_T^2(\theta).$$

However, neither form is particularly useful in the general case, because the lower bound depends on the estimator T through its bias; so it is not a "bound" in the usual sense.

In the class of *unbiased* estimators, the information equality does provide a meaningful bound, independent of the estimator; the bound is just the reciprocal of the information:

$$\text{m.s.e.} = \operatorname{var} T \geq \frac{1}{I(\theta)}, \qquad (T \text{ unbiased}).$$

If the lower bound is achieved, the estimator T is said to be *efficient*.[2] The (absolute) efficiency of an unbiased estimate T is defined as the ratio

$$e(T) = \frac{1/I(\theta)}{\text{var } T}.$$

Thus, $e(T) \leq 1$, and an efficient estimator has efficiency 1.

EXAMPLE 5–8. Consider the problem of estimating the variance θ of a normal population with *known* mean μ. The sample second moment about that known mean is sufficient and unbiased:

$$T \equiv \frac{1}{n} \sum (X_i - \mu)^2, \qquad E(T) = \theta.$$

The variance of T is

$$\begin{aligned} \text{var } T &= \frac{\text{var } (X - \mu)^2}{n} \\ &= \frac{1}{n}\left\{E[(X - \mu)^4] - (E[(X - \mu)^2])^2\right\} = \frac{2\theta^2}{n}. \end{aligned}$$

(Use was made here of the formula for moments in Section 3.3.1.) The information in the sample was computed in Example 4–35 to be

$$I_{\mathbf{X}}(\theta) = \frac{n}{2\theta^2} = \frac{1}{\text{var } T}.$$

Thus, T has efficiency 1—it is efficient.

EXAMPLE 5–9. Let X have a Poisson distribution with mean m so that $EX = \text{var } X = m$, and $EX^2 = m^2 + m$. Consider the problem of estimating the parameter $\theta = m^2$ using a single observation. Clearly, $T = X^2 - X$ is unbiased and its variance can be computed to be

$$\text{var } T = 4m^3 + 2m^2.$$

Now, in terms of θ the probability function of X is

$$f(x; \theta) = \theta^{x/2} \frac{e^{-\sqrt{\theta}}}{x!},$$

and

$$\log f(x; \theta) = -\theta^{1/2} + \frac{x}{2} \log \theta - \log (x!).$$

From this one easily finds

$$I(\theta) = \frac{1}{4\theta^{3/2}} = \frac{1}{4m^3}.$$

[2] This definition of efficiency, apparently due to Cramér, is open to criticism. It will be seen in the next section that the greatest lower bound of variances of unbiased estimators may be greater than the Cramér–Rao lower bound. This would mean that the unbiased estimator with minimum variance would not be "efficient"—even though it cannot be outdone by any other unbiased estimator.

The efficiency of T, therefore, is

$$e(T) = \frac{4m^3}{4m^3 + 2m^2} = \frac{2m}{2m + 1}.$$

Inasmuch as X is a natural estimate for $\sqrt{\theta}$, it might seem reasonable to try $U = X^2$ as an estimate of θ:

$$EU = \theta + \sqrt{\theta}, \qquad b_U(\theta) = \sqrt{\theta} = m,$$

and

$$\text{var } U = 4m^3 + 6m^2 + m.$$

The m.s.e. of this estimator, therefore, is larger than that of T. The Cramér–Rao inequality is strict:

$$\text{var } U > \frac{[1 + b'_U(\theta)]^2}{I(\theta)} = 4m^3 + 4m^2 + m.$$

The situation in which the lower bound in the information inequality is achieved proves interesting. The "inequality" is an equality if the correlation between V and T is $+1$ or -1. If this is the case, then either T is identically constant or V is a linear function of T with probability 1, and conversely. The coefficients in the linear relationship between V and T can be functions of θ (written for convenience as derivatives):

$$V = A'(\theta)T + B'(\theta),$$

where again the prime denotes differentiation with respect to θ. Integrating with respect to θ, one obtains

$$\log f(\mathbf{X}; \theta) = A(\theta)T + B(\theta) + K(\mathbf{X}),$$

where the "constant" of integration $K(\mathbf{X})$ does not depend on θ but may depend on $\mathbf{X}$. Thus, $f(x; \theta)$ is in the exponential family, and T is sufficient. (If T is unbiased, it is efficient.)

Conversely, if $f(x; \theta)$ is in the exponential family,

$$f(x; \theta) = B(\theta)h(x) \exp [Q(\theta)R(x)],$$

the statistic $\sum R(X_i)$ is sufficient, and the log-likelihood is a linear function of it. Hence, the lower bound is achieved, and among the class of estimators with the same bias, $\sum R(X_i)$ has minimum variance. If it is unbiased, it is efficient.

[*Note:* It has been stated that if T is a one-dimensional statistic that is sufficient for θ, then T has a distribution in the exponential family. One might be tempted to jump to the conclusion that if T is unbiased it would be efficient. However, for this to be true, the T (or corresponding dummy variable t) would have to enter into the exponent linearly: $tQ(\theta)$. If it does, efficiency can be asserted, but not otherwise.]

By *asymptotic efficiency* is meant the limit of the efficiency as the sample size becomes infinite. In order for this limit to be a finite positive number, it is necessary that the variance of the estimator behaves asymptotically as $1/n$. The limiting or asymptotic efficiency is then

$$\lim_{n\to\infty} e(T) = \lim_{n\to\infty} \frac{1/I(\theta)}{\text{var}\, T} = \frac{1}{c^2 E([(\partial/\partial\theta)\log f(X;\theta)]^2)},$$

where

$$c^2 = \lim_{n\to\infty} n \,\text{var}\, T.$$

When the asymptotic efficiency is 1, the estimator is said to be asymptotically efficient.

Asymptotic efficiency is sometimes defined even when the efficiency for finite samples is not defined for one reason or another. (The estimator may be biased or may not have moments.) If T is asymptotically normal with mean θ and variance c^2/n, the asymptotic efficiency of T is defined to be the expression given as the limit of $e(T)$.

5.1.4 Reduction of Variance

Suppose that U is being considered as an estimator for θ [or for any function $h(\theta)$], and that T is sufficient for θ. Then the conditional mean

$$g(t) \equiv E(U \mid T = t)$$

is a function of t alone, as the notation suggests. That is, $V \equiv g(T)$ is a *statistic*, and it has the same mean as U:

$$EV = E[E(U \mid T)] = EU.$$

Moreover, its variance is no greater than that of U. For, according to the relation between a conditional and unconditional variance in Section 2.3.1,

$$\text{var}\, U = \text{var}\, V + E[\text{var}\,(U \mid T)] \geq \text{var}\, V.$$

Also, because the bias in U is the same as the bias in V (no matter what is to be estimated), it follows that

$$\text{m.s.e.}\,(V) \leq \text{m.s.e.}\,(U).$$

So by means of conditioning on a sufficient statistic one may reduce the mean squared error, but in any case he would not increase it in the process. This justifies, in the case of estimation, the feeling that one may as well base statistical procedures on the value of a sufficient statistic.

The above inequality is referred to as the "Rao–Blackwell Theorem," and the construction of a new statistic by conditioning on a sufficient statistic has been termed "Rao–Blackwellization."

The inequality becomes an equality if and only if

$$P[\text{var}(U \mid T) = 0] = 1,$$

which is true if and only if U is essentially a function of T. For if U is a function of T, then $(U \mid T = t)$ is a constant, with zero variance, for all t. And, conversely, if the conditional distribution of U given T has zero variance, it is singular at its mean, $E(U \mid T = t) = g(t)$. This means that given $\mathbf{X} = \mathbf{x}$, with $t(\mathbf{x}) = t$, the value of U is $g(t) = g(t(\mathbf{x}))$ with probability 1. That is, U is a function of T. In summary, then, if U is already a function of the sufficient statistic T, Rao–Blackwellization gives back U, with no change in m.s.e. And if Rao–Blackwellization does *not* produce a decrease in m.s.e., then the U must have been already a function of T (with probability 1).

One might wonder whether he could find an estimator with the same mean as U but with a variance even smaller than that of $V = E(U \mid T)$. (Rao–Blackwelizing V would just produce V, but perhaps one could start with a different U.) If the family of distributions of the sufficient statistic T has the property of *completeness*, then there is a unique function of T with a given expectation. In such a case, V would be *the* function of T with minimum variance among those with the same expectation.

Definition. *The family of distributions of T, $\{F_T(t; \theta)\}$, is said to be* complete *if*

$$E[h(T)] = 0, \qquad \text{for all } \theta,$$

implies that $h(T) \equiv 0$ (with probability 1 for all θ).

If (to rephrase the definition) 0 is the only unbiased estimate of 0, then two functions of T with the same expectation must be identical, for their difference would be an unbiased estimate of 0.

It can be shown that if a population is in the one-parameter exponential family, the sufficient statistic $\sum R(X_i)$ has a distribution family that is complete. This completeness is often attributed, loosely, to the statistic itself, and it is said that "$\sum R(X_i)$ is complete."

EXAMPLE 5–10. Let $\mathbf{X}$ be a random sample, and $\mathbf{Y}$ the corresponding order statistic. The statistic X_1 (the first observation in the sample) is unbiased as an estimate of $\mu = EX$, and

$$P(X_1 = y_i \mid \mathbf{Y} = \mathbf{y}) = \frac{1}{n}.$$

(By symmetry, the first observation is just as likely to be the smallest as it is the second smallest, etc.) Therefore,

$$V = E(X_1 \mid \mathbf{Y}) = \frac{1}{n} \sum Y_i = \bar{X}.$$

Another unbiased estimator of μ, when the population is symmetric about μ, is the midrange, $U = \frac{1}{2}(Y_1 + Y_n)$. But $E(U \mid \mathbf{Y}) = U$, which is not surprising, since U is already a function of the order statistic.

If it is assumed that the population is normal with mean μ and unit variance, then $\sum X_i$ is minimal sufficient and

$$E(U \mid \sum X_i) = \bar{X}.$$

This might be shown directly, with some difficulty, but it follows from the completeness of $\bar{X}$ and the fact that $\bar{X}$ is unbiased in estimating μ. For the same reason,

$$E(X_1 \mid \sum X_i) = \bar{X}.$$

(Again, direct computation would yield this result; it will be called for in Problem 10–15.)

EXAMPLE 5–11. In Problem 5–9 it was found that the statistic $T = (n - 1)/\sum X_i$ is an unbiased estimate of λ in the case of an exponential population with mean $1/\lambda$. Since T is minimal sufficient, it cannot be improved by Rao–Blackwellization; and, indeed, it is the minimum variance unbiased estimator for λ, with variance (cf. Problem 5–9)

$$\text{var } T = \frac{\lambda^2}{n - 2}.$$

However, the information function of λ, as computed in Problem 4–51(c), is $I(\lambda) = n/\lambda^2$, and so the efficiency of T is less than 1:

$$e(T) = \frac{\lambda^2/n}{\lambda^2/(n - 2)} = 1 - \frac{2}{n}.$$

(This does approach 1 as n becomes infinite, so T is asymptotically efficient.) There is no efficient estimator here; and Problem 5–9 gave an estimator with smaller mean squared error than the minimum variance for unbiased estimators.

Showing that a statistic V is unbiased with minimum variance among unbiased estimators may not be accomplishing much. For one thing, there may be a biased statistic with smaller m.s.e. Moreover, as in the above example, the minimum variance unbiased estimator need not be efficient—or even close to it, as in the next example.

EXAMPLE 5–12. Suppose that it is desired to estimate $e^{-2\lambda}$ by using a single observation on a Poisson variable with mean λ. (The function $e^{-2\lambda}$ is the probability that a telephone switchboard, assuming a Poisson law for incoming calls, would have no calls in two time units if λ is the mean number per unit time; and the estimate of this probability is to be based on the number of calls received in a unit time.) Now,

$$f(x; \lambda) = \frac{\lambda^x}{x!} e^{-\lambda}, \qquad x = 0, 1, \ldots$$

and

$$E[h(X)] = \sum h(x) \frac{\lambda^x}{x!} e^{-\lambda}.$$

If this is to be equal to $e^{-2\lambda}$ so that $h(X)$ is unbiased, it must be that

$$\sum h(x)\frac{\lambda^x}{x!} = e^{-\lambda} = \sum (-1)^x \frac{\lambda^x}{x!}.$$

From the uniqueness of power series coefficients, it follows that the estimate $h(X)$ must be 1 or -1 according to whether X is even or odd. This is the *only* unbiased estimate, so it surely has the minimum variance property; but it is not a very appealing estimate. (However, as seen in Problem 5–13, the efficiency tends to 1 as λ tends to 0. This is not too surprising; for if the mean number of calls per unit time is very small, the value of X is apt to be 0, and the estimate 1 for the probability of no calls in two time units is not so bad.)

Problems

5–10. Show that the sample mean is an efficient estimate of the population mean for a population that is

(a) Normal with given variance. (b) Exponential.
(c) Bernoulli. (d) Geometric.

(See Problem 4–51.)

5–11. Show that if $\xi = a\theta + b$, with $a \neq 0$, and if T is an efficient estimate of θ, then $aT + b$ is an efficient estimate of ξ.

[*Hint:* Show first that $I(\xi) = I(\theta)/a^2$.]

5–12. Compute the asymptotic efficiency of the sample median in estimating the location parameter of a Cauchy distribution. (See Section 4.3.4 and Problem 4–52.)

5–13. Referring to Example 5–12 show the following, where $\theta = e^{-2\lambda}$:

(a) $\text{var } T = 1 - \theta^2$.
(b) $I(\theta) = -(2\theta^2 \log \theta)^{-1}$.
(c) $e(T) \to 0$ as $\theta \to 0$, and $e(T) \to 1$ as $\theta \to 1$.

5–14. Referring to Example 5–12, suppose that to estimate $e^{-2\lambda}$ one has an observation on Y, the number of calls arriving in four time units. Determine an unbiased estimate and compute its variance and efficiency.

5–15. Let T_1 and T_2 be each sufficient, and suppose that T_2 is a reduction of T_1, but not conversely. Let U be an unbiased estimate of θ, and $V_1 = E(U \mid T_1)$. Show that Rao–Blackwellizing V_1 using T_2 results in a statistic V_2 with $\text{var } V_2 < \text{var } V_1$.

5–16. Consider a random sample of size n from a normal population with unit variance. The estimator

$$U = \begin{cases} 1, & \text{if } X_1 \leq \lambda, \\ 0, & \text{if } X_1 > \lambda, \end{cases}$$

is unbiased in estimating $P(X \leq \lambda) = \Phi(\lambda - \mu)$, for a given λ. Use the sufficient statistic $\bar{X}$ to obtain an estimate with smaller variance than U. [Given: $X_1 \mid \bar{X} = t$ is normal with mean t and variance $1 - 1/n$, a fact that will be obtained in Chapter 10 (Problem 10-15).]

5–17. The mean $\bar{X}$ of a random sample is an unbiased estimate of the mean of a uniform population on $0 < x < \theta$, and so $2\bar{X}$ is an unbiased estimate of θ. Use the sufficient statistic $X_{(n)}$ to improve on $2\bar{X}$ by Rao–Blackwellization.

[*Hint:* $\bar{X}$ can be written as $[X_{(n)} + (n-1)\bar{X}']/n$, where $\bar{X}'$ is the mean of the $n-1$ X's that are not the largest among the n sample observations. Use the fact that $E[\bar{X}' \mid X_{(n)}] = X_{(n)}/2$; this is clear intuitively, but try to derive it.

5–18. In Example 5–9, the estimator $X^2 - X$ was proposed for the parameter m^2 of a Poisson distribution. It was found to have an efficiency less than 1. Show that it is the minimum variance unbiased estimator. (See the footnote on the definition of efficiency in Section 5.1.3.)

5.2 Deriving Estimators

The estimators considered so far have been more or less pulled out of the hat of intuition. It would be desirable to have some method or principle to follow that would always lead to good estimators. The closest to such a thing so far was the determination of an efficient estimate by inspection, in some instances of the exponential family; but sometimes this could be done and sometimes not. Then, too, the Rao–Blackwell theorem provides a means for improving an unbiased estimator by means of a sufficient statistic.

Unfortunately, there seems to be no generally applicable principle or method guaranteeing good estimators. This section will consider two techniques leading to estimators that often have good properties. (Another method will be given in Chapter 8.)

5.2.1 The Method of Moments

The oldest method of determining estimators (devised by K. Pearson in about 1894) is the method of moments. If there are k parameters to be estimated, the method consists of expressing the first k population moments in terms of these k parameters, equating them to the corresponding sample moments and taking the solutions of the resulting equations as estimates of the parameters. The method usually leads to relatively simple estimates.

The estimates obtained in this way are clearly functions of the sample moments. Since the sample moments are consistent estimates of population moments, the parameter estimates will generally be consistent.

Although the asymptotic efficiency of estimates obtained by the method of moments is often less than 1, such estimates may conveniently be used as first approximations from which more efficient estimates may be obtained by other means.

EXAMPLE 5–13. The estimate of μ^2 in any population having a mean would be the square of the sample mean, $\bar{X}^2$, according to the method of moments. This is biased but consistent. Efficiency could not be discussed without further assumptions as to the nature of the population.

EXAMPLE 5–14. The estimate of the parameter m in the Poisson family would be, according to the method of moments, the sample mean. For m is the population mean, and although m is also the population variance, the lowest order population moment is used.

5.2.2 Maximum Likelihood Estimation

The method of maximum likelihood estimation applies the maximum likelihood principle of Section 4.5.2 to the particular problem of estimation. The point estimate of θ consistent with the notion that the maximum likelihood state $\hat{\theta}$ provides the best explanation of observed data is clearly $\hat{\theta}$ itself. This maximum likelihood state is then defined to be the *maximum likelihood estimator* (m.l.e.).

Although, as seen earlier, the maximum likelihood state can occur at a value $\hat{\theta}$ at which the derivative of the likelihood function does not vanish or does not exist, the regular case is that in which $\hat{\theta}$ satisfies the *likelihood equation*:

$$\frac{\partial}{\partial\theta}\log L(\theta) = 0.$$

In the case of a vector parameter, $\boldsymbol{\theta} = (\theta_1, \ldots, \theta_k)$, the usual necessary condition for a maximum in the regular case is really k equations:

$$\begin{cases} \dfrac{\partial}{\partial\theta_1}\log L(\theta_1, \ldots, \theta_k) = 0 \\ \quad\vdots \qquad\qquad\qquad\qquad \vdots \\ \dfrac{\partial}{\partial\theta_k}\log L(\theta_1, \ldots, \theta_k) = 0. \end{cases}$$

A solution $(\hat{\theta}_1, \ldots, \hat{\theta}_k)$ of this system, assuming it corresponds to a maximum of L, is a maximum likelihood estimate of $(\theta_1, \ldots, \theta_k)$.

One might wonder whether an estimate obtained as the solution of the likelihood equation actually maximizes the likelihood function when the vanishing of $L'(\theta)$ does not by itself guarantee this. When there is any doubt on this point, it should be investigated. The usual situation is that the likelihood function (being a product of probabilities or densities) is bounded above and continuous in θ, and that the likelihood equation has only one solution, which then must maximize $L(\theta)$.

EXAMPLE 5–15. Consider a normal population with mean μ and variance θ. If μ is known, the likelihood is a function of θ alone, and

$$\log L(\theta) = -\frac{n}{2}\log(2\pi\theta) - \frac{1}{2\theta}\sum (X_i - \mu)^2,$$

with derivative

$$-\frac{n}{2\theta} + \frac{1}{2\theta^2}\sum (X_i - \mu)^2.$$

This vanishes at the value

$$\theta = \frac{1}{n}\sum (X_i - \mu)^2.$$

If *both* μ and θ are unknown, the likelihood equations are obtained by setting the partial derivatives with respect to μ and θ equal to 0:

$$\begin{cases} \dfrac{1}{\theta}\sum (X_i - \mu) = 0, \\ -\dfrac{n}{\theta} + \dfrac{1}{\theta^2}\sum (X_i - \mu)^2 = 0. \end{cases}$$

Solving the first equation for μ yields $\hat{\mu} = \bar{X}$, and this is then substituted in the second equation for θ to yield $\hat{\mu} = S^2$, the sample variance. Thus, the minimal sufficient $\bar{X}$ and S^2 are the (joint) maximum likelihood estimators for μ and θ, respectively.

In both cases the likelihood function is continuously differentiable and is bounded above. Therefore, the likelihood equations have unique solutions, and these must be values that maximize the likelihood.

EXAMPLE 5–16. Consider the basic experiment of a multinomial distribution, one that can result in any of k ways: $A_1, \ldots, A_k$. The parameters of the distribution are the corresponding probabilities $p_1, \ldots, p_k$, where $\sum p_i = 1$. Suppose that among n independent observations on this experiment there are f_i outcomes of type A_i, $i = 1, \ldots, k$. The probability of such a result is the required likelihood function

$$L(p_1, \ldots, p_k) = p_1^{f_1} \cdots p_k^{f_k},$$

with logarithm

$$\mathscr{L} = \log L(p_1, \ldots, p_k) = \sum_{i=1}^{k} f_i \log p_i.$$

In maximizing this, the probability vector $(p_1, \ldots, p_k)$ is restricted by the condition that $\sum p_i = 1$. Using the Lagrange method, one maximizes $\mathscr{L} - \lambda \sum p_i$, differentiating this with respect to each p_j:

$$\frac{\partial}{\partial p_j}(\mathscr{L} - \lambda \sum p_i) = \frac{f_j}{p_j} - \lambda.$$

These derivatives vanish for $j = 1, \ldots, k$ only if f_j/p_j is the same (equal to λ) for all j. That is, the maximum likelihood estimates must be proportional to the frequencies f_j. With the condition that $\sum p_j = 1$, this means that $\hat{p}_j = f_j/n$, the relative frequency of outcomes of type A_j.

If it is desired to estimate a function $g(\theta)$ of the parameter θ, the obvious value to announce if θ is assumed to be $\hat{\theta}$ is $g(\hat{\theta})$. So if $\hat{\theta}$ is the m.l.e. of θ, then $g(\hat{\theta})$ is the m.l.e. of $g(\theta)$.

EXAMPLE 5–17. The m.l.e. of the mean θ of an exponential population,

$$f(x; \theta) = \frac{1}{\theta} e^{-x/\theta}, \qquad \text{for } x > 0,$$

is the sample mean, $\bar{X}$. The m.l.e. of the parameter $\lambda = 1/\theta$, the mean number of "events" in a unit time interval, therefore, is $\hat{\lambda} = 1/\theta = 1/\bar{X}$. The m.l.e. of the variance $\sigma^2 = \theta^2$ is the square of the m.l.e. of θ, namely, $\bar{X}^2$.

The estimators produced by the method of maximum likelihood have some significant properties:

Property 1. *If T is sufficient for θ, a solution of the likelihood equation is a function of T.*

(This was demonstrated in Section 4.5.2, with a warning of possible exceptions when the solution is not unique.)

Property 2. *If there is an efficient estimate of θ, the method of maximum likelihood will produce it.*

This is seen as follows. If T is efficient, the score is a linear function of T:

$$V = \frac{\partial}{\partial \theta} \log f(\mathbf{X}; \theta) = C(\theta)T + D(\theta).$$

This vanishes at $\theta = \hat{\theta}$, the m.l.e.:

$$C(\hat{\theta})T + D(\hat{\theta}) = 0.$$

But the mean score is 0 for *all* θ:

$$EV = C(\theta)\theta + D(\theta) \equiv 0,$$

where ET has been evaluated as θ because an efficient estimator is unbiased. So, in particular, for $\theta = \hat{\theta}$:

$$C(\hat{\theta})\hat{\theta} + D(\hat{\theta}) = 0,$$

which implies that

$$T = -\frac{D(\hat{\theta})}{C(\hat{\theta})} = \hat{\theta}.$$

Property 3. *Under certain conditions of regularity, a maximum likelihood estimator is asymptotically efficient, having an asymptotically normal distribution with variance* $1/I(\theta)$.

(For a precise statement and proof see Cramér [5], p. 500, or Zacks [23], pp. 238 ff.)

EXAMPLE 5–18. The likelihood function for θ in a uniform distribution on $(\theta - \frac{1}{2}, \theta + \frac{1}{2})$ was seen in Example 4–30 to be

$$L(\theta) = \begin{cases} 1, & \text{if } X_{(n)} - \frac{1}{2} < \theta < X_{(1)} + \frac{1}{2}, \\ 0, & \text{otherwise.} \end{cases}$$

Any value of θ between $X_{(n)} - \frac{1}{2}$ and $X_{(1)} + \frac{1}{2}$ maximizes the likelihood function and is therefore a m.l.e. In particular, the midrange

$$A \equiv \tfrac{1}{2}[X_{(1)} + X_{(n)}]$$

is a m.l.e., and is unbiased, with mean θ and variance $[2(n + 2)(n + 1)]^{-1}$. Thus, A is consistent, but its asymptotic variance is of smaller order than the $1/n$ called for in the definition of asymptotic efficiency (and which, in regular cases, is the order of the asymptotic variance of an m.l.e.).

The statistic $X_{(n)} - \frac{1}{2}$, at the left end of the interval of m.l.e.'s, is itself a m.l.e. of θ, and is consistent, with mean $\theta - 1/(n + 1)$ converging to θ, and variance $n[(n + 1)^2(n + 2)]^{-1}$ converging to 0. The mean squared error comparison is as follows:

$$\text{m.s.e. } (A) = \text{var } A = \frac{1}{2(n + 2)(n + 1)},$$

$$\text{m.s.e. } (X_{(n)} - \tfrac{1}{2}) = \frac{n}{(n + 1)^2(n + 2)} + \left(\frac{1}{n + 1}\right)^2 = \frac{2}{(n + 2)(n + 1)}.$$

(It can be shown that A is the m.l.e. with smallest m.s.e.)

Problems

5–19. Use the method of moments to derive an estimator for the parameter $\theta = m^2$ in Example 5–9. What is the m.l.e. in this case?

5–20. Use the method of moments to obtain estimators for the population mean and variance in a normal population, based on a random sample.

5–21. Use the method of moments to estimate θ in a uniform population on $(0, \theta)$, using a random sample.

5–22. Explain why, in view of the results of Problem 5–10, the sample mean is the m.l.e. of the population mean for the following populations:

(a) Normal with known variance. (b) Exponential.
(c) Bernoulli. (d) Geometric.

5–23. A lot contains 10 articles, and a sample of 4 is drawn without replacement from the lot. Given that 1 of the 4 articles drawn is defective, what number of defectives in the lot would give the largest probability of this result?

5–24. Determine the maximum likelihood estimate of the parameter b in the density function

$$f(x; b) = \begin{cases} \exp(-|x - b|), & \text{for } x > b, \\ 0, & \text{for } x < b. \end{cases}$$

5–25. Show that the maximum likelihood estimates of p in a Bernoulli population based on a sequence of n independent trials, and based on a sequence of independent trials sufficient to obtain a specified number of successes agree (because the likelihood functions agree).

5–26. Determine the m.l.e. and the moment method estimator for $\theta = e^{-2\lambda}$ in Example 5–12.

5–27. Determine joint m.l.e.'s for θ_1 and θ_2 for a population that is uniform on (θ_1, θ_2). What is then the m.l.e. of the range $\theta_2 - \theta_1$?

5–28. Determine the m.l.e. of the mean of a uniform population on

(a) $0 < x < \theta$ (cf. Problem 4–46).
(b) $\theta_1 < x < \theta_2$ (cf. Problem 5–27).

5.3 Interval Estimates

In presenting an estimate of a population parameter—giving the value $\bar{X}$, for example, as an estimate of μ—no indication of the reliability of the estimate is found in the bare announcement of the estimated value. Giving the size of the sample used along with the estimate would be helpful but would assume that everyone who is to interpret the result is equipped to interpret sample size in terms of the "accuracy" of the estimate. Such is not ordinarily the case, and a more direct indication is desirable.

The standard error of estimate, given in Section 5.1.1, is one device used for indicating the reliability or precision of an estimation process. What is wanted, perhaps, is really a range of values within which the population parameter being estimated is almost sure to lie, in some sense. The notion of *confidence interval* for a parameter is commonly used to serve this purpose and will be introduced by means of an example.

EXAMPLE 5–19. Consider a sample of n from a normal population with unit variance. The interval from $\bar{X} - 2/\sqrt{n}$ to $\bar{X} + 2/\sqrt{n}$ is four standard deviations wide and ought to trap the actual population mean within it, one feels. What *can* be said is that the random interval from $\bar{X} - 2/\sqrt{n}$ to $\bar{X} + 2/\sqrt{n}$ has the property that

$$P\left(\bar{X} - \frac{2}{\sqrt{n}} < \mu < \bar{X} + \frac{2}{\sqrt{n}}\right) = P\left(-\frac{2}{\sqrt{n}} < \bar{X} - \mu < +\frac{2}{\sqrt{n}}\right) \doteq .95.$$

That is, an interval computed in this way will happen to cover the actual population mean 95 per cent of the time. Such a random interval is called a (95 per cent) confidence interval for the population mean.

Notice the effect of sample size in this example—the larger the sample, the narrower the confidence interval. Also, choosing a *confidence coefficient* other than the 95 per cent of the example would result in an interval of different width —narrower if the confidence coefficient is smaller and wider if the confidence coefficient is bigger. Thus, one can be "more confident" in the ability of a wider interval to trap the true value of the parameter.

Unfortunately, giving a confidence interval in an actual problem with actual numbers sounds like something different from what is really meant. Stating that a 90 per cent confidence interval for θ is (9.1, 9.6), for instance, would suggest that θ is a random variable that, with probability 0.90, lies between 9.1 and 9.6. In the analysis leading to a confidence interval, however, it is assumed that θ is a *constant* that, although not known, remains fixed throughout the sampling process. Nevertheless, a user of statistics (who is not a statistician) is likely to interpret a confidence interval statement as giving a probability for the "random variable" θ. A model in which such an interpretation is legitimate will be considered in Chapter 8.

In most of the situations to be encountered, the construction of a confidence interval begins with a quantity $p(T, \theta)$, called a *pivotal* quantity, whose distribution is independent of θ, and where $T = t(\mathbf{X})$ is a reasonable point estimate for θ. From the distribution of $p(T, \theta)$ one obtains points a and b such that

$$P_\theta(a < p(T, \theta) < b) = \gamma,$$

where γ is a specified confidence coefficient. Then, given T, the inequality is solved for θ to obtain a region of θ-values, often an interval, which is a confidence region for θ corresponding to the observed T-value.

EXAMPLE 5–20. If X is uniform on $0 < x < \theta$, the distribution of $X_{(n)}/\theta$, where $X_{(n)}$ is the largest of n independent observations, does not depend on θ:

$$F_{X_{(n)}/\theta}(u) = u^n, \qquad 0 < u < 1.$$

Thus, with $\gamma = .90$,

$$P_\theta\left(\sqrt[n]{.05} < \frac{X_{(n)}}{\theta} < \sqrt[n]{.95}\right) = .90,$$

or equivalently, for all θ,

$$P_\theta\left(\frac{X_{(n)}}{\sqrt[n]{.95}} < \theta < \frac{X_{(n)}}{\sqrt[n]{.05}}\right) = .90.$$

The inequality for θ in parentheses defines a 90 per cent confidence interval for θ.

A confidence interval can be constructed even when a pivotal quantity is not apparent. Again starting with an estimator T, if one chooses t_1 and t_2 depending on θ so that

$$\begin{cases} 1 - F_T(t_2; \theta) = \frac{1}{2}(1 - \gamma), \\ F_T(t_1; \theta) = \frac{1}{2}(1 - \gamma), \end{cases}$$

then

$$P_\theta[t_1(\theta) < T < t_2(\theta)] = F_T(t_2; \theta) - F_T(t_1; \theta) = \gamma,$$

for all θ, and the set

$$I_\gamma(T) \equiv \{\theta \mid t_1(\theta) < T < t_2(\theta)\}$$

has probability γ of covering the true value of θ, whatever it may be. This approach to constructing a confidence interval is not distinct from the earlier one, for the quantity $p(T, \theta) \equiv F_T(T; \theta)$ is pivotal.

There is a good deal of arbitrariness in the construction of a confidence interval—in the choice of the estimator T and in the allocation of the probability $1 - \gamma$ to the tails of the distribution of T. (The probability was allocated with equal amounts in each tail in the previous discussion but could have been allocated differently.) Presumably, it would be best to use an estimator based on a sufficient statistic.

Various notions of "good" confidence intervals have been introduced. The concept of "uniformly most accurate" involves essentially a mimimal probability of covering wrong parameter values. For two-sided intervals, arising usually from putting probability in both tails of T, the notion of a confidence interval of minimum length has been used to define optimality. (See Lehmann [16], p. 177, and Zacks [23], Chapter 10.)

Problems

5–29. It is known from long experience that the reliability of a certain chemical measurement is indicated by a standard deviation $\sigma = .005$ gm/ml. Determine a sample size n such that a 99 per cent confidence interval for μ has a width of .001 gm/ml. (Is it necessary to assume that the measurements are normal in order that the normal table be used?)

5–30. A series of measurements of a certain dimension of 25 parts has a mean 2.3 and a standard deviation .1. Construct an approximate confidence interval for the population mean, assuming that the sample size is large enough to make a normal approximation to the distribution of the sample mean and large enough so that using the sample standard deviation in place of the population standard deviation does not introduce much error. Use a 99 per cent confidence coefficient and then repeat with a 95 per cent confidence coefficient.

5–31. Construct a confidence interval for the mean of the population given the sample mean and variance of the preceding problem, but assuming a sample size of 400.

5–32. Construct the large-sample confidence interval for the mean of a variable with exponential density $(1/\theta)e^{-x/\theta}$, $x > 0$, in terms of the sample mean, using a 95 per cent confidence coefficient.

5–33. A sample of 60 observations from a Poisson population has a mean of 2.1. Construct the corresponding 90 per cent confidence interval for the population mean (using a normal approximation).

6

TESTING HYPOTHESES

An *hypothesis* is a statement or claim about the state of nature. Scientific investigators, industrial quality control engineers, market researchers, governmental decision makers, among others, will often have hypotheses about their particular facets of nature, hypotheses that need substantiation or verification—or rejection—for one purpose or another. To this end, they gather data and let the data support or cast doubt on an hypothesis. The process is termed the "testing of hypotheses" and is one of the main areas of statistical inference.

The testing problem is considered by some to be a matter of choosing between two actions—a decision problem; this approach will be taken up in Chapter 8. Others feel that the structure of decision theory is too restrictive in many situations. Most will agree that the rather classical theory to be presented in this chapter (or at least some of it) is pertinent in one context or another.

6.1 Basic Concepts

A statistical hypothesis may be very precise and specific, asserting that the state of nature is described by a certain completely defined probability model; this is a *simple* hypothesis. Or it may be more general, asserting that the state of nature has a certain property or satisfies a certain condition that is true of more than one specific probability model; this is called a *composite* hypothesis. A composite hypothesis is thought of as being composed of the various simple hypotheses that satisfy the condition defining it.

EXAMPLE 6–1. The statement that X is normally distributed is a composite hypothesis, since the property of normality does not completely define the distribution of X, the state of nature. It includes, for example, the simple hypothesis that X is normal with mean 2 and variance 8. The hypothesis that a random variable Y has mean 2 and variance 8 is composite, since these two moments do not alone define a distribution.

To "test" an hypothesis is to conduct an experiment of chance related to the state of nature and, on the basis of the outcome of the experiment, to decide whether the hypothesis can be "accepted" or should be "rejected." To accept an hypothesis does not mean the same thing to all investigators or in all situations. It does *not* mean, in a statistical experiment, that the hypothesis is "proved" in any rigorous sense, because the data in a sample give only incomplete information about a population and can easily be misleading.

To "accept" may mean to "believe," in the weak sense of being willing to act, or to proceed in some way, as though the hypothesis were true. For example, a lot of manufactured articles may be judged to be of acceptable quality (a statistical hypothesis) for sale or distribution, on the basis of the quality of a sample from the lot. Or an investigator may proceed with further research and experiment in a direction indicated by the hypothesis he accepts in a statistical test. Or a parent may become sufficiently convinced of the prophylactic value of a toothpaste to switch brands, on the basis of test results. In any of these situations, if the hypothesis being tested is not rejected, some are unwilling to say that it is accepted—fearing, perhaps, that their acceptance of it may be construed as conviction of its truth beyond any doubt. (See Section 6.1.3.) Here it will be assumed that an hypothesis is either accepted or (if not accepted) rejected.

A statistical test is based on data from an experiment of chance, and the essence of a test is a *rule* that tells the statistician whether the data he has collected call for accepting or rejecting his hypothesis. The term *test* will refer to this decision rule.

Definition. *A* test *of an hypothesis is a rule that assigns one of the inferences* accept the hypothesis *or* reject the hypothesis *to each forseeable result of an experiment.*

A test is usually described in terms of some statistic $T = t(\mathbf{X})$, a reduction of experimental data, called the *test statistic*. The range of values of T that, according to a given test, call for rejecting the hypothesis being tested is called the *critical region* of the test. The critical region defines the test and conversely.

A critical region C of T-values defines a region in the sample space of $\mathbf{X} = (X_1, \ldots, X_n)$, namely, the set of points $\mathbf{x}$ such that $t(\mathbf{x})$ falls in C. This $\mathbf{x}$-set is also called the critical region. That is, one can say either that a sample $\mathbf{X}$ falls in the critical region of $\mathbf{X}$-values, or that $t(\mathbf{X})$ falls in the critical region of T-values.

In devising tests, it is soon realized that it is difficult to construct a reasonable rule without some consideration, conscious or unconscious, of what *is* true about the model or population when the hypothesis being tested is *not* true. Specification of this alternative is usually part of the structure of the testing problem. The hypothesis to be tested is called the *null hypothesis*, and the set of *other* states of nature or models admitted as possible for a given experiment is called the *alternative hypothesis*. One says that he tests the null hypothesis "against" the alternative hypothesis. The null hypothesis will usually be denoted by H_0 and the alternative either by H_A or by H_1.

A null hypothesis is often rather more specific than the alternative; indeed, it is sometimes a simple hypothesis. Often, it is an hypothesis of "no difference" in the effects of two treatments, an application that perhaps gave rise to the adjective "null." The alternative hypothesis, although it can be and sometimes is a simple hypothesis, is usually more composite than the null hypothesis; and this may account for the asymmetry of the terminology. (It may also account for the attitude of not wanting to accept an hypothesis that is not rejected; for H_0 is often set up as overly precise, a specific model that is almost surely unbelievable.)

EXAMPLE 6–2. In testing the effectiveness of a new dentifrice, one would try it on a sample of children and compare the results with those of trying an old or standard type on a second sample (called the "control" group). The null hypothesis would be that there is no difference in dentifrices, made more precise, perhaps, in the hypothesis that the (population) mean number of new cavities is the same for each dentifrice, the alternative would be that the means are different (without saying how much different). In this instance, the null hypothesis is really too narrow; more likely, what one wants to know is whether he can act as though there is not an *appreciable* difference (rather than a difference of precisely zero), but the stated null hypothesis is easier to work with.

EXAMPLE 6–3. An archeological digging results in 13 skulls presumed to be either all from race A or all from race B, races whose mean skull diameters are known. It is desired to test the hypothesis that the mean μ of the population from which the unearthed skulls came is that of race A against the alternative that it is that of race B. Thus, one tests

$$H_0: \mu = \mu_A \qquad \text{against} \qquad H_1: \mu = \mu_B.$$

Here there is no reason to call one the null and the other the alternative hypothesis—it could have been reversed. If the skull diameter is assumed to be normally distributed with given variance, then both H_0 and H_1 are simple.

The application of a statistical test may lead to the wrong conclusion. Of course, there are good tests and bad tests; but even a good test may cause rejection of H_0 when it should be accepted, or acceptance of H_0 when it should be rejected. These wrong inferences are called *errors* of types I and II:

Type I error: Rejecting H_0 when H_0 is true.

Type II error: Accepting H_0 when H_0 is false.

Whether a test causes errors of one type or the other, when applied to a given set of data, is a matter of chance, depending on the particular sample that happens to be drawn. The performance of a test will be judged according to its tendency to lead to wrong conclusions.

6.1.1 Simple H_0 versus Simple H_A

The simplest problem to analyze is that in which both the null and alternative hypotheses are simple. In such a case, it is a matter of choosing between two models—two possible explanations for the data. If $f(\mathbf{x})$ denotes the density (or probability) function of the sample $\mathbf{X}$, the problem can be expressed as testing

$$H_0: f(\mathbf{x}) = f_0(\mathbf{x}) \qquad \text{versus} \qquad H_1: f(\mathbf{x}) = f_1(\mathbf{x})$$

where f_0 and f_1 are specific density functions (not depending on any parameter). *Error sizes* are defined as follows:

$$\alpha = \text{"size of the type I error"} = P_{H_0}\,(\text{reject } H_0)$$

and

$$\beta = \text{"size of the type II error"} = P_{H_1}\,(\text{accept } H_0).$$

In terms of the critical region C of values of $\mathbf{X}$ calling for rejection by the given test, one has (in the continuous case)

$$\alpha = P_{H_0}(C) = \int_C f_0(\mathbf{x})\, d\mathbf{x},$$

$$\beta = P_{H_1}(C^c) = \int_{C^c} f_1(\mathbf{x})\, d\mathbf{x} = 1 - \int_C f_1(\mathbf{x})\, d\mathbf{x}.$$

The type I error size can be made 0 by choosing $C = \varnothing$, that is, by adopting the rule of never rejecting H_0, no matter what the data turn out to be. But then $C^c = \Omega$, and $\beta = 1$. Likewise, if one adopts the critical region $C = \Omega$, then $\beta = 0$ but $\alpha = 1$. And if he tries to make α just close to 0, it will be found that this tends to make β large, and conversely, as is seen in the following example.

EXAMPLE 6–4. Let $\mathbf{X}$ denote a random sample of size 25 from a normal population with variance $\sigma^2 = 4$, whose mean is known to be either 0 or 1. The problem is to test

$$H_0: X \text{ is normal } (0, 4)$$

against

$$H_1: X \text{ is normal } (1, 4).$$

The sample mean is sufficient for this problem, and the family of normal distributions has a monotone likelihood ratio in terms of the sample mean. The intuitively

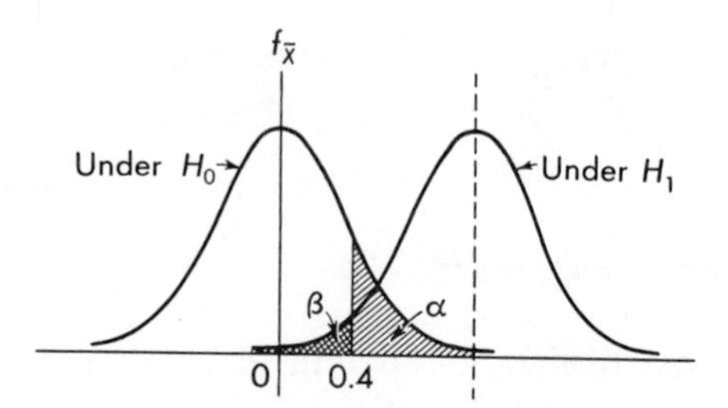

Figure 6–1. Error sizes for Example 6–4.

reasonable test would choose H_1 if $\bar{X}$ is large—larger than some constant, and otherwise choose H_0. The critical region of such a test would be of the form $\bar{X} > K$, and the error sizes for this test are as follows:[1]

$$\alpha = P(\bar{X} > K \mid \mu = 0) = 1 - \Phi\left(\frac{K - 0}{2/\sqrt{25}}\right),$$

$$\beta = P(\bar{X} < K \mid \mu = 1) = \Phi\left(\frac{K - 1}{2/\sqrt{25}}\right).$$

Figure 6–1 shows the density functions of the test statistic $\bar{X}$ under the two states of nature and exhibits areas equal to the error sizes corresponding to the particular test in which $K = .4$, that is, calling for rejection of H_0 if $\bar{X} > .4$. The values are $\alpha = .1587$ and $\beta = .0668$. It can be seen from the graph, by imagining the critical value K at different points along the horizontal scale, that increasing K decreases α but increases β, and that decreasing it increases α and decreases β. This "struggle" between α and β is also evident in a plot of the relation between α and β, given by the parametric equations above (expressing them in terms of the "parameter" K). This curve is shown in Figure 6–2 for the case $n = 4$, and the case $n = 25$, along with the corresponding curve for a different kind of critical region, $\bar{X} < K$. Notice the disastrous effect of using this latter, nonintuitive type of procedure. Notice also the

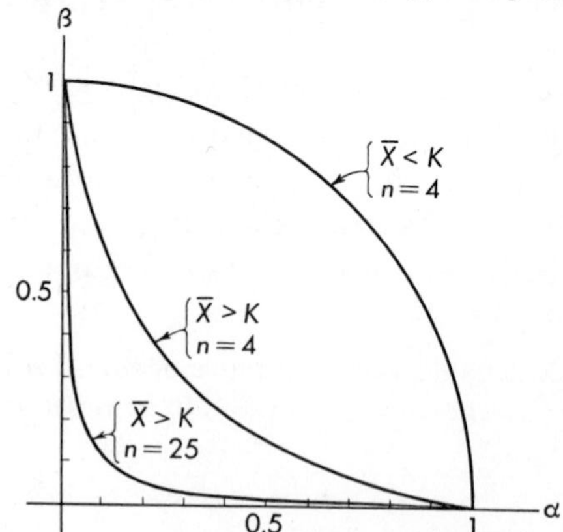

Figure 6–2. Plot of β versus α for Example 6–4.

[1] We take the liberty of using the vertical bar, usually indicating a conditional probability, to set off the state of nature assumed in a computation of probability. It will be seen in Chapter 8 that these uses are not really disparate.

effect of the larger sample size—the error sizes are generally smaller, and the curve is pulled in toward (0, 0).

In view of the inherent perversity of the error sizes—in that reducing one tends to increase the other—it is not clear how to choose a test, even if (as in Example 6–4) the proper *kind* of critical region to use is fairly clear. A number of approaches can be, and are, used.

One approach is that which considers the type I error the more serious. (This can be arranged by suitable labeling of the hypotheses involved.) An acceptable level α is prescribed, and the test is chosen to minimize β without letting the type I error size exceed the prescribed α. Even in this approach, however, the "acceptable" α is usually subjective. The investigator will have an intuitive feeling that it is all right to have a 1 per cent chance, say, of having H_0 rejected when it is true.

Another approach, to be developed in Chapter 8, is that which treats the state of nature as random and assigns probabilities to H_0 and H_1. If this be done, of course, one can compute a probability of error:

$$P(\text{test errors}) = \alpha P(H_0) + \beta P(H_1).$$

But better yet, as will be done in Chapter 8, costs can be attached to errors of each type so that the expected cost of using a test can be computed:

$$E(\text{cost}) = K_1 \alpha P(H_0) + K_2 \beta P(H_1),$$

where K_1 and K_2 are the respective costs of making errors of types I and II. The "best" test would then be one that minimizes the expected cost.

Sometimes it will be stated that a result that causes H_0 to be rejected by a 5 per cent test is "significant," and if rejected by a 1 per cent test, "highly significant." This kind of convention—whatever it means—is quite arbitrary. Nevertheless, tables of percentiles of test statistics are often given with the assumption that only a 1 per cent or a 5 per cent test will be used. (See Table IV in the Appendix, which perpetuates this phenomenon.)

It should be pointed out that whereas with a given α one can do only so well with β, when the sample size is fixed, it *is* usually possible to reduce both error sizes by taking more observations. This was apparent in Example 6–4. In the tests of that example one could achieve any specified α and β (except $\alpha = \beta = 0$) by choosing the critical boundary and sample size judiciously. (See Problem 6–5.)

Problems

6–1. Determine in each case whether the hypothesis given is simple or composite:

(a) A pair of dice is "straight."
(b) A pair of dice is "crooked."
(c) $E(X) = 3$.

(d) X is exponential with mean 3.
(e) X is uniformly distributed.
(f) A coin is biased.
(g) The distribution function of X is $1 - e^{-x}$, $x > 0$.
(h) The distribution function of X is not $1 - e^{-x}$, $x > 0$.

6–2. A bag contains five beads, some white and some not. Consider testing

$$\begin{cases} H_0\text{: At most one bead is white,} \\ H_1\text{: At least two beads are white,} \end{cases}$$

on the basis of a sample of two drawn without replacement from the bag.

(a) List all possible tests based on the sample of two—sensible or not.
(b) Determine the probability of a type I error for each simple hypothesis in H_0 and the probability of a type II error for each simple hypothesis in H_1, for the test that rejects H_0 if and only if any white beads are drawn.
(c) Is there a test with smaller type II error sizes than the test in (b)?

6–3. Suppose, in Example 6–4, that a type I error costs twice as much as a type II error, and that H_1 is twice as likely as H_0. Determine the best test among those of the form $\bar{X} > K$, when $n = 4$.

6–4. Referring to Example 6–4, compute the error sizes for the test that accepts H_0 if $0 < \bar{X} < 1$ but otherwise rejects H_0. [Notice where the point (α, β) falls in the plot of Figure 6–2.]

6–5. In testing $\mu = 0$ against $\mu = 1$ in a normal population with variance 4, express the error sizes in terms of n and K (cf. Example 6–4). Determine n and K:

(a) So that $\alpha = \beta = .01$. (b) So that $\alpha = .01$ and $\beta = .10$.

6–6. Let p denote the probability of "heads" in the toss of a coin. Construct all possible tests of the hypothesis that $p = \frac{1}{2}$ against the alternative that $p = 1$, based on the results of two tosses of the coin, and determine the error sizes for each test. (Since the number of heads in the two tosses is sufficient, use this statistic.)

6–7. It is desired to test the hypothesis that the expected number of arrivals in a 1-min period is $m = 1$ against the alternative that it is $m = 1.5$, assuming Poisson arrivals, and using Y, the number of arrivals in a 10-min period. Determine α and β for the test with critical region $Y > 12$.

6.1.2 The Power Function

When H_0 and H_A are not both simple, the error sizes are correspondingly not both defined. Thus if H_A is not simple, the expression $P_{H_A}(C^c)$ is not meaningful; for H_A is not a single, specific model, in which such a probability can be computed. However, the probability of the acceptance region C^c, or equivalently the probability of the critical region C, *can* be computed for each simple hypothesis in H_A and is a quantity of interest in judging performance. So instead of two error sizes, the necessary tool is a function—the *power function* of the test:

$$\pi(\theta) \equiv P_\theta(\text{reject } H_0) = P_\theta(C),$$

where θ denotes a particular model or simple hypothesis or state of nature—in either H_0 or H_A, and the subscript on P denotes the state of nature in which P is calculated.

The term *power* originated in describing the ability of a test to detect that H_0 is false when such is the case. This is the meaning of "power" in the term *power function* as it is defined for θ in H_A. However, the term has been carried over to denote the probability that a test rejects H_0 for θ in either H_0 or H_A.

Sometimes the probability that H_0 is *accepted* by a test is used in place of the power function. This probability, also a function of θ, is called the *operating characteristic* of the test:

$$OC(\theta) \equiv P_\theta(\text{accept } H_0) = P_\theta(C^c).$$

It is complementary to the power function,

$$OC(\theta) = 1 - \pi(\theta),$$

and it is a matter of taste which to use. In some areas of application, the operating characteristic function is more common, and in theoretical work one tends to find the power function as the descriptive vehicle.

For a test to be good it should not often make errors. Ideally, one would want a probability 1 of rejecting H_0 when it should be rejected—that is, for any θ on H_A, and a probability 0 of rejecting H_0 when it should be accepted—that is, for any θ in H_0:

$$\text{ideal power function} = \begin{cases} 0, & \text{for } \theta \in H_0, \\ 1, & \text{for } \theta \in H_A. \end{cases}$$

A test carried out on a finite amount of data cannot ordinarily achieve this ideal, and all one can ask is that $\pi(\theta)$ be small on H_0 and large on H_A.

EXAMPLE 6–5. Consider a Bernoulli experiment—success with probability p and failure with probability $1 - p$, and the following composite hypotheses:

$$\begin{cases} H_0: & 0 \le p \le .5, \\ H_1: & .5 < p \le 1. \end{cases}$$

Consider the particular test based on three observations that rejects H_0 if three successes occur. The power function for this test is as follows:

$$\pi(p) = P_p(3 \text{ successes in 3 trials}) = p^3.$$

For the test that rejects H_0 when either two or three successes occur in three trials, the power function is

$$\pi(p) = P_p(2 \text{ or 3 successes in 3 trials}) = 3p^2(1 - p) + p^3.$$

These two power functions are plotted on the same axes in Figure 6–3.

Notice that the division of [0, 1] into H_0 and H_1 does not enter the computation of $\pi(p)$, but of course it will enter in deciding whether a power function is acceptable or not. The power function for the second test is higher than that for the first test, not only on H_1 (which is desirable) but also on H_0 (which is not so desirable).

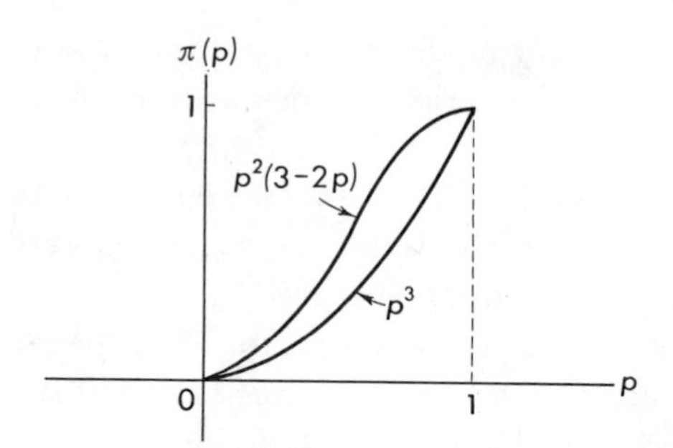

Figure 6–3. Power curves for Example 6–5.

EXAMPLE 6–6. Consider a normal population with variance 9; it is desired to test

$$H_0: \mu \leq 1 \qquad \text{against} \qquad H_1: \mu > 1,$$

using a critical region of the form $\bar{X} > K$, based on a random sample of size n from the population. The power function is

$$\begin{aligned}\pi(\mu) &= P_\mu(\mathbf{X} \text{ in } C)\\ &= P_\mu(\bar{X} > K) = 1 - \Phi\left(\frac{K - \mu}{3/\sqrt{n}}\right) = \Phi\left(\frac{\mu - K}{3/\sqrt{n}}\right).\end{aligned}$$

This function of μ looks like a cumulative normal distribution; it is shown in Figure 6–4 for $K = 1, n = 36$, and for $K = 1, n = 144$. Both functions are high on H_1 and not so high on H_0. The center of symmetry of the graphs is at $\mu = 1$, and it is clear from the computations above that a change in the boundary of the critical region from $K = 1$ to, say, $K = 2$ would shift the curve laterally so that the center of symmetry would be $\mu = 2$.

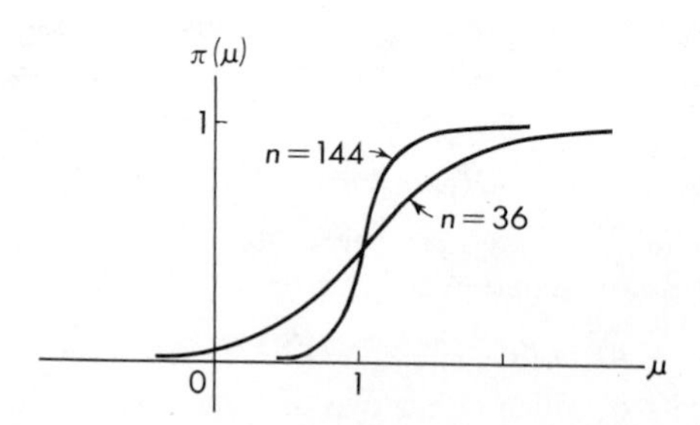

Figure 6–4. Power curves for Example 6–6.

The problems in these last two examples are what is called *one-sided*—the states of nature are indexed by a single real parameter, and all of the θ-values in H_0 lie to one side of those in H_A. As in the examples, it is often the case that a good test will involve a one-sided critical region of value of the test statistic.

(This will be investigated further in Section 6.2.) A *two-sided* problem, again for the case of a single real parameter, is one of the form

$$\begin{cases} H_0: & \theta_1 \le \theta \le \theta_2, \\ H_A: & \theta > \theta_2 \quad \text{or} \quad \theta < \theta_1. \end{cases}$$

This is often simplified to the form

$$\begin{cases} H_0: & \theta = \theta_0, \\ H_A: & \theta \ne \theta_0. \end{cases}$$

(Sometimes a problem is cast in this form when it is really a three-action problem—different actions being called for, when H_0 is not true, depending on whether θ is on one side or the other of H_0.) For such problems, if the likelihood ratio is monotone in terms of a statistic T, the type of critical region usually used (and intuitively plausible) is two-sided, excessively large or excessively small values of T both calling for rejection of H_0.

EXAMPLE 6–7. Let X have a Bernoulli distribution with unknown p, and let $S = X_1 + \cdots + X_5$ denote the number of successes in five independent trials. To test $p = .5$ against $p \ne .5$, it would seem reasonable to reject $p = .5$ for values of S that are either too small or too large. The test with critical region $S = 0, 1, 4$, or 5 has the power function

$$\begin{aligned} \pi(p) &= P_p[S = 0, 1, 4, \text{ or } 5] \\ &= 1 - \binom{5}{2} p^2(1-p)^3 - \binom{5}{3} p^3(1-p)^2 \\ &= \frac{3}{8} + 5(p - .5)^2 - 10(p - .5)^4, \end{aligned}$$

as shown in Figure 6–5. It is symmetrical about $p = .5$ and has a minimum at that value of p. Since H_0 is a simple hypothesis, α is uniquely defined as $\pi(.5) = \frac{3}{8}$.

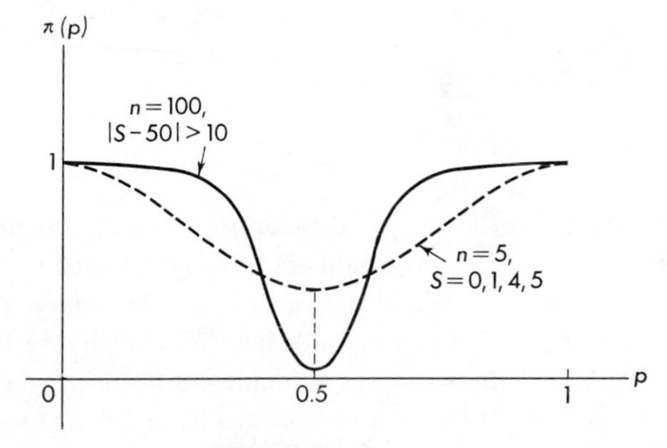

Figure 6–5

To see the effect of an increase in sample size, consider also the test under which H_0 is rejected if more than 60 or fewer than 40 successes occur in 100 independent trials. With this sample size, the sample sum S is approximately normal:

$$\pi(p) = P_p(|S - 50| > 10) \doteq 1 - \Phi\left(\frac{60.5 - 100p}{\sqrt{100pq}}\right) + \Phi\left(\frac{39.5 - 100p}{\sqrt{100pq}}\right).$$

The graph of this function is shown in Figure 6–5. Now α is $\pi(.5) = .036$.

Although the probability of rejecting H_0 when H_0 is true is not uniquely defined (depending as it does on the state θ), one might define a "size" for type I errors as the maximum (or supremum, if the maximum does not exist) of such probabilities:

$$\alpha = \max_{\theta \text{ in } H_0} P_\theta(\text{reject } H_0) = \max_{\theta \text{ in } H_0} \pi(\theta).$$

This is sometimes called the *significance level* of the test. The size for type II errors could similarly be taken to be

$$\beta = \max_{\theta \text{ in } H_1} [1 - \pi(\theta)].$$

In problems in which θ is a real parameter or a vector of real parameters and where $\pi(\theta)$ does not jump at the boundary between the region H_0 and the region H_1, the sizes defined in this way will add up to 1. They could not *both* be made small, no matter how closely the power function approximates the ideal function for θ not on the boundary. A typical situation might be that in Figure 6–6. Increasing the sample size, which might steepen the power curve, as in Example 6–6, could not reduce both α and β to acceptably small levels.

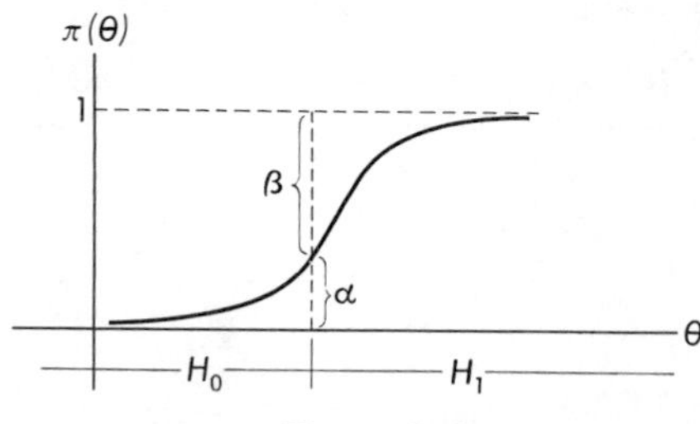

Figure 6–6

In such situations, where the division between H_0 and H_1 is a boundary point of each, a finite sampling experiment could not really be expected to discriminate well between H_0 and H_1 if the actual state is near that boundary point. Sometimes an "indifference zone" or "no man's land" including the boundary is defined, say, from θ' to θ'', in which one is indifferent to the action taken. The ideal power function is unrestricted there, as in Figure 6–7, for the one-sided case.

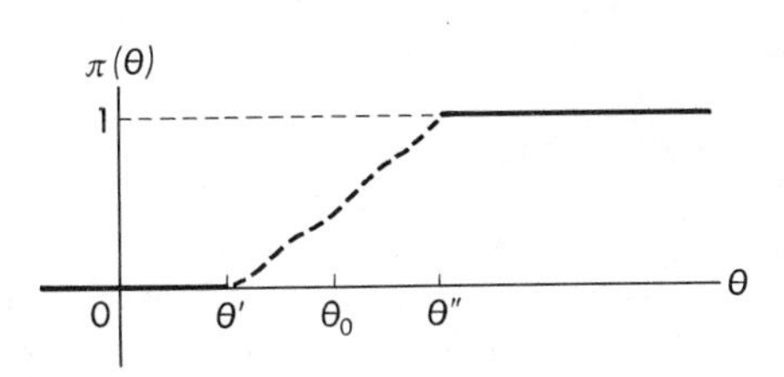

Figure 6–7

With such a modification, the error sizes could be defined to be

$$\alpha = \max_{H_0 - I} \pi(\theta) \qquad \text{and} \qquad \beta = \max_{H_1 - I} [1 - \pi(\theta)],$$

where I denotes the set of values of θ in the indifference zone. These are shown in Figure 6–8, for a problem with a monotone power function.

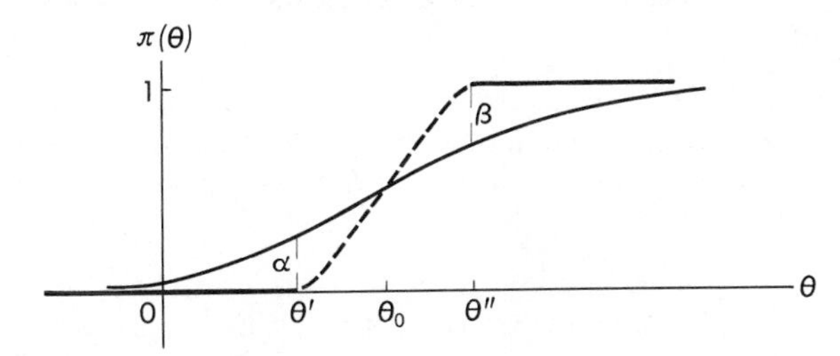

Figure 6–8. Modified error sizes.

Actually, the modification of a testing problem in this way amounts to replacing the given hypotheses with the simple hypotheses $\theta = \theta'$ and $\theta = \theta''$, and the error sizes are then those defined for this simpler problem. With the sample size and the critical boundary at one's disposal, he can select them to produce the specified α and β. However, having used this method as a practical means of constructing a test, one should realize that the test cannot be characterized by the two numbers α and β; the whole power function should be examined to see what protection is afforded at other values of θ.

EXAMPLE 6–8. Let X be normal with variance 9. It is decided that the probability of rejecting $\mu \leq 1$ when $\mu = .8$ should not exceed .05, and the probability of accepting $\mu \leq 1$ when $\mu = 1.2$ should not exceed .10. If the test used is $\bar{X} > K$, then

$$.05 = P_{\mu=.8}(\bar{X} > K) = 1 - \Phi\left(\frac{K - .8}{3/\sqrt{n}}\right)$$

and

$$.10 = P_{\mu=1.2}(\bar{X} < K) = \Phi\left(\frac{K - 1.2}{3/\sqrt{n}}\right).$$

These are readily solved with the aid of Table I (Appendix), giving the values of the standard normal distribution function $\Phi(z)$, to obtain $n = 481$ and $K = 1.025$.

Problems

6–8. For testing $p = \frac{1}{2}$ in a Bernoulli model, using three independent observations, a critical region with a large and small number of successes seems appropriate. Sketch the power function for the critical region $Y = 0$ or 3, and for the critical region $Y \neq 2$, where Y is the number of successes. Determine α for each test (note that it is uniquely defined), and compare the tests with respect to power on $H_A: p \neq \frac{1}{2}$.

6–9. Consider a test with critical region $\bar{X} > K$, where $\bar{X}$ is the mean of a random sample from a normal population with standard deviation 10. Select K and the sample size n so that the probability of the critical region is .05 when $\mu = 100$ and .98 when $\mu = 110$. Sketch the power function of the test. For what H_0 and H_1 is the test suited?

6–10. A lot contains 10 articles. To choose between accepting and rejecting the lot, a sample of 4 is drawn (without replacement). If more than 1 defective is found, the lot is rejected, but otherwise it is accepted. Determine the operating characteristic function of the test.

6–11. A test concerning the variance of a normal population has the critical region $S^2 > K$, where S^2 is the variance of a random sample of size 100. Use the asymptotic normality of the sample variance to determine K and the power function, given that the power is .98 at $\sigma^2 = .05$.

6–12. Determine the power function of the test with critical region $|\bar{X} - 1| > K$, where $\bar{X}$ is the mean of a random sample of size 25 from a normal population with variance 1, and K is determined so that in testing $\mu = 1$ against $\mu \neq 1$, the size of the type I error (which is uniquely defined) is .05.

6.1.3 Tests of Significance

The discussion of hypothesis testing so far has been in the tradition of J. Neyman and E. S. Pearson, who developed these ideas in the early 1930s. Prior to that time, hypotheses were subjected to "significance testing," a process vigorously championed by R. A. Fisher, who, though admitting that decision making is proper in industry, argued that it has no place in scientific inquiry. Belittling the Neyman–Pearson theory, he continued to be a "significance tester" along with many followers, who, even now, will reject but not accept an hypothesis.

In significance testing the investigator begins with a "null hypothesis"—an hypothesis that he would like to disprove on the basis of data he has gathered or will gather. It may be that the hypothesis has been developed according to some physical theory, and that preliminary observations suggest the possibility that the development of the theory may have been naïve or overly optimistic. Or it may be that the experimenter has a "treatment" that he thinks is apt to be effective in some way; he takes the hypothesis that the treatment has no effect as the null hypothesis and hopes that his data will disprove this.

Like the hypothesis tester of the previous discussion, the significance tester chooses a test statistic T whose extreme values would be thought to suggest the falsity of his null hypothesis. If the null hypothesis is of the form $\theta = \theta_0$, where θ is a real parameter, the statistic T would ordinarily be a good estimator of θ. The distribution of T under H_0 is calculated, and an observed value of T is judged according to this null distribution as being either reasonable (not unexpected), or so far out in the tail of the distribution (so "improbable") as to present a strong argument for the falsity of H_0. Such an extreme value is said to be "significant," and the null hypothesis is rejected when it is observed. If the rule "reject H_0 if $T > t_0$" is adopted, the *significance level* α of this rule or test is defined to be the probability of the condition $T > t_0$ under H_0 (as in the earlier discussion). Alternatively, one may compute the "P-value" corresponding to an observed value $T = t$ as the probability of observing $T = t$ or something more extreme. (This is the "α" at which the observed value $T = t$ is *just* significant.)

If T turns out to have a value that is not "significant," then no conclusion is drawn. Says R. A. Fisher ([10, p. 45]): "A test of hypothesis contains no criterion for 'accepting' a hypothesis. According to circumstances it may or may not influence its acceptability." Perhaps one reason for this is that there may be many other hypotheses or models that could equally well account for a value of T that is not significant. Indeed, when one asserts something as mathematically precise as $\mu = 17$ about a phenomenon of nature, he is almost sure to be wrong; and a consistent test will reject this assertion if he takes enough data.

EXAMPLE 6–9. Suppose a new fertilizer is to be tested for effectiveness as compared with a tried and true compound. The null hypothesis would be that there is no gain in mean yield with the new fertilizer. The test statistic T would likely be the average yield from several plots treated with the new fertilizer, and the null hypothesis would then be rejected at the 5 per cent significance level if $T > t'$, where the probability of $T > t'$ is .05 under H_0. If it should turn out that $T < t'$, no conclusion would be drawn. Indeed, there is almost sure to be some difference in mean yield with the new treatment, and the sample was just not big enough to detect the difference. On the other hand, the difference may be so slight that one would not want to reject H_0, having in mind that an effective treatment should really produce an *appreciable* difference in yield.

A significance tester, like a Neyman–Pearson tester, may make a mistake in his inference, owing to quirks in his data—but only one type of mistake, that of rejecting H_0 when it is true. The α, computed as the probability of rejecting H_0 when it is true, is not to be taken as the probability of an erroneous decision: "The calculation is based solely on a hypothesis, which, in the light of the evidence, is often not believed to be true at all, so that the actual probability of erroneous decision, supposing such a phrase to have any meaning, may be, for this reason only, much less than the frequency specifying the level of significance." (Fisher, [10], p. 45.) And again ([10], p. 47): "In general, tests of significance are based on hypothetical probabilities calculated from their null hypotheses. They do not generally lead to any probability statements about the real world, but to a rational and well-defined measure of reluctance to the acceptance of the hypotheses they test."

In any case, a significance level is arbitrary, a point well put by Fisher ([10], p. 45): "... in fact no scientific worker has a fixed level of significance at which from year to year, and in all circumstances, he rejects hypotheses; he rather gives his mind to each particular case in the light of his evidence and his ideas."

If the choice of α is arbitrary, even more so is the choice of critical region. Fisher seems to imply, and others perpetuate this notion, that improbability alone is evidence of incorrectness of the model in which it is calculated. But, for example, a coin that falls heads 50 times in 100 tosses is surely not rejected as unfair, even though this contingency is extremely unlikely (probability of .08 for a fair coin). In his intuitive choice of critical region, the significance tester can scarcely avoid consideration, conscious or unconscious, of the alternatives to the null hypothesis, and of the desire for high power on the alternatives.

6.1.4 Tests and Confidence Intervals

A test of $\theta = \theta_0$ can be constructed in a natural way using a confidence interval for θ. The rule defining the test is to reject θ_0 if it is not included in the confidence interval obtained from the observation $\mathbf{X}$. The size of the test (type I error size or significance level) is easily computed. For if $I_\gamma(\mathbf{X})$ is a confidence interval with confidence coefficient γ, then for all θ

$$P_\theta(\theta \in I_\gamma(\mathbf{X})) = \gamma,$$

and consequently the probability of rejecting a specific value θ_0 when θ_0 is correct is

$$\alpha = P_{\theta_0}(\theta_0 \notin I_\gamma(\mathbf{X})) = 1 - \gamma.$$

So, for instance, a 95 per cent confidence interval gives rise to a 5 per cent significance test.

Indeed, there is a duality here, because a family of tests can be used to construct a confidence interval. Suppose $C(\theta_0)$ is the critical region of a test of level α for $\theta = \theta_0$, defined for each θ_0:

$$P_{\theta_0}[\mathbf{X} \in C(\theta_0)] = \alpha.$$

Then define, for each sample point $\mathbf{x}$,

$$I(\mathbf{x}) = \{\theta \mid \mathbf{x} \notin C(\theta)\}.$$

There is a complete equivalence of the events

$$\{I(\mathbf{X}) \ni \theta\} \quad \text{and} \quad \{\mathbf{X} \notin C(\theta)\},$$

which means that their probabilities agree for any θ:

$$P_\theta(I(\mathbf{X}) \ni \theta) = 1 - P_\theta(\mathbf{X} \in C(\theta)) = 1 - \alpha.$$

So $I(\mathbf{X})$ is a confidence region for θ with confidence coefficient $1 - \alpha$.

The confidence region obtained as above from a family of tests may or may not be an interval. If one uses a two-sided critical region for a test statistic T that is linked closely to the parameter θ, then the confidence region will turn out to be an interval with finite limits, thought of as a two-sided confidence interval. If a one-sided critical region for T is used, then the confidence region turns out to be one-sided—an interval with one finite limit, extending to infinity in the other direction. The one limit is called a *confidence bound* (upper or lower, depending on which side of the interval it bounds) for θ.

EXAMPLE 6–10. Consider the family of normal distributions with unknown mean μ and variance $\sigma^2 = 1$, and the family of critical regions (one-sided) for given α:

$$C(\mu_0) = \{\mathbf{x} \mid \bar{x} > K\},$$

where K is determined so that

$$P_{\mu_0}(C) = P_{\mu_0}(\bar{X} > K) = 1 - \Phi\left(\frac{K - \mu_0}{1/\sqrt{n}}\right) = \alpha.$$

That is,

$$K = \mu_0 + \frac{z_{1-\alpha}}{\sqrt{n}},$$

where z_γ is the 100γ percentile of the standard normal distribution. Now define

$$\begin{aligned} I(X) &= \{\mu \mid \bar{X} < K\} \\ &= \left\{\mu \mid \bar{X} < \mu + \frac{z_{1-\alpha}}{\sqrt{n}}\right\} \\ &= \left\{\mu \mid \mu > \bar{X} - \frac{z_{1-\alpha}}{\sqrt{n}}\right\}. \end{aligned}$$

Thus the one-sided interval

$$\mu > \bar{X} - \frac{z_{1-\alpha}}{\sqrt{n}}$$

is a confidence region with confidence coefficient $1 - \alpha$, and $\bar{X} - z_{1-\alpha}/\sqrt{n}$ is a lower confidence bound for μ.

6.1.5 Large-Sample Tests for the Mean

Large-sample or asymptotic distributions of test statistics, which are often indifferent to the particular nature of the parent population, are usually simpler than small-sample distributions. A number of large-sample tests will be presented in subsequent chapters, but the Central Limit Theorem and the convergence of sample moments already provides the basis for large-sample testing of hypotheses about a population mean.

The distribution of $\bar{X}$ is approximtely normal with $E(\bar{X}) = EX$ and var $\bar{X} = \sigma^2/n$. So the statistic

$$Z = \frac{\bar{X} - \mu_0}{\sigma/\sqrt{n}}$$

(assuming σ to be known), which measures how far $\bar{X}$ is from μ_0, is approximately standard normal when $\mu = \mu_0$. And the power function of the critical region $Z > K$, or of the region $|Z| > K$, can be computed from Table I in the Appendix as a function of the true population mean.

When σ is not known, the value of

$$Z' = \frac{\bar{X} - \mu_0}{S/\sqrt{n}}$$

will be close to Z, inasmuch as S is (with high probability) close to σ when n is large. So critical regions based on Z' can be handled in the same way, with approximate power computable from Table I in the Appendix for $\Phi(z)$.

EXAMPLE 6–11. In large samples, the quantity

$$Z' = \frac{\bar{X} - \mu}{S/\sqrt{n}}$$

is approximately pivotal, and

$$P(|Z'| < 2) \doteq .95.$$

Hence, the interval

$$\bar{X} - \frac{2S}{\sqrt{n}} < \mu < \bar{X} + \frac{2S}{\sqrt{n}}$$

is an approximate 95 per cent confidence interval for μ, for any population with finite variance.

6.1.6 Randomized Tests

In the theory of testing, it is convenient to introduce the notion of a *randomized test*. Such a test is defined by a function ϕ on the sample space of the observations $\mathbf{X}$, a function with values on the range [0, 1], employed in testing as follows:

$$P(\text{reject } H_0 \text{ when } \mathbf{X} = \mathbf{x}) = \phi(\mathbf{x}).$$

That is, given the function $\phi(\cdot)$, one observes $\mathbf{X} = \mathbf{x}$, calculates $\phi(\mathbf{x})$, and then tosses a biased coin with $P(\text{heads}) = \phi(\mathbf{x})$ to decide between rejecting and accepting H_0.

In practice, one may never find statisticians tossing coins—bringing in extraneous experiments—to help in testing, yet the theory of tests is a little nicer if randomized tests are included to round out the range of possibilities.

A *non*randomized test, with critical region C, is included as a special case of a randomized test. The ϕ for a region C is just its indicator function:

$$\phi(\mathbf{x}) = \begin{cases} 1, & \text{if } \mathbf{x} \in C, \\ 0, & \text{if } \mathbf{x} \notin C. \end{cases}$$

The "coin" to be tossed is thus two-headed or two-tailed, and once $\mathbf{X} = \mathbf{x}$ is observed the decision is not random.

An important kind of randomized test is one in which the extra randomization really occurs only on the boundary of some region. The way in which this kind of test might be used to achieve a specified significance level in a problem in which the test statistic has a discrete distribution is illustrated in the following example.

EXAMPLE 6–12. Consider the problem of testing between $\lambda_0 = .1$ and $\lambda_1 = .2$ in a Poisson distribution for the number of events in an interval of 1 min, using tests with critical regions of the form $\sum X_i > K$, where X_i is the number of events in the ith minute in a series of 1-min intervals. Using 10 such intervals (that is, a sample of size 10), one finds the type I error sizes to be as follows:

Test	$\sum X_i > 0$	$\sum X_i > 1$	$\sum X_i > 2$	$\sum X_i > 3$
α	.6321	.2642	.0803	.0190

None of these gives exactly, say, .05. So if a test with $\alpha = .05$ is desired, it must be a randomized test. Consider, then, the randomized test that rejects λ_0 if $\sum X_i > 3$, accepts λ_0 if $\sum X_i < 3$, and rejects λ_0 with probability p if $\sum X_i = 3$. The α for this test is

$$\begin{aligned} P_{\lambda_0}(\text{reject } \lambda_0) &= P\big(\text{reject } \lambda_0 \mid \textstyle\sum X_i > 3\big)P_{\lambda_0}\big(\textstyle\sum X_i > 3\big) \\ &\quad + P\big(\text{reject } \lambda_0 \mid \textstyle\sum X_i = 3\big)P_{\lambda_0}\big(\textstyle\sum X_i = 3\big) \\ &\quad + P\big(\text{reject } \lambda_0 \mid \textstyle\sum X_i < 3\big)P_{\lambda_0}\big(\textstyle\sum X_i < 3\big) \\ &= 1 \times .0190 + p \times .0613 + 0 \times .9197. \end{aligned}$$

This can be made equal to .05 by taking $p = .506$.

The power function of a randomized test $\phi(\mathbf{x})$ is computed by averaging the conditional probability of rejecting H_0, given $\mathbf{X} = \mathbf{x}$ with respect to the distribution of the condition:

$$\begin{aligned} \pi_\phi(\theta) = P_\theta(\text{reject } H_0) &= E_\theta[P(\text{reject } H_0 \mid \mathbf{X} = \mathbf{x})] \\ &= E_\theta[\phi(\mathbf{X})]. \end{aligned}$$

Problems

6–13. In testing hypotheses about the mean of a normal population with unit variance, the mean of a sample of size 100 is found to be 2.15. In testing $\mu = 2$ against $\mu > 2$, what is the level at which this result would be just significant?

6–14. The mean and standard deviation blood platelet counts in a sample of 153 male cancer patients were found to be $\bar{X} = 359.9$ and $S = 169.9$ (measured in 1000 per mm^3).

(a) Construct a 95 per cent lower confidence bound for μ, the population mean platelet count.
(b) The mean for healthy males is assumed to be 235. Test the hypothesis that the mean μ of the cancer patient population is 235 against $\mu > 235$, at $\alpha = .01$.
(c) Determine the power function of the test in (b), (using the S from the sample in place of σ).

6–15. Consider the test $\bar{X} > K/\sqrt{n}$ for $\mu = 0$ against $\mu > 0$ in a population with $\sigma = 1$.

(a) Determine the power function.
(b) Determine n and K so that $\pi(.1) = .95$ and $\pi(-.1) = .05$.

6–16. Consider the two critical regions C_1 and C_2, and suppose a coin is tossed to decide whether to use C_1 or C_2.

(a) Express the power function of this procedure in terms of the power functions $\pi_{C_1}(\theta)$ and $\pi_{C_2}(\theta)$.
(b) Express the procedure as a randomized test by defining an equivalent $\phi(\mathbf{x})$.

6–17. A test based on a statistic T calls for rejecting H_0 when $T < 2$, accepting H_0 when $T > 2$, and tossing a (fair) coin when $T = 2$.

(a) Give the function $\phi(t)$ that defines this test.
(b) If T is discrete and H_0 is simple, with $P_{H_0}(T = t) = f_0(t)$, express α in terms of $f_0(t)$.

6–18. Determine a randomized test of level .10 for the problem of Example 6–12.

6.2 Evaluation and Construction of Tests

As in the problem of estimation there are two questions to be considered: (1) How well does a given test perform? and (2) How does one construct a good test? And again the answers are not altogether satisfactory, except in simple, usually unrealistic problems.

The performance of a test will be assessed in terms of its power function. But comparing functions always involves some measure of arbitrariness. Even in the case of simple H_0 and H_1, in which the power function is a two-point function, proper comparisons are not clear; one settles for a class of tests, each optimal in a sense, but with no obvious choice of best within the class.

There seems to be no analog of efficiency for the testing problem, but there are definitions for unbiasedness and consistency:

Definition. *A test is* unbiased *if its power on H_0 is always less than its power on H_A:*

$$\inf_{H_A} \pi(\theta) \geq \sup_{H_0} \pi(\theta).$$

Definition. *A sequence of tests $\{\phi_n\}$ each of given size α is* consistent *if their power functions approach unity for all θ in the alternative set:*

$$\pi_{\phi_n}(\theta) \to 1, \qquad \text{for } \theta \in H_A.$$

Whether these properties of tests are significant—good properties for a test to have—is a matter of taste. A general definition of unbiasedness that includes this definition for the case of testing, as well as the definition for estimators, will be given in Chapter 8. It will then be seen that the property of unbiasedness is not more significant here than it was in the case of estimation. A general definition of consistency will be given, but it will not quite include the above definition (although it will include the definition for estimators).

EXAMPLE 6–13. Let X be a random sample from a normal population with unit variance, and consider the critical region $\bar{X} > K_n$, with K_n so defined that, in testing $\mu = 0$,

$$P_{H_0}(\bar{X} > K_n) = 1 - \Phi(\sqrt{n}K_n) = \alpha,$$

or

$$K_n = \frac{z_{1-\alpha}}{\sqrt{n}}.$$

The power function of the test is

$$\pi(\mu) = P_\mu(\bar{X} > K_n) = \Phi\left(\frac{\mu - K_n}{1/\sqrt{n}}\right).$$

As n becomes infinite, $\pi(\mu)$ tends to 1, if $\mu > 0$. Moreover, for each n, $\pi(\mu)$ is an increasing function, with larger values on $\mu > 0$ than at $\mu = 0$. Hence, for testing $\mu = 0$ against $\mu > 0$, the region $\bar{X} > K_n$ is unbiased (for any n) and defines a consistent family of tests.

It is perhaps true that most tests used in statistical practice have been proposed on intuitive grounds, and, indeed, intuition does remarkably well in leading to tests that have good performance characteristics. Moreover, tests based on intuition usually are fairly simple to carry out, involving test statistics that are relatively easy to compute.

The next section will present a method, based on likelihood, for constructing optimal tests in the case of simple null and alternative hypotheses. And in

Section 6.2.2 a class of optimal tests will be defined for certain problems involving a single real parameter that are "one-sided." For more complicated problems, however, there seems to be no systematic approach or "principle" that will always lead to good tests. An approach based on a generalized maximum likelihood notion will be studied in Section 6.2.3, an approach that often gives good tests but is not foolproof.

It should be pointed out that a test may as well be based on a sufficient statistic. For, if T is sufficient and $\phi(\mathbf{x})$ is any test, then the test

$$\phi^*(t) \equiv E[\phi(\mathbf{X}) \mid T = t]$$

has the power function

$$\pi_{\phi^*}(\theta) = E\{E[\phi(\mathbf{X}) \mid T = t]\} = E[\phi(\mathbf{X})] = \pi_\phi(\theta),$$

exactly the same as the power function of ϕ. Thus, any test is equivalent to a randomized test based on a sufficient statistic.

6.2.1 Best Tests of Simple H_0 versus Simple H_1

When the null and alternative hypotheses are both simple, the error sizes α and β are uniquely defined (cf. Section 6.1.1). Comparing tests would involve comparing their characterizing pairs (α, β), but there is not a unique way of ordering such pairs—of ordering points in the plane. It could be said that (.1, .2) is better than (.2, .4), for instance, because when *both* error sizes for one test are smaller than the corresponding error sizes for a second test, the first is surely better. It is not clear, however, whether (.1, .3) is better or worse than (.3, .1).

To know what available tests are like in terms of their (α, β)'s, it will be useful to have the following result.

Theorem. *The set of* (α, β)*-representations of the randomized tests of a simple* H_0 *versus a simple* H_1 *is* convex. *Moreover, it is included in the closed unit square and contains the points* (0, 1) *and* (1, 0).

A set is *convex* if and only if it contains all of the line segment joining any two of its points. (It has no "dimples" or indentations.) Figure 6–9 shows a set that is convex and one that is not; in the latter case, a pair of points is indicated for which the line segment joining them is not in the set. Figure 6–10 shows possible (α, β)-sets; there is at least one test for each α $(0 \le \alpha \le 1)$, namely, the test $\phi(\mathbf{x}) \equiv \alpha$, and tests of this type (which ignore the data) are represented by points on the line $\alpha + \beta = 1$.

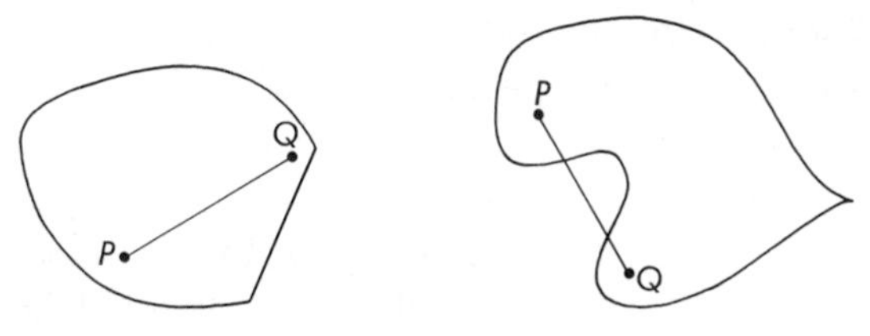

Figure 6–9

To show the convexity asserted in the theorem, let ϕ_1 and ϕ_2 denote any two tests, with corresponding error-size pairs (α_1, β_1) and (α_2, β_2). Let ϕ denote the randomized test

$$\phi(\mathbf{x}) = \gamma\phi_1(\mathbf{x}) + (1 - \gamma)\phi_2(\mathbf{x}),$$

where $0 \le \gamma \le 1$. The error sizes for this test are

$$\begin{aligned}\alpha = E_0[\phi(\mathbf{X})] &= \gamma E_0[\phi_1(\mathbf{X})] + (1 - \gamma)E_0[\phi_2(\mathbf{X})] \\ &= \gamma\alpha_1 + (1 - \gamma)\alpha_2\end{aligned}$$

and

$$\beta = E_1[1 - \phi(\mathbf{X})] = \gamma\beta_1 + (1 - \gamma)\beta_2.$$

But these relations are precisely the parametric equations for the line joining (α_1, β_1) to (α_2, β_2), being linear and yielding (α_2, β_2) for $\gamma = 0$ and (α_1, β_1) for $\gamma = 1$. That is, (α, β) lies on the line segment joining the two points. Conversely, every point (α, β) on the line segment has such a representation, for some γ on $0 \le \gamma \le 1$, and the test ϕ defined as above with that γ is a randomized test with error sizes (α, β).

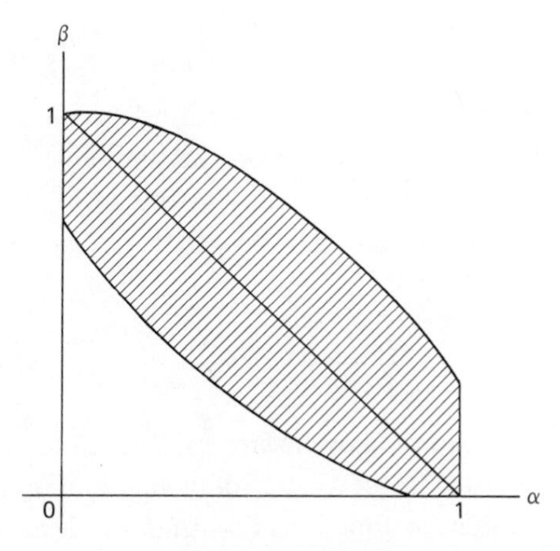

Figure 6–10. An (α, β)-set for simple H_0 versus simple H_1.

Although arbitrary points on the (α, β)-plot cannot always be compared, points that lie on a vertical line (or on a horizontal line) *can* be ordered. For clearly, if tests have the same α, the one with the smallest β (or greatest power) is best. The points with smallest β among all those having a given α lie on the bottom boundary of the (α, β)-set representing the collection of all tests (see Figure 6–10). Notice that such points also have the smallest α for given β, except possibly for those on the α-axis.

Now, in this case of simple H_0 and H_1, the term "power of the test" will mean power on H_1—the probability of "detecting" H_1 when this alternative is true. Thus, the power of a test is $1 - \beta$, and this language gives rise to the next definition.

Definition. *A test of simple H_0 versus simple H_1 that has the smallest β among tests with no larger α is called* most powerful.

Such tests can always be constructed, and the following reasoning suggests the method.

Let the sample observations have joint density $f_0(\mathbf{x})$ under H_0 and $f_1(\mathbf{x})$ under H_1. These functions attach numbers or measures to each point in the space of observation vectors $\mathbf{x}$. Imagine that f_0 assigns to each $\mathbf{x}$ a certain "cost" and that f_1 assigns a "return." Choosing a critical region C so as to have a certain size α amounts to putting into C enough points $\mathbf{x}$ so that the total cost of C is α:

$$\alpha = \int_C f_0(\mathbf{x})\, d\mathbf{x} = \text{total cost of } C.$$

Then, among those regions having the same total cost, look for one that has the smallest β or largest $1 - \beta$:

$$1 - \beta = \int_C f_1(\mathbf{x})\, d\mathbf{x} = \text{total return from } C.$$

It is now clear that the way to make this selection is to construct C by putting into it those points $\mathbf{x}$ having the largest return per unit cost: $f_1(\mathbf{x})/f_0(\mathbf{x})$. That is, points $\mathbf{x}$ are lined up according to the value of this ratio, and the points with the largest such values are put into C. The best critical region would then appear to be defined by

$$\frac{f_0(\mathbf{x})}{f_1(\mathbf{x})} < \text{const},$$

where the constant is chosen to make the size of C equal to α.

This line of reasoning then suggests the following lemma, in which again f_0 and f_1 denote either the density functions (continuous case) or the probability functions (discrete case) of the data to be used in testing.

Neyman–Pearson Lemma. *In testing $f_0(\mathbf{x})$ against $f_1(\mathbf{x})$, the critical region*

$$C_K \equiv \left\{\mathbf{x} \,\middle|\, \frac{f_0(\mathbf{x})}{f_1(\mathbf{x})} < K\right\}$$

is most powerful, for $K \geq 0$. That is, the power of any test $\phi(\mathbf{x})$ with an α no larger than that of C_K does not exceed the power of C_K:

$$\alpha_\phi \leq \alpha_{C_K} \qquad \textit{implies} \qquad \beta_\phi \geq \beta_{C_K}.$$

To prove this, it will be demonstrated that

$$(1 - \beta_{\phi_K}) - (1 - \beta_\phi) \geq 0,$$

where ϕ_K is the indicator function of C_K:

$$\phi_K(\mathbf{x}) = \begin{cases} 1, & \text{if } \mathbf{x} \in C_K, \\ 0, & \text{if } \mathbf{x} \notin C_K, \end{cases}$$

and ϕ is any other test such that

$$\alpha_{\phi_K} - \alpha_\phi = E_{H_0}(\phi_K) - E_{H_0}(\phi) = E_{H_0}(\phi_K - \phi) \geq 0.$$

(Throughout, it will be convenient to suppress the argument $\mathbf{x}$ in the various functions f and ϕ.) Observe first that for $\mathbf{x} \in C_K$,

$$\phi_K - \phi = 1 - \phi \geq 0 \qquad \text{and} \qquad f_1 > \frac{f_0}{K}$$

so that

$$(\phi_K - \phi)f_1 \geq \frac{1}{K}(\phi_K - \phi)f_0.$$

And for $\mathbf{x} \notin C_K$,

$$\phi_K - \phi = 0 - \phi \leq 0 \qquad \text{and} \qquad f_1 \leq \frac{f_0}{K}$$

so that again

$$(\phi_K - \phi)f_1 \geq \frac{1}{K}(\phi_K - \phi)f_0.$$

It follows then that

$$(1 - \beta_{\phi_K}) - (1 - \beta_\phi) = E_{H_1}(\phi_K - \phi) \geq \frac{1}{K} E_{H_0}(\phi_K - \phi) \geq 0,$$

as was to be proved.

EXAMPLE 6–14. Consider testing $H_0: \mu = 0$ against $H_1: \mu = 1$ in a normal population with variance 4, using a random sample of size n. The density function of the observation vector, under H_0, is

$$f_0(\mathbf{x}) = (8\pi)^{-n/2} \exp\left[-\tfrac{1}{8} \sum x_i^2\right],$$

and under H_1,

$$f_1(\mathbf{x}) = (8\pi)^{-n/2} \exp\left[-\tfrac{1}{8} \sum (x_i - 1)^2\right].$$

The likelihood ratio statistic is then

$$\frac{f_0(\mathbf{X})}{f_1(\mathbf{X})} = \exp\left[-\frac{1}{8} \sum X_i^2 + \frac{1}{8} \sum (X_i - 1)^2\right]$$

$$= \exp\left[-\frac{1}{8} \sum (2X_i - 1)\right].$$

This does not exceed the constant K', provided that

$$\bar{X} > \frac{1}{2} - \frac{4 \log K'}{n} \equiv K.$$

Thus, a critical region of the type $\bar{X} > K$ is best among all those having a given α. The value of K would be determined by that α. For example, if $\alpha = .1$ and $n = 25$,

$$.1 = P_{\mu=0}(\bar{X} > K) = 1 - \Phi\left[\frac{K - 0}{2/\sqrt{25}}\right],$$

from which it follows that $5K/2 = 1.28$ or $K = .512$.

As in this last example, the Neyman–Pearson critical region can often be expressed in terms of a test statistic that is easier to work with than the likelihood ratio itself. If the hypotheses being tested are defined by values θ_0 and θ_1 of a real parameter θ, and the family $\{f(x; \theta)\}$ has a monotone likelihood ratio with respect to the statistic T, then the Neyman–Pearson test is equivalent to the critical region $T > K$, when $\theta_0 < \theta_1$.

If $\mathbf{X}$ has a distribution in the one-parameter exponential family,

$$f(\mathbf{x}; \theta) = B(\theta)h(\mathbf{x}) \exp [Q(\theta)S(\mathbf{x})],$$

and H_0 and H_1 are defined by $\theta = \theta_0$ and $\theta = \theta_1$, respectively, then the likelihood ratio is a monotone increasing function of $S(\mathbf{X})$ if $Q(\theta_0) > Q(\theta_1)$, and the Neyman–Pearson test is of the form $S(\mathbf{X}) < K$. If $Q(\theta_0) < Q(\theta_1)$, it is of the form $S(\mathbf{X}) > K$. If (as in Example 6–14) the states θ_0 and θ_1 are particular values of a real parameter θ and the density $f(\mathbf{x}; \theta)$ is in the exponential family with a monotone increasing $Q(\theta)$, then the test $S(\mathbf{X}) > K$ is most powerful for any θ_0 against any θ_1 that is larger than θ_0.

EXAMPLE 6–15. Consider testing $p = \frac{1}{3}$ against $p = \frac{2}{3}$ in the Bernoulli model, using three independent observations. The Bernoulli distribution is in the exponential family with a monotone increasing $Q(p)$, and so the most powerful tests are of the form $S(X) = \sum X_i > K$. But (for $n = 3$) there are just five critical regions of this form, owing to the discrete character of the test statistic. A list of these, together with values of α and β for each, is given in the following table.

Critical Values of $\sum X_i$	α	β
—	0	1
3	$\frac{1}{27}$	$\frac{19}{27}$
3, 2	$\frac{7}{27}$	$\frac{7}{27}$
3, 2, 1	$\frac{19}{27}$	$\frac{1}{27}$
3, 2, 1, 0	1	0

The (α, β)-plot of these consists of just five points, and there is not a test of this type for every α. However, randomized tests can be constructed with (α, β)-representations on the line segments joining two points. For instance, let $C_1 = \{3\}$, and $C_2 = \{2, 3\}$, and let ϕ_1 and ϕ_2 denote the corresponding indicator functions. Then for any γ on $0 \le \gamma \le 1$, the following randomized test has an (α, β) on the line between $(\frac{1}{27}, \frac{19}{27})$ and $(\frac{7}{27}, \frac{7}{27})$:

$$\phi(X) \equiv \gamma\phi_2 + (1 - \gamma)\phi_1 = \begin{cases} 1, & \text{if } X_i > 2, \\ \gamma, & \text{if } X_i = 2, \\ 0, & \text{if } X_i < 2. \end{cases}$$

As will be discussed further, such tests will turn out to be most powerful and will collectively constitute the lower boundary of the (α, β) set representing all possible tests. Figure 6–11 gives the (α, β) plot for possible tests in this problem. The non-randomized Neyman–Pearson tests are the corner points along the bottom boundary, and the randomized ones are on the line segments between.

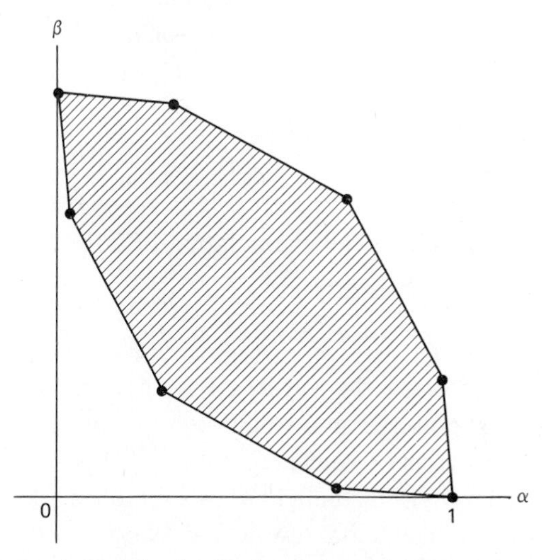

Figure 6–11. The (α, β)-plot for tests in Example 6–15.

As evident in the example, the Neyman–Pearson theory is more tidy when rounded out with randomized tests, and the following is a version of the main result that takes them into account.

Neyman–Pearson Lemma (Extended Version). *In testing $f_0(\mathbf{x})$ against $f_1(\mathbf{x})$, the following test based on $\Lambda \equiv f_0(\mathbf{X})/f_1(\mathbf{X})$ is most powerful (in the class of tests with no larger α):*

$$\phi^*(X) = \begin{cases} 1, & \text{if } \Lambda < K, \\ \gamma, & \text{if } \Lambda = K, \\ 0, & \text{if } \Lambda > K, \end{cases}$$

for any $K \geq 0$ and γ on $0 \leq \gamma \leq 1$. To achieve a specified α,

(a) *If there is a K such that $P(\Lambda < K) = \alpha$, then use that K and $\gamma = 0$.*
(b) *If not, then choose K so that*

$$P(\Lambda < K) < \alpha \leq P(\Lambda \leq K),$$

and let

$$\gamma = \frac{\alpha - P(\Lambda < K)}{P(\Lambda = K)}.$$

The proof of this version is an easy modification of the proof for the earlier version. In that proof, the essential inequality was

$$E_{H_1}(\phi^* - \phi) \geq \frac{1}{K} E_{H_0}(\phi^* - \phi),$$

evident after looking separately at the sets C_K and C_K^c, where C_K is the region $\Lambda < K$. Here there are three regions to look at—namely, $\Lambda < K$, $\Lambda > K$, and $\Lambda = K$. The first two are handled like C_K and C_K^c, and for points $\mathbf{x}$ such that $\Lambda = K$,

$$(\phi^* - \phi)f_1 = \frac{1}{K}(\phi^* - \phi)f_0,$$

so the inequality is again valid. The computation of α for the given K and γ is straightforward:

$$\alpha = E_{H_0}(\phi^*) = P(\Lambda < K) + \gamma P(\Lambda = K).$$

It is easily seen that tests like the Neyman–Pearson tests with the inequalities reversed would be *least* powerful and would be represented by (α, β)-points in the upper boundary of the set of points representing all possible tests.

A Neyman–Pearson test can be thought of as a modified maximum likelihood test. Thus, a maximum likelihood test would choose between H_0 and H_1 according to which density has the greater value at the observed value of X, and so would reject H_0 if $\Lambda < 1$. This is a particular Neyman–Pearson test and allows no flexibility in adjusting α and β according to importance or costs of the two types of error. The critical region $\Lambda < K$ is in the same spirit but does have, in the selection of K, the desired flexibility.

Problems

6–19. Let Z have one or the other of the two discrete distributions given in the following table.

Z	H_0	H_1
z_1	.2	.3
z_2	.3	.1
z_3	.1	.3
z_4	.3	.2
z_5	.1	.1

(a) Compute Λ for each Z-value and order the Z-values according to the value of Λ.
(b) Determine the most powerful tests (nonrandomized).
(c) Compare the Neyman–Pearson test of level $\alpha = .3$ with the test with critical region $Z = z_4$.

6–20. Show that the randomized test in Example 6–15 is equivalent to a procedure of this type: Toss a biased coin and follow test C_1 if the coin falls heads, C_2 if it falls tails.

6–21. Determine the most powerful critical regions for testing θ_0 against θ_1 (with $\theta_1 > \theta_0$) using a random sample of size n from an exponential population: $f(x; \theta) = \theta \exp(-\theta x)$, $x > 0$.

6–22. Obtain the most powerful tests for $\sigma = \sigma_0$ against $\sigma = \sigma_1$ using a random sample of size n from a normal population with known mean μ. Determine n and the critical boundary to achieve $\alpha = \beta = .01$. if $\sigma_0 = 2$ and $\sigma_1 = 3$.

6–23. Determine the best test at $\alpha = .10$ for $m = 1$ against $m = 2$ in a Poisson population with mean m, using a random sample of three observations.

6–24. A single observation is obtained from a Cauchy population:

$$f(x; \theta) = \frac{1/\pi}{1 + (x - \theta)^2},$$

and it is desired to test $\theta = 0$ against $\theta = 2$.

(a) Determine the maximum likelihood critical region for X, and compute the corresponding α and β.
(b) Express the critical region $\Lambda < 5$ in terms of X. Calculate α and β, and determine the test of the form $X > K$ having the same α. Compare the β's of the two tests.

(c) Express the critical region $\Lambda < \frac{17}{37}$ in terms of X, and comment on the nature of the critical regions for the Neyman–Pearson tests that have been considered.

6–25. In testing $p = \frac{1}{2}$ against $p \neq \frac{1}{2}$, show that the critical region $S = 0, 1, 3$, or 4 is unbiased, where S is the number of successes in four trials of a Bernoulli experiment. Is there any p_0 such that the region $S = 0, 1$, or 4 is unbiased for p_0 against $p \neq p_0$?

6–26. Show that for the family of Neyman–Pearson tests in a given problem, the relation between α and β is an inverse one—the larger the α the smaller the β.

6.2.2 Uniformly Most Powerful Tests

If H_0 is simple but H_A is composite, a test may be most powerful against one alternative in H_A but not against another. If it *is* most powerful against every simple alternative in H_A, it is said to be *uniformly most powerful* (UMP). In many of the parametric examples considered above, the Neyman–Pearson test for θ_0 against θ_1 had the same form no matter what value θ_1 greater than θ_0 was considered to be the alternative; any such test then would be UMP for θ_0 against H_A: $\theta > \theta_0$.

EXAMPLE 6–16. In Example 6–14 the region $\bar{X} > K$ was seen to be most powerful for testing $\mu = \mu_0$ against $\mu = \mu_1$, where $\mu_1 > \mu_0$, in a normal population with known variance. But the μ_1 can be *any* value of the population mean larger than μ_0, and so $\bar{X} > K$ is UMP for $\mu = \mu_0$ against $\mu > \mu_0$.

When H_0 is also composite, the size of the type I error is not uniquely defined. In such a case one defines the level α of a test to be the largest of the possible "sizes" of the type I error:

$$\alpha \equiv \sup_{H_0} P_{H_0}(\text{reject } H_0) = \sup_{\theta \in H_0} \pi(\theta).$$

(The "sup" or least upper bound is used in this definition to cover the cases in which a maximum is not actually achieved.) With this convention as to the meaning of the level or size of a test, one can define "UMP" in the general case:

Definition. *A test is said to be* uniformly most powerful (*UMP*) *when it is at least as powerful at every alternative as any other test of no greater size. That is,* ϕ^* *is UMP if and only if for any test* ϕ *such that*

$$\alpha_\phi = \sup_{H_0} \pi_\phi(\theta) \leq \sup_{H_0} \pi_{\phi^*}(\theta) = \alpha_{\phi^*},$$

it follows that for all θ *on* H_A,

$$\pi_{\phi^*}(\theta) \geq \pi_\phi(\theta).$$

A uniformly most powerful test is necessarily unbiased. For, given a UMP test of level α, the randomized test $\alpha \equiv \alpha$ has level α and has power α at each θ in H_A. The UMP test, therefore, must have power at least α at each θ in H_A and so is unbiased.

Uniformly most powerful tests do not always exist. They do, typically, in certain one-parameter problems in which the alternative is "one-sided." The UMP property of a test ϕ^* can sometimes be established by locating a value θ_0 for which

$$\pi_{\phi^*}(\theta_0) = \sup_{H_0} \pi_{\phi^*}(\theta)$$

and then showing the test to be UMP for $\theta = \theta_0$ against the given alternative. The class of tests ϕ with

$$\sup_{H_0} \pi_\phi(\theta) \leq \sup_{H_0} \pi_{\phi^*}(\theta) = \alpha$$

is contained in the class of tests with

$$\pi_\phi(\theta_0) \leq \pi_{\phi^*}(\theta_0) = \alpha,$$

so if ϕ^* has greatest power on H_A in this latter class of tests, it automatically has greatest power in the smaller class. Therefore, it is UMP for H_0 against H_A.

EXAMPLE 6–17. In testing $p \leq \frac{1}{2}$ against $p > \frac{1}{2}$ in the Bernoulli model using the number S of successes in five independent trials, the critical region $S = 4$ or 5 has the power function

$$\pi(p) = p^5 + 5p^4(1 - p).$$

This is monotonically increasing, so the power at $p = \frac{1}{2}$ is the level of the test:

$$\alpha = \sup_{p \leq 1/2} \pi(p) = \tfrac{3}{16}.$$

The test is a Neyman–Pearson test of $p = \frac{1}{2}$ against any single p greater than $\frac{1}{2}$; therefore, it has uniformly greatest power among all tests whose maximum type I error size does not exceed $\frac{3}{16}$ over all of $p \leq \frac{1}{2}$. It is UMP for $p \leq \frac{1}{2}$ against $p > \frac{1}{2}$.

In testing in a one-sided problem—a one-parameter family of possible states with H_A to one side of H_0—an approach used in Example 6–8 was to represent H_0 by a single state θ_0 and H_A by a single state θ_1, to construct a test for θ_0 against θ_1 with specified α and β, and then to study its performance for the original problem. If the test constructed for θ_0 against θ_1 is a Neyman–Pearson test, this will turn out to be UMP if the family of states has a monotone likelihood ratio.

Lemma. *Suppose that $\{f(\mathbf{x}; \theta)\}$ has a monotone likelihood ratio in terms of $T = t(\mathbf{X})$, and let C be the Neyman–Pearson critical region for θ_0 against θ_1, where $\theta_0 < \theta_1$. Then the power function of C is monotone increasing.*

To show the monotone character of C, let θ' and θ'' be any two values of θ with $\theta' < \theta''$. Then the condition $f(\mathbf{x}; \theta')/f(\mathbf{x}; \theta'') <$ const is equivalent to the condition $T >$ const, and this in turn is equivalent to $f(\mathbf{x}; \theta_0)/f(\mathbf{x}; \theta_1) <$ const. Thus, the Neyman–Pearson test of θ_0 against θ_1 is equivalent to a Neyman–Pearson test for θ' against θ'' and so is most powerful for θ' against θ''. So for any test ϕ with no greater α (at θ'), its power (at θ'') is greatest; and this is true in particular for the test $\phi \equiv \alpha_C$:

$$\pi_C(\theta'') \geq \pi_\phi(\theta'') = E_{\theta''}(\phi) = \alpha_C = \pi_C(\theta'),$$

which is precisely the inequality needed to establish the lemma.

Suppose now that C is the Neyman–Pearson critical region for θ_0 against θ_1 $(\theta_0 < \theta_1)$ in a monotone likelihood ratio family. This region is also most powerful for θ^* against any particular θ greater than θ^*, and so it is UMP for θ^* against $\theta > \theta^*$. Because $\pi_C(\theta)$ is monotone increasing, the level of the test in testing $\theta \leq \theta^*$ against $\theta > \theta^*$ is

$$\alpha_C = \sup_{\theta \leq \theta^*} \pi_C(\theta) = \pi_C(\theta^*),$$

and, therefore (as in Example 6–17 and the discussion preceding it), the region C is UMP for $\theta \leq \theta^*$ against $\theta > \theta^*$.

If the sample $\mathbf{X}$ has a distribution in the exponential family, with

$$f(\mathbf{x}; \theta) = B(\theta)h(\mathbf{x}) \exp [Q(\theta)S(\mathbf{x})],$$

and if $Q(\theta)$ is monotone increasing, then (as an application of the previous discussion) the critical region $S(\mathbf{X}) > K$ is uniformly most powerful for $\theta \leq \theta^*$ against $\theta > \theta^*$.

EXAMPLE 6–18. The joint density of a random sample $\mathbf{X}$ from an exponential population with mean θ is

$$f(\mathbf{x}; \theta) = \theta^{-n} \exp \left[-\frac{1}{\theta} \sum x_i\right], \qquad x_i > 0.$$

This is in the exponential family, with $Q(\theta) = -1/\theta$, a monotone increasing function of θ. Hence, the critical region $\bar{X} > K$ is UMP for $\theta \leq \theta^*$ against $\theta > \theta^*$.

As remarked earlier, a UMP test need not exist. A common situation of this kind is that of testing $\theta_1 \leq \theta \leq \theta_2$ against the alternative: $\theta > \theta_2$ or $\theta < \theta_1$. The possibility that a uniformly most powerful test may not exist in the case of a simple H_0 and a two-sided alternative can be seen as follows: A UMP test is unbiased, and its power function, therefore, has a minimum at H_0: $\theta = \theta_0$. On the other hand, a most powerful test of θ_0 against θ_1 where $\theta_1 > \theta_0$ would have a power function that in general would be increasing as θ passes through θ_0. For some θ's just to the right of θ_0, it would exceed the power function with a minimum at θ_0. This excess power on one side of θ_0 is paid for in considerably less power on the other. The following example illustrates this point.

EXAMPLE 6–19. In testing $p = .5$ against $p \neq .5$ in a Bernoulli population, consider the randomized test that calls for rejection of $p = .5$ when $S = 4$ or 5, and for rejection of $p = .5$ with probability .6 when $S = 3$, the statistic S again denoting the number of successes in five trials. The power function here is

$$\begin{aligned}\pi(p) &= 5p^4(1 - p) + p^5 + (.6)10p^3(1 - p)^2 \\ &= p^3(p - 2)(2p - 3).\end{aligned}$$

The graph of this power function is shown in Figure 6–12 along with the power function corresponding to the rejection region $S = 0, 1, 4,$ or 5. The randomization here was fixed up so that these two tests have the same α, but the one-sided test is more powerful than the symmetric one for $p > .5$. This extra power is achieved by sacrificing power on $p < .5$. At any rate, it is clear that the symmetric test is *not* UMP, although it is unbiased.

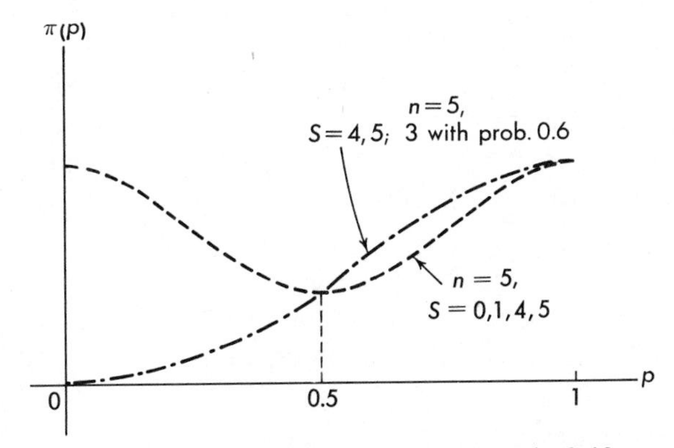

Figure 6–12. Power curves for Example 6–19.

6.2.3 Likelihood Ratio Tests

The maximum likelihood principal says to look for a state $\hat{\theta}$ that maximizes the likelihood function $L(\theta) = f(\mathbf{X}; \theta)$. It regards such state as providing the "best" explanation of the observed $\mathbf{X}$. In choosing between H_0 and H_A, then, it would be natural to compare the best explanation within H_0 with the best explanation within H_A and so reject H_0 if and only if

$$\sup_{H_0} f(\mathbf{x}; \theta) < \sup_{H_A} f(\mathbf{x}; \theta),$$

when $\mathbf{X} = \mathbf{x}$ is observed. Generalizing this idea to provide some flexibility in the level of the test, one would consider critical regions of the form

$$\Lambda^* \equiv \frac{\sup_{H_0} f(\mathbf{x}; \theta)}{\sup_{H_A} f(\mathbf{x}; \theta)} < \text{const.}$$

Such a critical region defines what is called a *likelihood ratio test.*

In many instances, it is convenient to consider the modified likelihood ratio in which the best θ in H_0 is compared with the best θ overall—that is, with the best θ in $H_0 + H_A$:

$$\Lambda = \frac{\sup_{H_0} L(\theta)}{\sup_{H_0+H_A} L(\theta)} < \text{const.}$$

Observe that if $\Lambda^* < 1$, then $\Lambda = \Lambda^*$, but if $\Lambda^* \geq 1$, then $\Lambda = 1$. The rejection region defined by $\Lambda < K$ is exactly the same as that defined by $\Lambda^* < K$, provided $K < 1$. In the problems in which we shall use the method, the least upper bound over $H_0 + H_A$ will be the same as the least upper bound over H_A, so that $\Lambda = \Lambda^*$.

For random samples it is again convenient to work with the logarithm of the likelihood function and to use $-\log \Lambda$, for which the critical region becomes $-\log \Lambda > \text{const.}$

If a test is to be based on the statistic Λ (or on $-\log \Lambda$), it is necessary to know the distribution of Λ. This is ordinarily very complicated, and it is useful to know that the asymptotic distribution of $-2 \log \Lambda$ is of the chi-square type (Table II, Appendix), under the null hypothesis H_0. The parameter of this chi-square distribution is the difference[2] in the dimension of $H_0 + H_A$ and the dimension of H_0.

Often the likelihood ratio test can be shown equivalent to a test involving a more natural or convenient statistic, whose distribution is known or readily derived. This is illustrated in the following example.

EXAMPLE 6–20. Consider testing $H_0: \mu = \mu_0$ against $H_A: \mu \neq \mu_0$ in a normal population with given variance v. The likelihood function is

$$L(\mu) = (2\pi v)^{-n/2} \exp\left[\frac{-1}{2v} \sum (X_i - \mu)^2\right].$$

This is a maximum over all μ for $\hat{\mu} = \bar{X}$, the maximum likelihood estimate of μ, and so the denominator of Λ is $L(\bar{X})$. The numerator is just $L(\mu_0)$, since H_0 consists of the single point μ_0. The ratio Λ is then $L(\mu_0)/L(\bar{X})$, and

$$\begin{aligned}\log \Lambda &= \log L(\mu_0) - \log L(\bar{X}) \\ &= -\frac{1}{2v}\left\{\sum (X_i - \mu_0)^2 - \sum (X_i - \bar{X})^2\right\} \\ &= -\frac{n}{2v}[\bar{X} - \mu_0]^2.\end{aligned}$$

Therefore, the critical region $\log \Lambda < \text{const}$ is equivalent to

$$|\bar{X} - \mu_0| > \text{const},$$

[2] See E. Lehmann [16], Section 7.13.

which is intuitively very appealing. (Incidentally, under H_0, $-2 \log \Lambda$ is not only *asymptotically* of the chi-square type, it *is* the square of a standard normal variable and, therefore, exactly of the chi-square type with one "degree of freedom.")

In this example, the likelihood ratio method led to a test that is consonant with intuition. It is not UMP, for no UMP test exists; however, it can be shown to be UMP in the class of unbiased tests (cf. Ferguson [9], p. 226). So the method sometimes works well, yielding reasonable tests; but sometimes it does not, as the next example shows.

EXAMPLE 6–21. It is desired to test $\theta = \theta_0$ against $H_A: 0 < \theta < 1$, using an observation Z whose distribution is discrete; the probability function is given in the following table.

	$f(z;\theta)$			
	$\theta = \theta_0$	$0 < \theta < 1$	$\sup f(z;\theta)$	Λ
z_1	$\frac{1}{12}$	$\frac{1}{3}\theta$	$\frac{1}{3}$	$\frac{1}{4}$
z_2	$\frac{1}{12}$	$\frac{1}{3}(1-\theta)$	$\frac{1}{3}$	$\frac{1}{4}$
z_3	$\frac{1}{6}$	$\frac{1}{2}$	$\frac{1}{2}$	$\frac{1}{3}$
z_4	$\frac{2}{3}$	$\frac{1}{6}$	$\frac{2}{3}$	1

The likelihood ratio test with $\alpha = \frac{1}{6}$ calls for rejecting H_0 if $Z = z_1$ or z_2, and the power function is

$$\pi(\theta) = \tfrac{1}{3}\theta + \tfrac{1}{3}(1-\theta) = \tfrac{1}{3}$$

for θ in H_A. However, the critical region $Z = z_3$ also has an α of $\frac{1}{6}$, and its power on H_A is $\frac{1}{2}$, uniformly greater than the power of the likelihood ratio test.

Problems

6–27. Determine the UMP test of $\theta \leq 1$ against $\theta > 1$ using a random sample of size 25 from a normal population with mean 0 and variance θ, with $\alpha = .05$.

6–28. An industrial sampling plan calls for accepting a lot of N articles, each classifiable as "good" or "defective," if a simple random sample of size n from the lot contains no more than c defectives. Show that this test is UMP for the hypothesis that the lot fraction defective does not exceed p_0 against the alternative that it does. (See Problem 4–50.)

6–29. Let X have a Cauchy distribution with unknown median θ. Show that the power function of the critical region $X > 1$ is monotone increasing. Can it be concluded from this that this region is UMP for $\theta < 0$ against $\theta > 0$?

6–30. Referring to Example 6–7, determine a test of $p = .5$ based on 100 tosses having the same α as in that example (.036), but which is UMP against $p > .5$. How does the power function of this test compare with that of the two-sided test in Example 6–7?

6–31. Construct the likelihood ratio test (using Λ^*) for $\mu = 0$ against $\mu = 1$ in a normal population, using a random sample of fixed size.

6–32. Determine the likelihood ratio test of $\theta = 0$ against $\theta \neq 0$ using a single observation from a Cauchy distribution with median θ. Show that the resulting test is unbiased. (Had the test been based on the mean of n observations, would you expect $-2 \log \Lambda$ to have the asymptotic distribution under H_0 that usually obtains?)

6–33. An observation Z takes on one of four values according to one of the three distributions or states of nature as shown in the following table of probabilities.

	z_1	z_2	z_3	z_4
θ_1	.2	.3	.1	.4
θ_2	.5	.1	.2	.2
θ_3	.3	0	.4	.3

Determine the critical regions and corresponding levels for the likelihood ratio tests $\theta = \theta_1$ against $\theta = \theta_2$ or θ_3.

6–34. Consider the problem of testing $v = v_0$ against $v \neq v_0$ in a normal population with *given* mean μ and unknown variance v. Discuss the nature of the likelihood ratio test in terms of the m.l.e. of $\hat{v}$.

6–35. Construct the likelihood ratio test for $\theta = \theta_0$ against $\theta \neq \theta_0$ in the density $\theta \exp(-\theta x)$, $x > 0$, obtaining as the critical region: $\bar{X} \exp(-\theta_0 \bar{X}) <$ constant. Sketch the graph of this function of $\bar{X}$ and indicate the corresponding critical region in terms of $\bar{X}$. Is this symmetrical about the population mean corresponding to θ_0? (In practice, one would probably take symmetrical tails for the rejection region, even though the right and left boundaries would correspond to different values of Λ, using the likelihood ratio notion only to justify the selection of a two-tailed critical region.)

6–36. Consider testing $p = .5$ against $p \neq .5$, using three independent observations on a Bernoulli variable. Construct the probability distribution of the statistic Λ under H_0.

6–37. Determine the likelihood ratio test of $v = v_0$ against $v = v_1$, using the variance of a sample from a normal population with unknown mean μ and variance v. Obtain also the power function of the test.

6.3 Sequential Tests

In testing a simple hypothesis against a simple alternative, the most powerful test was found to be given by a critical region of the form

$$\Lambda_n = \frac{f_0(X_1, \ldots, X_n)}{f_1(X_1, \ldots, X_n)} < K,$$

where $f_0(\mathbf{x})$ and $f_1(\mathbf{x})$ are the joint density functions of the observations (or probability functions in the discrete case) corresponding to H_0 and H_1, respectively. Given any two of the four quantities α, β, n, and K, the other two are determined by the relations

$$\alpha = P_{H_0}(\Lambda_n < K), \qquad \beta = P_{H_1}(\Lambda_n \geq K).$$

In carrying out such a test, one picks, say, α and β, and determines from them values of n and K, He then gathers the data, using that n as the size of the sample, computes Λ_n, and accepts or rejects H_0 according to whether $\Lambda_n \geq K$ or $\Lambda_n < K$. No provision is made to fulfill the natural desire to obtain more data when the nature of the sample is not sufficiently suggestive either of H_0 or of H_1 as to be really convincing. Nor, on the other hand, is there opportunity to cut the sampling short if a conclusion becomes obvious early in the process of obtaining the sample. Of course, nothing really *prevents* one from doing these things—gathering, according to how things go, more or less data than dictated by the choice of α and β; but it must be recognized that doing them alters the test and makes the given error sizes meaningless.

A procedure of "double sampling" would be somewhat more satisfying. For such a plan there is chosen a preliminary sample size m with corresponding constants C and D, and a second sample size $n - m$ with corresponding constant K. The procedure is then to draw a sample of size m, compute Λ_m, and

$$\begin{cases} \text{accept } H_0, & \text{if } \Lambda_m \geq D, \\ \text{reject } H_0, & \text{if } \Lambda_m \leq C, \\ \text{draw a sample of size } n - m, & \text{if } C < \Lambda_m < D. \end{cases}$$

If the second sample is called for, it is drawn, Λ_n computed, and

$$\begin{cases} H_0 \text{ accepted}, & \text{if } \Lambda_n \geq K, \\ H_0 \text{ rejected}, & \text{if } \Lambda_n < K. \end{cases}$$

This idea can clearly be extended to plans with more than two stages, at each stage a decision is made to reject, to accept, or (at stages prior to the last one) to continue sampling. The hope is that with such schemes one can get away with less sampling than would be necessary using a sample of fixed size, and such is usually the case. Of course, comparisons of sample sizes must be made on a probability basis, or on the average; for, the sample size in sequential plans is usually a random variable, since the decision to stop or to continue depends on the observations at hand.

In situations in which unknown parameters other than those of interest ("nuisance parameters") make planning difficult, an adaptive scheme of sampling in stages might be considered, in which results in the early stages furnish a guide to more intelligent planning of the future stages.

The only sequential procedure to be studied here is the *sequential probability ratio test* of Wald. An extensive treatment is given by Wald [21], and its optimality properties are considered in a decision theoretic framework in books by DeGroot [6] and Ferguson [9].

6.3.1 The Sequential Likelihood Ratio Test

Wald [20] first presented—and systematically studied—the following sequential test of a simple hypothesis against a simple alternative. Let H_0 denote the hypothesis that the population density (or probability function) is $f_0(x)$, and H_1 the hypothesis that it is $f_1(x)$. Numbers A and B are chosen, with $A < B$, and after each observation in a sequence the corresponding likelihood ratio is computed:

$$\Lambda_n = \frac{f_0(X_1, \ldots, X_n)}{f_1(X_1, \ldots, X_n)},$$

where the f_0 and f_1 refer to the joint distributions of $X_1, \ldots, X_n$ under H_0 and H_1, respectively. The procedure is then as follows: Reject H_0 if $\Lambda_n \leq A$, accept H_0 if $\Lambda_n \geq B$, and obtain another observation if $A < \Lambda_n < B$. It is conceivable that such a procedure might never terminate, but this contingency will be shown to have probability 0, in Section 6.3.2.

If the observations in the sequence $X_1, X_2, \ldots$ are independent random variables, the likelihood ratio is a product:

$$\Lambda_n = \prod_{i=1}^{n} \frac{f_0(X_i)}{f_1(X_i)},$$

and in such a case the logarithm is often easier to work with:

$$\log \Lambda_n = \sum_{i=1}^{n} \log \frac{f_0(X_i)}{f_1(X_i)} = Z_1 + \cdots + Z_n,$$

where

$$Z_i = \log \frac{f_0(X_i)}{f_1(X_i)}.$$

The sequence of Z's is again a sequence of independent random variables, if the X's are independent. The inequality for continuing sampling can be written in the form

$$\log A < \log \Lambda_n < \log B,$$

where usually $\log A < 0$ and $\log B > 0$. Figure 6–13 shows a plot of $(n, \log \Lambda_n)$ in a typical case, the points being connected by straight lines for ease in following the progress of the test. A decision is reached when the path first crosses a

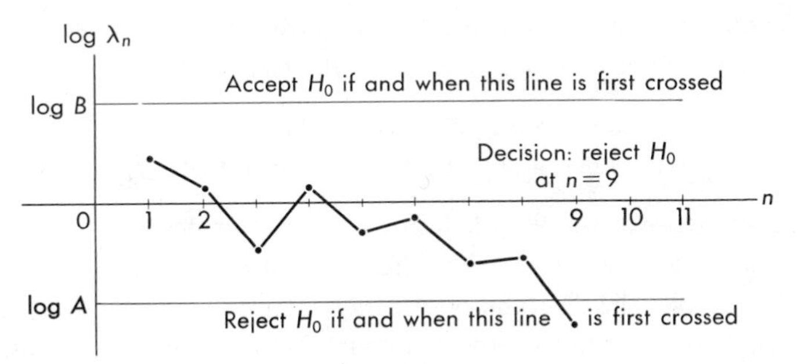

Figure 6–13

boundary; if that boundary is the upper one, H_0 is accepted, and if it is the lower one, H_0 is rejected. In either case, the sampling stops at that point.

In an actual situation the plot made would usually be neither Λ_n nor $\log \Lambda_n$, but some simpler statistic that arises naturally in the likelihood ratio. For instance, if the population distributions are in the exponential family, the log of the likelihood ratio for $\theta = \theta_0$ against $\theta = \theta_1$ is

$$\log \Lambda_n = n \log \frac{B(\theta_0)}{B(\theta_1)} + [Q(\theta_0) - Q(\theta_1)] \sum_{i=1}^{n} R(X_i),$$

and the inequality for continuing the sampling can be expressed in the form

$$C_1 + Dn < \sum_{i=1}^{n} R(X_i) < C_2 + Dn,$$

where the constants C_1, C_2, and D depend on θ_0 and θ_1. Thus, one can plot two parallel lines, corresponding to the linear functions of n in the extremes of this last inequality, and simply cumulate the $R(X_i)$ graphically as the sampling proceeds.

EXAMPLE 6–22. Consider testing $\theta = \theta_0$ against $\theta = \theta_1$, where $\theta_1 > \theta_0$, in a population with density $f(x; \theta) = \theta \exp(-\theta x)$. The log of the likelihood ratio for a single observation is

$$Z = \log \frac{f(X; \theta_0)}{f(X; \theta_1)} = \log \frac{\theta_0}{\theta_1} - (\theta_0 - \theta_1)X,$$

and so

$$\log \Lambda_n = Z_1 + \cdots + Z_n = n \log \frac{\theta_0}{\theta_1} - (\theta_0 - \theta_1) \sum X_i.$$

The inequality for continuing sampling after the nth observation is then

$$\frac{1}{\theta_1 - \theta_0} \left[\log A + n \log \frac{\theta_1}{\theta_0}\right] < \sum X_i < \frac{1}{\theta_1 - \theta_0} \left[\log B + n \log \frac{\theta_1}{\theta_0}\right].$$

Error sizes for the sequential likelihood ratio test are easily expressed, formally, in terms of the numbers A and B that define the test:

$$\begin{cases} \alpha = P_{H_0}(\Lambda_1 \leq A) + P_{H_0}(A < \Lambda_1 < B \text{ and } \Lambda_2 \leq A) + \cdots, \\ \beta = P_{H_1}(\Lambda_1 \geq B) + P_{H_1}(A < \Lambda_1 < B \text{ and } \Lambda_2 \geq B) + \cdots. \end{cases}$$

Although easy to write down, these expressions are by no means easily computed. Moreover, one could not hope to *solve* these equations for A and B in terms of given α and β, despite the desirability of being able to do so in setting up a test to meet specified protection. Another approach is fortunately more fruitful.

The probability in terms of which α and β are defined is actually a probability on a sample space of infinite sequences, and in this probability measure, the set of sequences leading to a decision at a finite stage has been asserted to have probability 1. Moreover, the probability of a set of sequences leading to rejection, say, at stage k is really a marginal distribution in the space of values of the first k observations, $\mathbf{X} = (X_1, \ldots, X_k)$, inasmuch as the decision to terminate the sampling comes from those observations. Thus, if E_k denotes the set of $(x_1, \ldots, x_k)$ that calls for termination at stage k with a decision to reject H_0, and F_k the set calling for accepting H_0 at that stage, then

$$\alpha = P_{H_0}\,(\text{reject } H_0) = \sum P_{H_0}(E_k)$$

and

$$1 - \beta = P_{H_1}\,(\text{reject } H_0) = \sum P_{H_1}(E_k).$$

But according to the rule that determines the decision, the inequality $f_0(\mathbf{x}) \leq Af_1(\mathbf{x})$ holds at each point of E_k so that

$$P_{H_0}(E_k) = \int_{E_k} f_0(\mathbf{x})\, d\mathbf{x} \leq A \int_{E_k} f_1(\mathbf{x})\, d\mathbf{x} = AP_{H_1}(E_k).$$

Summing on k yields

$$\alpha = \sum P_{H_0}(E_k) \leq A(1 - \beta).$$

A similar argument results in the inequality

$$\beta \leq \frac{1 - \alpha}{B}.$$

Dividing these inequalities by, respectively, $1 - \beta$ and $1 - \alpha$ (which, for practical purposes can be assumed positive, since $\alpha = 1$ implies $\beta = 0$ and $\beta = 1$ implies $\alpha = 0$), one obtains

$$\frac{\alpha}{1 - \beta} \leq A \quad \text{and} \quad \frac{\beta}{1 - \alpha} \leq \frac{1}{B}.$$

These relations satisfied by the α and the β of the test defined by a given A and B do not exactly yield α and β, but they almost do.

Since *in*equalities are obtained in this derivation simply because Λ_n does not usually attain *exactly* the value A or the value B, the inequalities are almost equalities. Indeed, in practice, A and B are taken to be *equal* to $\alpha/(1-\beta)$ and $(1-\alpha)/\beta$, respectively. Doing so, of course, means that the test actually carried out has error sizes somewhat different from those specified. Let α' and β' denote the actual sizes of the type I and type II errors of the test defined by the limits $A = \alpha/(1-\beta)$ and $B = (1-\alpha)/\beta$. According to the inequalities, then,

$$\frac{\beta'}{1-\alpha'} \leq \frac{1}{B} = \frac{\beta}{1-\alpha} \quad \text{and} \quad \frac{\alpha'}{1-\beta'} \leq A = \frac{\alpha}{1-\beta}.$$

Multiplying these through to eliminate denominators and adding the results, one obtains

$$\alpha' + \beta' \leq \alpha + \beta.$$

It could not be, therefore, that *both* $\alpha' > \alpha$ and $\beta' > \beta$; so, at most one of the error sizes will be larger than specified when using the approximate formulas for A and B. Further, it follows from the inequalities obtained that for small α and β,

$$\alpha' \leq \frac{\alpha'}{1-\beta'} \leq \frac{\alpha}{1-\beta} \doteq \alpha(1+\beta)$$

and

$$\beta' \leq \frac{\beta'}{1-\alpha'} \leq \frac{\beta}{1-\alpha} \doteq \beta(1+\alpha).$$

Thus, the one error size that does increase does not increase by more than a factor of about $(1+\alpha)$ or $(1+\beta)$. For example, if both α and β are .05, the α' and β' actually achieved by using A and B as specified are bounded by about .0525.

Although using the approximate values of A and B results in error sizes that are not appreciably larger than specified, it is possible that they are *smaller* than specified. This would only be disturbing in the sense that the statistician could get away with less sampling and still be within the desired α and β. This effect should be as slight, as is the increase in α and β, since both are caused by the discontinuity of the sample number necessary to reach a decision. If one could imagine a continuous sample number, the formulas for A and B would be correct. Wald ([20], pp. 65–69) shows that the change in sample size is at most slight; in certain common cases it amounts to at most one or two in small samples and around 1 or 2 per cent in larger samples.

EXAMPLE 6–23. It is desired to test $H_0: p = .5$ against $H_1: p = .2$ in a Bernoulli population, using a sequential test with $\alpha = .1$ and $\beta = .2$. The likelihood ratio after n independent observations is

$$\Lambda_n = \frac{(.5)^f(.5)^{n-f}}{(.2)^f(.8)^{n-f}},$$

where f is the number of successes in n independent trials. Thus,

$$\Lambda_{n+1} = \begin{cases} \frac{5}{2}\Lambda_n, & \text{if success occurs in the } (n+1)\text{st trial,} \\ \frac{5}{8}\Lambda_n, & \text{if failure occurs in the } (n+1)\text{st trial.} \end{cases}$$

A success raises the value of Λ_n, and a failure lowers it. If Λ_n falls below the value $A = .1/(1 - .2) = 1/8$, H_0 is rejected in favor of the lower value $p = .2$. If Λ_n reaches $B = (1 - .1)/.2 = 9/2$, H_0 is accepted. Otherwise, sampling is continued.

If a sequence of observations turns out to be $T, H, H, T, H, T, H, H, T, T, \ldots,$ the corresponding Λ_n's are

$$.625,\quad 1.56,\quad 3.91,\quad 2.44,\quad 6.10,\quad 3.82,\quad 9.53,\quad 23.8,\quad 14.9,\quad 9.30,\quad \ldots.$$

At the fifth observation, H_0 is accepted and the sampling stopped (despite the return at the sixth observation to the region between A and B, since this observation would not have been obtained in practice).

6.3.2 Finite Termination of the Test

It will be demonstrated here that although one might obtain a sequence of observations that called for continuing the sampling at each stage, with no decision reached, the probability of such an occurrence is 0. As before, let the hypothesis being tested be that the population density (or probability) function is $f_0(x)$ and the alternative that it is $f_1(x)$. The logarithm of the likelihood ratio for any single observation,

$$Z = \log \frac{f_0(X)}{f_1(X)},$$

will be 0 if and only if f_0 and f_1 agree at that point. If they are to define essentially different distributions, then the set of points at which Z is not 0 would have to have positive probability under any state of nature. Assuming, then, that $P(Z = 0) < 1$, it must follow that there is a $d > 0$ and a $q > 0$ so that either $P(Z > d) \geq q$ or $P(Z < -d) \geq q$. In the former case, choose an integer r such that $rd > \log(B/A)$, and break up the sum of Z's that gives Λ_n for $n = kr$ into k sections of length r:

$$Z_1 + \cdots + Z_n = (Z_1 + \cdots + Z_r) + \cdots + (Z_{(k-1)r+1} + \cdots + Z_{kr}).$$

For the first group of terms,

$$P\left\{\left|\sum_1^r Z_i\right| > \log \frac{B}{A}\right\} \geq P\left(\sum_1^r Z_i > \log \frac{B}{A}\right) \geq P\left(\sum_1^r Z_i > rd\right)$$
$$\geq P(Z_i > d, i = 1, \ldots, r) \geq q^r > 0.$$

If, on the other hand, $P(Z < -d) = q$, the integer r is chosen such that $-rd < -\log(B/A)$, with the same result, namely,

$$P\left[\left|\sum_1^r Z_j\right| > \log\frac{B}{A}\right] \geq q^r.$$

Moreover, the same result would be true for each group of r Z's in the sum of kr Z's, since the Z's are identically distributed.

Now let N denote the (random) sample size necessary to reach a decision. The events $\{N > n\}$ form a descending sequence whose intersection is the event that N is not finite, and then

$$P(N \text{ not finite}) = \lim_{n\to\infty} P(N > n) = \lim_{k\to\infty} P(N > kr),$$

where r is the positive integer defined before. The last equality follows from the fact that the sequence of probabilities $P(N > n)$ is monotone, bounded below, and so has a limit; this limit is also the limit of any subsequence—such as the sequence consisting of every rth element. But if $N > kr$, then no decision is reached through stage kr. Hence,

$$\log A < \log \Lambda_n < \log B, \qquad \text{for } n \leq kr,$$

which means that none of the groups of Z's can exceed $\log(B/A)$:

$$|Z_{(j-1)r+1} + \cdots + Z_{jr}| < \log\frac{B}{A}, \qquad j = 1, 2, \ldots, k$$

(for otherwise $\log \Lambda_n$ could not remain between $\log A$ and $\log B$ for $n \leq kr$). Hence,

$$P(N > kr) \leq P\left\{|Z_{(j-1)r+1} + \cdots + Z_{jr}| < \log\frac{B}{A}, j = 1, 2, \ldots, k\right\}$$

$$= \left\{P\left(|Z_1 + \cdots + Z_r| < \log\frac{B}{A}\right)\right\}^k \leq (1 - q^r)^k.$$

Taking the limit as k becomes infinite, one finds

$$P(N \text{ not finite}) \leq \lim_{k\to\infty} (1 - q^r)^k = 0$$

because $q > 0$ and, therefore, $1 - q^r < 1$.

Problems

6–38. Construct and carry out a sequential test for $p = \frac{1}{2}$ against the alternative $p = \frac{1}{3}$, with $\alpha = .1$ and $\beta = .2$, using an ordinary coin.

6–39. Construct the sequential likelihood ratio test for $\mu = \mu_0$ against $\mu = \mu_1$ in a normal population with given variance σ^2. Express the inequality for continuing sampling in terms of the sample sum, and determine the nature of the curves bounding the region in which this inequality is satisfied. Sketch these curves for the particular values $\mu_0 = 4$, $\mu_1 = 6$, $\alpha = \beta = .05$, and $\sigma^2 = 9$. Observe the change in the curves corresponding to a change in error sizes to $\alpha = .1$ and $\beta = .2$, and also corresponding to a change in μ_1 to 10. What are the curves like if one considers the sample *mean* instead of the sample sum?

6–40. Obtain the sequential test, in terms of the sum of a sequence of observations, $m = m_0$ against $m = m_1$ with $m_1 > m_0$, where m is the mean of a Poisson distribution. Use $\alpha = \beta = .05$.

6–41. Fill in the details showing the asserted inequality, in Section 6.3.2, in the case $P(Z < -d) \geq q$.

6.3.3 The Operating Characteristic

The operating characteristic (probability of accepting H_0) of a sequential likelihood ratio test can be obtained, somewhat indirectly, as follows: Consider a test defined by given constants A and B, with $A < 1 < B$, to test $H_0: f_0(x)$ against $H_1: f_1(x)$. If one considered f_0 and f_1 to be the only possible states of nature, there would be little point to an "operating characteristic." However, the population is frequently one of a family $f(x; \theta)$, with $f_1(x) = f(x; \theta_1)$ and $f_0(x) = f(x; \theta_0)$ used as convenient states for defining the test, the interest lying in distinguishing $\theta \leq \theta^*$ from $\theta > \theta^*$. In such a case one *is* interested in the operating characteristic as a function of all possible values of θ.

Let θ be fixed and determine as a function of that θ a value of h other than 0 for which

$$E_\theta\left\{\left(\frac{f(X; \theta_0)}{f(X; \theta_1)}\right)^h\right\} = 1.$$

This expectation is 1 when $h = 0$, but there is often just one other value of h for which it is also 1. Observe in particular that $h = 1$ does the trick if $\theta = \theta_1$ and that $h = -1$ works for $\theta = \theta_0$.

From the integral expression for the expected value:

$$\int_{-\infty}^{\infty} \left\{\frac{f(x; \theta_0)}{f(x; \theta_1)}\right\}^h f(x; \theta)\, dx = 1,$$

it follows that the integrand function can be thought of as a density function:

$$f^*(x; \theta) = \left\{\frac{f(x; \theta_0)}{f(x; \theta_1)}\right\}^h f(x; \theta).$$

Consider now the auxiliary problem of testing

$$H^*: f^*(x; \theta) \qquad \text{against} \qquad H: f(x; \theta),$$

which are simple hypotheses for fixed h and θ. Using the constants A^h and B^h in the inequality for continuing sampling in testing H^* against H,

$$A^h < \prod \frac{f^*(X_i;\theta)}{f(X_i;\theta)} = \prod \left\{\frac{f(X_i;\theta_0)}{f(X_i;\theta_1)}\right\}^h < B^h,$$

one finds, upon taking the $1/h$ power (assume $h > 0$), the same inequality as that used for continuing sampling in testing H_0 against H_1. (If $h < 0$, the lower boundary could have been taken as B^h and the upper one as A^h, with the same net result.) Consequently,

$$P_\theta(\text{accept } H_0) = P_\theta(\text{accept } H^*) = P_H(\text{accept } H^*) = \beta^*,$$

where β^* is the size of the type II error for the auxiliary problem. This can be expressed in terms of the cease-sampling limits for the *auxiliary* problem:

$$OC(\theta) = P_\theta(\text{accept } H_0) = \beta^* \doteq \frac{1 - A^h}{B^h - A^h}.$$

Thus, the operating characteristic is expressed in terms of h. But θ, too, depends on h, and these two relations define the operating characteristic curve $OC(\theta)$ parametrically: Each choice of h determines a θ and a value of $P(\text{accept } H_0)$, that is, a point on the OC curve.

The expression relating h and θ does not define θ for $h = 0$, since it is automatically satisfied for any θ; however, the point on the operating characteristic curve corresponding to $h = 0$ can be determined by passing to the limit as $h \to 0$. Similarly, the expression for $OC(\theta)$ is indeterminate for $h = 0$, but it can be evaluated by means of l'Hospital's rule, with the result:

$$OC(\theta)|_{h=0} = \frac{-\log A}{\log B - \log A}.$$

The values $h = 1$ and $h = -1$ have been seen to correspond to $\theta = \theta_1$ and $\theta = \theta_0$, respectively, and of course $OC(\theta_1) = \beta$, and $OC(\theta_0) = 1 - \alpha$. As $h \to \infty$, the operating characteristic clearly tends to 0, and as $h \to -\infty$, it tends to 1. Thus, there are five convenient points, which frequently suffice to furnish an adequate sketch of the OC curve:

h	$-\infty$	-1	0	1	∞
θ		θ_0		θ_1	
OC	1	$1 - \alpha$	$\dfrac{-\log A}{\log B - \log A}$	β	0

EXAMPLE 6–24. Consider testing $p = p_0$ against $p = p_1 > p_0$ in a Bernoulli population. Here $f(x;p) = p^x(1-p)^{1-x}$, $x = 0, 1$, so that

$$E_p\left\{\left[\frac{f(X;p_0)}{f(X;p_1)}\right]^h\right\} = \left[\frac{1-p_0}{1-p_1}\right]^h(1-p) + \left(\frac{p_0}{p_1}\right)^h p.$$

Upon setting this equal to 1 and solving for p, the expression for p as a function of h is found to be

$$p = \frac{1 - [(1 - p_0)/(1 - p_1)]^h}{(p_0/p_1)^h - [(1 - p_0)/(1 - p_1)]^h}.$$

As $h \to 0$, this becomes

$$\frac{-\log [(1 - p_0)/(1 - p_1)]}{\log (p_0/p_1) - \log [(1 - p_0)/(1 - p_1)]}.$$

Taking the particular values $p_0 = .5$, $p_1 = .9$, $\alpha = \beta = .05$, one obtains these points on the operating characteristic curve, shown in Figure 6–14.

h	$-\infty$	-1	0	1	∞
p	0	.5	.733	.9	1
OC	1	.95	.5	.05	0

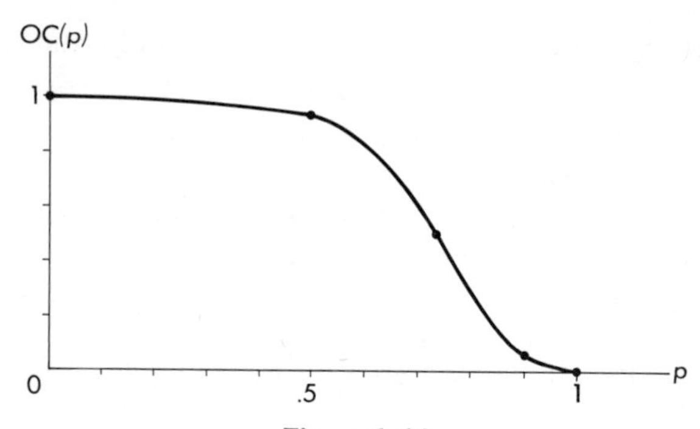

Figure 6–14

6.3.4 Required Sample Size

The number of observations required to reach a decision, in a given sequential likelihood ratio test, is a random variable. Let it be denoted by N. Its distribution depends on the state of nature that obtains during the sampling process. In Section 6.3.2, it was seen that $N < \infty$ with probability 1 so that N has values 1, 2, 3, . . . with probabilities $p_1, p_2, \ldots,$ where $\sum p_i = 1$.

The moments of N are not easily computed but can be shown to be finite. With an integer r chosen as in Section 6.3.2, consider the terms defining $E(N^i)$ grouped in bunches of r:

$$\begin{aligned} E(N^i) &= [1^i p_1 + \cdots + r^i p_r] + [(r+1)^i p_{r+1} + \cdots + (2r)^i p_{2r}] + \cdots \\ &\leq r^i(p_1 + \cdots + p_r) + (2r)^i(p_{r+1} + \cdots + p_{2r}) + \cdots. \end{aligned}$$

Use of the inequalities

$$p_1 + \cdots + p_r \leq 1, \qquad p_{r+1} + \cdots + p_{2r} \leq 1 - q^r,$$
$$p_{2r+1} + \cdots + p_{3r} \leq (1 - q^r)^2, \qquad \text{etc.},$$

which follow from Section 6.3.2, yields

$$E(N^i) \leq r^i + (2r)^i(1 - q^r) + (3r)^i(1 - q^r)^2 + \cdots$$
$$= r^i[1 + 2^i(1 - q^r) + 3^i(1 - q^r)^2 + \cdots].$$

The series in brackets can be shown to be finite by the standard ratio test, given (as is the case) that $0 < q \leq 1$.

Suppose for the moment that $Z, Z_1, Z_2, \ldots$ are independent, identically distributed random variables, and that N is a random variable with values $1, 2, \ldots$ such that the event $\{N \geq i\}$ is independent of any events involving $Z_i, Z_{i+1}, \ldots$. Let Y_i be 0 if $N < i$ and 1 if $N \geq i$. Then

$$E(Z_1 + \cdots + Z_N) = E\left\{\sum_{i=1}^{\infty} Y_i Z_i\right\} = \sum_{i=1}^{\infty} E(Y_i Z_i) = E(Z) \sum_{i=1}^{\infty} E(Y_i),$$

provided that it is permissible to interchange summation and expectation. Now,

$$\sum_{i=1}^{\infty} E(Y_i) = \sum_{i=1}^{\infty} P(N \geq i) = \sum_{i=1}^{\infty} \sum_{j=i}^{\infty} P(N = j) = \sum_{j=1}^{\infty} \sum_{i=1}^{j} P(N = j) = E(N),$$

which gives the desired expression for $E(\log \lambda_N)$, namely, $E(Z)E(N)$. The interchange employed is valid if the following series is convergent:

$$\sum_{i=1}^{\infty} |E(Y_i Z_i)| \leq \sum_{i=1}^{\infty} E(|Y_i|)E(|Z_i|) = E(|Z|)E(N).$$

It will be convergent if $E(|Z|)$ is finite and $E(N)$ is finite.

Application of this last result to the sequence $Z_i = \log [f_0(X_i)/f_1(X_i)]$ provides an expression for the value of $E(N)$:

$$E(N) = \frac{E(\log \Lambda_N)}{E(Z)},$$

which can be computed approximately. For, the random variable Λ_N is approximately a Bernoulli variable, with the value $\log A$ if the decision reached is to reject H_0 and the value $\log B$ if the decision reached is to accept H_0:

$$E_\theta(\log \Lambda_N) \doteq (\log A)\pi(\theta) + (\log B)[1 - \pi(\theta)],$$

where $\pi(\theta)$ is the power function or probability that the given test rejects H_0 when the state is actually θ. In particular, then,

$$E(N \mid H_0) = \frac{1}{E(Z \mid H_0)} [\alpha \log A + (1 - \alpha) \log B]$$

and

$$E(N \mid H_1) = \frac{1}{E(Z \mid H_1)} [(1 - \beta) \log A + \beta \log B].$$

EXAMPLE 6–25. To test $\mu = 0$ against $\mu = 1$ in a normal population with unit variance with $\alpha = \beta = .01$, one uses constants $A = \frac{1}{99}$ and $B = 99$. Then $\log A = -\log 99 = -\log B = -4.595$, and

$$E_{H_0}(\log \Lambda_N) = \alpha \log A + (1 - \alpha) \log B = (1 - 2\alpha) \log 99$$
$$= 4.5031.$$

The log of the likelihood ratio is

$$Z = \log \frac{f_0(X)}{f_1(X)} = \log \left\{ \frac{\exp(-\frac{1}{2}X^2)}{\exp[-\frac{1}{2}(X - 1)^2]} \right\} = \frac{1}{2} - X,$$

and then $E(Z) = E(\frac{1}{2} - X) = \frac{1}{2} - \mu$. If $\mu = 0$,

$$E(N) = \frac{E(\log \Lambda_N)}{E(Z)} = \frac{4.5031}{1/2 - 0} \doteq 9.$$

Similarly, if $\mu = 1$,

$$E(N) = \frac{(1 - \beta) \log A + \beta \log B}{1/2 - 1} \doteq 9.$$

In a likelihood ratio test of fixed sample size, the conditions

$$.01 = \alpha = P(\bar{X} > k \mid \mu = 0),$$
$$.01 = \beta = P(\bar{X} < k \mid \mu = 1)$$

lead to $n \doteq 22$, much larger than the expected sample size in the sequential test.

The expected sample size can be computed for other states than the two used to define the test, if H_0 and H_1 correspond to particular values θ_0 and θ_1 of a parameter θ:

$$E(N \mid \theta) = \frac{1}{E(Z \mid \theta)} \{(\log A)\pi(\theta) + (\log B)[1 - \pi(\theta)]\}.$$

(Notice that logs are to the base e if the log in Z is to that base.)

The expected value of B is usually a function of θ. One annoying feature of the sequential likelihood ratio test, when it is used to test a one-sided situation such as $H_0: \theta \leq \theta'$ against $H_1: \theta > \theta'$ is the following: The test would be set up by choosing θ_0 in H_0 (i.e., $\theta_0 < \theta'$) and θ_1 in H_1 ($\theta_1 > \theta'$), the zone between θ_0 and θ_1 being an "indifference zone." And yet, if the actual population is described by a θ in this indifference zone (i.e., near θ'), $E(N)$ tends to be largest. Thus the test tends to take longer to reach a decision when θ is near the point θ' where there is perhaps little concern as to which way the test turns out. It is certainly intuitively reasonable that it should not take so long to discover that a population is definitely of one kind or the other as it does to discover which kind it is when it is near the borderline. What is annoying, then, is that in the borderline case, wrong decisions are likely not to be very costly anyway, and it seems a shame to invest the effort or cost in the large sample required to reach a decision. Ways of altering the sequential testing procedure have been proposed but these are not taken up here.

EXAMPLE 6–26. In testing p_0 against p_1, possible values of a Bernoulli parameter p, it was found in Example 6–24 that

$$p = \frac{1 - (q_0/q_1)^h}{(p_0/p_1)^h - (q_0/q_1)^h},$$

where $q_i = 1 - p_i$ and h is such that $E(e^{hZ}) = 1$. The power in terms of h is

$$\pi[p(h)] = \frac{B^h - 1}{B^h - A^h}.$$

Now, using again $p_0 = .5$, $p_1 = .9$, and $\alpha = \beta = .05$, one obtains

$$E(Z) = p \log \frac{p_0 q_1}{q_0 p_1} + \log \frac{q_0}{q_1} = -p \log 9 + \log 5.$$

At $h = 0$ both EZ and $E(\log \Lambda_N)$ are 0, but their ratio can be evaluated by computing its limit as $h \to 0$, the result given in the following table.

h	$-\infty$	-1	0	1	∞
p	0	.5	.733	.9	1
$E(N \mid p)$	1.83	5.19	9.17	7.2	5.01

The entry for $h = 0$ involves evaluation of an indeterminate form. This is perhaps most easily accomplished by using series expansions. For instance, since $B^h = \exp[h \log B]$,

$$B^h - A^h = e^{h \log B} - e^{h \log A} = h(\log B - \log A) + \frac{h^2}{2}(\log^2 B - \log^2 A) + \cdots.$$

Similar expansions for other ingredients of $E(N)$ yield

$$E(N)|_{h=0} = \frac{\log A \log B}{\log 5 \log (5/9)} = 9.17.$$

Although finite, the required N can be very large in any single experiment—much larger than one might want to tolerate. In practice, one establishes a bound n_0 beyond which he refuses to go; if no decision is reached by the (n_0)th stage, sampling is stopped anyway, with H_0 accepted if $\Lambda_{n_0} > 1$ and rejected if $\Lambda_{n_0} < 1$. The effect of such a modification, which surely alters the error sizes of the test is slight if n_0 is large (see Wald [21]).

Graphically, the topping rule just described amounts to altering the decision lines from those shown in Figure 6–13 to those shown in Figure 6–15.

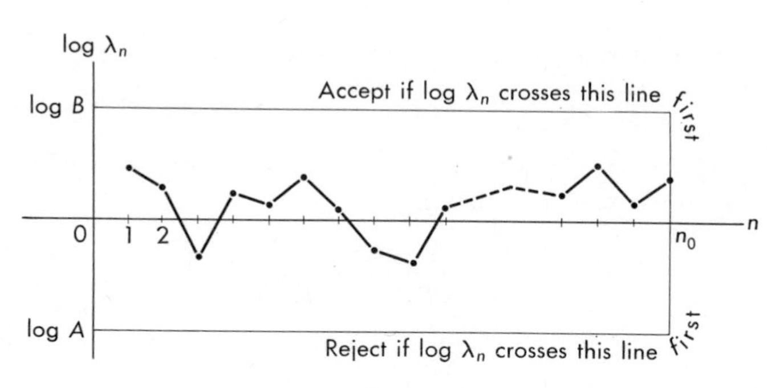

Figure 6–15

Problems

6–42. Determine the relation between μ and h in a sequential test for $\mu = 0$ against $\mu = 1$ in a normal population with unit variance. Plot the power function of the test with $\alpha = \beta = .01$.

6–43. Use the result of Problem 6–42 to sketch (using 5 points) the graph of the expected sample size for the test of that problem.

6–44. Show that in testing $m = m_0$ against $m = m_1$ in a Poisson population, the relation between m and h is

$$m = \frac{(m_1 - m_0)^h}{1 - (m_0/m_1)^h}.$$

6–45. Obtain the graphs of $OC(\theta)$ and $E(N \mid \theta)$ in a sequential test of the density $\theta_0 e^{-\theta_0 x}$ against $\theta_1 e^{-\theta_1 x}$ $(x > 0)$ using $\theta_0 = 2$ and $\theta_1 = 1$, $\alpha = .05$, $\beta = .10$. Observe that for one of the five points used, $E(N) = 0$. In view of the fact that $N \geq 1$, this result must be wrong. Explain.

7

UNIVARIATE NORMAL INFERENCE

Both the techniques of inference and the distribution theory that gives those techniques a mathematical basis have been most highly developed in the case of a normal population. This is partly because the form of the normal density makes it particularly amenable to such a development; partly because—although perhaps never *exactly* the right model—the normal model is often sufficiently close to correct for practical purposes; partly because the very scope of the methods and theory of the normal model provide a prototype for such development; and partly because the methods developed for normal populations often turn out to work reasonably well when the population is moderately different from normal.

7.1 Distribution Theory

In this section the distributions of the minimal sufficient statistic $(\bar{X}, S^2)$ will be derived, as well as the distributions of various other statistics encountered in problems of inference in univariate normal populations. Large-sample distributions have already been given and used; to be taken up now are the distributions for "small" samples—samples of a size not large enough for asymptotic results to give answers that are sufficiently precise.

7.1.1 Gamma Distributions

Integrals of the form

$$\int_0^\infty x^{t-1}e^{-\lambda x}\,dx$$

have been encountered earlier for the case of integer t (in studying sums of exponential variables) and in the case of t equal to half an integer (in the chi-square distribution). The integral is an improper one, for the range of integration is infinite; and if $t < 1$, the point $x = 0$ is a point of discontinuity of the integrand. It can be shown by standard methods that the integral converges for all real, positive values of t (assuming $\lambda > 0$), and so it defines a function of t on that range; when $\lambda = 1$ it is called the *gamma function*:

$$\Gamma(t) = \int_0^\infty x^{t-1}e^{-x}\,dx.$$

Certain other integrals can be reduced, by appropriate change of variable, to gamma functions. In particular, these are useful:

$$\int_0^\infty x^{t-1}e^{-\lambda x}\,dx = \frac{1}{\lambda^t}\,\Gamma(t), \qquad \text{for } \lambda > 0,$$

and

$$\int_0^\infty x^{2t-1}e^{-x^2/2}\,dx = 2^{t-1}\Gamma(t).$$

The latter formula gives expressions for the even order moments of a standard normal distribution:

$$E(X^{2k}) = \frac{1}{\sqrt{2\pi}}\int_{-\infty}^\infty x^{2k}e^{-x^2/2}\,dx$$
$$= \sqrt{\frac{2}{\pi}}\int_0^\infty x^{2k}e^{-x^2/2}\,dx = \frac{2^k}{\sqrt{\pi}}\,\Gamma\left(k + \frac{1}{2}\right).$$

These moments were obtained earlier from the moment generating function:

$$E(X^{2k}) = (2k - 1)(2k - 3)\cdots 5\cdot 3\cdot 1,$$

and equating the two expressions yields a formula for $\Gamma(k + \frac{1}{2})$. In particular, for $k = 0$, $\Gamma(\frac{1}{2}) = \sqrt{\pi}$.

One interesting property of the gamma function is that it can be considered as a "factorial" function—its value at a positive integer is a factorial. To see how this comes about, integrate by parts (recalling that t is a positive real number):

$$\Gamma(t) = \int_0^\infty e^{-x}x^{t-1}\,dx = \frac{e^{-x}x^t}{t}\Big|_0^\infty + \int_0^\infty \frac{x^t}{t}e^{-x}\,dx$$
$$= 0 + \frac{1}{t}\,\Gamma(t + 1),$$

or

$$\Gamma(t + 1) = t\Gamma(t).$$

This relationship automatically gives the values of $\Gamma(t)$ on any interval $n < t \le n + 1$ in terms of its values on the interval $0 < t \le 1$. On the latter range the point $t = 1$ is easy to handle:

$$\Gamma(1) = \int_0^\infty e^{-x}\,dx = 1,$$

whence

$$\Gamma(2) = \Gamma(1) = 1,$$
$$\Gamma(3) = 2\Gamma(2) = 2 \cdot 1,$$

and by a simple induction,

$$\Gamma(n) = (n - 1)!$$

This result is the reason for calling $\Gamma(t)$ a *generalized factorial* function—a function defined for all positive t that has the value of an ordinary factorial at the integers.

The value of $\Gamma(t)$ at the points halfway between the integers is obtainable from the value $\Gamma(\frac{1}{2}) = \sqrt{\pi}$, given previously. Again from the recurrence relation $\Gamma(t + 1) = t\Gamma(t)$ there follows:

$$\Gamma\left(\frac{3}{2}\right) = \frac{1}{2}\Gamma\left(\frac{1}{2}\right) = \frac{1}{2}\sqrt{\pi},$$

$$\Gamma\left(\frac{5}{2}\right) = \frac{3}{2}\Gamma\left(\frac{3}{2}\right) = \frac{3}{2}\cdot\frac{1}{2}\sqrt{\pi},$$

and in general (by induction),

$$\Gamma\left(k + \frac{1}{2}\right) = (2k - 1)(2k - 3)\cdots 5 \cdot 3 \cdot 1\frac{\sqrt{\pi}}{2^k},$$

which is the same result as that obtained earlier by equating expressions for the moments of even order of a standard normal distribution.

It can be shown that $\Gamma(t)$ is continuous, is concave up ($\Gamma''(t) > 0$), and tends to $+\infty$ as $t \to 0+$ and as $t \to \infty$. With these facts and the values of $\Gamma(t)$ for t a multiple of $\frac{1}{2}$, one can sketch $\Gamma(t)$, as in Figure 7–1.

The integration formula

$$\int_0^\infty x^{\alpha - 1}e^{-\lambda x}\,dx = \frac{\Gamma(\alpha)}{\lambda^\alpha}, \qquad \alpha, \gamma > 0,$$

gives the proper multiplying constant to make the integrand (which is a non-negative function) a density:

$$f(x; \alpha, \lambda) = \frac{\lambda^\alpha}{\Gamma(\alpha)}x^{\alpha-1}e^{-\lambda x}, \qquad \text{for } x > 0.$$

Distributions defined by such a density are said to be *gamma distributions*. The characteristic function is

$$\phi(t) = \frac{\lambda^\alpha}{\Gamma(\alpha)}\int_0^\infty e^{-x(\lambda - it)}x^{\alpha - 1}\,dx$$

$$= \left(1 - \frac{it}{\lambda}\right)^{-\alpha}.$$

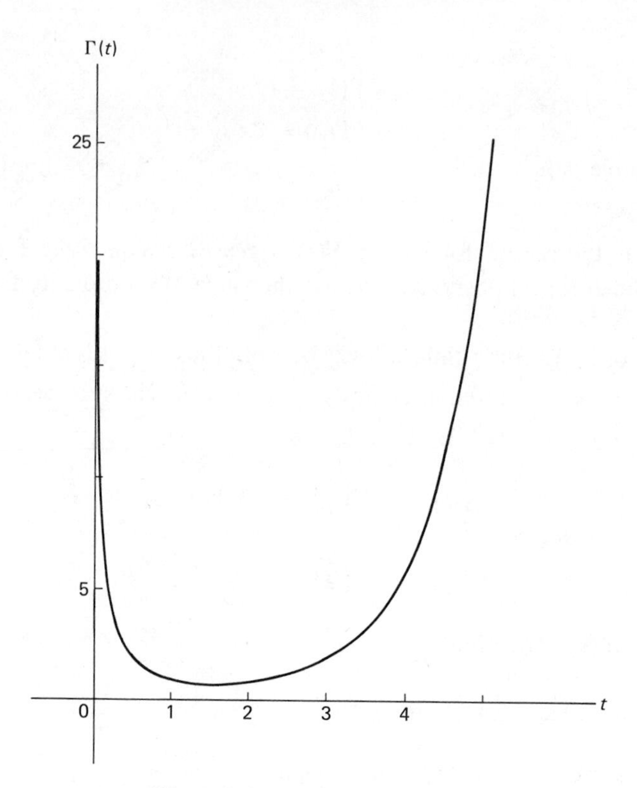

Figure 7–1. The gamma function.

Moments of the distribution can be computed from the coefficients in the expansion of $\phi(t)$, or more directly as expected values that are expressible as gamma functions.

7.1.2 Beta Distributions

The "beta function" arises in considering the product of two gamma functions [using the form involving $\exp(-\frac{1}{2}x^2)$]:

$$\begin{aligned}\Gamma(s)\Gamma(t) &= \int_0^\infty \int_0^\infty \left(\frac{u^2}{2}\right)^{s-1} \left(\frac{v^2}{2}\right)^{t-1} \exp\left[-\frac{1}{2}(u^2+v^2)\right] uv\, du\, dv \\ &= \int_0^{\pi/2} \int_0^\infty \left(\frac{r^2}{2}\right)^{s+t-1} 2(\cos\theta)^{2s-1}(\sin\theta)^{2t-1} \exp\left(-\frac{1}{2}r^2\right) r\, dr\, d\theta \\ &= \int_0^\infty w^{s+t-1}e^{-w}\, dw \int_0^{\pi/2} 2(\cos\theta)^{2s-1}(\sin\theta)^{2t-1}\, d\theta \\ &= \Gamma(s+t)B(s,t),\end{aligned}$$

where, then,

$$B(s, t) = 2\int_0^{\pi/2} \cos^{2s-1}\theta \sin^{2t-1}\theta\, d\theta = \frac{\Gamma(s)\Gamma(t)}{\Gamma(s+t)}$$

$$= \int_0^1 x^{s-1}(1-x)^{t-1}\, dx = \int_0^1 (1-y)^{s-1}y^{t-1}\, dy = B(t, s).$$

This symmetric function of s and t is called the *beta function* (the B is a capital "beta"). This derivation of $B(s, t)$ is along the same lines as the calculation of the integral of the normal density, and the latter is a special case of the present formula, with $s = t = \frac{1}{2}$:

$$\Gamma(\tfrac{1}{2})\Gamma(\tfrac{1}{2}) = B(\tfrac{1}{2}, \tfrac{1}{2})\Gamma(1)$$

$$= 2\int_0^{\pi/2} \cos^0\theta \sin^0\theta\, d\theta = \pi.$$

The family of *beta distributions* is defined by the density

$$f(x) = \frac{1}{B(r, s)} x^{r-1}(1-x)^{s-1}, \qquad \text{for } 0 < x < 1,$$

where r and s are positive parameters. The kth moment about zero is readily computed:

$$\frac{1}{B(r, s)}\int_0^1 x^k x^{r-1}(1-x)^{s-1}\, dx = \frac{B(r+k, s)}{B(r, s)}.$$

In particular, the expected value is obtained by setting $k = 1$:

$$\frac{B(r+1, s)}{B(r, s)} = \frac{r}{r+s}.$$

The family of beta distributions—shown in Problem 3–77 to belong to the two-parameter exponential family—provides a versatile and useful set of models for distributions on the unit interval $0 < x < 1$. For $r > 1$ and $s > 1$ they are unimodal, with maximum density at the value $x = (r-1)/(r+s-2)$, having various degrees of peakedness depending on the size of $r + s$, and various degrees of skewness depending on the ratio of r to s. (For $r = s$ they are symmetric.) For $r = s = 1$ the density is uniform on $(0, 1)$, for $r < 1$ it is infinite at $x = 0$, and for $s < 1$ it is infinite at $x = 1$. When $r > 1$ the graph passes through the origin like x^{r-1}, and when $s > 1$ it passes through $(1, 0)$ like $(1-x)^{s-1}$, the order of tangency increasing with r and s, respectively. Several members of the beta family of density functions are shown in Figure 7–2.

A related family of distributions, on the positive real axis $0 < x < \infty$, is suggested by a form of the beta function obtained by making the change of variable $x = u/(1+u)$ in the earlier integral formula; the result of the change is

$$B(r, s) = \int_0^\infty \frac{u^{r-1}\, du}{(1+u)^{r+s}}.$$

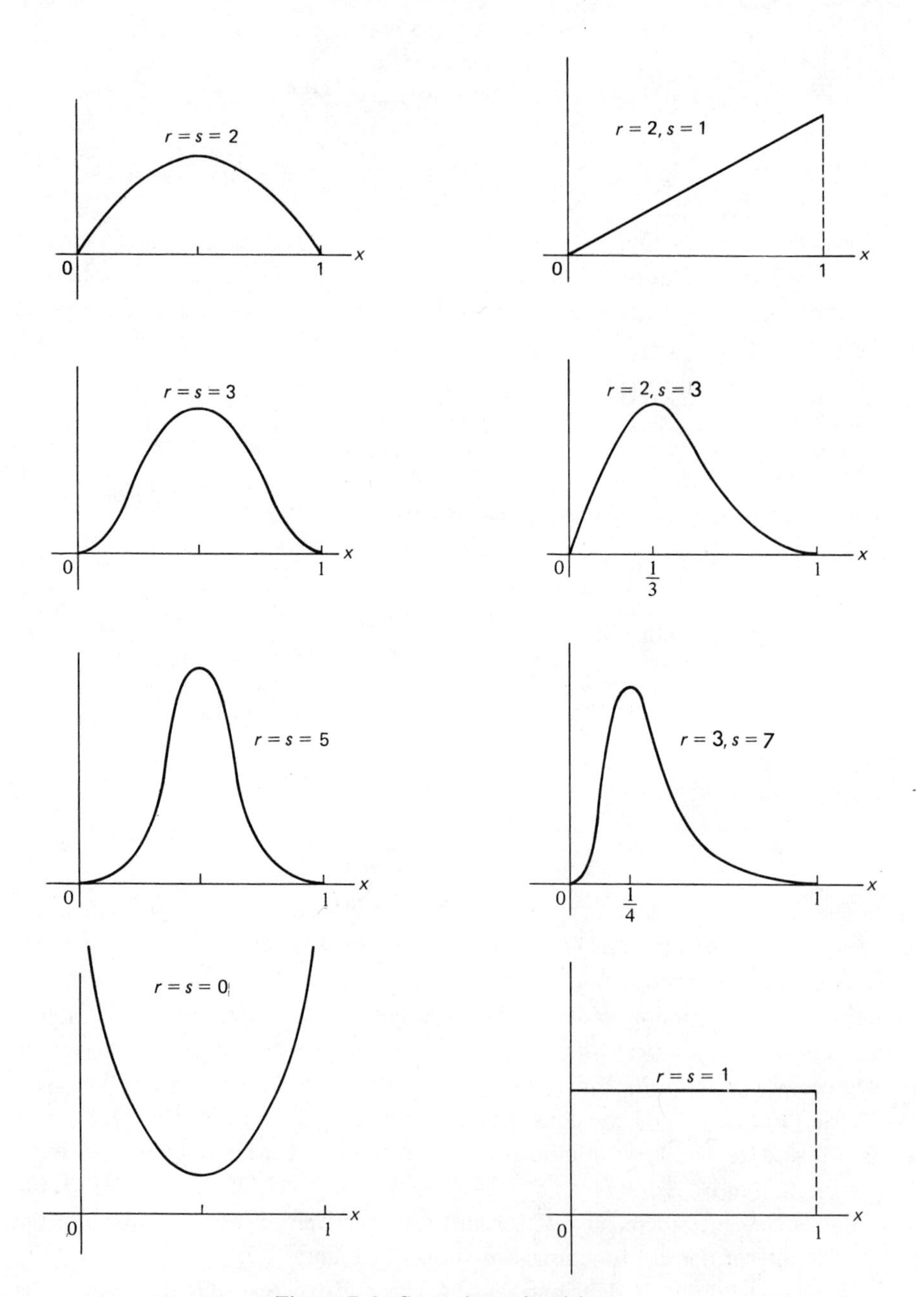

Figure 7–2. Some beta densities.

The integrand, divided by $B(r, s)$, is a density function on $(0, \infty)$; it will be encountered shortly in the study of the ratio of chi-square variables.

Problems

7–1. Compute the following:

(a) $\Gamma(6)$.

(b) $\Gamma(\frac{11}{2})$.

(c) $B(2, \frac{3}{2})$.

(d) $\int_0^1 (-\log_e u)^{3/2} u\, du$. (Let $u = e^{-x}$.)

(e) $\int_0^\infty e^{-3x} x^4\, dx$.

(f) $\int_0^1 x^5 (1 - x)^9\, dx$.

(g) $\int_0^{\pi/2} \cos^4 \theta \sin^6 \theta\, d\theta$.

(h) $\int_0^\infty \frac{u^2\, du}{(1 + u)^5}$.

7–2. Given that X has a gamma distribution (α, λ), evaluate EX and EX^2 by recognizing the integrals involved as gamma functions. From these obtain var X.

7–3. Calculate the variance of a beta distribution.

7–4. Let W have the density

$$f_W(w) = \frac{1}{B(r, s)} \frac{w^{r-1}}{(1 + w)^{r+s}}, \qquad \text{for } w > 0.$$

Calculate EW and EW^2 by recognizing the integrals involved as beta functions.

7–5. Let W have the density of Problem 7–4, and let $U = 1/W$. Obtain the density of U, and show that U is a variable of the same family—all that happens is that r and s are interchanged.

7.1.3 The Chi-Square Distribution

The chi-square distribution was defined in Section 3.3.4 as the distribution of a sum of squares of independent, standard normal random variables. The number of terms in the sum is called the number of *degrees of freedom*. The characteristic function of the chi-square distribution with k degrees of freedom was seen to be

$$\phi(t) = (1 - 2it)^{-k/2}.$$

This is the characteristic function of a gamma distribution (cf. Section 7.1.1) with $\alpha = k/2$ and $\lambda = \frac{1}{2}$, and so the density function of the distribution is

$$f(x) = \frac{1}{2^{k/2}\Gamma(k/2)} x^{k/2-1} e^{-x/2}, \qquad x > 0.$$

The mean of the distribution is k, the number of degrees of freedom, and the variance is $2k$.

The chi-square distribution is often encountered in large-sample theory as an asymptotic distribution of a test statistic. For instance, it was given in Section

6.2.3 as the asymptotic distribution of $-2 \log \Lambda$, where Λ is the likelihood ratio statistic; and it will provide the asymptotic distribution of a goodness-of-fit statistic and of certain contingency table statistics in Chapter 9. In small-sample theory, the chi-square distribution is what is needed in studying the sample variance in the case of a normal population and various estimates of variance in the analysis of designed experiments involving normal models (Chapter 12).

The "standardized" observations $(X_i - \mu)/\sigma$ in a random sample from a normal population with mean μ and variance σ^2 are independent standard normal variates. The sum of their squares then has the chi-square distribution with n degrees of freedom. This fact is the starting point in problems in the "analysis of variance" (Chapter 12).

The chi-square distribution obeys an addition law, to be stated formally along with some important consequences:

Theorem 1. *If X and Y have independent chi-square distributions with k and m degrees of freedom, respectively, then $X + Y$ is chi-square with $k + m$ degrees of freedom.*

The proof of this was called for in Problem 3–68. It follows easily using characteristic functions, but perhaps more fundamentally it is the result of the structure of a chi-square variate as the sum of squares of independent standard normal variates. Adding two such variates together clearly yields (if they are independent) a variate having the same structure, the parameters adding together to yield the parameter of the sum.

Corollary. *If $X_1, \ldots, X_n$ are independent and X_i is chi-square with k_i degrees of freedom $(i = 1, \ldots, n)$, then $X_1 + \cdots + X_n$ is chi-square with $k_1 + \cdots + k_n$ degrees of freedom.*

This follows from Theorem 1 by induction on n.

Theorem 2. *If X and Y are independent, and if X is chi-square with k degrees of freedom and $X + Y$ is chi-square with m degrees of freedom, then Y is chi-square with $m - k$ degrees of freedom.*

Again the proof uses characteristic functions:

$$\phi_Y(t) = \frac{\phi_{X+Y}(t)}{\phi_X(t)} = \frac{(1 - 2it)^{-m/2}}{(1 - 2it)^{-k/2}} = (1 - 2it)^{-(m-k)/2}.$$

7.1.4 The Distribution of $\bar{X}$ and S^2

The mean $\bar{X}$ of a random sample from a normal population is already known (Problem 4–13) to be normally distributed with mean μ and variance σ^2/n, where μ and σ^2 are the population parameters as usual. The distribution of the sample variance S^2 will now be obtained, following a demonstration of the fact that $\bar{X}$ and S^2 are *independent* random variables.

Lemma. *If a bivariate characteristic function factors so that one factor is a marginal characteristic function:*

$$\phi_{U,V}(s, t) = \phi_U(s)h(t),$$

then the other factor is the other marginal characteristic function. That is, $h(t)$ is the characteristic function of V, and U and V are independent. (Either or both U and V can be vectors.)

This follows immediately from the fact that $\phi_V(t) = \phi_{U,V}(0, t)$.

Theorem 3. *Given a random sample* $\mathbf{X}$ *from a normal population, the statistic $\bar{X}$ and the vector $(X_1 - \bar{X}, \ldots, X_n - \bar{X})$ are independent.*

To prove this, write out the joint characteristic function of the $n + 1$ quantities involved:

$$\phi_{\bar{X},X_1-\bar{X},\ldots,X_n-\bar{X}}(t, t_1, \ldots, t_n) = E\{\exp(i[t\bar{X} + t_1(X_1 - \bar{X}) + \cdots + t_n(\bar{X}_n - \bar{X})])\}.$$

But observe that the bracketed quantity in the exponent can be rewritten:

$$t\bar{X} + \sum t_j(X_j - \bar{X}) = \sum \left[\frac{t}{n} + (t_j - \bar{t})\right] X_j \equiv \sum a_j X_j,$$

which is a linear combination of independent, normal variables, with coefficients

$$a_j \equiv \frac{t}{n} + (t_j - \bar{t})$$

having the properties

$$\sum a_j = t \quad \text{and} \quad \sum a_j^2 = \frac{t^2}{n} + \sum (t_j - \bar{t})^2.$$

The joint characteristic function of $\bar{X}$ and the vector $(X_1 - \bar{X}, \ldots, X_n - \bar{X})$ can then be written as

$$\begin{aligned} E\left[\exp\left(\sum ia_jX_j\right)\right] &= \prod \phi_{X_j}(a_j) = \prod \exp\left(i\mu a_j - \frac{\sigma^2}{2} a_j^2\right) \\ &= \exp\left(i\mu \sum a_j - \frac{\sigma^2}{2} \sum a_j^2\right) \\ &= \exp\left[i\mu t - \frac{\sigma^2 t^2}{2n}\right] \cdot \exp\left[-\frac{\sigma^2}{2} \sum (t_j - \bar{t})^2\right]. \end{aligned}$$

This last is of the form

$$\phi_{\bar{X}}(t)h(t_1, \ldots, t_n),$$

and by the lemma it follows that $h(t_1, \ldots, t_n)$ is the characteristic function of $(X_1 - \bar{X}, \ldots, X_n - \bar{X})$; moreover, this vector is independent of $\bar{X}$.

Corollary. *The mean $\bar{X}$ and the variance S^2 of a random sample from a normal population are independent random variables.*

This follows from Theorem 3 because S^2 is a function of the vector $(X_1 - \bar{X}, \ldots, X_n - \bar{X})$, and because functions of independent quantities are independent.

Theorem 4. *If X is a random sample from a normal population, then nS^2/σ^2 has a chi-square distribution with $n - 1$ degrees of freedom.*

This follows from the parallel axis theorem:

$$\sum \left(\frac{X_i - \mu}{\sigma}\right)^2 = \frac{nS^2}{\sigma^2} + \frac{(\bar{X} - \mu)^2}{\sigma^2/n}.$$

The left-hand side has a chi-square distribution with n degrees of freedom because it is the sum of squares of n independent standard normal variables, and the second term on the right is chi-square with one degree of freedom because it is the square of the standard normal variable $(\bar{X} - \mu)/\sqrt{\sigma^2/n}$. Hence, by the corollary to Theorem 3 (which gives independence of the two terms on the right in the equation just given) and by Theorem 2 of the preceding section, the first term on the right has a chi-square distribution with $n - 1$ degrees of freedom.

Problems

7–6. Obtain the formula $2k$ for the variance of a chi-square distribution with k degrees of freedom by computing it as the variance of a sum.

7–7. Calculate $E(1/U)$, where U has a chi-square distribution with k degrees of freedom.

7–8. Compute the mean and variance of S^2, the sample variance of a random sample from a normal population.

7–9. Show that in the case of a random sample from a normal population, the sample mean and the sample mean deviation, $\sum |X_i - \bar{X}|$, are independent.

7–10. Show that the distribution of $\sum Z_i^2$, given $\sum Z_i = 0$, is chi-square with $n - 1$ degrees of freedom if the Z's are independent standard normal variables.

[*Hint:* Express the sum of squares in terms of the mean and variance, and then exploit their independence.]

7–11. Let **X** be a random sample from a normal population with mean μ. Compute the expected value of the quantity $(\bar{X} - \mu)/S$. (Observe that the independence of numerator and denominator is crucial.)

7.1.5 *F* and *t* Distributions

In the problem of comparing normal populations with respect to their variances, as well as in a variety of other problems, it will be necessary to know the distribution of the ratio of two chi-square random variables.

Let U and V denote independent chi-square variables with, respectively, m and n degrees of freedom. The joint distribution of U and V is defined by the joint density, which is obtained by multiplying together the densities of U and V:

$$f_{U,V}(u, v) = \frac{2^{-(m+n)/2}}{\Gamma(m/2)\Gamma(n/2)} u^{m/2-1}v^{n/2-1}e^{-(u+v)/2},$$

which holds for $u > 0$ and $v > 0$. Consider then the ratio $W = U/V$: This random variable has the distribution function (for $w > 0$)

$$\begin{aligned} F_W(w) &= P(W < w) = P\left(\frac{U}{V} < w\right) \\ &= \iint\limits_{u/v<w,\,u>0} f_{U,V}(u, v)\,du\,dv \\ &= \int_0^\infty \int_0^{vw} f_{U,V}(u, v)\,du\,dv \\ &= \frac{2^{-(m+n)/2}}{\Gamma(m/2)\Gamma(n/2)} \int_0^\infty e^{-v/2}v^{n/2-1} \int_0^{vw} e^{-u/2}u^{n/2-1}\,du\,dv. \end{aligned}$$

The density function is then

$$f_W(w) = \frac{d}{dw} F_W(w) = \frac{2^{-(m+n)/2}w^{m/2-1}}{\Gamma(m/2)\Gamma(n/2)} \int_0^\infty e^{-(v+vw)/2}v^{n/2+m/2-1}\,dv.$$

And since

$$\int_0^\infty e^{-(1+w)v/2}v^{(m+n)/2-1}\,dv = \frac{\Gamma[(m+n)/2]2^{(m+n)/2}}{(1+w)^{(m+n)/2}},$$

the density of W reduces to

$$f_W(w) = \frac{1}{B(m/2, n/2)} w^{m/2-1}(1+w)^{-(m+n)/2},$$

for $w > 0$. Thus, W has a distribution of the form given in Problem 7–4, obtainable from a beta distribution by a change of variable $(w/(1+w) = x)$.

The F distribution is now defined as the distribution of the ratio of two independent chi-square variables, each divided by the corresponding number of degrees of freedom:

$$F = \frac{U/m}{V/n} = \frac{n}{m} W.$$

The density function of F is obtained at once from the density of W:

$$\begin{aligned} f_F(x) &= \frac{m}{n} f_W\left(\frac{m}{n} x\right). \\ &= \frac{(m/n)^{m/2}}{B(m/2, n/2)} x^{m/2-1}\left(1 + \frac{m}{n} x\right)^{-(m+n)/2}, \end{aligned}$$

for $x > 0$. The expected value of F exists for $n > 2$ and is readily obtained from the expected value of W (Problem 7–4):

$$E(F) = E\left(\frac{n}{m} W\right) = \frac{n}{m} E(W) = \frac{n}{n-2}.$$

The percentiles of the F distribution are available in tables (see Table IV, Appendix), but it is customary to give only the lower or the higher percentiles, not both. For instance, if $F_{.05}$ is given, then $F_{.95}$ need not be listed because it is contained elsewhere in the table of 5th percentiles. For,

$$\begin{aligned} .05 = P(F < F_{.05}) &= 1 - P(F > F_{.05}) \\ &= 1 - P\left(\frac{1}{F} < \frac{1}{F_{.05}}\right) \end{aligned}$$

and, therefore, $1/F_{.05}$ is the 95th percentile of the distribution of $1/F$. But $1/F$ is again a random variable with an F distribution, n degrees of freedom in the numerator, and m in the denominator (if F itself has m degrees of freedom in the numerator and n in the denominator).

The *t distribution* with n degrees of freedom can be defined as that of a random variable symmetrically distributed about 0 whose square has the F distribution with 1 and n degrees of freedom in numerator and denominator, respectively. Let T denote such a random variable so that T^2 has the F density

$$f_{T^2}(x) = \frac{\Gamma((1+n)/2)}{n\Gamma(1/2)\Gamma(n/2)}\left(\frac{x}{n}\right)^{1/2-1}\left(1 + \frac{x}{n}\right)^{-(1+n)/2}, \qquad x > 0.$$

Then, for $x > 0$,

$$f_{|T|}(x) = \frac{d}{dx} P(|T| < x) = \frac{d}{dx} P(T^2 < x^2) = 2x f_{T^2}(x^2).$$

But since T is symmetrically distributed, its distribution density is obtained from that of $|T|$ as follows:

$$f_T(\pm x) = \frac{1}{2} f_{|T|}(|x|) = |x| f_{T^2}(x^2)$$
$$= \frac{\Gamma((n+1)/2)}{\sqrt{n\pi}\,\Gamma(n/2)} \left(1 + \frac{x^2}{n}\right)^{-(n+1)/2}.$$

This is the desired density function of a t distribution.

The symmetry of the t distribution about $x = 0$ implies that the mean value of T, if it exists, must be 0. The integral defining the mean is clearly absolutely convergent for $n > 1$, and so for those values of n, $E(T) = 0$. For $n = 1$, the density of T reduces to what has been called the *Cauchy density*, having no absolute moments of any integral order. For $n > 2$, the integral defining $E(T^2)$ is absolutely convergent, and so for those values of n, $\operatorname{var} T = E(T^2) = n/(n-2)$, as derived above for the F distribution.

As $n \to \infty$, the density function of T approaches that of a standard normal variate. For,

$$\left(1 + \frac{x^2}{n}\right)^{-(n+1)/2} = \left\{\left(1 + \frac{x^2}{n}\right)^{n/x^2}\right\}^{-x^2/2} \left(1 + \frac{x^2}{n}\right)^{-1/2} \to \exp\left(-\frac{x^2}{2}\right).$$

Moreover, the constant factor tends to $1/\sqrt{2\pi}$. This can be seen using "Stirling's formula":[1]

$$\Gamma(p) \sim \sqrt{\frac{2\pi}{p}} \left(\frac{p}{e}\right)^p, \qquad \text{(large } p\text{)},$$

which implies that

$$\frac{\Gamma(p+h)}{\Gamma(p)} \sim p^h, \qquad \text{(large } p\text{)}.$$

This approach to normality accounts for the fact that in t tables (such as Table III, Appendix), there is often a sequence of entries for $n = \infty$, which are simply the corresponding points on a standard normal distribution.

7.1.6 Noncentral Chi-Square Distribution

A sum of squares of independent, normal variables, each having unit variance but with possibly nonzero means is said to have a *noncentral chi-square* distribution. That is, if $Z_1, \ldots, Z_k$ are independent, and $Z_i \overset{d}{=} \mathcal{N}(\mu_i, 1)$, then

$$\chi'^2 = Z_1^2 + \cdots + Z_k^2$$

[1] See H. Cramér [5], p. 128.

has the noncentral chi-square distribution with k degrees of freedom. This distribution would seem to depend on the k parameters $\mu_1, \ldots, \mu_k$, but the following development will show that the dependence is only through the value of the *noncentrality parameter*:

$$\lambda \equiv \tfrac{1}{2}(\mu_1^2 + \cdots + \mu_k^2).$$

The moment generating function of the noncentral χ'^2 is computed as follows. If Z is a normal variate with unit variance and mean μ, then

$$\psi_{Z^2}(t) = \frac{1}{\sqrt{2\pi}} \int_{-\infty}^{\infty} \exp\left[tz^2 - \frac{(z-\mu)^2}{2}\right] dz = (1-2t)^{-1/2} \exp\left(\frac{\mu^2 t}{1-2t}\right).$$

The moment generating function of χ'^2 can then be obtained as a product of such functions:

$$\psi_{\chi'^2}(t) = \prod \psi_{Z_i^2}(t) = (1-2t)^{-k/2} \exp\left(\frac{2\lambda t}{1-2t}\right).$$

It is evident from this that the distribution depends on the μ_i's through the value of λ. It is also clear that with $\lambda = 0$ (or $\mu_i = 0$ for all i) the distribution becomes the *central* chi-square, or ordinary chi-square distribution defined earlier.

The corresponding noncentral chi-square density function does not have a closed form, but various tables of the distribution function are available.[2]

"Noncentral" F and t distributions are required for power functions of certain tests concerning normal populations. A ratio of chi-square variates, each divided by the corresponding number of degrees of freedom, has a noncentral F distribution if the numerator has a noncentral chi-square distribution that is independent of the central chi-square distribution in the denominator. The noncentral t-distribution is the distribution of the ratio of a normal random variable with mean μ and variance 1 to the square root of a central chi-square variable divided by its number of degrees of freedom.[3]

Problems

7–12. Compute the variance of the F-distribution. (See Problem 7–4.)

7–13. Determine the following:

(a) The 1st percentile of an F-distribution with 15 and 8 degrees of freedom in the numerator and denominator, respectively.

[2] For example, E. Fix, "Tables of Noncentral χ^2," *Univ. Calif. Publ. Statistics* **1**, 15–19 (1949).

[3] Tables of these distributions are available: G. Lieberman and G. Resnikoff, *Tables of the Non-central t-Distribution* (Stanford Calif.: Stanford University Press, 1957; see also M. Fox, "Charts of the Power of the F-Test," *Ann. Math. Stat.* **28**, 484–97 (1956).

(b) A relationship between certain columns of the F-tables and certain columns of the t-table.

7–14. Show that the density of $F(1, n)$ converges to the density of $\chi^2(1)$ as n becomes infinite and account for this phenomenon in terms of t.

7–15. Show that if $\bar{X}$ and S^2 are the mean and variance, respectively, of a random sample from a normal population, then the statistic

$$T \equiv \sqrt{n-1}\,\frac{\bar{X}-\mu}{S}$$

has a t-distribution with $n - 1$ degrees of freedom.

7–16. Use the idea of Problem 7–11 and the result of Problem 7–7 to calculate the expected value of an F-distribution, and so check the formula in the text.

7–17. Calculate the mean and variance of a noncentral chi-square variable with k degrees of freedom and noncentrality parameter λ, exploiting its structure as a sum of squares.

7.2 One-Sample Problems

The joint density functions of the observations in a random sample $\mathbf{X}$ from a normal population is

$$f(\mathbf{x}; \mu, \sigma^2) = (2\pi\sigma^2)^{-n/2} \exp\left(-\sum \frac{(x_i - \mu)^2}{2\sigma^2}\right).$$

A minimal sufficient statistic for this family (which is included in the two-parameter exponential family) is given by the sum and the sum of squares of the observations: $(\sum X_i, \sum X_i^2)$, or equivalently, by the sample mean and sample variance (see Example 4–22). The sample mean and sample variance are joint maximum likelihood estimates of the population mean and population variance, and they are jointly asymptotically efficient (an extension to the two-parameter case of the notion of asymptotic efficiency of Chapter 5).

The statistics $\bar{X}$ and S^2 were seen in Section 7.1.4 to be *independent* random variables, $\bar{X}$ being $\mathcal{N}(\mu, \sigma^2/n)$ and nS^2/σ^2 having a chi-square distribution with $n - 1$ degrees of freedom.

7.2.1 Dispersion

The theory for inference about σ^2 is simplest when the mean μ is known. In this case the m.l.e. of σ^2 is

$$\hat{\sigma}^2 = \frac{1}{n}\sum (X_i - \mu)^2,$$

which is unbiased, consistent, and efficient. The normal family with known μ has a monotone likelihood ratio in terms of $\hat{\sigma}^2$, so the test $\hat{\sigma}^2 >$ const is UMP for $\sigma^2 \leq \sigma_0^2$ against $\sigma^2 > \sigma_0^2$. The distribution of $n\hat{\sigma}^2/\sigma^2$ is chi-square with n degrees of freedom.

The usual situation, however, is that the population mean is not known, and a second moment that is to provide information about variability must be taken about some sample quantity. As pointed out, $(\bar{X}, S^2)$ is minimal sufficient, and so the pertinent second moment is that about the sample mean.

The variance S^2 has a distribution that depends only on σ^2 and the sample size. More precisely, nS^2/σ^2 has a chi-square distribution with $n - 1$ degrees of freedom. Consequently, the expectation and variance of S^2 are

$$E(S^2) = \frac{n-1}{n}\sigma^2$$

and

$$\text{var } S^2 = \frac{2(n-1)}{n^2}\sigma^4.$$

As an estimate of σ^2, the sample variance S^2 is consistent. Since it is biased, its efficiency for a finite sample is not defined. It was seen earlier (Section 5.1.1) that $nS^2/(n+1)$ has a smaller m.s.e. than S^2; but, of course, their asymptotic properties are the same.

It may be more to the point to estimate the standard deviation σ, a parameter of dispersion equivalent to σ^2 but measured on a more intuitively meaningful scale, that of the observations themselves. The estimate of σ obtained either by the method of maximum likelihood or the method of moments is S, the sample standard deviation, which is therefore consistent and asymptotically normal. The distribution of S/σ (being the square root of S^2/σ^2) is independent of σ, involving only the sample size. Thus,

$$E(S) = \alpha_n \sigma \qquad \text{and} \qquad \text{var } S = \beta_n^2 \sigma^2,$$

where α_n and β_n^2 can be computed from the distribution of nS^2/σ^2. Indeed, the rth moment can easily be computed:

$$\begin{aligned} E(S^r) &= E\left[\left(\frac{nS^2}{\sigma^2}\right)^{r/2}\right]\frac{\sigma^r}{n^{r/2}} \\ &= \frac{\sigma^r}{n^{r/2}2^{(n-1)/2}\Gamma[1/2](n-1)]}\int_0^\infty x^{r/2}x^{(n-3)/2}e^{-x/2}\,dx \\ &= \frac{\Gamma[(n+r-1)/2]}{\Gamma[\frac{1}{2}(n-2)]}\left(\frac{2}{n}\right)^{r/2}\sigma^r. \end{aligned}$$

Putting $r = 1$ and $r = 2$, one obtains

$$\alpha_n = \sqrt{\frac{2}{n}} \frac{\Gamma(n/2)}{\Gamma[(n-1)/2]}, \qquad \beta_n^2 = 1 - \frac{1}{n} - \alpha_n^2.$$

Although an unbiased estimate of σ is obtained by dividing S by α_n, which is then the minimum variance unbiased estimate (based on a sufficient statistic for a complete family), the estimator S has a smaller mean squared error, namely:

$$\left[2(1 - \alpha_n) - \frac{1}{n}\right]\sigma^2.$$

Indeed, the estimator

$$T = \frac{n}{n-1} \alpha_n S$$

has the smallest m.s.e. among multiples of S (see Problem 7–20). For all three estimators, an asymptotic formula for the m.s.e. is $\sigma^2/(2n)$, and an estimate of the square root of this might be termed the standard error of estimate:

$$\text{standard error} = \frac{.707}{\sqrt{n}} S.$$

The sample range—another measure of sample dispersion—also can be used in making inferences about σ or about σ^2. It is not based on a sufficient statistic, so the fact that it is generally less effective than the sample variance should come as no surprise. Nevertheless, the range is useful in certain situations (e.g., industrial) where its arithmetic simplicity is an advantage.

If data X are transformed to Y, where $Y_i = (X_i - \mu)/\sigma$, then the *standardized range* $W \equiv R/\sigma$ is the range of the Y's—the range of a random sample from a standard normal population. The distribution of W depends only on n, and, therefore, the distribution of R depends only on n and σ.

A formula for the density function of W can be obtained from the general expression for the density of a range as derived in Section 4.3.3, upon substitution of the appropriate population density and c.d.f. The formula can be handled numerically, and this has been done. Table V of the Appendix gives the mean, variance, and certain percentiles of the distribution of W, for small-sample sizes. (The range is ordinarily not used for large samples because of its low efficiency.)

In terms of the notation

$$E(W) = E\left(\frac{R}{\sigma}\right) = a_n, \qquad \text{var } W = b_n^2,$$

the estimator R/a_n is an unbiased estimate of σ with m.s.e. given by its variance:

$$\text{m.s.e.} = \text{var}\left(\frac{R}{a_n}\right) = \frac{b_n^2}{a_n^2}\sigma^2.$$

Since both R/a_n and S/α_n are unbiased estimates of σ, their relative efficiency is defined as

$$\frac{\text{var}\,(S/\alpha_n)}{\text{var}\,(R/a_n)} = \frac{\beta_n^2/\alpha_n^2}{b_n^2/a_n^2}.$$

The distributions of S/σ and R/σ being independent of σ (as well as of μ), it is an easy matter to construct confidence intervals for σ based on S and based on R. Thus, given the 2.5 and 97.5 percentiles of W from Table V,

$$P\left(W_{.025} < \frac{R}{\sigma} < W_{.975}\right) = .95,$$

and inversion of the inequalities yields the 95 per cent confidence interval for σ:

$$\frac{R}{W_{.975}} < \sigma < \frac{R}{W_{.025}}.$$

The same kind of manipulation, beginning with the chi-square distribution of nS^2/σ^2, results in a 95 per cent confidence interval for σ^2:

$$\frac{nS^2}{\chi^2_{.975}} < \sigma^2 < \frac{nS^2}{\chi^2_{.025}}.$$

(Taking square roots would give a corresponding interval for σ.)

EXAMPLE 7–1. The following are nine readings of an inlet oil temperature, assumed to be normally distributed:

99, 93, 99, 97, 90, 96, 93, 88, 89.

The range is $R = 11$, and the standard deviation is $S = 3.97$. For $n = 9$, Table V in the Appendix gives $a_n = 2.97$, and a simple computation yields $\alpha_n = .9139$. Using the range leads to the point estimate $11/2.97 = 3.71$ and the 95 per cent confidence interval $2.34 < \sigma < 7.10$ (where, for instance, $2.34 = R/W_{.975}$).

The estimates S, S/α_n, and $n\alpha_n S/(n-1)$ for σ are, respectively, 3.97, 4.34, and 4.08, with corresponding mean squared errors of $.0611\sigma^2$, $.0643\sigma^2$, and $.0604\sigma^2$. (The r.m.s. error of about 25 per cent of the quantity being estimated is that large because of the rather small-sample size.) The 95 per cent confidence interval based on S is $2.85 < \sigma < 8.07$ (where $2.85 = [nS^2/\chi^2_{.975}]^{1/2}$).

Testing between two specific values of σ^2, say, v_0 and $v_1 > v_0$, is *not* a problem of one simple hypothesis against another simple hypothesis, since specification of the variance does not completely determine a normal population. A likelihood ratio test can be constructed:

$$\Lambda^* = \frac{L(\bar{X}, v_0)}{L(\bar{X}, v_1)} = \left(\frac{v_1}{v_0}\right)^{n/2} \exp\left[-\frac{1}{2}nS^2\left(\frac{1}{v_0} - \frac{1}{v_1}\right)\right] < \text{const.}$$

This is equivalent to $S^2 >$ constant. The power function of this test is obtained by using the fact that nS^2/σ^2 has a chi-square distribution with $n - 1$ degrees of freedom:

$$\pi(\sigma^2) = P(S^2 > c) = P\left(\frac{nS^2}{\sigma^2} > \frac{nc}{\sigma^2}\right)$$

$$= 1 - F\left(\frac{nc}{\sigma^2}\right),$$

where F denotes the chi-square distribution function with $n-1$ degrees of freedom. This is clearly an increasing function of σ^2, and so the test is at least unbiased for a problem in which the σ^2 in H_0 are to the left of those in H_1.

A corresponding test based on sample range is given by the critical region $R > c$. The power function for this test is

$$\pi(\sigma) = P_\sigma(R > c) = 1 - F_W\left(\frac{c}{\sigma}\right),$$

where $W = R/\sigma$, and F_W is given in Table V, Appendix.

EXAMPLE 7–2. In Figure 7–3 are plotted power functions for the test $S^2 > 4.32$ and for the test $R > 7.66$, based on a sample size 15. These were constructed to have the same power of .30 at $\sigma^2 = 4$. (If H_0 were $\sigma^2 \leq 4$, these tests would have the same $\alpha = .30$. Notice that the test based on S is slightly more powerful than that based on R. Indeed, the former strictly dominates the latter, but the degree of dominance is not overwhelming.)

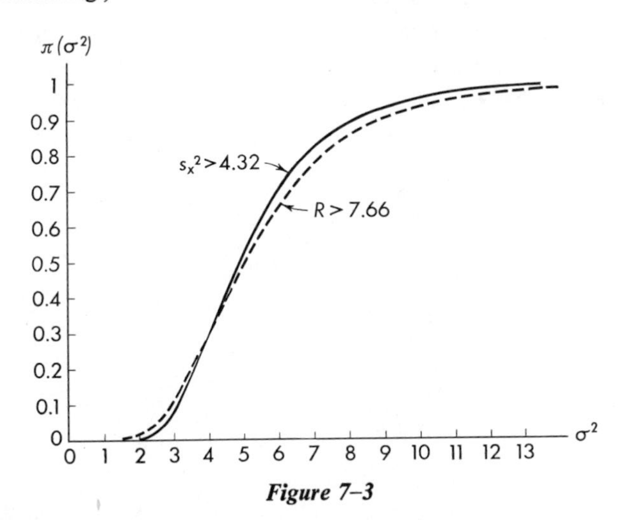

Figure 7–3

7.2.2 The Mean

The mean $\bar{X}$ of a random sample is a good estimator for the mean of a normal population from many points of view. It is consistent, and in terms of quadratic loss it is the best linear unbiased function of the observations (BLUE). It is efficient if σ^2 is known, and if σ^2 is unknown it is asymptotically efficient, jointly with S^2 for σ^2. Its distribution is readily available, being $\mathcal{N}(\mu, \sigma^2/n)$; but the dependence of its variance on an unrelated parameter σ^2 is sometimes awkward.

The sample median is another candidate for estimating μ, which is also the population median. Its distribution (cf. Section 4.3.4) is asymptotically $\mathcal{N}(\mu, \pi\sigma^2/(2n))$, so its asymptotic efficiency is $2/\pi$, or about 64 per cent.

The midrange is another easily computed statistic used to estimate μ:

$$A \equiv \tfrac{1}{2}[X_{(1)} + X_{(n)}].$$

If the data are standardized by the transformation $Y_i = (X_i - \mu)/\sigma$, the midrange of the Y's is the midrange of a random sample from a standard normal population, related to A as follows:

$$B \equiv \frac{1}{2}[Y_{(1)} + Y_{(n)}] = \frac{A - \mu}{\sigma}.$$

Since $EB = 0$, it follows that $EA = \mu$, and A is unbiased. Its efficiency relative to the sample mean could be computed as

$$\frac{\operatorname{var} \bar{X}}{\operatorname{var} A} = \frac{1/n}{\operatorname{var} B},$$

where var B is a function of n alone that is obtainable numerically from the distribution of B given in Section 4.3.3.

A likelihood ratio test can be constructed for testing the hypothesis $\mu = \mu_0$ against the two-sided alternative $\mu \neq \mu_0$. The numerator of the likelihood ratio Λ is the maximum of $L(\mu, \sigma^2)$ over σ^2 holding $\mu = \mu_0$, achieved when σ^2 is $\sum (X_i - \mu_0)^2/n$. The denominator is $L(\bar{X}, S^2)$, the maximum when both μ and σ^2 are allowed to vary; and, therefore, the likelihood ratio is

$$\Lambda = \frac{L(\mu_0, \sum (X_i - \mu_0)^2/n)}{L(\bar{X}, S^2)} = \left[1 + \left(\frac{\bar{X} - \mu_0}{S}\right)^2\right]^{-n/2}.$$

The critical region $\Lambda <$ const is clearly equivalent to

$$T^2 = \frac{(n-1)(\bar{X} - \mu_0)^2}{S^2} > \text{const},$$

where

$$T = \sqrt{n-1}\,\frac{\bar{X} - \mu_0}{S} = \frac{\sqrt{n}(\bar{X} - \mu_0)/\sigma}{\sqrt{nS^2/(n-1)\sigma^2}}$$

has the t-distribution with $n - 1$ degrees of freedom (cf. Problem 7–15), when $\mu = \mu_0$ (i.e., under H_0). The likelihood ratio test for $\mu = \mu_0$ against $\mu \neq \mu_0$, then, is of the form $|T| > K$, where K is determined from a given α as the $100(1 - \alpha/2)$ percentile of $t(n - 1)$. [Alternatively, K^2 is the $100(1 - \alpha)$ percentile of $F(1, n - 1)$.]

A natural test for the one-sided $\mu \leq \mu_0$ against $\mu > \mu_0$ is provided by the critical region $T > K$. The power function of this test, and that of the test $|T| > K$ for the two-sided problem, can be given in terms of the noncentral t-distribution; for, if $\mu \neq \mu_0$, T has a noncentral t-distribution with noncentrality parameter $n(\mu - \mu_0)^2/(2\sigma^2)$.

A two-sided confidence interval for μ can be constructed using T^2, with μ_0 replaced by a general μ. For instance, a 95 per cent confidence interval is obtained from the F-distribution with $(1, n - 1)$ degrees of freedom:

$$P(T^2 < F_{.95}) = .95,$$

or

$$P\left\{-\sqrt{F_{.95}} < \sqrt{n-1}\,\frac{\bar{X} - \mu}{S} < +\sqrt{F_{.95}}\right\} = .95.$$

Since $\sqrt{F_{.95}} = t_{.975}$ and $-\sqrt{F_{.95}} = t_{.025}$ (percentiles of the t distribution with $n - 1$ degrees of freedom), the confidence interval can be written in the form

$$\bar{X} - t_{.975}\frac{S}{\sqrt{n-1}} < \mu < \bar{X} + t_{.975}\frac{S}{\sqrt{n-1}}.$$

7.2.3 Confidence Regions for Mean and Variance

A region of (μ, σ^2) values can be constructed from the $(\bar{X}, S^2)$ of a given sample, which with a specified probability will cover the actual (μ, σ^2), as follows: Let z_p and χ_p^2 denote percentiles of the standard normal distribution and of the chi-square distribution with $n - 1$ degrees of freedom, respectively. Given a confidence coefficient η, choose δ and ϵ such that $\eta = (1 - 2\delta)(1 - 2\epsilon)$. Then

$$\begin{aligned} P&\left\{\left(\frac{\bar{X} - \mu}{\sigma/\sqrt{n}}\right)^2 < z^2_{1-\sigma} \text{ and } \chi^2_\epsilon < \frac{nS^2}{\sigma^2} < \chi^2_{1-\epsilon}\right\} \\ &= P\left\{\left(\frac{\bar{X} - \mu}{\sigma/\sqrt{n}}\right)^2 < z^2_{1-\delta}\right\} P\left\{\chi^2_\epsilon < \frac{nS^2}{\sigma^2} < \chi^2_{1-\epsilon}\right\} \\ &= (1 - 2\delta)(1 - 2\epsilon) = \eta. \end{aligned}$$

The inequalities whose probability is given here define a region of the (μ, σ^2) plane, depending on $(\bar{X}, S^2)$, which is the desired confidence region.

This confidence region is not just the rectangle defined by the confidence intervals constructed for μ alone and for σ alone. Strictly speaking, these confidence intervals could not be combined by intersection to obtain a confidence rectangle whose level is the product of the individual levels, since the statistics T and S^2 on which the intervals are based are not independent. The figure in the following example shows, however, that the rectangular region would be a reasonable approximation for the actual confidence region.

EXAMPLE 7–3. Suppose that it is desired to construct a 90 per cent confidence region on the basis of a sample of size 20, in which $\bar{X} = 11.2$ and $S^2 = 6.8$. Since $.90 \doteq .95 \times .95$, let $2\delta = 2\epsilon = .05$. The inequalities defining the confidence region are then

$$(11.2 - \mu)^2 < (1.96)^2 \frac{\sigma^2}{20} \qquad \text{and} \qquad 8.91 < \frac{136}{\sigma^2} < 32.9.$$

The boundaries of this region are the curves defined by making the inequalities into equalities, and are the parabola and horizontal lines shown in Figure 7–4. Also shown in that figure are the 95 per cent confidence intervals for μ and σ^2 alone.

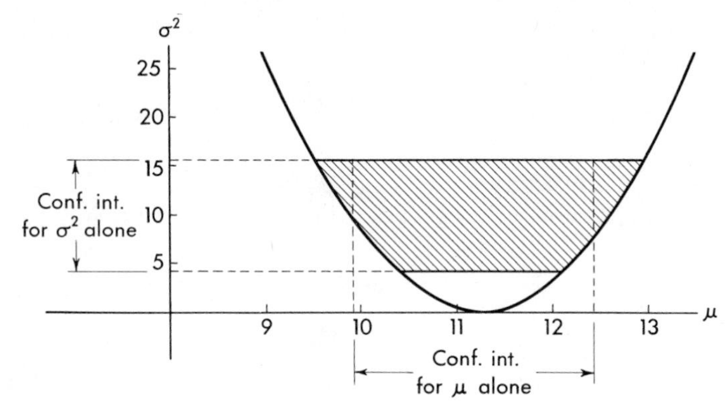

Figure 7–4

Problems

7–18. Refer to Example 7–1.

(a) Determine rejection limits K_1 and K_2 so that the tests $S > K_1$ and $R > K_2$ have the same power of 0.2 at $\sigma = 3$.

(b) Compare their powers at $\sigma = 5$.

7–19. Outside diameters of 70 supposedly identical parts are given in the following frequency table:

Diameter (inches)	Frequency
1.0950	2
1.0945	2
1.0940	9
1.0935	15
1.0930	26
1.0925	10
1.0920	6

(a) Determine a 95 per cent confidence interval for σ^2 using S^2.

(b) Determine a 95 per cent confidence interval for μ.

(c) Determine a 90 per cent confidence region for (μ, σ^2).

7–20. Determine k such that the estimate kS of σ in a normal population has minimum mean squared error, and compute that smallest m.s.e.

7–21. Compute $\alpha_n = E(S/\sigma)$ for $n = 10$, and compute the m.s.e. for each of the four estimators S/α_n, S, $n\alpha_n S/(n-1)$, and R/a_n, where a_n is given in Table V in the Appendix. [Compare the m.s.e.'s for the first three with the asymptotic value $\sigma^2/(2n)$.]

7–22. Given these observations from a normal population:

$$\begin{array}{cccccc} 4.28, & 4.32, & 4.32, & 4.29, & 4.31, & 4.35, \\ 4.32, & 4.33, & 4.28, & 4.27, & 4.38, & 4.28, \end{array}$$

give estimates of μ and σ based on the midrange and range, respectively, and determine a standard error for the estimate of σ.

7–23. Assuming σ^2 to be known, determine (in terms of K, σ^2, and n) the large sample power function $\pi(\mu)$ of the test having as a critical region $m > K$, where m is the sample median.

7–24. Construct the likelihood ratio Λ^* for testing $\mu = \mu_0$ against $\mu = \mu_1$. Show that the test $\Lambda^* < 1$ is equivalent to $\bar{X} > (\mu_0 + \mu_1)/2$, if $\mu_1 > \mu_0$. Does $(\Lambda^*)^{2/n}$ have a distribution that is familiar? Obtain an expression for the power function of the test $\Lambda^* < 1$.

7.3 Comparisons

A common statistical problem is that of comparing two populations with respect to some characteristic. These populations may arise or exist naturally, such as the populations of males and of females, of Democrats and Republicans, of left-handed and right-handed persons, and of front wheel and rear wheel tires. In other instances, one population is "treated" in some way and the comparison with an untreated population is for the purpose of judging the effectiveness of the treatment: for example, a population is compared with itself before and after some educational process; or a population of sober drivers is compared with that of intoxicated drivers; or patients given some medical treatment are compared with untreated patients.

In treating a problem statistically, comparisons are made by comparing samples from the two populations. These may be independent random samples, or they may be related by "pairing." Thus, the "before" and "after" scores for a student are naturally paired, as are the wear on a front tire and the wear on a rear tire on a given vehicle. Pairing may also be introduced as a device to reduce variability.

Comparison of some numerical characteristic of the two populations may be made with respect to location or with respect to dispersion, to mention the more common cases. This is often done by computing statistics of location or dispersion for a sample from each population and combining them in some way such as taking their difference or ratio. A comparison of the populations can be made intelligently only if the distribution of the comparison statistic is known. This is generally simple for large samples (e.g., the difference of two sample means). For small samples the distribution and corresponding procedure are determined by the nature of the population, and in this section procedures will be given for the case of normal populations, first for independent samples and then for observations related by pairing.

Comparisons of categorical populations will be taken up in Chapter 9, and some small-sample tests to use with numerical populations of unknown form will be given in Chapter 11. Comparisons of more than two populations will be considered (in the normal case) in Chapter 12.

7.3.1 Comparing Variances

Populations 1 and 2 will be assumed normal with means μ_1 and μ_2 and variances σ_1^2 and σ_2^2, respectively. Independent random samples are obtained, one from population 1 with mean and variance $(\bar{X}_1, S_1^2)$, and one from population 2 with mean and variance $(\bar{X}_2, S_2^2)$. These sample statistics are minimal sufficient, as is evident from the likelihood function, whose logarithm is as follows:

$$\log L(\mu_1, \mu_2; \sigma_1^2, \sigma_2^2) = -\frac{1}{2}(n_1 + n_2)\log(2\pi) - \frac{1}{2}n_1 \log \sigma_1^2 - \frac{1}{2}n_2 \log \sigma_2^2 - \frac{n_1}{2\sigma_1^2}[S_1^2 + (\mu_1 - \bar{X}_1)^2] - \frac{n_2}{2\sigma_2^2}[S_2^2 + (\mu_2 - \bar{X}_2)^2].$$

To construct a likelihood ratio test of $\sigma_1^2 = \sigma_2^2 = v$ against $\sigma_1^2 \neq \sigma_2^2$, the maximum likelihood estimates are needed:

$$\text{In } H_0\text{:} \quad \hat{\mu}_1 = \bar{X}_1, \quad \hat{\mu}_2 = \bar{X}_2, \quad \hat{v} = \frac{n_1 S_1^2 + n_2 S_2^2}{n_1 + n_2}.$$

$$\text{In } H_0 + H_1\text{:} \quad \hat{\mu}_1 = \bar{X}_1, \quad \hat{\mu}_2 = \bar{X}_2, \quad \hat{\sigma}_1^2 = S_1^2, \quad \hat{\sigma}_2^2 = S_2^2.$$

Then

$$\Lambda^2 = \left(\frac{L(\bar{X}_1, \bar{X}_2; \hat{v}, \hat{v})}{L(\bar{X}_1, \bar{X}_2; S_1^2, S_2^2)}\right)^2 = (\text{const})\frac{Z^{n_1}}{(1 + Z)^{n_1 + n_2}},$$

where Z is the ratio of sums of squared deviations about sample means:

$$Z = \frac{n_1 S_1^2}{n_2 S_2^2}.$$

The critical region $\Lambda < K$ defines a critical region for Z that calls for rejecting equality of variances if Z is either too large or too small. This is seen in Figure 7–5, in which the distribution of Z is sketched (typically) beneath the sketch of the functional relationship between Λ and Z. The two shaded regions, whose total area is the probability of rejecting H_0 with the test $\Lambda < \Lambda_0$, are not necessarily of equal area, and it is convenient to choose modified limits for Z such that they would be of equal area. Such a modification yields a test that is not the same as the likelihood ratio test—but which is not radically different. In practice, it is the ratio of the unbiased variance estimates,

$$F = \frac{n_2 - 1}{n_1 - 1} Z = \frac{\tilde{S}_1^2}{\tilde{S}_2^2},$$

which is used rather than Z. This ratio has a distribution that is $F(n_1 - 1, n_2 - 1)$ under H_0; for, upon division of numerator and denominator each by the common variance v, it is the ratio of independent chi-squared variates, each divided by its number of degrees of freedom. (See Section 7.1.5.)

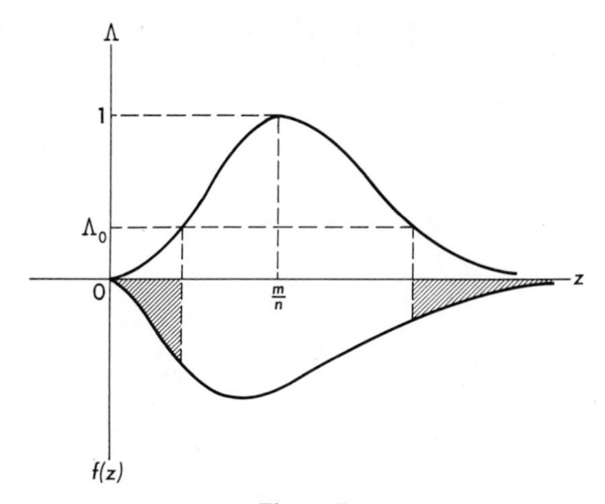

Figure 7–5

EXAMPLE 7–4. The 12 observations in Problem 7–22 are determinations by an alcohol method of the per cent of nickel in solution, with mean 4.311 and variance 9.91×10^{-4}. The following 10 observations were obtained by an aqueous method:

4.27, 4.32, 4.29, 4.30, 4.31, 4.30, 4.30, 4.32, 4.28, 4.32,

with mean 4.301 and variance 2.69×10^{-4}. The pertinent F-ratio for comparing variances is

$$F = \frac{2.69 \times 10/9}{9.91 \times 12/11} = .28.$$

Table IV (Appendix) of F-distribution percentiles is not constructed so that .28 can be interpreted as a percentile. However, the 5th percentile of $F(9, 13)$ is .33, so the observed F-value suggests, but not strongly, that perhaps the population variances are not equal. Further testing may be desirable.

The power function of a test employing the test ratio F, whether the test be one-sided or two-sided, can be obtained from the F-distribution, since if $\sigma_1^2 \neq \sigma_2^2$,

$$P(F < K) = P\left(\frac{\tilde{S}_1^2/\sigma_1^2}{\tilde{S}_2^2/\sigma_2^2} < \frac{\sigma_2^2}{\sigma_1^2} K\right).$$

The probability on the right is the value of the c.d.f. of $F(n_1 - 1, n_2 - 1)$ at the point $\sigma_2^2 K/\sigma_1^2$.

7.3.2 Comparing Means

The comparison of the means of two normal populations is complicated by the dependence of the distribution of the sample means on the population variances. The simplest case, to be considered first, is that in which the population variances can be assumed *equal*: $\sigma_1^2 = \sigma_2^2 = v$. (The notation of the preceding section will be continued here.)

A likelihood ratio test can be constructed for the null hypothesis of equal means: $\mu_1 = \mu_2 = \mu$, against the two-sided alternative $\mu_1 \neq \mu_2$, with $\sigma_1^2 = \sigma_2^2 = v$, based on independent random samples from the two populations. The necessary maximum likelihood estimates are as follows:

In H_0: $\hat{\mu} = \dfrac{n_1\bar{X}_1 + n_2\bar{X}_2}{n_1 + n_2}$,

$$\hat{v}_0 = \frac{1}{n_1 + n_2}\{n_1 S_1^2 + n_1(\bar{X}_1 - \hat{\mu})^2 + n_2 S_2^2 + n_2(\bar{X}_2 - \hat{\mu})^2\}.$$

In $H_0 + H_1$: $\hat{\mu}_1 = \bar{X}_1$, $\hat{\mu}_2 = \bar{X}_2$, and $\hat{v} = \dfrac{n_1 S_1^2 + n_2 S_2^2}{n_1 + n_2}$.

Then

$$\max_{H_0+H_1} L(\mu_1, \mu_2 \colon v) = L(\bar{X}_1, \bar{X}_2; \hat{v}) = (2\pi\hat{v})^{-(n_1+n_2)/2} e^{-(n_1+n_2)/2}$$

and

$$\max_{H_0} L(\mu, \mu; v) = L(\hat{\mu}, \hat{\mu}; \hat{v}_0) = (2\pi\hat{v}_0)^{-(n_1+n_2)/2} e^{-(n_1+n_2)/2}$$

so that

$$\Lambda^{-2/(n_1+n_2)} = \frac{\hat{v}_0}{\hat{v}} = 1 + \frac{n_1(\bar{X}_1 - \hat{\mu})^2 + n_2(\bar{X}_2 - \hat{\mu})^2}{n_1 S_1^2 + n_2 S_2^2}.$$

A little elementary algebra yields

$$n_1(\bar{X}_1 - \hat{\mu})^2 + n_2(\bar{X}_2 - \hat{\mu})^2 = \frac{n_1 n_2}{n_1 + n_2}(\bar{X}_1 - \bar{X}_2)^2,$$

and upon substitution one obtains

$$\Lambda^{-2/(n_1+n_2)} = 1 + \frac{T^2}{n_1 + n_2 - 2},$$

where

$$T = \frac{\bar{X}_1 - \bar{X}_2}{\sqrt{\dfrac{n_1 S_1^2 + n_2 S_2^2}{n_1 + n_2 - 2}\left(\dfrac{1}{n_1} + \dfrac{1}{n_2}\right)}} = \frac{Z}{\sqrt{U}}$$

and

$$Z = \frac{\bar{X}_1 - \bar{X}_2}{\sigma\sqrt{1/n_1 + 1/n_2}}, \qquad U = \frac{n_1 S_1^2 + n_2 S_2^2}{\sigma^2(n_1 + n_2 - 2)}.$$

Now, under H_0, Z is standard normal and U is a chi-square variate divided by its number of degrees of freedom; because they are independent, the null distribution of T is a t-distribution with $n_1 + n_2 - 2$ degrees of freedom (and T^2 is $F(1, n_1 + n_2 - 2)$). The critical region $\Lambda <$ const is equivalent to one of the form $|T| > K$, and the critical value K is then determined from a given significance level by means of the t-table.

The statistic T, resulting from the above manipulation of the likelihood ratio, is intuitively appealing as the thing to compute in judging whether $\mu_1 = \mu_2$ or not. Its numerator is just the difference between sample means, and the denominator is an estimate of the standard deviation of that difference—the usual kind of measuring stick to use in deciding whether an observed difference is reasonably explained by sampling fluctuations under the null hypothesis or is so large as to suggest a real difference in population means.

Although T^2 was derived for use with a two-sided alternative, it is clear that a one-sided critical region for T would be appropriate when the alternative is one-sided—say, $\mu_1 > \mu_2$. In either case, the power function would come from distribution of T, which is noncentral t when the means are not equal.

A confidence interval for the difference $\mu_1 - \mu_2$ can be constructed by using the fact that the random variable T^2, defined as before except that $(\bar{X}_1 - \bar{X}_2)^2$ is replaced by $[(\bar{X}_1 - \bar{X}_2) - (\mu_1 - \mu_2)]^2$, has an F-distribution. Thus,

$$P(T^2 < t^2_{.975}(n_1 + n_2 - 2)) = .95,$$

and the 95 per cent confidence limits for $\mu_1 - \mu_2$ are

$$\bar{X}_1 - \bar{X}_2 \pm t_{.975}\sqrt{\frac{n_1 S_1^2 + n_2 S_2^2}{n_1 + n_2 - 2}\left(\frac{1}{n_1} + \frac{1}{n_2}\right)}.$$

EXAMPLE 7–5. In a test to compare men drivers with women drivers with respect to economical driving, five drivers from each group drove a car over a measured course through city traffic. Gasoline consumption figures were as follows:

Men: 0.94, 1.20, 1.00, 1.06, 1.02.
Women: 1.40, 0.98, 1.22, 1.16, 1.34.

The means are 1.044 and 1.220, with a difference of 0.176. The variances are .007584 and .0216, so the estimate of the common variance (assuming the variances of the populations to be equal) is

$$S^2 = \frac{5 \times .007584 + 5 \times .0216}{5 + 5 - 2} = (.135)^2.$$

The estimate of the standard deviation of the difference in sample means is then

$$.135\sqrt{\tfrac{1}{5} + \tfrac{1}{5}} = .0853.$$

Using a multiplier of $1.86 = t_{.95}(8)$, one obtains the following confidence limits for the difference in population means:

$$.176 \pm 1.86 \times .0853.$$

Since the interval (.017, .335) does not contain 0, one would reject the hypothesis of no difference at $\alpha = .10$ in a two-sided test. [One might argue that a one-sided test is in order, in which case he would calculate T and compare with the 90th percentile of $t(8)$.]

When the population variances cannot be assumed equal, the problem of comparing means is called the *Behrens–Fisher* problem. After an extensive history of proposed solutions and controversy, the problem does not have a universally accepted solution. But various solutions are good enough for practical applications.

One test for $\mu_1 = \mu_2$ uses a critical region of the form

$$\frac{|\bar{X}_1 - \bar{X}_2|}{\sqrt{S_1^2/n_1 + S_2^2/n_2}} > k,$$

in which the test statistic, quite reasonably, is a measure of difference scaled according to an estimate of the standard deviation of the difference in sample means. The difficulty is that the probability of this critical region depends on the unknown values of the population variances, or more precisely, on the ratio of the variances. But it can be shown that if k is chosen as the $100(1 - \alpha/2)$ percentile of $t(n_0 - 1)$, where $n_0 = \min(n_1, n_2)$, the probability of the critical region under H_0 is at most α; indeed, α is the least upper bound of this probability and so is what was called earlier the "size" of a test with a composite null hypothesis.

[A survey of various solutions of the Behrens–Fisher problem and a study of their power characteristics is given in H. Scheffe, "Practical Solutions of the Behrens–Fisher Problem," *J. Am. Stat. Assn.* **65**, 1501 ff. (1970).]

EXAMPLE 7–6. For the determination of per cent nickel by two methods given in Example 7–4, the data were summarized as follows:

Alcohol: $n_1 = 12$, $\bar{X}_1 = 4.311$, $S_1^2 = .000991$.
Aqueous: $n_2 = 10$, $\bar{X}_2 = 4.301$, $S_2^2 = .000269$.

From these one computes

$$\frac{S_1^2}{n_1} + \frac{S_2^2}{n_2} = .0001905,$$

and the test statistic is

$$\frac{.010}{\sqrt{.000352}} = .956.$$

The 97.5 percentile of $t(9)$ is 2.26, and rejecting equality for a value of the test statistic exceeding 2.26 in magnitude has an α of at most .05. The given data do not call for rejection at this level.

7.3.3 Paired Comparison of Means

When, as discussed earlier, observations from one population are naturally paired with observations from another population, yielding pairs $(X_1, Y_1), \ldots, (X_n, Y_n)$, there is ordinarily a relationship between the observations in a pair. Thus, in comparing a dimension of left hands versus right hands, the pair of observations on the left and right hands of a given person are related, simply because the hands belong to the same person. If a person is large, both hands tend to be large, although not usually the same. So the X's and Y's could not be thought of as comprising two independent samples. And without knowing the nature of the relationship beween X_i and Y_i, one may as well consider that he has simply a sample of n differences, $d_i \equiv X_i - Y_i$, whose mean happens to be the difference of means $\bar{X} - \bar{Y}$. The problem is then a one-sample problem, and a t-test based on $\bar{d}$ and on an estimate of $\sigma_{\bar{d}} = \sigma_d/\sqrt{n}$ would be used—if the differences can be assumed to be normally distributed and independent.

EXAMPLE 7–7. Experimentation suggests that it may be possible for a person to control certain body phenomena previously thought to be beyond his control if he is trained to do so in a program of "biofeedback" exercises. The following data and blood pressure measurements (in millimeters of mercury) before and after the training or conditioning, for each of six subjects:

Subject	Before	After	Difference
1	136.9	130.2	−6.7
2	201.4	180.7	−20.7
3	166.8	149.6	−17.2
4	150.0	153.2	3.2
5	173.2	162.6	−10.6
6	169.3	160.1	−9.2

The mean and standard deviation are -10.2 and 7.662 so that

$$T = \sqrt{n-1}\,\frac{\bar{d}}{S_d} = -2.98.$$

This suggests ($t_{.95}(5) = 2.01$), a real reduction in blood pressure achieved by self-control as learned in the training program.

It should be observed that the variation in blood pressure reading from person to person, either before or after, is considerable, and the difference in means of two independent samples would have much greater variability than that observed in the above d's.

Paired data are sometimes obtained by design. For example, to study the effect of a new diet on pig growth one could select a random sample of young pigs for the "treatment" group and another random sample for a "control" group. The treatment group would get the new diet and the control group a standard diet. However, the variability in the difference of mean weight gains can be reduced significantly by choosing, instead, a pair of pigs from each of n litters, randomly assigning one of the pair to the new diet and one to the standard diet. Pigs in the same litter tend to be more nearly alike in their response to diet than are pigs generally, and this means that the differences between the weight gains in a pair is less variable than the difference between weight gains of two randomly chosen pigs. Even when there is nothing like a "litter" to use in pairing, it can often be effectively done by selecting pairs of subjects so that the two in a pair are nearly alike in many of the ways that may contribute to variability of response. In whatever way the pairing comes about, the n paired differences are used in a (one-sample) t-test if normality can be assumed.

The point of reducing variability is to increase the sensitivity of the test—the power to detect the presence of a treatment effect of given magnitude. The power of a two-sided t-test is $P(T^2 > K)$, the complement of the c.d.f. of a noncentral F. The noncentrality parameters here are as follows:

Two-sample test: $\dfrac{(\mu_1 - \mu_2)^2/2}{2\sigma^2/n}$.

Paired comparison: $\dfrac{(\mu_1 - \mu_2)^2/2}{\sigma_d^2/n}$.

The two-sample tests employs independent samples of size n from populations with a common variance σ^2, and the paired comparison uses the differences d in n pairs. For a given difference $\mu_1 - \mu_2$ the noncentrality parameters will be equal, and the power of the tests would be the same, provided $2\sigma^2/n = \sigma_d^2/n$. This would be the case when the X and Y in a pair are independent, and nothing would be gained in that case by pairing. But if pairs can be arranged so that $\sigma_d^2 < 2\sigma^2$, then the paired t-test is more sensitive. For instance, if $\sigma_d^2 = \sigma^2/2$, the power of the paired t-test is the same at $\mu_1 - \mu_2 = \delta$ as the power of the two-sample t-test at $\mu_1 - \mu_2 = 2\delta$.

Small-sample tests for paired comparisons that do not require the assumption of normality will be taken up in Chapter 11.

Problems

7–25. Given the data in Example 7–5, apply the F-test for equality of variances. Use a two-sided test at $\alpha = .10$.

7–26. Using the F-distribution of $(\sigma_2^2 \tilde{S}_1^2)/(\sigma_1^2 \tilde{S}_2^2)$, assuming independent random samples from normal populations, obtain a 90 per cent confidence interval for the ratio of population variances,

(a) As a general formula.
(b) In the special case of the data of Example 7–6.

7–27. Obtain the power as a function of σ_1^2/σ_2^2 for a two-sided test of equality of variances. (Use equal tail areas.)

7–28. When n_1 and n_2 are large, the statistic

$$Z = \frac{\bar{X}_1 - \bar{X}_2}{\sqrt{S_1^2/n_1 + S_2^2/n_2}}$$

is approximately standard normal. (Are assumptions of normality and equal population variances required for this?) In a three-year study of the effectiveness of sodium monofluorophospate (MFP) as a decay-fighting agent in toothpaste,[4] results reported were as shown in the following table.

Dentrifrice	Number of Children	Average Number of Cavities per Child	Standard Deviation
MFP	208	19.98	10.61
Control	201	22.39	11.96

Test the hypothesis of no difference in means at the 5 per cent level.

7–29. The following are determinations of rate of flow of a certain gas through two different soil types:

Soil A: 21, 28, 19, 24, 22, 30.
Soil B: 21, 29, 34, 32, 26, 28, 36, 26.

(a) Assuming equal population variances (and normal populations), test the hypothesis of no difference in means.
(b) If equal variances cannot be assumed, would this alter the conclusion?

[4] S. N. Frankl and J. E. Alman, *J. Oral Therapeut. Pharmacol.* **4** (1968), 443–49.

7–30. Pairs of pigs are selected from each of 10 litters, one of each pair being fed a new diet and the other a standard diet. Weight gains over a given period are as shown in the following table.

Pair	1	2	3	4	5	6	7	8	9	10
New	36.0	32.7	39.2	37.6	32.0	40.2	34.4	30.7	36.4	37.2
Standard	35.2	30.0	36.5	38.1	29.4	36.0	31.3	31.6	31.1	34.0

Test the hypothesis of no difference at $\alpha = .05$ (one-sided).

7–31. Given $\sigma_1^2 = \sigma_2^2 = \sigma^2$, in the notation of Section 7.3.3, obtain the value of the correlation coefficient ρ implied by the condition $\sigma_d^2 = \sigma^2/2$.

8

STATISTICAL DECISION THEORY

Although statistical problems are quite varied in nature, some of them, notably in business and industry, are clearly *decision* problems. In these, the state of nature is unknown, but a decision must be made—a decision whose consequences depend on the unknown state of nature. Such a problem is a *statistical* decision problem when there are data that give partial information about the unknown state.

It is perhaps not too surprising, following the parallel development in the 1930s of the Neyman–Pearson theory of tests and the von Neumann theory of games, that A. Wald was led in the 1940s to formulate the statistical decision problem in the setting of game theory, and so to provide a unifying structure in which most statistical problems can be profitably studied. Statisticians are not all agreed that the decision theoretic approach is the definitive one for treating statistical problems; indeed, R. A. Fisher did not hide his contempt for it, in regard to the use of statistics as a scientific method. But some feel that it provides the only rational approach.

8.1 Preliminaries

Statistical decision theory views an inference problem as a two-person game, with the statistician as one player and "nature" as the other. The statistician chooses a "strategy" not knowing what "strategy" nature has chosen, and the "payoff" is the consequence of these choices.

Section 8.1.2 takes up the measurement of consequences on a meaningful scale; and Section 8.1.3 discusses the notion of probability as a measure of belief, needed when thinking of the state of nature as random—as a model for the uncertainty one has about nature.

8.1.1 Convex Combinations

To prepare for the representation of actions available to a decision-maker by points in a vector space, a review of certain geometrical notions is in order. Euclidean k-space will be the space most frequently referred to, and in particular that in which $k = 2$ (the Cartesian plane) will lend itself most readily to graphical methods and to visualization.

It will be recalled that in a vector space there is defined an addition, $P_1 + P_2$, and multiplication by a scalar, aP, these operations satisfying the commutative, associative, and distributive properties that suggest adoption of the symbols for arithmetic operations:

$$P_1 + P_2 = P_2 + P_1,$$
$$P_1 + (P_2 + P_3) = (P_1 + P_2) + P_3,$$
$$a(P_1 + P_2) = aP_1 + aP_2.$$

In the case of the plane, where P is a point (x, y), also written as a vector $\begin{pmatrix} x \\ y \end{pmatrix}$, the operations are performed coordinate-wise:

$$\text{Addition:} \quad \begin{pmatrix} x_1 \\ y_1 \end{pmatrix} + \begin{pmatrix} x_2 \\ y_2 \end{pmatrix} = \begin{pmatrix} x_1 + x_2 \\ y_1 + y_2 \end{pmatrix}.$$

$$\text{Multiplication by a scalar:} \quad a\begin{pmatrix} x \\ y \end{pmatrix} = \begin{pmatrix} ax \\ ay \end{pmatrix}.$$

A *linear combination* of vectors $P_1, \ldots, P_m$ is one of the form

$$a_1 P_1 + \cdots + a_m P_m = \sum a_i P_i.$$

A linear combination of two points in which the coefficients sum to 1 is a point on the *line* through the two points:

$$P = \gamma P_1 + (1 - \gamma) P_2.$$

In the $k = 2$ case, this represents the two equations,

$$\begin{cases} x = \gamma x_1 + (1 - \gamma) x_2, \\ y = \gamma y_1 + (1 - \gamma) y_2, \end{cases}$$

which are the parametric equations of the line through (x_1, y_1) and (x_2, y_2), being linear and satisfied by P_1 for $\gamma = 1$ and by P_2 for $\gamma = 0$.

A linear combination $\sum a_i P_i$ in which the coefficients are probabilities: $\sum a_i = 1$ and $a_i \geq 0$, is called a *convex combination*. Convex combinations will be encountered in decision problems in representing a randomized choice of action. They arise naturally in mechanics (for $k = 2$ or 3) in determining the center of gravity of a set of mass points. Given a mass m_1 at $P_1, \ldots$, and a mass m_n at P_n, the center of gravity is given by the formula

$$\text{c.g.} = \sum \frac{m_i}{M} P_i, \qquad \text{where } M = \sum m_i.$$

But since $m_i/M \geq 0$ and $\sum m_i/M = 1$, this is precisely what is defined to be a convex combination of $P_1, \ldots, P_n$. This application to mechanics helps in visualizing the nature of the set of all convex combinations of a set of points, because any convex combination can be interpreted as a center of gravity of a mass distribution.

In the case of two points in 2- or 3-space, the center of gravity of a system of two point masses must lie on the line segment joining the points. In general, the *line segment* joining P_1 and P_2 is defined as the set of all convex combinations:

$$P = \gamma P_1 + (1 - \gamma)P_2, \qquad \text{for } 0 \leq \gamma \leq 1.$$

Each such point P lies "between" P_1 and P_2, because each coordinate of P is between the corresponding coordinates of P_1 and P_2.

A set is called *convex* if it contains the entire line segment joining any pair of its points (as already defined for $k = 2$ in Section 6.2.1). For any given collection of points $P_1, \ldots, P_m$, the set of all convex combinations is a convex set. For if P and Q are two such convex combinations:

$$P = \sum \alpha_i P_i, \qquad \sum \alpha_i = 1, \qquad \alpha_i \geq 0,$$
$$Q = \sum \beta_i P_i, \qquad \sum \beta_i = 1, \qquad \beta_i \geq 0,$$

then each point on the line segment joining P and Q is of the form

$$\begin{aligned} \gamma P + (1 - \gamma)Q &= \gamma \sum \alpha_i P_i + (1 - \gamma) \sum \beta_i P_i \\ &= \sum [\gamma\alpha_i + (1 - \gamma)\beta_i] P_i, \end{aligned}$$

for some γ on $0 \leq \gamma \leq 1$. This is again a convex combination of the P_i's because

$$\gamma\alpha_i + (1 - \gamma)\beta_i \geq 0$$

and

$$\sum [\gamma\alpha_i + (1 - \gamma)\beta_i] = \gamma \sum \alpha_i + (1 - \gamma) \sum \beta_i = 1.$$

The collection of convex combinations of $P_1, \ldots, P_m$ is called the convex set *generated* by those points. The smallest convex set containing any given collection of points is called their *convex hull*, and the convex set generated by a finite number of points is their convex hull. In the case of $k = 2$, the convex set generated by m points is a polygon, which can be thought of as formed by a rubber band stretched to include all of the points and released to rest on pegs inserted at the various points. The convex set generated by m points in 3-space is a polyhedron.

EXAMPLE 8–1. Consider the five points (1, 1), (1, 2), (2, 2), (2, 4), and (4, 0), plotted in Figure 8–1 as $P_1, \ldots, P_5$. The convex set generated by these points—the set of possible centers of gravity—is the quadrilateral shown shaded in the figure. The point P_3 is inside the set, and so it can be represented as a convex combination

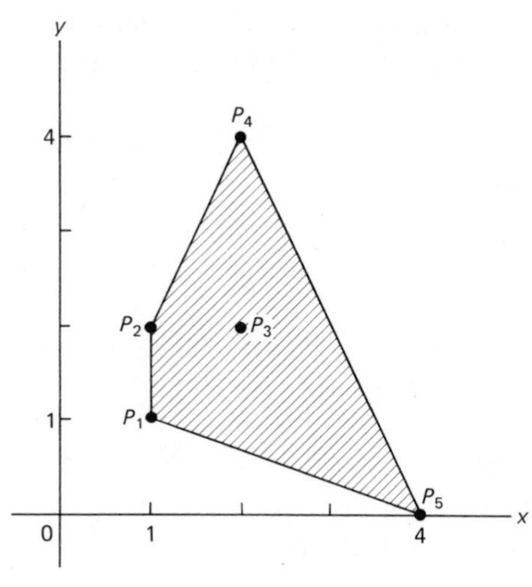

Figure 8–1

of the other four points. (Indeed, any convex combination of the five points is equivalent to a convex combination of P_1, P_2, P_4, and P_5.) No point outside the quadrilateral could be a convex combination, for a negative mass at one or more of the given points would be required to produce such a center of gravity.

8.1.2 Utility

Loss is not necessarily monetary. It can be as vague as loss of prestige, comfort, or goodwill. But to be useful, loss must be measured on at least an ordered, if not a numerical, scale. The term *utility* has come to be used to denote a kind of measure of gain (or negative loss) which appears to be appropriate for decision problems.

Specifically, utility denotes a real-valued function on prospects with which one is faced that (a) is bounded, (b) orders prospects according to preference, and (c) is computed as an expected value for prospects that are random. In terms of the notation $u(\mathscr{P})$ for the utility of a prospect $\mathscr{P}$, property (b) says that if $\mathscr{P}_1$ is preferred to $\mathscr{P}_2$, then $u(\mathscr{P}_1) \geq u(\mathscr{P}_2)$. The essence of property (c) is that if $\mathscr{P}$ is the random propsect of facing prospect $\mathscr{P}_1$ with probability α and prospect $\mathscr{P}_2$ with probability $1 - \alpha$, then $\mathscr{P}$'s utility is

$$u(\mathscr{P}) = \alpha u(\mathscr{P}_1) + (1 - \alpha)u(\mathscr{P}_2),$$

or the weighted average that defines the expected value of the random variable $u(\mathscr{P})$. The boundedness of property (a) is a condition that avoids certain paradoxes, like that of the following example.

EXAMPLE 8–2. The following proposition is offered, for a fee: You will be given a reward of 2^k dollars if, in repeated tosses of a coin, the first heads turns up on the kth toss. How large a fee would *you* pay to enter this arrangement?

Ordinarily entry fees (apart from the overhead for the "house") are determined to be the expected monetary gain—at least, such a determination makes the game "fair." Here the expected gain is

$$E(2^X) = \sum_1^\infty 2^k\left(\frac{1}{2^k}\right) = \infty,$$

which not only would be prohibitive as an entry fee, but also is not even approached by what people are actually willing to pay to play this game. The paradox is resolved by the fact that people's attitudes are not determined solely by dollar amounts—utility in this situation is not proportional to amount of money.

It is not easy for one to determine whether he actually has a utility function, satisfying the above properties, and it is usually even harder to determine what the function is, given that there is one. The existence of a utility function can be shown to follow as a consequence of certain reasonable axioms for preference patterns.[1]

Utility is clearly a personal matter, but it is not uncommon that two people have similar utility functions. If they do, a reasonable theory of decisions based on utility as a measure of gain should lead them to similar behavior rules in a given problem. On the other hand, if they have quite different utility functions, they might well be led to opposite behaviors, and still both be considered rational.

The notion of utility helps to explain why people sometimes enter into an unfair betting situation. A bet resulting in a new total capital M_W with probability p and with new total capital M_L with probability $1 - p$ is said to be *fair* if the expected total capital after the bet is equal to the initial capital M_0:

$$M_0 = pM_W + (1 - p)M_L.$$

The utility U_B of the random prospect consisting of taking the bet is computed by taking the expected value of the possible utilities after the bet:

$$U_B = pu(M_W) + (1 - p)u(M_L).$$

Putting these two equations together in the form

$$\begin{pmatrix} M_0 \\ U_B \end{pmatrix} = p\begin{pmatrix} M_W \\ u(M_W) \end{pmatrix} + (1 - p)\begin{pmatrix} M_L \\ u(M_L) \end{pmatrix},$$

one finds that the quantity U_B can be interpreted as the ordinate of that point whose abscissa is M_0 and which is a convex combination of $[M_W, u(M_W)]$ and $[M_L, u(M_L)]$, lying, therefore, on the line segment joining these two points. The

[1] See M. H. DeGroot [6], Chapter 7.

bet will be worthwhile, then, if the utility curve at M_0 is below that line segment, as in Figure 8–2. That is, if $U_B > u(M_0)$, one is better off taking the bet than not taking it.

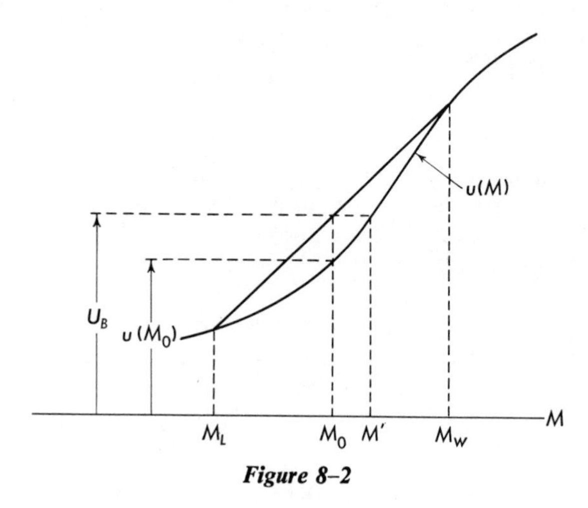

Figure 8–2

The bet is monetarily disadvantageous (and advantageous to the one offering the bet) if

$$M_0 > pM_W + (1 - p)M_L,$$

which implies that the expected net gain is negative:

$$p(M_W - M_0) + (1 - p)(M_L - M_0) < 0.$$

Inspection of the graph shows how much to the right of its present value one could move M_0 and still keep $u(M_0) < U_B$ so that the bet would be still worthwhile. (Note that U_B, depending only on M_L, M_W, and p, does not change as one considers larger M_0's or more disadvantageous bets.) The largest M_0 for which the bet is still worthwhile is indicated on the graph in Figure 8–2 as M'.

8.1.3 Personal Probability

In Chapter 1, probability was defined axiomatically as a mathematical model for experiments of chance. Such models were constructed, a priori, for certain familiar experiments whose symmetries and long histories of repetition make the construction rather objective—a matter of wide agreement. Other experiments, repeatable but without features that suggest particular models, a priori, are thought of as having objective models which, although not known in complete detail, could be revealed through sufficient repetition of the experiment. And because the axioms for probability are simply idealizations of properties that hold for relative frequencies in repeated trials, it is not surprising that the mathematical structure built on them does well in representing repeatable chance phenomena.

Probability models are also useful in describing an individual's beliefs about an unknown state of affairs, or about a phenomenon yet to take place, even when these are hard to conceive of as repeatable experiments. Probabilities are termed *personal* or *subjective* in such applications. On the basis of personal opinions, hunches, information, and beliefs, one may be willing to bet with certain odds that, say, Congress will cut income taxes in its next session; these odds define his personal probability of the event. Or, to cite a more relevant example, one may have certain prior convictions or feelings about the value of a parameter θ in an ordinary probability model—beliefs that can be represented by a distribution over the space of values of θ. This distribution would be concentrated heavily in regions of θ-values where he feels θ is "most likely" to be, according to his beliefs. It can be thought of as a probability distribution, if properly normalized, a distribution of personal probability for θ-values.

To assign a probability distribution to θ-values is not necessarily to believe that the true value of the parameter came about as the result of an experiment of chance. That is, it is not necessary to think of "nature" as a gambler, or to think that a given state of affairs was just the result of "chance." Nevertheless, it will be convenient to refer to the "random variable" Θ, a mathematical entity having the distribution that represents one's uncertainties about θ.

Two questions arise: (1) If the phenomenon being modeled is not repeatable, how does one establish any features of the distribution? (2) Are one's beliefs really susceptible of such modeling in the first place? (That is, are they sufficiently orderly and consistent to permit mathematical representation?) The answer to the second question is that the existence of a model for personal probability must be assumed—at some level or other. For example, it can be shown that if a notion of "at least as likely as" is assumed as a primitive relation, satisfying certain simple axioms, then there is a unique probability model agreeing with that relation. (See DeGroot [6], Chapter 6, and the references given there.)

The determination of personal probabilities, assuming their existence, can be approached by the device of comparison with objective experiments having known probabilities. To take a simple example, if a person is indifferent to (a) the prospect of receiving a reward if a fair coin falls heads, and (b) the prospect of receiving the same reward if an event E occurs, then his personal probability for E is $\frac{1}{2}$. Constructive definitions of probability based on such ideas as this have been considered by Ramsey, de Finetti, Savage, and many others. A useful bibliography on subjective probability is included in a book by Kyburg and Smokler.[2]

[2] H. E. Kyburg, Jr., and H. E. Smokler (eds.), *Studies in Subjective Probability* (New York: John Wiley & Sons, 1964).

Problems

8–1. Determine which of the following restrictions on (x, y) define convex sets in the plane:

(a) $x^2 + y^2 \leq 1$.
(b) $x^2 + y^2 \geq 1$.
(c) $y > x^2$.
(d) $y < x^2$
(e) $xy = 0$.
(f) $xy > 0$.
(g) $\begin{cases} x + 2y = 3 \\ 0 < x < 1. \end{cases}$
(h) $x + 2y \geq 3$.
(i) $x + y \leq 1$.
(j) $\begin{cases} x^2 + y^2 \leq 4 \\ y > 1. \end{cases}$

8–2. Show (algebraically, rather than geometrically) that the plane set defined by $x + y \leq 1$, $y \geq 0$, and $x \geq 0$ is convex.

8–3. Determine the convex combination of (0, 4, 0), (3, 1, 0), (0, 0, 4), and (4, 3, 5) in which the line $x = y = z$ issuing from the origin first strikes the convex set generated by the four points.

8–4. Show that the set of linear combinations of $P_1, P_2, \ldots, P_m$ is the smallest convex set containing those points. (That is, show that each convex combination of the P's is included in any convex set containing them all.)

8–5. A man needs \$6 to buy a ticket to a Saturday football game he very much wants to see, but he can only scrape up \$3 around the house. Are the numbers of dollars good representatives of his utility? Devise a graph that would better indicate his utility (on that day) as a function of how much money he has. In terms of this utility scale, determine his utility in tossing a coin with a friend for \$3, double or nothing. Is his utility improved by taking this bet?

8–6. Determine the expected utility in playing the game of Example 8–2 (winning 2^k dollars if the first heads in repeated tosses of a coin turns up on the kth toss) for a person whose utility for money (x in dollars) is

$$u(x) = \begin{cases} x/2^b, & \text{if } x \leq 2^b, \\ 1, & \text{if } x > 2^b. \end{cases}$$

What entry fee in dollars ought he to be willing to pay?

8–7. A man is considering a bet in which he will end up with \$10 if he wins and with nothing if he loses. His utility for money is

$$u(x) = \begin{cases} 0, & \text{if } x < 1, \\ \frac{1}{9}(x - 1), & \text{if } x \geq 1, \end{cases}$$

over the range of values of x (in dollars) of interest. Determine the utility in his taking the bet. Is there any probability of success so unfavorable that he should not take the bet, assuming that his initial capital is \$1? Assuming that his initial capital is \$2?

8–8. Given the bet of the preceding problem, what probability of winning makes it a "fair" bet if the man has initially \$1? Given this probability of winning, what is the most that the man could start with and still consider the bet worthwhile?

8.2 The No-Data Problem

Although statistical decision problems specifically involve the use of data as an aid in decision making, the problem of making a decision in the absence of data will be considered first. Not only are such problems simpler, but the approach to treating problems involving data is to reduce their solution to the solving of a no-data problem.

A decision-maker is faced with a set of possible actions; his "decision" is to choose one of them. The problem lies in the fact that for each action he might choose, the consequences depend on the state of nature—which is unknown to him. The set of possible actions will be referred to as the *action space* and denoted by $\mathscr{A}$; individual actions will be denoted by a or a_i. The set of possible states of nature, or *state space*, will be denoted[3] by Ω and the individual states by θ or θ_j.

8.2.1 Loss and Regret

It will be assumed that for each combination of state θ and action a, there is a loss $\ell(\theta, a)$, giving the negative utility of the consequences of taking action a when nature is in state θ. This *loss function* must generally be assumed to be known in order for the decision problem to be tackled in a rational way, although sometimes one can make good decisions knowing only certain aspects of the loss function.

If the state space Ω is finite, consisting of states $\theta_1, \ldots, \theta_k$, then for each action a, the losses $\ell(\theta_i, a)$ can be interpreted as coordinates of a point $(L_1, \ldots, L_k)$, where $L_i \equiv \ell(\theta_i, a)$. These points, one for each of the various actions in $\mathscr{A}$, provide a geometrical representation of the action space.

EXAMPLE 8–3. Consider a problem in which there are two states of nature and three possible actions. To have a concrete (though possibly fanciful) illustration in mind, think of the states of nature as "rain" (θ_1) and "no rain" (θ_2), and the actions as "stay at home" (a_1), "go out without an umbrella" (a_2), and "go out with an umbrella," (a_3). That is, one has an errand to do and must decide whether to take an umbrella and face the possible mortification of carrying it in the sunshine, leave the umbrella home and possibly get soaked, or just give up on the errand. The following table gives losses (negative utilities) that represent the dilemma which

		States of Nature	
		θ_1 (rain)	θ_2 (no rain)
	a_1 (stay home)	4	4
Actions	a_2 (go, no umbrella)	5	0
	a_3 (go with umbrella)	2	5

[3] The name Ω was used earlier for a probability space, but there is no longer need for this meaning of the symbol. The capital theta will be needed for the state of nature when thought of as a random variable.

constitutes the decision problem. The points in the plane representing the actions a_1, a_2, and a_3 are, respectively, (4, 4), (5, 0), and (2, 5), shown in Figure 8–3.

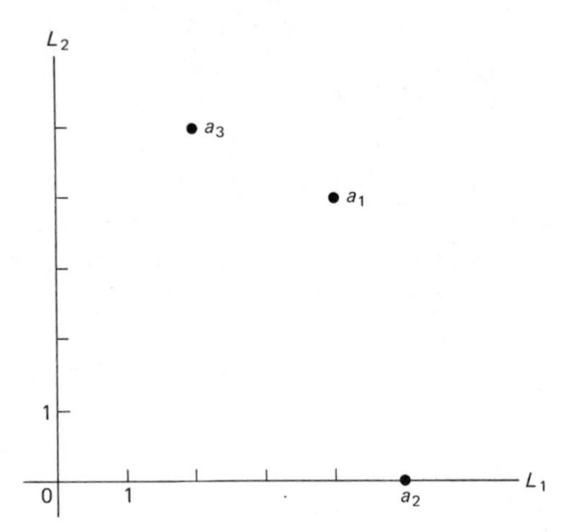

Figure 8–3. Loss plot for actions in Example 8–3.

It is felt by some that actions should be based on what is called *regret* instead of on *loss*. The regret function $r(\theta, a)$ is defined as the result of subtracting from the loss $\ell(\theta, a)$ the minimum loss for the given θ:

$$r(\theta, a) = \ell(\theta, a) - \min_a \ell(\theta, a).$$

That is, for each state of nature, one determines the smallest loss he could get by with if that state of nature were known to be the true state; this is a contribution to loss that even a *good* decision cannot avoid, and so, to obtain a quantity more appropriate to the decision process, it is subtracted from the loss. The difference $r(\theta, a)$ represents the loss that could have been avoided had the state of nature been known—hence the term *regret.*

A regret function has a minimum value of 0 for each θ. Conversely, a loss function $\ell(\theta, a)$ that has a minimum of 0 for each θ is equal to the corresponding $r(\theta, a)$; such a loss function is already a regret function, with no modification necessary.

EXAMPLE 8–4. The loss table of Example 8–3 is repeated here with a row giving the minimum loss for each state:

	θ_1	θ_2
a_1	4	4
a_2	5	0
a_3	2	5
Minimum loss	2	0

Subtracting these minima from the losses yields the regrets, which is shown in the following regret table:

	θ_1	θ_2
a_1	2	4
a_2	3	0
a_3	0	5

If nature is known to be in state θ_1, the correct action is a_3, which produces the smallest loss, 2. This loss is not the fault of the decision-maker. However, had he chosen action a_2, the loss would be 5, of which 3 units could have been avoided (if only the state of nature had been known to be θ_1) and so is regretted. The graphical representation of the actions according to regret is shown in Figure 8–4. The geometrical effect of "regretizing" is to translate the action points so that, with all in the first quadrant, at least one is on each of the coordinate axes.

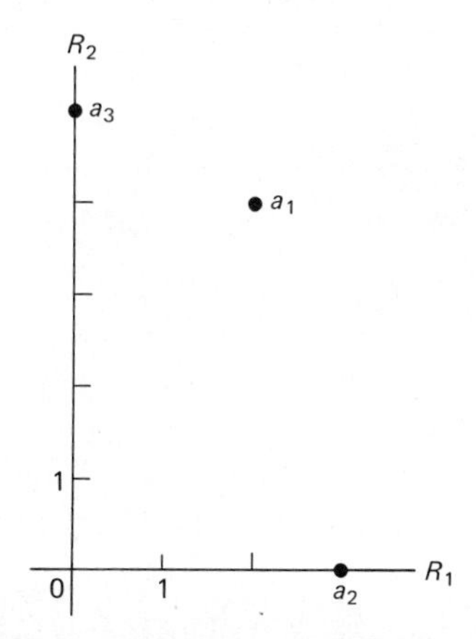

Figure 8–4. Regret plot for actions in Example 8–4.

Regret sometimes goes under the name of *opportunity loss*, a term that is appropriate when thinking in terms of gain rather than loss. The gain resulting from action a under state θ is $-\ell(\theta, a)$, and so the minimum loss is the negative of the maximum gain. The regret is then

$$r(\theta, a) = \ell(\theta, a) - \min_a \ell(\theta, a) = \max_a g(\theta, a) - g(\theta, a),$$

where $g = -\ell$. This represents the maximum that could have been gained if the state of nature were known, minus the amount actually gained, hence the "opportunity" loss.

In studying a decision problem, the question of whether it would make any difference if one used regret instead of loss naturally arises. In the subsequent discussion it will become evident that in some cases it does make a difference and in others it does not. The classical treatment of statistical problems, when viewed as problems in decision theory, frequently amounts to assuming a loss function that is already a regret function. This is surely defensible if the method used is one for which replacing actual losses by opportunity losses does not alter the solution. At any rate, the decision-maker, or the statistician, if he himself does not incur the loss, would presumably be interested in regrets—the amounts of loss he could not avoid even with the decision that is best when the state of nature is known. His client, however, would suffer the losses, not the regrets (if they are different).

8.2.2 Mixed Actions

In game theory, it is found that it pays to consider *mixed actions*—procedures that employ an extraneous device to make a random choice of action. To choose a mixed action is to choose a particular random device. The original actions are then referred to in contrast as *pure* actions and are special cases of mixed actions in which the random devices are really deterministic.

A mixed action is defined by a probability distribution on the action space; and for each particular mixed action that might be employed, the loss is a random variable (for a given state θ). But the utility of a random prospect is computed as an expected value; and so when a mixed action is employed, the loss incurred (for given θ) is the expected value of the loss function with respect to the distribution over $\mathscr{A}$ that defines the mixed action.

In a problem with finitely many states, $\theta_1, \ldots, \theta_m$, and finitely many actions, $a_1, \ldots, a_k$, the m losses characterizing a given mixed action $(p_1, \ldots, p_k)$ are as follows:

$$\begin{cases} L_1 \equiv E[\ell(\theta_1, a)] = \ell(\theta_1, a_1)p_1 + \cdots + \ell(\theta_1, a_k)p_k \\ \vdots \\ L_m \equiv E[\ell(\theta_m, a)] = \ell(\theta_m, a_1)p_1 + \cdots + \ell(\theta_m, a_k)p_k \end{cases}$$

or in matrix notation,

$$\begin{pmatrix} L_1 \\ \vdots \\ L_m \end{pmatrix} = p_1 \begin{pmatrix} \ell(\theta_1, a_1) \\ \vdots \\ \ell(\theta_m, a_1) \end{pmatrix} + \cdots + p_k \begin{pmatrix} \ell(\theta_1, a_k) \\ \vdots \\ \ell(\theta_m, a_k) \end{pmatrix}.$$

Thus, the point in m-space representing the mixed action $(p_1, \ldots, p_k)$ is a convex combination of the points that represent the pure actions. And the set of all mixed actions is the convex set generated by the pure actions.

EXAMPLE 8–5. For the three-action problem of Examples 8–3 and 8–4, a mixed action is a vector of three probabilities: (p_1, p_2, p_3). The pure actions are the probability vectors (1, 0, 0), (0, 1, 0), and (0, 0, 1). For a given mixed action **p**, the losses L_1 and L_2 under θ_1 and θ_2, respectively, are given by

$$\begin{pmatrix} L_1 \\ L_2 \end{pmatrix} = p_1 \begin{pmatrix} 4 \\ 4 \end{pmatrix} + p_2 \begin{pmatrix} 5 \\ 0 \end{pmatrix} + p_3 \begin{pmatrix} 2 \\ 5 \end{pmatrix}.$$

The set of mixed actions is represented, therefore, by the convex combinations of the pure actions, as shown in Figure 8–5.

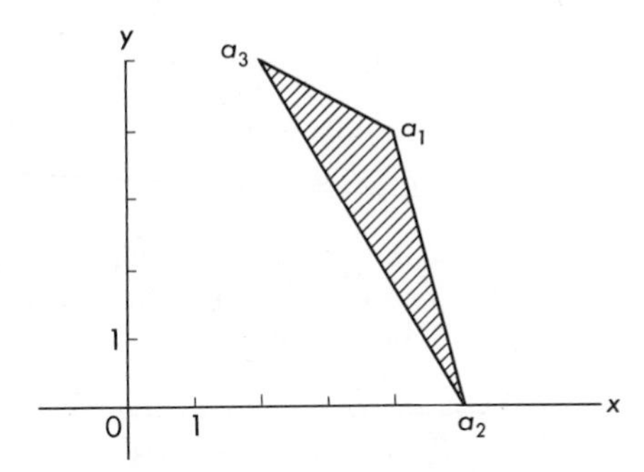

Figure 8–5. Set of mixed actions in Example 8–5.

When losses are converted to regrets, the mixed action points are translated along with the pure actions. For, the minimum subtracted from each loss is independent of a, and is equal to its expectation with respect to the mixing distribution:

$$E[r(\theta, a)] = E[\ell(\theta, a)] - \min_a \ell(\theta, a).$$

8.2.3 The Minimax Principle

It has been seen that even in the simplest problems, the desire to choose an action to minimize loss is not an adequate guide, since the action that minimizes the loss for one state of nature does not necessarily do so for another. It should come as no surprise, then, that no single method or principle for selecting an action has been universally accepted as "best." Two principles are considered in this and the following section; each principle proposes a way of assigning a single number to each action, and so *ordering* them according to this number. The one assigned the smallest number is presumed to be "best" and is the action called for by that particular ordering.

The *minimax principle* is that one should expect the worst and prepare for it. That is, for each action determine the maximum possible loss that might be incurred under the various states of nature. This number is assigned to that action and serves to order the actions. The one with the smallest (or minimum) maximum loss is the one taken.

EXAMPLE 8–6. Consider again the loss table of Example 8–3; a column has been added, listing the maximum loss for each action:

	θ_1	θ_2	$\max \ell(\theta, a)$
a_1	4	4	4
a_2	5	0	5
a_3	2	5	5

If action a_1 is selected, the maximum loss is 4, incurred for either state of nature. This maximum is smaller than the maximum of 5 encountered for a_2 or a_3, and so a_1 is the minimax action.

The minimax principle can also be applied to the regret function. The regret table corresponding to the above loss table is the following, in which a column of maximum regrets for the various actions is given:

	θ_1	θ_2	$\max r(\theta, a)$
a_1	2	4	4
a_2	3	0	3
a_3	0	5	5

The minimum of the maximum regrets is 3, achieved for action a_2. The minimax regret action is not the same as the minimax loss action in this example.

A graphical representation is enlightening. As in Example 8–3, the actions are represented by the points (L_1, L_2), with $L_i = \ell(\theta_i, a)$. Now, for any point (x, y) above the line $x = y$, the second coordinate (y) is the maximum one; for a point below the line $x = y$, the first coordinate (x) is maximum. To choose between two actions *both above* $x = y$, take the one with the smaller y—the lower one, or the one first met by a horizontal line moving upward. To choose between two actions *both below* $x = y$, take the one first met by a vertical line moving to the right. For points some of which lie above, some on, and some below the line $x = y$, take (for the minimax action) the one first hit in moving together a horizontal line up *and* a vertical line to the right—the point first met by a wedge moving with its vertex on $x = y$, as shown in Figure 8–6.

This example illustrates the fact that the minimax principle may lead to different actions, depending on whether it is applied to losses or to regrets.

In choosing from the set of mixed actions, the minimax procedure is applied to *expected* losses (or to expected regrets). The graphical procedure for the case of two states of nature, developed above for choosing a pure action, can also be applied in the case of mixed actions.

EXAMPLE 8–7. The set of mixed actions for the problem of the preceding few examples was shown in Figure 8–5. It is repeated in Figure 8–7, along with the moving wedge used in obtaining the minimax point. From the graph it is evident that the

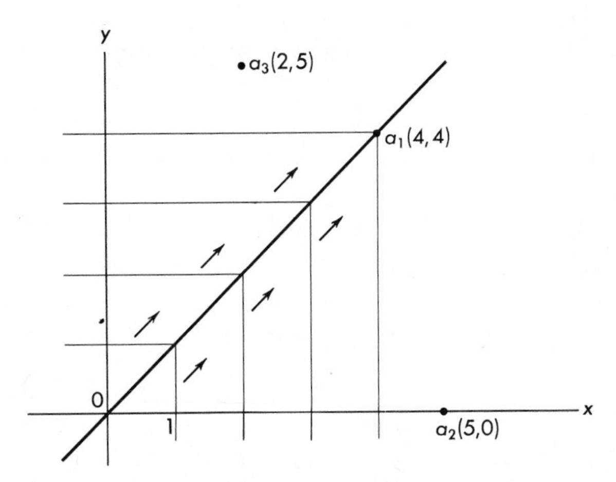

Figure 8–6. Determining a minimax action (Example 8–6).

wedge strikes the set of mixed actions first at the point where the line $x = y$ strikes it, a point that is a mixture of a_2 and a_3:

$$\gamma\binom{5}{0} + (1 - \gamma)\binom{2}{5}.$$

Equating the components $5\gamma + 2(1 - \gamma)$ and $5(1 - \gamma)$ one obtains $\gamma = \frac{3}{8}$, corresponding to the point $(\frac{25}{8}, \frac{25}{8})$. Thus, the minimax expected loss is $\frac{25}{8}$, obtained by the mixed action $(0, \frac{3}{8}, \frac{5}{8})$. (Note that as in the case of pure actions, the minimax mixed action would be different if regrets had been used rather than losses.)

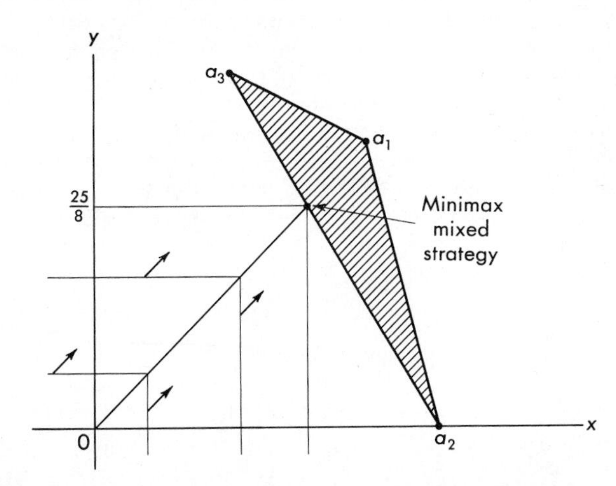

Figure 8–7. Graphical determination of minimax mixed action.

It is clear from the graphical procedure for the case of two states of nature that only boundary points of the set of mixed actions could be minimax actions. Moreover, only those points that are on, loosely speaking, the lower left portion of the boundary can be minimax. Indeed, there will only be one minimax action unless the wedge first strikes the set of mixed actions in a portion of the boundary that is parallel to one of the axes.

The graphical method for finding a minimax action given above is not general. Minimax mixed actions in the more general problem with finitely many actions and states can be located by algebraic methods; but these will not be developed here because the minimax principle is not a major tool for determining good statistical procedures.

There is one other important case in which a graphical procedure yields the minimax action, namely, that in which there are just two actions. This procedure will be illustrated in the next example. It has some importancein that the problem of testing hypotheses will be treated as a two-action decision problem in Section 8.3.3.

EXAMPLE 8–8. A decision problem with two actions and four states is defined by the following loss table, which also gives the expected losses for each state, assuming a mixed action $(p, 1 - p)$. The expected losses, as functions of p, are

	a_1	a_2	Expected Loss
θ_1	4	0	$4p$
θ_2	2	-1	$3p - 1$
θ_3	1	5	$5 - 4p$
θ_4	-1	2	$2 - 3p$

plotted in Figure 8–8. For each p, the maximum expected loss is the ordinate on the line that is highest at that p. The heavy line in the graph is a plot of the maximum expected loss as a function of p, and the minimum maximum occurs at the lowest point on the maximum plot. Equating $4p$ and $5 - 4p$ (the functions whose graphs intersect at the desired point) yields $p = \frac{5}{8}$, and so the minimax mixed action is the probability vector $(\frac{5}{8}, \frac{3}{8})$.

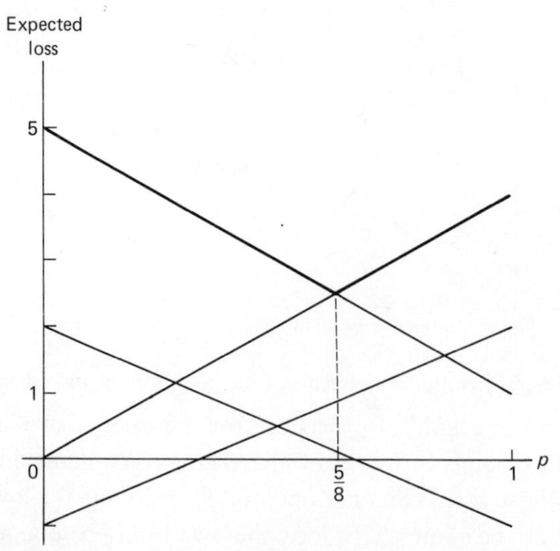

Figure 8–8. Graphical determination of minimax mixed action.

Minimax actions are optimal in the theory of two-person, zero-sum games, in the sense that by using such a strategy, one can be assured of no more loss than the minimum maximum—no matter what the other player (nature, in decision problems) does. Moreover, the other player has a strategy, following the same principle, that is optimal; and with it he can be assured that the first player (the decision-maker) does no better than he can do by using the minimax strategy and incurring the minimax loss. (But to assume that nature is trying to win the "game" and knows that her minimax action is optimal is perhaps unduly pessimistic.)

Problems

8–9. Given the following loss table determine the minimax loss (pure) action and also the minimax regret (pure) action.

	θ_1	θ_2	θ_3
a_1	2	−3	−1
a_2	4	0	5
a_3	1	1	−1
a_4	0	2	−2

8–10. Investigate the minimax strategies (pure and mixed, using loss and then regret) for the following loss table.

	θ_1	θ_2
a_1	10	10
a_2	15	0

8–11. Given the following loss table, plot points corresponding to the various actions and determine the minimax action (mixed). Determine the corresponding regret table and determine the minimax regret action.

	a_1	a_2	a_3	a_4	a_5
θ_1	−1	1	0	2	5
θ_2	3	2	5	3	2

8–12. The state of nature is defined by θ, where $0 \le \theta \le 1$, and an action is to be chosen that is a number a on the range $0 \le a \le 1$. If the loss function is $(\theta - a)^2$, determine the minimax action. (Would it be different using regret?)

8–13. A lot of five articles is to be accepted or rejected. The state of nature is defined by the number M of defective articles present among the five; thus, $M = 0, 1, 2, 3, 4,$ or 5. Assuming losses equal to twice the number of defectives in a lot if it is passed, and equal to the number of good articles in a lot if it is rejected, determine the minimax mixed action.

8.2.4 Bayes Actions

Anticipating the worst state of affairs imaginable, as the minimax principle does, may not be a rational thing to do—especially if one is convinced that the worst state has only a remote chance of being the actual state. Adopting personal probability as a model for one's beliefs or convictions about an unknown state of affairs permits him to think of nature as random, and one would then average the prospective losses for each action with respect to the weighting in the model for this randomness.

Let the state of nature be denoted by Θ. If the state space is discrete, with (personal) probability distribution given by

$$g(\theta) = P(\Theta = \theta),$$

then the random loss $\ell(\Theta, a)$ has expectation

$$B(a) \equiv E[\ell(\Theta, a)] = \sum \ell(\theta, a) g(\theta).$$

If θ is a numerical parameter, Θ is a random variable; if its distribution function is $G(\theta)$, then

$$B(a) = E[\ell(\Theta, a)] = \int \ell(\theta, a)\, dG(\theta),$$

with corresponding sum and integral formulas for the discrete and continuous cases.

The average $B(a)$, called the *Bayes loss* for action a, provides a numerical ordering of the available actions. A *Bayes action* is an action a_B that minimizes the Bayes loss:[4]

$$B(a_B) = \min B(a).$$

Thus, in comparing actions, if an extremely large loss can occur for a given action when nature is in a state believed by the decision-maker to be quite unlikely, the impact of the extreme loss is minimized by weighting only slightly the state that would produce it.

The distribution of Θ is often called a *prior* or an *a priori distribution* (often shortened to "prior"), and the motivation for the term will become apparent after consideration of the problem of making decisions using data. Indeed, it will be only then that the connection between a "Bayes action" and Bayes' theorem will become evident.

Again, in the case of two states of nature, a graphical procedure can be employed to locate the Bayes action, as the following example will illustrate.

[4] The term *Bayes loss* is sometimes used to refer to the minimum value, $B(a_B)$. But it seems convenient to have a name for $B(a)$, the function of a that is being minimized. (It could be called simply "expected loss," but this term is rather ambiguous.)

EXAMPLE 8–9. The loss table for the rain problem (Example 8–3 and others) is repeated here with an additional column giving prior probabilities and a row giving expected losses computed with respect to the prior distribution. The prior probabilities are *assumed* to be given by

$$P(\Theta = \theta_1) = .4 = 1 - P(\Theta = \theta_2),$$

reflecting a belief that on the date in question, it is a little less likely to rain than not.

	a_1	a_2	a_3	$g(\theta)$
θ_1	4	5	2	.4
θ_2	4	0	5	.6
$B(a)$	4	2.0	3.8	

An example of the computation of $B(a)$ is

$$B(a_3) = 2 \times .4 + 5 \times .6 = 3.8.$$

The action with the smallest Bayes loss is a_2, and so a_2 is the Bayes action corresponding to the assumed prior.

The Bayes solution can be determined graphically, using the plot of (L_1, L_2) introduced in Example 8–5. The Bayes loss for an action with losses (L_1, L_2) is

$$B(a) = .4L_1 + .6L_2.$$

All actions with a given Bayes loss K would satisfy the condition

$$.4L_1 + .6L_2 = K,$$

and so be represented by points on the line having this equation. For various K's, the equation defines a *family* of lines with slope $-.4/.6$, higher lines corresponding to larger K's. (The intercept on the vertical axis is $K/.6$.) Thus, to compare two actions, it is only necessary to determine which action lies on the lower line in this family of lines. Or, if there are several actions to compare, move a line with slope $-.4/.6$ (which is the slope common to all of the lines $.4L_1 + .6L_2 = K$) from below all of the points representing actions until it first hits one of those points. The corresponding action is the Bayes solution. Figure 8–9 illustrates this, and examination of that figure also shows that for different prior weights, which would give families of lines with slopes other than $-.4/.6$ (but always nonpositive), the Bayes solution would be always either a_2 or a_3. (For *one* choice of prior weights, the line would hit a_2 and a_3 simultaneously, in which case it does not matter which of these actions is taken.)

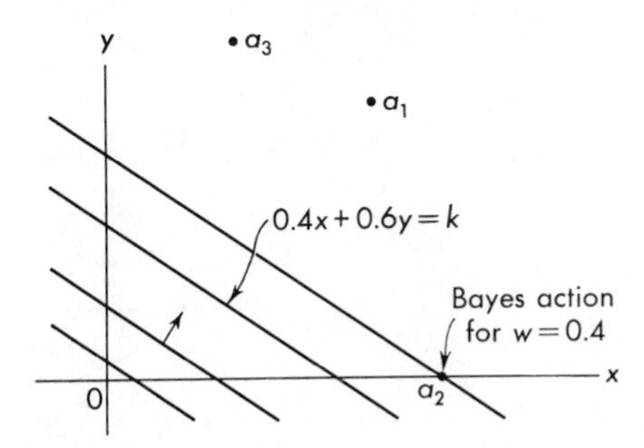

Figure 8–9. Graphical determination of Bayes action.

If the decision-maker wants to consider *mixed* actions, the Bayes method can be applied to the expected losses. Thus, if A denotes the random action defined by a mixture or distribution over the action space $\mathscr{A}$, one computes the Bayes loss as an average of the expected losses,

$$\text{Bayes loss} = E^{\Omega}[E^{\mathscr{A}}\ell(\Theta, A)],$$

and then choose a distribution for A that minimizes this Bayes loss. When the action space and the state space are discrete, with corresponding distributions defined by probability functions $p(a)$ and $g(\theta)$, respectively, the formula for Bayes loss is

$$\sum_i E^{\mathscr{A}}[\ell(\theta_i, A]g(\theta_i) = \sum_i \sum_j \ell(\theta_i, a_j)p(a_j)g(\theta_i)$$
$$= \sum_j \left\{\sum_i \ell(\theta_i, a_j)g(\theta_i)\right\}p(a_j),$$

a convex combination of the Bayes losses corresponding to the pure actions.

EXAMPLE 8–10. If mixed actions are permitted in the problem of the preceding example, the same graphical method can be applied to the set of mixed actions. Figure 8–10 shows the set of mixed actions and lines with various slopes labeled according to the prior probability w for θ_1, at points where the line families meet the action set. Observe that the Bayes mixed action is really a pure action, except for the prior defined by $w = \frac{5}{8}$, which defines a family of lines parallel to the line through a_2 and a_3. In that exceptional case, any mixture of a_2 and a_3 is a Bayes action; but there is no gain in using a mixture in preference to either a_2 or a_3—the Bayes loss would be the same.

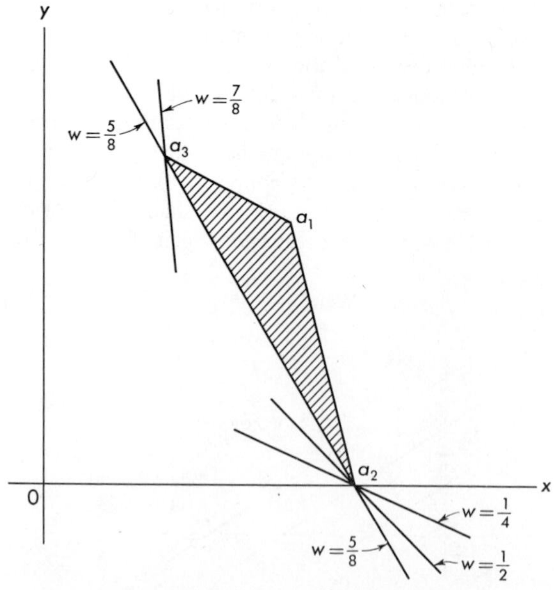

Figure 8–10. Bayes solutions (Example 8–10).

This example illustrates a general fact. As far as Bayes solutions are concerned, there is no gain in using mixed actions in preference to pure actions. That is,

$$\min_{\mathscr{A}} E^{\Omega}\ell(\theta, a) = \min E^{\mathscr{A}}[E^{\Omega}\ell(\theta, A)],$$

where the minimum on the right is taken over competing mixtures—distributions for A. To establish this equality, it must be shown that

$$\min_{\mathscr{A}} B(a) = \min E^{\mathscr{A}}[B(A)],$$

where $B(a)$ is the Bayes loss using the action a and the minimum on the right is again taken over distributions on $\mathscr{A}$. But the average of any random variable is always between its minimum and maximum, so $E[B(A)]$ is no smaller than the minimum on the left—and the same is then true of the miniumm of $E[B(A)]$. On the other hand, the minimum on the left can be thought of as taken over the class of degenerate distributions—those that deposit the entire probability mass at a single action; and because this class is included in the class of all distributions on $\mathscr{A}$, the left-hand side is no smaller on the right. Hence, they are equal.

In obtaining Bayes solutions, it does not matter whether the technique is applied to the loss function or to the regret function. This is apparent in the case of two states of nature from the graphical approach involving families of parallel lines; for the method does not involve the location of the origin, and the translation of the action set that corresponds to regretizing the losses does not affect the solution. It is also true in general, because the expected regret corresponding to any prior with c.d.f. $W(\theta)$ is

$$\int r(\theta, a)\, dW(\theta) = \int \ell(\theta, a)\, dW(\theta) - \int \min_{\mathscr{A}} \ell(\theta, a)\, dW(\theta),$$

and (because the second term on the right is independent of a) the action that minimizes the expected loss minimizes the expected regret.

There is another kind of graphical Bayes solution of the decision problem with two states of nature. It not only serves to point out a relationship between the minimax and Bayes principles but will also be a useful tool in the later study of Bayes sequential testing. The following example illustrates the method.

EXAMPLE 8–11. Consider again the decision problem of the preceding examples and assume prior probabilities w and $1 - w$ for θ_1 and θ_2, respectively. The Bayes losses for the various possible actions are as follows:

$$B(a_1) = 4w + 4(1 - w) = 4.$$
$$B(a_2) = 5w + 0(1 - w) = 5w.$$
$$B(a_3) = 2w + 5(1 - w) = 5 - 3w.$$

These functions of w are plotted in Figure 8–11. It is seen there that for any w less than $\frac{5}{8}$, action a_2 has the smallest Bayes loss; and for $w > \frac{5}{8}$ the Bayes action is a_3. For $w = \frac{5}{8}$, either a_2 or a_3 is Bayes; and the minimum Bayes loss in that case is $\frac{25}{8}$, the largest such loss for any prior. The prior distribution $(\frac{5}{8}, \frac{3}{8})$ is, in that sense, *least favorable* to the decision-maker. Also, observe that the minimum Bayes loss for the least favorable prior is precisely the minimax loss (cf. Example 8–7).

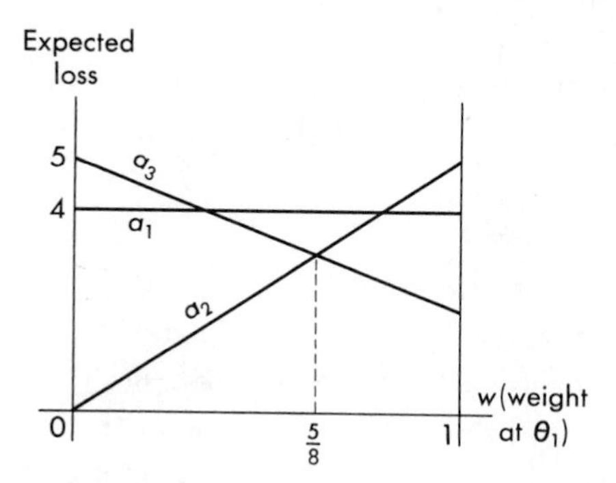

Figure 8–11. Graphical determination of Bayes action.

In considering the decision problem as a zero-sum game, one in which one player's loss is the other player's gain, the losses for the decision-maker are the negative losses for nature. Thus, by inverting the graph in Figure 8–11 one obtains the graph of losses as a function of a mixed action for nature. And so, determining the prior that maximizes the minimum Bayes loss is equivalent to determining (by the graphical technique of Example 8–8) a mixed action for nature that minimizes her maximum loss—that is, her minimax mixed action. A theorem in game theory states that by taking his minimax mixed action, each player is acting optimally, and the minimax loss for one player is just the negative of the minimax loss for the other. This is why, in the example just given, the maximum Bayes loss is equal to the minimax loss and is incurred when the prior is least favorable.

In viewing the decision problem as just that, and not as a game against a "malevolent nature," the Bayes and minimax principles are just devices for ordering the actions so that one can be picked as "best." The minimax principle, though, does have the game-theoretic feature of preparing against all conceivable states of nature (and in particular, against the worst), whereas in the Bayes approach it is assumed that nature's "strategy" is known—at least as a mixed action, or a distribution for Θ.

A frequent objection to the Bayesian approach is that it is difficult to determine a prior distribution. Even among Bayesian disciples there has long been an unresolved question as to how to represent a state of complete ignorance about a parameter. Bayes (1763) considered, without really adopting, the notion that ignorance could be represented by a uniform distribution; Laplace's apparent espousal of the idea led to so many controversial applications that the Bayesian approach fell into general disfavor. One basic difficulty in representing ignorance by uniformity is that if someone is ignorant of θ, he is also ignorant of θ^2 or $\log \theta$ or any other parameter in a reparametrization; but a uniform distribution for θ is not generally uniform for $g(\theta)$.

Another difficulty with the uniform prior is that if the parameter space is not bounded, a constant density does not define a probability distribution. A distribution that cannot at least be normalized to have total mass 1 is said to be *improper*. Mathematically, even an improper prior can provide the ordering of actions needed to select a "best" one, although the "Bayes loss" computed therefrom will not be measured on the same scale as the loss.

An improper uniform prior can be approximated by a sequence of proper priors. For instance, if $g(\theta)$ is uniform on $[0, M]$, the action that minimizes

$$B_M(a) = \frac{1}{M} \int_0^M \ell(\theta, a)\, d\theta$$

would also minimize $MB_M(a)$, whose limit would define the "Bayes loss" for the improper uniform prior. Similarly, the limit of the proper density $Me^{-M\theta}$ (for $\theta > 0$) as M approaches 0 would be uniform on $[0, \infty]$. And the Bayes action for given M may in either case approach the action that minimizes the improper Bayes loss as $M \to 0$.

The difficulty and resulting imprecision in the specification of a prior does not bother some Bayesian statisticians.[5] He will usually have enough data so that the influence of the prior on his conclusions is minimal (see Section 8.4). In the case of a Bayesian decision-maker, such as a plant manager who must make decisions with limited data, the fact that different people will have different priors is not bothersome. For, either the prior is not crucial and they all come to the same conclusion anyway, or it is crucial and they *should* come to different conclusions. In the latter case, they would either have to gather more data or to resolve their differences, a process that is healthy in provoking thought and reevaluation.

8.2.5 Admissibility

It would seem intuitively obvious that if taking action a_1 produces losses that are always less than the losses incurred by taking action a_2, then action a_2 should not be considered—or at least, nothing is lost by excluding it from consideration in solving a decision problem.

Definition. *Action a_1 is said to* dominate *action a_2 if for all θ, $\ell(a_1, \theta) \leq \ell(a_2, \theta)$. The domination is said to be* strict *if the inequality is strict for some θ.*

Definition. *An action is called* admissible *if it is not strictly dominated by any other action.*

[5] Compare G. E. P. Box and G. C. Tiao [3], Chapter 1.

These definitions, of course, apply also to mixed actions, by making comparisons in terms of expected losses.

EXAMPLE 8–12. Consider a no-data problem with two states of nature and four actions, in which losses are those in the following table. Action a_4 is strictly dominated by action a_3 and so is inadmissible; for the loss in taking a_4 is always at least as

		Actions			
		a_1	a_2	a_3	a_4
States	θ_1	4	5	2	3
	θ_2	4	0	5	5

large as the loss in taking a_3, and it is actually larger for state θ_1. There is no dominance relation among a_1, a_2, and a_3, but if mixed actions are allowed, some mixtures of a_2 and a_3 dominate a_1 strictly. For example, the mixed action (0, .5, .5, 0) does. In Figure 8–12, the heavy line indicates the mixtures of a_2 and a_3 that dominate a_1. In the set of mixed actions, only mixtures of a_2 and a_3 are admissible.

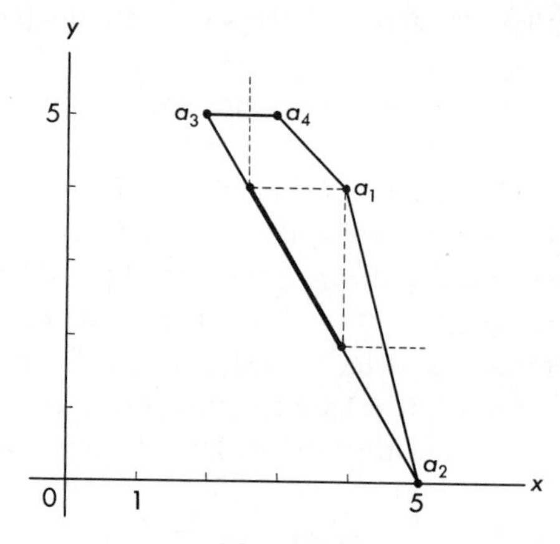

Figure 8–12

Admissibility appears to be a good property, since it would seem pointless to consider one procedure if another has no larger losses and actually a smaller loss for some states. However, an "inadmissible" procedure that is only a little bit inadmissible might still be very useful. In Figure 8–12, for instance, there are clearly actions that are arbitrarily close to being admissible. As loss functions are usually far from precise, one might find it hard to argue that a slight amount of inadmissibility should condemn a procedure.

The idea that in searching for good procedures one can restrict his attention to a relatively small class of procedures that are optimal in some sense gives rise to the next definition.

Definition. *A set of actions is called a* complete class *if any action not in the class is strictly dominated by one that is. A* minimal complete class *is a complete class that does not include a complete proper subclass.*

A complete class includes all admissible actions. The class of admissible actions may not be complete, but if it is, then it is minimal complete. If a minimal complete class exists, it is identical with the class of admissible actions.

Figure 8–13 shows a set of possible actions as a convex set (shaded). The set of Bayes actions (the heavy boundary $\widehat{ABC}$) is a complete class. The admissible actions are on the heavy arc $\widehat{BC}$ and constitute a minimal complete class.

The two-dimensional picture suggests that (with obvious exceptions) Bayes actions are admissible, and that admissible actions are Bayes for some prior. Precise theorems of this type and related results are given in Chapter 2 of Ferguson [9].

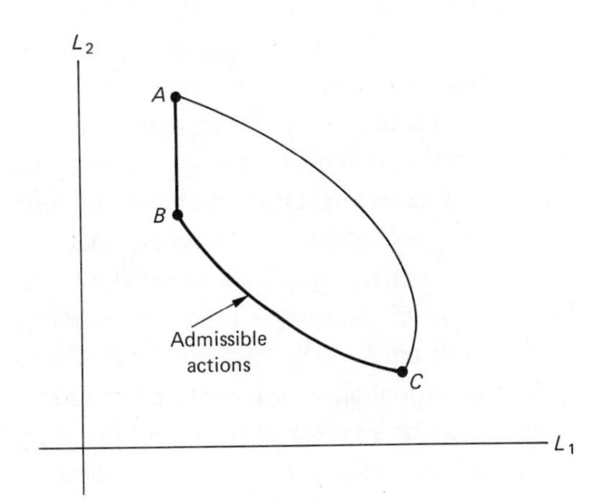

Figure 8–13

Problems

8–14. Determine the Bayes action for the loss table in Problem 8–9 given prior probabilities $(\frac{1}{6}, \frac{1}{3}, \frac{1}{2})$ for $(\theta_1, \theta_2, \theta_3)$.

8–15. Determine the Bayes action (as a function of w) for the loss table of Problem 8–11 given prior probabilities $g(\theta_1) = w$, $g(\theta_2) = 1 - w$.

8–16. Consider the decision problem given in Problem 8–13.

(a) Determine the Bayes action for a uniform prior.

(b) Determine the Bayes action if the prior probability of M defectives in the lot is binomial with $n = 5$ and $p = \frac{1}{5}$.

(c) Determine the Bayes action if the prior distribution of M is binomial with $n = 5$ and unknown p, as a function of p.

+**8–17.** In the decision problem of Problem 8–12, where θ and a are numbers on [0, 1] and the loss function is $(\theta - a)^2$,

(a) Determine the Bayes action if $g(\theta) = 1$.
(b) Determine the Bayes action if $g(\theta) = 6\theta(1 - \theta)$.
(c) Show that, in general, the Bayes action is $E(\Theta)$.

8–18. Show that if a is inadmissible, there is a Bayes solution that does not involve a.

8–19. Look for dominance relations in the loss table of Problem 8–9 and show that a_2 and a_3 are not admissible in the class of mixed actions.

8.3 Using Data in Decisions

Looking at data resulting from performing the experiment whose model is under study can be thought of as a form of "spying" on nature. Such spying usually costs something so that the amount of spying permitted or desired is not necessarily infinite. How much should be done depends on the extent to which spying increases one's information about nature; if he is spying in the wrong way—doing the wrong experiment—a large amount of it may be futile. Before a balance of cost of experimentation and gain of information can be made, it is necessary to see what information is to be gained by experimenting and, for a given amount of experimenting, how to get the most out of the data. For the present, then, the size of the sample used will be considered fixed.

The general discussion will be given in terms of "data" Z, where Z may be the value of a single observation on a univariate distribution or on a multivariate distribution; or it may be a vector of the values in a sample, or the value of some statistic computed from the sample. In any case, there is a corresponding distribution of Z.

8.3.1 The Risk Function

A decision rule or strategy that is to lead to a decision based on the data Z must take into account that Z can have many values. A rule for choosing an action is not complete until it prescribes an action a for each conceivable value z of Z. Such a rule is called a *statistical decision function*:

$$a = d(z).$$

This is a mapping from the space of possible data points to the space of actions. The statistician's task is then the selection of a suitable decision rule, or decision function, $d(z)$.

The action taken is random, since the data to which the decision rule is applied are random. Hence, the loss is random: $\ell(\theta, d(Z))$. It is customary, and consistent with rules for utility, to base the analysis on the *expected value* of the loss, called the *risk function*:

$$R(\theta, d) = E_\theta[\ell(\theta, d(Z))].$$

Notice that there is now a third possible meaning of the phrase, "expected loss." It was used first to mean an average with respect to a mixed action; then it was used to mean an average with respect to a prior distribution (mixed action for nature); and here the average is computed with respect to the distribution of the data. This distribution of Z depends on θ, so the dependence of $R(\theta, d)$ on θ enters through the θ in $\ell(\theta, a)$ and also through the θ in the distribution of Z. The expected loss $R(\theta, d)$ is called the *risk function.*

Some use the term *risk function* to denote the expected value of the *regret* function. If the loss function used is already a regret function, as will be the case in many of the important statistical applications to be discussed, there is no difference in usage. It is interesting that the expected regret is the same as what would be obtained by applying the "regret-izing" process to the risk function:

$$E_\theta[r(\theta, d(Z))] = R(\theta, d) - \min_d R(\theta, d).$$

(To show this is an exercise in inequalities and minima.)

The data have thus been used to change the problem of the statistician from one of selecting an action in view of a certain loss function, to one of selecting a decision function d in view of a certain risk function. Interpreting an "action" as a *decision function* and "loss" as *risk*, the problem is now like the original no-data problem, and the concepts developed for no-data problems can be applied to the present situation.

In particular, minimax and Bayes decision functions can be defined. The decision function d^* that minimizes

$$M(d) \equiv \max_\theta R(\theta, d)$$

is the minimax rule, and the minimum maximum $M(d^*)$ can be compared with the minimum maximum loss for no-data rules to determine the worth of the data. Similarly, the function d^{**} that minimizes the *Bayes risk*,

$$B(d) \equiv E[R(\Theta, d)]$$

is the Bayes rule, and the minimum Bayes risk $B(d^{**})$ can be compared with the minimum Bayes loss for no-data rules to determine the worth of the data in the Bayesian approach (for a given prior).

EXAMPLE 8–13. Consider the rain problem of Example 8–3, with loss function as follows:

		θ_1 (rain)	θ_2 (no rain)
(Stay home)	a_1	4	4
(Go, no umbrella)	a_2	5	0
(Go with umbrella)	a_3	2	5

A single observation of a rain indicator (or a weather report) will make up the "datum" Z. If this datum is to be useful, one must know the way in which the distribution of the datum depends on the state of nature. For this problem, assume the following, in which z_1 indicates an observation (or prediction) of "rain," and z_2, of "no rain":

	θ_1 (rain)	θ_2 (no rain)
z_1 (rain)	.8	.1
z_2 (no rain)	.2	.9

That is, assume as known from past experience that when it is going to rain, the probability is .8 that the indicator will show "rain," and .2 that it will show "no rain."

Since there are finitely many possible actions and possible values of Z, the number of decision functions is finite. Indeed, it is $3^2 = 9$, since there are three actions and two values of Z (a decision function assigns an action to each value of Z). These nine decision functions are listed as follows:

	d_1	d_2	d_3	d_4	d_5	d_6	d_7	d_8	d_9
If outcome is z_1, take action	a_1	a_2	a_3	a_1	a_2	a_1	a_3	a_2	a_3
If outcome is z_2, take action	a_1	a_2	a_3	a_2	a_1	a_3	a_1	a_3	a_2

The computation of the risk function $R(\theta, d)$ proceeds as follows: For $\theta = \theta_1$ and $d = d_5$,

$$\begin{aligned} R(\theta_1, d_5) &= E_{\theta_1}[\ell(\theta_1, d_5(Z))] \\ &= \ell(\theta_1, d_5(z_1))p_{\theta_1}(z_1) + \ell(\theta_1, d_5(z_2))p_{\theta_1}(z_2) \\ &= \ell(\theta_1, a_2) \times .8 + \ell(\theta_1, a_1) \times .2 = 5 \times .8 + 4 \times .2 = 4.8. \end{aligned}$$

In similar fashion, one can compute $R(\theta, d)$ for each combination of decision function d and state θ. The results are tabulated:

	θ_1	θ_2
d_1	4	4
d_2	5	0
d_3	2	5
d_4	4.2	0.4
d_5	4.8	3.6
d_6	3.6	4.9
d_7	2.4	4.1
d_8	4.4	4.5
d_9	2.6	0.5

This array can now be attacked as though it gave the loss function in a no-data problem. For instance, d_1, d_5, d_6, and d_8, are *inadmissible*, even in the set of pure actions, and can be eliminated from the competition. Similarly, d_4 and d_7 are dominated by mixtures of other actions, but there is no clear choice among d_2, d_3, and d_9.

Figure 8–14 illustrates the effect of the data. Points (R_1, R_2) are plotted for each decision function, where $R_i = R(\theta_i, d)$. The original no-data rules appear as before, but called here d_1, d_2, and d_3; the set of no-data mixtures is shown as the shaded triangle. The mixed decision rules available when using the data Z are represented by the points in the set outlined by the larger polygon. These rules include some that dominate most of the no-data actions, and some that are worse—because they use the data in a contrary fashion. Improving the loss picture through the use of data is possible here because the data are related to the state of nature. (See Problem 8–20.)

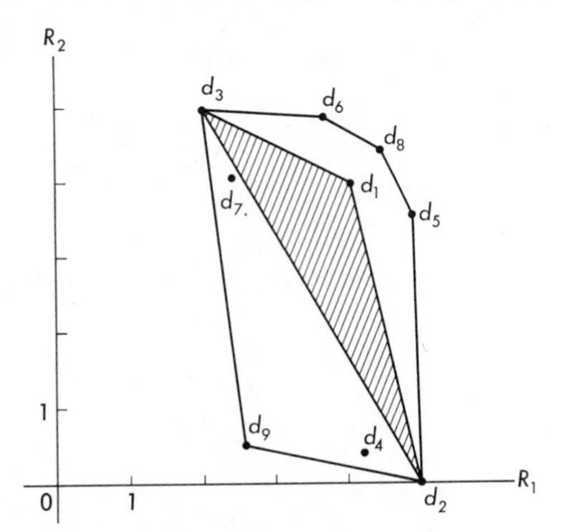

Figure 8–14. Decision rules for Example 8–13.

The minimax rule is the boundary point first met by the line $R_1 = R_2$, seen in Figure 8–14 to be the mixture $(\frac{7}{17}, \frac{10}{17})$ of the rules d_3 and d_9. The minimax risk is $\frac{40}{17}$, as compared with the minimax loss of $\frac{25}{8}$ using no-data rules (cf. Example 8–7). The data Z is "worth" $\frac{25}{8} - \frac{40}{17}$ in using the minimax approach.

The Bayes rule for the prior $(\frac{1}{2}, \frac{1}{2})$, for example, is readily seen from Figure 8–14 to be d_9, for which the Bayes risk is $(2.6 + .5)/2 = 1.55$. The no-data Bayes rule for the same prior is d_2, with Bayes loss 2.5; so the data would be worth 2.5–1.55 in this case.

EXAMPLE 8–14. The problem is to estimate the value of the parameter θ in an exponential density with expected value θ. That is, it is assumed that nature is in one of the states described by this family of distributions. The actions here are the various possible choices of an estimated value of θ, so the action space is the same as the state space. For purposes of this example, assume that the loss function is quadratic and has a zero minimum for each action:

$$\ell(\theta_0, a) = (a - \theta_0)^2,$$

the loss incurred upon choosing a as the estimate when the state is θ_0.

A decision procedure is to be based on the result of n performances of the basic exponential experiment, a sample of size n from the population. As illustrations of the computation of risks, consider the following (arbitrarily selected) decision procedures:

$$d_1(\mathbf{X}) = \bar{X}, \qquad d_2(\mathbf{X}) = \bar{X} + 1, \qquad d_3(\mathbf{X}) = X_n.$$

The risk functions for these procedures are as follows:

$$R(\theta, d_1) = E_\theta(\bar{X} - \theta)^2 = \text{var } \bar{X} = \frac{\theta^2}{n}.$$

$$R(\theta, d_2) = E_\theta(\bar{X} + 1 - \theta)^2 = \text{var } \bar{X} + 1 = \frac{\theta^2}{n} + 1.$$

$$R(\theta, d_3) = E_\theta(X_n - \theta)^2 = \text{var } X = \theta^2.$$

These are plotted as functions of θ in Figure 8–15. It is clear from the figure that d_1 dominates both d_2 and d_3 strictly but that d_2 and d_3 are not comparable with each other. It is perhaps a little surprising (cf. Problem 8–21) that d_1 is itself not the best possible estimator, even though it is efficient.

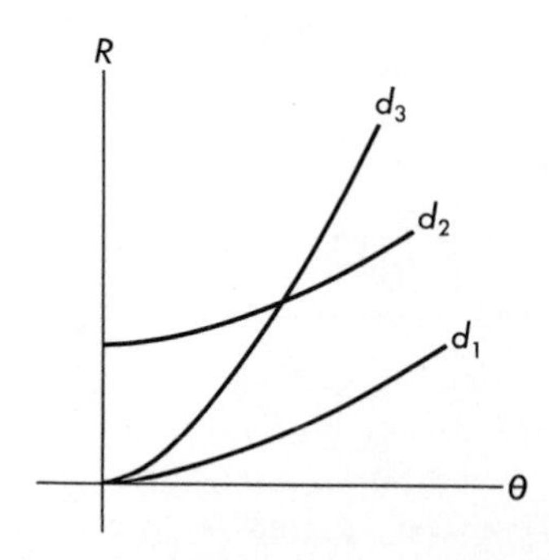

Figure 8–15. Risk functions for Example 8–14.

EXAMPLE 8–15. Flash bulbs come in "lots" of ten, and one is taken from each lot and tested (a process that ruins it). The remaining nine are either to be sold at 15 cents each, with a double your money back guarantee, or to be junked at a cost of 10 cents for the lot. Thus, one of two actions is to be selected: Sell the lot or junk it. The state of nature is described by the number k of defective bulbs among the ten.

Since the experiment has only two outcomes, assigning one of the two actions to each outcome can be done in four distinct ways; these are the possible decision functions:

d_1: Sell the lot if the test bulb is good; junk if defective.
d_2: Junk the lot if the test bulb is good; sell if defective.
d_3: Sell the lot if the test bulb is good; sell if defective.
d_4: Junk the lot if the test bulb is good; junk if defective.

Consider first $d_1(X)$, where X denotes the outcome of the test. The loss is

$$\ell(k, d_1(X)) = \begin{cases} -1.35 + (.30)k, & \text{if } X = \text{good}, \\ .10, & \text{if } X = \text{defective}, \end{cases}$$

and the risk function is then

$$\begin{aligned} R(k, d_1) &= E_k[\ell(k, d_1(X))] \\ &= (-1.35 + .3k)\left(1 - \frac{k}{10}\right) + .10\left(\frac{k}{10}\right) \\ &= -.01(3k^2 - 44.5k + 135). \end{aligned}$$

For $d_2(X)$,

$$\ell(k, d_2(X)) = \begin{cases} .10, & \text{if } X = \text{good}, \\ -1.35 + .30(k - 1), & \text{if } X = \text{defective}, \end{cases}$$

and

$$R(k, d_2) = .01(3k^2 - 17.5k + 10).$$

Similarly,

$$\begin{aligned} R(k, d_3) &= -.09(15 - 3k), \\ R(k, d_4) &= .10. \end{aligned}$$

These four risk functions are plotted (as though k were continuous, for ease in following the graphs) in Figure 8–16. For some states the best decision function is d_3; for others it is d_2; and for still others it is d_4. And d_1, which is never best, is not far from being best at all states.

From Figure 8–16 it is apparent that d_4 has the risk function with smallest maximum value, .10. When no data are taken, the losses are

$$\begin{aligned} \ell(k, \text{sell}) &= .30k - 1.50, \\ \ell(k, \text{junk}) &= .10, \end{aligned}$$

and the minimum maximum loss is again .10, so the data do not reduce the minimax loss and are pointless. Moreover, it would be an unusual manufacturer who would make flash bulbs and then junk everything made for fear of the large loss he might incur if the lot fraction defective were quite high. Rather, the manufacturer would develop a process so that a lot with more than, say, half-defective bulbs

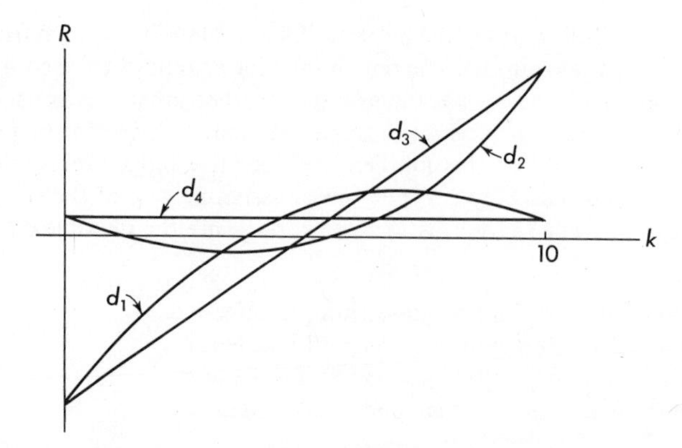

Figure 8–16. Risk functions for Example 8–15.

would "practically" never occur. More realistically, then, one might consider the maximum risk over small values of k, in which case the smallest maximum would be achieved for d_3. But this focusing of the attention on certain states deemed practically possible is perhaps better handled using the Bayes approach.

Suppose a prior distribution is assumed that is uniform on the three values $k = 0, 1, 2$. The Bayes risk using d_1 is

$$B(d_1) = -.01 \sum_{k=0}^{2} \tfrac{1}{3}(3k^2 - 44.5k + 135) = .955,$$

and one finds similarly,

$$B(d_2) = -.025, \qquad B(d_3) = -1.08, \qquad \text{and} \qquad B(d_4) = .10.$$

The minimum of these Bayes risks is -1.08, achieved by using decision rule d_3. The Bayes losses for the no-data actions are

$$B(\text{sell}) = E(.30k - 1.50) = -1.20,$$
$$B(\text{junk}) = E(.10) = .10,$$

and the minimum is -1.20. Since this is actually less than the minimum risk using the data rules, it is cheaper not to make the test observation with the prior that was assumed.

In this situation, it would make sense to think of the population parameter k as a binomial variable—the number of defectives in a sample of 10 flash bulbs from a production line that is "in control," putting out bulbs with a constant probability p_0 of a defective. With a binomial prior for k, with parameters $n = 10$ and $p = p_0$, the Bayes risks are easily calculated using $E(k) = 10p_0$ and $E(k^2) = \text{var}\, k + [E(k)]^2 = 10p_0(1 - p_0) + (10p_0)^2$:

$$\begin{aligned} B(d_1) &= -.01E(3k^2 - 44.5k + 135) \\ &= -.05(54p_0^2 - 83p_0 + 27). \end{aligned}$$

Similarly,

$$B(d_2) = .05(54p_0^2 - 29p_0 + 2),$$
$$B(d_3) = -1.35(1 - 2p_0),$$
$$B(d_4) = .10.$$

If p_0 is not large (or even just less than $\frac{1}{2}$), the Bayes rule is d_3 and it does not pay to test. However, if k is uniform on $k = 0, 1, \ldots, 10$, the Bayes rule is d_1, and it does pay to test (Problem 8–24).

Problems

8–20. Calculate the risk table and plot the corresponding points, for the rain problem of Example 8–13 when there is available

(a) An observation Y, with two possible values, that is perfectly related to the state of nature:

$$P(Y = y_1 \mid \theta_1) = P(Y = y_2 \mid \theta_2) = 1.$$

(Conclusion?)

(b) An observation X that has a distribution independent of θ: $f(x; \theta_1) = f(x; \theta_2) = p$. (Can this result be generalized?)

8–21. Refer to Example 8–14 and show that $n\bar{X}/(n + 1)$ has a risk function that is uniformly smaller (on $0 < \theta < \infty$) than the risk function for $\bar{X}$.

8–22. Suppose θ and a are numerical variables on [0, 1]. Given the loss function $(\theta - a)^2$, determine the risk of each of the following rules when Z is binomial with parameters $n = 2$ and $p = \theta$.

(a) $d_1 = \dfrac{Z}{2}$. (b) $d_2 = \dfrac{(Z + 1)}{4}$.

Which rule is better? Which rule is better if one takes into account his beliefs by a uniform distribution for θ?

8–23. Given losses in this table:

	θ_1	θ_2	θ_3
a_1	0	4	5
a_2	3	2	6
a_3	4	0	3

and an observation Z with distributions as follows:

	θ_1	θ_2	θ_3
z_1	.2	.4	.7
z_2	.8	.6	.3

(a) Determine the Bayes rule and the value of the data assuming a uniform prior for the three states.

(b) Determine the minimax (pure) decision function.

(c) Determine the maximum likelihood rule.

8–24. In Example 8–15, verify the claim at the end, that the Bayes rule for a uniform prior is d_1. How much is the observation worth?

★8.3.2 Randomized Decision Rules

An extraneous experiment of chance has been introduced, for mathematical convenience, both into the problem of testing hypotheses—as a device for choosing an action after making an observation, and into the decision problem—as a device for choosing randomly among competing decision functions. Because the testing of hypotheses can be considered as a two-action decision problem (a_1: accept H_0, a_2: reject H_0), it is natural to wonder if there is a connection between the two ways of introducing randomness. As reflection on Problem 6–20 might suggest, there is sometimes an equivalence.

Definition. *A* behavioral decision rule *is an assignment of a probability distribution on the action space $\mathscr{A}$ to each outcome z of the data Z. It is carried out by sampling to obtain $Z = z$ and then performing the experiment in $\mathscr{A}$ assigned to that z-value by the rule, to see what action to take.*

Definition. *A* mixed decision rule *is a distribution on the space of decision functions $d(z)$. It is carried out by performing an experiment with that distribution to obtain a decision rule $d(\cdot)$, obtaining data $Z = z$, and then taking the action assigned to z by the rule $d(\cdot)$.*

EXAMPLE 8–16. Consider a decision problem with two available actions, a_1 and a_2, and an observation Z with three values: z_1, z_2, and z_3. The eight (2^3) decision rules are given in the following table.

	d_1	d_2	d_3	d_4	d_5	d_6	d_7	d_8
z_1	a_1	a_2	a_1	a_2	a_2	a_1	a_1	a_2
z_2	a_1	a_2	a_2	a_1	a_2	a_1	a_2	a_1
z_3	a_1	a_2	a_2	a_2	a_1	a_2	a_1	a_1

A mixed decision function is defined by a probability vector $(q_1, q_2, \ldots, q_8)$, where q_i is the probability assigned to the rule d_i. Given such a mixed rule, one can make the following calculations:

$$P(a_1 \mid z_1) = q_1 + q_3 + q_6 + q_7 \equiv \phi_1(z_1).$$
$$P(a_1 \mid z_2) = q_1 + q_4 + q_6 + q_8 \equiv \phi_1(z_2).$$
$$P(a_1 \mid z_3) = q_1 + q_5 + q_7 + q_8 \equiv \phi_1(z_3).$$
$$P(a_2 \mid z_1) = q_2 + q_4 + q_5 + q_8 \equiv \phi_2(z_1).$$
$$P(a_2 \mid z_2) = q_2 + q_3 + q_5 + q_7 \equiv \phi_2(z_2).$$
$$P(a_2 \mid z_3) = q_2 + q_3 + q_4 + q_6 \equiv \phi_2(z_3).$$

The functions ϕ_1 and ϕ_2 so defined, with $\phi_1 + \phi_2 = 1$, define a behavioral rule.

Conversely, suppose that a behavioral rule is given. To be specific, let it be defined as follows:

z	z_1	z_2	z_3
$\phi_1(z)$	.2	.4	.7
$\phi_2(z)$	.8	.6	.3

If there is to be a mixed rule $\mathbf{q}$ equivalent to this, it must have components satisfying the six equations obtained by putting the specified ϕ-values into the six equations above:

$$q_1 + q_3 + q_6 + q_7 = .2, \qquad \text{(etc.)}$$

These six equations in eight unknowns have many solutions. For instance, if one arbitrarily sets $q_1 = q_2 = 0$, he finds

$$q_3 = q_6 = q_7 = \frac{.2}{3}, \qquad q_8 = q_4 = \frac{.5}{3}, \qquad q_5 = \frac{1.4}{3}.$$

More systematically, a solution can be constructed as a mixture $d(z)$ such that $d(z_1)$, $d(z_2)$, and $d(z_3)$ are independent:

$$\begin{cases} q_1 = P(a_1 \mid z_1)P(a_1 \mid z_2)P(a_1 \mid z_3) = .2 \times .4 \times .7 = .056 \\ \vdots \\ q_8 = P(a_2 \mid z_1)P(a_1 \mid z_2)P(a_1 \mid z_3) = .8 \times .4 \times .7 = .224. \end{cases}$$

It can be verified that the mixture obtained in this way:

$$(.056, .144, .036, .096, .336, .024, .084, .224)$$

satisfies the six equations in the eight q's given.

Equivalence of a behavioral and a mixed decision rule would mean that the risk function is the same. For a behavioral decision rule, the risk function is the expected loss:

$$E^{\mathscr{Z}|\theta}\{E^{\mathscr{A}|Z}[\ell(\theta, A)]\},$$

where the superscript $\mathscr{A} \mid Z$ means that for given Z the average is computed with respect to the distribution in $\mathscr{A}$ that defines the rule; and then this function of the random variable Z is averaged with respect to the distribution in the sample space defined by the parameter value θ.

For a mixed decision rule the risk, or expected loss, is

$$E^{\mathscr{D}}\{E^{\mathscr{Z}|\theta}[R(\theta, d(Z))]\},$$

where the superscript $\mathscr{Z} \mid \theta$ means that for a given rule $d(\cdot)$ the loss is averaged over the sample space. The result is then averaged over the space $\mathscr{D}$ of decision functions. The notion of a distribution on a function space is beyond our scope, so the rather (but not completely) general equivalence of behavioral and mixed rules will not be proved. But in the case of finite sample spaces and finitely many actions, the equivalence can be demonstrated much as in the example above.

Suppose that $\mathscr{A} = \{a_1, \ldots, a_m\}$ and $\mathscr{Z} = \{z_1, \ldots, z_k\}$, in which case the number of decision functions is $m^k \equiv r$. A *mixed* decision rule is then defined by a probability vector $(q_1, \ldots, q_r)$. For a given rule $d_i(z)$, define

$$d_{ij}(z) = \begin{cases} 0, & \text{if } d_i(z) \neq a_j, \\ 1, & \text{if } d_i(z) = a_j \end{cases}$$

and observe that

$$\phi_j(z) \equiv P(A = a_j \mid z) = \sum_i d_{ij}(z) q_i,$$

for $j = 1, \ldots, m$. These equations for the mk ϕ's in terms of the m^k q's define a behavioral rule (that is, they define the ϕ's). The risk for the mixed rule $\mathbf{q}$ is

$$\begin{aligned} R(\theta, \mathbf{q}) &= \sum_i \sum_z \ell(\theta, d_i(z)) f(z \mid \theta) q_i \\ &= \sum_i \sum_j \ell(\theta, a_j) \left\{ \sum f(z \mid \theta) \right\} q_i, \end{aligned}$$

where the inner sum is taken over the set S_{ij} of z's such that $d_i(z) = a_j$:

$$\sum_{S_{ij}} f(z \mid \theta) = \sum_z d_{ij}(z) f(z \mid \theta).$$

Interchanging the order of summations yields

$$\begin{aligned} R(\theta, \mathbf{q}) &= \sum_j \ell(\theta, a_j) \left\{ \sum_z \left(\sum_i d_{ij}(z) q_i \right) f(z \mid \theta) \right\} \\ &= \sum_j \ell(\theta, a_j) \sum_z \phi_j(z) f(z \mid \theta), \end{aligned}$$

which is the risk function corresponding to the behavioral rule defined by the ϕ's.

To define a mixed rule equivalent to a given behavioral rule $(\phi_1, \ldots, \phi_k)$, let the mixed rule be identified by a vector subscript $\mathbf{j} = (j_1, \ldots, j_k)$, where each component is one of the integers $1, 2, \ldots, m$. There are $m^k = r$ such vector subscripts, each identifying one of the r possible mixed rules, as follows:

$$d_{\mathbf{j}}(z) = \begin{cases} a_{j_1}, & \text{if } z = z_1 \\ \vdots & \\ a_{j_k}, & \text{if } z = z_k. \end{cases}$$

Then define the mixture as a probability vector $\mathbf{q}$ with components

$$q_{\mathbf{j}} \equiv \phi_{j_1}(z_1) \cdots \phi_{j_k}(z_k).$$

It can be checked that these add up to 1 and that, starting with this mixture, one is led back to the given $\phi_v(z)$ as the (marginal) probability $P(A = a_v \mid z)$.

8.3.3 Estimation and Testing as Special Cases

As suggested in Example 8–14, a problem of estimation can be thought of as a decision problem in which the state space Ω is identical with the action space $\mathscr{A}$. That is, the "action" chosen is a value from the parameter space, a value that is announced as the estimate of the true parameter value. Any of various loss functions might be adopted, but it seems natural to require that $\ell(\theta, \theta) = 0$, and that for each θ the loss is an increasing function of the error, $|\theta - a|$. Simple functions of this type are $|\theta - a|$ itself, or powers of this error, and in particular the second power, called *quadratic loss*:

$$\ell(\theta, a) \equiv (\theta - a)^2.$$

(More generally, one might want to permit a multiplicative function of θ in front of the quadratic.) With this loss, the risk function is just the mean squared error:

$$R(\theta, T) = E(\theta - T)^2 = \text{m.s.e.}$$

Strictly speaking, quadratic loss cannot be measured in units of utility if utility is required to be bounded. The maximum risk would be infinite for all T, and the minimax principle yields nothing; for such considerations, the loss would have to be modified, but otherwise, the unboundedness of quadratic loss does not seem to be troublesome.

An excuse often given for using quadratic loss in estimation—essentially assessing estimators according to variance—is that the mathematical manipulations involved are simpler than with other loss functions. This excuse only carries weight if there are other reasons as well. One such reason will now be given. It applies only to the case of *normal, unbiased* estimates. It will be shown that for such estimators and a wide class of loss functions, a comparison of estimators on the basis of risk computed with quadratic loss is equivalent to a comparison based on risk computed with other loss functions.

Assume, then, that the loss function is of the type in which the loss is an increasing function of $|\theta - a|$ with the property $\ell(\theta, \theta) = 0$. It is then a regret function; let the risk based on this regret be denoted $R(\theta, T)$. The risk based on quadratic regret is just var T, when T is unbiased. Consider a normal estimator T and another normal estimator T' that is not as good as T in terms of variance:

$$\text{var } T' > \text{var } T.$$

Then

$$R(\theta, T) = E[\ell(\theta, T)] = \int_{-\infty}^{\infty} \ell(\theta, t)\, dF_T(t)$$

$$= \int_0^1 \ell(\theta, F_T^{-1}(u))\, du,$$

the last integral being obtained by a change of variable, $F_T(t) = u$. Similarly,

$$R(\theta, T') = \int_0^1 \ell(\theta, F_{T'}^{-1}(u))\, du.$$

Now if T and T' are both unbiased and normal, the inverse functions both have the value θ at $u = \frac{1}{2}$; and if var $T' >$ var T, these inverse functions are as pictured in Figure 8–17. Observe in this figure that

$$F_{T'}^{-1}(u) < F_T^{-1}(u) < \theta, \qquad \text{for } u < \tfrac{1}{2},$$

and

$$F_{T'}^{-1}(u) > F_T^{-1}(u) > \theta, \qquad \text{for } u > \tfrac{1}{2}.$$

In either case, and hence for the whole interval $0 < u < 1$,

$$\ell(\theta, F_T^{-1}) < \ell(\theta, F_{T'}^{-1}),$$

from which it follows that $R(\theta, T) < R(\theta, T')$. That is, the estimator T' which was assumed worse than T using quadratic regret is also worse than T when compared in terms of the more general loss function.

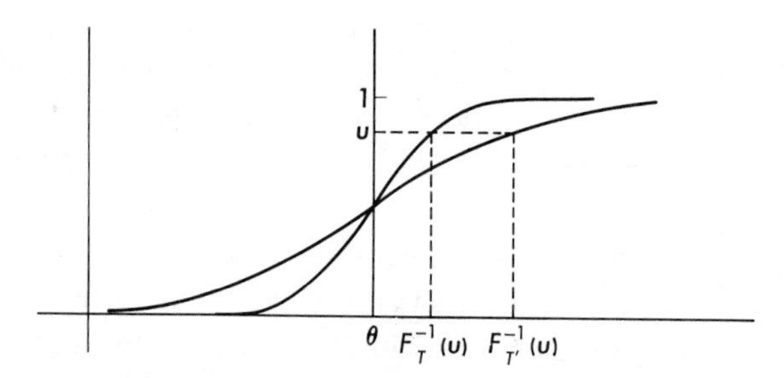

Figure 8–17

An hypothesis-testing problem can be viewed as a decision problem in which the available actions are A: accept H_0, and B: reject H_0—or perhaps more specific actions that would be appropriate when H_0 is known to be true (A), or known to be false (B). Conversely, a two-action problem with actions A and B defines hypotheses H_0 and H_1 according to a given loss structure. The corresponding regret function is 0 for (at least) one action for each state θ; that action is then correct if the state is θ. So the state space Ω is divided into two subsets—the set of states for which A is correct, called H_0, and the set of states for which B is correct, called H_1. Thus, the regret function is of this form:

$$r(\theta, A) = \begin{cases} 0, & \text{for } \theta \text{ in } H_0, \\ b(\theta), & \text{for } \theta \text{ in } H_1, \end{cases}$$

$$r(\theta, B) = \begin{cases} a(\theta), & \text{for } \theta \text{ in } H_0, \\ 0, & \text{for } \theta \text{ in } H_1, \end{cases}$$

where $a(\theta)$ and $b(\theta)$ are nonnegative functions.

For given data Z, a decision function assigns to each Z-value either the action A or the action B. The critical region C (as before) is the set of Z-values assigned the action B (reject H_0), and the risk function or expected regret is then computed as follows:

$$R(\theta, C) = \begin{cases} a(\theta)\pi_C(\theta), & \text{for } \theta \text{ in } H_0, \\ b(\theta)[1 - \pi_C(\theta)], & \text{for } \theta \text{ in } H_1, \end{cases}$$

where (as in Chapter 6)

$$\pi_C(\theta) = P_\theta(C) = P_\theta(\text{reject } H_0).$$

EXAMPLE 8–17. Consider a two-action (A and B) testing problem with losses (see Figure 8–18)

$$\ell(\theta, B) = \begin{cases} 4 - 8\theta, & \text{for } \theta \le .5, \\ 0, & \text{for } \theta > .5, \end{cases}$$

$$\ell(\theta, A) = \begin{cases} 0, & \text{for } 0 \le .5, \\ 4\theta - 2, & \text{for } \theta > .5. \end{cases}$$

This amounts to testing $H_0: \theta \le \frac{1}{2}$ against $H_1: \theta > \frac{1}{2}$. If θ is the mean of a normal population with unit variance and the test is based on a random sample of size 25, the natural type of test to use is of the form $\bar{X} > K$. In particular, if C_1 is $\bar{X} > .5$ and C_2 is $\bar{X} > .6$, then

$$\pi_{C_1}(\theta) = \Phi(5\theta - 2.5)$$

and

$$\pi_{C_2}(\theta) = \Phi(5\theta - 3).$$

The risk functions, computed from the formula just preceding this example, are shown in Figure 8–19. Observe that neither critical region dominates the other.

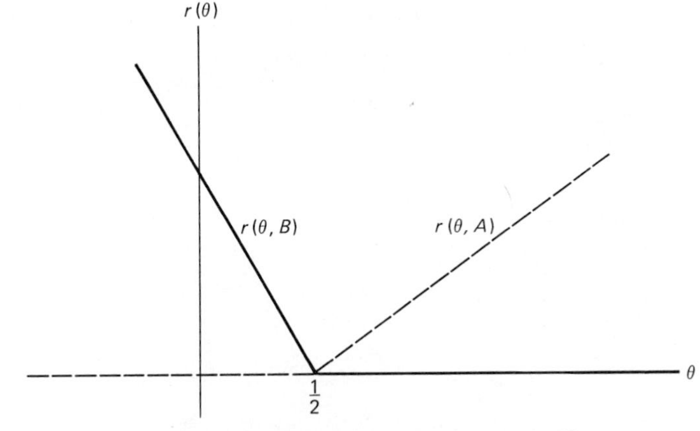

Figure 8–18. Losses for Example 8–17.

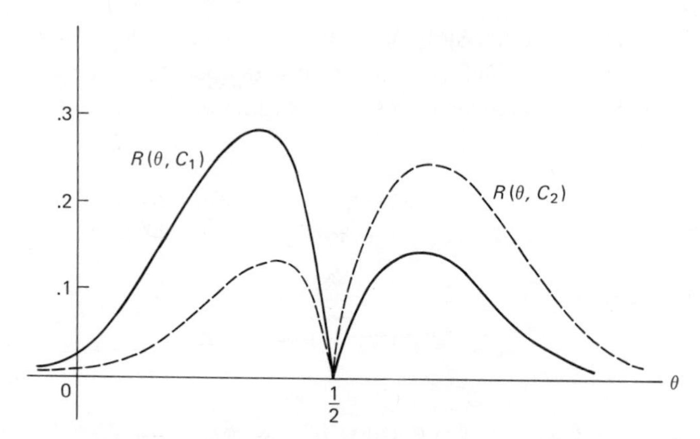

Figure 8–19. Risk functions for Example 8–17.

In the case of simple H_0 and simple H_1, the functions $a(\theta)$ and $b(\theta)$ defining the regret are just constants, and the risk function for the critical region C is

$$R(\theta, C) = \begin{cases} a\alpha, & \text{if } \theta = H_0, \\ b\beta, & \text{if } \theta = H_1. \end{cases}$$

Thus the risks are just constants times the error sizes, and the graphical representation of decision rules C in the two-state case by points (R_0, R_1), where $R_i = R(H_i, C)$, is similar to the representation of tests by means of the (α, β)-plot. The axes are simply stretched by the factors a and b that give the costs of wrong decisions—that is, or errors of types I and II, respectively.

EXAMPLE 8–18. Consider testing $\mu = 0$ against $\mu = 1$ in a normal population with variance 4, using four observations. The (R_0, R_1)-plots of the most powerful tests $\bar{X} > K$ are shown in Figure 8–20, assuming losses $a = 3$ and $b = 1$. Marked on the graph are the points representing the minimax test (given by $K = .98$, with $3\alpha = \beta = .49$) and the Bayes test corresponding to the prior probabilities (.2, .8), (given by $K = .213$, with $\alpha = .416$ and $\beta = .216$).

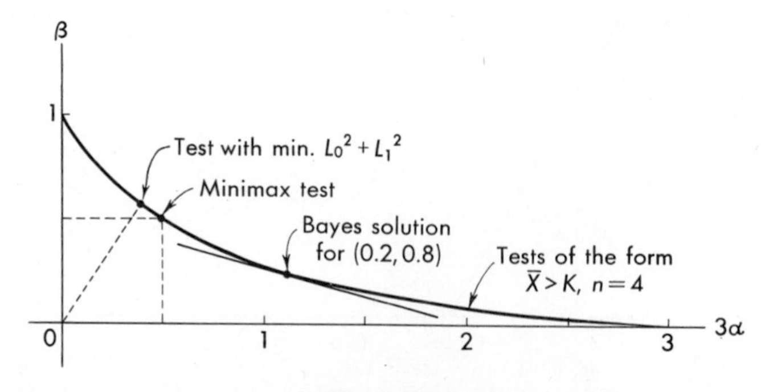

Figure 8–20

8.3.4 Properties of Decision Rules

Since estimation and the testing of hypotheses can be considered as decision problems, one might expect to generalize the various properties of procedures encountered in these special cases. First, let it be noted that the definition of admissibility given earlier for the no-data problem applies in the general case with data, when it is given in terms of the risk function. Thus, a rule or procedure d dominates the rule d_1, provided that

$$R(\theta, d_1) \geq R(\theta, d), \qquad \text{for all } \theta.$$

A procedure would be called *inadmissible* if it were strictly dominated by some other procedure. However, whether this is bad or not depends on the degree to which it is dominated, as discussed in Section 8.2.5.

If two procedures d and d' are both based on samples of a given size, and if d dominates d' strictly, d can be thought of as more *efficient* than d', since in general it would take more observations using d' to equalize the risks. The relative efficiency of d' with respect to d might be defined as the ratio of the risk of d to that of d' for a given sample size, or as the ratio of sample sizes needed to make the risks equal, if this is possible. (The ratios need not agree and would usually depend on the state θ.) Further, the risk may have a lower bound or minimum achievable risk (in restricted classes of procedures) such as was used as the basis of a measure of "absolute" efficiency in Chapter 5.

A procedure d is said to be *consistent* if the expected regret tends to zero as the sample size becomes infinite. This implies that the "procedure" includes instructions as to how to apply it to samples of different sizes. That is, consistency applies to a *sequence* of procedures, one for each sample size. Presumably, if the expected regret tends to 0, it will be small for "large" samples; this is of little comfort if a large sample is expensive.

The mean square consistency of an estimator as defined in Chapter 5 is a special case of the above general definition when the loss function is quadratic, because in such a case the risk function is just the mean squared error. However, the definition of a consistent test of hypotheses as given in Chapter 6 is not a special case, since it calls for convergence to 0 of the risk only for alternative hypotheses, with α held fixed.

A procedure based on the statistic Z is said to be *unbiased* if for all states of nature θ and θ',

$$E_\theta[\ell(\theta', d(Z))] \geq E_\theta[\ell(\theta, d(Z))],$$

where again the subscript on E denotes the state to be used in computing the expectation. This can be interpreted to sound very plausible: Thinking of

$\ell(\theta, a)$ as related to how close the action a comes to the action that should correspond to θ, the relation states that the decision procedure d comes closer *on the average* to the correct decision than to an incorrect one. Yet, despite the appealing sound of this statement, fulfillment of the condition of unbiasedness by no means guarantees a good procedure.

This general definition of unbiasedness includes as special cases the definitions given previously for estimation and testing. In the case of estimation, with the assumption of a quadratic loss $(T - \theta)^2$ for an estimator T of θ, the condition for unbiasedness becomes

$$E_\theta[(T - \theta_1)^2] \geq E_\theta[(T - \theta)^2], \qquad \text{for all } \theta, \theta_1.$$

The condition is satisfied by an estimator T whose mean is θ, since the right member is then the variance of T, which is indeed the smallest second moment of T. Conversely, if the condition is satisfied, it follows (upon expanding the squares, averaging term by term, canceling, and forming squares again) that

$$[E_\theta(T) - \theta_1]^2 \geq [E_\theta(T) - \theta]^2, \qquad \text{all } \theta, \theta_1;$$

but then it follows, substituting $\theta_1 = E_\theta(T)$, that the right side must be zero, which implies that $E_\theta(T) = \theta$. That is, under the assumption of a quadratic loss function, the property of unbiasedness becomes equivalent to the property that the distribution of the estimator T be "centered" at the parameter θ in the sense of expectation.

Using the loss function $|T - \theta|$, one would find that the condition of unbiasedness is equivalent to the condition that the distribution of the estimator T be centered at θ in the sense of the median, that is, that the median of T be θ. An estimator with this property is sometimes called *median unbiased.*

In the case of hypothesis testing, the special definition of unbiasedness can be included in the general one if the losses are assumed to be constant on H_0 and on H_1: $a(\theta) \equiv a$ and $b(\theta) \equiv b$. Unbiasedness of a procedure $d(\mathbf{X})$ would imply that if θ is in H_0 and θ' is in H_0,

$$E_\theta[\ell(\theta', d(\mathbf{X}))] = a\pi(\theta) \geq E_\theta[\ell(\theta, d(\mathbf{X}))] = a\pi(\theta).$$

For θ in H_0 and θ' in H_1, it implies that

$$E_\theta[\ell(\theta', d(\mathbf{X}))] = b[1 - \pi(\theta)] \geq E_\theta[\ell(\theta, d(\mathbf{X}))] = a\pi(\theta).$$

The first of these conditions is not a restriction, and so for θ in H_0,

$$b[1 - \pi(\theta)] \geq a\pi(\theta).$$

Similarly, for θ in H_1, unbiasedness requires that

$$a\pi(\theta) \geq b[1 - \pi(\theta)].$$

That is, it must be that

$$\begin{cases} \pi(\theta) \le \dfrac{b}{a+b}, & \text{for } \theta \text{ in } H_0, \\ \pi(\theta) \ge \dfrac{b}{a+b}, & \text{for } \theta \text{ in } H_1. \end{cases}$$

The power function, therefore, must be smaller for any θ in H_0 than for any θ in H_1. This last condition is the usual definition of an unbiased test; if it is fulfilled, then there are losses a and b such that $b/(a+b)$ is an upper bound for $\pi(\theta)$ on H_0 and a lower bound for $\pi(\theta)$ on H_1, in which case the general definition of an unbiased procedure is fulfilled.

★8.3.5 Monotone Problems and Procedures

Many statistical problems are concerned with families of distributions indexed by a *single* real parameter. Of these, some are of a type called *monotone* —problems in which the loss function has a monotone character and in which the family of distributions has a monotone likelihood ratio.

A regret function $r(\theta, a)$ defines a function of θ for each action—a family of regret curves each identified by a particular action. These curves will lie above the θ-axis, except that some will be 0 for certain states θ. When a regret is 0, the corresponding action is optimal or best for that state. The set of θ-values for which a given regret curve is 0 will be called the *optimality set* for that action. A regret function $r(\theta, a)$ will be called a monotone regret function if the regret curve it defines satisfies the following conditions:

1. The optimality sets for the various actions are intervals (finite, infinite, single point, or empty) which constitute a partition of the parameter space.
2. For each action, the regret curve does not decrease as θ moves away from the optimality set in either direction.
3. Any two regret curves intersect at most once. [That is, the difference $r(\theta, a_1) - r(\theta, a_2)$ changes sign at most once for a given pair of actions a_1 and a_2.]

Figure 8–21 gives three examples of a monotone regret function. In Figure 8–21 (a) a quadratic regret $(\theta - a)^2$ is shown for the estimation problem, although curves are drawn for only three values of a. For each action a, there is just one parameter value ($\theta = a$) for which a is the correct action, and these points of zero regret fill out the θ axis; it is also clear that the other two conditions are fulfilled. Figures 8–21 (b) and (c) show monotone regret functions for the two- and three-action cases, respectively.

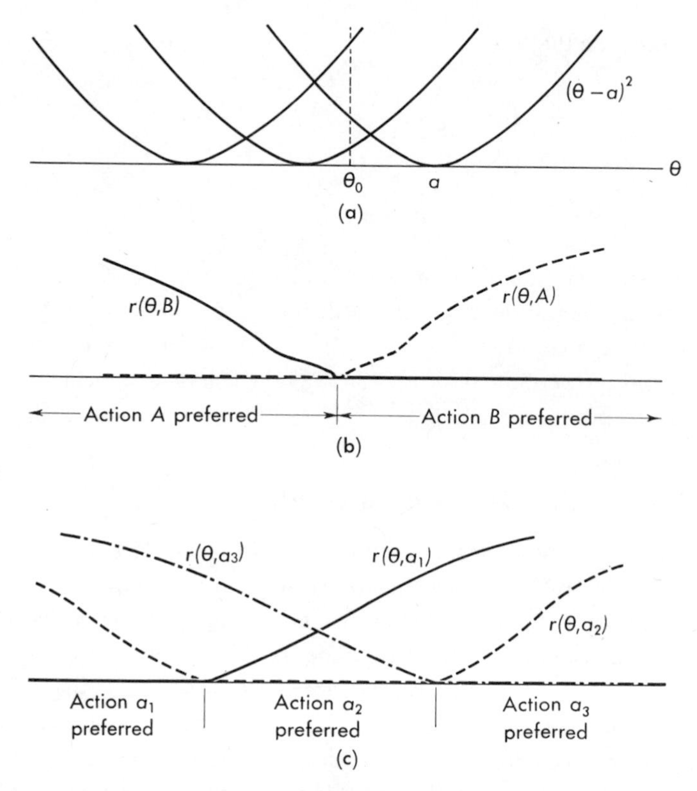

Figure 8–21. Some monotone regret functions.

In Figure 8–22 regret functions are shown that are *not* monotone. In Figure 8–22(a), the first two conditions are fulfilled but condition (3) is not, since $r(\theta, a_1) - r(\theta, a_2)$ changes sign twice. In Figure 8–22 (b), all three conditions are violated.

EXAMPLE 8–19. Lots of a certain product (say, eggs) are to be classified into three grades according to the average lot quality θ. That is, if $\theta > \theta_2$, the lot is grade A. If $\theta_1 < \theta < \theta_2$, it is grade B; if $\theta < \theta_1$, it is grade C. The classification is to be done on the basis of a sample from each lot. Suppose that the penalties for misclassifications are those given in the following regret table.

		Lot Quality (state of nature, θ)		
		$\theta < \theta_1$	$\theta_1 < \theta < \theta_2$	$\theta > \theta_2$
Action (grade assigned)	A	2	1	0
	B	1	0	1
	C	0	1	2

This regret function satisfies conditions (1) through (3) and so is monotone.

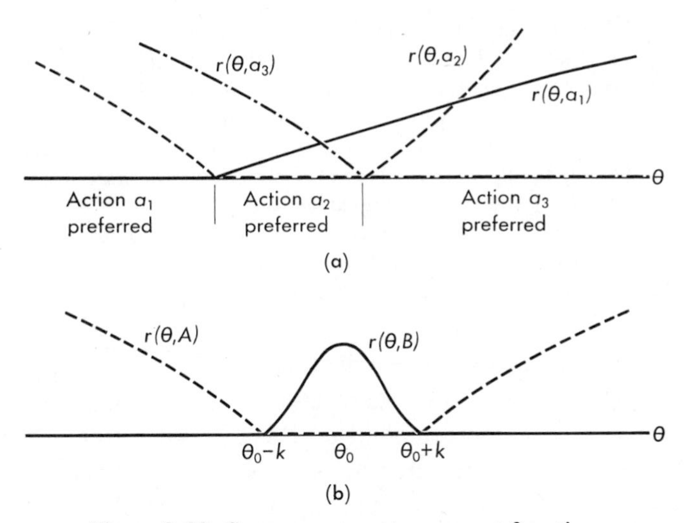

Figure 8–22. Some nonmonotone regret functions.

To describe the known results concerning monotone problems, it is necessary to define the notion of a *monotone procedure*. In a monotone decision problem, in which the likelihood ratio is monotone in terms of the statistic $T = t(\mathbf{X})$, a decision function $d(\mathbf{X})$ is said to be monotone provided that the inequality $t(\mathbf{X}) \leq t(\mathbf{Y})$, for any two given samples $\mathbf{X}$ and $\mathbf{Y}$, is equivalent to the statement that the optimality set for the action $d(\mathbf{X})$ is not to the right of the optimality set for $d(\mathbf{Y})$. This means simply that the axis of values of $t(\mathbf{X})$ is partitioned by the decision rule in a manner that corresponds to the partition of the parameter axis by the optimality sets. Thus, if there are just *two* actions, the parameter axis is partitioned by the regret function into two sets, $\theta < \theta^*$ and $\theta > \theta^*$ (with $\theta = \theta^*$ put into either one); and a monotone procedure is defined by a value t^* that divides the T-axis into $T \leq t^*$ and $T > t^*$, the former inequality calling for the action corresponding to $\theta < \theta^*$, and the latter for the action corresponding to $\theta > \theta^*$. If there are *three* available actions, a monotone rule is defined by points t_1 and t_2 such that $T < t_1$ calls for the action that is best for small θ, $T > t_2$ calls for the action that is best for large θ, and $t_1 < T < t_2$ calls for the action that is best for intermediate θ. And so on, for any finite number of actions. For an estimation problem, a monotone procedure announces as an estimate of θ the value of a statistic that is a monotone increasing function of $t(\mathbf{X})$.

The action corresponding to one of the $n - 1$ boundary points can be given by putting the boundary point into one or the other of the contiguous sets, either by specifying this at the outset or by the toss of a (biased) coin. The rule in the latter case would be a behavioral decision rule, and would be defined by a set of probabilities, one for each boundary point (together with the specification of those points).

EXAMPLE 8–20. Consider again the (egg) classification problem of Example 8–19, and suppose that the classification is to be made on the basis of the weight of a single egg taken from the lot. If this weight X is (for practical purposes) normally distributed with given variance and mean μ, the family of distributions of X (as indexed by μ) has a monotone likelihood ratio in terms of X, since these distributions are in the exponential family with $R(x) = x$. A *monotone procedure* is then defined by four constants: x_1, x_2, p_1, and p_2, and carried out as follows: If $X < x_1$, classify the lot as C; if $x_1 < X < x_2$, classify it as B; if $X > x_2$, classify it as A; if $X = x_1$, classify it as C or B according to the toss of a coin with $p = p_1$; and if $X = x_2$ classify it as B or A according to the toss of a coin with $p = p_2$.

The class of all monotone procedures has been found to be *essentially complete* for monotone problems with finitely many actions, and for the monotone estimation problem.[6] That is, for any procedure *not* in the class there is one *in* the class that is at least as good (as measured by risk) for all θ. It can also be shown, subject to increasingly less general conditions as n increases, that the monotone procedures for a monotone problem with n actions are admissible.[7] (That is, not only do the monotone procedures include essentially all of the admissible ones, they include nothing but admissible ones.) These matters will not be investigated further here. Indeed, although the existence of such results should give some assurance that restricting candidate procedures for monotone problems to monotone procedures is justifiable, the concepts of this section will be used in the remaining chapters only to obtain uniformly most powerful tests for certain testing problems.

Problems

8–25. Determine the Bayes test for Example 8–18 corresponding to the prior $g(H_0) = g(H_1) = .5$, and compute the error sizes of the test.

8–26. Determine the risk function for the test $S > 3$, where S is the number of successes in five independent Bernoulli trials with parameter p, given loss 2 if H_0 is rejected when $p \leq \frac{1}{2}$ and loss 1 if H_0 is accepted when $p > \frac{1}{2}$ (and 0 otherwise).

8–27. Consider an estimation problem with loss function

$$L(\theta, a) = \begin{cases} 0, & \text{if } |a - \theta| \leq 1, \\ 1, & \text{if } |a - \theta| > 1. \end{cases}$$

Determine the risk function and show that if a sequence of estimators $\{T_n\}$ is consistent in mean square, it is also consistent with respect to the above loss function $L(\theta, a)$. (Would the reverse implication be true?)

[6] These results are obtained by S. Karlin and H. Rubin in "The Theory of Decision Procedures for Distributions with Monotone Likelihood Ratios," *Ann. Math. Stat.* **27**, 272–99 (1956). They discuss the results in a paper at a somewhat more elementary level in *J. Amer. Stat. Assn.* **1**, 637–43 (1956).

[7] Ibid.

8–28. Consider the test $X > K$, where X is an observation from a Cauchy population with location parameter θ.

(a) Show that the power function is monotone. Can you use this to deduce that the test is UMP for $\theta \leq \theta^*$ versus $\theta > \theta^*$?
(b) Show that the test is unbiased for $\theta \leq \theta^*$ versus $\theta > \theta^*$.
(c) Is the test admissible for $\theta = 0$ versus $\theta = 2$? (See Problem 6–24.)
★(d) Is any of the likelihood ratio (Neyman–Pearson) tests for $\theta = 0$ versus $\theta = 2$ in Problem 6–24 a monotone procedure?

★8–29. Refer to the decision problem in Example 8–13.

(a) Express the minimax (mixed) decision function obtained there as a behavioral decision rule.
(b) Express the following behavioral rule as a mixture of the nonrandomized decision functions:

If $Z = z_1$, toss two coins and take action a_3 unless two heads turn up, in which case take action a_2.
If $Z = z_2$, toss two coins and take action a_3 if two heads turn up and action a_2 otherwise.

★8–30. Suppose that nature is in one of the states θ_1, θ_2, or θ_3, and that an observation X is made having a normal distribution with unit variance and mean -2 under θ_1, mean 0 under θ_2, and mean $+2$ under θ_3. Let the loss function be as given in the following table:

	θ_1	θ_2	θ_3
A	0	1	2
Action B	1	0	1
C	2	1	0

Consider the following decision rules:

$d_1(X) = A$, if $X < -1$; B, if $-1 < X < 1$; C, if $X > 1$.
$d_2(X) = A$, if $X < -1.5$; B, if $-1.5 < X < 1.5$; C, if $X > 1.5$.
$d_3(X) = A$, if $X > 0$; C, if $X < 0$.
$d_4(X) = B$, if $X < -1$; C, if $-1 < X < 1$; A, if $X > 1$.

Calculate the risk function for each rule. Which rules are monotone?

8.4 Using Bayes' Theorem

The basic assumption in the solution of the decision problem termed *Bayesian* is that one's beliefs about nature can be represented by the model of a probability distribution. This distribution has been called (in Section 8.2.4) a prior distribution, to suggest that it represents the statistician's beliefs as they exist prior to the collection and assimilation of any data.

Treating the state of nature as a random variable Θ means that the probability function or density function $f(z; \theta)$ is actually a *conditional* probability or density, the condition being that nature is in state θ. Thus, when using a probability distribution for Θ, it is more appropriate to use the notation $f(z \mid \theta)$ for the probability or density function of the data Z. From the conditional distribution of Z and the given prior distribution for Θ, one can calculate the joint distribution of (Z, Θ):

$$f_{Z,\Theta}(z, \theta) = f(z \mid \theta)g(\theta),$$

and the corresponding marginal or absolute distribution of Z, with density or probability[8]

$$f_Z(z) = E_g[f(z \mid \Theta)] = \begin{cases} \int f(z \mid \theta)g(\theta)\, d\theta, & (\Theta \text{ continuous}), \\ \sum f(z \mid \theta_i)g(\theta_i), & (\Theta \text{ discrete}). \end{cases}$$

Bayes' theorem (Section 1.3.2) can now be put to use to give the conditional distribution of the state of nature, given the data $Z = z$, according to its density or probability function:

$$h(\theta \mid z) = \frac{f(z \mid \theta)g(\theta)}{f_Z(z)}.$$

(Here f and f_Z are both probability functions or both density functions, according as Z is discrete or continuous, and h and g are likewise both probability functions or both density functions, according as Θ is discrete or continuous.)

8.4.1 The Posterior Distribution

The distribution for Θ defined by $h(\theta \mid z)$ is called the *posterior* distribution and represents the statisticians (new) beliefs about Θ, conditional on the observed data $Z = z$. The notion of using Bayes' theorem to modify prior beliefs by incorporating the information from a sample to obtain better educated beliefs is the essence of "Bayesian" statistics.

The posterior density (or probability function, if Θ is discrete) is essentially the product of the likelihood function for the observed Z-value and the prior density (or probability function). The denominator in the formula for h is just a proportionality constant—just what is needed to make the total posterior probability equal to 1.

[8] A word about subscripts is in order. The spirit of our use is that the subscript indicates, in one way or another, the distribution with respect to which the averaging is to be carried out. Thus $E_\theta[\ ]$ denoted an average with respect to the distribution (for X, as the context made clear) indexed by the parameter θ and (usually) defined by a density or probability function $f(x; \theta)$. Here $E_g[\ ]$ denotes an average with respect to the distribution defined by g, that is, by $g(\theta)$, and the context makes clear that Θ is the random quantity being averaged out.

EXAMPLE 8–21. The probability table giving the conditional distribution of the data in Example 8–13 (the rain problem again) is repeated here, with a row of prior probabilities (.7, .3) assumed to represent beliefs based on past experience or hunches.

	$f(z \mid \theta_1)$	$f(z \mid \theta_2)$	$f_Z(z)$
z_1 (rain)	.8	.1	$.8 \times .7 + .1 \times .3 = .59$
z_2 (no rain)	.2	.9	$.2 \times .7 + .9 \times .3 = .41$
$g(\theta)$	.7	.3	

The last column in this table gives the computation of the unconditional or absolute probability of each value of Z, as averages of the conditional probabilities with respect to the prior probabilities. The posterior probabilities are now ratios of the terms composing the sum to the value of the sum, in each case:

$$h(\theta_1 \mid z_1) = \frac{.8 \times .7}{.59} = \frac{56}{59}, \qquad h(\theta_2 \mid z_1) = \frac{.1 \times .3}{.59} = \frac{3}{59},$$

$$h(\theta_1 \mid z_2) = \frac{.2 \times .7}{.41} = \frac{14}{41}, \qquad h(\theta_2 \mid z_2) = \frac{.9 \times .3}{.41} = \frac{27}{41}.$$

(Of course, the posterior probabilities, for a given z, must add up to 1, as they do here.) Thus, an indication of "rain" (z_1) alters the odds from 7 to 3 in favor of rain (prior) to 56 to 3 in favor of rain (posterior). Also, an indication of "no rain" (z_2) alters the odds to 14 to 27 against rain. The amount by which the odds are altered depends on the strength of the relationship between the data and the state of nature—on the quality of the data. (See also Problem 8–33.)

EXAMPLE 8–22. Consider the estimation of p in the Bernoulli probability function $p^x(1 - p)^{1-x}$, $x = 0$ or 1, using the statistic k, the number of "successes" in n independent trials. Given p, the probability function for k is binomial:

$$f(k \mid p) = \binom{n}{k} p^k(1 - p)^{n-k}, \qquad k = 0, 1, \ldots, n.$$

Assuming a uniform prior distribution for p on [0, 1], one finds[9] the following absolute probability function for k:

$$f(k) = \binom{n}{k} \int_0^1 p^k(1 - p)^{n-k}\, dp = \frac{1}{n + 1}, \qquad k = 0, 1, \ldots, n.$$

The posterior distribution of p, given k, is then the integrand of $f(k)$ divided by $f(k)$:

$$h(p \mid k) = \frac{1 \cdot f(k \mid p)}{f(k)} = \binom{n}{k}(n + 1)p^k(1 - p)^{n-k}.$$

For instance, if the n trials all result in "success," then $k = n$, and the distribution function for p is

$$H(p \mid k = n) = p^{n+1}, \qquad 0 \le p \le 1.$$

[9] The integration is performed using the following formula derived in Chapter 7:

$$\int_0^1 x^r(1 - x)^s\, dx = \frac{r!\,s!}{(r + s + 1)!},$$

where r and s are nonnegative integers.

For large n, a sequence of n successes would put most of the weight near $p = 1$ in the posterior distribution.

It is of interest to observe that if the posterior distribution is used as a prior distribution for a second experiment, the resulting new posterior distribution based on the outcome of the second experiment is the same as the posterior distribution that would have been obtained based on the *combined* data from the two experiments and the initial prior distribution. Let the outcomes of the experiments be (X, Y), with joint density

$$f(x, y) \mid \theta) = f(x \mid \theta) f(y \mid x, \theta).$$

The posterior density of θ, given $X = x$ and assuming a prior density $g(\theta)$, is

$$h(\theta \mid x) = \frac{g(\theta) f(x \mid \theta)}{\int g(\theta) f(x \mid \theta) \, d\theta},$$

and using this as a prior distribution together with the result $Y = y$, one obtains the posterior distribution:

$$\begin{aligned} k(\theta \mid x, y) &= \frac{h(\theta \mid x) f(y \mid x, \theta)}{\int h(\theta \mid x) f(y \mid x, \theta) \, d\theta} \\ &= \frac{g(\theta) f(x \mid \theta) f(y \mid x, \theta)}{\int g(\theta) f(x \mid \theta) f(y \mid x, \theta) \, d\theta} = \frac{g(\theta) f(x, y \mid \theta)}{\int g(\theta) f(x, y \mid \theta) \, d\theta}, \end{aligned}$$

which is what would have been obtained by using the combined result $X = x$ and $Y = y$ together with the prior distribution given by $g(\theta)$. This property of posterior distributions is illustrated in Problem 8–31.

8.4.2 Solving the Decision Problem

The posterior distribution can be used as a tool in computing Bayes decision rules. The method is to compute the posterior expected loss:

$$E_h[\ell(\Theta, a)] = \int \ell(\theta, a) h(\theta \mid z) \, d\theta,$$

and then to select the action a (for the given z) that minimizes this quantity. The dependence of the chosen a on the given z provides a decision function $a = d^*(z)$, which is precisely the Bayes decision function for the given prior distribution.

To establish this claim, the expression for the Bayes risk corresponding to a decision function $d(z)$ is manipulated as follows:

$$\begin{aligned} B(d) = E_g[R(\Theta, d)] &= \int R(\theta, d)g(\theta)\,d\theta \\ &= \int\!\!\int \ell(\theta, d(z))f(z \mid \theta)g(\theta)\,dz\,d\theta \\ &= \int f(z)\left\{\int \ell(\theta, d(z))h(\theta \mid z)\,d\theta\right\} dz \\ &= \int f(z)E_h[\ell(\Theta, d(z))]\,dz. \end{aligned}$$

From the last representation of $B(d)$, it is apparent that $B(d)$ is minimized by a function $d(z)$, whose value a for each z is selected to minimize the expected posterior loss, $E_h[\ell(\Theta, a)]$.

The method, then, is equivalent to the following: Determine the posterior distribution, given $Z = z$, and use it as a distribution for Θ in a no-data problem with the original loss function. The Bayes solution for this no-data problem is the desired Bayes action for the given problem.

The significance of this method of determining $d(z)$ is that it is only necessary to compute the value of $d(z)$ for the z actually obtained as the result of experimentation. Computing $d(z)$ as a Bayes decision function from the basic definition involves a determination of the whole function—the value of $d(z)$ for each foreseeable z. Mathematically, a problem in the calculus of variations (minimizing over a set of possible functions) is replaced by a problem in the calculus (minimizing over a set of numbers). It may prove helpful to consider the schematic diagrams for computation of Bayes strategies according to the two methods, as shown in Figures 8–23 and 8–24.

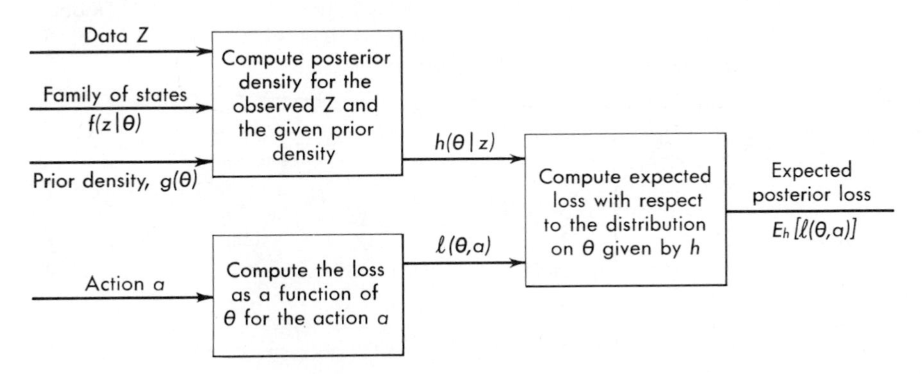

Figure 8–23

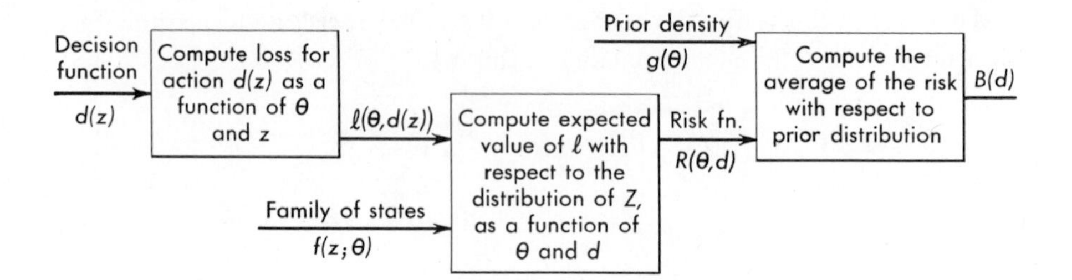

Bayes decision function is the $d(z)$ that yields the minimum $B(d)$

Figure 8–24

EXAMPLE 8–23. The two methods of computing a Bayes strategy will now be applied to the rain problem, with data Z and prior probabilities as in Example 8–21. It was seen earlier (Example 8–13) that the only admissible strategies are

d_2: Go without umbrella, no matter what Z is.
d_3: Go with the umbrella, no matter what Z is.
d_9: Go with it if Z = "rain," go without it if Z = "no rain."

The risk table for these three decision rules, together with a column giving the Bayes risk for each computed by weighting the risks with prior probabilities, is as follows:

	θ_1	θ_2	$B(d)$
d_2	5	0	3.5
d_3	2	5	2.9
d_9	2.6	.5	1.97
$g(\theta)$	.7	.3	

The Bayes rule is d_9, since this produces the smallest $B(d)$. The alternative computation is to apply posterior probabilities to the original loss table. This is done in the following table, for each of the possible observations:

	θ_1	θ_2	$B(a \mid z_1)$	$B(a \mid z_2)$
a_1	4	4	$\frac{236}{59}$	$\frac{164}{41}$
a_2	5	0	$\frac{280}{59}$	$\frac{70}{41}$
a_3	2	5	$\frac{127}{59}$	$\frac{163}{41}$
$h(\theta \mid z_1)$	$\frac{56}{59}$	$\frac{3}{59}$		
$h(\theta \mid z_2)$	$\frac{14}{41}$	$\frac{27}{41}$		

To illustrate the computations, the Bayes loss $B(a_3 \mid z_2)$ using the posterior probabilities given $Z = z_2$ is

$$B(a_3 \mid z_2) = 2 \times \tfrac{14}{41} + 5 \times \tfrac{27}{41} = \tfrac{163}{41}.$$

The Bayes action given $Z = z_1$ is a_3 (which gives the minimum Bayes loss of $\frac{127}{59}$), and the Bayes action given $Z = z_2$ is a_2 (which gives the minimum Bayes loss of $\frac{70}{41}$). These results—take a_3 (go with umbrella) if $Z = z_1$ (rain), and take a_2 (go without umbrella) if $Z = z_2$ (no rain)—constitute what was identified as decision rule d_9, the Bayes rule.

Exploiting the computations of previous examples has perhaps served to mask the comparison in the amounts of computation needed for each method. In applying prior probabilities to risks, the risk for each of the nine decision rules must be obtained, whereas in applying posterior probabilities to losses, only the posterior probabilities corresponding to the observed value of Z need be computed. Making the entire set of necessary computations in each case "from scratch" would make the comparison more convincing.

EXAMPLE 8–24. Consider the estimation of θ in an exponential density $\theta e^{-\theta x}$ (for $x > 0$, $\theta > 0$), using a single observation, with loss function

$$\ell(\theta, a) = |\theta - a|.$$

If one assumes the improper prior $g(\theta) \equiv 1$ for $\theta > 0$, the posterior density becomes

$$h(\theta \mid x) = x^2\theta e^{-\theta x},$$

which is a proper density for $\theta > 0$. The estimated value a is then chosen to minimize the expected posterior loss:

$$E_h|\theta - a| = \int_0^\infty |\theta - a| x^2 \theta e^{-\theta x}\, d\theta.$$

According to the result of Problem 2–98, this absolute first moment of Θ is minimized by choosing a to be the median of the distribution—the number m such that

$$\int_0^m x^2\theta e^{-x\theta}\, d\theta = \tfrac{1}{2}.$$

This can be found by numerical methods to be about $1.7/x$. (The maximum likelihood estimate, by way of contrast, is $1/x$.)

In view of the simplification afforded by introduction of the posterior distribution, the Bayesian statistician might consider his task to be that of reporting the posterior distribution resulting from his data and his assumed prior. Then anyone can apply that posterior to his own problem and loss function and so finish the job of inference or decision. However, because the posterior depends on the prior, which is also a personal matter, it may be more fitting for the statistician simply to report the likelihood function. Then each consumer of the report can multiply in his own personal prior to obtain the appropriate posterior for application to his loss function. In this sense, the Bayesian agrees with the "likelihood statistician" that the relevant information contained in data is best summarized and reported as a likelihood function.

Problems

8–31. Suppose, in the rain problem of Example 8–13 and others, that in addition to the data Z of Example 8–21 one has an independent observation Y, whose distribution is given here along with that of Z:

	θ_1	θ_2
z_1	.8	.1
z_2	.2	.9

	θ_1	θ_2
y_1	.6	.3
y_2	.4	.7

Assume equal prior probabilities for θ_1 and θ_2 to compute the posterior, given $(Y, Z) = (y_1, z_1)$. Do it in two ways: first by computing a posterior given $Y = y_1$ and then a new posterior given $Z = z_1$ (with the first posterior as prior); and second by a single modification of the prior using the *joint* distribution of (Y, Z).

+ **8–32.** Suppose a Bernoulli parameter p is given the improper density $(pq)^{-1}$, $0 < p < 1$. Obtain the corresponding posterior, given k successes in n independent trials, as a beta distribution. Use results of Section 7.1.2 to determine the Bayes estimate of p (in terms of k and n) given the quadratic loss $(p - a)^2$.

8–33. (a) Show that if all of the prior probability is concentrated on a single state, then the posterior distribution is the same. That is, no amount of data can convince a statistician whose mind is already made up. (There is one exceptional case—what is it?)

(b) Show that the posterior distribution is the same as the prior if the distribution of the data is independent of the state of nature.

+ (c) Show that if T is sufficient for $f(x; \theta)$, the posterior distribution depends on $\mathbf{X}$ only through the values of T.

8–34. A lot of N articles contain M defective. One observation is obtained by testing one article drawn at random. Assuming a prior distribution for M that is binomial (N, p), determine the posterior distribution, given that the tested article is defective.

8–35. Determine the posterior distribution of the mean μ of a normal population with known variance, given a random sample $(X_1, \ldots, X_n)$ and a uniform (improper) prior for μ.

8–36. In Example 8–23, compute the minimum Bayes risk in two ways: as the minimum $B(d)$ for the Bayes rule d_B, and as the weighted sum of the expected posterior losses $B(a \mid z_i)$ using absolute probabilities of the z_i's as weights.

8.4.3 Conjugate Families

It is often the case that there is a family of distributions for Θ, corresponding to data $\mathbf{X}$ with distribution indexed by θ, such that if the prior belongs to the family, so also does the posterior—for any sample size and any particular $\mathbf{X} = \mathbf{x}$. Such a family is called a *conjugate family of distributions*. It is a useful thing to have if in addition to being fairly simple, it is sufficiently rich in distributions to enable one at least to approximate by members of the family priors of the sort he might expect to encounter.

A conjugate family can be constructed by finding a family that includes the likelihood function (as a function of θ) and that is closed under multiplication. It can be shown that this is possible when there is a sufficient statistic of fixed dimension for any sample size, as is the case for the exponential family (cf. Ferguson [9], p. 163).

EXAMPLE 8–25. Consider a random sample from a normal population with known variance σ^2. The likelihood function can be written

$$L(\mu) = \exp\left[-\frac{n}{2\sigma^2}(\mu - \bar{X})^2\right],$$

which (with suitable normalizing constant) is a member of the normal family of distributions. Assuming a prior for μ that is normal, say with mean ν and variance τ^2, results in a posterior density whose essential part (the part depending on μ) is

$$\exp\left[-\frac{n}{2\sigma^2}(\mu^2 - 2\mu\bar{X} + \bar{X}^2) - \frac{1}{2\tau^2}(\mu^2 - 2\mu\nu + \nu^2)\right]$$

$$= \exp\left\{-\frac{1}{2}\left(\frac{n}{\sigma^2} + \frac{1}{\tau^2}\right)\left[\mu^2 - 2\mu\frac{\bar{X}n\tau^2 + \nu\sigma^2}{n\tau^2 + \sigma^2} + \cdots\right]\right\},$$

where the terms and factors not written out do not involve μ. The exponent being quadratic in μ, this is again a normal density; and so the normal distributions constitute a conjugate family for the given problem. The mean and variance of the posterior distribution for μ can be read from the above exponent:

Posterior mean: $\dfrac{\bar{X}n\tau^2 + \nu\sigma^2}{n\tau^2 + \sigma^2}$.

Posterior variance: $\dfrac{\sigma^2\tau^2}{n\tau^2 + \sigma^2} = \dfrac{1}{n/\sigma^2 + 1/\tau^2}$.

These can perhaps be better interpreted in terms of *precision*, or the reciprocal of the variance:

Precision in sample: $\pi_s \equiv \dfrac{n}{\sigma^2}$.

Precision of prior: $\pi_{pr} \equiv \dfrac{1}{\tau^2}$.

Precision of posterior: $\pi_{po} \equiv \dfrac{n}{\sigma^2} + \dfrac{1}{\tau^2}$

$$\equiv \pi_s + \pi_{pr}.$$

Thus, the precision in the posterior is the precision in the sample plus the precision of the prior. Moreover, the posterior mean is the weighted average of the sample mean and the prior mean in which the weights are the precisions:

$$E_h(\mu) = \frac{\pi_s\bar{X} + \pi_{pr}\nu}{\pi_{po}}.$$

The following table on p. 412 gives some examples of conjugate families. In each case the prior is simply a function of θ of the form of the likelihood function, but with arbitrary parameters, so that the family will be closed under multiplications, and also so that one can represent a variety of reasonable priors by a member of the family. (Normalizing factors have been omitted in giving densities.)

Likelihood	Prior	Posterior
$\theta^{\Sigma X_i}(1-\theta)^{n-\Sigma X_i}$ (binomial or negative binomial)	$\theta^{\alpha-1}(1-\theta)^{\beta-1}$ (beta)	$\theta^{\alpha+\Sigma X_i-1}(1-\theta)^{\beta+n-1-\Sigma X_i}$ (beta)
$\exp\left[-\frac{n}{2\sigma^2}(\theta-\bar{X})^2\right]$ (normal population, given σ^2)	$\mathcal{N}(\nu, \tau^2)$	$\mathcal{N}\left(\frac{\bar{X}n\tau^2+\nu\sigma^2}{n\tau^2+\sigma^2}, \frac{\sigma^2\tau^2}{n\tau^2+\sigma^2}\right)$
$e^{-n\theta}\theta^{\Sigma X_i}$ (Poisson)	$\theta^{\alpha-1}e^{-\lambda\theta}$ (gamma)	$\theta^{\alpha-1-\Sigma X_i}e^{-\theta(\lambda+n)}$ (gamma)
$\theta^n e^{-\theta\Sigma X_i}$ (exponential population)	$\theta^{\alpha-1}e^{-\lambda\theta}$ (gamma)	$\theta^{\alpha+n-1}e^{-\theta(\lambda+\Sigma X_i)}$ (gamma)

8.4.4 Estimation and Testing

In any decision problem, the Bayes decision rule is to choose an estimated value that minimizes the expected loss with respect to the posterior distribution given the data Z. When the problem is that of estimation and the loss is quadratic, the Bayes estimate is the value a that minimizes

$$E_h[\ell(\Theta, a)] = E_h[(\Theta - a)^2].$$

But the second moment of any random variable is smallest when taken about the mean, so the Bayes estimate of θ is the mean of the posterior distribution of Θ, namely, $\theta_B \equiv E_h\Theta$.

It might be remarked that for a uniform prior the posterior density is proportional to the likelihood function, and so a maximum likelihood estimate can be interpreted as the *mode* of that posterior density.

EXAMPLE 8–26. If a quadratic loss is assumed in estimating the parameter p of a Bernoulli distribution, and the prior is a beta distribution (α, β), then the Bayes estimate is the mean of a beta distribution with parameters $(\alpha + k, \beta + n - k)$, where k is the observed number of successes in n trials. The mean is

$$p_B = E_h(p) = \frac{\alpha + k}{\alpha + \beta + n}.$$

For $\alpha = \beta = 1$, the prior is uniform, and $p_B = (k + 1)/(n + 2)$. The m.l.e. k/n can be obtained by setting $\alpha = \beta = 0$, which defines an improper prior.

EXAMPLE 8–27. If a normal prior (ν, τ^2) is assumed for the mean μ of a normal population with known variance σ^2, the posterior (cf. Example 8–25) has mean

$$\mu_B = \frac{\pi_s \bar{X} + \pi_{pr}\nu}{\pi_{po}},$$

where the π's are the precisions (reciprocal variances) of the sample mean, the prior, and the posterior. Thus, a prior of low precision—rather flat and uninformative results in a Bayes estimate close to the sample mean. And a precise prior—that is, as compared with the precision in the sample, would yield an estimate close to the prior mean.

In testing a simple hypothesis $f_0(x)$ against a simple alternative $f_1(x)$, the posterior probabilities are

$$h_0 = \frac{g_0 f_0}{g_0 f_0 + g_1 f_1} \quad \text{and} \quad h_1 = \frac{g_1 f_1}{g_0 f_0 + g_1 f_1},$$

where g_0 and g_1 are prior probabilities for f_0 and f_1. Applying the probabilities h_0 and h_1 to the usual regret table,

	H_0	H_1
Accept H_0	0	b
Reject H_0	a	0

one obtains these expected losses:

$$\text{Accept } H_0\text{: } \frac{b g_1 f_1}{f_0 g_0 + f_1 g_1}.$$

$$\text{Reject } H_0\text{: } \frac{a g_0 f_0}{f_0 g_0 + f_1 g_1}.$$

The Bayes action is the one corresponding to the smaller of these losses. Equivalently, H_0 is rejected if and only if

$$\frac{a f_0 g_0}{b f_1 g_1} < 1,$$

or

$$\frac{f_0(x)}{f_1(x)} < \frac{b g_1}{a g_0}.$$

This is a Neyman–Pearson test. And conversely, given any Neyman–Pearson test, one can find losses and prior probabilities such that it is a Bayes test.

In the case of composite hypotheses, suppose that the losses for wrong decisions in a testing problem are constant:

	θ in H_0	θ in H_1
Accept H_0	0	b
Reject H_0	a	0

The expected loss with respect to a posterior density h is then given by

$$E_h[\ell(\Theta, a)] = \begin{cases} bP_h(H_1), & \text{if } H_0 \text{ is accepted,} \\ aP_h(H_0), & \text{if } H_0 \text{ is rejected.} \end{cases}$$

The Bayes action is to reject H_0 if the posterior distribution puts much weight on H_1. More precisely, H_0 is rejected if and only if

$$P_h(H_0) < \frac{b}{b + a}.$$

EXAMPLE 8–28. Assume the constant loss structure as above in testing $\mu \leq \mu_0$ against $\mu > \mu_0$, where μ is the mean of a normal population with known variance. If the prior is assumed to be normal, the posterior is normal, and

$$P_h(\mu \leq \mu_0) = \Phi\left(\frac{\mu_0 - E_h\mu}{1/\sqrt{\pi_{po}}}\right),$$

where π_{po} is the posterior precision and $E_h\mu$ is a linear function of $\bar{X}$. The inequality $P_h(H_0) <$ const is then equivalent to $\bar{X} >$ const so that the Bayes test is given by this UMP critical region.

A probability interval for θ would be an interval (L_1, L_2) such that for given γ,

$$P_h(L_1 < \Theta < L_2) = \gamma.$$

For any posterior h there would be many such intervals, but it would be perhaps most natural to take the shortest one as an interval estimate for θ.

EXAMPLE 8–29. Assume again a normal population with known variance and a normal prior for the mean. The posterior, being normal, is symmetric about its mean, and a shortest interval of given probability γ would be centered at the mean. Thus,

$$P_h\left(z_{.025} < \frac{\mu - E_h\mu}{\sqrt{1/\pi_{po}}} < z_{.975}\right) = .95,$$

where z refers to the standard normal distribution. So a 95 per cent probability interval for μ has limits

$$E_h\mu \pm z_{.975}\sqrt{1/\pi_{po}}.$$

If (and only if) the prior is flat ($\pi_{pr} = 0$), this reduces to the usual 95 per cent confidence interval for μ.

Problems

8–37. Determine a conjugate family for the normal population with mean 0 and "precision" θ (reciprocal variance), assuming a random sample of size n. Calculate the posterior mean corresponding to an arbitrary prior from the conjugate family.

8–38. Given a random sample of 10 from a Bernoulli population, suppose that there are 8 successes.

(a) Compute the posterior probability of the interval $\frac{1}{2} < p < 1$, if the prior is in the conjugate family with $\alpha = \beta = 1$.
(b) What is the limit of the Bayes estimate of p as α and β tend to infinity, with $\alpha = \beta$?

8–39. Assume a prior from the conjugate family for the exponential population $\theta e^{-\theta x}$, $x > 0$, and consider a random sample of size n.

(a) Obtain the Bayes estimate of θ.
(b) Examine the cases (1) $\alpha = 1$, $\lambda = 0$ and (2) $\alpha = 0$ and $\lambda = 0$. (Are these proper priors?) Observe that one of these gives the m.l.e.
(c) Examine the case $n \to \infty$ with α and λ fixed.

★8.5 Sequential Procedures

A stream of potential observations $\mathbf{X} = (X_1, X_2, \ldots)$ is assumed, and the distribution of each finite segment $\mathbf{X}_n \equiv (X_1, \ldots, X_n)$ is assumed to be known, depending on the state of nature θ. In the simplest case, the X's would be independent and identically distributed with

$$f_n(x_1, \ldots, x_n \mid \theta) = \prod_1^n f(x_i \mid \theta)$$

defining the distribution of $\mathbf{X}_n$ in terms of a population density $f(x \mid \theta)$. The cost of obtaining observations will generally depend on θ, on the number of observations used, and possibly on the observations themselves. For simplicity, the discussion will be given for the important case in which the cost is proportional to the number of observations:

$$[\text{cost of obtaining } \mathbf{X}_n] = cn.$$

The most general decision procedure involves a stopping rule $\mathbf{s}$ to say when to stop sampling, and a terminal decision rule $\mathbf{d}$ to say what action to take when the sampling is stopped. The *terminal decision rule* can be thought of as a vector of decision rules, one for each sample size:

$$\mathbf{d} = [d_0, d_1(x_1), d_2(x_1, x_2), \ldots],$$

where d_0 is one of the available actions and $d_n(x_1, \ldots, x_n)$ is an assignment of an action to the observation vector value $(x_1, \ldots, x_n)$.

A *stopping rule* is characterized by a family of functions of the form

$$\boldsymbol{\phi} = [\phi_0, \phi_1(x_1), \phi_2(x_1, x_2), \ldots],$$

where (in the nonrandomized case)

$$\phi_0 = \begin{cases} 0, & \text{if the rule says take the first observation,} \\ 1, & \text{if the rule says take no observations} \end{cases}$$

and

$$\phi(x_1, \ldots, x_n) = \begin{cases} 0, & \text{if the rule says take another observation, given } \mathbf{x}_n, \\ 1, & \text{if the rule says take no more observations, given } \mathbf{x}_n. \end{cases}$$

Given the functions ϕ_k, one can determine (for any given stream of observations) when to stop; and if the rule is given in other terms, the functions ϕ can be determined from that rule.

A stopping rule can also be characterized by the family of functions

$$s = [s_0, s_1(x_1), s_2(x_1, x_2), \ldots],$$

where $s_0 \equiv \phi_0$, and for $n \geq 1$,

$$\begin{aligned} s_n(\mathbf{x}_n) &= (1 - \phi_0)[1 - \phi_1(x_1)]\cdots[1 - \phi_{n-1}(x_{n-1})]\phi_n(x_n) \\ &= \begin{cases} 1, & \text{if the rule calls for stopping for the first time at the } n\text{th observation, given } \mathbf{X} = \mathbf{x}, \\ 0, & \text{otherwise.} \end{cases} \end{aligned}$$

(The "otherwise" here would mean either that the rule called for stopping or says to go on after the nth observation.)

The number of observations that are actually obtained when following a given rule $\mathbf{s}$ is a random variable, since it depends on the values of the observations obtained up to the time of stopping; call it N. Given $\mathbf{X} = \mathbf{x}$, the functions s_n give conditional probabilities for N:

$$P(N = n \mid \mathbf{X} = \mathbf{x}) = s_n(\mathbf{x}_n).$$

The unconditional distribution for N (but assuming a given θ) is obtained by averaging:

$$P(N = n) = E[s_n(X_1, \ldots, X_n)],$$

a quantity that depends on the rule $\mathbf{s}$ and on θ.

EXAMPLE 8–30. The sequential likelihood ratio test of Section 6.3.1 is defined by the following rule $(\mathbf{s}, \mathbf{d})$. Let A, B, and M be given constants, with $A < M < B$, and let Λ_n denote the likelihood ratio for testing $f_0(x)$ against $f_1(x)$ based on n independent observations. Then define $\phi_0 = 0$ and, for $n \geq 0$,

$$\phi_n(\mathbf{X}_n) = \begin{cases} 0, & \text{if } A < \Lambda_n < B, \\ 1, & \text{if } \Lambda_n \leq A \text{ or } \Lambda_n \geq B. \end{cases}$$

And define, for $n > 0$,

$$d_n(\mathbf{X}_n) = \begin{cases} \text{reject } H_0, & \text{if } \Lambda_n < M, \\ \text{accept } H_0, & \text{if } \Lambda_n \geq M. \end{cases}$$

With these ϕ's defining **s**, the rule (**s**, **d**) is a sequential decision procedure for testing f_0 against f_1.

The determination of optimal procedures is generally a rather difficult task. In Section 8.5.1 the decision component **d** of a Bayes procedure (**s**, **d**) will be derived. It is simply a fixed sample size Bayes decision rule applied whenever the sampling stops. In Section 8.5.2 the nature of the stopping component **s** of a Bayes procedure will be determined for the special case of testing one simple hypothesis against another. It is essentially a sequential likelihood ratio test.

8.5.1 The Bayes Decision Rule

Given a loss function $\ell(\theta, a)$ the total loss incurred in using a procedure (**s**, **d**) is

$$\ell(\theta, d_N(\mathbf{X}_N)) + cN,$$

and the risk function is the expected value of this random quantity:

$$R(\theta; \mathbf{s}, \mathbf{d}) = E[\ell(\theta, d_N(\mathbf{X}_N)) + cN],$$

where the expectation is taken with respect to the distribution of (N, **X**), given θ. This can be calculated by first averaging with respect to N, given **X**, as a sum:

$$\begin{aligned} R(\theta; \mathbf{s}, \mathbf{d}) &= E^{\mathscr{X}|\theta}\left[\sum_n [\ell(\theta, d_n(\mathbf{X}_n)) + cn]P(N = n \mid \mathbf{X})\right] \\ &= \sum_n E^{\mathscr{X}|\theta}\{s(\mathbf{X}_n)[\ell(\theta, d_n(\mathbf{X}_n)) + cn]\}, \end{aligned}$$

where the superscript indicates that the expectation is taken over the sample space of **X** using the distribution defined by $\Theta = \theta$.

For a given prior on the space Ω of θ's, the Bayes risk is the average of the above with respect to that prior:

$$\begin{aligned} B(\mathbf{s}, \mathbf{d}) &= \sum_n E^{\Omega}E^{\mathscr{X}|\Theta}\{s_n(\mathbf{X}_n)[\ell(\Theta, d_n(\mathbf{X}_n)) + cn]\} \\ &= \sum_n E\{s_n(\mathbf{X}_n)[\ell(\Theta, d_n(\mathbf{X}_n)) + cn]\}. \end{aligned}$$

To minimize this risk, it will suffice first to minimize over all decision rules for a given stopping rule, and then to minimize the result over the class of all stopping rules. The following theorem shows how to do the first part of the job.

Theorem. *Given a stopping rule* **s** *and a prior distribution for* Θ*, the Bayes risk* $B(\mathbf{s}, \mathbf{d})$ *is minimized by the decision rule* $d_B \equiv [d_{0B}, d_{1B}, d_{2B}, \ldots]$*, where* d_{iB} *is the Bayes decision rule for the first* i *observations considered as a sample of given size* i.

To establish this, observe first that the Bayes risk can be written using an expectation with respect to the absolute distribution of $\mathbf{X}$ of the conditional expectation over Ω given $\mathbf{X}$:

$$\begin{aligned} B(\mathbf{s}, \mathbf{d}) &= \sum_n E^{\mathscr{X}}\{E^{\Omega|\mathbf{X}}[s_n(\mathbf{X}_n)\{\ell(\Theta, d_n(\mathbf{X}_n)) + cn\}]\} \\ &= \sum_n E^{\mathscr{X}}\{s_n(\mathbf{X}_n)[E^{\Omega|\mathbf{X}}\{\ell(\Theta, d_n(\mathbf{X}_n))\} + cn]\}. \end{aligned}$$

Now, because s_n and cn are nonnegative, this will be minimized if in each term the expected posterior loss:

$$E^{\Omega|\mathbf{X}}[\ell(\Theta, d_n(\mathbf{X}_n))]$$

is minimized by a suitable choice of d_n. But this is accomplished if for given $\mathbf{X}_n$ one chooses $d_n(\mathbf{X}_n)$ to be the Bayes rule for the sample $\mathbf{X}_n$ of fixed size n, that is, for $d_n = d_{nB}$, as asserted.

8.5.2 A Sequential Bayes Test

The stopping rule will now be developed for a Bayes sequential test of a simple H_0 against a simple H_1. As before, the losses in this testing situation are assumed to be 0 for correct decisions and $\ell(H_0, \text{reject}) = a \geq 0, \ell(H_1, \text{accept}) = b \geq 0$, where the actions "reject" and "accept" are, respectively, the rejection and acceptance of H_0. A constant cost c per observation is assumed. Let f_0 and f_1 denote the population densities under H_0 and H_1, respectively.

Consider the first stage in the sequential sampling. The choice between making a decision with no data and taking one or more observations will be made by comparing the minimum Bayes risks for these alternatives, on the basis of an assumed initial prior. Let the prior probabilities for H_0 and H_1 be w and $1 - w$, respectively.

If the sampling is not to begin at all, the action to be taken (according to the theorem of the preceding section) is the Bayes action determined by calculating Bayes losses using the given prior probabilities and basic loss function:

$$\begin{cases} B(\text{reject}) = wa, \\ B(\text{accept}) = (1 - w)b. \end{cases}$$

These functions are shown in Figure 8–25, the locus of the minimum as a function of w being drawn with a heavy line. This minimum Bayes loss, with no sampling, will be denoted by $\rho_0(w)$. It is achieved by rejecting H_0 if $w < b/(a + b)$ and accepting H_0 if $w > b/(a + b)$.

Next consider the class $\mathscr{C}_1$ of rules $(\mathbf{s}, \mathbf{d})$ that call for taking at least one observation. As above, N is the number of observations actually taken, and $\mathbf{X}_n$ is the

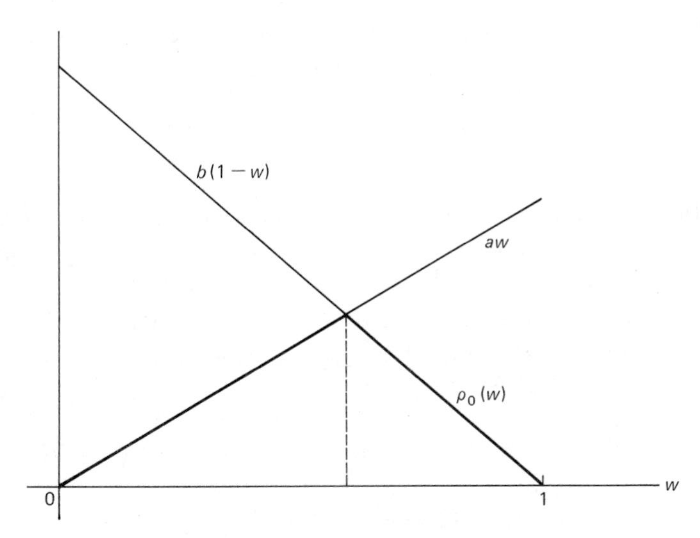

Figure 8–25

truncated observation sequence. If $N \geq 1$, the risk function (including sampling costs) has the values

$$\begin{cases} R_0 = a\alpha + cE(N \mid H_0), & \text{if } \theta = H_0, \\ R_1 = b\beta + cE(N \mid H_1), & \text{if } \theta = H_1. \end{cases}$$

The Bayes risk is then

$$B(\mathbf{s}, \mathbf{d}) = wR_0 + (1 - w)R_1.$$

Let $\rho_1(w)$ denote the minimum Bayes risk in the class of procedures $(\mathbf{s}, \mathbf{d})$ that call for at least one observation:

$$\rho_1(w) \equiv \min_{\mathscr{C}_1} [wR_0 + (1 - w)R_1].$$

This function has certain properties that permit one to draw useful conclusions about the nature of the stopping rule:

Theorem

1. $c \leq \rho_1(w) \leq c + \rho_0(w)$.
2. $\rho_1(0) = \rho_1(1) = c$.
3. $\rho_1(w)$ *is concave down.*

The first part of (1) follows when it is noted that $EN \geq 1$ and, therefore, that R_0 and R_1 must each be as large as c, since a, b, and c are nonnegative. The second inequality says that the minimum risk does not exceed the cost of one observation plus the minimum risk with no additional observations, where this sum is the minimum risk in the class of procedures calling for exactly one observation (a subclass of $\mathscr{C}_1$). To obtain (2), observe that

$$\rho_1(0) = \min_{\mathscr{C}_1} R_1 = \min_{\mathscr{C}_1} [b\beta + cE(N \mid H_0)],$$

which is clearly minimized by taking $N = 1$ and rejecting H_0, no matter what the data say, so that $\beta = 0$. The minimum value is c. A similar argument yields $\rho_1(1) = c$.

The concavity claimed in (3) in the theorem is established as follows. Let w' and w'' be numbers between 0 and 1, and consider the value of $\rho_1(w)$ at the point $w = \lambda w' + (1 - \lambda)w''$, where $0 < \lambda < 1$:

$$\begin{aligned}\rho_1(w) &= \min_{\mathscr{C}_1} \{[\lambda w' + (1 - \lambda)w'']R_0 + [1 - \lambda w' - (1 - \lambda)w'']R_1\} \\ &= \min_{\mathscr{C}_1} \{\lambda[w'R_0 + (1 - w')R_1] + (1 - \lambda)[w''R_0 + (1 - w'')R_1]\} \\ &\geq \lambda\rho_1(w') + (1 - \lambda)\rho_1(w'').\end{aligned}$$

Thus, the graph of ρ_1 at the intermediate point w is higher than the ordinate on the chord between the points on ρ_1 corresponding to w' and w''. So ρ_1 is concave down. This can also be seen intuitively, since $\rho_1(w)$ is the lower envelope of a family of line segments joining pairs of points $(0, R_1)$ and $(1, R_0)$.

According to the characteristics of $\rho_1(w)$ given in the theorem, there are two possible configurations of ρ_1 with respect to ρ_0, shown in Figure 8–26. If the cost c per observation is high, then $\rho_0 \leq \rho_1$, and no matter what the prior distribution is, no observations should be taken. If c is small enough, then there are two numbers, C and D, such that the Bayes stopping rule is as follows:

(i) If $C < w < D$, take at least one observation.

(ii) If $w < C$, take no observations and reject H_0.

(iii) If $w > D$, take no observations and accept H_0.

That is, this is the portion of the stopping rule that applies to the first stage.

Suppose next that n observations ($n \geq 1$) have been taken, and the question is whether to stop and make a decision or to take still more observations. This choice can be handled in excactly the same way as was the choice faced at the first stage. For, the function $\rho_1(w)$, now defined in terms of the class of rules calling for at least one *more* observation, has the same structure as before. This is because the distribution of $(X_{n+1}, X_{n+2}, \ldots)$ is the same as the distribution of $(X_1, X_2, \ldots)$ and the loss structure is unchanged. However, in defining $\rho_1(w)$, the variable N would denote the number of *additional* observations (after

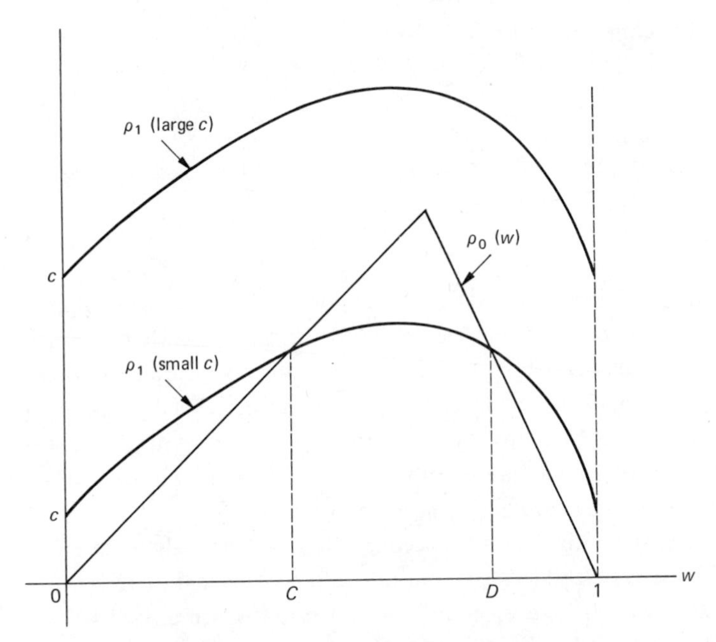

Figure 8–26

the n already at hand), and in place of the prior probability w one would use the posterior probability h_n based on the first n observations:

$$h_n \equiv \frac{w \prod_1^n f_0(X_i)}{w \prod_1^n f_0(X_i) + (1 - w) \prod_1^n f_1(X_i)}$$

$$= \frac{\Lambda_n}{\Lambda_n + (1 - w)/w},$$

where Λ_n is the likelihood ratio statistic based on the first n observations. (Note that the cost of the observations already at hand is not a factor in a decision as to whether to obtain more data.)

The function $\rho_0(\cdot)$ is also the same as before, but now the choice between ρ_0 and ρ_1 is made in terms of the posterior distribution defined by h_n. So the Bayes rule at the nth stage is as follows:

(i) Continue sampling if

$$C < \frac{\Lambda_n}{\Lambda_n + (1 - w)/w} < D,$$

or equivalently, if

$$\frac{C}{1 - C} \frac{1 - w}{w} < \Lambda_n < \frac{D}{1 - D} \frac{1 - w}{w}.$$

(ii) Stop sampling and reject H_0 if

$$\Lambda_n < \frac{C}{1 - C} \frac{1 - w}{w}.$$

(iii) Stop sampling and accept H_0 if

$$\Lambda_n > \frac{D}{1 - D} \frac{1 - w}{w}.$$

Although the above analysis gives the nature of the Bayes stopping rule, the constants C and D could only be determined from a knowledge of the function $\rho_1(w)$. This is not easy to come by; but it is possible to write a functional equation for $\rho_1(w)$, one that can be handled by numerical methods, by making use of the principle that upon breaking into an optimal procedure at any stage, the remaining portion of the procedure is optimal when considered in its own right. (The problem is one in the area of *dynamic programming*.)

The Bayes sequential testing procedure discussed strongly resembles a sequential likelihood ratio test (SLRT), but one difference is that there is no "no-data" stage in the latter. However, it can be shown that given any SLRT there exists a loss structure and a sampling cost c per observation such that the Bayes sequential test for some prior is the given SLRT.

9

ANALYSIS OF CATEGORICAL DATA

Experiments will be considered here in which the outcomes are assigned to categories, rather than measured on numerical scales. The simpler problems are "univariate," involving a single scheme of classification with a finite number of categories. More generally, one may classify an experimental unit (or population member) according to each of several schemes, the interest being in any possible relationships among the schemes. In the univariate case, one might have two (or more) populations, and the problem would be to compare the populations with regard to a given classification scheme. It will be seen that there is an intimate relation between the problem of comparing univariate categorical populations and that of studying the relationship of the two classification schemes in a bivariate model.

For more thorough and extensive treatments, reference is made to the books of Kendall and Stuart [15], Vol. 2; and Bishop, Fienberg, and Holland [2].

9.1 Goodness of Fit

Suppose that each experimental unit is classified into one of k categories, $A_1, \ldots, A_k$. A probability model is defined by a probability vector $\mathbf{p} = (p_1, \ldots, p_k)$, where p_i is the probability of category A_i. A random sample of size n has the likelihood function

$$L(\mathbf{p}) = p_1^{f_1} \cdots p_k^{f_k},$$

where f_i is the frequency of A_i among the n observations, and (clearly) $\sum f_i = n$. The vector of frequencies $(f_1, \ldots, f_k)$ is minimal sufficient and has a multinomial distribution. (See Sections 3.4 and 4.4.5.)

The *goodness of fit* problem is that of testing a given set of specific probabilities:

$$H_0\colon \mathbf{p} = \boldsymbol{\pi} = (\pi_1, \ldots, \pi_k)$$

against the alternative that the correct model is not given by the vector $\boldsymbol{\pi}$.

9.1.1 Pearson's Chi-Square Statistic

It seems natural to measure the goodness of fit of a given set of data to the model $\mathbf{p} = \boldsymbol{\pi}$ in terms of the differences $f_i - n\pi_i$ between the observed and expected cell frequencies. In 1900, K. Pearson introduced the following statistic:

$$\chi^2 = \sum_{i=1}^{k} \frac{(f_i - n\pi_i)^2}{n\pi_i},$$

which has the value 0 for a "perfect" fit and is large when the fit is bad. The critical region is of the form $\chi^2 > K$.

The distribution of χ^2 depends on the model being tested, but a redeeming feature is that under H_0 the asymptotic distribution (large n) is independent of the model being tested. So a test based on χ^2 is asymptotically "distribution free." The asymptotic distribution is chi-square with $k - 1$ degrees of freedom under the null hypothesis (cf. Cramér [5], pp. 417 ff.). But there remains the question as to how small a sample can be and yet be large enough so that the asymptotic distribution is useful. No simple answer can be given, but it has been found that when the sample size is, say, four or five times the number of cells, the approximation is rather good, even if some of the expected frequencies $n\pi_i$ are quite small (as small as 1, or smaller). (Sometimes a rule of thumb is given that each $n\pi_i$ should be at least 5, but this is unnecessarily conservative.)

EXAMPLE 9–1. To test the equal likelihood of the six faces of a die, the die was cast 120 times with results follows:

Face	1	2	3	4	5	6
Frequency	18	23	16	21	18	24

The expected frequency of each face in 120 tosses is 20, so the test statistic has the value

$$\chi^2 = \frac{(18 - 20)^2}{20} + \frac{(23 - 20)^2}{20} + \cdots + \frac{(24 - 20)^2}{20} = 2.5.$$

The 95th percentile of the chi-square distribution with 5 degrees of freedom is 11.1; therefore, in a test with $\alpha = .05$ one would accept the null hypothesis.

If a sample is not large enough for one to get by with the asymptotic distribution, it may be that the exact distribution of χ^2 under H_0 can be calculated for a particular problem. (See Problem 9–4.)

The power of a chi-square test is analytically rather unmanageable, because of the extremely composite nature of the alternative hypothesis. It can be shown that the power tends to 1 for any alternative, as n becomes infinite; that is, the test is consistent. And it is sometimes pointed out (as in the case when H_0 is simple and some alternatives are arbitrarily "close" to H_0) that this means almost sure rejection of H_0 if one only takes a large enough sample. This is because one is unlikely to specify H_0 accurately enough to represent precisely the true distribution.

The power of the chi-square test has been studied empirically for certain classes of alternatives, mainly for its application to numerical populations (see Section 11.2 for discussion and references). This is done by artificial sampling—what is called a "Monte Carlo" method. A calculation of power is called for in Problem 9–4 for a small-sample test.

The Pearson chi-square test can be modified to test a distribution in which the expected frequencies of the categories depend on an unknown parameter, $\boldsymbol{\theta} = (\theta_1, \ldots, \theta_r)$. The statistic used is

$$\chi^2 = \sum_{i=1}^{k} \frac{[f_i - n\pi_i(\hat{\boldsymbol{\theta}})]^2}{n\pi_i(\hat{\boldsymbol{\theta}})},$$

where $\pi_i(\boldsymbol{\theta})$ is the probability of category i under H_0 and $\hat{\boldsymbol{\theta}}$ is an estimate of $\boldsymbol{\theta}$ obtained from the sample. It can be shown that if $\hat{\boldsymbol{\theta}}$ is consistent and asymptotically normal and efficient (as are maximum likelihood estimators), then the limiting distribution of the test statistic, under H_0, is again of the chi-square type but with $k - 1 - r$ degrees of freedom. Reduction of the number of degrees of freedom by the number of parameters being estimated shifts the critical boundary, so that χ^2 has to be smaller for acceptance at a given level. This is not surprising, inasmuch as the fit is bound to be better when the values from the sample are used for the parameters. Indeed, a *perfect* fit ($\chi^2 = 0$) can be achieved by estimating each cell probability, as a population parameter, by the corresponding sample relative frequency; this would mean that $r = k - 1$, and no degrees of freedom are available to test goodness of fit. But as long as there are at least two more cells than unknown parameters, there is residual information in the sample with which to test fit after estimating the parameters.

EXAMPLE 9–2. Five "coins" with identical but unknown values of $p = P(\text{heads})$ are tossed together 100 times to test the hypothesis that the number of heads per toss follows a binomial distribution. (Perhaps some kind of dependence is introduced in the tossing process.) The results are given as follows:

Number of heads	0	1	2	3	4	5
Frequency	3	16	36	32	11	2

The maximum likelihood estimate of p is the mean number of heads per five coins divided by 5, which turns out to be .476. Using this to calculate the cell probabilities by the binomial formula, one obtains the following expected frequencies:

4.0, 17.9, 32.6, 29.6, 13.5, 2.4.

The value of χ^2 is then found to be

$$\chi^2 = \frac{(3-4)^2}{4} + \cdots + \frac{(2-2.4)^2}{2.4} \doteq 1.53.$$

The 5 per cent rejection limit would be the 95th percentile of the chi-square distribution with $6 - 1 - 1 = 4$ degrees of freedom, which is 9.49. Since $1.53 < 9.49$, the null hypothesis is accepted.

A point needs to be mentioned in connection with using the chi-square test in the unspecified parameter case, when the model actually tested is an approximation to the one it is desired to test. This situation arises when categories are combined to increase expected cell frequencies or (as in the Poisson model) to reduce an infinite set of categories to a finite set. The question is how to estimate the parameters—from the reduced data (i.e., from the frequency tabulation actually used in the construction of χ^2) or from the original data. Thus, if the parameter to be estimated is the population mean, how should the sample mean be computed? The answer is that the asymptotic distribution claimed for χ^2 may not be correct in either case;[1] it *is* correct if one uses the estimate $\hat{\theta}$ that maximizes the likelihood function:

$$L(\theta) = \prod_1^k [p_i(\theta)]^{f_i},$$

where $p_i(\theta)$ is the probability of the ith of the k class intervals used in forming χ^2. The maximizing $\hat{\theta}$ may be awkward to obtain, but unfortunately the naïve estimates may cause one to operate at a significance level higher than he intended. In some cases, including the Poisson, the error is not serious.

9.1.2 The Likelihood-Ratio Test

The maximum of the likelihood function, when $\mathbf{p}$ is not restricted (i.e., under $H_0 + H_1$), has been seen in Example 5–16 to be attained for $p_i = f_i/n$. Hence, in testing $\mathbf{p} = \boldsymbol{\pi}$ against $\mathbf{p} \neq \boldsymbol{\pi}$, the likelihood ratio is

$$\Lambda = \frac{L(\boldsymbol{\pi})}{L(\hat{\mathbf{p}})} = \prod_1^k \frac{\pi_i^{f_i}}{(f_i/n)^{f_i}},$$

The asymptotic distribution of $-2 \log \Lambda$ is chi-square with $k - 1$ degrees of freedom under H_0, where $k - 1$ is the difference between $k - 1$, the number of free parameters in the denominator, and 0, the number in the numerator. So the approximate significance level of the critical region $\Lambda < K$ can be determined from the chi-square table.

[1] See H. Chernoff and E. L. Lehmann, "The Use of the Maximum Likelihood Estimates in χ^2 Tests of Goodness of Fit," *Ann. Math. Stat.* **25**, 579 (1954).

It is of interest to note that under H_0 Pearson's χ^2 is asymptotically equal (in the sense of limit in probability) to $-2 \log \Lambda$. The following steps could be "rigorized":

$$\log \pi_i - \log \frac{f_i}{n} \doteq \frac{1}{f_i/n}\left(\pi_i - \frac{f_i}{n}\right) - \frac{1}{(f_i/n)^2}\frac{(\pi_i - f_i/n)^2}{2},$$

from which

$$\begin{aligned} -2 \log \Lambda &= -2 \sum f_i\left(\log \pi_i - \log \frac{f_i}{n}\right) \\ &\doteq -2 \sum n\left(\pi_i - \frac{f_i}{n}\right) + \sum \frac{n^2}{f_i}\left(\pi_i - \frac{f_i}{n}\right)^2 \\ &\doteq 0 + \sum \frac{(n\pi_i - f_i)^2}{n\pi_i}. \end{aligned}$$

The asymptotic equivalence of $-2 \log \Lambda$ and χ^2 can be thought of as justifying, in a sense, the intuitive choice of a one-tail critical region for χ^2. For, the region $\chi^2 >$ const is asymptotically equivalent to $\Lambda <$ const, which, as a likelihood ratio test, has certain optimality properties (see Wilks [22], p. 413).

EXAMPLE 9–3. Consider again the die-tossing data of Example 9–1:

Face	1	2	3	4	5	6
Frequency	18	23	16	21	18	24

The test is defined in terms of the statistic

$$\begin{aligned} -2 \log \Lambda &= -2n \log n - 2 \sum f_i(\log \pi_i - \log f_i) \\ &= -2[120 \log 120 + 120 \log \tfrac{1}{6} - 18 \log 18 - \cdots - 24 \log 24] = 2.503. \end{aligned}$$

(Recall, in computing this, that "log" means natural log.) As was to be expected, this is close to the χ^2 value of 2.5 found in Example 9–1, and the inference is the same.

If the probabilities π_i being tested are defined in terms of an unknown parameter vector $\boldsymbol{\theta}$, the likelihood in the numerator of Λ is maximized over $\boldsymbol{\theta}$:

$$\Lambda = \frac{L(\pi(\hat{\boldsymbol{\theta}}))}{L(\hat{\mathbf{p}})},$$

and the asymptotic chi-square distribution of $-2 \log \Lambda$ (under H_0) has $k - 1 - r$ degrees of freedom, r again being the dimension of $\boldsymbol{\theta}$. This adds, perhaps, to the plausibility of $k - 1 - r$ as the parameter of the asymptotic distribution of χ^2 when r parameters are estimated to obtain expected cell frequencies. For, although the argument suggested for relating Λ and χ^2 was given for the simpler case ($r = 0$), it is equally convincing more generally, since both $\pi(\hat{\boldsymbol{\theta}})$ and f_i/n converge to $\pi(\boldsymbol{\theta})$, if $\pi(\cdot)$ is a reasonable function.

In the case of small samples the null distribution of the likelihood ratio is not simple and depends on H_0, as in the case of χ^2. The statistics χ^2 and Λ are not always equivalent in their ordering of frequency vectors, and it is not clear which is the better one to use. (See Problem 9–4.)

Problems

9–1. Record the last digit in 50 successive entries of a four-place log table (Table XV in the Appendix), and test the hypothesis that the digits are uniformly distributed among 0, 1, . . ., 9.

9–2. In 30 settings of gill nets in a lake, the following counts of northern pike were noted:

Number of fish in a set	0	1	2	3	4
Frequency	12	10	6	0	2

Test the hypothesis that the number of pike per set has a Poisson distribution.

9–3. Show that if $Y_1, \ldots, Y_k$ are independent Poisson variates with $EY_i = np_i$, then their conditional joint distribution given $\sum Y_i = n$ is multinomial with parameters $(n; p_1, \ldots, p_k)$.

9–4. Consider testing $p_1 = p_2 = p_3$ in a discrete population with three possible outcomes, using six independent trials.

(a) Determine the distribution of χ^2 under H_0.
(b) Determine the distribution of Λ under H_0.
(c) Sketch the chi-square c.d.f. with two degrees of freedom and superpose a sketch of the c.d.f. of χ^2.
(d) Calculate the power of the test $\chi^2 > 3$ for the particular alternative $p_2 = p_3 = .5$.

9.2 The Bernoulli Model

Problems of inference for a categorical population with two categories—a Bernoulli population—have been taken up in earlier chapters to illustrate general concepts of inference for a one-parameter family of distributions. Estimation of the Bernoulli parameter p was encountered in Chapter 5 (Problems 5–8 and 5–10, and Example 5–16), and Bayesian estimation of p was considered in Chapter 8. The problem of testing based on a random sample of fixed size was handled in Chapter 6, (Examples 6–5, 6–7, 6–17, and 6–19, and Problems 6–8 and 6–25), as were the cases of sequential sampling (Examples 6–23, 6–24, and 6–26) and sampling without replacement (Problems 6–10 and 6–28).

A large-sample test for $p = p_0$ can be constructed on the basis of the asymptotic normality of the sample relative frequency of success. If there are f successes in n independent trials, then

$$Z = \frac{f/n - p_0}{\sqrt{p_0(1 - p_0)/n}}$$

is asymptotically standard normal, so a critical region of the form $f/n - p_0 > K$ is related to a given significance level by the standard normal table. The quantity Z^2 has an asymptotic distribution that is chi-square with one degree of freedom. It is shown in Problem 9–5 that Z^2 is precisely the same as χ^2, the Pearson goodness-of-fit statistic computed for the present problem—that of testing the probabilities $(p_0, 1 - p_0)$ for the two categories of success and failure. (This shows, incidentally, that the limiting distribution of χ^2 is indeed chi-square when $k = 2$.)

A large-sample confidence interval for p is obtained by starting with the statement,

$$P\left(\frac{|f/n - p|}{\sqrt{p(1 - p)/n}} < z_{1-\alpha/2}\right) = 1 - \alpha,$$

and solving the inequality for p:

$$\frac{(f/n - p)^2}{(p - p^2)/n} < k^2, \qquad \text{where } k = z_{1-\alpha/2},$$

or

$$\left(\frac{f}{n}\right)^2 - 2p\frac{f}{n} + p^2 < p\frac{k^2}{n} - p^2\frac{k^2}{n}.$$

This quadratic inequality in p is satisfied for p between the limits

$$\frac{1}{1 + k^2/n}\left(\frac{f}{n} + \frac{k^2}{2n} \pm k\sqrt{\frac{(f/n)(1 - f/n)}{n} + \frac{k^2}{4n^2}}\right),$$

which are then the $1 - \alpha$ confidence limits. Observe that if n is quite large, one can neglect the terms k^2/n and $k^2/(4n^2)$ and obtain the approximate limits

$$\frac{f}{n} \pm k\sqrt{\frac{(f/n)(1 - f/n)}{n}}.$$

This result can be thought of as a special case of the large-sample confidence limits $\bar{X} \pm kS/\sqrt{n}$ for a population mean. The success of the approximation in simpler limits depends on the value of p, but generally speaking, it requires a larger n than is needed for a normal approximation to be adequate. If p is neither close to 0 nor close to 1, a sample size of 25 or so is usually adequate for the normal approximation; but it should be 100 or more to use the cruder confidence limits.

A large-sample test for comparing two Bernoulli proportions p_1 and p_2 employing independent random samples is constructed as follows. Let f_i be the frequency of success among n_i observations on population i. Then if n_1 and n_2 are large, $f_1/n_1 - f_2/n_2$ is approximately normal with mean $p_1 - p_2$ and variance

$$\text{var}\left(\frac{f_1}{n_1} - \frac{f_2}{n_2}\right) = \frac{p_1(1 - p_1)}{n_1} + \frac{p_2(1 - p_2)}{n_2}.$$

Under the null hypothesis of no difference between populations, $p_1 = p_2 \equiv p$, this becomes

$$p(1 - p)\left(\frac{1}{n_1} + \frac{1}{n_2}\right).$$

The m.l.e. of p is the proportion of successes in the combined sample: $\hat{p} = (f_1 + f_2)/(n_1 + n_2)$, and substitution of this for p yields a measure of variability computable from the sample. The statistic

$$Z = \frac{f_1/n_1 - f_2/n_2}{\sqrt{\dfrac{f_1 + f_2}{n_1 + n_2}\left(1 - \dfrac{f_1 + f_2}{n_1 + n_2}\right)\left(\dfrac{1}{n_1} + \dfrac{1}{n_2}\right)}}$$

is then approximately standard normal. The hypothesis $p_1 = p_2$ would be rejected if Z is too far from 0 in a one-sided or two-sided critical region according to the nature of the alternative hypothesis.

EXAMPLE 9–4. Suppose a new drug is found to give relief by 111 of 300 people who try it, and that 39 out of 150 given a placebo for the same purpose find relief. Is the drug effective? A test for the hypothesis of no difference in probability of relief using the drug and using the placebo is based on the statistic

$$Z = \frac{\frac{111}{300} - \frac{39}{150}}{\sqrt{\frac{150}{450}\frac{300}{450}\left(\frac{1}{300} + \frac{1}{150}\right)}} \doteq 2.33.$$

This exceeds the critical value in a 5 per cent test (one-sided) and at that level one would infer a real difference. Whether that difference is great enough to call the drug "effective" is perhaps not for the statistician to say.

Problems

9–5. Show that in testing the probabilities $(p_0, 1 - p_0)$ for a Bernoulli model, Pearson's χ^2 is equal to $Z^2 \equiv (f/n - p_0)^2/[p_0(1 - p_0)/n]$, where f is the number of successes in n independent trials.

9–6. Compute approximate 95 per cent confidence intervals (use $k = 2$) for p in the following cases: (a) $n = 50, f = 5$; and (b) $n = 100, f = 50$. (Observe the effect of a small f/n in the accuracy of the cruder approximation.)

9–7. Discuss the accuracy in a poll asking a question with just two responses, when it is based on a random sample of 2000 individuals.

[*Note:* Many opinion polls use samples of between 1000 and 2000, but their sampling techniques (stratification, for instance) result in errors somewhat less than that inherent in ordinary random sampling.]

9–8. One poll shows 960 out of 2000 (48 per cent) in favor of a candidate, and another shows 765 out of 1500 (51 per cent) in favor. Is there evidence that they are based on samples from different populations? (Assume independent random samples.)

9.3 Two-Way Contingency Tables

A set of categories into which experimental outcomes are classed has been referred to as a "scheme of classification." The "categories" are possible states of something that *varies* from one outcome to another, and so the simpler term *variable* will be used—despite the earlier convention that a "variable" is numerical. Thus hair color, sex, political preference, and habitat, for example, are variables according to which experimental units may be classified.

When each outcome is classified according to more than one variable, data can be presented in what is called a *contingency table*—an array of frequencies, one for each cell in the array defined by a combination of choices of a category for each variable. When just two variables are considered, the table is called a *two-way* contingency table.

For example, a person may be classed according to marital status as M (married), S (single), or D (previously married, now living alone), and he may also be classed according to highest educational level as E (elementary school,) H (high school), C (college), G (graduate work).[2] The table of frequencies can be laid out in the following 3×4 array:

	E	H	C	G	
M	n_{11}	n_{12}	n_{13}	n_{14}	n_{1+}
S	n_{21}	n_{22}	n_{23}	n_{24}	n_{2+}
D	n_{31}	n_{32}	n_{33}	n_{34}	n_{3+}
	n_{+1}	n_{+2}	n_{+3}	n_{+4}	n

Row and column totals are given in the margin, the symbol + indicating a subscript that has been "summed out." These marginal entries are the frequencies of the categories of one variable or the other when considered by itself.

The discussion will employ the following notation. Variable A has r categories: $A_1, \ldots, A_r$, and variable B has c categories: $B_1, \ldots, B_c$. The contingency table

[2] Classifications have their practical problems, and it is not always possible to be clear cut. "College" may mean college degree or just some college attendance.

then has r rows and c columns, and (of course) rc cells. The frequency of the combination (A_i, B_j) is denoted by n_{ij} for $i = 1, \ldots, r$, $j = 1, \ldots, c$. As above, marginal frequencies are given by

$$n_{i+} = \sum_j n_{ij}, \qquad n_{+j} = \sum_i n_{ij}, \qquad n \equiv n_{++} = \sum \sum n_{ij}.$$

It will be assumed throughout that the individual outcomes are the results of independent, identical trials, with the implication that it suffices to summarize the results of sampling in the cell frequencies.

The *probability* of the *ij*-cell the probability that an outcome falls in category A_i of A and in category B_j of B will be denoted by p_{ij}. Marginal probabilities—the probabilities of the categories for one of the two variables by itself—are obtained by summing over one subscript or the other:

$$p_{i+} = \sum_j p_{ij}, \qquad p_{+j} = \sum_i p_{ij}, \qquad \sum p_{i+} = \sum p_{+j} = 1.$$

9.3.1 Independence and Homogeneity

There are two approaches to the statistical study of a relationship between two categorical variables, each involving its own model and mode of sampling. The approach suggested in the above introduction is to consider (A, B) as bivariate, defined on each member of a population (or for each outcome of an experiment), A random sample of size n is drawn and the sample outcomes are put into the appropriate cells and counted, the number of occurrences of the pair (A_i, B_j) being n_{ij}. These cell frequencies, as well as the marginal cell frequencies computed from them, are all random variables. The distribution of the cell frequencies is multinomial:

$$P(n_{11}, \ldots, n_{rc}) = \frac{n!}{\prod \prod n_{ij}!} \prod \prod p_{ij}^{n_{ij}}.$$

(Indexing symbols and their ranges, for products and sums, will be omitted when clearly implied by the context. Here the double products are over i and j with $i = 1, \ldots, r$ and $j = 1, \ldots, c$.)

An alternative approach is to consider that one variable, say B, defines c populations corresponding to the categories of B: $B_1, \ldots, B_c$. These may or may not be thought of as subpopulations of an encompassing population. A random sample of size n_{+j} is drawn from population B_j, for $j = 1, \ldots, c$. In each sample, the outcomes are classified according to A, and n_{ij} will denote the number of occurrences of A_i in the sample of n_{+j} outcomes from B_j. Again, the frequencies n_{ij} define a two-way table, but now the column totals (the sample sizes n_{+j}) are constants. The cell frequencies n_{ij}, for $i = 1, \ldots, r$, are the frequencies in a sample of size n_{+j} from a multinomial (r-nomial, to be precise) population; and the joint probability function of all the cell frequencies is the product of multinomial probabilities:

$$P(n_{11}, \ldots, n_{rc} \mid n_{+1}, \ldots, n_{+c}) = \prod_j \frac{n_{+j}!}{\prod_i n_{ij}!} \prod_i p_{i|j}{}^{n_{ij}},$$

where $p_{i|j}$ denotes the probability of A_i in subpopulation B_j. (The vertical bar here and in the left member of the above equality, can be interpreted in the usual sense of indicating a condition, as will be seen; but it can also mean that what follows is a parameter that would be too messy to attach to the notation in some other way.)

EXAMPLE 9–5. Suppose data on students are reported as shown in the following contingency table. The data could have been obtained by determining categories for

	Liberal Arts	Technology	Agriculture	Education
Smoker	30	25	15	20
Nonsmoker	20	25	35	30

each of the variables (college, smoker or not) for each student in a sample of 200 drawn at random from the students at large in a given school. Or they could have been obtained by drawing samples of 50 from each of the four colleges involved and determining the smoker status of each student. In either case, the interest would lie in a possible relationship between smoking and educational category; but the models and sampling methods are different. The fact that each of the column totals is 50 suggests strongly that the sampling was done from individual colleges. This type of sampling is feasible, since each college would presumably have a list of the students enrolled. However, such subpopulations are not always easy to isolate in order to sample from them.

The particular situation at hand will usually suggest or even dictate that one method of obtaining data be used in preference to the other. In either case the reason for gathering data is to learn about a possible relationship between variables A and B. When sampling is done at large from the bivariate population, the hypothesis of *independence* expresses a lack of relationship:

$$H_{\text{ind}}: p_{ij} = p_{i+}p_{+j}, \qquad \text{all } i, j.$$

When independent samples are obtained from the univariate (or conditional) populations $B_1, \ldots, B_c$, the hypothesis of *homogeneity* expresses a lack of relationship:

$$H_{\text{hom}}: p_{i|j} \text{ is a function of } i \text{ alone.}$$

Now, these apparently different models are both intended to describe the phenomenon of "no relationship," so it is perhaps not surprising to find that they are essentially equivalent:

Theorem 1. *Variables A and B are independent if and only if the variables $A \mid B_1, \ldots, A \mid B_c$ are identically distributed.*

Notice that if only the c univariate populations $B_1, \ldots, B_c$ with identical categories $(A_1, \ldots, A_r)$ are given, a bivariate population (A, B) can be constructed by using *any* distribution $(p_{+1}, \ldots, p_{+c})$ to form the joint probabilities:

$$p_{ij} \equiv p_{+j} p_{i|j},$$

and the theorem then applies.

It may be instructive to formalize the essence of the proof in a result about two-way tables of numbers:

Lemma. *In the matrix (n_{ij}) the following conditions are equivalent:*

1. *Rows are proportional.*
2. *Columns are proportional.*
3. $n_{ij} = n_{i+} n_{+j} / n_{++}$.

Conditions (1) and (2) are both expressed in the factorization $n_{ij} = a_i b_j$. That is, as a function of j the n_{ij} in a row are a constant multiple (depending on the row) of some numbers $b_1, \ldots, b_c$; and as a function of i, the n_{ij} in a column are a constant multiple (depending on the column) of numbers $a_1, \ldots, a_r$. Condition (3) clearly implies the other two; and if $n_{ij} = a_i b_j$, then $n_{i+} = a_i \sum b_j$, $n_{+j} = b_j \sum a_i$, and $n_{++} = \sum a_i \sum b_j$, and (3) follows.

Returning to Theorem 1, then, the condition $p_{ij} = p_{i+} p_{+j}$, which defines H_{ind} holds if and only if the columns of (p_{ij}) are proportional—that is, if and only if the conditional probabilities for A are identical for all categories of the conditioning variable B.

Upon substitution of the conditions defining H_{ind} and H_{hom} into the probability functions given earlier for the cell frequencies, one obtains the following results.

Theorem 2. *Under the hypothesis of* independence*:*

1. *The marginal totals for A and the marginal totals for B are independent multinomial vectors, with probability functions (p.f.'s)*

$$f_1(n_{1+}, \ldots, n_{r+}) = \frac{n!}{\prod n_{i+}!} \prod p_{i+}^{\,n_{i+}}$$

and

$$f_2(n_{+1}, \ldots, n_{+c}) = \frac{n!}{\prod n_{+j}!} \prod p_{+j}^{n_{+j}}.$$

2. *The joint p.f. of the cell frequencies is*

$$f_{H_{\text{ind}}}(n_{11}, \ldots, n_{rc}) = \frac{n!}{\prod\prod n_{ij}!} \prod p_{i+}^{n_{i+}} \prod p_{+j}^{n_{+j}}$$
$$= f_1 f_2 f_3,$$

where f_1 and f_2 are the p.f.'s of marginal totals as in (1), *and*

$$f_3 = \frac{\prod n_{i+}! \prod n_{+j}!}{n! \prod\prod n_{ij}!}.$$

3. *The function f_3 in* (2) *is the conditional p.f. of the cell frequencies given all marginal totals:*

$$f_3 = f_{H_{\text{ind}}}(n_{11}, \ldots, n_{rc} \mid n_{1+}, \ldots, n_{+c}).$$

Theorem 3. *Under the hypothesis of* homogeneity, *given samples of size n_{+1} from $B_1, \ldots, n_{+c}$ from B_c:*

1. *The marginal totals $(n_{1+}, \ldots, n_{r+})$ are multinomial with p.f. f_1 defined as in* (1) *of Theorem* 2.
2. *The joint p.f. of the cell frequencies is*

$$f_{H_{\text{hom}}}(n_{11}, \ldots, n_{rc}) = \prod_j \frac{n_{+j}!}{\prod_i n_{ij}!} \prod p_{i+}^{n_{i+}} = f_1 f_3,$$

where f_1 and f_3 are as in (1) *of Theorem* 2.

3. *The conditional p.f. of the cell frequencies given all marginal totals is again f_3:*

$$f_{H_{\text{hom}}}(n_{11}, \ldots, n_{rc} \mid n_{1+}, \ldots, n_{+c}) = \frac{\prod n_{i+}! \prod n_{+j}!}{n! \prod\prod n_{ij}!}.$$

The various p.f.'s above that involve the marginal probabilities as parameters are in the exponential family, a fact that can be exploited (using extensions of results quoted in Section 5.1.4) to establish the following:

Theorem 4. *Under H_{ind} the marginal totals $(n_{1+}, \ldots, n_{+c})$ are complete and sufficient for $(p_{1+}, \ldots, p_{+c})$. Under H_{hom}, the marginal row totals $(n_{1+}, \ldots, n_{r+})$ are complete and sufficient for $(p_{1+}, \ldots, p_{r+})$.*

The structure of "unrelated" rows and columns of numbers as given in the lemma to Theorem 1 is present in the matrix of cell means, under independence, unconditionally (clearly), and also conditional on the marginal totals:

Theorem 5. *Under H_{ind} (or H_{hom}) the expected cell frequencies, given all marginal totals, are obtained by multiplying marginal frequencies as follows:*

$$E(n_{ij} \mid n_{1+}, \ldots, n_{+c}) = \frac{n_{i+}n_{+j}}{n}.$$

This follows from the fact (Theorem 4) that the marginal totals (MT) are complete and sufficient. For, the statistic n_{ij} is an unbiased estimate of $np_{i+}p_{+j}$, and so the conditional mean $E(n_{ij} \mid MT)$ is the *only* function of the marginal totals that is an unbiased estimate of $np_{i+}p_{+j}$. But $n_{i+}n_{+j}/n$ is also a function of the marginal totals and is an unbiased estimate of $np_{i+}p_{+j}$. Therefore, it must be identical with $E(n_{ij} \mid MT)$, as was to be shown.

Maximum likelihood estimates are needed in the construction of likelihood ratio statistics:

Theorem 6. *Under H_{ind} the m.l.e.'s of the marginal probabilities are the corresponding marginal relative frequencies:*

$$\hat{p}_{i+} = \frac{n_{i+}}{n}, \qquad \hat{p}_{+j} = \frac{n_{+j}}{n}.$$

Under H_{hom} the m.l.e.'s of the common conditional probabilities $p_{i|j} = p_{i+}$ are the corresponding row total relative frequencies:

$$\hat{p}_{i|j} = \frac{n_{i+}}{n}.$$

Under either H_{ind} or H_{hom}, the expected cell frequencies e_{ij} have m.l.e.'s ($n\hat{p}_{ij}$ or $n_{+j}\hat{p}_{i|j}$, respectively)

$$\hat{e}_{ij}' = \frac{n_{i+}n_{+j}}{n}.$$

Proofs of the assertions in Theorem 6 are left as exercises for the reader.

The likelihood ratio technique can be used to devise a test for independence or for homogeneity. Not surprisingly, the likelihood ratio statistics for these two statistics situations are identical:

Theorem 7. *The likelihood ratio statistic Λ_{ind} used in testing H_{ind} against the alternative that the marginal variables are not independent is identical with the likelihood ratio statistic Λ_{hom} used in testing H_{hom} against the alternative that the populations $B_1, \ldots, B_c$ are not identically distributed.*

In testing H_{hom}, the likelihood function is

$$L = \prod \prod (p_{i|j})^{n_{ij}},$$

maximized at $p_{i|j} = n_{ij}/n_{+j}$. Under the null hypothesis, the likelihood becomes

$$L = \prod \prod (p_{i+})^{n_{ij}} = \prod p_{i+}{}^{n_{i+}},$$

maximized at $\hat{p}_{i+} = n_{i+}/n$. The likelihood ratio is then

$$\frac{\prod \prod (n_{i+}/n)^{n_{ij}}}{\prod \prod (n_{ij}/n_{+j})^{n_{ij}}} = \prod \prod \left(\frac{n_{i+}n_{+j}/n}{n_{ij}}\right)^{n_{ij}}.$$

The maximization in the numerator was over a space of dimension $r - 1$, and in the denominator over a space of dimension $c(r - 1)$, since there are $r - 1$ free parameters in each of the c columns. The difference in these dimensions, needed for the asymptotic distribution, is

$$c(r - 1) - (r - 1) = (r - 1)(c - 1).$$

It will be left to Problem 9–12 to show that under H_{ind} the likelihood ratio is the same and that the number of degrees of freedom is the same.

Problems

9–9. Suppose four observations on a bivariate categorical variable yield the following marginal totals:

			2
			2
1	2	1	4

(a) Determine the table of expected frequencies, given the marginal totals, under H_{ind}.

(b) Write out each of the possible tables with the given marginal totals, and for each one determine (i) the value of Λ and (ii) the probability of that table given the marginal totals. (In computing Λ, define $x^x = 1$ when $x = 0$.)

9–10. Show that the m.l.e. of p_{ij} under H_{ind} is $n_{i+}n_{+j}/n^2$. (Use the method of Lagrange's multipliers.)

9–11. Obtain the m.l.e. of $p_{i|j}$ under H_{hom}, and from it the m.l.e. of $E(n_{ij}) = n_{+j}p_{i|j}$.

9–12. Obtain the likelihood ratio, as a function of cell and marginal frequencies, for the problem of testing H_{ind} under the general alternative that H_{ind} is false. (Discuss degrees of freedom.)

9–13. Consider the case of two rows and two columns: $r = c = 2$. Observe that, given the marginal totals, all cell frequencies are determined when just one of them is specified. Show that the conditional distribution of that one, under either H_{ind} or H_{hom}, is hypergeometric. Verify the result of Theorem 5 by direct computation.

9.3.2 Large-Sample Tests

The likelihood ratio technique provides a large-sample test for independence and also a large-sample test for homogeneity. As seen in Theorem 7 of the preceding section, the test statistic is the same in both cases:

$$\Lambda = \frac{\prod n_{i+}{}^{n_{i+}} \prod n_{+j}{}^{n_{+j}}}{n^n \prod\prod n_{ij}{}^{n_{ij}}} = \prod\prod \left[\frac{(n_{i+}n_{+j})/n}{n_{ij}}\right]^{n_{ij}}.$$

The distribution of $-2 \log \Lambda$, under either null hypothesis, is approximately chi-square with $(r - 1)(c - 1)$ degrees of freedom, so the critical region $-2 \log \Lambda > K$ has size α if the critical value K is taken to be the $100(1 - \alpha)$ percentile of that chi-square distribution.

EXAMPLE 9–6. The following contingency table was encountered in Example 9–5:

30	25	15	20	90
20	25	35	30	110
50	50	50	50	

Carrying out the computation

$$-\log \Lambda = n \log n + \sum\sum n_{ij} \log n_{ij} - \sum n_{i+} \log n_{i+} - \sum n_{+j} \log n_{+j},$$

one obtains $-2 \log \Lambda = 10.25$. For 3 degrees of freedom, $\chi^2_{.95} = 7.81$ and $\chi^2_{.99} = 11.3$, so the observed result is significant at the 5 per cent level but not at 1 per cent. There is evidence that smoking (the row variable) and college of enrollment (the column variable) are related somewhat. (Whether the statistical relationship is the consequence of a causal relationship, though an important question, is not answerable by statistical analysis alone.)

Even though the hypotheses of independence and homogeneity are essentially the same, and the likelihood ratio test is the same in each case (for a given α), considerations of power would be different in the two models and corresponding methods of sampling. For instance, in the Example 9–6, a given set of unequal probabilities, say (.4, .5, .5, .5), for $p_{1|j}$ would correspond to a degree of independence different from that defined by (.5, .5, .5, .4) if the subpopulations B_1 and B_4 are of different size (so that $p_{+1} \neq p_{+4}$).

The testing of independence can be treated by considering it to be a problem of testing goodness of fit to the model $p_{ij} = p_{i+}p_{+j}$, in which there are $r + c - 2$ unspecified parameters (p_{i+} and p_{+j} for $i = 1, \ldots, r - 1$ and $j = 1, \ldots, c - 1$). The Pearson chi-square statistic is constructed as usual, using the m.l.e.'s of the mean cell frequencies (see Theorem 6):

$$\chi^2 = \sum\sum \frac{(n_{ij} - n_{i+}n_{+j}/n)^2}{n_{i+}n_{+j}/n}.$$

Under H_{ind}, the asymptotic distribution of χ^2 is chi-square with

$$rc - 1 - (r + c - 2) = (r - 1)(c - 1)$$

degrees of freedom, and this is consistent with the fact that (under H_{ind}) χ^2 is asymptotically the same as $-2 \log \Lambda$. The critical region for χ^2 would be $\chi^2 > K$, using the same critical constant K as in the case of $-2 \log \Lambda$.

The chi-square statistic just defined can be formally calculated for any contingency table, even if it summarizes data obtained for testing H_{hom}—namely, if it gives the frequencies of A_i in independent samples from populations $B_1, \ldots, B_c$, each with the categories of A. Now, given p_{i+}, it is clear that under H_{hom} the quantity

$$\sum_j \left\{ \sum_i \frac{(n_{ij} - n_{+j}p_{i+})^2}{n_{+j}p_{i+}} \right\}$$

is asymptotically a sum of c independent chi-square variates and so is chi-square with $c(r - 1)$ degrees of freedom. But when the estimates n_{i+}/n are substituted for p_{i+} in each of the c terms, it is not immediately obvious that the result is asymptotically chi-square with $c(r - 1) - (r - 1)$ degrees of freedom. Nevertheless, it can be shown that such is the case—that the limiting distribution of χ^2 under H_{hom}, as each sample size n_{+j} becomes infinite (with no n_{+j}/n approaching 0), is chi-square with $(r - 1)(c - 1)$ degrees of freedom. So the large sample chi-square test used for testing independence can also be used as a test for homogeneity. In setting the critical value for a given significance level, it is not necessary to know which of the two sampling methods was used in obtaining the data.

EXAMPLE 9–7. Again consider the table of Examples 9–5 and 9–6:

30	25	15	20	90
20	25	35	30	110
50	50	50	50	200

As mentioned earlier, the equal marginal totals at the bottom suggest that the data result from independent samples of size 50, obtained for testing homogeneity. Whether that is the case, or whether 200 observations on a bivariate population were obtained (with the column totals just happening to come out equal), the chi-square statistic provides a test for the presence of a relationship. The expected values in the first row are $50 \times 90/200$ and in the second, $50 \times 110/200$. Using these one obtains $\chi^2 = 10.10$ (as compared with 10.25 for $-2 \log \Lambda$ obtained earlier). This exceeds 7.81, the critical value corresponding to $\alpha = .05$, so at that level homogeneity is rejected.

9.3.3 Small-Sample Tests

A small-sample test for either independence or homogeneity—called *Fisher's exact test*—can be constructed as follows, using the conditional distribution of the cell frequencies given the marginal totals. From among all possible tables that have given totals, one selects certain ones for inclusion in a critical set, a set of probability (in the conditional distribution) not exceeding a specified α. Having chosen such a critical set for each set of marginal totals, one obtains the sample in the form of a contingency table, notes the marginal totals, and determines whether the sample table is one of those in the critical set for those marginal totals. If it is, then the hypothesis of independence (or homogeneity) is rejected. The size of the test carried out in this way does not exceed α:

$$P_{H_0}(\text{reject } H_0) = E[P_{H_0}(\text{reject } H_0 \mid \text{marginal totals})] \le E(\alpha) = \alpha.$$

For given marginal totals, the critical set should include tables that suggest lack of independence, and this can be achieved by using either χ^2 or Λ with which to order the tables—although these variables may not always yield the same ordering of tables. Tables with smallest Λ-values (or largest χ^2 values) would be put first into the critical set.

Lest the prospect of preparing all those critical sets—one for each set of marginal totals—appear prohibitively tedious, it should be pointed out that there is no need to make any probability computations other than those that apply to the marginal totals actually observed.

The probabilities needed in constructing a critical set of size not to exceed α are the conditional probabilities f_3 given in Theorems 2 and 3 of the preceding section:

$$f_3 = P_{H_0}(n_{11}, \ldots, n_{rc} \mid n_{1+}, \ldots, n_{+c}) = \frac{\prod n_{i+}! \prod n_{+j}!}{n! \prod\prod n_{ij}!},$$

where H_0 may mean either H_{ind} or H_{hom}. In the simplest case, namely, $r = c = 2$, the probabilities define a hypergeometric distribution:

$$f_3 = \frac{n_{1+}!}{n_{11}!\, n_{12}!} \cdot \frac{n_{2+}!}{n_{21}!\, n_{22}!} \cdot \frac{n_{+1}!\, n_{+2}!}{n!} = \frac{\binom{n_{1+}}{n_{11}}\binom{n_{2+}}{n_{21}}}{\binom{n}{n_{+1}}}.$$

This is a univariate distribution—the distribution of n_{11}, say, since n_{12}, n_{21}, and n_{22} are determined by it and the marginal totals. More generally, the distribution would be that of $(r - 1)(c - 1)$ free variables, the rest of the rc cell frequencies being determined by the marginal totals. (See Problem 9–9.)

EXAMPLE 9–8. To test $p_1 = p_2$ in two Bernoulli populations [$p_i = P$(success) in population i], the following data are obtained:

	Success	Failure
From population 1	3	3
From population 2	7	2
	10	5

There are six 2 × 2 tables with the same marginal totals:

$$\begin{array}{cc|} 6 & 0 \\ 4 & 5 \\ \hline \end{array} \quad \begin{array}{cc|} 5 & 1 \\ 5 & 4 \\ \hline \end{array} \quad \begin{array}{cc|} 4 & 2 \\ 6 & 3 \\ \hline \end{array} \quad \begin{array}{cc|} 3 & 3 \\ 7 & 2 \\ \hline \end{array} \quad \begin{array}{cc|} 2 & 4 \\ 8 & 1 \\ \hline \end{array} \quad \begin{array}{cc|} 1 & 5 \\ 9 & 0 \\ \hline \end{array}$$

Identifying each table with the value of n_{12}, one finds the following probabilities and values of χ^2 and $-2 \log \Lambda$:

n_{12}	χ^2	$-2 \log \Lambda$	$P_{H_0}(n_{12} \mid MT)$
0	5	6.73	.042
1	1.25	1.323	.252
2	0	0	.42
3	1.25	1.243	.24
4	5	5.178	.045
5	11.25	13.689	.002

A test for homogeneity at $\alpha \leq .10$ would be carried out (for the observed marginal totals) by rejecting H_{hom} for $n_{12} = 0, 4$, or 5. With the given data one accepts H_{hom} at the 10 per cent level.

Fisher's exact test is an example of what is called a *conditional test*, in which a critical region is defined for the data given the value of some statistic T. Such a test may have optimality properties if the statistic T is not informative concerning the truth or falsity of the hypothesis being tested. In the case of a 2 × 2 table, the matter seems not to be resolved to everyone's satisfaction. There may be a little information in the marginal totals—but only a little, so that practically speaking the conditional test is good enough, even though it may not be strictly optimal.

9.3.4 Measures of Association

Upon dividing the value of Pearson's χ^2, computed for testing independence, by the sample size n, one obtains a quantity that depends only on the sample *proportions* in the various cells:

$$\frac{\chi^2}{n} = \sum\sum \frac{\left(\frac{n_{ij}}{n} - \frac{n_{i+}}{n} \cdot \frac{n_{+j}}{n}\right)^2}{\frac{n_{i+}}{n} \cdot \frac{n_{+j}}{n}}.$$

A similar computation in terms of *population* proportions, or cell probabilities, defines a population parameter:

$$\theta^2 \equiv \sum \sum \frac{(p_{ij} - p_{i+}p_{+j})^2}{p_{i+}p_{+j}}.$$

This has the value 0 when and only when $p_{ij} = p_{i+}p_{+j}$, for all i, j. The value of θ is sometimes used as a measure of departure from independence. A quantity proportional to θ:

$$\gamma \equiv \left[\frac{\theta^2}{\min(r-1, c-1)}\right]^{1/2}$$

has the property that $0 \leq \gamma \leq 1$, and that $\gamma = 1$ if and only if there is exactly one nonzero element in each row of the matrix (p_{ij}) if $r \geq c$ (or in each column if $r < c$).

The sample version of γ, namely,

$$C \equiv \left[\frac{\chi^2}{n \min(r-1, c-1)}\right]^{1/2}$$

measures association in the sample and would be used to estimate γ. As in the case of γ, $C = 0$ would mean that the rows (and columns) of the matrix (n_{ij}) are proportional, and $C = 1$ would imply that there is exactly one nonzero n_{ij} in each row (if $r \geq c$). The variance of C is known but is very complicated. An approximation sometimes proposed for large n is

$$\text{var } C \doteq [n \min(r-1, c-1)]^{-1},$$

but this may not be very good except under H_{ind}. It has been argued that except in the 2×2 case, the measure C is not readily interpretable, and in particular it does not permit computation of such things as a conditional probability of correctly classifying a population member according to variable A given his category in variable B. [Various measures of association are discussed in an important paper by L. A. Goodman and W. H. Kruskal, "Measures of Association for Cross Classifications," *J. Am. Stat. Assn.* **49**, 733–63 (1954).]

EXAMPLE 9–9. Schoolchildren were classified according to standard of clothing and according to intelligence with results as follows:[3]

	Intelligence Class						
	1	2	3	4	5	6	Totals
1	33	48	113	209	194	39	636
Clothing 2	41	100	202	255	138	15	751
Standard 3	39	58	70	61	33	4	265
4	17	13	22	10	10	1	73
Total	130	219	407	535	375	59	1725

[3] Given by M. G. Kendall and A. Stuart [15], quoting from a *Biometrika* article (Vol. 8, p. 94) by W. H. Gilby.

Calculations show $\chi^2 = 174.82$ and $-2 \log \Lambda = 164.19$. And then $C = [\chi^2/(3 \times 1725)]^{1/2} = .184$, with $\sigma_C \doteq .014$. An hypothesis of independence would be rejected, because the evidence for association is very strong. It might be pointed out that this would account for the discrepancy between χ^2 and $-2 \log \Lambda$, since one expects them to be approximately equal only for large n *and* under the null hypothesis of independence.

Problems

9–14. Compute χ^2 and test for independence in the following 2×2 table:

32	88	120
68	112	180
100	200	300

9–15. Given the following 2×2 table:

200	0
0	600

(a) Calculate the measure of association C.
(b) Calculate $-2 \log \Lambda$ and compare with χ^2. (Is the discrepancy surprising?)

9–16. Calculate C and its standard deviation, for the data of Example 9–7.

9–17. (Calculator desirable.) An opinion research group asked this question of 2023 persons: "Do you read newspapers less, more, or about the same as you did a few years ago?" The results, together with information on the level of education, are given in the table below. Test for independence of response to the question and level of education at $\alpha = .01$.

	Some College	High School	Less
Read more	209	298	227
Read less	87	131	202
Same	180	284	348
Don't know	10	15	32

9–18. Determine the critical set for testing independence with $\alpha \le .30$ given the following marginal totals:

		4
		8
9	3	

9–19. Four independent trials are carried out in each of three Bernoulli populations, with results as follows:

	Population 1	2	3
Success	1	3	0
Failure	3	1	4

Carry out the conditional exact test for homogeneity—that is, for the hypothesis that the probability of success is the same in all three populations.

9–20. For the case $r \leq c$, show that if there is exactly one nonzero frequency in each column of a contingency table, then $C = 1$.

9.3.5 A Loglinear Model

The model for a two-way contingency table can be recast in terms of new parameters—the logarithms of the cell probabilities, a formulation whose extension to tables of higher dimension (Section 9.4) seems to provide useful insights and realistic models.

The multinomial distribution—the model for cell counts when independent observations are cross-classified—is in the general exponential family, with log-likelihood

$$\log L = \sum \sum n_{ij} \log p_{ij} = \sum \sum n_{ij} \log e_{ij} - n \log n,$$

where e_{ij} is the expected cell frequency among n observations: $e_{ij} = np_{ij}$. It turns out to be fruitful to work with the "natural" parameters $\log p_{ij}$ or equivalently $\log e_{ij}$ in the construction of models.[4]

The condition $p_{ij} = p_{i+}p_{+j}$ that defines independence of the marginal variables becomes a condition of additivity in the structure of $\log e_{ij}$:

$$H_{\text{ind}}: \quad \log e_{ij} = \log e_{i+} + \log e_{+j} - \log n,$$

where $e_{i+} = np_{i+}$ and $e_{+j} = np_{+j}$. Thus, under the hypothesis of independence, the expected cell frequency has a logarithm that is the sum of a term depending on i and a term depending on j (and a constant). Such a sum can always be written in the form

$$\log e_{ij} = \alpha_i + \beta_j + \lambda,$$

where α_i and β_j are defined so that

$$\sum \alpha_i = \sum \beta_j = 0.$$

Moreover, if $\log e_{ij}$ has this additive structure, then independence obtains. For if

$$np_{ij} = e_{ij} = e^{\alpha_i + \beta_j + \lambda},$$

then

$$np_{i+} = e^{\alpha_i + \lambda} \sum e^{\beta_j},$$

$$np_{+j} = e^{\beta_j + \lambda} \sum e^{\alpha_i},$$

[4] When the term in the exponent of an exponential family density that mixes parameter and observation is of the form $\sum \theta_k R_k(x)$, the parameters θ are called *natural.*

and

$$n = e^{\lambda} \sum e^{\alpha_i} \sum e^{\beta_j}.$$

Multiplication then shows that $n^2 p_{i+} p_{+j} = n^2 p_{ij}$.

The significance of independence is that it corresponds to the lack of an "interaction" term— a term involving both subscripts—in the expression for the log expected frequency. The resulting additive model is similar to some that will be encountered in studying factorial experiments in Chapter 12.

9.4 Three-Way Classifications

Independent observations are classified according to three variables—variable A with categories A_i $(i = 1, \ldots, r)$, variable B with categories B_j $(j = 1, \ldots, s)$, and variable C with categories C_k $(k = 1, \ldots, t)$. The probability of the cell (A_i, B_j, C_k) will be called p_{ijk}, and the frequency of occurrence of this cell among n observations, n_{ijk}. As before, a " + " replacing a subscript will indicate that the subscript has been summed out. And in particular, $p_{+++} = \sum\sum\sum p_{ijk} = 1$, and $n_{+++} = n$.

The contingency table for the data n_{ijk} is three-dimensional. The third dimension, corresponding to variable C, defines what will be referred to as *layers* of the table. Thus a $2 \times 2 \times 2$ table will be given as follows:

n_{111}	n_{121}	n_{1+1}
n_{211}	n_{221}	n_{2+1}
n_{+11}	n_{+21}	n_{++1}

(Layer 1)

n_{112}	n_{122}	n_{1+2}
n_{212}	n_{222}	n_{2+2}
n_{+12}	n_{+22}	n_{++2}

(Layer 2)

n_{11+}	n_{12+}	n_{1++}
n_{21+}	n_{22+}	n_{2++}
n_{+1+}	n_{+2+}	n

(Marginal layer)

The most general model will have p_{ijk} unrestricted except for the condition $p_{+++} = 1$. The likelihood function in this model is given by the multinomial formula:

$$L = \prod\prod\prod p_{ijk}{}^{n_{ijk}},$$

with logarithm

$$\log L = \sum\sum\sum n_{ijk} \log e_{ijk} - n \log n,$$

where $e_{ijk} = np_{ijk}$ is the expected cell frequency. The likelihood is maximized when the cell probability has the value $\hat{p}_{ijk} = n_{ijk}/n$, the parameter space of the maximization having dimension $rst - 1$. These m.l.e. values for p_{ijk} fit the data perfectly, in the sense that the corresponding estimates of expected cell frequencies equal the observed frequencies.

Restricted models for the expected cell frequency will be assumed to have the form

$$\log e_{ijk} = \alpha_i + \beta_j + \gamma_k + \xi_{jk} + \eta_{ki} + \zeta_{ij} + \lambda,$$

where

$$\sum \alpha_i = \sum \beta_j = \sum \gamma_k = 0$$

and

$$\sum_j \xi_{jk} = \sum_k \xi_{jk} = \sum_k \eta_{ki} = \sum_i \eta_{ki} = \sum_i \zeta_{ij} = \sum_j \zeta_{ij} = 0.$$

By successively dropping, first the third-order "interaction" term to obtain this model, and then the second-order terms one by one, a hierarchy of models for $\log e_{ijk}$ is defined. This can be done in six ways, one of which results in the following hierarchy:

$$H_1: \quad \alpha_i + \beta_j + \gamma_k + \xi_{jk} + \eta_{ki} + \zeta_{ij} + \lambda,$$
$$H_2: \quad \alpha_i + \beta_j + \gamma_k + \xi_{jk} + \eta_{ki} + \lambda,$$
$$H_3: \quad \alpha_i + \beta_j + \gamma_k + \xi_{jk} + \lambda,$$
$$H_4: \quad \alpha_i + \beta_j + \gamma_k + \lambda.$$

The unrestricted model p_{ijk} would have to have a third-order interaction term; it will be denoted H_0.

A test for H_a in the framework of H_b, that is, for H_a against the alternative $H_b - H_a$ can be constructed using the likelihood ratio:

$$\log \Lambda_{ab} = \log \frac{\sup_{H_a} L(p)}{\sup_{H_b} L(p)}$$
$$= \sum \sum \sum n_{ijk}(\log \hat{e}_a - \log \hat{e}_b),$$

where $\hat{e}_a$ is the m.l.e. of the expected cell frequency under H_a and $\hat{e}_b$ is the m.l.e. under H_b. The variable $-2 \log \Lambda_{ab}$ has an asymptotic distribution under H_b that is chi-square with degrees of freedom given by the dimension of H_b minus the dimension of H_a.

9.4.1 Complete Independence

Model H_4 is the model for complete independence of the three variables of classification:

$$p_{ijk} = p_{i++}p_{+j+}p_{++k}.$$

For, if this condition is satisfied, then $\log e_{ijk}$ is clearly of the form of model H_4; and conversely, if model H_4 is assumed, then this multiplicative condition for independence is satisfied (see Problem 9–21).

Under H_4, the likelihood function has logarithm

$$\log L = \sum \sum \sum n_{ijk} \log (p_{i++}p_{+j+}p_{++k})$$
$$= \sum n_{i++} \log p_{i++} + \sum n_{+j+} \log p_{+j+} + \sum n_{++k} \log p_{++k}.$$

The maximum over H_4, a space of $r + s + t - 3$ dimensions, is achieved for

$$\hat{p}_{i++} = \frac{n_{i++}}{n}, \qquad \hat{p}_{+j+} = \frac{n_{+j+}}{n}, \qquad \hat{p}_{++k} = \frac{n_{++k}}{n},$$

and these m.l.e.'s are sufficient. The corresponding m.l.e.'s of the expected cell frequencies are

$$\hat{e}_{ijk} = \frac{n_{i++}n_{+j+}n_{++k}}{n^2}.$$

To test the hypothesis of independence against the general alternative (H_0) that they are not independent, one uses the statistic

$$\begin{aligned} -2 \log \Lambda_{40} &= -2 \sum\sum\sum n_{ijk}(\log \hat{e}_4 - \log \hat{e}_0) \\ &= -2\left\{\sum\sum\sum n_{ijk}[\log (n_{i++}n_{+j+}n_{++k}) - \log n_{ijk}] - 2n \log n\right\}, \end{aligned}$$

which is asymptotically chi-square, under H_4, with

$$\begin{aligned} &(rst - 1) - (r + s + t - 3) \\ &\quad = (r-1)(s-t)(t-1) + (r-1)(s-1) + (s-1)(t-1) + (t-1)(r-1) \end{aligned}$$

degrees of freedom.

EXAMPLE 9–10. One hundred observations are classified according to variables A, B, and C, each with two categories, the results being summarized in the following $2 \times 2 \times 2$ table:

Layer 1			Layer 2			Marginal layer		
6	14	20	6	4	10	12	18	30
4	16	20	34	16	50	38	32	70
10	30	40	40	20	60	50	50	100

(Layer 1) (Layer 2) (Marginal layer)

Here, $n_{1++} = 30$, $n_{2++} = 70$, $n_{+1+} = n_{+2+} = 50$, $n_{++1} = 40$, and $n_{++2} = 60$; and these determine the expected cell frequencies, under H_{ind}, as follows:

6	6	9	9	15	15
14	14	21	21	35	35

Using these to calculate $-2 \log \Lambda_{40}$ one obtains the value 30.685, well beyond the 99 per cent point, say, on the chi-square distribution with 4 degrees of freedom. Independence would be rejected—but one would naturally wonder if there may be any relationship somewhere between complete independence and the general model that simply denies independence.

9.4.2 Higher-Order Models

Adding the second-order interaction term ξ_{jk} to the model for $\log e_{ijk}$ that defines independence yields model H_3:

$$\log e_{ijk} = \alpha_i + \beta_j + \gamma_k + \xi_{jk} + \lambda.$$

It can be shown (along the lines of the demonstration in Problem 9–22) that if H_3 holds, then

$$p_{ijk} = p_{i++}p_{+jk},$$

or equivalently,

$$P(A_iB_jC_k) = P(A_i)P(B_jC_k).$$

Variable A is then said to be jointly independent of B and C. Under this model the log-likelihood is

$$\log L = \sum n_{i++} \log p_{i++} + \sum\sum n_{+jk} \log p_{+jk},$$

which is maximized by the m.l.e.'s

$$\hat{p}_{i++} = \frac{n_{i++}}{n}, \qquad \hat{p}_{+jk} = \frac{n_{+jk}}{n}.$$

These are sufficient for H_3 and define the expected cell frequencies under H_3:

$$\hat{e}_{ijk} = \frac{n_{i++}n_{+jk}}{n}.$$

The parameter space of H_3 has dimension $(r - 1) + (st - 1)$.

Adding another second-order term η_{ki} to the model for log-expected frequency yields H_2,

$$\log e_{ijk} = \alpha_i + \beta_j + \gamma_k + \xi_{jk} + \eta_{ki} + \lambda.$$

This can be shown to be equivalent to the following condition on cell probabilities:

$$p_{ijk} = \frac{p_{i+k}p_{+jk}}{p_{++k}}, \qquad \text{all } i, j, k.$$

In terms of the variables A, B, and C, this can be written

$$P(A_iB_j \mid C_k) = P(A_i \mid C_k)P(B_j \mid C_k), \qquad \text{all } i, j, k.$$

That is, A and B are independent conditional on C—they are independent in layers. In this model the m.l.e.'s of the sufficient cell probabilities, p_{i+k}, p_{+jk}, p_{++k} yield the following m.l.e. of the expected cell frequency:

$$n\hat{p}_{ijk} = \frac{n_{i+k}n_{+jk}}{n_{++k}}.$$

To obtain this, it was necessary to maximize over a space of dimension

$$(rt - t) + (st - t) + (t - 1) = rt + st - t - 1.$$

The model called H_1, with all three second-order interaction terms present, does not correspond to any simple interpretation in terms of independence. It can be shown, however, that in the $2 \times 2 \times 2$ case it is equivalent to the constancy of the *cross-product ratio* across layers:

$$\frac{p_{111}p_{221}}{p_{211}p_{121}} = \frac{p_{112}p_{222}}{p_{212}p_{122}}.$$

The marginal totals n_{ij+}, n_{i+k}, n_{+jk} are sufficient, but it is not possible to give direct formulas, in terms of them, for the m.l.e.'s of the expected cell frequencies. However, these m.l.e.'s can be obtained by an iterative process that will be demonstrated in Example 9–11. The dimension of the parameter space for H_1 is

$$rs + st + tr - (r + s + t).$$

9.4.3 Selecting a Model

There are six hierarchies of models, corresponding to the six orders in which one adds second-order interaction terms one by one. Consider again the hierarchy defined earlier, consisting of models designated $H_0, \ldots, H_4$. With the likelihood ratio Λ_{ab} defined as before, it is apparent that

$$-2 \log \Lambda_{04} = -2[\log \Lambda_{01} + \log \Lambda_{12} + \log \Lambda_{23} + \log \Lambda_{34}],$$

a decomposition in which the terms correspond to the decreasing complexity of the model as one proceeds down the hierarchy from H_0 to H_4. Degrees of freedom are computed as follows:

a	Dimension of H_a
0	$rst - 1$
1	$rs + st + tr - (r + s + t)$
2	$rt + st - t - 1$
3	$r + st - 2$
4	$r + s + t - 3$

ab	Degree of Freedom, H_a versus H_b
10	$(r - 1)(s - 1)(t - 1)$
21	$(r - 1)(s - 1)$
32	$(r - 1)(t - 1)$
43	$(s - 1)(t - 1)$

Although the degrees of freedom add, it should be remembered that the terms in the decomposition of $\log \Lambda$ are not always independent statistics.

For the given hierarchy, one may compute the component likelihood ratios and take them as measures of degradation of the fit as the model is successively simplified. No hard and fast rule will be given here (if indeed one can be given, in view of the subjective nature of the problem) for deciding how much simplification is warranted.

It should be pointed out that the hierarchy used here is but one of six, and that using different hierarchies may lead to different conclusions. So the hierarchy should, if possible, be chosen according to the particular field of application by deciding which interactions might be reasonably dropped first.

EXAMPLE 9–11. Consider again the data given in Example 9–10:

6	14	20	6	4	10	12	18	30
4	16	20	34	16	50	38	32	70
10	30	40	40	20	60	50	50	100

The m.l.e.'s of expected cell frequencies under H_4 were given in Example 9–10. Under H_3 and H_2, they are as follows (in each case the marginal totals that determine the m.l.e.'s are circled):

H_3 [A independent of (B, C)]

3	9	12	12	6	18	15	15	(30)
7	21	28	28	14	42	35	35	70
(10)	30	(40)	(40)	20	(60)	50	50	100

H_2 (independent in layers)

5	15	(20)	$\frac{40}{6}$	$\frac{20}{6}$	(10)
5	15	20	$\frac{200}{6}$	$\frac{100}{6}$	50
10	30	(40)	(40)	20	(60)

These were computed by the formulas given in the preceding section. In the case of H_1, the iterative process mentioned there yielded the following results (approximate). The process itself will be explained below.

H_1 (all three interactions present)

5.17	14.82	19.99	6.83	3.18	10.01
4.84	15.17	20.01	33.16	16.83	49.99
10.01	29.99	40	39.99	2.01	60

(The marginal layer totals agree exactly with those of the data.) Inserting these various estimates in the appropriate likelihood formula to obtain the needed maxima, one obtains the following values for $-2 \log \Lambda_{ab}$:

a	b	$-2 \log \Lambda_{ab}$	Degree of Freedom
4	3	17.261	1
3	2	12.654	1
2	1	.029	1
1	0	.742	1
4	0	30.685	4

It seems rather clear that simplification beyond model H_2 is not called for, but that this model of independence in layers fits the data rather well.

The iterative scheme for the estimate of expected cell frequencies under H_1 is as follows. Starting with 1's in place of all of the n_{ijk} multiply each row by a factor that makes the rows total correct—that is, "correct" according to the sufficient marginal totals:

		20			10		12	18
		20			50		38	32
10	30		40	20				

Adjusting entries of 1's:

1	1	2	1	1	2
1	1	2	1	1	2

so that the row totals are correct gives:

10	10	20	5	5	10
10	10	20	25	25	50
20	20		30	30	

Next the columns are multiplied by a factor that makes the column totals (which are wrong so far) correct:

5	15	20	20/3	10/3	10	35/3	55/3
5	15	20	100/3	50/3	50	115/3	95/3
10	30		40	20			

Now the row and column sums are correct but the sums across the layers, shown at the far right, are not. So the next adjustment, multiplying by an appropriate factor across layers, makes them correct:

36/7	162/11	48/7	36/11	12	18
114/23	288/19	760/23	320/19	30	32

But this has made the row and column totals a bit off, so the whole cycle is repeated. One more repetition yields the results given earlier in the example, in which the row and column totals are within .01 of the right values. (This is usually close enough for purposes of computing Λ.)

The iterative scheme demonstrated in the above example can be shown to give the correct m.l.e.'s, not only in the case of H_1, but (appropriately modified) in other cases as well. And in those instances in which the m.l.e.'s can be computed by formulas from the relevant marginal totals, the process quickly yields the exact answers. These matters and details of this admittedly sketchy account of three-way tables, as well as a thorough study of higher-dimensional tables, are taken up in Bishop, Fienberg, and Holland [2].

Problems

9–21. Show that if Model H_4 is assumed, the cell probabilities in a three-way table satisfy the multiplicative condition for independence.

9–22. Show that if log e_{ijk} has just one second-order interaction term, then one variable of classification is jointly independent of the other two, as claimed at the beginning of Section 9.4.2.

9–23. Derive the m.l.e.'s of the probabilities p_{i++} and p_{+jk} under H_3. (The method of Lagrange multipliers is useful.)

9–24. Derive the expected cell frequency estimates under H_3 by means of an iterative scheme. (One cycle suffices.)

9–25. Verify the computation of at least one of the log-likelihood ratio values in Example 9–11.

10

MULTIVARIATE DISTRIBUTIONS

Multivariate distributions—joint distributions of several random variables—have been encountered in earlier chapters. Basic definitions were given in Sections 2.1.7–2.1.10, 2.3.3, and 2.4.5. The multinomial model (Section 3.4 and Chapter 9) is multivariate, as is the joint distribution of the observations in a sample and the joint distribution of statistics (e.g., $\bar{X}$ and S^2) defined on a sample.

This chapter presents a further discussion of multivariate distributions as needed in preparation for Chapters 11 and 12. It includes also a brief discussion of correlation and prediction in bivariate populations. Inference for multivariate populations generally constitutes the subject of multivariate analysis, surveyed in several books (e.g., Anderson [1]).

10.1 Transformations

A review of certain aspects of transformations in one dimension will perhaps serve to point out the similarities between transformations in one dimension and transformations in several dimensions and make the latter seem more natural. Consider, then, a transformation defined by the continuously differentiable function $y = g(x)$, which maps the region R of x-values into the region S of y-values. If the derivative is continuous and does not vanish in R:

$$\frac{dy}{dx} = g'(x) \neq 0, \qquad \text{all } x \text{ in } R,$$

the transformation is monotonic. The graph of $y = g(x)$ is then steadily increasing or steadily decreasing. For each y in S there is a unique x in R, called $g^{-1}(y)$, such that $g[g^{-1}(y)] = y$. The function $g^{-1}(y)$ is the *inverse* function, and this is differentiable with derivative

$$\frac{dg^{-1}(y)}{dy} = \frac{dx}{dy} = \frac{1}{dy/dx} = \frac{1}{g'(x)}.$$

The differential coefficient dy/dx provides a change-in-length factor from dx to $dy = (dy/dx)\,dx$, and dx/dy similarly provides the change factor in going from dy to $dx = (dx/dy)\,dy$. A definite integral is transformed as follows:

$$\int_R h(x)\,dx = \int_S h(g^{-1}(y))\,\frac{dx}{dy}\,dy,$$

in which, of course, dx/dy represents the function of y obtained by differentiating $g^{-1}(y)$. If X is a random variable with density function $f_X(x)$, the random variable $Y = g(X)$ has the density function

$$f_Y(y) = f_X(g^{-1}(y))\left|\frac{dx}{dy}\right|.$$

10.1.1 Bivariate Transformations

In two dimensions, a transformation

$$\begin{cases} u = g(x, y), \\ v = h(x, y) \end{cases}$$

maps a region R of points in the xy plane into a region S of points in the uv-plane. It is assumed that g and h are continuously differentiable. The quantity that plays the role of the derivative is now the *Jacobian* of the transformation:

$$\frac{\partial(u, v)}{\partial(x, y)} = \begin{vmatrix} \dfrac{\partial u}{\partial x} & \dfrac{\partial u}{\partial y} \\ \dfrac{\partial v}{\partial x} & \dfrac{\partial v}{\partial y} \end{vmatrix}.$$

Suppose further that there is an inverse transformation:

$$\begin{cases} x = G(u, v), \\ y = H(u, v), \end{cases}$$

which takes each point (u, v) of S into a unique point (x, y) in R such that

$$\begin{cases} g[G(u, v), H(u, v)] = u, \\ h[G(u, v), H(u, v)] = v, \end{cases}$$

these being identities in (u, v). Then, if the Jacobian of the transformation is nonvanishing[1] in R, the Jacobian of the inverse transformation is defined in S as the reciprocal of the Jacobian of the direct transformation:

$$\frac{\partial(x, y)}{\partial(u, v)} = \left(\frac{\partial(u, v)}{\partial(x, y)}\right)^{-1},$$

[1] The nonvanishing of the Jacobian does not (as in R_1) guarantee the existence of a unique inverse globally, even though an inverse would be defined locally at each point.

and is the factor needed for conversion of area elements. Thus, the change of variables in a double integral is accomplished as follows:

$$\iint_R f(x, y)\, dx\, dy = \iint_S f(G(u, v), H(u, v)) \left| \frac{\partial(x, y)}{\partial(u, v)} \right| du\, dv,$$

where S is the image of the region R under the transformation.

If (x, y) is a possible value of the random vector (X, Y), the transformation being considered defines the new random vector (U, V) with

$$\begin{cases} U = g(X, Y), \\ V = h(X, Y). \end{cases}$$

This transformation induces a probability distribution in the uv-plane as follows: If S is a set in the uv-plane and R is the set of all points in the xy-plane that have "images" in S under the transformation, then

$$P[(U, V) \text{ in } S] = P[(X, Y) \text{ in } R].$$

If the distribution of (X, Y) is of the continuous type, the probability of S can be expressed as an integral over R:

$$\begin{aligned} P[(U, V) \text{ in } S] &= \iint_R f_{X,Y}(x, y)\, dx\, dy \\ &= \iint_S f_{X,Y}(G(u, v), H(u, v)) \left| \frac{\partial(x, y)}{\partial(u, v)} \right| du\, dv, \end{aligned}$$

where $f_{X,Y}(x, y)$ is the joint density function of (X, Y). Since this relation holds for each event S in the uv plane, the density of (U, V) is the integrand function of the uv-integral:

$$f_{U,V}(u, v) = f_{X,Y}(G(u, v), H(u, v)) \left| \frac{\partial(x, y)}{\partial(u, y)} \right|.$$

That is, to obtain the density of (U, V), solve the transformation equations for x and y in terms of u and v, substitute for x and y in the joint density of X and Y, and multiply by the absolute value of the Jacobian of (x, y) with respect to (u, v). And then, similarly, the density of (X, Y) as obtained from that of (U, V) is

$$f_{X,Y}(x, y) = f_{U,V}(g(x, y), h(x, y)) \left| \frac{\partial(u, v)}{\partial(x, y)} \right|.$$

EXAMPLE 10–1. Let R denote the rectangular region in the plane of (λ, μ) defined by the inequalities $0 < \lambda < \pi/2$, $0 \leq \mu < 2\pi$, and consider the transformation to the (x, y) plane defined by

$$x = \sin \lambda \cos \mu, \qquad \lambda = \sin \lambda \sin \mu.$$

The region R is mapped into the interior of the unit circle in the xy plane, which is then the region S shown in Figure 10–1. The random vector (θ, ϕ) defined on R with density $(\sin \lambda)/2\pi$ (as discussed in Example 2–23) is transformed into (X, Y), where

$$X = \sin \theta \cos \phi, \qquad Y = \sin \theta \sin \phi.$$

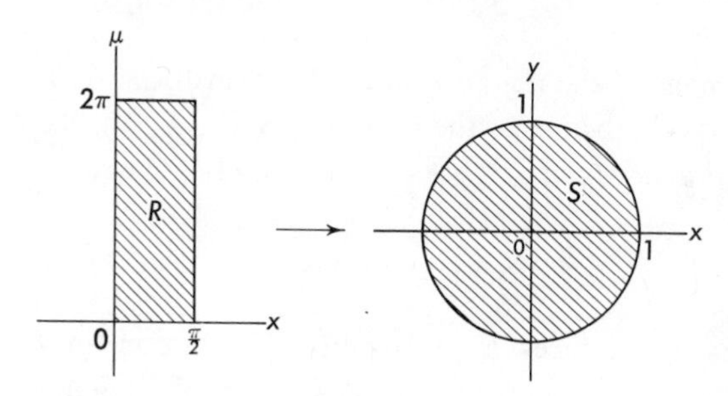

Figure 10–1

The density function of (X, Y) is easily written down, using the inverse transformation,

$$\begin{cases} \lambda = \arcsin (x^2 + y^2)^{1/2}, \\ \mu = \arctan \dfrac{y}{x}. \end{cases}$$

It is

$$f_{X,Y}(x, y) = f_{\theta,\phi}\left(\arcsin (x^2 + y^2)^{1/2}, \arctan \frac{y}{x}\right) \frac{1}{\sin \lambda \cos \lambda}$$

$$= \frac{1/(2\pi)}{[1 - (x^2 + y^2)]^{1/2}},$$

which holds for $x^2 + y^2 < 1$, or in S. The factor $[\sin \lambda \cos \lambda]^{-1}$ is the Jacobian of (λ, μ) with respect to (x, y), or the reciprocal of

$$\frac{\partial(x, y)}{\partial(\lambda, \mu)} = \begin{vmatrix} \cos \lambda \cos \mu & -\sin \lambda \sin \mu \\ \cos \lambda \sin \mu & \sin \lambda \cos \mu \end{vmatrix}.$$

It is sometimes convenient to derive the distribution of a function of random variables as the marginal distribution of a transformed bivariate distribution. The following example illustrates the technique.

EXAMPLE 10–2. Let X and Y be independent, standard normal variables, and define

$$U = \frac{X}{Y}, \qquad V = Y.$$

The inverse transformation is defined (except on a set of probability 0) by

$$X = UV, \qquad Y = V,$$

with Jacobian

$$\frac{\partial(x, y)}{\partial(u, v)} = \begin{vmatrix} v & u \\ 0 & 1 \end{vmatrix} = v$$

(which is nonvanishing except on the set mentioned, $v = y = 0$). Then,

$$f_{U,V}(u, v) = f_{X,Y}(uv, v)\left|\frac{\partial(x, y)}{\partial(u, v)}\right|$$

$$= \frac{|v|}{2\pi} \exp\left(-\frac{1}{2}(u^2v^2 + v^2)\right).$$

The marginal density of U is obtained by integrating out v:

$$f_{X/Y}(u) = \frac{1}{2\pi}\int_{-\infty}^{\infty} |v|e^{-(u^2+1)v^2/2}\, dv$$

$$= \frac{1}{\pi}\int_0^{\infty} ve^{-(u^2+1)v^2/2}\, dv = \frac{1/\pi}{1 + u^2}.$$

This is the density of a Cauchy distribution with median 0.

The results discussed above will now be specialized to the particular type of transformation called *linear*:

$$\begin{cases} x = au + bv, \\ y = cu + dv. \end{cases}$$

The Jacobian of this transformation is just the determinant of the coefficients:

$$\frac{\partial(x, y)}{\partial(u, v)} = \begin{vmatrix} a & b \\ c & d \end{vmatrix} = ad - bc \equiv \Delta.$$

If Δ vanishes, the coefficients a and b are proportional to c and d, so that y is a multiple of x. This degenerate case is ruled out by the assumption that the Jacobian of the transformation does not vanish, in which case there is a unique inverse:

$$\begin{cases} u = Ax + By, \\ v = Cx + Dy, \end{cases}$$

where $A = d/\Delta$, $B = -b/\Delta$, $C = -c/\Delta$, and $D = a/\Delta$. The Jacobian of the inverse transformation is

$$\frac{\partial(u, v)}{\partial(x, y)} = \begin{vmatrix} A & B \\ C & D \end{vmatrix} = AD - BC = \frac{1}{\Delta}.$$

Suppose now that the transformation is applied to define a random vector (X, Y) from a given random vector (U, V):

$$\begin{cases} X = aU + bV, \\ Y = cU + dV. \end{cases}$$

If (U, V) has a continuous distribution with joint density function $f_{U,V}(u, v)$, the random vector (X, Y) has the joint density

$$f_{X,Y}(x, y) = \frac{1}{|\Delta|} f_{U,V}(Ax + By, Cx + Dy),$$

in the nonsingular case ($\Delta \neq 0$).

EXAMPLE 10–3. Let U and V have independent, exponential distributions with mean 1, and define

$$\begin{cases} X = U + V, \\ Y = U - V. \end{cases}$$

The joint density of (U, V) is the product of the marginal densities:

$$f_{U,V}(u, v) = e^{-u-v}, \qquad u > 0 \text{ and } v > 0.$$

The Jacobian of the transformation is

$$\frac{\partial(x, y)}{\partial(u, v)} = \begin{vmatrix} 1 & 1 \\ 1 & -1 \end{vmatrix} = \Delta = -2,$$

and the inverse transformation is readily seen (upon solving for U and V) to be

$$\begin{cases} U = \frac{1}{2}(X + Y), \\ V = \frac{1}{2}(X - Y). \end{cases}$$

The first quadrant of the uv-plane, which carries the distribution of (U, V), is mapped into the region X between $x = y$ and $x = -y$, as shown in Figure 10–2. The joint density function of (X, Y) is

$$f_{X,Y}(x, y) = \tfrac{1}{2} f_{U,V}(\tfrac{1}{2}[x + y], \tfrac{1}{2}[x - y]) = \tfrac{1}{2}e^{-x}, \qquad \text{for } (x, y) \text{ in } S.$$

From this, incidentally, one can determine the density of the sum $U + V$ as the marginal density of X, by integrating out the y:

$$\begin{aligned} f_{U+V}(x) = f_X(x) &= \int_{-\infty}^{\infty} f_{X,Y}(x, y)\, dy \\ &= \tfrac{1}{2}\int_{-x}^{x} e^{-x}\, dy = xe^{-x}, \qquad x > 0. \end{aligned}$$

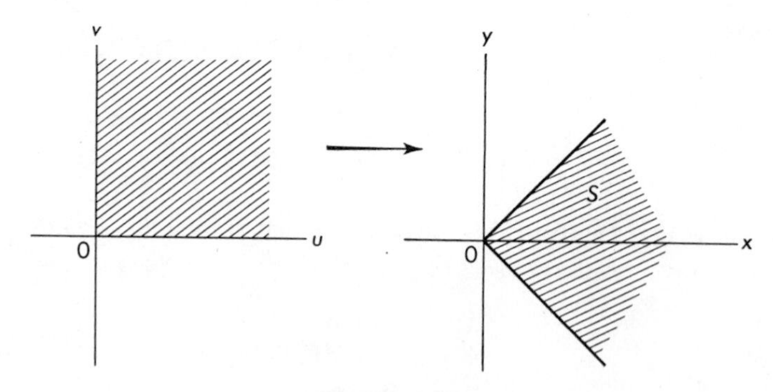

Figure 10–2

A transformation of the special type

$$\begin{cases} u = (\cos\theta)x - (\sin\theta)y, \\ v = (\sin\theta)x + (\cos\theta)y \end{cases}$$

is a *rotation*, having the property that $u^2 + v^2 = x^2 + y^2$; that is, distances from the origin are unchanged. The inverse of a rotation is also a rotation, and the Jacobian in each case is 1:

$$\begin{cases} x = (\cos\theta)u + (\sin\theta)v, \\ y = (-\sin\theta)u + (\cos\theta)v. \end{cases}$$

The following theorem will be useful in studying bivariate normal distributions.

Theorem. *Given any bivariate distribution, there is a rotation transformation that results in a distribution with uncorrelated components.*

To see this, consider a rotation through an angle θ, to be determined,

$$\begin{cases} Z = (\cos\theta)X - (\sin\theta)Y, \\ W = (\sin\theta)X + (\cos\theta)Y, \end{cases}$$

and calculate the covariance of Z and W:

$$\operatorname{cov}(Z, W) = (\cos\theta)(\sin\theta)(\sigma_X^2 - \sigma_Y^2) + (\cos^2\theta - \sin^2\theta)\sigma_{X,Y}.$$

This vanishes if θ is chosen so that

$$\cot 2\theta = \frac{\cos^2\theta - \sin^2\theta}{2\sin\theta\cos\theta} = \frac{\sigma_Y^2 - \sigma_X^2}{2\sigma_{X,Y}}.$$

Corollary. *If X and Y are not perfectly correlated, there is a linear transformation such that the new variables are uncorrelated and have unit variances. Moreover, this transformation is nonsingular (has an inverse).*

This follows by setting $U = Z/\sigma_Z$ and $V = W/\sigma_W$, which can be done provided neither σ_Z nor σ_W is zero. But if $\rho^2 < 1$, X and Y are not linearly related, so *no* linear combination can have zero variance (i.e., be a constant). The determinant of the combined transformation (from (X, Y) to (U, V)) is just $(\sigma_Z\sigma_W)^{-1}$, so it is nonsingular.

10.1.2 Multivariate Transformations

Given the outcome $(x_1, \ldots, x_n)$, one may wish to define the related quantities:

$$\begin{cases} u_1 = g_1(x_1, \ldots, x_n) \\ \vdots \qquad \vdots \\ u_n = g_n(x_1, \ldots, x_n), \end{cases}$$

for given functions $g_1, \ldots, g_n$. For inverses and integral transformations, the crucial quantity is again the *Jacobian* of the transformation:

$$\frac{\partial(u_1, \ldots, u_n)}{\partial(x_1, \ldots, x_n)} = \begin{vmatrix} \frac{\partial u_1}{\partial x_1} & \cdots & \frac{\partial u_n}{\partial x_1} \\ \vdots & & \vdots \\ \frac{\partial u_1}{\partial x_n} & \cdots & \frac{\partial u_n}{\partial x_n} \end{vmatrix}.$$

If this does not vanish in a region R, and there is an inverse transformation defined on the image of R:

$$\begin{cases} x_1 = G_1(u_1, \ldots, u_n) \\ \vdots \qquad \vdots \\ x_n = G_n(u_1, \ldots, u_n), \end{cases}$$

whose Jacobian is the reciprocal of that of the forward transformation, then the transformation from $\mathbf{x}$ to $\mathbf{u}$ in a multiple integral over R is accomplished by substituting $\mathbf{x}$ from the inverse transformation and transforming the volume element as follows:

$$dx_1 \cdots dx_n \to \left| \frac{\partial(x_1, \ldots, x_n)}{\partial(u_1, \ldots, u_n)} \right| du_1 \cdots du_n,$$

the integral in $\mathbf{u}$ then extending over the image of R.

If $(x_1, \ldots, x_n)$ is a possible value of the random vector $(X_1, \ldots, X_n)$, the transformation from $\mathbf{x}$ to $\mathbf{u}$ defines a random vector $\mathbf{U}$ with components

$$\begin{cases} U_1 = g_1(X_1, \ldots, X_n) \\ \vdots \qquad \vdots \\ U_n = g_n(X_1, \ldots, X_n). \end{cases}$$

If S denotes a set of points $(u_1, \ldots, u_n)$ and R the set of all points $(x_1, \ldots, x_n)$ having images in S, then

$$\begin{aligned} P[(U_1, \ldots, U_n) \text{ in } S] &= P[(g_1(X_1, \ldots, X_n), \ldots, g_n(X_1, \ldots, X_n)) \text{ in } S] \\ &= P[(X_1, \ldots, X_n) \text{ in } R]. \end{aligned}$$

And in the continuous case, this last probability can be expressed as an integral of the joint density function of X's:

$$\begin{aligned} P(S) &= \int \underset{R}{\cdots} \int f(x_1, \ldots, x_n)\, dx_1 \cdots dx_n \\ &= \int \underset{S}{\cdots} \int f(G_1(u_1, \ldots, u_n), \ldots, G_n(u_1, \ldots, u_n)) \left| \frac{\partial(x_1, \ldots, x_n)}{\partial(u_1, \ldots, u_n)} \right| du_1 \cdots du_n. \end{aligned}$$

The integrand of this $\mathbf{u}$-integral is then the joint density functions of the U's.

So far, the transformed variable has been assumed to be of the same dimension as the X's, but it is possible that one may be interested in a smaller number of U's than X's:

$$\begin{cases} U_1 = g_1(X_1, \ldots, X_n) \\ \vdots \qquad \vdots \\ U_m = g_m(X_1, \ldots, X_n), \end{cases}$$

where $m \leq n$. (If m were greater than n, the U's would be "overdetermined," and if not incompatible, then not all necessary.) One could supply $n - m$ additional compatible relations and then determine the distribution of the mU's of interest as a marginal distribution of that of the nU's. However, the transformation to $(U_1, \ldots, U_m)$ and the distribution in the $\mathbf{X}$ space determine a probability distribution directly in the m-dimensional $\mathbf{U}$ space, in the usual way:

$$P[(U_1, \ldots, U_m) \text{ in } T] = \int \cdots_R \int f(x_1, \cdots, x_n)\, dx_1 \cdots dx_n,$$

where R is the set of $(x_1, \ldots, x_n)$ with image points in T. Taking T to be a semi-infinite rectangle and differentiating yields, as usual, the density function of $(U_1, \ldots, U_m)$.

EXAMPLE 10–4. Let $X_1, \ldots, X_n$ be successive interarrival times in a Poisson process—independent random variables with the exponential density e^{-x}, $x > 0$. The times to the successive arrivals measured from time zero, the reference point of X_1, are the sums:

$$\begin{cases} Y_1 = X_1 \\ Y_2 = X_1 + X_2 \\ \quad \vdots \\ Y_n = X_1 + \cdots + X_n. \end{cases}$$

The inverse of this transformation of X's to Y's is given by

$$\begin{cases} X_1 = Y_1 \\ X_2 = Y_2 - Y_1 \\ \quad \vdots \\ X_n = Y_n - Y_{n-1}, \end{cases}$$

and the Jacobian of the transformation (either way) is 1. The joint density of the Y's is, therefore,

$$\begin{aligned} f_{\mathbf{Y}}(\mathbf{y}) &= f_{\mathbf{X}}(y_1, y_2 - y_1, \ldots, y_n - y_{n-1}) \\ &= \exp\left[-y_1 - (y_2 - y_1) - \cdots - (y_n - y_1)\right] = e^{-y_n}, \end{aligned}$$

a formula that holds for $0 < y_1 < y_2 < \cdots < y_n$. The marginal density of $Y_n = \sum X_i$ can then be obtained by integrating out $y_1, \ldots, y_{n-1}$:

$$\begin{aligned} f_{Y_n}(y_n) &= e^{-y_n} \int_0^{y_n} \int_0^{y_{n-1}} \cdots \int_0^{y_2} dy_1\, dy_2 \cdots dy_{n-1} \\ &= \frac{y_n^{n-1}}{(n-1)!} e^{-y_n}, \qquad \text{for } 0 < y_n. \end{aligned}$$

Problems

10–1. Discuss the geometric significance of the transformation $x' = x + h$, $y' = y + k$, for given constants h, k. Determine the joint density function of (X', Y') induced by this transformation on the variables (X, Y) with density $f(x, y)$.

10–2. Obtain the joint density of $2X - Y$ and $X + Y$ if X and Y are independent standard normal variables.

10–3. Show that a rotation transformation does not change the form of a distribution which is

(a) That of (X, Y) in Problem 10–2.
(b) Uniform on the region $x^2 + y^2 = 1$.

(Generalize?)

10–4. Let X and Y be independent observations on a uniform distribution on [0, 1]. Let $Z = XY$, $W = Y$, and obtain the density of $Z = XY$ as a marginal density of (Z, W).

10–5. Show that if X and Y have independent gamma distributions with $\lambda = 1$ (see Section 7.1.1), then $X/(X + Y)$ has a beta distribution.
[*Hint:* Let $Z = X + Y$, $W = X/(X + Y)$.]

10–6. Let X and Y have independent chi-square distributions, and let $U = X/Y$, $V = X + Y$. Derive the joint density of U and V and the marginal density of U. [The distribution of U could have been obtained by using, with it, $V = Y$, but doing as suggested yields the independence of X/Y and $X + Y$; verify this.]

10–7. Suppose (X, Y) has a bivariate distribution with $\mu_X = \mu_Y = 1$, $\sigma_X^2 = \sigma_Y^2 = 2$, and $\sigma_{X,Y} = 1$. Obtain new variables (U, V) as a linear transformation on (X, Y) so that $\sigma_U = \sigma_V = 1$, $\mu_U = \mu_V = 0$, $\sigma_{U,V} = 0$.

10–8. Let $\mathbf{U}$ have a density that is uniform, $f_{\mathbf{U}}(\mathbf{u}) = n!$, on the region $0 < u_1 < u_2 < \cdots < u_n < 1$. Let $X_1 = U_1$, $X_2 = U_2 - U_1, \ldots, X_n = U_n - U_{n-1}$, and obtain the density of $\mathbf{X}$.

10–9. Let X_i and Y_i be defined as in Example 10–4, and define $Z_1 = Y_1/Y_2, \ldots, Z_{n-1} = Y_{n-1}/Y_n$, and $Z_n = Y_n$. Obtain the joint density function of the Z's, and in this way show that they are independent.

10.1.3 The General Linear Transformation

In studying linear transformations, it is convenient to use matrix notation and methods.[2] The vectors $(x_1, \ldots, x_n)$ and $(u_1, \ldots, u_n)$ are written as column matrices:

$$\mathbf{x} = \begin{pmatrix} x_1 \\ \vdots \\ x_n \end{pmatrix}, \qquad \mathbf{u} = \begin{pmatrix} u_1 \\ \vdots \\ u_m \end{pmatrix}.$$

[2] Along with the basic operations of addition, multiplication, transposition, and inversion, the discussion (here and in Section 10.2.2) will include the notion of rank and the processes of diagonalizing square matrices and factoring nonnegative definite matrices.

The linear transformation from $\mathbf{x}$ to $\mathbf{u}$ defined by the equations

$$\begin{cases} u_1 &= a_{11}x_1 + \cdots + a_{1n}x_n \\ \vdots & \quad \vdots \qquad\qquad\quad \vdots \\ u_m &= a_{m1}x_1 + \cdots + a_{mn}x_n \end{cases}$$

is then written

$$\mathbf{u} = \mathbf{A}\mathbf{x},$$

where $\mathbf{A}$ is the matrix of the transformation:

$$\mathbf{A} = (a_{ij}) = \begin{pmatrix} a_{11} & \cdots & a_{1n} \\ \vdots & & \\ a_{m1} & \cdots & a_{mn} \end{pmatrix}.$$

If $m = n$ and if the Jacobian of the transformation, which is then the determinant of the square matrix $\mathbf{A}$ (written $\det \mathbf{A}$), does not vanish, the inverse transformation exists. It is given by

$$\mathbf{x} = \mathbf{A}^{-1}\mathbf{u},$$

with

$$\mathbf{A}^{-1} = \frac{1}{\det \mathbf{A}} \begin{pmatrix} A_{11} & \cdots & A_{n1} \\ \vdots & & \\ A_{1n} & \cdots & A_{nn} \end{pmatrix},$$

where A_{ij} is $(-1)^{i+j}$ times the determinant of order $n - 1$ obtained by striking from $\det \mathbf{A}$ the row and column containing a_{ij}. (This reduced determinant associated with a_{ij} is its "minor," and A_{ij} is its "cofactor.") The inverse matrix $\mathbf{A}^{-1}$ has the property that $\mathbf{A}\mathbf{A}^{-1} = \mathbf{A}^{-1}\mathbf{A} = \mathbf{I}$, the "identity matrix" of order n, which has 0's in every position except along the diagonal where $i = j$, and 1's along that diagonal.

Let $\mathbf{V}$ denote an m by n matrix whose elements are random variables V_{ij}. Define

$$E(\mathbf{V}) = \begin{pmatrix} E(V_{11}) & \cdots & E(V_{1n}) \\ \vdots & & \vdots \\ E(V_{m1}) & \cdots & E(V_{mn}) \end{pmatrix}$$

and observe that for any constant matrices $\mathbf{A}$ and $\mathbf{B}$ (of dimensions such that the products are defined):

$$E(\mathbf{AVB}) = \mathbf{A}E(\mathbf{V})\mathbf{B}.$$

Consider now the random vector $(X_1, \ldots, X_n)$ written as a column matrix $\mathbf{X}$. The column matrix of mean values is

$$\mu = E(\mathbf{X}) = \begin{pmatrix} E(X_1) \\ \vdots \\ E(X_n) \end{pmatrix},$$

and the *covariance matrix* $\mathbf{M}$, defined to be the matrix of second moments, about the means, is

$$\mathbf{M} = E[(\mathbf{X} - \mu)(\mathbf{X} - \mu)'],$$

where the prime denotes transposition or interchange of rows and columns. The element in the ith row and jth column of $\mathbf{M}$ is m_{ij}, the covariance of X_i and X_j. Clearly, $\mathbf{M}$ is *symmetric*: $\mathbf{M} = \mathbf{M}'$, since $m_{ij} = m_{ji}$. If $(X_1, \ldots, X_n)$ are independent, the covariance matrix is diagonal (nonzero elements occur only on the diagonal $i = j$), all covariances being 0 for $i \neq j$.

A covariance matrix is *nonnegative definite*. This means that it is symmetric and that for any constants $(c_1, \ldots, c_n)$ the quadratic form

$$\mathbf{c}'\mathbf{M}\mathbf{c} = \sum_{i=1}^{n}\sum_{j=1}^{n} c_i c_j m_{ij}$$

is nonnegative. This is seen as follows:

$$\begin{aligned}\mathbf{c}'\mathbf{M}\mathbf{c} = \mathbf{c}'E[(\mathbf{X} - \boldsymbol{\mu})(\mathbf{X} - \boldsymbol{\mu})']\mathbf{c} &= E[\mathbf{c}'(\mathbf{X} - \boldsymbol{\mu})(\mathbf{X} - \boldsymbol{\mu})'\mathbf{c}] \\ &= E([\mathbf{c}'(\mathbf{X} - \boldsymbol{\mu})]^2) \geq 0.\end{aligned}$$

Moreover, if at least one c_i is not 0, the vanishing of $\mathbf{c}'\mathbf{M}\mathbf{c}$ would imply that the linear combination $\mathbf{c}'(\mathbf{X} - \boldsymbol{\mu})$ is 0 with probability 1, and some $X_i - \mu_i$ is a linear combination of the other X's with probability 1. This would mean that the distribution of $\mathbf{X}$ is *singular*; that is, all of the probability mass is concentrated in a subspace of lower dimension, and the n-variate density does not exist. If $\mathbf{X}$ is *not* singular, then $\mathbf{c}'\mathbf{M}\mathbf{c}$ is actually positive unless $c_1 = \cdots = c_n = 0$; in this case $\mathbf{M}$ is said to be *positive definite*.

The determinant of $\mathbf{M}$ is positive or zero according to whether M is positive definite or just nonnegative definite. For if $\mathbf{c}'\mathbf{M}\mathbf{c} = 0$ for some $\mathbf{c} \neq \mathbf{0}$, then $\mathbf{c}'\mathbf{M}\mathbf{c} = E[\mathbf{c}'(\mathbf{X} - \boldsymbol{\mu})]^2 = 0$, and so $\mathbf{c}'(\mathbf{X} - \boldsymbol{\mu}) = 0$ with probability 1. But then $\mathbf{c}'\mathbf{M} = \mathbf{0}$, which says that the rows of $\mathbf{M}$ are linearly dependent, and $\det \mathbf{M} = 0$. Conversely, if $\mathbf{c}'\mathbf{M} = \mathbf{0}$ for some $\mathbf{c} \neq \mathbf{0}$, it follows that $\mathbf{c}'\mathbf{M}\mathbf{c} = 0$.

If $r < n$, where r is the rank of $\mathbf{M}$, then there is a subset of r of the X's whose r-variate distribution is not singular. For there is then a (principal) $r \times r$ submatrix of $\mathbf{M}$ which is positive definite, and this is the covariance matrix of a corresponding set of r of the X's. Moreover, since any larger square submatrix is singular, it follows that the remaining X's can be expressed as linear combination of the r that are linearly independent.

The joint characteristic function of $(X_1, \ldots, X_n)$ is conveniently expressed in matrix notation:

$$\phi_{\mathbf{X}}(\mathbf{t}) = E[\exp(it_1X_1 + \cdots + it_nX_n)] = E(e^{i\mathbf{t}'\mathbf{X}}).$$

Suppose now that **Y** is an m-dimensional random vector (written as a column matrix) defined by a linear transformation on **X**:

$$\mathbf{Y} = \mathbf{A}\mathbf{X},$$

where $m \leq n$. Then

$$\boldsymbol{\mu}_{\mathbf{Y}} = E(\mathbf{Y}) = E(\mathbf{A}\mathbf{X}) = \mathbf{A}E(\mathbf{X}) = \mathbf{A}\boldsymbol{\mu}_{\mathbf{X}},$$

and the matrix of second moments of **Y** is

$$\begin{aligned}\mathbf{M}_{\mathbf{Y}} &= E[(\mathbf{Y} - \boldsymbol{\mu}_{\mathbf{Y}})(\mathbf{Y} - \boldsymbol{\mu}_{\mathbf{Y}})'] = E[\mathbf{A}(\mathbf{X} - \boldsymbol{\mu}_{\mathbf{X}})(\mathbf{X} - \boldsymbol{\mu}_{\mathbf{X}})'\mathbf{A}'] \\ &= \mathbf{A}E[(\mathbf{X} - \boldsymbol{\mu}_{\mathbf{X}})(\mathbf{X} - \boldsymbol{\mu}_{\mathbf{X}})']\mathbf{A}' = \mathbf{A}\mathbf{M}_{\mathbf{X}}\mathbf{A}'.\end{aligned}$$

The characteristic function of **Y** is also easily obtained in terms of the characteristic function of **X**:

$$\phi_{\mathbf{Y}}(\mathbf{t}) = E(e^{i\mathbf{t}'\mathbf{Y}}) = E(e^{i\mathbf{t}'\mathbf{A}\mathbf{X}}) = E(e^{i[\mathbf{A}'\mathbf{t}]'\mathbf{X}}) = \phi_{\mathbf{X}}(\mathbf{A}'\mathbf{t}).$$

If **Y** is n-dimensional, so that **A** is square, and if **A** is nonsingular (that is, if $\det \mathbf{A} \neq 0$), there is an inverse transformation, $\mathbf{X} = \mathbf{A}^{-1}\mathbf{Y}$. If **X** has the density function $f_{\mathbf{X}}(\mathbf{x})$, the corresponding density function for **Y** is

$$f_{\mathbf{Y}}(\mathbf{y}) = \frac{1}{|\det \mathbf{A}|} f_{\mathbf{X}}(\mathbf{A}^{-1}\mathbf{y}).$$

The assumption of nonsingularity of **A** simply means that the random variables **Y** are really n distinct random variables; that is, one is not just a linear combination of the others, in which case the distribution of **Y** would be singular (concentrated in the hyperplane of the linear relation).

10.2 Normal Distributions

It was seen in Problem 3–65 that every linear combination of independent normal variables is normally distributed. The term *normal* will now be applied to any multivariate distribution the linear combinations of whose component variables are always normal.

10.2.1 Bivariate Normal Distributions

The case of two variables will be considered first, without the aid of matrix notation and methods. The treatment of the general multivariate normal distribution using matrices, given in the next section, will include all of the results given here as special cases; however, the easier visualization in the case of two dimensions may help in acquiring a feel for what is going on in the general case.

Definition. *The random vector (X, Y) will be said to have a* bivariate normal *distribution if and only if the combination $aX + bY$ is normal for every a and b.*

The characteristic function of a bivariate normal distribution can be derived from the characteristic function of the univariate normal distribution. Thus if (X, Y) is bivariate normal, then for any a and b

$$\begin{aligned}\phi_{aX+bY}(t) = E[e^{it(aX+bY)}] &= \phi_{X,Y}(at, bt) \\ &= \exp[it\mu - \tfrac{1}{2}\sigma^2 t^2],\end{aligned}$$

where

$$\begin{cases}\mu = a\mu_X + b\mu_Y, \\ \sigma^2 = a^2\sigma_X^2 + b^2\sigma_Y^2 + 2ab\sigma_{X,Y}.\end{cases}$$

Upon substitution of these expressions one obtains

$$\phi_{X,Y}(at, bt) = \exp\{i[(at)\mu_X + (bt)\mu_Y] - \tfrac{1}{2}[(at)^2\sigma_X^2 + 2(at)(bt)\sigma_{X,Y} + (bt)^2\sigma_Y^2]\}$$

so that, finally,

$$\phi_{X,Y}(r, s) = \exp\{i[r\mu_X + s\mu_Y] - \tfrac{1}{2}[r^2\sigma_X^2 + 2rs\sigma_{X,Y} + s^2\sigma_Y^2]\}.$$

Theorem 1. *If (X, Y) is bivariate normal with $\rho_{X,Y} = 0$, then X and Y are independent.*

This follows because the characteristic function then factors into the product of two univariate normal characteristic functions.

Theorem 2. *The marginal distributions of a bivariate normal distribution are normal.*

The marginal variables X and Y are each linear combinations of X and Y. [Their normality would also be evident upon setting $r = 0$ or $s = 0$ in $\phi_{X,Y}(r, s)$.]

Theorem 3. *If (U, V) is obtained from the bivariate normal (X, Y) by linear transformation, then (U, V) is bivariate normal.*

Every linear combination $aU + bV$ of the linear combinations U and V is again a linear combination of X and Y, and so is normal.

Theorem 4. *If (X, Y) is bivariate normal with $\rho^2 < 1$, the distribution is non-singular, with density*

$$\begin{aligned}f(x, y) = \frac{1}{2\pi\sigma_X\sigma_Y\sqrt{1-\rho^2}} \exp\Bigg\{\frac{-1}{2(1-\rho^2)}\Bigg[\left(\frac{x-\mu_X}{\sigma_X}\right)^2 \\ - 2\rho\left(\frac{x-\mu_X}{\sigma_X}\right)\left(\frac{y-\mu_Y}{\sigma_Y}\right) + \left(\frac{y-\mu_Y}{\sigma_Y}\right)^2\Bigg]\Bigg\}.\end{aligned}$$

First, suppose that $\mu_X = \mu_Y = 0$. According to the corollary of Section 10.1.1, there is a nonsingular linear transformation

$$\begin{cases} U = aX + bY, \\ V = cX + dY \end{cases}$$

such that $\sigma_U^2 = \sigma_V^2 = 1$ and $\sigma_{U,V} = 0$, which implies here that U and V are independent standard normal variables (Theorem 1). The inverse transformation:

$$\begin{cases} X = \dfrac{dU - bV}{\Delta}, \\ Y = \dfrac{-cU + aV}{\Delta}, \end{cases}$$

exists because the determinant is not 0:

$$\Delta \equiv ad - bc \neq 0.$$

Now,

$$\sigma_X^2 = \frac{d^2 + b^2}{\Delta^2}, \qquad \sigma_Y^2 = \frac{a^2 + c^2}{\Delta^2}, \qquad \sigma_{X,Y} = \frac{-(ab + cd)}{\Delta^2},$$

and

$$\sigma_X^2\sigma_Y^2(1 - \rho^2) = \sigma_X^2\sigma_Y^2 - \sigma_{X,Y}^2 = \frac{1}{\Delta^2}.$$

The density of (X, Y) is obtained from that of (U, V):

$$f_{X,Y}(x, y) = f_{U,V}(ax + by, cx + dy)\left|\frac{\partial(u, v)}{\partial(x, y)}\right|,$$

where, because U and V are independent standard normal:

$$f_{U,V}(u, v) = \frac{1}{2\pi}\exp\left[-\frac{1}{2}(u^2 + v^2)\right].$$

The required Jacobian is

$$\frac{\partial(u, v)}{\partial(x, y)} = \Delta = \frac{1}{\sigma_X\sigma_Y\sqrt{1 - \rho^2}},$$

and

$$\begin{aligned} u^2 + v^2 &= (ax + by)^2 + (cx + dy)^2 \\ &= (a^2 + c^2)x^2 + (b^2 + d^2)y^2 + 2xy(ab + cd). \end{aligned}$$

Putting all of this together in the formula for $f_{X,Y}$ yields the asserted result when the means are both 0; the more general case is obtained by a simple translation.

The cross sections of a function $f(x, y)$ parallel to the xy-plane are called *level curves* (curves of constant "altitude") or *contour curves.* In the case of a bivariate normal distribution, the density function is constant if the quadratic in the exponent is constant:

$$\left(\frac{x - \mu_X}{\sigma_X}\right)^2 - 2\rho\left(\frac{x - \mu_X}{\sigma_X}\right)\left(\frac{y - \mu_Y}{\sigma_Y}\right) + \left(\frac{y - \mu_Y}{\sigma_Y}\right)^2 = \text{const.}$$

Such an equation defines an ellipse (see Figure 10–3), provided that the discriminant is negative, or

$$\left(\frac{\rho}{\sigma_X\sigma_Y}\right)^2 - \frac{1}{\sigma_X^2}\frac{1}{\sigma_Y^2} = \frac{\rho^2 - 1}{\sigma_X^2\sigma_Y^2} < 0,$$

and this is indeed the case when the distribution is nonsingular:

Theorem 5. *The level curves of a bivariate normal density for (X, Y) are ellipses centered at (μ_X, μ_Y). The axes are parallel to the coordinate axes if and only if X and Y are independent.*

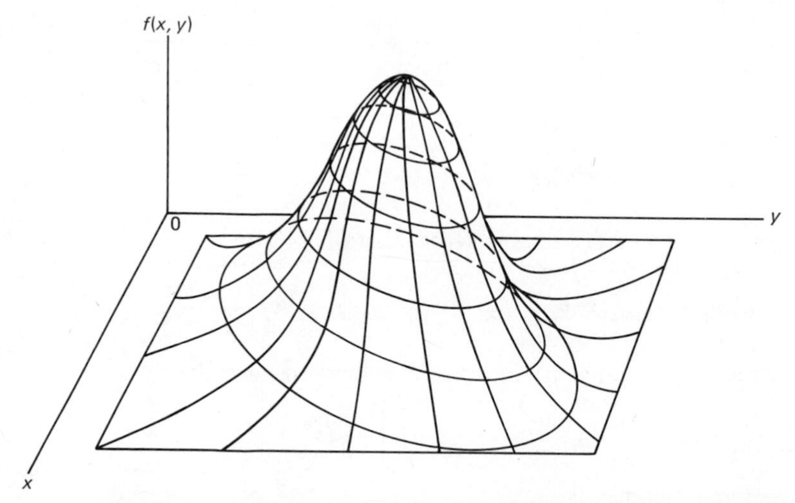

Figure 10–3. Level curves of a bivariate normal density.

With a proper choice of multiplicative constant, any function of the form

$$f(x, y) = (\text{const}) \exp\left[-\tfrac{1}{2}(ax^2 + 2bxy + cy^2)\right]$$

can serve as the density of a bivariate normal distribution if $a > 0$ and $ac > b^2$ (so that the level curves are ellipses). It is simply necessary to identify the coefficients in the quadratic with corresponding coefficients in the bivariate normal density. Doing so yields

$$\sigma_X^2 = \frac{c}{ac - b^2}, \qquad \sigma_Y^2 = \frac{a}{ac - b^2}, \qquad \rho = \frac{-b}{\sqrt{ac}}.$$

EXAMPLE 10–5. Suppose that (X, Y) has the density

$$f(x, y) = (\text{const}) \exp\left[-\tfrac{1}{2}x^2 + xy - y^2 - x + 2y\right].$$

The linear terms can be eliminated by a translation—set $x = x' + h, y = y' + k$ and deduce $h = 0, k = 1$. The discriminant of the quadratic terms (unchanged by the translation) is negative:

$$b^2 - ac = -1,$$

so the level curves are ellipses. From the preceding formulas, one obtains $\sigma_X^2 = 2$, $\sigma_Y^2 = 1$, and $\rho = 1/\sqrt{2}$. Alternatively, one may proceed thus:

$$\begin{aligned} -\tfrac{1}{2}[x^2 - 2xy + 2y^2 + 2x - 4y] &= -\tfrac{1}{2}[(x - y)^2 + y^2 + 2(x - y) - 2y] \\ &= -\tfrac{1}{2}[(x - y + 1)^2 + (y - 1)^2 - 2], \end{aligned}$$

and the linear transformation

$$\begin{cases} U = X - Y + 1, \\ V = Y - 1, \end{cases} \quad \text{or} \quad \begin{cases} X = U + V, \\ Y = V + 1 \end{cases}$$

evidently reduces the density to the product of standard normal densities. The parameters of (X, Y) can be calculated from those of the independent, standard normal U and V.

The density of the conditional variable $Y \mid X = x$ is proportional to the joint density in the cross section $X = x$:

$$f_{Y|x_0}(y) = \frac{f(x_0, y)}{f_X(x_0)}$$

and so in the case of a normal distribution the conditional density is an exponential function with a quadratic in the exponent. With

$$x' = \frac{x - \mu_X}{\sigma_X}, \qquad y' = \frac{y - \mu_Y}{\sigma_Y},$$

it is

$$f_{Y|x}(y) = (\text{const}) \exp\left\{-\frac{1}{2(1 - \rho^2)}\left[y'^2 - 2\rho x'y' + (\text{term in } x)\right]\right\}.$$

As a univariate density, this is normal (see Figure 10–4); completing the square in y' puts the parameters on display:

$$\begin{aligned} f_{Y|x}(y) &= (\text{const}) \exp\left\{-\frac{1}{2(1 - \rho^2)}\left[(y' - \rho x')^2 + (\text{terms in } x)\right]\right\} \\ &= (\text{const}) \exp\left\{-\frac{1}{2\sigma_Y^2(1 - \rho^2)}\left[(y - \mu_Y - \rho\sigma_Y x')^2 + (\text{terms in } x)\right]\right\}. \end{aligned}$$

The result is summarized as follows.

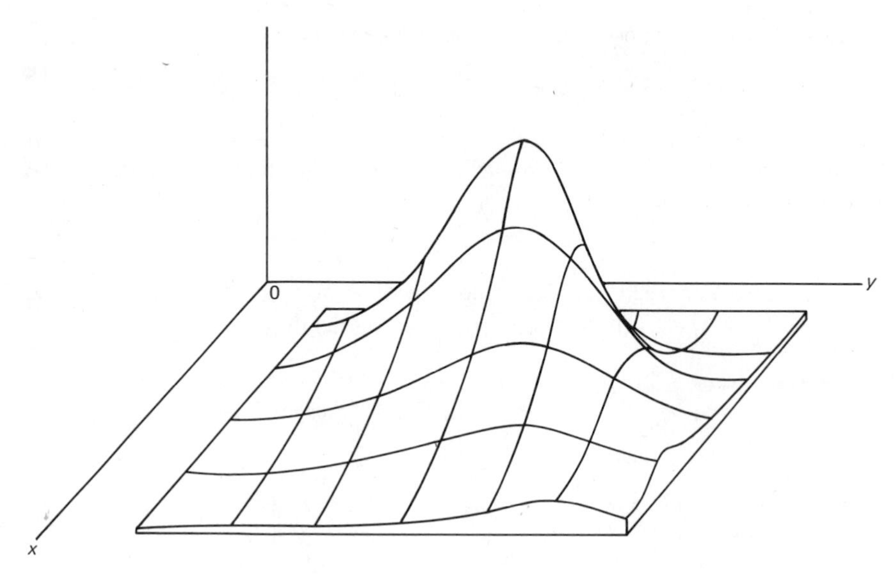

Figure 10–4. Conditional densities as cross sections.

Theorem 6. *In a bivariate normal distribution the conditional distributions of $X \mid y$ and $Y \mid x$ are (univariate) normal. The regression functions are linear and the conditional variances are constant:*

$$E(Y \mid x) = \mu_Y + \frac{\sigma_{X,Y}}{\sigma_X^2}(x - \mu_X),$$

$$\text{var}\,(Y \mid x) = \sigma_Y^2(1 - \rho^2),$$

with analogous formulas for $X \mid y$.

Problems

10–10. Write out the density function of a bivariate normal distribution for (X, Y) with $EX = 0$, $EY = 4$, var $X = 1$, var $Y = 9$, and cov $(X, Y) = 2$.

10–11. Given that the following is a bivariate normal density, determine the first and second moments of the distribution:

$$f(x, y) = (\text{const}) \exp\,(-x^2 + xy - 2y^2).$$

10–12. Let X and Y have independent standard normal distributions.

(a) What are the level curves of the joint density function?
(b) Show that variables obtained by a rotation transformation are again independent standard normal variables.
(c) Obtain the distribution of $U = \exp\,[-\frac{1}{2}(X^2 + Y^2)]$ from that of $X^2 + Y^2$.
(d) Deduce the distribution of $V = \arctan\,(Y/X)$ by interpreting it geometrically.
(e) Obtain the joint distribution of U and V in (c) and (d).
(f) Use the result in (e) to explain how one might use a "random number generator" (which generates a stream of independent uniform observations) to obtain a stream of independent standard normal observations.

10–13. Given the bivariate normal density of Problem 10–10,

(a) Determine a rotation transformation such that the transformed variables are independent.
(b) Use the method in the latter part of Example 10–5 to obtain a linear transformation to (U, V) in which $V = Y$, such that U and V are independent. (This is said to be a *shearing* transformation. Study its effect on the rectangular region bounded by $x = 0$, $x = a$, $y = 0$, $y = b$.)

10–14. Use characteristic functions to show that if $\mathbf{X}$ is a random sample from a normal population, then two linear combinations $\sum a_i X_i$ and $\sum b_i X_i$ have a bivariate normal distribution. (Without loss of generality, assume $E\mathbf{X} = 0$.)

10–15. Use the result of Problem 10–14 to show that if $\mathbf{X}$ is a random sample from a normal population, then $E(X_1 \mid \bar{X}) = \bar{X}$. (This is a Rao–Blackwellization of X_1 using the sufficient statistic $\bar{X}$. See Example 5–10.)

10–16. Let U be the deviation of X about its mean and V the deviation of Y from the regression line of Y on X:

$$U = X - \mu_X, \qquad V = Y - E(Y \mid X).$$

Show that U and V are independent if $f(X, Y)$ is bivariate normal.

10–17. Suppose that the regression function of a bivariate distribution is linear: $E(Y \mid x) = a + bx$.

(a) Use the fact that $E[g(X, Y) \mid X = x] = E[g(x, Y) \mid X = x]$ to show that $\sigma_{X,Y} = b\sigma_X^2$.
(b) Assume further that the conditional variance is constant: $\text{var}\,(Y \mid x) = k$. Use (a) and the relation of Section 2.3.1 between variance and conditional variance to show that $k = \sigma_Y^2(1 - \rho^2)$.

10–18. Let (X, Y) be bivariate normal with $\text{var}\, X = \text{var}\, Y = 1$. Show that the correlation between X^2 and Y^2 is ρ^2, the square of the correlation between X and Y,

(a) By using brute-force integration, calculating $E(X^2 Y^2)$, and so forth.
(b) By making a linear transformation to eliminate dependence and computing $E(X^2 Y^2)$ from moments of the new variables.

[*Hint:* A rotation will work, as will also a transformation of the form $U = aX + bY$, $V = Y$, with a and b chosen to make $\text{var}\, U = 1$, $\text{cov}\,(U, V) = 0$, as in 10–13(b).]

10.2.2 Multivariate Normal Distributions

The random p-vector $\mathbf{X}$ is said to be *multivariate normal* if and only if the linear combination

$$\mathbf{a}'\mathbf{X} = a_1 X_1 + \cdots + a_p X_p$$

is normal for all $\mathbf{a}$. That such distributions exist is clear from the fact (Problem 3–65) that any linear combination of independent normal variables is normal.

The characteristic function of a multivariate normal distribution can be derived from the univariate normal characteristic function. Thus, if $\boldsymbol{\mu}$ and $\mathbf{M}$ are, respectively, the mean vector and covariance matrix of $\mathbf{X}$, then

$$\mu_{\mathbf{a}'\mathbf{X}} = \mathbf{a}'\boldsymbol{\mu},$$
$$\sigma^2_{\mathbf{a}'\mathbf{X}} = \mathbf{a}'\mathbf{M}\mathbf{a},$$

and

$$\phi_{\mathbf{a}'\mathbf{X}}(t) = E[e^{i\mathbf{a}'\mathbf{X}t}] = \phi_{\mathbf{X}}(t\mathbf{a})$$
$$= \exp\left[i(\mathbf{a}'\boldsymbol{\mu})t - \tfrac{1}{2}(\mathbf{a}'\mathbf{M}\mathbf{a})t^2\right.,$$

so that (with $\mathbf{s} \equiv t\mathbf{a}$)

$$\phi_{\mathbf{X}}(\mathbf{s}) = \exp[i\mathbf{s}'\boldsymbol{\mu} - \tfrac{1}{2}\mathbf{s}'\mathbf{M}\mathbf{s}].$$

Theorem 1. *If* $\mathbf{X}$ *is multivariate normal and its second moment matrix* $\mathbf{M}$ *is diagonal—all covariances equal to 0—then the components of* $\mathbf{X}$ *are independent.*

This follows because the characteristic function then factors into the product of univariate normal characteristic functions.

Theorem 2. *The marginal distributions of a multivariate normal distribution are all multivariate normal.*

The characteristic function of a marginal distribution is obtained by setting equal to 0 the t's corresponding to the other components in the joint characteristic function, and the result is a characteristic function that is again of the multivariate normal type.

Theorem 3. *If* $\mathbf{X}$ *is multivariate normal, then* $\mathbf{AX}$ *is multivariate normal for any (compatible) matrix* $\mathbf{A}$.

This is evident from the fact that for any vector $\mathbf{a}$ of suitable dimension, $\mathbf{a}'(\mathbf{AX}) = (\mathbf{A}'\mathbf{a})'\mathbf{X}$, which must be univariate normal.

Again, in the multivariate case, the definition of "normal" includes singular distributions. The nonsingular ones correspond to nonsingular second moment matrices and have densities given in the following theorem.

Theorem 4. *If* $\mathbf{X}$ *is p-variate normal with mean* $\boldsymbol{\mu}$ *and has a nonsingular second moment matrix* $\mathbf{M}$, *then* $\mathbf{X}$ *has a density function, given by*

$$f_{\mathbf{X}}(\mathbf{x}) = \frac{1}{(2\pi)^{p/2}\sqrt{\det \mathbf{M}}} \exp\left[-\frac{1}{2}(\mathbf{x} - \boldsymbol{\mu})'\mathbf{M}^{-1}(\mathbf{x} - \boldsymbol{\mu})\right].$$

According to a theorem of matrix theory, there is a nonsingular matrix $\mathbf{A}$ such that $\mathbf{AA}' = \mathbf{M}$. Let $\mathbf{Y} = \mathbf{A}^{-1}(\mathbf{X} - \boldsymbol{\mu})$, which is multivariate normal with second moment matrix

$$\mathbf{M_Y} = \mathbf{A}^{-1}\mathbf{M}(\mathbf{A}^{-1})' = \mathbf{I},$$

so that

$$f_{\mathbf{Y}}(\mathbf{y}) = (2\pi)^{-p/2}e^{-\mathbf{y}'\mathbf{y}/2}.$$

The density of $\mathbf{X}$ is then

$$\begin{aligned} f_{\mathbf{X}}(\mathbf{x}) &= f_{\mathbf{Y}}(\mathbf{A}^{-1}(\mathbf{x} - \boldsymbol{\mu}))\,|\det(\mathbf{A}^{-1})| \\ &= \frac{1}{(2\pi)^{p/2}\,|\det \mathbf{A}|}\exp\left[-\frac{1}{2}(\mathbf{x} - \boldsymbol{\mu})'(\mathbf{A}^{-1})'\mathbf{A}^{-1}(\mathbf{x} - \boldsymbol{\mu})\right], \end{aligned}$$

and since $(\mathbf{A}^{-1})'\mathbf{A}^{-1} = (\mathbf{AA}')^{-1} = \mathbf{M}^{-1}$ and $|\det \mathbf{A}| = \sqrt{\det \mathbf{M}}$, the asserted result follows.

Corollary. *If $\mathbf{X}$ is multivariate normal and its second-moment matrix $\mathbf{M}$ has rank r, then there is a $\mathbf{C}$ and a vector $\mathbf{d}$ such that $\mathbf{X} = \mathbf{CY} + \mathbf{d}$, where $\mathbf{Y}$ is r-variate normal with zero mean vector and identity covariance matrix. (That is, the Y's are r independent, standard normal variables.)*

It can be assumed without loss of generality that the first r rows of $\mathbf{M}$ are linearly independent, so that the submatrix $\mathbf{M}_r$ of second moments of $(X_1, \ldots X_r)$ is nonsingular. Then, according to Theorem 4, there is a linear transformation on independent standard normal variables $(Y_1, \ldots, Y_r)$ that yields $(X_1, \ldots, X_r)$. Since the remaining X's can be expressed as linear combination of the first r X's it follows that $(X_1, \ldots, X_p)$ is a linear transformation on $(Y_1, \ldots, Y_r)$, as asserted.

EXAMPLE 10–6. Let Y_1, Y_2, and Y_3 be independent, standard normal random variables. Consider $\mathbf{X}$ defined by

$$\begin{cases} X_1 = 2Y_1 - Y_2 + Y_3, \\ X_2 = Y_1 - 3Y_2, \end{cases}$$

or

$$\mathbf{X} = \begin{pmatrix} 2 & -1 & 1 \\ 1 & -3 & 0 \end{pmatrix}\mathbf{Y}.$$

The covariance matrix of $\mathbf{X}$ is

$$\mathbf{M} = \begin{pmatrix} 2 & -1 & 1 \\ 1 & -3 & 0 \end{pmatrix}\begin{pmatrix} 2 & 1 \\ -1 & -3 \\ 1 & 0 \end{pmatrix} = \begin{pmatrix} 6 & 5 \\ 5 & 10 \end{pmatrix},$$

with det $\mathbf{M} = 35$ and

$$\mathbf{M}^{-1} = \frac{1}{35}\begin{pmatrix} 10 & -5 \\ -5 & 6 \end{pmatrix}.$$

With these one can write out the characteristic function of $\mathbf{X}$:

$$\exp\left[-\frac{1}{2}(6t_1^2 + 10t_1t_2 + 10t_2^2)\right];$$

and the density function:

$$\frac{1}{2\pi\sqrt{35}}\exp\left[-\frac{1}{70}(10x_1^2 - 10x_1x_2 + 6x_2)^2\right].$$

Suppose next that $\mathbf{Z}$ is defined as

$$\mathbf{Z} = \begin{pmatrix} 1 & -1 \\ 2 & 1 \\ 1 & 2 \end{pmatrix}\mathbf{X}.$$

The distribution of $\mathbf{Z}$ must be singular, since there are more $\mathbf{Z}$'s than $\mathbf{X}$'s; indeed, $Z_3 = Z_2 - Z_1$, so the distribution of $\mathbf{Z}$ is concentrated in the plane $z_1 - z_2 + z_3 = 0$. The covariance matrix of $\mathbf{Z}$ is

$$\mathbf{M_Z} = \begin{pmatrix} 1 & -1 \\ 2 & 1 \\ 1 & 2 \end{pmatrix}\begin{pmatrix} 6 & 5 \\ 5 & 10 \end{pmatrix}\begin{pmatrix} 1 & 2 & 1 \\ -1 & 1 & 2 \end{pmatrix} = \begin{pmatrix} 6 & -3 & -9 \\ -3 & 54 & 57 \\ -9 & 57 & 66 \end{pmatrix},$$

which is singular.

EXAMPLE 10–7. Consider this matrix of second moments of $\mathbf{X}$:

$$\mathbf{M} = \begin{pmatrix} 5 & -6 \\ -6 & 9 \end{pmatrix},$$

which is positive definite (all principal minor determinants are positive). It is apparent that adding $\frac{6}{5}$ of the first row (or column) to the second row (column) puts a 0 in the second row (column). Thus,

$$\begin{pmatrix} 1 & 0 \\ \frac{6}{5} & 1 \end{pmatrix}\begin{pmatrix} 5 & -6 \\ -6 & 9 \end{pmatrix}\begin{pmatrix} 1 & \frac{6}{5} \\ 0 & 1 \end{pmatrix} = \begin{pmatrix} 5 & 0 \\ 0 & \frac{9}{5} \end{pmatrix}.$$

The resulting diagonal entries can be made 1's by pre- and postmultiplying by the reciprocal of the square root; so let

$$\mathbf{C} = \begin{pmatrix} 1/\sqrt{5} & 0 \\ 0 & \sqrt{5}/3 \end{pmatrix}\begin{pmatrix} 1 & 0 \\ \frac{6}{5} & 1 \end{pmatrix} = \begin{pmatrix} 1/\sqrt{5} & 0 \\ 2/\sqrt{5} & \sqrt{5}/3 \end{pmatrix},$$

a matrix such that $\mathbf{CMC'} = \mathbf{I}$. But then $\mathbf{M} = \mathbf{C}^{-1}(\mathbf{C}^{-1})'$:

$$\begin{pmatrix} 5 & -6 \\ -6 & 9 \end{pmatrix} = \begin{pmatrix} \sqrt{5} & 0 \\ -6/\sqrt{5} & 3/\sqrt{5} \end{pmatrix}\begin{pmatrix} \sqrt{5} & -6/\sqrt{5} \\ 0 & 3/\sqrt{5} \end{pmatrix}$$

and $\mathbf{X}$ can be written

$$\mathbf{X} - \boldsymbol{\mu} = \begin{pmatrix} \sqrt{5} & 0 \\ -6/\sqrt{5} & 3/\sqrt{5} \end{pmatrix}\mathbf{Y},$$

where **Y** has independent standard normal components. (This representation is not unique. It can be checked, for example, that **X** can also be given in terms of such **Y** by the transformation $X_1 - \mu_1 = Y_1 + 2Y_2$, $X_2 - \mu_2 = -3Y_2$.)

Theorem 6 in the preceding section can be extended to the general multivariate case (cf. Rao [19], p. 522); that is, the conditional distribution of a subset of components, given values of the remaining components, is multivariate normal.

Problems

10–19. Let **X** have a multivariate normal distribution with mean and covariance matrices

$$\boldsymbol{\mu} = \begin{pmatrix} 3 \\ 3 \\ 0 \\ 0 \end{pmatrix}, \qquad \mathbf{M} = \begin{pmatrix} 2 & 0 & 2 & 0 \\ 0 & 1 & 1 & 0 \\ 2 & 1 & 5 & 1 \\ 0 & 0 & 1 & 1 \end{pmatrix}.$$

(a) Show that **M** is nonsingular.
(b) Compute the correlation between X_1 and X_3.
(c) Let $\mathbf{Y} = \mathbf{CX} + \mathbf{d}$, with

$$\mathbf{C} = \begin{pmatrix} 1 & 1 & 1 & -1 \\ 1 & -1 & 1 & 1 \\ 1 & 0 & 1 & 0 \end{pmatrix}, \qquad \mathbf{d} = \begin{pmatrix} 2 \\ 0 \\ -1 \end{pmatrix}.$$

Compute the mean and covariance matrices of **Y**.
(d) Obtain the characteristic function and density function of Y_1, Y_2 in (c).
(e) Determine the mean and variance of Y_3 directly, using $E(\cdot)$ computations.

10–20. Obtain the density function of $\mathbf{X} = \mathbf{C}^{-1}\mathbf{Y}$ in Example 10–7 (let $\boldsymbol{\mu} = \mathbf{0}$) using the **C** derived in the example, and also using the transformation mentioned parenthetically at the end. (They should agree.)

10–21. Let **X** be p-variate normal with mean vector **0** and second-moment matrix **M**. show that the quadratic form $Q = \mathbf{X}'\mathbf{M}^{-1}\mathbf{X}$ has a chi-square distribution with p degrees of freedom.

[*Hint:* Write the characteristic function of Q as an expectation and substitute $\mathbf{M} = \mathbf{AA}'$, where $\mathbf{A}^{-1}\mathbf{X}$ is a vector of independent standard normal variables.]

10–22. In Problem 10–14, it was seen that two linear combinations $\mathbf{a}'\mathbf{X}$ and $\mathbf{b}'\mathbf{X}$ of independent normal variables $X_1, \ldots, X_p$ have a bivariate normal distribution. Show that this is so if the X's have a p-variate normal distribution (not necessarily independent), and show that $\mathbf{a}'\mathbf{X}$ and $\mathbf{b}'\mathbf{X}$ are independent if and only if $\mathbf{aM_Xb} = 0$.

10.3 Correlation and Prediction

In the study of multivariate models one is naturally very interested in relationships among the variables—in their existence and strength, in their pointing toward possible causal relationships, and in the possibility of exploiting a relationship (even if only stochastic) for predicting one variable from a knowledge of others. The matter of causal relationships is not statistical and can only be examined in a given field of application.

10.3.1 Prediction in a Bivariate Model

The predicting of the value of a random variable has its element of risk. If one predicts the value of Y to be k, the *prediction error* is defined to be $Y - k$. In the spirit of estimation theory, we use the *mean squared prediction error* (m.s.p.e.) as a criterion for success:

$$\text{m.s.p.e.} \equiv E(Y - k)^2.$$

This second moment of Y is clearly minimized when taken about the mean, so the value $k = EY$ gives the best prediction in the sense of mean squared error. The minimum of this m.s.p.e. is the variance σ_Y^2.

When Y is one component of a random pair (X, Y) and it is possible to observe X before making the prediction of Y, it would be expected that unless X and Y are independent one could improve his prediction through his knowledge of X. Using a function $h(X)$ as a predictor of Y results in the mean squared prediction error

$$\text{m.s.p.e.} = E[Y - h(X)]^2 = E\{E[(Y - h(x))^2 \mid X]\},$$

and this is minimized by choosing (for each x) $h(x)$ to be the conditional mean of Y at $X = x$:

$$\text{best predictor} = E(Y \mid x).$$

For $X = x$, the minimum m.s.p.e. is the conditional variance, so unconditionally one has

$$\begin{aligned}(\text{m.s.p.e.})_{\min} &= E[\text{var}\,(Y \mid X)]\\ &= \text{var}\,Y - \text{var}\,[E(Y \mid X)].\end{aligned}$$

Thus, the m.s.p.e can be reduced from its value (var Y) when X is not used at all, provided that $E(Y \mid X)$ is not constant in X.

Sometimes for practical reasons, one may be content with an approximate, *linear* prediction—a function of the form $a + bX$ or (for more convenient manipulation) $a + b(X - \mu_X)$. With this function, the m.s.p.e. is

$$\begin{aligned}E[Y - a - b(X - \mu_X)]^2 &= a^2 - 2a\mu_Y + b^2\sigma_X^2 - 2b\sigma_{XY} + E(Y^2)\\ &= (a - \mu_Y)^2 + \sigma_X^2\left(b - \frac{\sigma_{XY}}{\sigma_X^2}\right)^2 + \sigma_Y^2 - \frac{\sigma^2{}_{XY}}{\sigma_X^2}.\end{aligned}$$

This is clearly a minimum when

$$a = \mu_Y, \qquad b = \frac{\sigma_{XY}}{\sigma_X^2}.$$

That is, the best linear predictor for Y in terms of X is

$$\mu_Y + \frac{\sigma_{XY}}{\sigma_X^2}(X - \mu_X),$$

and the corresponding m.s.p.e. is

$$(\text{m.s.p.e.})_{\min} = \sigma_Y^2(1 - \rho^2).$$

So a linear predictor reduces the r.m.s. prediction error from its value σ_Y when X is not used at all to $\sigma_Y\sqrt{1 - \rho^2}$. This dependence of ρ is not surprising when it is recalled that ρ is a coefficient of *linear* correlation.

EXAMPLE 10–8. In educational circles it is found that variables (such as high-school rank and placement scores) used in predicting college performance are seldom more highly correlated with college performance than about $\rho = .6$. Suppose that a predictor variable X and a performance variable Y are scaled so that $\mu_X = \mu_Y = 70$, $\sigma_X = \sigma_Y = 10$, and assume $\rho = .6$. The best linear predictor is then given by

$$Y = 70 + .6(X - 70),$$

and the r.m.s.p.e. is $\sigma_Y\sqrt{1 - \rho^2} = 8$. This is not a great reduction from the r.m.s. error of 10 that would be involved in predicting performance to be $\mu_Y = 70$, regardless of X.

If the regression function happens to *be* linear, then the best linear predictor must be the best predictor, with m.s.p.e. given by the formula $\sigma_Y^2(1 - \rho^2)$. (This was obtained as a formula for conditional variance in Problem 10–17. It is the same for each value x of the conditioning variable X.

When (X, Y) is bivariate *normal*, then the regression function is linear and is given by the same formula as gives the best linear predictor. (See Theorem 6 of Section 10.2.1.)

This discussion seems to pretend that the constants defining the bivariate model are known. And indeed, there are situations (e.g., educational testing) where the wealth of data and experience is practically as good as knowing the precise model. However, if the model is not known, one must sample it, obtaining data for the form $(X_1, Y_1), \ldots, (X_n, Y_n)$, and then use this to estimate the parameters of the model. From these estimates one can make approximate predictions of Y for some other X. This will be considered in Section 12.2.8.

10.3.2 Correlation in the Bivariate Normal Case

Given a random sample $(X_1, Y_1), \ldots, (X_n, Y_n)$ from a bivariate normal population, the likelihood function has logarithm

$$\log L(\mu_X, \mu_Y; \sigma_X^2, \sigma_Y^2; \rho)$$

$$= -\frac{n}{2}[\log(2\pi) + \log\sigma_X^2 + \log\sigma_Y^2 + \log(1 - \rho^2)]$$

$$-\frac{1}{2(1 - \rho^2)}\sum\left\{\left(\frac{X_i - \mu_X}{\sigma_X}\right)^2 - 2\rho\left(\frac{X_i - \mu_X}{\sigma_X}\right)\left(\frac{Y_i - \mu_Y}{\sigma_Y}\right) + \left(\frac{Y_i - \mu_Y}{\sigma_Y}\right)^2\right\}.$$

It can be shown (with persistence) that the joint m.l.e. of the parameters is given by the minimal sufficient statistic $(\bar{X}, \bar{Y}; S_X^2, S_Y^2; r)$, where r is the sample correlation coefficient:

$$r = \frac{S_{XY}}{S_X S_Y} = \frac{\sum (X_i - \bar{X})(Y_i - \bar{Y})}{[\sum (X_i - \bar{X})^2 \sum (Y_i - \bar{Y})^2]^{1/2}}.$$

Like the sample moments, r can be thought of as the correlation in a probability-type distribution that puts equal masses at the n data points; thus, properties of ρ are also properties of r. In particular, $r^2 \leq 1$, and $r^2 = 1$ if and only if the data points are collinear.

The statistic r approaches ρ in probability, and it is natural to use it in large-sample inferences about ρ. To do this, one must know the distribution of r, and this was obtained by Fisher (see Kendall and Stuart [15], Vol. 1, p. 387). It is a little too complicated to be useful. In the important special case of testing $\rho = 0$, however, the null distribution is simpler. It will be demonstrated in Chapter 12 (see Problem 12–10) that the statistic

$$T = \sqrt{n-2}\,\frac{r}{\sqrt{1-r^2}}$$

has a $t(n-2)$ distribution when $\rho = 0$.

EXAMPLE 10–9. If 27 observations on (X, Y) yield $r = .2$, the corresponding value of T is

$$T = \sqrt{25}\,\frac{.2}{\sqrt{1-.04}} \doteq 1.02.$$

This is just under the 85th percentile of $t(25)$, so a two-sided test for $\rho = 0$ at $\alpha = .3$ or less would call for accepting $\rho = 0$.

Fisher proposed a transformation that can be used in large-sample tests for other values of ρ:

$$Z = \frac{1}{2}\log\frac{1+r}{1-r}.$$

He showed (see Kendall and Stuart, Vol. 1, [15], p. 391) that Z is approximately normal (for $n > 50$, say) with

$$E(Z) \doteq \frac{1}{2}\log\frac{1+\rho}{1-\rho} \qquad \text{and} \qquad \operatorname{var} Z \doteq \frac{1}{n-3}.$$

This large-sample distribution can also be used in constructing confidence intervals.

EXAMPLE 10–10. Suppose that in a sample of 67 pairs, one finds $r = .2$. Construction of a 95 per cent confidence interval for ρ proceeds as follows:

$$P\left\{\left|\frac{1}{2}\log\frac{1+r}{1-r} - \frac{1}{2}\log\frac{1+\rho}{1-\rho}\right| < \frac{2}{\sqrt{n-3}}\right\} \doteq .95,$$

and for the given data the inequality becomes

$$\left|\log\left(\frac{3}{2}\frac{1-\rho}{1+\rho}\right)\right| < \frac{1}{2}.$$

Solving this for ρ yields $-.047 < \rho < .424$, which is the desired confidence interval. Notice that it includes $\rho = 0$, so that if one were testing independence at $\alpha = .05$, he should accept it on the basis of the given data.

Problems

10–23. Two contest judges turned in ratings as follows:

	Contestants							
Judges	1	2	3	4	5	6	7	8
A	92	90	85	96	92	88	96	88
B	89	90	88	93	90	85	95	90

(a) Compute the correlation coefficient of the ratings of the two judges.
(b) If the sample statistics found in (a) are taken as population parameters, what linear predictor is best for Judge A's rating of a contestant given Judge B's rating? What predictor would be used if Judge B's rating is not known, and what reduction in r.m.s. prediction error is achieved by using it when it is known?

10–24. Given the data in Example 10-10, test the hypothesis that $\rho = .4$ against the one-sided alternative $\rho < .4$, using $\alpha = .05$.

11

NONPARAMETRIC INFERENCE

In most problems considered so far, the class of distributions or states of nature assumed as possible models have been defined by a density or probability function of given form but depending on a finite number of real parameters. Such problems are said to be *parametric*, and others are called *nonparametric*. The distinction, however, is not clear-cut. For example, a large-sample test for the mean using $\bar{X}$ does not require that a particular form of density be specified, but it is usually considered a parametric problem. Contingency tables are sometimes included and sometimes not included under the heading of "nonparametric." On the other hand, other attempts at defining the term "nonparametric" appear to be no more successful.

Despite the absence of an indexing parameter θ, there are some problems of estimation that are nonparametric—estimating a median, for example, and estimating the probability $P(X > Y)$ in a bivariate population. However, this chapter considers mainly some problems of testing hypotheses. In them, the notion of power "function" is difficult to manage without the parametric structure, but it is sometimes possible to study power for a particular class of alternatives. The approach here will be to consider various intuitively contructed tests and the appropriate null distributions—in the spirit of significance testing but with some reference to large-sample performance as measured by asymptotic relative efficiency.

For a more complete discussion of nonparametric inference the reader is referred to the book by Gibbons [12].

11.1 Order Statistics and Related Distributions

Much of nonparametric inference deals with continuous populations, for which the order statistic based on a random sample is defined and is sufficient (cf. Section 4.4.3). Some aspects of its distribution were considered in Section 4.3.2; here the joint distribution will be derived. Incidentally, it will be shown that the various permutations of the sample observations are equally likely to have given rise to a given order statistic, a fact on which a number of nonparametric procedures are based.

The next two sections will derive this result and the formula for the density of the order statistic. (They may be omitted if these results are accepted without proof.)

11.1.1 Distribution of the Order Statistic

Let **X** denote a random sample from a continuous population. Let $\boldsymbol{\nu}$ denote one of the $n!$ possible permutations of the integers $1, 2, \ldots, n$:

$$\boldsymbol{\nu} = (\nu_1, \nu_2, \ldots, \nu_n).$$

Each of these permutations defines a region of sample points **x**:

$$R_{\boldsymbol{\nu}} = \{\text{set of } \mathbf{x} \text{ such that } x_{\nu_1} < x_{\nu_2} < \cdots < x_{\nu_n}\}.$$

Let R denote the region $R_{\boldsymbol{\nu}}$ in which $\boldsymbol{\nu}$ is the particular permutation $(1, 2, \ldots, n)$:

$$R = \{\text{set of } \mathbf{x} \text{ such that } x_1 < x_2 < \cdots < x_n\}.$$

The $n!$ sets $\{R_{\boldsymbol{\nu}}\}$ together make up the space of sample points **x**, except for the boundaries, which have probability 0 under the assumption of a continuous population.

Each permutation $\boldsymbol{\nu}$ defines a linear transformation $\mathbf{y} = T_{\boldsymbol{\nu}}\mathbf{x}$, where $y_k = x_{\nu_k}$. Such a transformation has the property that $\det T_{\boldsymbol{\nu}} = \pm 1$, and the inverse is just another permutation transformation. Under such a transformation, the region $R_{\boldsymbol{\nu}}$ of points **x** such that $x_{\nu_1} < \cdots < x_{\nu_n}$ is carried into the region R of points **y** such that $y_1 < \cdots < y_n$. Further, since the joint density of the random sample **X** is a *symmetric* function of the coordinates of **x** it is unchanged by $\mathbf{y} = T_{\boldsymbol{\nu}}\mathbf{x}$:

$$f(\mathbf{x}) \equiv \prod f(x_k) = \prod f(x_{\nu_k}) = \prod f(y_k) = f(\mathbf{y}).$$

Therefore,

$$P(R_{\boldsymbol{\nu}}) = \int_{R_{\boldsymbol{\nu}}} f(\mathbf{x})\, d\mathbf{x} = \int_R f(\mathbf{y})\, d\mathbf{y} = P(R),$$

which means that each $R_{\boldsymbol{\nu}}$ has the probability $1/n!$.

Consider now the order statistic

$$\mathbf{Y} = t(X_1, \ldots, X_n) = (X_{(1)}, \ldots, X_{(n)}).$$

This function $t(\mathbf{x})$ carries each **x** into a point in R, and so induces a measure in R which is the probability distribution of the ordered observations. This distribution has zero density outside R, but for any set A in R, the probability that **Y** is in A can be computed. Let $A_{\boldsymbol{\nu}}$ denote the set of those points in $R_{\boldsymbol{\nu}}$ whose coordinates when ordered yield a point in A. Then

$$P(\mathbf{Y} \text{ in } A) = P(\mathbf{X} \text{ in } \sum_{\boldsymbol{\nu}} A_{\boldsymbol{\nu}}) = \sum_{\boldsymbol{\nu}} P(\mathbf{X} \text{ in } A_{\boldsymbol{\nu}}) = \sum_{\boldsymbol{\nu}} \int_{A_{\boldsymbol{\nu}}} f(\mathbf{x})\, d\mathbf{x}$$

$$= \sum_{\boldsymbol{\nu}} \int_A f(\mathbf{y})\, d\mathbf{y} = n!\, P(\mathbf{X} \text{ in } A).$$

Therefore,

$$f_{\mathbf{Y}}(\mathbf{y}) = \begin{cases} n!f(\mathbf{y}), & \text{if } \mathbf{y} \text{ is in } R, \\ 0, & \text{otherwise.} \end{cases}$$

That is, the joint density is essentially unchanged in R, being only multiplied by $n!$ to take into account the requirement that the total probability be 1.

EXAMPLE 11–1. It is perhaps helpful to write out some of these relations for the simple case $n = 2$. There are just two permutations of (1, 2); call them **0**: (1, 2), and **1**: (2, 1). The regions $R = R_0$ and R_1 are the half-planes in which $x_1 < x_2$ and $x_2 < x_1$, respectively, as shown in Figure 11–1. Shown also is a set A of points in R and the set A_1 of points in R_1, which are carried into A under the transformation $y_1 = x_2$, $y_2 = x_1$. The order statistic $(X_{(1)}, X_{(2)})$ has "values" only in R and takes each point (where $x_1 \neq x_2$) into a point in R. For instance, both (4, 1) and (1, 4) would be carried into (1, 4), which is in R; and

$$\begin{aligned} P[X_{(1)}, X_{(2)}) \text{ in } A] &= P(\mathbf{X} \text{ in } A_0) + P(\mathbf{X} \text{ in } A_1) \\ &= 2P(\mathbf{X} \text{ in } A_0) = \int_{A_0} 2f(x_1)f(x_2)\,dx_1\,dx_2. \end{aligned}$$

The joint density of $X_{(1)}$ and $X_{(2)}$ is 0 at any point in R_1 and is just twice the joint density of X_1 and X_2 in R_0.

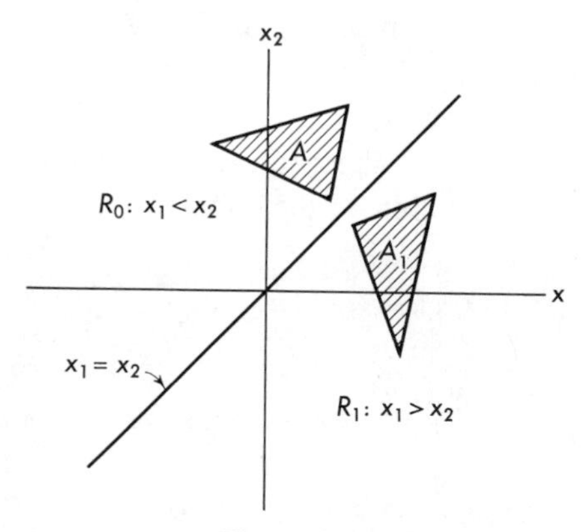

Figure 11–1

11.1.2 Conditional Distribution Given the Order Statistic

The order statistic determined from a random sample **X** is easily seen to be sufficient for any family of distributions by the factorization criterion. For the joint density at $(x_1, \ldots, x_n)$ is just

$$f_{\mathbf{X}}(\mathbf{x}) = \prod f(x_i) = g(\mathbf{y}),$$

where $f(x)$ is the population density and $\mathbf{y} = (x_{(1)}, \ldots, x_{(n)})$ is the order statistic value corresponding to **x**. That is, given the ordered coordinates $x_{(1)}, \ldots, x_{(n)}$ one can determine $f_{\mathbf{X}}(\mathbf{x})$, which is then a function (g) of the ordered coordinates.

The sufficiency can also be seen by computing the conditional distribution of **X** given the order statistic, which turns out to be independent of the population distribution. This conditional distribution is of interest in itself, so (even though the sufficiency has been quickly established above) it will now be derived. Specifically, it will be shown that this conditional distribution assigns probability $1/n!$ to each of the $n!$ sample points whose coordinates are permutations of a given point in R.

Consider a set B of sample points and a set A of possible "values" of the order statistic **Y**; that is, A lies in R, the set of **x** such that $x_1 < \cdots < x_n$. Then

$$\begin{aligned} P(\mathbf{Y} \text{ in } A \text{ and } \mathbf{X} \text{ in } B) &= P(\mathbf{X} \text{ in } \tilde{A}B) \\ &= \int_{\tilde{A}} \phi_B(\mathbf{x}) f(\mathbf{x})\, d\mathbf{x}, \end{aligned}$$

where $\tilde{A}$ denotes the union of all A_ν as defined earlier, and $\phi_B(x)$ is the "indicator function" of B, the function that is 1 for **x** in B and 0 for **x** in B^c. Since a permutation transformation T_ν would not alter the set $\tilde{A}$ (which already contains all permutations of any point in it), such a change of variable in the integral above yields

$$P(\mathbf{Y} \text{ in } A \text{ and } \mathbf{X} \text{ in } B) = \int_{\tilde{A}} \phi_B(T_\nu \mathbf{x}) f(\mathbf{x})\, d\mathbf{x}.$$

Since this is the same for each permutation $\boldsymbol{\nu}$, it is the same as the average over all such permutations:

$$P(\mathbf{Y} \text{ in } A \text{ and } \mathbf{X} \text{ in } B) = \int_{\tilde{A}} \frac{1}{n!} \sum_{\boldsymbol{\nu}} \phi_B(T_\nu \mathbf{x}) f(\mathbf{x})\, d\mathbf{x}.$$

Now the integrand has the same values in any A_ν as in A, and, therefore, this integral over $\tilde{A}$ can be written as $n!$ times the integral over A:

$$P(\mathbf{Y} \text{ in } A \text{ and } \mathbf{X} \text{ in } B) = \int_A \left[\frac{1}{n!} \sum_{\boldsymbol{\nu}} \phi_B(T_\nu \mathbf{y})\right] n!\, f(\mathbf{y})\, d\mathbf{y}.$$

This can be interpreted as the integral with respect to the distribution of **Y** of the conditional probability of **X** given **Y** = **y**. That is,

$$P(\mathbf{X} \text{ in } B \mid \mathbf{Y} = \mathbf{y}) = \frac{1}{n!} \sum_{\boldsymbol{\nu}} \phi_B(T_\nu \mathbf{y}).$$

Since $\phi_B(\mathbf{x}) = 1$ for **x** in B, and 0 otherwise, the sum here is just the number of permutations of **y** that are in B. In particular, if B is a single point that is one of the permutations of **y** in R, the conditional probability is $1/n!$, as claimed. That is, the various permutations that could have led to a given order statistic are equally likely to have been the original sample point.

11.1.3 The Transformation $F(X)$

Consider a continuous population with c.d.f. $F(x)$, and let x_u denote the maximum of the possible inverses $F^{-1}(u)$. Then if $U \equiv F(X)$, and for $0 \le u \le 1$,

$$F_U(u) = P[F(X) \le u] = P(X \le x_u) = F(x_u) = u.$$

For $u < 0$, $F_U(u) = 0$, and for $u > 1$, $F_U(u) = 1$. Thus U is uniform on $[0, 1]$.

Now consider a random sample $\mathbf{X}$ and let $U_1 = F(X_1), \ldots, U_n = F(X_n)$. These U's are independent, being functions of independent random variables, and each is uniform on $[0, 1]$. Because $F(x)$ is monotone increasing, the smallest U is the transform of the smallest X, and so forth, so that the ordered U's are, respectively, the transforms of the ordered X's:

$$V_i \equiv U_{(i)} = F(X_{(i)}).$$

The distribution of $(V_1, \ldots, V_n)$ is, therefore, the distribution of the order statistic of a random sample from the uniform population on $[0, 1]$:

$$f_{\mathbf{V}}(\mathbf{v}) = \begin{cases} n!, & \text{for } 0 < v_1 < \cdots < v_n < 1, \\ 0, & \text{elsewhere,} \end{cases}$$

which is simply a uniform density on the set of values of $\mathbf{V}$.

The random variable $V_k = F(X_{(k)})$ is the area under the population density to the left of the kth smallest observation. Its distribution is the distribution of the kth smallest observation in a sample of n from a uniform distribution on $[0, 1]$, with density (from Section 4.3.2)

$$f_{V_k}(v) = n\binom{n-1}{k-1} v^{k-1}(1-v)^{n-k},$$

for $0 < v < 1$. The mean value of this involves a beta function (Section 7.1.2):

$$E(V_k) = \int_0^1 n\binom{n-1}{k-1} v^k(1-v)^{n-k}\, du = \frac{k}{n+1}.$$

A useful statistic is the area between two successive ordered observations:

$$V_k - V_{k-1} = F(X_{(k)}) - F(X_{(k-1)}),$$

with mean

$$E(V_k) - E(V_{k-1}) = \frac{k}{n+1} - \frac{k-1}{n+1} = \frac{1}{n+1}.$$

This says that the expected area under the population density between two successive ordered observations is $1/(n+1)$; the ordered observations thus tend to divide the area under the density curve into $n + 1$ equal areas.

The total area between the smallest and largest observations:

$$Z \equiv V_n - V_1 = F(X_{(n)}) - F(X_{(1)})$$

is used in the construction of *tolerance intervals*. The difference $V_n - V_1$ is, of course, simply the range of a random sample from a uniform population, having density (according to Section 4.3.2)

$$f_Z(z) = n(n-1)(1-z)z^{n-2}, \qquad \text{for } 0 < z < 1.$$

This is independent of the population distribution, $F(x)$. The following example illustrates the notion of a tolerance interval.

EXAMPLE 11–2. Suppose that it is desired to find a sample size n such that at least 99 per cent of a certain population, with probability .95, will lie between the smallest and largest sample observations. That is, n is to be chosen so that

$$\begin{aligned} .95 &= P[F(X_{(n)}) - F(X_{(1)}) > .99] \\ &= P(V_n - V_1 > .99) = n(n-1)\int_{.99}^{1} (1-z)z^{n-2}\,dz \\ &= 1 - (.99)^{n-1}(.01n + .99). \end{aligned}$$

This yields an n of about 475. [Note $(.99)^{n-1} \doteq e^{-.01(n-1)}$.] In about 95 per cent of samples of size 475, the extreme values of the sample will include 99 per cent of the population, or more.

In a later section, it will be necessary to transform the sample distribution function, defined by

$$F_n(x) = \frac{k}{n}, \qquad \text{for } X_{(k)} \le x < X_{(k+1)},$$

with $F_n(x) = 0$ for $x < X_{(1)}$ and 1 for $x \ge X_{(n)}$. Now consider the function

$$F_n^*(y) \equiv F_n(F^{-1}(y)) = \frac{k}{n}, \qquad \text{for } X_{(k)} \le F^{-1}(y) < X_{(k+1)}.$$

This last inequality is equivalent to $F(X_{(k)}) \le y < F(X_{(k+1)})$, since $F(x)$ is increasing, and so

$$F_n^*(y) = \frac{k}{n}, \qquad \text{for } V_k \le y < V_{k+1},$$

where the V's are the ordered observations in a random sample from a uniform population (no matter what $F(x)$ is). That is, $F_n^*(y)$ is the sample distribution function of a random sample from the uniform distribution on [0, 1]; this can be used to show that such functionals as

$$\int_{-\infty}^{\infty} g[F_n(x), F(x)]\,dF(x) \qquad \text{and} \qquad \sup_x g[F_n(x), F(x)]$$

are independent of the population c.d.f. $F(x)$, a fact needed in Section 11.2. For, making the change of variable $u = F(x)$ reduces these to

$$\int_0^1 g[F_n^*(u), u]\,du \qquad \text{and} \qquad \sup_{0<u<1} g[F_n^*(u), u],$$

respectively.

Problems

11–1. Let (X_1, X_2, X_3) be a random sample from a population with density $f(x) = e^{-x}$, $x > 0$. Find the probability of the region $x_2 < x_3 < x_1$ by carrying out the integration of the joint density.

11–2. Obtain the joint density of the successive areas under the population density between the observations:

$$Y_1 = F(X_{(1)}), \quad Y_2 = F(X_{(2)}) - F(X_{(1)}), \ldots, \quad Y_n = F(X_{(n)}) - F(X_{(n-1)}).$$

[Why not include $Y_{n+1} = 1 - F(X_{(n)})$?]

11–3. Determine the joint density of the smallest and next smallest of four independent observations from a uniform distribution on (0, 1) by integrating out unwanted variables from the joint density function of the order statistic.

11–4. Let **X** be a random sample from a population that is uniform on $[0, \theta]$, and let **Y** be the corresponding order statistic. Obtain the joint density of $(Y_1, \ldots, Y_{n-1})$ given Y_n and observe that it does not depend on θ. [The distribution of **Y** given Y_n is singular; but it is again independent of θ, as is the distribution of **X** given **Y**. To conclude that the distribution of **X** given Y_n does not involve θ (which would establish sufficiency of Y_n) seems justifiable, but rigorous verification is beyond our scope.]

★11–5. Let X be continuous with $F(0) = 0$. Find the density of the vector of spacings of the n observations in a random sample: $Z_1 = X_{(1)}$, $Z_2 = X_{(2)} - X_{(1)}, \ldots$, $Z_n = X_{(n)} - X_{(n-1)}$.

(a) Integrate out extra z's to obtain the joint density of Z_1, Z_2, and Z_3.

(b) Transform to the density of (nZ_1, nZ_2, nZ_3) and pass to the limit as n becomes infinite, to show that the nZ's are, asymptotically, independent exponential variables with parameter $\lambda = f(0)$. (The result can clearly be generalized with a fixed m in place of 3 and a in place of 0.)

11.2 Goodness of Fit

The problem of testing the "fit" of observed data to an hypothesized model was considered in Chapter 9 for discrete populations. Here the problem will be studied in the case of continuous univariate populations, in which the ordering in the sample space seems to call for different methods. It should be noted, however, that the Pearson chi-square test can be applied in the continuous case, and people often do this. In order to apply it, the continuous model to be tested must be approximated by a discrete one obtained by round-off, replacing the given sample space by a finite system of class intervals. Whether by reason of the necessary approximation or its failure to use or recognize order in the sample space, the chi-square test appears to be generally less satisfactory for continuous models than those to be taken up here.

11.2.1 The Kolmogorov–Smirnov Test

The sample distribution function:

$$F_n(x) = \frac{j}{n}, \qquad \text{for } X_{(j)} \le x < X_{(j+1)}, \quad j = 0, \ldots, n,$$

(with $X_{(0)} = -\infty$ and $X_{(n+1)} = \infty$) will generally differ from the population distribution function. But if it differs from an assumed distribution $F(x)$ by "too much," this may serve as grounds to reject the hypothesis that $F(x)$ is the population distribution function. That is, the amount of the difference between the empirical and assumed distribution functions should be a useful statistic in determining whether or not to accept the assumed distribution function as correct.

It is the actual numerical difference $|F_n(x) - F(x)|$ that is used in the Kolmogorov–Smirnov test. More precisely, since this difference depends on x, the Kolmogorov–Smirnov statistic is taken to be the least upper bound of such differences:

$$D_n = \sup_{\text{all } x} |F_n(x) - F(x)|.$$

This statistic has a distribution, under H_0, which is independent of the c.d.f. $F(x)$ that defines H_0. Such a statistic is called *distribution-free*. (The chi-square statistic is asymptotically distribution-free.) That D_n is distribution-free was demonstrated in Section 11.1.3.

For a given n, a single table is required for the distribution function of D_n and can be used for any $F(x)$. This table can be computed through the use of recursion formulas, and has been computed for various sample sizes (Table VI, Appendix). Asymptotic percentiles can be computed from the limiting distribution:

$$\begin{aligned}\lim_{n\to\infty} P\left(D_n < \frac{z}{n}\right) &= 1 - 2\sum_{j=1}^{\infty} (-1)^{j-1} \exp(-2j^2z^2) \\ &\doteq 1 - 2\exp(-2z^2).\end{aligned}$$

The statistic D_n is essentially "two-sided," involving the *absolute* difference of $F(x)$ and $F_n(x)$. The critical region $D_n >$ const is used to test $F(x)$ against the alternative that the c.d.f. is *not* $F(x)$.

EXAMPLE 11–3. Consider testing the hypothesis that a distribution is normal with mean 32 and variance 3.24, using the 10 observations: 31.0, 31.4, 33.3, 33.4, 33.5, 33.7, 34.4, 34.9, 36.2, and 37.0. The sample distribution function and the population distribution function being tested are sketched in Figure 11–2. The maximum deviation is about .56.

According to Table VI, Appendix, the 95th percentile of the distribution of D_{10} is .409. Since .56 > .409, the distribution being tested is rejected at the 5 per cent level.

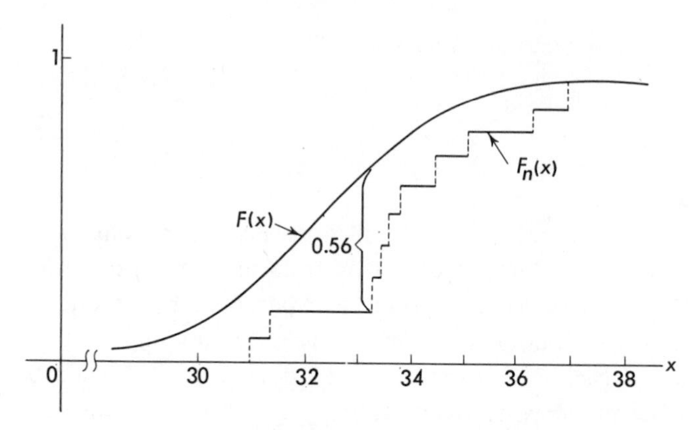

Figure 11–2

It is not possible to talk about a power "function," since the alternatives are too numerous to be indexed by even a finite number of parameters. One known result concerning the power of the Kolmogorov–Smirnov test is shown in Figure 11–3. For each value of Δ, there is given a lower bound on the power among the class of alternatives $F_1(x)$ for which

$$\Delta = \sup_{\text{all}\,x} |F_1(x) - F_0(x)|.$$

One curve gives this bound for the test $D_n > K$, where K is chosen so that $\alpha = .05$, and the other curve corresponds to $\alpha = .01$. Experiments described in the article from which the curves are taken indicate that the lower bounds given are usually quite conservative. It is also shown in that article that the Kolmogorov–Smirnov test is consistent and biased.

It has been assumed so far that the populations in question are continuous. However, it happens that in the case of a population that is not of the continuous type, the probability of rejecting H_0 is no larger (under H_0) than that given in the table constructed for use in the continuous case. That is, using Table VI, Appendix, to determine the rejection limit for a given α results in a test with an α that is no larger than that given.

EXAMPLE 11–4. Four coins are tossed 160 times, with results given in the following tabulation. Also shown in this tabulation are the values of the binomial distribution function with $n = 4$ and $p = .5$ multiplied by 160, and the sample cumulative frequency function. Examination of the differences, also shown in the table, yields the value of D_n: $\frac{7}{160} \doteq .044$.

Number of Heads	0	1	2	3	4
Frequency	10	33	61	43	13
Sample cumulative frequency	10	43	104	147	160
Cumulative frequency under H_0	10	50	110	150	160
Differences	0	7	6	3	0

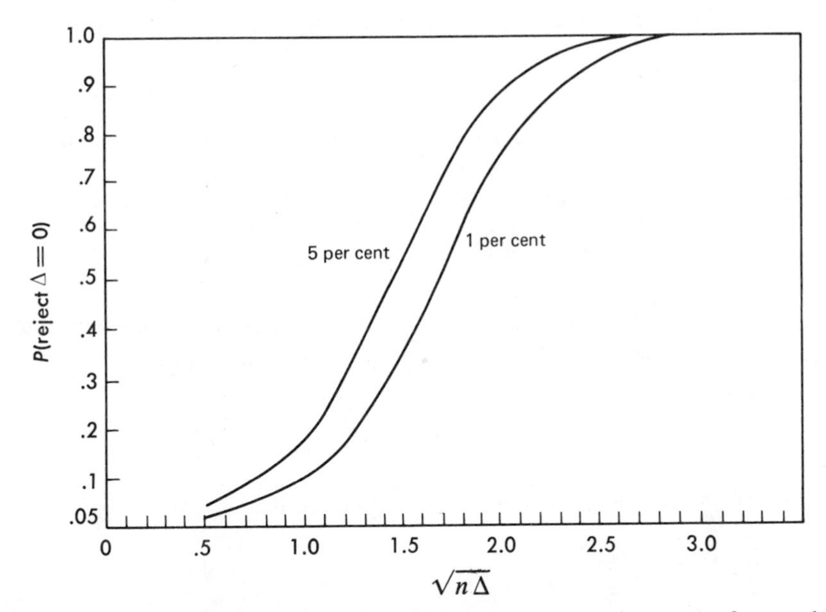

Figure 11–3 From F. J. Massey, Jr., "The Kolmogorov–Smirnov test for goodness of fit," *J. Am. Stat. Assn.* **46**, 68–78 (1951).

The 20 per cent rejection limit for a continuous population is

$$\frac{1.07}{\sqrt{160}} \doteq .085.$$

The test $D_n > .085$ in the present problem has then an α that does not exceed .20. Thus, at a level not exceeding .20, the null hypothesis is accepted, since D_n turned out to be .044, which does not exceed the rejection limit .085.

Since D_n is based on the notion of a distribution function, the values of the random variable X and their ordering play an essential role, as opposed to the circumstances of a chi-square test, in which any ordering of outcomes is not taken into account. In order to apply the Kolmogorov–Smirnov test to an experiment in which the outcomes are not inherently numerical, one would first have to set up an arbitrary coding of outcomes with numbers. This coding is not unique, and maximum deviations D_n can be different for different codings, as seen in the following example. This phenomenon suggests that the Kolmogorov–Smirnov test should not be used in such cases.

EXAMPLE 11–5. To test the equal likelihood of the six faces of a die, it is cast 120 times, with the following resulting frequencies of the six faces: 18, 23, 16, 21, 18, 24. The sample distribution function has step heights proportional to 0, 18, 41, 57, 78, 96, 120, as compared with the step heights of the H_0 distribution, which are proportional to 0, 20, 40, 60, 80, 100, 120. The maximum discrepancy is $\frac{4}{120} \doteq .033$, which is smaller than the 10 per cent rejection limit $1.22/\sqrt{120} = .111$. Thus, acceptance of H_0 is called for.

Suppose, however, that the following code is used: one dot = 2, two dots = 5, three dots = 1, four dots = 4, five dots = 3, and six dots = 6. Using the same sample of 120 tosses as above, we now find that the sample distribution function has step heights above the horizontal proportional to 16, 34, 52, 73, 96, 120. The maximum departure from the null distribution function is now $\frac{8}{120} \doteq .067$, considerably closer to the rejection limit than with the original coding, even though the same experimental results have been used.

A one-sided Kolmogorov–Smirnov statistic is sometimes employed to test $F_0(x)$ against the one-sided alternative $F(x) \geq F_0(x)$, with $F(x) \not\equiv F_0(x)$. The statistic used is

$$D_n^+ = \sup_{\text{all } x} [F_n(x) - F_0(x)].$$

The distribution of this statistic under the null hypothesis is defined by[1]

$$P(D_{n+} > u) = \sum_{0 \leq k \leq n(1-u)} \binom{n}{k} \left(u + \frac{k}{n}\right)^{k-1} u \left(1 - u - \frac{k}{n}\right)^{n-k}.$$

The asymptotic formula

$$P\left(D_n^+ < \frac{x}{\sqrt{n}}\right) \doteq 1 - \exp(-2x^2)$$

was obtained by Smirnov.

An exposition of Kolmogorov–Smirnov tests and of the Cramér–von Mises tests to be considered in the next section is given in a paper by Darling,[2] in which their power is also discussed. For some alternatives the Kolmogorov–Smirnov test has a power advantage over the chi-square test. For testing a specific normal population against the alternative of a mean shifted to one side, the power of the Kolmogorov–Smirnov does not compare favorably with that of the usual uniformly most powerful test. Chapman[3] considers the relative powers of various one-sided tests including that based on D_n^+.

An important feature of the Kolmogorov–Smirnov statistic is that, with it, one can easily construct a confidence band for the population c.d.f. (See Problem 11–9.) Another feature is that the test based on this statistic is an "exact" test—the exact distribution of the test statistic is known, even for small samples, and, equally important, it is distribution-free. On the other hand, the chi-square

[1] See Z. W. Birnbaum and F. H. Tingey, "One-Sided Confidence Contours for Distribution Functions," in *Ann. Math. Stat.* **22**, 592 (1951).

[2] D. A. Darling, "The Kolmogorov–Smirnov, Cramér–von Mises Tests," *Ann. Math. Stat.* **28**, 823 (1957).

[3] D. G. Chapman, "A Comparative Study of Several One-Sided Goodness-of-Fit Tests," *Ann. Math. Stat.* **29**, 655 (1959).

statistic, sometimes used (as mentioned earlier) for testing a continuous model is only asymptotically chi-square, independent of the population; it is a large-sample test.

Pearson's chi-square test for goodness of fit does have the important flexibility of being adaptable to testing a composite hypothesis, one consisting of a family of models indexed by a real parameter. It would be desirable to be able to use the Kolmogorov–Smirnov approach for testing a family of continuous models, but no analytic treatment of such problems is available. Lilliefors has done a Monte Carlo study[4] in which he obtains rejection limits for the particular problem of testing for *normality* of a population distribution.

The null hypothesis in this problem is that X is normal with unknown parameters. If $\bar{X}$ and S^2 from a random sample are used as parameters for a normal distribution, and if the c.d.f. of this distribution is called $F_0(x)$, then the statistic D_n, defined as before, has a distribution (under H_0) whose percentiles are given by Lilliefors, including the following:

n	10	11	12	13	14	15	16	17	18	19	20	25
95th percentile	.258	.249	.242	.234	.227	.220	.213	.206	.200	.195	.190	.173

For $n > 25$ he proposes the asymptotic value $.886/\sqrt{n}$.

Lilliefors also found, again by the Monte Carlo technique, that this modification of the Kolmogorov–Smirnov test has greater power than the chi-square test in the situations he considered.

11.2.2 Other Tests

Other measures of discrepancy between two distribution functions lead to corresponding statistics for testing goodness of fit. Given a nonnegative weight function $G(y)$, the statistic

$$n\int_{-\infty}^{\infty} [F_n(x) - F(x)]^2 G[F(x)]\, dF(x),$$

like the Kolmogorov–Smirnov statistic, is distribution-free—its distribution does not depend on $F(x)$. Using such a statistic was proposed by Cramér and by von Mises; the particular form with weighting function $G \equiv 1$ was proposed by Smirnov, who then also obtained an expression for its limiting distribution. With $G \equiv 1$, the statistic reduces to

$$n\omega_n^2 = n\int_{-\infty}^{\infty} [F_n(x) - F(x)]^2\, dF(x) = n\int_0^1 [F_n^*(y) - y]^2\, dy.$$

[4] H. W. Lilliefors, "On the Kolmogorov–Smirnov Test for Normality with Mean and Variance Unknown," *J. Am. Stat. Assn.* **64**, 399–402 (1967).

Some of the percentiles of its asymptotic distribution are as follows:[5]

$P(n\omega_n^2 \leq x)$	.80	.85	.90	.95	.98	.99
x	.241	.284	.347	.461	.620	.743

For computations in samples that are not too large, the following formula is useful (it is obtained by carrying out the integration indicated in $n\omega_n^2$):

$$n\omega_n^2 = \frac{1}{12n} + \sum_{j=1}^{n} \left(\frac{2j-1}{2n} - F(X_{(j)})\right)^2,$$

where $X_{(1)}, \ldots, X_{(n)}$ are the ordered observations and $F(x)$ the distribution function of the null hypothesis.

EXAMPLE 11–6. Consider the 10 observations of Example 11–3 for use in testing (as in that example) a normal distribution with mean 32 and standard deviation 1.8. For $X_{(1)}$,

$$F(X_{(1)}) = F(31) = \Phi\left(\frac{31 - 32}{1.8}\right) \doteq .29$$

and

$$\left\{\frac{2-1}{20} - .29\right\}^2 \doteq .057.$$

Proceeding in this fashion, one finds $n\omega_n^2$ to be about .88, whereas for $\alpha = .05$, the rejection limit is .461 (from the table of the asymptotic distribution given above). The null hypothesis that the distribution is normal with mean 32 and standard deviation 1.8 is rejected at the 5 per cent level.

A test of the Cramér–von Mises type for the null hypothesis $F(x; \theta)$, where θ is an unknown parameter and estimated from the sample, has been constructed by Darling.[6] The test statistic is again $n\omega_n^2$, except that the value of θ used is a sample estimate of θ. If this estimate is properly chosen, the asymptotic distribution of the test statistic can be determined under H_0 and for certain problems is distribution-free.

The effect of grouping data in the Kolmogorov–Smirnov and Cramér–von Mises tests has been studied in a series of Russian papers.[7]

[5] These entries come from an article by Anderson and Darling, "Asymptotic Theory of Certain 'Goodness of Fit' Criteria Based on Stochastic Processes," *Ann. Math. Stat.* **23**, 193 (1952). Computations by A. W. Marshall [*Ann. Math. Stat.* **29**, 307 (1958)], indicate that the convergence to the asymptotic distribution is very rapid, the asymptotic expressions being useful even for n as small as 3 or 4.

[6] D. A. Darling, "The Cramér–Smirnov Test in the Parametric Case," *Ann. Math. Stat.* **26**, 1 (1955).

[7] For references, see the survey paper by D. A. Darling, "The Kolmogorov–Smirnov, Cramér–von Mises Tests," *Ann. Math. Stat.* **28**, 823 (1957).

A somewhat different measure of discrepancy between theoretical and empirical distribution functions has been proposed by Sherman:[8]

$$\sum_{i=1}^{n+1} \left| F(X_{(i)}) - F(X_{(i-1)}) - \frac{1}{n+1} \right|,$$

where $X_{(0)} = -\infty$ and $X_{(n+1)} = \infty$. This statistic is suggested by the fact that the expected area under the population density curve between a pair of successive ordered observations is $1/(n + 1)$, as shown in Section 11.1.3. The exact distribution is given by Sherman, who also shows that it is asymptotically normal.

Another type of goodness of fit statistic for testing normality has been proposed by Shapiro and Wilk;[9] the statistic is approximately

$$W = r^2_{\mathbf{Y},\mathbf{m}},$$

where $Y_i = X_{(i)}$, $\mathbf{X}$ is the random sample being used for the test, m_i is the expected value of the ith order statistic corresponding to a random sample from a standard normal population,[10] and $r_{\mathbf{Y},\mathbf{m}}$ is the correlation of the vectors $\mathbf{Y}$ and $\mathbf{m}$. (Observe that because the correlation coefficient is unchanged by a linear transformation of either X or Z, the statistic W has to do only with normality, and not with a particular normal population.) The critical region used is of the form $W < K$. A table of critical values for sample sizes up to 20 is given in the referenced paper, as well as the results of an empirical study of power characteristics for various alternatives—results suggesting that this W-test will outperform the Kolmogorov–Smirnov test and the chi-square test.[11]

A relatively simple test for normality due to Geary (1935), particularly useful in testing against symmetric alternatives, employs the ratio of mean deviation to standard deviation as a test statistic. This test and others are compared with the Shapiro–Wilk test in a pair of recent articles, and the problem continues to receive attention.[12]

[8] B. Sherman, "A Random Variable Related to the Spacing of Sample Values," *Ann. Math. Stat.* **21**, 339 (1950).

[9] S. Shapiro and M. Wilk, "An Analysis of Variance Test for Normality," *Biometrika* **52**, 591–611 (1965).

[10] The statistic actually proposed involves a set of constants $\mathbf{a}$ for which $2\mathbf{m}$ is an approximation.

[11] In using the chi-square test for normality, which requires an approximation involving finitely many class intervals, the point made just prior to Section 9.1.2 is particularly relevant—the case of a normal population is one in which the error in significance level mentioned there may be significant.

[12] R. B. D'Agostino and B. Rosman, "The Power of Geary's Test of Normality," p. 181, and A. R. Dyer, "Comparisons of Tests for Normality with a Cautionary Note," p. 185, both in *Biometrika* **61** (1974).

11.2.3 Comparison of Distributions

Different populations have different distribution functions, and it is expected that samples from these different populations will have sample distribution functions that differ. Of course, random sampling fluctuations can introduce a difference in sample distribution functions even though the samples be from the same population, but a very large discrepancy between sample distribution functions might reasonably serve as the basis for an inference that the populations are different.

Tests based on the discrepancy between sample distribution functions are ordinarily employed when the alternatives to the null hypothesis of identical populations include all ways in which the populations can differ, as opposed to problems in which the alternatives are that one population has simply "shifted" away from the other. Different ways of measuring discrepancies between sample distribution functions lead to different tests. The *two-sample Kolmogorov–Smirnov test* (also called the Smirnov test) is based on the statistic D, defined as follows:

$$D = \sup_{\text{all } x} |F_m(x) - G_n(x)|,$$

where $F_m(x)$ is the sample distribution function of a sample of size m from X and $G_n(x)$ is the sample distribution function of a sample of size n from Y. Smirnov has shown that for large samples of sizes m and n:

$$P_{H_0}\left[D > z\left(\frac{1}{m} + \frac{1}{n}\right)^{1/2}\right] \doteq 2 \sum_{k=1}^{\infty} (-1)^{k-1} \exp\left[-2k^2z^2\right],$$

the same function appearing on the right as showed up in the one-sample Kolmogorov–Smirnov test. Again the first-term approximation is very good and on the safe side. Using this, the above probability is seen to be approximately α if $z = [-\frac{1}{2} \log (\alpha/2)]^{1/2}$, and so a test with level α is obtained by rejecting H_0 for

$$D > \left[-\frac{1}{2}\left(\frac{1}{m} + \frac{1}{n}\right) \log \frac{\alpha}{2}\right]^{1/2}.$$

Values of this asymptotic rejection limit for $\alpha = .05$ and $\alpha = .01$ are given in Table VII, Appendix, along with small-sample rejection limits for these two levels.

EXAMPLE 11–7. Consider the two samples of size 60 given in the accompanying frequency tabulations. The maximum absolute difference in cumulative relative frequencies is .1 and the 5 per cent rejection limit from Table VII (see Appendix) is about .25. The null hypothesis of identical distributions is accepted.

x_i	Frequencies		Cumulative Frequencies	
	Sample 1	Sample 2	Sample 1	Sample 2
0	5	6	5	6
1	9	2	14	8
2	4	5	18	13
3	5	7	23	20
4	7	8	30	28
5	2	4	32	32
6	8	4	40	36
7	4	9	44	45
8	9	6	53	51
9	7	9	60	60

The test just described is two-sided, since either positive or negative differences are counted. A one-sided test is provided by the one-sided statistic

$$D^+ = \sup_{\text{all } x} [F_m(x) - G_n(x)],$$

which would intuitively be appropriate for a one-sided class of alternatives, say, $F_Y(\lambda) < F_X(\lambda)$. The large-sample distribution of D^+ was shown by Smirnov to be defined by

$$P\left[D^+ > z\left(\frac{1}{m} + \frac{1}{n}\right)^{1/2}\right] = \exp(-2z^2), \qquad z \geq 0.$$

From this it is evident that the large-sample distribution of

$$\frac{4(D^+)^2mn}{m + n}$$

is given by

$$P\left[\frac{4(D^+)^2mn}{m + n} < u\right] = 1 - \exp\left(-\frac{u}{2}\right), \qquad u \geq 0,$$

which is exponential and happens also to be the chi-square distribution function with 2 degrees of freedom.

EXAMPLE 11–8. Had the data been gathered as in Example 11–7, but to test instead the hypothesis that the population distributions are identical against the alternative that population 2 is shifted to the right, the one-sided D^+ should be used. Since for the data obtained all the differences between the sample distribution functions are positive, $D^+ = D = .1$. Then, since

$$\frac{4(D^+)^2mn}{m + n} = 1.2 < 5.99,$$

where 5.99 is the 95th percentile of the chi-square distribution with 2 degrees of freedom, the null hypothesis would be accepted at the 5 per cent level.

Problems

11–6. A medical facility has determined, from long experience, that the platelet count in the blood of healthy males is normal with mean 235,000 per mm^3 and standard deviation 44,600 per mm^3. The platelet counts of 25 lung cancer patients were obtained as follows (in units of 1000 per mm^3):

173, 189, 196, 207, 215, 237, 275, 282, 293, 300,
305, 316, 346, 382, 395, 399, 401, 437, 480, 504,
524, 634, 682, 882, 999.

(a) Use the Kolmogorov–Smirnov test to decide, at $\alpha = .01$, whether these observations can be considered as coming from the population of counts for healthy males.
(b) Use the data to test for normality of the population from which they are drawn, by means of the Lilliefors modification of the Kolmogorov–Smirnov test. (Use $\alpha = .05$ so that the rejection limits given in the text can be applied.) Given: $\bar{X} = 402.1$ and $S = 210.7$.

11–7. Use the Kolmogorov–Smirnov statistic to test the hypothesis that X has a binomial distribution with $n = 4$ and $p = \frac{1}{2}$ on the basis of the following results:

x_i	0	1	2	3	4
f_i	6	38	58	47	11

11–8. Compute the power of the test $|\bar{X} - \mu_0| > K$, with K chosen so that $\alpha = .05$ and $n = 25$, at the alternative $\mu = \mu_0 + .6$ in a problem involving normal populations with known variance $\sigma^2 = 1$. Compare this with the lower bound given by the curve in Figure 11–3 for the power of the Kolmogorov–Smirnov test.

11–9. Discuss the construction of a confidence band about $F_n(x)$ based on the distribution of the Kolmogorov–Smirnov statistic. [A confidence band at level α would be defined by a pair of functions $A(x)$ and $B(x)$ such that $P[A(x) < F(x) < B(x)] = 1 - \alpha$, where A and B are determined by $F_n(x)$.]

11–10. Carry out the details in Example 11–6.

11–11. Determine a 5 per cent critical region in terms of D_n^+ for testing $F(x) = F_0(x)$ against the hypothesis $F(x) \geq F_0(x)$ for a sample of size 25. (Use the Smirnov asymptotic formula.)

11–12. (a) Discuss the first-term approximation and the statement that it is on the safe side, for the distribution of D in Section 11.2.3.
(b) Verify the claim in Section 11.2.3 concerning the large-sample distribution of $4(D^+)^2mn/(m + n)$.

(c) Measurements of viscosity for a certain substance were made with the following results:

First day	37.0,	31.4,	34.4,	33.3,	34.9,	36.2,	31.0,	33.5,
	33.7,	33.4,	34.8,	30.8,	32.9,	34.3,	33.3.	
Second day	28.4,	31.3,	28.7,	32.1,	31.9,			
	32.8,	30.2,	30.2,	32.4,	30.7.			

Would you say that the population has changed from one day to the next?

11.3 Randomness

A random sample was defined in Chapter 4 as a sequence of identically distributed, independent random variables—a sequence of independent observations on a certain "population." If there is any reason to suspect that the process of obtaining a "random sample" might *not* fit the mathematical model of a random sample, it would be in order to test the hypothesis

$$H_0\colon\ F(x_1, \ldots, x_n) = F(x_1)\cdots F(x_n),$$

where $F(x_1, \ldots, x_n)$ is the joint distribution function of the sample observations $(X_1, \ldots, X_n)$ and $F(x)$ is a fixed (but unknown) univariate distribution function, namely, the population distribution function.

The statistical problem is not completely posed until there is specified an alternative to H_0, and herein lies one of the major obstacles to an adequate treatment of this problem of "randomness." The most inclusive alternative hypothesis would be that the observations are *not* independent replicas of a population variable—in other words, that H_0 is not true. This class of alternatives is rather unwieldy. For instance, given *any* test with a critical region of specified size under H_0, there would be many states not in H_0 for which the power of that test is zero. (Any multivariate distribution with zero mass on the critical region and dependent marginals would serve as such a state.) Thus, it is better to consider a more restricted class of alternative states that are especially feared, and then to look for a test that has reasonably high power for this restricted class, not worrying about *all* conceivable alternatives to randomness.

One way to restrict the class of alternatives is to keep the assumption of independence of the observations, allowing the distribution function to vary from observation to observation. In particular, a *trend* alternative is one in which the joint distribution function of the observations is assumed to be of the form

$$F(x_1, \ldots, x_n) = \prod_{i=1}^{n} F[x_i - g(i)].$$

With this assumption the distributions for the various observations are identical in shape but shifted in location, as would be the case if the mean of a population with density of given shape were gradually changing as the observations were taken.

Even with such restriction of alternatives it is not apparent how to "derive" suitable tests from basic principles, so the tests that have been proposed have only an intuitive basis. Power is again difficult to manage, and the discussion here will include at most a derivation of the distribution of the test statistic under the null hypothesis.

11.3.1 Run Tests

A *run* in a sequence of symbols is a group of consecutive symbols of one kind preceded and followed by (if anything) symbols of another kind. For example, in the sequence

$$+ + + - + + - - - - + + - -$$

the runs can be exhibited by putting vertical bars at the changes of symbol:

$$+ + + \mid - \mid + + \mid - - - - \mid + + \mid - -$$

Here there are a run of three $+$'s, a run of one $-$, a run of two $+$'s, and so on. There are altogether six runs, three runs of $+$'s and three of $-$'s.

Consider now a sequence of observations from a continuous population, and let each observation be assigned the letter "a" if it is above the median of the observations and the letter "b" if it is below. The sample sequence then defines a sequence of a's and b's. If, to simplify the discussion, it is agreed to ignore the median observation when the number of observations is odd, the sequence of a's and b's will have an even number of terms—say, $2m$ terms, of which then m are a's and m are b's. In this sequence there will be a certain number of runs of a's and a certain number of runs of b's; let these numbers be denoted by r_a and r_b, respectively, and let $r = r_a + r_b$ denote the total number of runs. Run tests are then based on the intuitive notion that an unusually large or an unusually small value of r would suggest lack of randomness.

For instance, if a downward trend is present, the a's will tend to come at the beginning and the b's at the end of the sequence of a's and b's, resulting in a relatively small number of runs. Also, one could imagine certain kinds of dependences among the observations which would result in a systematic bouncing back and forth from one side of the median to the other, producing an unusually large number of runs.

The distributions of r, r_a, and r_b can be readily computed under the null hypothesis of randomness. Under this hypothesis the $(2m)!$ arrangements of the observations are equally likely to have produced a given order statistic (see Section 11.1.2). Each arrangement leads to a sequence of a's and b's, and the distinct arrangements of m a's and m b's are then also equally likely since each one comes from $(m!)^2$ arrangements of the observations. So the probability of a given run configuration can be computed as the ratio of the number of the arrangements of a's and b's having that configuration to the total number.

To compute the joint probability that $r_a = x$ and $r_b = y$, then, it is only necessary to count the number of arrangements of a's and b's having this property and divide by $\binom{2m}{m}$. In counting arrangements for the numerator there are three cases to be considered: either (i) $x = y + 1$, (ii) $x = y - 1$, or (iii) $x = y$. The probability that r_a differs from r_b by *more* than 1 is zero, so all possibilities are covered in these three cases.

In case (i) the sequence of a's and b's begins with an a and ends with an a. To divide the a's into x separated groups, slots for the b's are inserted in $x - 1$ places selected from the $m - 1$ spaces between the m a's. This can be done in $\binom{m-1}{x-1}$ ways. Having decided where to put the b's, one partitions them into y (or $x - 1$) groups, just as the a's were partitioned, in one of $\binom{m-1}{y-1}$ ways, and puts the groups of b's into the prepared slots. The arrangement, then, has x runs of a's and y runs of b's and could have been accomplished in $\binom{m-1}{x-1}\binom{m-1}{y-1}$ ways.

Case (ii) is exactly analogous to case (i), and the number of arrangements is the same. In case (iii), in which $x = y$, the sequences begin with a and end with b, or else they begin with b and end with a. In either instance, the number of ways, computed as for case (i), is again $\binom{m-1}{x-1}\binom{m-1}{y-1}$; the total for case (iii) is twice that number. Finally, then, given the order statistic one has

$$P(r_a = x \text{ and } r_b = y) = \begin{cases} 2\binom{m-1}{x-1}\binom{m-1}{y-1}\Big/\binom{2m}{m}, & \text{if } x = y, \\ \binom{m-1}{x-1}\binom{m-1}{y-1}\Big/\binom{2m}{m}, & \text{if } x = y \pm 1, \\ 0, & \text{if } |x - y| > 1. \end{cases}$$

Since this is the same for any order statistic, it is also the absolute probability of $r_a = x$ and $r_b = y$, and the condition has been omitted from the notation.

The probability function for the total number of runs can be computed from the above joint probability function as follows:

$$P(r = z) = \sum_{x+y=z} P(r_a = x \text{ and } r_b = y).$$

If z is an even number, $z = 2k$, the sum has only one term—that in which $x = y = k$. If z is odd, $z = 2k + 1$, there are two terms in the sum, one in which $x = k$ and $y = k + 1$ and one in which $x = k + 1$ and $y = k$. Hence,

$$P(r = z) = \begin{cases} 2\binom{m-1}{k-1}\binom{m-1}{k-1}\Big/\binom{2m}{m}, & \text{if } z = 2k, \\ 2\binom{m-1}{k}\binom{m-1}{k-1}\Big/\binom{2m}{m}, & \text{if } z = 2k + 1, \end{cases}$$

for $z = 2, 3, \ldots, 2m$. The corresponding cumulative distribution function is readily computed from this and is available in tables.[13]

EXAMPLE 11–9. In samples of size six from a continuous population, there will be three observations above and three observations below the sample median. That is, $m = 3$. The number of runs can vary from two to six, and the distribution function of this number r is as follows (computed from the formulas given above):

z	2	3	4	5	6
$F_r(z)$	.1	.3	.7	.9	1.0

It is evident that a test which calls for rejecting randomness when fewer than three runs occur has a significance level

$$\alpha = P_{H_0}(\text{reject } H_0) = P_{H_0}(2 \text{ runs}) = .1.$$

It can be shown that the mean and variance of the distribution of r are

$$E(r) = m + 1, \qquad \text{var } r = \frac{m(m-1)}{2m-1},$$

and that for large samples, r is approximately normally distributed.[14] Using a continuity correction and the approximation

$$\text{var } r \doteq \tfrac{1}{4}(2m - 1),$$

one finds the percentiles of r to be obtained from those of the standard normal distribution as follows:

$$r_p = \tfrac{1}{2}(2m + 2 + z_p\sqrt{2m - 1}),$$

where r_p and z_p denote, respectively, the $100p$ percentiles of the distribution of r and of a standard normal variate.

Another type of run test employs the signs of the differences of successive observations. If the population mean has a rising trend, for instance, there

[13] F. Swed and C. Eisenhart, "Tables for Testing Randomness of Grouping in a Sequence of Alternatives," *Ann. Math. Stat.* **14**, 66–87 (1943).

[14] For these results see A. M. Mood, "The Distribution Theory of Runs," *Ann. Math. Stat.* **11**, 367–92 (1940).

would be a tendency for observations to increase from one observation to the next, and +'s would more often occur in groups than with no trend. A given sequence of n observations defines a sequence of $n - 1$ signs of differences of successive pairs of observations. Let s denote the total number or runs of +'s and −'s in such a sequence. Various aspects of the distribution of s under the null hypothesis of randomness have been studied.[15] The mean and variance of s are

$$E(s) = \frac{2n-1}{3}, \qquad \text{var } s = \frac{16n-29}{90},$$

and the large-sample distribution of s is known to be approximately normal.

EXAMPLE 11–10. In samples of size 50, the mean and variance of the total number of runs up and down, under H_0, are

$$E(s) = 33 \qquad \text{and} \qquad \text{var } s \doteq 8.57.$$

The 5th percentile of the distribution of s under H_0 is then

$$s_{.05} \doteq 33 - 1.64\sqrt{8.57} \doteq 27.2.$$

Thus, 27 or fewer runs would be considered significant in testing against the presence of a trend, at the 5 per cent level, and would call for rejection of randomness.

11.3.2 Other Tests for Randomness

Another test for randomness is based on the statistic

$$r = \frac{d^2}{S^2},$$

where d^2 denotes the mean square successive difference:[16]

$$d^2 = \frac{1}{2(n-1)} \sum_{i=1}^{n-1} (X_{i+1} - X_i)^2,$$

and S^2 is the sample variance (unbiased version). For each i the difference $X_{i+1} - X_i$ has mean zero and variance $2\sigma^2$ under the null hypothesis that $(X_1, \ldots, X_n)$ is a random sample from a population with variance σ^2. The expected value of d^2 is then σ^2 under H_0. If a trend is present, d^2 is not altered nearly so much as the variance estimate S^2, which increases greatly. Thus, the critical region $r <$ const is employed in testing against the alternative of a trend.

[15] See J. Wolfowitz and H. Levene, *Ann. Math. Stat.* **15**, 58, 153 (1944).
[16] See B. I. Hart, "Significance Levels for the Ratio of the Mean Square Successive Difference to the Variance," *Ann. of Math. Stat.* **13**, 445–47 (1942), and references given there.

In order to use this test, of course, it is necessary to know the distribution of r. It can be shown that in the case of a *normal* population

$$E(r) = 1 \qquad \text{and} \qquad \operatorname{var} r = \frac{1}{n+1}\left(1 - \frac{1}{n-1}\right),$$

and that r is approximately normal for large samples (say, $n > 20$).

EXAMPLE 11–11. Consider the following observations from a normal population:

39, 42, 38, 53, 51, 30, 40, 40, 28, 43, 46, 53, 55, 29, 24,
34, 53, 66, 43, 42, 38, 34, 57, 26, 33,

listed in the order of observation. From these can be computed the values $d^2 \doteq 97$ and $S^2 \doteq 111.5$, so that $r \doteq .87$. The 5th percentile of r (for $n = 25$) is approximately

$$\mu_r - 1.64\sigma_r \doteq 1 - 1.64\left[\frac{1}{26}\left(1 - \frac{1}{24}\right)\right]^{1/2} \doteq .685,$$

and randomness would be accepted at the 5 per cent level, since $.87 > .685$.

Whereas the test based on the mean square successive difference requires a knowledge of the population distribution under H_0 even to evaluate the significance level, a "randomization" technique (due to R. A. Fisher) provides a test whose level is independent of the distribution assumed to be common to the observations under H_0. A test based on this technique is called a *permutation test* and is constructed as follows in the problem at hand.

If $(X_1, \ldots, X_n)$ is a random sample from a population with distribution function $F(x)$, the conditional probabilities of the $n!$ permutations of the observations, given the order statistic, are each $1/n!$. Then, if for each order statistic there is defined a critical region consisting of k of the $n!$ permutations, the corresponding significance level is $\alpha = k/n!$ The test defined in this way is a conditional test, and (as when conditional tests were introduced in Section 9.3.3) one need determine the critical set of permutations only for the order statistic actually observed.[17]

One choice of statistic to use in arranging permutations of a given order statistic according to evidence for or against randomness is the mean square successive difference d^2 defined above. (The variance S^2 in the denominator of the ratio r used there is the same for each permutation and can be disregarded here.) Permutations with small values of d^2 would be put first into a critical set.

[17] There seems to be little concern over the propriety of using a conditional test in this situation, since the conditioning statistic would appear intuitively to have no information about randomness.

More generally (see Problem 11–18), one might use a *serial correlation*, a statistic of the form

$$R_h = \sum_{1}^{n-h} X_i X_{i+h}$$

for some fixed h. The critical values of R_h, and perhaps h itself, would be chosen in light of the alternatives to randomness.

Such tests as these suffer the disadvantage of requiring computation of the values of the statistic used in forming the critical region for each permutation of the order statistic—for each new problem, after the order statistic has been obtained. No tables can be prepared in advance. But this is not the case if one uses a statistic based on ranks.

The rank t_i of the observation X_i is its position in the order statistic (that is, one more than the number of observations less than X_i in the sample). In the case of a continuous population each sample $(X_1, \ldots, X_n)$ defines a sequence of ranks $(t_1, \ldots, t_n)$ which is one of the $n!$ permutations of the integers $(1, 2, \ldots, n)$. Under the null hypothesis of randomness these are equally likely, and again a critical region is constructed by putting into it rank sequences ordered according to some function of the ranks, $T(t_1, \ldots, t_n)$. Two of the many rank-order statistics that have been proposed are

$$\sum_{i=1}^{n} it_i, \quad \text{and} \quad \sum_{i=1}^{n} E[Z_{(i)}]t_i,$$

where $Z_{(i)}$ denotes the ith smallest among n observations from a standard normal population.[18] (Table X gives values of $E[Z_{(i)}]$.)

Another test for randomness is based on the number of inequalities $X_j < X_k$ for $j < k$ (for downward trend alternatives).[19] This shares with the run, permutation, and rank tests the property of being "distribution-free," the distribution of the statistic under the null hypothesis being independent of the population distribution.

EXAMPLE 11–12. Suppose that in a sample of five from a continuous population the results are (1, 3, 9, 4, 7). There are 5! or 120 permutations of these numbers. Some of them, together with values of $T = \sum_{i=1}^{n-1} (X_{i+1} - X_i)^2$, the corresponding rank sequence, and the values of $\sum it_i$ are listed in the following table. [Of course,

[18] Such statistics are studied by A. Stuart, "The Asymptotic Relative Efficiencies of Distribution-free Tests of Randomness Against Normal Alternatives," *J. Am. Stat. Assn.* **49**, 147–57 (1954); see also I. R. Savage, "Contributions to the theory of Rank Order Statistics—the 'Trend' Case," *Ann. Math. Stat.* **27**, 968–77 (1957); and in Reference [16], p. 258.

[19] H. B. Mann, "Nonparametric Tests Against Trend," *Econometrica* **13**, 245 (1945).

for each entry listed, the permutation with numbers reversed would have the same T-value; for instance, $T(9, 7, 4, 3, 1) = 18$.]

Permutation	T	Ranks	$\sum it_i$
1 3 4 7 9	18	1 2 3 4 5	55
3 1 4 7 9	26	2 1 3 4 5	54
1 4 3 7 9	30	1 3 2 4 5	54
4 1 3 7 9	33	3 1 2 4 5	52
1 3 4 9 7	34	1 2 3 5 4	54
3 1 4 9 7	42	2 1 3 5 4	53
4 3 1 7 9	45	3 2 1 4 5	51
1 3 7 9 4	49	1 2 4 5 3	52
3 4 1 7 9	50	2 3 1 4 5	52
1 4 3 9 7	50	1 3 2 5 4	53
1 3 9 7 4	53	1 2 5 4 3	51
4 1 3 9 7	53	3 1 2 5 4	51
1 3 7 4 9	54	1 2 4 3 5	54
1 4 9 7 3	54	1 3 5 4 2	48

These are listed in order of increasing T-values, but observe that they would be ordered differently using the rank statistic $\sum it_i$. For a test based on T with $\alpha = .1$ the 12 permutations with smallest T-values would make up the critical region; that is, randomness would be rejected for $T \leq 42$; and the observed sequence given at the outset, for which $T = 74$, would call for accepting randomness. Observe, as commented earlier, that the T-values would change if the order statistic were, say (1, 3, 4, 7, 10), whereas the values of $\sum it_i$ (and so its distribution) would not.

Problems

11–13. Verify the distribution of the total number r of runs in samples of size six, given in Example 11–9: (a) by substituting in the formula for $P(r = z)$, and (b) by writing the 20 arrangements of three a's and three b's and counting the number of runs in each. Also, verify the formulas given for the mean and variance of the number of runs.

11–14. The following 30 observations are taken from a table of "random sampling numbers" (which should exhibit randomness if anything should):

15, 77, 01, 64, 69, 69, 58, 40, 81, 16, 60, 20, 00, 84, 22,
28, 26, 46, 66, 36, 86, 66, 17, 34, 49, 85, 40, 51, 40, 10.

Would you accept randomness on the basis of the number of runs above and below the median? On the basis of runs up and down?

11–15. Add $i - 1$ to the ith observation ($i = 1, \ldots, 25$) given in Example 11–11 and test the resulting sequence of numbers for randomness: (a) using the number of runs above and below the median, (b) the number of runs up and down, and (c) the mean square successive difference statistic.

11–16. Show that permuting two successive ranks in the rank sequence defined by a particular permutation of ranks changes the value of $\sum it_i$ by the amount of the difference between those two ranks. Use this fact to determine all rank permutations for which $\sum it_i = 53$ in Example 11–12 from those given with the value 54, and so conclude that the critical region of size $\alpha = .1$ for this statistic is different from that given in terms of the T in the example.

11–17. Given var $d^2 = (3n - 4)\sigma^4/(n - 1)^2$ in the case of a normal population, determine the efficiency of d^2 relative to S^2 as an estimate of σ^2.

11–18. Show that using the statistic d^2 is almost equivalent to using the serial correlation R_h with $h = 1$. Compute R_1 for the permutations of Example 11–12.

11–19. Compute the value of the statistic $\sum t_i E[Z_{(i)}]$ for the permutations given in Example 11–12.

11.4 One-Sample Location Tests

Location is frequently measured, in nonparametric inference, by the median m. It will be assumed that population distributions are continuous, so that for any observation X,

$$P(X > m) = P(X < m) = \tfrac{1}{2}.$$

The tests to be considered here for the population median can also be applied to the problem of comparison when observations are paired. (See Section 7.3.3.) Thus, if X and Y are from the same population,

$$P(X > Y) = P(X < Y) = \tfrac{1}{2},$$

so that 0 is the median of the population of differences.

11.4.1 The Sign Test

Suppose that it is desired to test the hypothesis that $m = m_0$. If m_0 is actually the median, approximately half of the observations in a random sample would be expected to lie on either side of m_0. This suggests using the number of sample observations to the right of m_0 (or equivalently, the number to the left) as the test statistic. Too many or too few observations on one side of m_0 would be used as a basis for rejecting $m = m_0$.

The distribution of the number of observations to the right of m_0 is binomial with parameter

$$p = P(X > m_0).$$

Under H_0: $m = m_0$, the value of p is $\frac{1}{2}$, and the expected number of observations on each side is $n/2$. The power (probability of rejecting H_0) would be the same for all alternative states having a given p, and this $\pi(p)$ can be computed from the binomial distribution or from the normal distribution if n is large.

EXAMPLE 11–13. Consider a test that rejects $m = m_0$ when more than 15 or fewer than 5 of 20 observations in a random sample fall to the right of m_0. The α for this test is the probability of the critical region under $p = \frac{1}{2}$:

$$\left[\binom{20}{0} + \cdots + \binom{20}{4} + \binom{20}{16} + \cdots + \binom{20}{20}\right] \Big/ 2^{20} \doteq .012.$$

The power function is given approximately by

$$\begin{aligned}\pi(p) &= 1 - P_p(5, 6, \ldots, \text{or } 15 \text{ observations to the right of } m_0)\\ &\doteq 1 - \left[\Phi\left(\frac{15.5 - 20p}{\sqrt{20pq}}\right) - \Phi\left(\frac{4.5 - 20p}{\sqrt{20pq}}\right)\right].\end{aligned}$$

Given the following 20 observations:

37.0, 31.4, 34.4, 33.3, 34.9, 31.6, 31.3, 34.6, 32.6, 31.6,
36.2, 31.0, 33.5, 33.7, 33.4, 34.4, 32.1, 33.3, 32.7, 31.5,

to test $m = 32$ against $m \neq 32$, the following pattern of signs yields the desired statistic, + indicating an observation above and − an observation below the value 32:

+ − + + + − − + + −
+ − + + + + + + + −

There are 14 plus signs, not enough for rejection of $m = 32$ at the level .012.

Suppose next that the observations in a random sample from X of size n and those in a random sample from Y of the same size n are paired according to the order of observation: (X_i, Y_i), $i = 1, \ldots, n$. For each pair a plus sign (+) or a minus sign (−) is recorded according as the Y exceeds or is exceeded by the X. Assume for the present that the distribution of (X, Y) is continuous so that ties have probability zero. If the probability that Y exceeds X is called p: $p = P(Y > X)$, the null hypothesis that X and Y are identically distributed yields the value $p = \frac{1}{2}$. Of course, p can be $\frac{1}{2}$ even though X and Y are not identically distributed, and it is really the hypothesis $p = \frac{1}{2}$ that the sign test is designed for.

The test statistic used is the number of + signs or the number of pairs of observations in which Y exceeds X. Because each pair (X_i, Y_i) is a trial of a Bernoulli experiment, the statistic is simply the number of successes in n independent trials, which has been seen to be minimal sufficient for problems involving p. The statistic has the binomial distribution with parameters (n, p).

Pairing the observations makes the two-sample problem into a one-sample problem—indeed, into a parametric one-sample problem, which has already been studied. The potential disadvantage in the requirement of equal sample sizes is offset by the fact that X and Y do not have to be independent; the test can be used in comparing twins, or right- and left-hand characteristics in individuals, or two sides of a tire, and so forth. In fact, all that is required is that the probability $P(Y_i > X_i)$ remain fixed.

Even though the populations be assumed continuous, ties *do* occur in practice owing to round-off. Rather than make half of the ties + and half −, or to assign + and − by the toss of a coin, the best procedure for handling ties appears to be to ignore them.[20] That is, the test is applied, using only those pairs in which there is not a tie. Using this procedure results in a significance level that is at most as large as that used in determining the critical region based on the assumption of continuous populations. For with the critical region C_m determined for each m so that under the null hypothesis

$$P(C_m \mid m \text{ nonzero differences}) \leq \alpha,$$

it follows that the probability of rejecting H_0 is

$$\sum_{m=0}^{n} P_{H_0}(C_m \mid m \text{ nonzero differences}) P_{H_0}(n \text{ nonzero differences}) \leq \sum_{m=0}^{n} \alpha P_{H_0}(m \text{ nonzero differences}) = \alpha.$$

The sign test can be applied to discrete populations to test the hypothesis $P(X > Y) = P(X < Y)$, by applying it to the signs of $X - Y$ in those pairs where there is a difference. However, the following example points out that in this discrete case there must be some thought prior to the discarding of tied observations.

EXAMPLE 11–14. On the basis of road tests, 50 drivers are asked to compare two makes of cars with respect to ease of handling. Suppose the results are as follows:

Prefer	Frequency
M	7
C	3
Neither	40

Should one really discard the ties and make a comparison on the basis of the 10 drivers who reported a difference? Is there no information in the 40 ties?

The population might be defined as follows:

Outcome	Probability
$M > C$	p_+
$M < C$	p_-
$M = C$	p_0

[20] See E. Lehmann [16], Section 4.7, and J. Hemelryk, "A Theorem on the Sign Test When Ties Are Present," *Proc. Kon. Ned. Akad. Sect. Sci.* **A55**, 322 (1952).

Ignoring the ties does provide a reasonable basis for testing $p_+ = p_-$, for there is no information about this hypothesis in the tied data. On the other hand, it is perhaps not clear as to what hypothesis about the model corresponds to what is meant when it is said that there is "no appreciable difference in handling between makes." For instance, does $p_+ > p_-$ mean that M is really better when $p_0 = .80$? The buyer of make M might want to conclude that it is better and to test $p_+ \leq \frac{1}{2}$ against $p_+ > \frac{1}{2}$, in which case all the evidence is relevant.

It should be noted that in Example 11–14, only comparisons of the form "Car M handles better than car C" were made. Even though the characteristic thought of as "handling" might be conceived of as quantitative, it was not necessary to adopt a scale on which to measure an "X" and a "Y," since only the sign of the difference is used. Without such a scale, of course, the t-test could not be used; and when scaling is awkward and arbitrary, the sign test is especially useful.

NO

11.4.2 The Signed-Rank Test

Suppose Y is continuous and symmetric about 0. It follows then (cf. Problem 11–26) that $|Y|$ is independent of the sign of Y.

Let $\mathbf{Y} = (Y_1, \ldots, Y_n)$ be a sample from such a population, let $|\mathbf{Y}|$ denote the vector of magnitudes: $(|Y_1|, \ldots, |Y_n|)$, and let

$$S_i = \begin{cases} +, & \text{if } Y_i > 0, \\ -, & \text{if } Y_i < 0. \end{cases}$$

The S's are independent, and $P(S_i = +) = \frac{1}{2}$, so that the probability of any sequence $\mathbf{a}$ of plus and minus signs is

$$P(\mathbf{S} = \mathbf{a}) = \frac{1}{2^n}.$$

Now let $t_{|\mathbf{Y}|}(\mathbf{a})$ denote the permutation transformation that orders the a_i's according to the magnitudes $|Y_i|$, and define $\mathbf{S}^* = t_{|\mathbf{Y}|}(\mathbf{S})$. Thus S_1^* is the sign of the Y_i with smallest magnitude, and so on. [For instance, if $\mathbf{Y} = (2, -1, 3)$, then $\mathbf{S} = (+, -, +)$, and $\mathbf{S}^* = (-, +, +)$.] Given any sequence $\mathbf{a}$ of plus and minus signs,

$$\begin{aligned} P(\mathbf{S}^* = \mathbf{a}) &= P(\mathbf{S} = t_{|\mathbf{Y}|}^{-1}(\mathbf{a})) \\ &= \sum_{\mathbf{b}} P(\mathbf{S} = \mathbf{b},\, t_{|\mathbf{Y}|}^{-1}(\mathbf{a}) = \mathbf{b}), \end{aligned}$$

where $\mathbf{b}$ is a possible "value" of a sequence of signs, ranging over the $n!$ such distinct sequences. But inasmuch as $\mathbf{S}$ and $|\mathbf{Y}|$ are independent, the last probability factors:

$$P(\mathbf{S}^* = \mathbf{a}) = \sum P(\mathbf{S} = \mathbf{b}) P(t_{|\mathbf{Y}|}^{-1}(\mathbf{a}) = \mathbf{b}) = \frac{1}{2^n},$$

because the first factor is $1/2^n$ in each term and the second factors sum to 1. So $\mathbf{S}^*$ has the same distribution as $\mathbf{S}$, namely, that of independent observations on $\{+, -\}$ with $P(+) = \frac{1}{2}$. The result is summarized as follows:

Theorem. *If the observations in a random sample from a population symmetric about* 0 *are ordered according to magnitude, the* 2^n *sequences of signs of the ordered observations are equally likely.*

Suppose that it is desired to test $m = m_0$, where m is the median of a population X that is symmetric about that median. Given a random sample $\mathbf{X}$, the deviations $Y_i \equiv X_i - m_0$ are symmetric about 0 if $m = m_0$, and the first step in constructing the signed-rank statistic is to put the deviations Y_i (like the previous Y's) in order according to the magnitude $|Y_i|$. *Wilcoxon's signed-rank statistic* is then defined as the sum of the ranks of the positive elements in this sequence of deviations. Equivalently, it is the sum of the ranks of the $+$'s in the sequence of signs $\mathbf{S}^*$ that results from ordering the Y's according to $|Y|$:

$$W_+ \equiv \sum iI_X(i),$$

where $I_X(i)$ is 1 or 0 according to whether S_i^* is $+$ or $-$. It can also be written as a sum of independent variables:

$$W_+ = \sum V_i, \qquad \text{where } V_i = \begin{cases} i, & \text{if } S_i^* = +, \\ 0, & \text{if } S_i^* = -, \end{cases}$$

the sequences $\mathbf{S}^*$ having a uniform distribution among the 2^n possible sequences of $+$'s and $-$'s.

Clearly, W_+ is symmetrically distributed on the integers $0, 1, \ldots, n(n+1)/2$ so that

$$E(W_+) = \tfrac{1}{4}n(n+1).$$

Its variance is readily computed using the independence of the V's and the fact that $\operatorname{var} V_i = i^2/12$:

$$\operatorname{var} W_+ = \sum \operatorname{var} V_i = \tfrac{1}{24}n(n+1)(2n+1).$$

Exact cumulative probabilities for W_+ for small samples are given in Table IX, Appendix. Large-sample probabilities can be calculated approximately from the fact that W_+ is asymptotically normally distributed. (Although W_+ is a sum of independent variables, these are not identically distributed, and a form of "Central Limit Theorem" more general than the one given in Chapter 3 is used in establishing the asserted result.)

EXAMPLE 11–15. For $n = 4$, 8 of the 16 patterns of plus and minus signs are listed in the following table, together with the corresponding value of W_+. The other 8 would be obtained by interchanging the roles of plus and minus and subtracting W_+ from 10.

S^*	W_+
+ + + +	10
− + + +	9
+ − + +	8
+ + − +	7
+ + + −	6
− − + +	7
− + − +	6
+ − − +	5

The resulting distribution of W_+ is as follows:

$$P(W_+ = k) = \begin{cases} \frac{1}{16}, & \text{for } k = 0, 1, 2, 8, 9, 10, \\ \frac{2}{16}, & \text{for } k = 3, 4, 5, 6, 7. \end{cases}$$

The value of W_+ will be small if there are either (i) few positive differences $X_i - m_0$ or (ii) if the positive differences are not very large in magnitude; and in either case one would tend to reject H_0: $m = m_0$ in favor of the alternative $m < m_0$. Large values of W_+ would suggest $m > m_0$, and a two-sided critical region for W_+ would be appropriate for the alternative $m \neq m_0$.

The signed-rank test can clearly be used as a paired comparison test of the hypothesis of no population difference by applying it to the differences $D_i = Y_i - X_i$, given a random sample of pairs (X_i, Y_i), provided that D_i is symmetric about 0. In Problem 11–27, it is shown that D *is* symmetric about 0 if X and Y are independent and identically distributed, but this is not a necessary condition.

EXAMPLE 11–16. The data given in Example 7–7 consisted of before and after measurements of blood pressure, with difference as follows:

Subject	Difference (after − before)
1	−6.7
2	−20.7
3	−17.2
4	3.2
5	−10.6
6	−9.2

The sequence of differences ordered by magnitude is

$$(3.2, -6.7, -9.2, -10.6, -17.2, -20.7),$$

so that

$$S^* = (+, -, -, -, -, -)$$

and $W_+ = 1$. Here one can easily compute:

$$P(W_+ = 0 \text{ or } 1) = \frac{1}{2^6} + \frac{1}{2^6} \doteq 0.03$$

(which can be verified in Table IX, Appendix), so at $\alpha = .05$ one would reject the hypothesis of no difference and infer that the special biofeedback training is effective.

The sign test can also be applied in this situation, and because

$$P(\text{at most } 1 +) = \frac{1}{2^6} + \frac{6}{2^6} = \frac{7}{64} \doteq 0.11,$$

the null hypothesis would not be rejected, even at $\alpha = .10$. It is apparent that the signed-rank test took into account that the one positive difference is a very small one; it is more sensitive or powerful than the sign test. (The t-test, applied in Example 7–7, uses not only the fact that 3.2 is the smallest difference but also the amounts by which it is smaller than the others. One would expect it to be even more powerful in the normal case to which it applies.)

11.4.3 Asymptotic Relative Efficiency

The remarks at the end of Example 11–16 were based on intuition and reflected the fact that tests are ordinarily compared with respect to power. In the case of nonparametric tests, and for the wide classes of alternatives for which they are devised, exact power calculations for finite samples are laborious and overly specific; it is nearly impossible to make general statements.

A large-sample approach to comparing tests was introduced by E. J. G. Pitman (unpublished Columbia Lecture Notes, 1948). Most tests in common use are consistent—the power approaches 1 as n becomes infinite, so to compare powers, Pitman considered a sequence of alternatives approaching the null hypothesis. Then, rather than compare powers, he compares sample sizes necessary to yield the same power. Thus, the *asymptotic relative efficiency* (ARE) of one test with respect to another test of the same size α is defined as the limiting ratio of sample sizes required in order that they have the same power on a sequence of alternatives tending to the null hypothesis. This limiting ratio often turns out to be independent of α but may well depend on the particular class of alternatives considered.

The computation of ARE for the sign test relative to the t-test in the case of normal alternatives (where the t-test should shine) is fairly straightforward. As mentioned earlier, the power of the sign test is a binomial probability; that of the t-test is a noncentral t, but both become normal in the limit. The details are as follows.

Let S_n denote the number of positive observations in a random sample of size n. This is binomially distributed with $ES_n = np$ and var $S_n = np(1 - p)$, where $p = P(+)$. A large-sample size test of approximate size α uses the critical region $S_n > K$, where

$$K = \frac{n}{2} + z_{1-\alpha}\sqrt{\frac{n}{4}}.$$

Now consider an alternative for which

$$p = P(+) = \frac{1}{2} + \frac{a}{2\sqrt{n}}.$$

Then

$$E(S_n) = \frac{n}{2}\left(1 + \frac{a}{\sqrt{n}}\right), \qquad \text{var } S_n = \frac{n}{4}\left(1 - \frac{a^2}{n}\right),$$

and the power of the test $S_n > K$ is

$$P(S_n > K) = 1 - \Phi\left(\frac{K - E(S_n)}{\sqrt{\text{var } S_n}}\right) = 1 - \Phi\left(\frac{z_{1-\alpha}\sqrt{\frac{1}{4}n} - \frac{1}{2}na/\sqrt{n}}{\sqrt{\frac{1}{4}n(1 - a^2/n)}}\right),$$

which tends to $1 - \Phi(z_{1-\alpha} - a)$ as n becomes infinite. And the same limit is achieved if, in place of a, we had used a sequence $\{a_n\}$ such that $a_n \to a$.

Consider, in particular, a sequence of normal alternatives approaching the null hypothesis that the median of X is 0:

$$H_n\colon \quad X \text{ is normal with } EX = b\sigma/\sqrt{n} \text{ and } \sigma_X = \sigma.$$

For such alternatives, the "p" can be explicitly calculated:

$$p = P(+) = \frac{1}{2} + \int_0^{b/\sqrt{n}} \frac{e^{-z^2/2}}{\sqrt{2\pi}}\, dz = \frac{1}{2} + \frac{a_n}{2\sqrt{n}},$$

where $a_n \to 2b/\sqrt{2\pi}$. Thus, the power approaches

$$1 - \Phi\left(z_{1-\alpha} - \frac{2b}{\sqrt{2\pi}}\right)$$

as n becomes infinite.

A t-test with the same approximate α is given by the critical region $T_n > z_{1-\alpha}$, where

$$T_n = \frac{\bar{X} - 0}{S}\sqrt{n - 1},$$

since T_n is asymptotically normal (0, 1). Moreover, writing T_n in the form

$$T_n = \frac{\bar{X}/(\sigma/\sqrt{n})}{\sqrt{S^2/\sigma^2}}\sqrt{\frac{n-1}{n}},$$

in which the denominator has limit in probability 1, it follows that T_n is asymptotically normal with mean b' and variance 1 when X is normal with mean $b'\sigma\sqrt{n'}$ and variance σ^2. The power of the test $T_n > z_{1-\alpha}$ is then

$$P(T_n > z_{1-\alpha}) \doteq 1 - \Phi(z_{1-\alpha} - b').$$

This will be the same as the asymptotic power of the sign test if $b' = 2b/\sqrt{2\pi}$, and the normal alternatives will agree if $b/\sqrt{n} = b'/\sqrt{n'}$, or if $n'/n = 2/\pi$. This ratio of sample sizes is the desired ARE.

It can be shown that the ARE of the signed-rank test relative to the t-test is $3/\pi \doteq .955$ against normal alternatives. The ARE of the sign test for the two-sample problem with shift alternatives: $F_X(\lambda) = F_Y(\lambda - \theta)$, relative to the t-test, was shown by Pitman to be $4\sigma^2 f^2(0)$, where $f(x) = F'(x)$, when $X - Y$ is symmetrical about 0. (This reduces to $2/\pi$ in the normal case.) In this two-sample case, it is understood that both sample sizes become infinite in a fixed ratio.

11.4.4 Confidence Intervals

A confidence interval for the population median m can be obtained as follows. Let $X_{(1)}, \ldots, X_{(n)}$ denote the observations in a random sample arranged in numerical order. Then if $r < s$,

$$P[X_{(r)} < m < X_{(s)}] = P(\text{exactly } r, r + 1, \ldots, \text{ or } s - 1 \text{ observations} < m)$$

$$= \sum_{k=r}^{s-1} \binom{n}{k}\left(\frac{1}{2}\right)^n.$$

The random interval from the rth smallest to the sth smallest observation is, therefore, a confidence interval for m with confidence coefficient given by the above sum of binomial probabilities.

EXAMPLE 11–17. In the case of 20 observations, the interval from the fourth smallest to the sixteenth smallest observation will include m with probability .993, according to the calculation of Example 11–13. For the 20 observations in that example, the fourth and sixteenth observations (in order of size) are 31.5 and 34.6. These are then 99.3 per cent confidence limits.

Confidence intervals for the median can also be constructed using the signed-rank statistic. (Recall the discussion in Section 6.1.4 of the duality between tests and confidence intervals.) The reader is referred to the book by Gibbons [12] for a treatment of two procedures, one a trial-and-error process leading non-uniquely to a confidence interval, and the other a method of G. Noether that is easier to apply. J. Tukey has given a graphical method for the latter, as follows: Define an integer h such that

$$P(W_+ \le h) = \frac{1 - \eta}{2},$$

where η is the desired confidence coefficient or something close to it. Construct an isosceles triangle whose base is the line segment from $X_{(1)}$ to $X_{(n)}$ and draw lines through the other observations parallel to the equal sides of the triangle, as shown for nine observations in Figure 11–4. Then draw vertical lines through the $(h + 1)$st intersection point from each end—that is, so that there are h intersections to the left of the left line and h to the right of the right line. These lines meet the horizontal axis at the desired confidence limits.

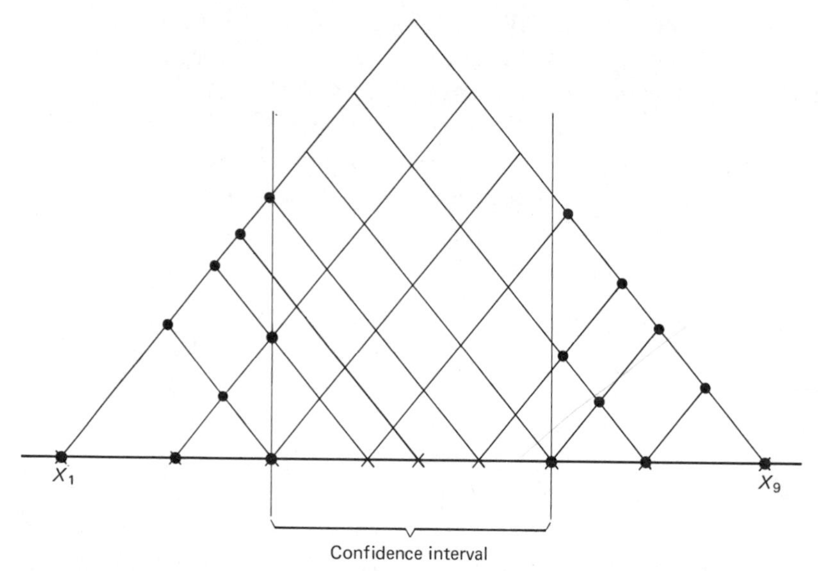

Figure 11–4 Confidence interval of 90.2 per cent for the median, $n = 9$.

Problems

11–20. In a table of "random normal numbers" giving observations from a supposedly standard normal population, 88 out of 150 entries are found to be positive. Would you accept the hypothesis that the mean is 0?

11–21. Determine a one-sided critical region for a sign test with an α of approximately 10 per cent if $n = 8$. Sketch the power of the test as a function of $p = P(+)$. Could you achieve an α of *exactly* 10 per cent?

11–22. Thirty pairs of puppies are selected, each pair from a single litter, and one puppy from each pair is fed diet A, the rest getting diet B. If, as judged after a given period of time, diet A puppies are found to have fared better (according to some criterion) in 16 of the 30 pairs and to have fared worse in 10 pairs, does the evidence point to the superiority of diet A, as was expected might be the case?

11–23. Determine the probability that the population median is included between the fortieth and sixtieth observations (when in numerical order) of a random sample of 100 observations from a continuous population.

11–24. To decide whether to switch from type A tomato plant to type B, I planted five pairs, one of each type in each pair, matching the pairs according to appearance and according to location of planting. The yields were as follows:

Pair number	1	2	3	4	5
Type A	2.6	2.2	2.4	2.8	1.9
Type B	3.6	2.8	2.3	3.0	2.3

Test the hypothesis that there is no reason for one to switch types, at $\alpha = .10$, using the signed-rank test. Also carry out a t-test (what assumptions are needed?).

11–25. Twelve mineral specimens from the Black Forest were dated by a potassium–argon method with the following results (age in millions of years);

249, 254, 243, 268, 253, 269, 287, 241, 273, 306, 303, 280.

Use the Tukey graphical method to determine a confidence interval for the median age of minerals in the area, with a confidence coefficient of approximately 90 per cent.

11–26. Show that if Y is continuous and symmetric about 0, the magnitude and sign of Y are independent.

11–27. Show that if X and Y are independent and identically distributed with a continuous distribution, the difference $X - Y$ is symmetrically distributed about 0.

11–28. Show that the ARE of the sign test of $\theta = 0$ relative to the t-test for the case of a uniform population on $[\theta - \frac{1}{2}, \theta + \frac{1}{2}]$ is 1/3. Use a sequence of alternatives with $\theta_n = b/\sqrt{n}$ to carry out computations parallel to those given in the normal case, and also check the result with the formula $4\sigma^2 f^2(0)$.

11.5 Two-Sample Location Problems

Populations X and Y are to be compared with regard to location by testing the hypothesis

$$H_0: \quad F_X(\lambda) \equiv F_Y(\lambda)$$

against shift alternatives:

$$H_A: \quad F_X(\lambda) \equiv F_Y(\lambda - \theta), \qquad \text{for } \theta \neq 0.$$

Again, tests will be proposed on an intuitive basis with this alternative in mind. The distribution of the test statistic under H_0 will be given, to permit the setting of critical values corresponding to given significance levels. The only mention of performance will be in terms of asymptotic relative efficiency (ARE).

The tests to be presented are essentially permutation tests, conditional tests given the order statistic, based on the fact that each permutation of a random sample has the same chance of having produced a given order statistic. Independent random samples of size m from X and of size n from Y can be considered as a single random sample of size $m + n$ if H_0 is true. Thus, given the combined order statistic (obtained by ordering $X_1, \ldots, X_m, Y_1, \ldots, Y_n$), each permutation thereof has probability $1/(m + n)!$ under the null hypothesis. If, for each possible order statistic, one puts k permutations of it into a critical region, the resulting test has the significance level $k/(m + n)!$, as follows:

$$P_{H0}(\text{reject } H_0) = E[P_{H_0}(\text{reject } H_0 \mid \text{order statistic})]$$
$$= E\left[\frac{k}{(m + n)!}\right] = \frac{k}{(m + n)!}$$

The selection of permutations to use in the rejection region for a given order statistic can be made according to any criterion thought to be appropriate for the alternatives against which protection is desired.

EXAMPLE 11–18. Although the statistic $\bar{X} - \bar{Y}$ is certainly not distribution-free, it can be used to determine which permutations are appropriate for the rejection of the hypothesis of identical distributions of X and Y against shift alternatives Actually, given the order statistic, one is also given the sum $\sum X_i + \sum Y_j$, so one could use $\sum X_i$ or $\sum Y_j$ equally well. Moreover, since the ordering among the X's and the ordering among the Y's does not play a role in comparing values of the sum used, only the $\binom{m + n}{n}$ distinct assignments of m of the $m + n$ observations to X (the remaining n being Y's) need be considered. And these are equally likely.

To illustrate, suppose that three observations from X and two from Y yield the combined order statistic (4, 6, 7, 8, 9). Of the 120 arrangements of these numbers that might have been the combined sample $(X_1, X_2, X_3, Y_1, Y_2)$, the 12 in which the X's are (4, 6, 7) in some order and the Y's are (8, 9) all give the same value of $\sum Y_j$, namely, 17. The 10 selections of three of the five numbers in the order statistic for X's are shown in the following table, together with a corresponding value of $\sum Y_j$ in each case. Each has conditional probability $\frac{1}{10}$ (given the order statistic), and in devising a test at $\alpha = .2$ against the hypothesis that the Y-population is shifted to the right, the two permutations with $\sum Y_j = 17$ and $\sum Y_j = 16$ would make up the critical region.

X's	Y's	$\sum Y_j$
4, 6, 7	8, 9	17
4, 6, 8	7, 9	16
4, 7, 8	6, 9	15
4, 6, 9	7, 8	15
4, 7, 9	6, 8	14
6, 7, 8	4, 9	13
4, 8, 9	6, 7	13
6, 7, 9	4, 8	12
6, 8, 9	4, 7	11
7, 8, 9	4, 6	10

In the following sections, tests for comparing locations are given that are permutation tests based on the ranks of the X's and the ranks of the Y's in the combined order statistic. Knowing these is equivalent to knowing the sequence of X's and Y's indicating the population of origin of the successive elements of the order statistic. For instance, if the Y-ranks are (3, 5, 6, 7) in a sequence of four Y's and three X's, then that sequence is (X, X, Y, X, Y, Y, Y). Each sequence of X's and Y's and so each sequence of Y-ranks (or of X-ranks) comes from $m!\,n!$ of the $(m + n)!$ equally likely permutations of the order statistic and so has probability $1\Big/\binom{m+n}{m}$. That is, the $\binom{m+n}{m}$ distinct sequences of m X's and n Y's are equally likely.

11.5.1 The Median Test

For simplicity, it is assumed throughout this section that $m + n = 2k$, so that there are k observations on either side of the median in the order statistic. A statistic computed from the sequence of X's and Y's determined by a given order statistic that is indicative of a shift (or of the absence thereof) is the number of X's to the left of the median of the combined sample. This statistic determines and is determined by the following two-way table with fixed marginal totals:

	Left of Median	Right of Median	
From X	L	$m - L$	m
From Y	$k - L$	$k - m + L$	n
	k	k	

Since the L X's can appear in $\binom{k}{L}$ ways to the left of the median and the $m - L$ X's in $\binom{k}{m-L}$ ways to the right, the conditional distribution (given the order statistic) under the null hypothesis is given by

$$P(L = \ell) = \frac{\binom{k}{\ell}\binom{k}{m-\ell}}{\binom{2k}{m}},$$

which is independent of the condition and hence is also the unconditional probability. Notice that it is precisely the same as the conditional probability of a two-way table, given marginal totals, in testing for independence or homogeneity.

Large or small values of L would suggest a shift of Y from X, one way or the other, respectively; and the critical point would be determined from a given α by means of the hypergeometric distribution—in the small-sample case. For large samples, one may use the fact that

$$Z \equiv \frac{L - km/(m+n)}{\sqrt{\frac{1}{4}mn/(m+n)}}$$

is approximately standard normal as m and n become infinite in a fixed (or asymptotically fixed) proportion. A two-sided test of the form $\chi^2 >$ const, where χ^2 is the usual contingency table statistic, would be equivalent to $Z^2 >$ const.

In the small-sample case, it would be necessary to randomize to achieve a given α. It is known that such a test is most powerful for a one-sided alternative. The ARE of the median test relative to the t-test for normal shift alternatives is the same as that of the sign test, namely, $2/\pi$ or about 0.63.

11.5.2 The Wilcoxon–Mann-Whitney Test

A test that uses the *ranks* of the X's (say) in the combined order statistic should be more sensitive to shifts than the median test, which uses only the information as to whether an X is larger or smaller than the median. F. Wilcoxon introduced the statistic R_X, defined to be the sum of the ranks of the X's in the order statistic formed from two independent random samples X and Y. When there is a shift of X to the right, the X's tend to be in the higher ranking positions in the order statistic and will lead to larger value of R_X than might be expected if the X's and Y's are interspersed—as they tend to be when taken from the same population.

It might be remarked that if R_Y denotes the sum of the Y-ranks, then

$$R_X + R_Y = \sum_1^N i = \tfrac{1}{2}N(N+1),$$

where $N = n + m$. Knowing R_X is equivalent to knowing R_Y.

For given sample sizes, the distribution of R_X under the null hypothesis can be obtained by brute-force calculation, using the fact that the $\binom{N}{m}$ rank sequences are equally likely, as in the following example. Tables of probabilities can be constructed using a recurrence relation, and Table VIII in the Appendix gives tail probabilities for certain small values of m and n.

EXAMPLE 11–19. If $m = 3$ and $n = 5$ there are $\binom{8}{3} = 56$ sequences of X's and Y's, each defining a rank sequence. Thus, in the sequence (X, Y, X, X, Y, Y, Y, Y), the sum of the X-ranks is $1 + 3 + 4 = 8$. and the sum of the Y-ranks is $8 \cdot 9/2 - 8 = 28$. The probabilities for R_X are obtained by listing the various X-rank sequences and computing R_X for each:

X-Ranks	R_X
1, 2, 3	6
1, 2, 4	7
1, 3, 4	8
1, 2, 5	8
2, 3, 4	9
1, 3, 5	9
1, 2, 6	9
⋮	⋮

The distribution of R_X is then as follows, in part:

R_X	6	7	8	9	10	...
Probability	$\frac{1}{56}$	$\frac{1}{56}$	$\frac{2}{56}$	$\frac{3}{56}$	$\frac{4}{56}$	...

A test for no difference in X and Y against the alternative of a shift of X to the left would use small values of R_X. Taking $R_X = 6, 7, 8$ as critical values would define a test with $\alpha = \frac{4}{56}$. The critical region (6, 7, 20, 21) would be appropriate for a two-sided shift alternative and would have an α of $\frac{4}{56}$. (Check these in Table VIII, Appendix.)

The Mann–Whitney statistic is based on "inversions." Consider the sequence of observations as they appear in the order statistic, noting only which population an observation comes from. For each Y observation count the number of X observations preceding it, and let U_Y denote the total of these numbers. That is, for each pair of observations X_i and Y_j define

$$Z_{ij} = \begin{cases} 1, & \text{if } X_i < Y_j, \\ 0, & \text{if } X_i > Y_j. \end{cases}$$

The sum of these Z's is the statistic U_Y:

$$U_Y = \sum_{i=1}^{m} \sum_{j=1}^{n} Z_{ij}.$$

A similar sum with 1 and 0 reversed can be used in defining U_X as the total number of X inversions; that is, the total of the numbers of Y observations preceding the X observations. Since there are mn terms in the summation of Z_{ij}, and since interchanging 0's and 1's in U_Y produces U_X, it is evident that $U_X + U_Y = mn$.

Proposed independently, the Wilcoxon and Mann–Whitney statistics turn out to be simply related:

$$R_Y = U_Y + \tfrac{1}{2}n(n + 1).$$

For, if the rank of the smallest Y is r_1, there are $r_1 - 1$ inversions for that Y; if the rank of the next smallest Y is r_2, there are $r_2 - 2$ corresponding inversions; and so on. Adding these numbers of inversions yields the relationship given.

Since the statistic R_Y is a constant plus the statistic U_Y, procedures based on one can be interpreted as procedures based on the other, with the same kinds of critical region. Rejection limits are tabulated in Table VIII (Appendix) only for R_Y or R_Y.

EXAMPLE 11–20. Suppose that the observations from X are 13, 17, 11, 14, and that the observations from Y are 10, 16, 15. The combined order statistic is then *10*, 11, 13, 14, *15*, *16*, 17, the italicized ones coming from Y. In terms of which populations the observations come from, the order statistic can be represented this way: Y, X, X, X, Y, Y, X. In this sequence the first X, the second X, and the third X are each preceded by one Y; the fourth X is preceded by three Y's. Therefore, $U_X = 6$. Similarly, the first Y has no X's ahead of it, and the other two Y's are each preceded by three X's; so $U_Y = 6$, and $U_X + U_Y = 12 = 4 \cdot 3$. Observe that $R_Y = 1 + 5 + 6 = U_Y + 3 \times \frac{4}{2}$.

Under the null hypothesis that the observations are independent random variables with $P(X > Y) = P(X < Y) = \frac{1}{2}$, the mean and variance of U_Y can be computed. Since $E(Z_{ij}) = \frac{1}{2}$ for all i and j,

$$E(U_Y) = E\left\{\sum_i \sum_j Z_{ij}\right\} = \sum_i \sum_j E(Z_{ij}) = \frac{mn}{2}.$$

The variance computation is not quite so simple, since the terms in the sum representing U_Y are not independent random variables:

$$\text{var } U_Y = \sum_i \sum_j \sum_h \sum_k \text{cov}\,(Z_{ij}, Z_{hk}).$$

Now,

$$Z_{ij}Z_{hk} = \begin{cases} 1, & \text{if } X_i < Y_j \text{ and } X_h < Y_k, \\ 0, & \text{otherwise} \end{cases}$$

so that

$$\begin{aligned} E(Z_{ij}Z_{hk}) &= P(X_i < Y_j \text{ and } X_h < Y_k) \\ &= \begin{cases} \frac{1}{2}, & i = h \text{ and } j = k, \\ \frac{1}{4}, & i \neq h \text{ and } j \neq k, \\ \frac{1}{3}, & i = h \text{ and } j \neq k, \text{ or } j = k \text{ and } i \neq h. \end{cases} \end{aligned}$$

Hence,

$$\text{cov}\,(Z_{ij}, Z_{hk}) = \begin{cases} 0, & i \neq h \text{ and } j \neq k, \\ \frac{1}{4}, & i = h \text{ and } j = k, \\ \frac{1}{12}, & i = h \text{ and } j \neq k, \text{ or } j = k \text{ and } i \neq h. \end{cases}$$

To complete the computation of var U_Y, it is only necessary to count the number of terms in each case. There are mn terms in which $i = h$ and $j = k$, and m^2n terms in which $j = k$. Of the latter, however, mn have also $i = h$, leaving $m^2n - mn = mn(m - 1)$ in which $j = k$ and $i \neq h$, Similarly, there are $mn(n - 1)$ terms in which $i = h$ and $j \neq k$. Finally, then,

$$\text{var } U_Y = \tfrac{1}{4}mn + \tfrac{1}{12}[mn(m - 1) + mn(n - 1)] = \tfrac{1}{12}mn(m + n + 1).$$

With formulas for mean and variance at hand, a large sample test can be constructed using the fact that the test statistic (either U or R) is asymptotically normal,[21] even when the null hypothesis is false.[22] Thus, the statistic

$$Z = \frac{U_X + \frac{1}{2} - mn/2}{[mn(m + n + 1)/12]^{1/2}},$$

which includes a continuity correction term, is approximately standard normal.

The asymptotic efficiency of the Wilcoxon–Mann–Whitney test relative to the t test for shift alternatives was shown by Pitman to be

$$12\sigma^2\left(\int f^2(x)\, dx\right)^2,$$

where f is the density function and σ^2 the variance. (The integral of f^2 and the variance would be the same for each alternative.) In the normal case, this reduces to $3/\pi$, or about .96, and Hodges and Lehmann have shown that it is always at least .864.

11.5.3 The Fisher–Yates Test

The Fisher–Yates c_1 test calls for putting arrangements of m X's and n Y's ($N = m + n$) into the critical region according to the size of

$$c_1 = \frac{1}{n}\sum_{i=1}^{n} E(Z_{s_i}),$$

where $Z_1, \ldots, Z_N$ are the ordered observations in a random sample from a standard normal population, and $s_1, \ldots, s_n$ is the sequence of Y ranks in the combined order statistic of X's and Y's. For one-sided shift alternatives the critical region is $c_1 >$ constant (or $c_1 <$ constant, depending on the direction of the shift), and it can be shown that this test is most powerful in the case of normal alternatives for sufficiently small shifts.[23] It is also known that the

[21] See E. Fix and J. Hodges, "Significance Probabilities of the Wilcoxon Test," *Ann. Math. Stat.* **26**, 301–12 (1955).

[22] E. Lehmann, "Consistency and Unbiasedness of Nonparametric Tests," *Ann. Math. Stat.* **22**, 301–12 (1955).

[23] See E. Lehmann [16], pp. 236 ff.

asymptotic efficiency of the c_1 test relative to the t test is always at least 1 for shift alternatives and is equal to 1 only in the normal case.[24] Further, c_1 is asymptotically normal as m and n become infinite with a nonzero, finite limiting ratio with mean zero and variance

$$\frac{m}{nN(N-1)} \sum_{i=1}^{N} a_{N,i}^{\;2},$$

where[25] $a_{N,i} = E(X_i)$.

EXAMPLE 11–21. Consider the c_1 test based on samples of sizes $m = 4$ and $n = 6$. The values of $a_{10,i}$ found in Table X (Appendix), are as follows: -1.539, -1.001, $-.656$, $-.376$, $-.123$, .123, .376, .656, 1.001, and 1.539. The sequences having largest c_1 values are then found to be those listed:

Sequence	Y-Ranks	$6c_1 = \sum_{i=1}^{6} a_{10,s_i}$
X X X X Y Y Y Y Y Y	5 6 7 8 9 10	3.57
X X X Y X Y Y Y Y Y	4 6 7 8 9 10	3.32
X X X Y Y X Y Y Y Y	4 5 7 8 9 10	3.07
X X Y X X Y Y Y Y Y	3 6 7 8 9 10	3.04
X X X Y Y Y X Y Y Y	4 5 6 8 9 10	2.82
X X Y X Y X Y Y Y Y	3 5 7 8 9 10	2.79
X Y X X X Y Y Y Y Y	2 6 7 8 9 10	2.69
X X Y X Y Y X Y Y Y	3 5 6 8 9 10	2.54
X X Y Y X X Y Y Y Y	3 4 7 8 9 10	2.54
X X X Y Y Y Y X Y Y	4 5 6 7 9 10	2.54
X Y X X Y X Y Y Y Y	2 5 7 8 9 10	2.45
X X Y Y X Y X Y Y Y	3 4 6 8 9 10	2.30
X X Y X Y Y Y X Y Y	3 5 6 7 9 10	2.26

For $\alpha = .05$, there would be $.05 \times \binom{10}{4} = 10.5$ points in the critical region; rather, $\theta = 0$ would be accepted for $6c_1 \leq 2.30$, rejected for $6c_1 \geq 2.54$, and rejected or accepted according to the toss of a suitably chosen "coin" if $6c_1 = 2.45$, the value that puts α just over the desired .05.

If the result of sampling is as follows: 22, 25, 27, 30, from X, and 24, 29, 32, 33, 35, 36, from Y, the corresponding sequence of X's and Y's in the combined order statistic is $X\,Y\,X\,X\,Y\,X\,Y\,Y\,Y\,Y$, with $6c_1 = 2.45$. This is the boundary point of the 5 per cent critical region.

[24] See H. Chernoff and I. R. Savage, "Asymptotic Normality and Efficiency of Certain Nonparametric Test Statistics," *Ann. Math. Stat.* **29**, 972 (1958).

[25] Tables of $a_{N,i}$ are available in *Biometrika Tables for Statisticians*, Vol. 1. Cambridge, England: Cambridge Univ. Press, 1954; and in a paper by D. Teichrow, *Ann. Math. Stat.* **27**, 410 (1956). A short version appears here in Table X, Appendix.

It is interesting to compute the critical value of c_1 that comes from the asymptotic distribution; the approximate variance is .587, and for $\alpha = .05$ the critical value would be $1.645\sqrt{.0587} = .4$, or $6c_1 = 2.4$, which is approximately what was obtained. Similarly, for $\alpha = .025$, $1.96\sqrt{.0587} = .475$, or $6c_1 = 2.85$. This would put the first four points into the critical region, with $\alpha = 4\Big/\binom{10}{4} = .019$. The agreement is fairly good, considering the small sample sizes.

Problems

11–29. Data are obtained as follows:

From X: 12.9, 12.4, 14.7, 13.8, 15.4, 11.8.
From Y: 14.3, 13.4, 14.0.

Apply (a) the Wilcoxon test and (b) the c_1-test of no difference in populations, at $\alpha = .10$, against the alternative that there has been a shift in location (two-sided).

11–30. Rank tests can be applied if experimental units can be ordered, without the necessity of establishing a numerical scale for measuring the characteristic of interest. Six samples of flint from area X were compared with five samples from area Y and ordered (by a series of pairwise comparisons) according to hardness, as follows:

$$X, X, Y, X, X, Y, X, X, Y, Y, Y.$$

(The listing gives just the area of origin.) Apply the Wilcoxon test for the hypothesis that the hardness is the same in both areas against a two-sided alternative at $\alpha = .10$.

11–31. Given data as follows:

From X: 18, 16, 21, 18, 14, 19.
From Y: 14, 12, 17, 15.

Test the hypothesis of no difference in a one-sided test at an α of about 5 per cent, (a) using the median test, (b) using the Wilcoxon test, and (c) using the c_1-test.

11–32. Refer to Example 11–18 and demonstrate the stated equivalence of rankings, according to $\bar{X} - \bar{Y}$ and according to $\sum Y_i$.

11–33. Check the accuracy of the large-sample normal distribution for the Wilcoxon statistic in the case $m = n = 10$, say for the table entry for $c = 79$.

11–34. Compute the Pitman asymptotic efficiency of the Wilcoxon test relative to the t-test for the case of shift alternatives and a uniform population on $[\theta - \frac{1}{2}, \theta + \frac{1}{2}]$.

11–35. Verify the entries in Table VIII of the Appendix for the cases $m = n = 3$, and $m = 3$, $n = 5$.

12

LINEAR MODELS AND ANALYSIS OF VARIANCE

The problems to be studied in this chapter deal with the means of several normal populations and so generalize the comparison problem of Section 7.3.2. Here, too, the null hypothesis will often be that the means are all equal, but, of course, the more general problem will raise new questions. In some cases, it is not expected that the means will be equal, and the interest lies in determining the functional dependence of the means on a numerical index of the populations. In others, the means may depend on several nonnumerical factors, and it is required to test the effect of one factor in the presence of the others. In the models to be considered, the dependence on the parameters that define the mean will be linear in those parameters. Such *linear models* can be treated in a quite general framework, using a matrix formulation, but only a brief introduction to this general model will be included.

The material to be presented is but a brief beginning in a subject to which complete books are devoted (Scheffé [20], Graybill [13], and Kempthorne [14], to mention a few). And even for those models and methods that are given, the matter of optimality will not be considered (see Lehmann [16], Chapter 7).

12.1 Partitioning Chi-Square

In studying the variance of a sample from a normal population, the following decomposition of a sum of squares was encountered:

$$\sum\left(\frac{X_i - \mu}{\sigma}\right)^2 = \sum\left(\frac{X_i - \bar{X}}{\sigma}\right)^2 + \left(\frac{\bar{X} - \mu}{\sigma/\sqrt{n}}\right)^2,$$

which is of the form $Q = Q_1 + Q_2$, each Q being a quadratic expression in standard normal variables.The distribution of Q is clearly $\chi^2(n)$ and that of Q_2 is almost as clearly $\chi^2(1)$. If Q_1 and Q_2 are found to be independent, then (according to Theorem 2 of Section 7.1.3) it can be concluded that Q_1 is $\chi^2(n-1)$; and this was done in Section 7.1.4.

In this chapter, in the *analysis of variance* (*ANOVA*), it will be necessary to have a generalization of these ideas to handle the decomposition of sums of squares into several quadratic forms, and Cochran's theorem is one answer to this need.

12.1.1 Cochran's Theorem

Consider the linear transformation from $(u_1, \ldots, u_n)$ to $(L_1, \ldots, L_m)$:

$$\begin{cases} L_1 = a_{11}u_1 + \cdots + a_{1n}u_n \\ \quad\vdots \\ L_m = a_{m1}u_1 + \cdots + a_{mn}u_n, \end{cases}$$

and let Q denote the sum of squares of these L's:

$$Q = L_1^2 + \cdots + L_m^2.$$

The *rank* of Q is defined to be the largest number of L's among which there is no linear relation. (It is the rank of the matrix $\mathbf{A}$ of the transformation.)

Cochran's Theorem. *Let $U_1, \ldots, U_n$ be independent, standard normal variables, and suppose that one can write an identity of the form*

$$\sum_1^n U_i^2 = Q_1 + \cdots + Q_k,$$

where each Q_i is a sum of squares of linear combinations of the U's. Let r_i be the rank of Q_i. Then if

$$r_1 + \cdots + r_k = n,$$

it follows that the Q_i's have independent chi-square distributions, each with a number of degrees of freedom given by its rank as a quadratic form.[1]

This result will be applied to situations in which $X_1, \ldots, X_n$ are independent normal variables with $E(X_i) = \mu$ and var $X_i = \sigma^2$, and

$$\sum_{i=1}^n \left(\frac{X_i - \mu}{\sigma}\right)^2 = Q_1 + \cdots + Q_k,$$

each Q being a sum of squares of linear combinations of X's. Of course, a linear combination

$$L_i = a_{i1}X_1 + \cdots + a_{in}X_n$$

[1] Cochran's theorem is frequently given in a slightly different but equivalent form, in which each Q is simply assumed to be a quadratic form in the $\mathbf{U}$'s: $Q = \mathbf{U'BU}$, rather than a sum of squares of linear combinations. In the version given here, the $\mathbf{B}$ is $\mathbf{A'A}$, and it can be shown that the rank of $\mathbf{A}$ is the same as that of $\mathbf{A'A}$. Therefore, the rank condition can be investigated for either $\mathbf{A}$ or $\mathbf{B}$. In the applications to be made in this book, the Q will arise as a sum of squares of linear combinations; hence, the given version seems simpler to use.

is automatically a linear function of U's, with $U_i = (X_i - \mu)/\sigma$:

$$L_i = \sigma(a_{i1}U_1 + \cdots + a_{in}U_n) + c_i$$

and is homogeneous if $c_i = 0$. But since $E(L_i) = c_i$, it is only necessary to verify that $E(L_i) = 0$ and examine the rank in the linear dependence of the L's on the X's. If there are j linear relations among the m L's, the rank of the transformation is at most $m - j$. Then the fact that the rank of a sum of quadratic forms is no larger than the sum of the ranks[2] can often be exploited to conclude that the rank is exactly $m - j$. This will be illustrated in the following example.

EXAMPLE 12–1. Let $X_1, \ldots, X_n$ be independent and normally distributed with common mean μ and common variance σ^2. Since for each i the variable $(X_i - \mu)/\sigma$ is standard normal, the following sum of squares has a chi-square distribution with n degrees of freedom:

$$\sum \left(\frac{X_i - \mu}{\sigma}\right)^2 = \sum \left(\frac{X_i - \bar{X}}{\sigma}\right)^2 + n\left(\frac{\bar{X} - \mu}{\sigma}\right)^2$$
$$= Q_1 + Q_2,$$

where the partitioning into the two Q's is essentially the parallel axis theorem. The term Q_2 is a square of a single linear combination of the standard normal variables $(X_i - \mu)/\sigma$, and the rank is just 1. The term Q_1 is the sum of squares of $L_1, \ldots, L_n$, where

$$L_i = \frac{X_i - \bar{X}}{\sigma} = \frac{X_i - \mu}{\sigma} - \frac{1}{n}\sum \frac{X_j - \mu}{\sigma}.$$

The sum of these L's is 0, and so the rank of Q_1 does not exceed $n - 1$. But the rank of the sum is n, which is no larger than 1 plus the rank of Q_1: that is, the rank of Q_1 is at least $n - 1$. Therefore, it is exactly $n - 1$, and the ranks add up as in the condition of Cochran's theorem. The conclusion then follows that Q_1 and Q_2 have independent chi-square distributions with $n - 1$ and 1 degree of freedom, respectively.

★12.1.2 Proof of the Theorem

As before, let $\mathbf{L} = \mathbf{Au}$, where $\mathbf{A}$ is of rank r, and $Q = L_1^2 + \cdots + L_m^2$. If it is assumed that the nonsingular matrix of rank r comes from the first r rows of $\mathbf{A}$ (which can be arranged by renumbering, if necessary), the quantities $L_{r+1}, \ldots, L_m$ can each be expressed as linear combinations of the first r L's in the sum of squares Q, resulting in a quadratic expression involving just the first r L's:

$$Q = \sum_{j=1}^{r}\sum_{i=1}^{r} b_{ij}L_iL_j = \mathbf{L'BL},$$

[2] See F. Hohn, *Elementary Matrix Algebra*, 2nd ed. (New York: Macmillan, 1964), p. 240.

where **L** is the column matrix with entries $L_1, \ldots, L_r$, and **B** is the symmetric matrix of b_{ij}'s. Since Q is clearly positive definite (as a quadratic form in the L's), **B** is nonsingular.

Now, the crux of the whole argument is that Q is expressible as a sum of exactly r squares of linear combinations of the u's, and this fact is obtained as follows: As in the two-dimensional case in which a rotation of axes eliminates the cross-product term, so in the n-dimensional case there is an orthogonal transformation of coordinates that will eliminate the cross-product terms in a quadratic form, leaving only the squared terms. The matrix of the new form is diagonal and can be written as $\mathbf{P}'\mathbf{B}\mathbf{P}$, where **B** is the original matrix and **P** is the transformation matrix. After elimination of the cross-product terms, a further change of scale along the various axes will give each squared term a coefficient 1, so that the matrix of the new form is the identity matrix. By combining the two transformations, then, it is possible to find a (nonsingular) matrix **P** with the property that $\mathbf{P}'\mathbf{B}\mathbf{P}$ is the identity matrix.

With the matrix **P** determined in this way, so that $\mathbf{P}'\mathbf{B}\mathbf{P} = \mathbf{I}$, let $\mathbf{v} = \mathbf{P}^{-1}\mathbf{L}$, or $\mathbf{L} = \mathbf{P}\mathbf{v}$, and substitute this in Q:

$$Q = \mathbf{L}'\mathbf{B}\mathbf{L} = \mathbf{v}'\mathbf{P}'\mathbf{B}\mathbf{P}\mathbf{v} = \mathbf{v}'\mathbf{v} = v_1^2 + \cdots + v_r^2,$$

which represents Q as the sum of r linear combinations of u's, as was claimed to be possible. That is, the v's, being linear combinations of L's, are then linear combinations of the u's.

This result will now be applied. Suppose that $U_1, \ldots, U_n$ are independent, standard, normal random variables, and that

$$\sum_{i=1}^{n} U_i^2 = Q_1 + Q_2 + \cdots + Q_k,$$

where *each* Q_i is a sum of squares of linear combinations of the U's. Let r_i denote the number of degrees of freedom (rank), as already defined for Q, of the term Q_i. It is now known that Q_i is expressible as the sum of exactly r_i squares of linear combinations of U's:

$$\begin{cases} Q_1 = W_1^2 + \cdots + W_{r_1}^2 \\ Q_2 = W_{r_1+1}^2 + \cdots + W_{r_1+r_2}^2 \\ \vdots \end{cases}$$

where there are, all-in-all, $r_1 + r_2 + \cdots + r_k$ W's or linear combinations of U's involved. If, then, this sum of the ranks is n, one has

$$\sum_{i=1}^{n} U_i^2 = \mathbf{U}'\mathbf{U} = W_1^2 + \cdots + W_n^2 = \mathbf{W}'\mathbf{W}.$$

But, since each W_i is a linear combination of U's, $\mathbf{W} = \mathbf{CU}$; and

$$\mathbf{U}'\mathbf{U} = (\mathbf{CU})'(\mathbf{CU}) = \mathbf{U}'\mathbf{C}'\mathbf{CU}.$$

Clearly, $\mathbf{C}'\mathbf{C} = \mathbf{I}$, and since $\mathbf{C}$ is square and $\mathbf{C}^{-1} = \mathbf{C}'$, then also $\mathbf{CC}' = \mathbf{I}$. But, if $\mathbf{U}$ is multivariate normal with mean vector $(0, 0, \ldots, 0)$ and covariance matrix $\mathbf{I}$, $\mathbf{W} = \mathbf{CU}$ is multivariate normal, with mean vector $(0, 0, \ldots, 0)$ and covariance matrix $\mathbf{CIC}' = \mathbf{I}$. That is, $W_1, W_2, \ldots,$ and W_n are independent, standard, normal variables, which means that $Q_1, \ldots, Q_k$ are independent chi-square variables, with $r_1, \ldots, r_k$, degrees of freedom, respectively, as asserted by Cochran's theorem.

12.2 Regression

In many fields of application, one finds instances in which a response Y depends in some way on a "controlled variable" x, a quantity that can be prescribed and fixed by setting a dial or giving a dose of medicine. The interest here is in those situations in which the variable Y is a random variable, either because it includes a random measurement error ϵ along with a deterministic function of the controlled variable,

$$Y_x = g(x) + \epsilon,$$

or because different subjects to which a given treatment is applied respond randomly. In the latter case, one is interested in the average response and how it varies with x, so that if one defines $E(Y_x) = g(x)$, the above equation again applies with ϵ interpreted as the deviation from the mean. Some examples are the increase in blood pressure induced by a certain drug, the yield in a chemical reactor obtained at a given temperature, the sales resulting from various amounts of advertising, the hardness of an alloy determined by the amount of carbon, the reaction time of a subject as affected by the amount of alcohol, and so on.

Another context in which a relation of the form $Y_x = g(x) + \epsilon$ arises is that in which X and Y are both random, but one is concerned with the variation of Y given the value $X = x$. Some examples are the height of a father (X) and the height of his son (Y), and a student's high-school rank (X) and his college grade point average (Y). In such instances, the function $g(x)$ is the conditional mean:

$$g(x) = E(Y \mid X = x),$$

termed the *regression function* (of Y on X). This name for $g(x)$ has been carried over to the cases in which x is a controlled variable, and the study of $g(x)$ in any case is called *regression analysis*. When the variable x has more than one component, the problem is said to be one of *multiple* regression.

The regression function is often assumed to have a given form, such as:

$$\alpha + \beta x, \quad a + bx + cx^2 + dx^3, \quad ke^{ax}, \quad ax^p, \quad b \sin t, \quad a_1x_1 + a_2x_2 + a_3x_3, \ldots,$$

defined by a finite set of parameters which are then the object of study. The simplest cases are those in which the dependence on the parameters is linear, although sometimes (ke^{ax}, ax^p) a logarithmic transformation will yield such a function and be just as tractable.

The data for a regression analysis will ordinarily consist of independent pairs $(x_1, Y_1), \ldots, (x_n, Y_n)$, in which either the x's are present values of a controlled variable or they are the particular values of $X_1, \ldots, X_n$ that occur in n bivariate observations. Errors ϵ_i are defined by

$$Y_i = g(x_i) + \epsilon_i, \qquad i = 1, \ldots, n,$$

where $g(x)$ is the regression function. It is assumed, in the simplest models, that the ϵ's are independent random variables with zero mean and constant variance:

$$E(\epsilon_i) = 0, \qquad \text{var } \epsilon_i = \sigma^2.$$

In the bivariate case, the deviation ϵ about the conditional mean always does have mean 0, so the assumption is just that the conditional variance is constant. When a distributional assumption is needed, the one usually made is that the Y's (and so the ϵ's) are normal. In the bivariate case, an assumption that (X, Y) is bivariate normal implies that the conditional distribution of Y given $X = x$ is normal with constant variance (and linear mean function).

12.2.1 Least Squares

The problem of fitting a function of specified form with a finite number of parameters to a given set of points $(x_1, Y_1), \ldots, (x_n, Y_n)$ has no solution if an exact fit is required, and not a unique solution if an approximate fit is acceptable, unless the number of parameters at least equals the number of points to be fitted. Gauss' method of least squares, yielding a good approximate fit, is traditional, analytically and practically tractable, and therefore useful.

Since the Y-value is thought of as a random variable distributed about $g(x)$ along the vertical line corresponding to a given x, it makes sense to measure the closeness of fit of given data to a curve $g(x)$, proposed as an approximation to the true $g(x)$, in terms of *vertical deviations*, also called *residuals*:

$$d_i \equiv Y_i - g(x_i).$$

The fit will be good if the sum of squared deviations

$$Q \equiv \sum d_i^2$$

is small, and the method of least squares is to choose $g(x)$ from the class of functions being considered so as to minimize the measure Q.

The method will be carried out first for the simplest case, that of a *linear* regression function. For reasons that will emerge, it will be convenient to parametrize the function in terms of $\bar{x}$, the mean of the controlled variable values:

$$g(x) = a + b(x - \bar{x}),$$

where then b is the slope and a is the height of the line at $x = \bar{x}$. For a trial line defined by given values of a and b, the mean squared deviation is a function of those values: $Q = Q(a, b)$, and this function is then to be minimized by suitable choice of a and b.

The minimization can be accomplished either algebraically or by means of calculus. The algebraic method, based on completing the square, will be given first. By expanding the square and collecting terms, it is readily verified that

$$\begin{aligned} Q(a, b) &= \sum [Y_i - a - b(x_i - \bar{x})]^2 \\ &= \sum Y_i^2 + na^2 - 2a \sum Y_i + b^2 \sum (x_i - \bar{x})^2 - 2b \sum (x_i - \bar{x}) Y_i. \end{aligned}$$

Completion of the square in a and b yields (if $S_x^2 \neq 0$)

$$\begin{aligned} \frac{1}{n} Q &= S_Y^2 + (a - \bar{Y})^2 + S_x^2\left(b - \frac{S_{xY}}{S_x^2}\right)^2 - \frac{S_{xY}^2}{S_x^2} \\ &= S_Y^2 (1 - r^2) + (a - \bar{Y})^2 + S_x^2 \left(b - \frac{S_{xY}}{S_x^2}\right)^2, \end{aligned}$$

where the earlier notation S^2, S_{xY}, and r^2 has been employed even though the x's are fixed. (The use of a lowercase x in the subscripts is intended to serve as a reminder that x is not random.) It is at once clear that Q is minimized by choosing the values

$$\hat{\alpha} = \bar{Y}, \qquad \hat{\beta} = \frac{S_{xY}}{S_x^2}$$

for a and b, respectively. [The notation $\hat{\alpha}$ and $\hat{\beta}$ anticipates the use of these quantities as estimators for the true parameters α and β when the regression function is assumed to be linear—of the form $\alpha + \beta(x - \bar{x})$. The condition $S_x^2 \neq 0$ simply requires that the x's are not all the same; if they *were* all the same, there would be no point in trying to fit a function of the given form.] It is also apparent that the minimum value of Q/n is $S_Y^2 (1 - r^2)$.

EXAMPLE 12–2. An experiment was designed to study the effect of dissolved sulfur on the surface tension of liquid copper. The decrease in surface tension was assumed to be a linear function of the logarithm of the percentage of sulfur. Data were obtained as follows:

x = Log Per Cent Sulfur	Y = Decrease in Surface Tension (deg/cm)
−3.38	308
−2.38	426
−1.20	590
−.92	624
−.49	649
−.19	727

Calculations of slope and intercept according to the formulas for $\hat{\beta}$ and $\hat{\alpha}$, respectively, yielded the following empirical regression function:

$$y = 736 + 127.6x.$$

The correlation coefficient is $r = .996$, so $1 - r^2 = .008$ and

$$S_Y^2(1 - r^2) \doteq 166 \doteq (12.9)^2.$$

The data are plotted along with the fitted line in Figure 12–1.

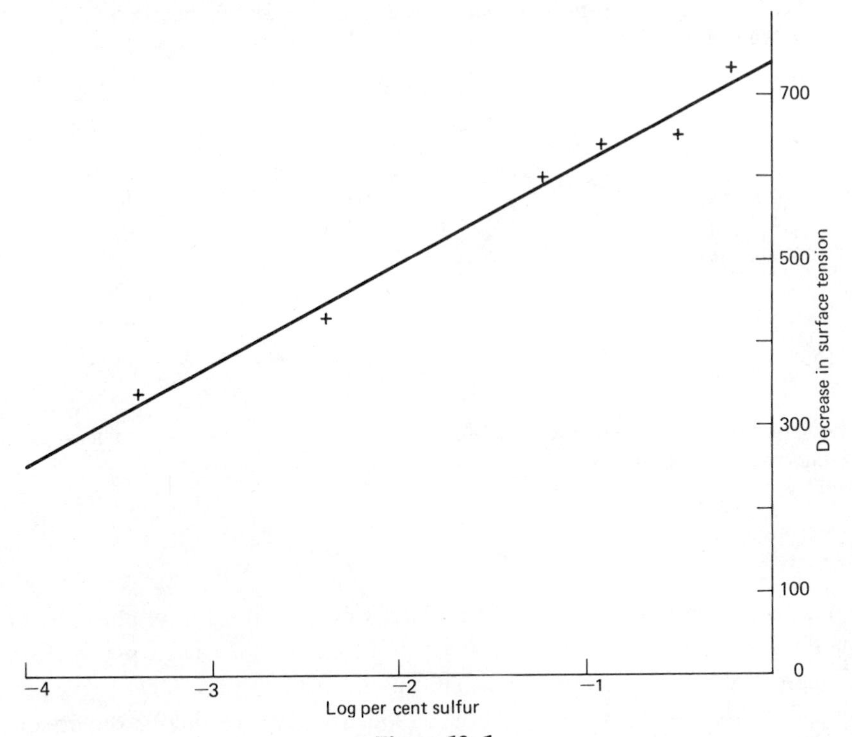

Figure 12–1

The method of least squares can be used to handle any assumed regression function. If the dependence on the parameters $\beta_1, \ldots, \beta_p$ is linear, the minimization of Q, the sum of squared residuals, is particularly simple and does not require numerical methods. Thus, the first-order partial derivatives of $Q = \sum d_i^2$ with respect to the parameters must all vanish at a minimum point, Q being everywhere differentiable:

$$\frac{\partial Q}{\partial \beta_j} = 2 \sum d_i \frac{\partial d_i}{\partial \beta_j} = 0, \qquad \text{for } j = 1, \ldots, p.$$

When the residual d_i is linear in β_j, the partial derivative $\partial d_i/\partial \beta_j$ does not depend on the parameters, so the equations $\partial Q/\partial \beta_j = 0$ are all *linear equations* in the parameters. These equations are called (even in nonlinear cases) the *normal equations* of the least squares process, and in the linear case their solution exists and is unique unless the experiment is designed with ill-chosen x-values (or not enough data for the kind of model assumed).

A unique solution of the normal equations must yield the minimum value of Q. For Q is bounded below and so *has* a minimum value, and the normal equations must be satisfied at a minimum.

EXAMPLE 12–3. In fitting the quadratic $a + bx + cx^2$ by the method of least squares, the sum of squared residuals is

$$Q = \sum (Y_i - a - bx_i - cx_i^2)^2,$$

and

$$\frac{\partial Q}{\partial a} = -2 \sum d_i, \qquad \frac{\partial Q}{\partial b} = -2 \sum x_i\, d_i, \qquad \frac{\partial Q}{\partial c} = -2 \sum x_i^2\, d_i.$$

So the normal equations are these:

$$\begin{cases} na + b \sum x_i + c \sum x_i^2 = \sum Y_i, \\ a \sum x_i + b \sum x_i^2 + c \sum x_i^3 = \sum Y_i x_i, \\ a \sum x_i^2 + b \sum x_i^3 + c \sum x_i^4 = \sum Y_i x_i^2. \end{cases}$$

The solution of this system of equations will be unique if there are at least three distinct x's.

12.2.2 Estimation of the Parameters

The least squares solutions to the problem of fit constitute a set of estimators for the parameters in a regression function of given form. Their properties as estimators will now be considered, but only in the case of a linear regression: $g(x) = \alpha + \beta(x - \bar{x})$, the assumption of which makes reasonable the attempt to fit a straight line to the data.

The least squares estimators $\hat{\alpha}$ and $\hat{\beta}$ are *linear* functions of the observed Y's:

$$\hat{\alpha} = \frac{1}{n}\sum Y_i, \qquad \hat{\beta} = \frac{1}{nS_x^2}\sum (x_i - \bar{x})Y_i.$$

Moreover, the Y's are assumed independent, so one can easily calculate first and second moments, recalling that $EY_i = g(x_i)$ and $\text{var } Y_i = \sigma^2$:

$$E(\hat{\alpha}) = \alpha + \frac{\beta}{n}\sum (x_i - \bar{x}) = \alpha.$$

$$E(\hat{\beta}) = \frac{1}{nS_x^2}\sum (x_i - \bar{x})E(Y_i) = \beta.$$

$$\text{var } \hat{\alpha} = \frac{\sigma^2}{n}.$$

$$\text{var } \hat{\beta} = \frac{1}{n^2S_x^4}\sum (x_i - \bar{x})^2 \text{ var } Y_i = \frac{\sigma^2}{nS_x^2}.$$

$$\text{cov}(\hat{\alpha}, \hat{\beta}) = \frac{1}{n^2S_x^2}\text{cov}\left[\sum Y_i, \sum (x_j - \bar{x})Y_j\right]$$

$$= \frac{1}{n^2S_x^2}\sum (x_j - \bar{x})\sigma^2 = 0.$$

So $\hat{\alpha}$ and $\hat{\beta}$ are unbiased estimates of α and β; and they are uncorrelated, which happens because of the particular way in which α was defined.

Recall that the minimum value of Q/n, the value at $a = \hat{\alpha}$ and $b = \hat{\beta}$, is given by

$$\hat{\sigma}^2 \equiv S_Y^2(1 - r^2) = \frac{1}{n}\sum [Y_i - \hat{\alpha} - \hat{\beta}(x_i - \bar{x})]^2.$$

It has the structure of a variance estimate—the average of squared deviations of the Y's, each about an estimate of its mean. Using it as an estimate of the (conditional) variance σ^2 provides "standard error" formulas for $\hat{\alpha}$ and $\hat{\beta}$:

$$\text{s.e. }(\hat{\alpha}) = \frac{\hat{\sigma}}{\sqrt{n}}, \qquad \text{s.e. }(\hat{\beta}) = \frac{\hat{\sigma}}{\sqrt{nS_x^2}}.$$

Observe that $\hat{\sigma}^2$ is 0 when $r^2 = 1$, that is, when the data lie on a straight line.

An interesting property of the least squares estimators is that they provide a linear unbiased estimate of the regression function with minimum variance. This is a special case of the *Gauss–Markov theorem.* It can be proved as follows. Let U be a linear combination of the Y's:

$$U = c_1Y_1 + \cdots + c_nY_n,$$

and let $b_1, \ldots, b_n$ be the coefficients of the Y's in the difference between U and the least squares regression function:

$$U = \sum b_iY_i + \hat{\alpha} + \hat{\beta}(x - \bar{x}).$$

In order for U to be unbiased in estimating $\alpha + \beta(x - \bar{x})$, it must be that

$$E(U) = \sum b_i[\alpha + \beta(x_i - \bar{x})] + \alpha + \beta(x - \bar{x}) = \alpha + \beta(x - \bar{x})$$

for all α and β so that

$$\sum b_i = 0 \qquad \text{and} \qquad \sum b_i(x_i - \bar{x}) = 0.$$

These conditions imply (as the reader should verify) that $\sum b_i Y_i$ is uncorrelated with $\hat{\alpha}$ and with $\hat{\beta}$ so that

$$\text{var } U = \sigma^2 \sum b_i^2 + \frac{\sigma^2}{n} + \frac{\sigma^2}{nS_x^2}(x - \bar{x})^2.$$

This is clearly a minimum if $b_1 = \cdots = b_n = 0$, as was to be shown.

When the Y's (and so the ϵ's) are *normally* distributed, the estimators $\hat{\alpha}$ and $\hat{\beta}$, being linear functions of the Y's, are normal. Indeed, the pair $(\hat{\alpha}, \hat{\beta})$ is bivariate normal, and because $\text{cov}(\hat{\alpha}, \hat{\beta}) = 0$ it follows that $\hat{\alpha}$ and $\hat{\beta}$ are independent.

Assumption of the normal model permits calculation of maximum likelihood estimates of the parameters. The likelihood function has logarithm

$$\log L(\alpha, \beta; \sigma^2) = -\frac{n}{2}\log(2\pi\sigma^2) - \frac{1}{2\sigma^2}\sum [Y_i - \alpha - \beta(x_i - \bar{x})]^2.$$

For this to be a maximum, it must be a maximum in (α, β) for fixed σ^2 and conversely. Thus, the $(\hat{\alpha}, \hat{\beta})$ in the maximizing $(\hat{\alpha}, \hat{\beta}, \hat{\sigma}^2)$ must minimize the sum of squared residuals, and so the maximum likelihood estimates are precisely the least squares estimates. These, together with the equation obtained by setting the partial derivative with respect to σ^2 equal to 0,

$$-\frac{n}{2\sigma^2} + \frac{1}{2\sigma^4}\sum [Y_i - \hat{\alpha} - \hat{\beta}(x_i - \bar{x})]^2 = 0,$$

determine the m.l.e. of σ^2:

$$\hat{\sigma}^2 = \frac{1}{n}\sum [Y_i - \hat{\alpha} - \hat{\beta}(x_i - \bar{x})]^2 = \frac{1}{n} Q_{\min}.$$

This expression for $\hat{\sigma}^2$ is just the minimum average squared deviation of the data about the least squares regression line. (It was in anticipation of these identifications that the circumflex notation was used in the least squares method.)

The distribution of the m.l.e. $\hat{\sigma}^2$ can be given just about as easily in the general case as in the case of a straight line regression. If the regression function is linear in the parameters $\beta_1, \ldots, \beta_p$:

$$g(x) = g_1(x)\beta_1 + \cdots + g_p(x)\beta_p,$$

the normal equations are (as seen earlier) linear in the β's:

$$\sum [Y_i - \hat{g}(x_i)]g_k(x_i) = 0, \qquad k = 1, \ldots, p,$$

where $\hat{g}(x)$ is the least squares regression function, $\sum g_j(x)\hat{\beta}_j$. Moreover, the solutions $\hat{\beta}_j$ are linear in the observations $Y_1, \ldots, Y_n$. And just as in the straight line case, the m.l.e. of σ^2 is the minimum of the average squared residual about the least squares regression function:

$$\hat{\sigma}^2 = \frac{1}{n}\sum [Y_i - \hat{g}(x_i)]^2.$$

Theorem. *Under the normal model, with the mean function $g(x)$ depending linearly on p parameters, the variable $n\hat{\sigma}^2/\sigma^2$ has a chi-square distribution with $n - p$ degrees of freedom.*

This is established with the aid of Cochran's theorem. Starting with the identity

$$Y_i - g(x_i) = [Y_i - \hat{g}(x_i)] + [\hat{g}(x_i) - g(x_i)],$$

squaring and adding yields the following:

$$\sum [Y_i - g(x_i)]^2 = \sum [Y_i - \hat{g}(x_i)]^2 + \sum [\hat{g}(x_i) - g(x_i)]^2.$$

The cross-product term has disappeared upon summation by virtue of the normal equations given above. The first term on the right is at most of rank $n - p$, since there are p linear relations among the $\hat{g}$'s (namely, the normal equations). The second term on the right is

$$\sum_i \Big[\sum_j g_j(x_i)(\hat{\beta}_j - \beta_j)\Big]^2,$$

which is of rank at most p, because it is a quadratic form in just p linear functions of the Y's. Thus, the rank on the right is both at most and at least n, so (according to Cochran's theorem) the two terms on the right, after division by σ^2, have independent chi-square distributions with degrees of freedom $n - p$ and p, respectively.

It should be remarked that the distribution of $\hat{\sigma}^2$, the estimate of the "error variance" σ^2 based on deviations about the least squares estimate of the regression function, is independent of the values of any of the regression parameters.

Problems

+ **12–1.** Determine the line that best fits the following points in the sense of least squares: (0, 2), (1, 1), (4, 3), (5, 2).

12–2. Obtain the least squares estimates of the regression parameter or parameters if

(a) $g(x) = \gamma x^2$.
(b) $g(x, z) = \alpha x + \beta z$.

12–3. Given the data of Example 12–2.

(a) Write the empirical regression function in the form $\hat{\alpha} + \hat{\beta}(x - \bar{x})$.
(b) Calculate the standard errors of $\hat{\alpha}$ and $\hat{\beta}$.
(c) Construct a 90 per cent confidence interval for σ^2.

12–4. Obtain $\hat{\alpha}$ and compute cov $(\hat{\alpha}, \hat{\beta})$ when $g(x) = \alpha + \beta x$.

12–5. Obtain a minimal sufficient statistic for the family of distributions of $Y_1, \ldots, Y_n$ when the regression is linear and the error variance σ^2 is constant.

12–6. It is sometimes said (and written) that the least squares estimates $\hat{\alpha}$ and $\hat{\beta}$ are consistent, without any specification as to how observations are to be taken as n becomes infinite. Suppose that $x_1 = 0$ and that $x_2 = x_3 = \cdots = 1$. Calculate the variance of $\hat{\alpha}$ and of $\hat{\beta}$. Does $\hat{\beta}$ converge to β?

12–7. In studying the reaction rate of a certain synthetase of a bovine lens, it is found that the reciprocal of this rate appears to be linearly related to the reciprocal of the substrate concentration. Obtain the least squares estimate of this linear relationship, given the following data:

Reciprocal of Substrate Concentration (nanomoles of product formed per minute)	Reciprocal of Reaction Rate (micromoles per liter)
24	.429, .444
20	.293, .293
16	.251, .268
12	.207, .216
8	.239, .218
6	.180, .199
2	.156, .167

12.2.3 Testing Hypotheses

Analyzing tests of hypotheses about a regression function requires a distributional assumption, and it will be assumed here that the independent observations or responses $Y_1, \ldots, Y_n$ are *normally* distributed. The following result will be often useful:

Lemma. *Under the normal model, the maximum value of the likelihood function $L(g(x), \sigma^2)$ is given by*

$$[2\pi\hat{\sigma}^2]^{-n/2}e^{-n/2},$$

where $\hat{\sigma}^2$ is the average squared residual about the least squares regression function—that is, the m.l.e. of σ^2.

This follows from the fact that $\hat{\sigma}^2$ is obtained by substituting the m.l.e.'s of the regression parameters into the regression function in the exponent of the normal density. This exponent then becomes

$$-\frac{1}{2\hat{\sigma}^2}\sum [Y_i - \hat{g}(x_i)]^2 = -\frac{n}{2},$$

as asserted in the lemma.

Consider now a problem of testing the hypothesis $\alpha = \alpha_0$ against $\alpha \neq \alpha_0$, with β unspecified, when it is assumed that the regression function is $\alpha + \beta(x - \bar{x})$. The likelihood ratio statistic is

$$\Lambda = \sup_{(\beta,\sigma^2)} L(\alpha_0, \beta; \sigma^2) / \sup_{(\alpha,\beta,\sigma^2)} L(\alpha, \beta; \sigma^2)$$

$$= \left(\frac{\hat{\sigma}_0^2}{\hat{\sigma}^2}\right)^{-n/2},$$

everything but the variance estimates having canceled out (see the lemma above). Here $\hat{\sigma}_0^2$ is the m.l.e. of σ^2 under the null hypothesis:

$$\hat{\sigma}_0^2 = \frac{1}{n}\sum [Y_i - \alpha_0 - \hat{\beta}(x_i - \bar{x})]^2,$$

where $\hat{\beta}$ (as should be verified by the reader) is the same whether obtained as the m.l.e. when $\alpha = \alpha_0$ or with α unrestricted: $\hat{\beta} = S_{xY}/S_x^2$. By adding and subtracting $\hat{\alpha}$ inside the bracket before squaring, one obtains

$$\hat{\sigma}_0^2 = \hat{\sigma}^2 + (\hat{\alpha} - \alpha_0)^2.$$

Thus,

$$\Lambda = \left[1 + \frac{(\hat{\alpha} - \alpha_0)^2}{\hat{\sigma}^2}\right]^{-n/2},$$

so the likelihood ratio critical region is equivalent to

$$T^2 \equiv (n - 2)\frac{(\hat{\alpha} - \alpha_0)^2}{\hat{\sigma}^2} > \text{const.}$$

The notation anticipates the fact that T^2 is $F(1, n - 2)$ under H_0, and that $\sqrt{n - 2}\,(\hat{\alpha} - \alpha_0)/\hat{\sigma}$ is $t(n - 2)$. It is already known that $n(\hat{\alpha} - \alpha_0)^2/\sigma^2$ is $\chi^2(1)$ and that $n\hat{\sigma}^2/\sigma^2$ is $\chi^2(n - 2)$, but independence is needed. This can be deduced with the aid of Cochran's theorem.

The theorem of the preceding section that gave the distribution of $\hat{\sigma}^2$ involved a decomposition of the sum of squares of the Y's about the true $g(x)$ into two terms: (a) the sum of squares of the Y's about the least squares estimate $\hat{g}(x)$, and (b) the sum of squares of the deviations of the $\hat{g}(x_i)$'s from the $g(x_i)$'s. This second term is as follows, in the present context:

$$\sum [\hat{\alpha} + \hat{\beta}(x_i - \bar{x}) - \alpha - \beta(x_i - \bar{x})]^2 = n(\hat{\alpha} - \alpha)^2 + nS_x^2(\hat{\beta} - \beta)^2,$$

where the cross-product term has vanished because $\sum (x_i - \bar{x}) = 0$. Thus, Cochran's theorem says that when divided by σ^2, the two terms on the right (each being of rank 1) have independent $\chi^2(1)$ distributions—which is known from other considerations, and that these terms and $\hat{\sigma}^2$ are independent. It then follows that when $\alpha = \alpha_0$,

$$T_{\alpha_0}^2 = \left(\frac{\hat{\alpha} - \alpha_0}{\sigma/\sqrt{n}}\right)^2 \bigg/ \frac{n\hat{\sigma}^2}{(n-2)\sigma^2}$$

has the F-distribution claimed. Similarly (and this is usually of more interest), the test statistic

$$T_{\beta_0} \equiv \sqrt{n-2}\,\frac{\hat{\beta} - \beta_0}{\hat{\sigma}/S_x}$$

is $t(n-2)$ under the hypothesis $\beta = \beta_0$. (When $\beta \neq \beta_0$, the distribution would be noncentral t, and power functions are obtainable from that.)

Confidence intervals can be constructed for α and for β. Let $-k$ denote the $100\gamma/2$ percentile of the t distribution with $n-2$ degrees of freedom, or equivalently, $k^2 = F_{1-\gamma}(1, n-2) = t^2_{1-\gamma/2}(n-2)$. Then

$$P\left\{(n-2)\frac{(\hat{\alpha} - \alpha)^2}{\hat{\sigma}^2} < k^2\right\} = 1 - \gamma,$$

or

$$P\left(\hat{\alpha} - \frac{k\hat{\sigma}}{\sqrt{n-2}} < \alpha < \hat{\alpha} + \frac{k\hat{\sigma}}{\sqrt{n-2}}\right) = 1 - \gamma.$$

Similarly, a confidence interval for β is found to have the limits

$$\hat{\beta} \pm \frac{k\hat{\sigma}}{\sqrt{(n-2)S_x^2}}.$$

EXAMPLE 12–4. Suppose that calculations for a given set of 100 points (x_i, Y_i) yield $S_x^2 = 9.7$, $\hat{\alpha} = 1.1$, $\hat{\beta} = .02$, and $\hat{\sigma}^2 = .0036$. The normal distribution can be used to approximate the t distribution with 98 degrees of freedom, so that for $\gamma = .10$, $k = 1.645$. The 90 per cent confidence intervals for α and β are, respectively,

$$1.1 \pm 1.645 \times \frac{.06}{\sqrt{98}} \quad \text{or} \quad (1.0900, 1.1010)$$

and

$$.02 \pm 1.645 \times \frac{.06}{\sqrt{98 \times 9.7}} \quad \text{or} \quad (.0168, .0232).$$

The likelihood ratio test of $\beta = \beta_0$ is equivalent to the rule of rejecting β_0 if the confidence interval for β does not include it. According to this test, the value $\beta = 0$ (equality of means) would be rejected because the confidence interval obtained does not include it.

It should be pointed out that although inferences can be made concerning α without any assumptions on β, and vice versa, the statistics used for testing these two parameters are not independent, and care should be taken in making statements about α and β jointly. For instance, it cannot be asserted that both α and β lie in the corresponding 90 per cent confidence intervals with probability $.90 \times .90$. On the other hand, a *simultaneous confidence region* for (α, β) can be constructed, taking the ratio of the *sum* of the second and third terms in the chi-square partitioning to the first term; the numerator would be chi-square with 2 degrees of freedom, and the ratio would then have the F distribution with $(2, n - 2)$ degrees of freedom. Thus, given a confidence coefficient $1 - \gamma$:

$$P\left\{\frac{(\hat{\alpha} - \alpha)^2 + S_x^2(\hat{\beta} - \beta)^2}{2\hat{\sigma}^2/(n-2)} < F_{1-\gamma}(2, n-2)\right\} = 1 - \gamma.$$

The inequality is satisfied (for given $\hat{\alpha}$, $\hat{\beta}$, and $\hat{\sigma}^2$) by the points (α, β) in an ellipse. This elliptical region, which depends on the observations and is therefore random, is the confidence region.

12.2.4 Testing Linearity

So far, in the analysis of regression functions of one variable, the values $x_1, \ldots, x_n$ of the controlled variable were labeled as if distinct, but at no point in the discussion was it required that they all be distinct. (It was necessary, however, to assume that they were not all the same.) If some x's are the same, this simply means that several observations are taken at the same setting of the controlled variable, and the data can be thought of as comprising samples from each of several populations, one at each of the distinct x-values. The notation will be as follows.

Controlled Variable	Sample Size	Observations	Sample Mean	Population Mean
x_1	n_1	$Y_{11}, \ldots, Y_{1n_1}$	$\bar{Y}_1$	$g(x_1)$
$\vdots$	$\vdots$	$\vdots$	$\vdots$	$\vdots$
x_k	n_k	$Y_{k1}, \ldots, Y_{kn_k}$	$\bar{Y}_k$	$g(x_k)$

Altogether there are $n = \sum n_i$ observations, and the corresponding likelihood function depends on $k + 1$ parameters:

$$L(\gamma_1, \ldots, \gamma_k; \sigma^2) = (2\pi\sigma^2)^{-n/2} \exp\left[-\tfrac{1}{2}\sum\sum(Y_{ij} - \gamma_i)^2\right],$$

where $\gamma_i \equiv g(x_i)$, double sums extending over $j = 1, \ldots, n_i$ and $i = 1, \ldots, k$. A minimal sufficient statistic is readily seen to be $(\bar{Y}_1, \ldots, \bar{Y}_k; \hat{\sigma}^2)$, where $\hat{\sigma}^2$

is the mean squared deviation of the observations about the corresponding sample means:

$$\hat{\sigma}^2 = \frac{1}{n} \sum \sum (Y_{ij} - \bar{Y}_i)^2.$$

This "within samples" estimate of variance divided by σ^2/n is clearly the sum of k independent chi-square variables and so is itself chi-square with $(n_1 - 1) + \cdots + (n_k - 1) = n - k$ degrees of freedom—no matter what the regression function may be.

Thus, the existence of several observations at a single x, or at each of several x's, provides an estimate of variance that permits testing linearity. More precisely, one tests

$$H_0: \quad \gamma_i = \alpha + \beta(x_i - \bar{x}).$$
$$H_A: \quad (x_i, \gamma_i) \text{ are not collinear.}$$

The maximization of the likelihood under H_0 is just as before, when a linear regression was assumed as the basic model; but with the new notation the formulas are now as follows:

$$\hat{\alpha} = \bar{Y} = \frac{1}{n} \sum n_i \bar{Y}_i, \qquad \hat{\beta} = \frac{S_{xY}}{S_x^2},$$

$$S_x^2 = \frac{1}{n} \sum n_i (x_i - \bar{x})^2, \qquad S_{xY} = \frac{1}{n} \sum n_i (x_i - \bar{x}) \bar{Y}_i,$$

$$\hat{\sigma}_0^2 = \frac{1}{n} \sum \sum (Y_{ij} - \tilde{Y}_i)^2, \qquad \text{where } \tilde{Y}_i = \hat{\alpha} + \hat{\beta}(x_i - \bar{x}).$$

Since (in a now-familiar decomposition)

$$\hat{\sigma}_0^2 = \hat{\sigma}^2 + \frac{1}{n} \sum n_i (\bar{Y}_i - \tilde{Y}_i)^2,$$

the critical region is defined by

$$F = \frac{\sum n_i (\bar{Y}_i - \tilde{Y}_i)^2 / (k - 2)}{\sum \sum (Y_{ij} - \bar{Y}_i)^2 / (n - k)} > \text{const.}$$

That is, if the sample means differ too much from the linear function fitted by least squares, in comparison with the variance estimate obtained from within the samples, the hypothesis of linearity is rejected.

Again we have been led to an "analysis of variance"—a decomposition of variability into components identified with specific sources of variation. And Cochran's theorem can again be used (Problem 12–11) to show that the terms in the decomposition have (upon division by σ^2) independent chi-square distributions with degrees of freedom as shown in the following ANOVA table:

Source	Sum of Squares	Degrees of Freedom
Nonlinearity	$\sum n_i(\bar{Y}_i - \tilde{Y}_i)^2$	$k - 2$
Error	$\sum\sum (Y_{ij} - \bar{Y}_i)^2$	$n - k$
Total	$\sum\sum (Y_{ij} - \tilde{Y}_i)^2$	$n - 2$

It follows that the test ratio F has an $F(k - 2, n - k)$ distribution under the null hypothesis, which would then be used to relate the critical value of F to a significance level. (Power is again given by a noncentral F distribution.)

EXAMPLE 12–5. Some simple artificial data, made up to look nonlinear, will demonstrate the technique and the rationale:

Controlled Variable	Observations
1	3, 4, 5
2	1, 3,
3	3, 4, 5

It is apparent that the best linear fit is given by a horizontal line: $y = \bar{Y} = \frac{7}{2}$. The necessary computations may be summarized in the following table:

x_i	n_i	Y_{ij}	$\bar{Y}_i$	$\tilde{Y}_i$	$n_i(\bar{Y}_i - \tilde{Y}_i)^2$	$\sum (Y_{ij} - \bar{Y}_i)^2$	$\sum (Y_{ij} - \tilde{Y}_i)^2$
1	3	3, 4, 5	4	$\frac{7}{2}$	$3(\frac{1}{4})$	2	$\frac{11}{4}$
2	2	1, 3	2	$\frac{7}{2}$	$2(\frac{9}{4})$	2	$\frac{26}{4}$
3	3	3, 4, 5	4	$\frac{7}{2}$	$3(\frac{1}{4})$	2	$\frac{11}{4}$
Total	8	28	—	—	6	6	12

The ANOVA table can now be constructed:

Source	Sum of Squares	Degrees of Freedom
Nonlinearity	6	1
Error	6	5
Total	12	6

The test ratio is

$$F = \frac{6/1}{6/5} = 5.$$

The rejection limit for $\alpha = .05$ is $F_{.95}(1, 5) = 6.61$, and the 10 per cent limit is 4.06. Thus, the null hypothesis would be rejected at $\alpha = .10$ but accepted at $\alpha = .05$.

12.2.5 Selecting a Model

The testing of $\beta = 0$, for example, in the regression model $\alpha + \beta x$ amounts to testing the simple model $g(x) = \alpha$ against one that is not so simple and includes the simpler one as a special case—whereas, in fact, neither one may be the true model. And unless a priori considerations in the field of application suggest that the linear model or some other particular model is the right one, an aim of a regression analysis may be to find at least an adequate if not the true model.

In seeking a good regression model, it is natural to want to find one that is as simple as possible and still fits the data well. To do this, it seems reasonable to try something simple such as $g(x) = \beta_1$ and then to add terms one at a time, each with a new parameter. Thus, one is led to consider a *hierarchy* of models—a chain in which the successively simpler models are obtained from the most complex by successive restrictions on the parameters. For example, a hierarchy of polynomial models would be as follows:

$$\begin{aligned} H_1&: \; g(x) = \beta_1, \\ H_2&: \; g(x) = \beta_1 + \beta_2 x, \\ &\;\vdots \\ H_k&: \; g(x) = \beta_1 + \beta_2 x + \cdots + \beta_k x^{k-1}. \end{aligned}$$

An analysis of variance suggests how to test one model in a hierarchy against a more complicated one.

First suppose that one wants to test the null hypothesis that the regression function is given by "model 1" with m_1 parameters, a model that is embedded in or included as a special case of "model 2" with m_2 parameters. The alternative would be that the true model is included in model 2, but is not as simple as model 1. Let it be assumed in both models that observations are independent and normal with constant variance σ^2. The likelihood ratio is

$$\Lambda = \frac{\sup L(\beta_1, \ldots, \beta_{m_1}; \sigma^2)}{\sup L(\beta_1, \ldots, \beta_{m_1}, \ldots, \beta_{m_2}; \sigma^2)} = \left(\frac{\hat{\sigma}_1^2}{\hat{\sigma}_2^2}\right)^{-n/2},$$

in which

$$\begin{aligned} n\hat{\sigma}_1^2 &= \sum [Y_i - \hat{g}_1(x_i)]^2 \equiv Q_1, \\ n\hat{\sigma}_2^2 &= \sum [Y_i - \hat{g}_2(x_i)]^2 \equiv Q_2, \end{aligned}$$

where $\hat{g}_1$ and $\hat{g}_2$ are the least squares estimates of the regression function under models 1 and 2, respectively. According to an earlier theorem, Q_1/σ^2 and Q_2/σ^2 are chi-square, with $n - m_1$ and $n - m_2$ degrees of freedom, under model 1 and model 2, respectively.

Now, Q_1 can be decomposed according to the identity

$$Q_1 = (Q_1 - Q_2) + Q_2.$$

Such decompositions have been encountered several times in special cases, so it should not be hard to believe these results from a general theory of linear models:

1. $(Q_1 - Q_2)/\sigma^2$ is $\chi^2(m_2 - m_1)$ under model 1.
2. Q_2/σ^2 is $\chi^2(n - m_2)$ under model 2.
3. Q_2 and $Q_1 - Q_2$ are independent under model 2.

The likelihood ratio test statistic

$$F \equiv \frac{(Q_1 - Q_2)/(m_2 - m_1)}{Q_2/(n - m_2)}$$

has an $F(m_2 - m_1; n - m_2)$ distribution under the null hypothesis.

Similar results can be obtained for a more general hierarchy:

$$H_1 \subset H_2 \subset \cdots \subset H_k,$$

where $A \subset B$ means that model A is obtained from model B as a special case by setting certain parameters equal to zero. Let m_i denote the number of independent parameters in H_i. The ANOVA table is as follows:

Source	Sum of Squares	Degrees of Freedom
Fitting H_2 after H_1	$Q_1 - Q_2$	$m_2 - m_1$
Fitting H_3 after H_2	$Q_2 - Q_3$	$m_3 - m_2$
$\vdots$		
Fitting H_k after H_{k-1}	$Q_{k-1} - Q_k$	$m_k - m_{k-1}$
Error	Q_k	$n - m_k$
Total	Q_1	$n - m_1$

Typically, adding parameters to the model reduces the average sum of squares (which is an estimate of error variance); adding parameters that do not reduce it by much may be of little value.

EXAMPLE 12–6. Consider the following artificial data, for simplicity of computation, given together with some columns useful in computing estimates:

x	Y	$x - \bar{x}$	$Y - \bar{Y}$	$Y(x - \bar{x})$	$\tilde{Y}$	$Y - \tilde{Y}$
1	0	-1	-2	0	$-\frac{1}{2}$	$\frac{1}{2}$
2	1	0	-1	0	2	-1
3	5	1	3	5	$\frac{9}{2}$	$\frac{1}{2}$

Suppose that the following hierarchy of polynomial regression models is proposed:

H_1: $g(x) = \alpha$.
H_2: $g(x) = \alpha + \beta x$.
H_3: $g(x) = \alpha + \beta x + \gamma x^2$.

Under H_1, $\hat{\alpha} = \bar{Y} = 2$, and $Q_1 = \sum (Y_i - \bar{Y})^2 = 14$. Under H_2, the linear regression model,

$$\hat{\beta} = \frac{5}{2}, \qquad \hat{\alpha} = \bar{Y} - \bar{x} = -3, \qquad \tilde{Y}_i = -3 + \frac{5x}{2},$$

and

$$Q_2 = \sum (Y_i - \tilde{Y}_i)^2 = \frac{3}{2}.$$

Under H_3 the normal equations are

$$\begin{cases} 3\alpha + 6\beta + 14\gamma = 6, \\ 6\alpha + 14\beta + 36\gamma = 17, \\ 14\alpha + 36\beta + 98\gamma = 49, \end{cases}$$

with solution $\hat{\alpha} = 2$, $\hat{\beta} = -\frac{7}{2}$, $\hat{\gamma} = \frac{3}{2}$. And then

$$Q_3 = \sum (Y_i - \hat{\alpha} - \hat{\beta}x_i - \hat{\gamma}x_i^2)^2 = 0.$$

The ANOVA table is as follows:

Source	Sum of Squares Component	Degrees of Freedom	Mean
Fitting β after α	$\frac{25}{2}$	1	$\frac{25}{2}$
Fitting γ after α, β	$\frac{3}{2}$	1	$\frac{3}{2}$
Error	0	$n - 3 = 0$	—
Totals	14	$n - 1 = 2$	7

In each line the "sum of squares" is a component of the total sum attributable to the "source" indicated. If it is large, this indicates that a significantly better fit has been achieved; for the total of the components in the remaining lines gives an error sum of squares for the model fitted at that point. (In this example, with only three points to fit, the error after fitting a parabola is 0.)

Problems

12–8. The amounts of a certain chemical that dissolved in a given amount of water at various temperatures were determined as follows:

Amount dissolved (gm)	8	12	21	31	39	58
Temperature (°C)	0	10	20	30	40	60

Assuming a linear regression function, obtain an 80 per cent confidence interval for the slope.

12–9. Given the following data, assume independent, normal observations with common variance σ^2:

x	Y
1	2, 5, 5
2	4, 6, 5
3	6, 4, 8

(a) Test for linearity of the regression function.
(b) Assuming linearity, give estimates of the parameters.
(c) Assuming linearity, determine a 95 per cent confidence region for the pair (α, β).

12–10. Show that in testing $\beta = 0$ in the model $g(x) = \alpha + \beta x$, the test statistic can be written in the form

$$t = \sqrt{n-2}\,\frac{r}{\sqrt{1-r^2}},$$

where r is the correlation coefficient of the x's and Y's. Infer then that the conditional distribution given the x's (t with $n - 2$ degrees of freedom) is also the unconditional distribution. (This fact was used in testing $\rho = 0$ in Section 10.3.)

12–11. Apply Cochran's theorem, as suggested in Section 12.2.4, to establish the distribution of the test statistic for linearity.

12–12. Given these data points: $(-1, 1)$, $(0, 0)$, $(1, 1)$ and $(2, 3)$, set up the ANOVA table corresponding to the hierarchy:

H_1: $g(x) = \gamma x^2$.
H_2: $g(x) = \gamma x^2 + \beta x$.
H_3: $g(x) = \gamma x^2 + \beta x + \alpha$.

12.2.6 Designing the Experiment

The design of a regression experiment includes the choosing of values of the independent variable at which to take observations and the specification of how many observations to take at each value. The problem of determining optimum designs has been studied by Elfving and others, but is given only brief mention here.[3]

Different regression problems will lead to different definitions of "optimum." Thus, in testing linearity, it is desirable to have k (the number of distinct x-values) large. But for a good estimate of slope, given that the regression function is linear, it is desirable to make S_x^2 as large as possible. Ordinarily, there are practical upper and lower limits to the value that x may be given, so to make S_x^2 large, it suffices to put all of the observations at one end or the other of this feasible range. If a proportion p is put at x_1, and $q = 1 - p$ at x_2, then

$$nS_x^2 = px_1^2 + qx_2^2 - (px_1 + qx_2)^2 = pq(x_1 - x_2)^2.$$

This is maximized for $p = q = \frac{1}{2}$. Elfving showed that for estimating any single linear function of the parameters in a general regression problem, exactly two x-values should be used, and he gave a scheme for determining the proper allocation between these two points.

12.2.7 Prediction

In the model $Y = g(x) + \epsilon$, the best prediction of the value of Y at $x = x_0$ in terms of mean squared error is given by the mean $g(x_0)$. When $g(x)$ is unknown, it is natural to use the estimate $\hat{g}(x_0)$, and the corresponding error of prediction is

$$e = \hat{g}(x_0) - Y_0,$$

where Y_0 is the value observed at x_0. This error is normal with mean 0, if $g(x)$ is linear in its parameters, and variance

$$\sigma_e^2 = \sigma^2 + \text{var}\, \hat{g}(x_0),$$

since Y_0 is independent of the observations Y_i that define the estimate $\hat{g}$. In the case of a linear regression: $g(x) = \alpha + \beta(x - \bar{x})$, the mean squared prediction error is

$$\begin{aligned}\sigma_e^2 &= \sigma^2 + \text{var}\,[\hat{\alpha} + \hat{\beta}(x_0 - \bar{x})] \\ &= \sigma^2\left[1 + \frac{1}{n} + \frac{(x_0 - \bar{x})^2}{nS_x^2}\right].\end{aligned}$$

[3] See G. Elfving, "Optimum Allocation in Linear Regression Theory," *Ann. Math. Stat.* **23**, 255 (1952); also J. Kiefer and J. Wolfowitz, "Optimum Designs in Regression Problems," *Ann. Math. Stat.* **30**, 271 (1959), and references given there.

Observe that this increases with $|x_0 - \bar{x}|$, as would be expected; for the effect on the prediction of an error in slope is magnified as x_0 moves away from the center of the x's.

A confidence interval for Y_0, better described as a *prediction interval*, can be constructed using the fact that $n\hat{\sigma}^2/\sigma^2$ is $\chi^2(n - m)$, where m is the number of independent parameters in $g(x)$. For, e^2/σ_e^2 is $\chi^2(1)$ and is independent of $\hat{\sigma}^2$ so that, if $s^2 = n\hat{\sigma}^2/(n - m)$, the ratio

$$F = \frac{e^2/\sigma_e^2}{s^2/\sigma^2}$$

has a distribution that is $F(1, n - m)$. And then

$$P\left\{\frac{|Y_0 - \hat{g}(x_0)|}{s(\sigma_e/\sigma)} < t_{1-\eta/2}\right\} = 1 - \eta,$$

where t_γ is the 100γ percentile of $t(n - m)$, or

$$P\left[\hat{g}(x_0) - t_{1-\eta/2}s\left(\frac{\sigma_e}{\sigma}\right) < Y_0 < \hat{g}(x_0) + t_{1-\eta/2}s\left(\frac{\sigma_e}{\sigma}\right)\right] = 1 - \eta.$$

In the linear case,

$$\frac{\sigma_e}{\sigma} = \left[1 + \frac{1}{n} + \frac{(x_0 - \bar{x})^2}{nS_x^2}\right]^{1/2},$$

and the extremes of the above inequality are calculable from the sample. They define the desired prediction interval.

A kind of inverse prediction problem, sometimes termed *discrimination*, and encountered in bioassay work, is that of determining the value x_0 at which a given set of observations $Y_1', \ldots, Y_m'$ is recorded, given the usual data for a regression of Y on x. The m.l.e. of x_0 is

$$\hat{x}_0 = \frac{\bar{Y}' - \hat{\alpha}}{\hat{\beta}} + \bar{x}.$$

A confidence interval for x_0 is readily constructed using the fact that the quantity $\bar{Y}' - \hat{\alpha} - \hat{\beta}(x_0 - \bar{x})$ is normal with mean 0 and variance

$$\sigma^2\left[\frac{1}{m} + \frac{1}{n} + \frac{(x_0 - \bar{x})^2}{ns_x^2}\right],$$

and is independent of the estimate $\hat{\sigma}^2$.

12.2.8 The Bivariate Normal Model

Up to this point, the development has been for those problems in which x is a nonrandom controlled variable. However, as discussed at the outset, the simple regression assumptions made (independent, normal observations with constant variance) are sometimes fulfilled by the conditional variable $Y \mid X = x$ in a bivariate model for (X, Y). In particular, if (X, Y) is bivariate normal, the regression function $E(Y \mid x)$ is linear in x. So the various results obtained for a linear regression function are valid in this bivariate normal case, conditional on given values of x.

In the case of the t-statistic for testing $\beta = \beta_0$,

$$T = \sqrt{n-2}\,\frac{\hat{\beta} - \beta_0}{\hat{\sigma}/S_x},$$

the distribution (when $\beta = \beta_0$) was shown to be $t(n - 2)$; but in the bivariate normal case, this would be conditional on the X-values used. On the other hand, since the density of T does not involve those values (depending only on n), it follows that the unconditional distribution of T is also $t(n - 2)$.

The prediction of Y given a value of X is perhaps the most frequent regression problem in the bivariate model. It has been seen that the prediction error in the linear case, when the empirical regression function is used for prediction, is given by

$$\sigma_e^2 = \sigma^2\left[1 + \frac{1}{n} + \frac{(x_0 - \bar{x})^2}{nS_x^2}\right].$$

The σ^2 in this expression is the conditional variance of Y, constant in the normal model:

$$\sigma^2 = \operatorname{var}(Y \mid X = x) = \sigma_Y^2(1 - \rho^2),$$

and, of course, σ_e^2 is a conditional mean squared error, given the X-values used in estimating the regression line. One could calculate the unconditional variance, but it can be argued (as R. A. Fisher did) that it is the conditional mean squared error that is relevant in the prediction process.

EXAMPLE 12–7. The following midquarter and final examination scores were achieved by 16 students in a statistics class:

MQ	Final	MQ	Final
60	136	85	116
62	128	81	124
77	157	68	137
95	186	60	142
44	90	61	54
66	92	42	97
62	59	80	129
79	163	80	148

The least squares regression was found to be $y = 17.34 + 1.525x$, so the final exam score predicted for a midquarter score of 80 would be $17.34 + 1.525 \times 80 = 139$. The 90 per cent prediction interval has limits

$$139 \pm t_{.95}(16) \times (27.87) \times \left[1 + \frac{1}{14} + \frac{11.125^2}{3170}\right]^{1/2}$$

or 139 ± 51.5. The 27.87 was calculated as the value of $\hat{\sigma} = S_Y(1 - r^2)^{1/2}$, with $S_Y = 35.19$ and $r^2 = .373$.

12.2.9 A Matrix Formulation

The model for a regression problem with a regression function that is linear in the parameters can be written in matrix notation as follows:

$$\mathbf{Y} = \mathbf{X}\boldsymbol{\beta} + \boldsymbol{\epsilon},$$

where $\boldsymbol{\epsilon}$ is an n-vector of independent errors with mean 0 and constant variance σ^2, $\boldsymbol{\beta}$ is a p-vector of parameters, and $\mathbf{X}$ is an $n \times p$ matrix defining the design of the experiment. In the case $g(x) = \alpha + \beta x$, $E\mathbf{Y}$ is given by

$$\mathbf{X}\boldsymbol{\beta} = \begin{pmatrix} 1 & x_1 \\ 1 & x_2 \\ & \vdots \\ 1 & x_n \end{pmatrix} \begin{pmatrix} \alpha \\ \beta \end{pmatrix}.$$

More generally, with $g(x) = \sum g_j(x)\beta_j$, the matrix $\mathbf{X}$ is defined by $(X)_{ij} = g_j(x_i)$.

The least squares estimates of the parameters can be found by minimizing the sum of squared residuals of the Y's about the regression function, and this can be accomplished by a generalization of the process of completing the square. It is readily verified that

$$\begin{aligned} Q \equiv (\mathbf{Y} - \mathbf{X}\boldsymbol{\beta})'(\mathbf{Y} - \mathbf{X}\boldsymbol{\beta}) &= \boldsymbol{\beta}'\mathbf{S}\boldsymbol{\beta} - 2\boldsymbol{\beta}'\mathbf{X}'\mathbf{Y} + \mathbf{Y}'\mathbf{Y} \\ &= (\boldsymbol{\beta} - \mathbf{S}^{-1}\mathbf{X}'\mathbf{Y})'\mathbf{S}(\boldsymbol{\beta} - \mathbf{S}^{-1}\mathbf{X}'\mathbf{Y}) + \mathbf{Y}'\mathbf{A}\mathbf{Y}, \end{aligned}$$

where $\mathbf{S} \equiv \mathbf{X}'\mathbf{X}$ and $\mathbf{A} \equiv \mathbf{I} - \mathbf{X}\mathbf{S}^{-1}\mathbf{X}'$, provided, of course, $\mathbf{S}$ is of rank p ("full rank") and so has an inverse. This is the usual case in regression problems and will be assumed to hold in what follows. From the last expression for Q above, it is apparent that Q is minimized by

$$\hat{\boldsymbol{\beta}} = \mathbf{S}^{-1}\mathbf{X}'\mathbf{Y}.$$

This least squares estimate of $\boldsymbol{\beta}$ is a linear transformation of $\mathbf{Y}$ and so is p-variate normal with

$$\begin{aligned} E(\hat{\boldsymbol{\beta}}) &= \mathbf{S}^{-1}\mathbf{X}'(\mathbf{X}\boldsymbol{\beta}) = \boldsymbol{\beta}, \\ \mathbf{M}_{\hat{\boldsymbol{\beta}}} &= \mathbf{S}^{-1}\mathbf{X}'(\sigma^2\mathbf{I})\mathbf{X}\mathbf{S}^{-1} = \sigma^2\mathbf{S}^{-1}. \end{aligned}$$

The Gauss–Markov Theorem states that the least squares estimate $\hat{\boldsymbol{\beta}}$ has a property of minimum variance in the class of unbiased, linear estimates, namely, that the variances of its components are simultaneously smallest. This is seen as follows. Consider an arbitrary linear estimate $\boldsymbol{\beta}^* = \mathbf{CY}$, and define $\mathbf{B}$ by

$$\boldsymbol{\beta}^* - \hat{\boldsymbol{\beta}} = (\mathbf{C} - \mathbf{S}^{-1}\mathbf{X}')\mathbf{Y} \equiv \mathbf{BY}.$$

The requirement of unbiasedness,

$$E(\boldsymbol{\beta}^*) = E[(\mathbf{B} + \mathbf{S}^{-1}\mathbf{X}')\mathbf{Y}] = \mathbf{BX}\boldsymbol{\beta} + \mathbf{S}^{-1}\mathbf{X}'\mathbf{X}\boldsymbol{\beta} = \mathbf{BX}\boldsymbol{\beta} + \boldsymbol{\beta} = \boldsymbol{\beta},$$

implies that $\mathbf{BX}$ must be $\mathbf{0}$. The second moment matrix is then

$$\mathbf{M}_{\boldsymbol{\beta}^*} = \sigma^2(\mathbf{B} + \mathbf{S}^{-1}\mathbf{X}')(\mathbf{B} + \mathbf{S}^{-1}\mathbf{X}')' = \sigma^2(\mathbf{BB}' + \mathbf{S}^{-1}).$$

Since $\mathbf{S}^{-1}$ is constant and the diagonal elements of $\mathbf{BB}'$ are

$$(\mathbf{BB}')_{ii} = b_{i1}^2 + \cdots + b_{in}^2,$$

the smallest variances of the β_i are achieved for $\mathbf{B} = \mathbf{0}$ (which satisfies $\mathbf{BX} = \mathbf{0}$); that is, $\boldsymbol{\beta}^* = \hat{\boldsymbol{\beta}}$.

The m.l.e. of σ^2 is given as usual by Q_E/n, where Q_E is the minimum mean squared residual about $\mathbf{X}\hat{\boldsymbol{\beta}}$:

$$Q_E \equiv (\mathbf{Y} - \mathbf{X}\hat{\boldsymbol{\beta}})'(\mathbf{Y} - \mathbf{X}\hat{\boldsymbol{\beta}}) = \mathbf{Y}'\mathbf{AY},$$

with $\mathbf{A}$ defined as before. (It can be verified that $\mathbf{A}$ is symmetric and idempotent: $\mathbf{A}^2 = \mathbf{A}$.) Since the quantities $\mathbf{Y} - \mathbf{X}\hat{\boldsymbol{\beta}}$ satisfy p linear relations: $\mathbf{X}'(\mathbf{Y} - \mathbf{X}\hat{\boldsymbol{\beta}}) = 0$, by virtue of the definition of $\hat{\boldsymbol{\beta}}$, the rank of $\mathbf{A}$ is at most $n - p$.

The sum of squared deviations about the true regression function can be decomposed as follows (adding and subtracting $\mathbf{X}\hat{\boldsymbol{\beta}}$ before expanding):

$$\begin{aligned}(\mathbf{Y} - \mathbf{X}\boldsymbol{\beta})'(\mathbf{Y} - \mathbf{X}\boldsymbol{\beta}) &= (\mathbf{Y} - \mathbf{X}\hat{\boldsymbol{\beta}})'(\mathbf{Y} - \mathbf{X}\hat{\boldsymbol{\beta}}) + (\hat{\boldsymbol{\beta}} - \boldsymbol{\beta})'\mathbf{S}(\hat{\boldsymbol{\beta}} - \boldsymbol{\beta}) \\ &= Q_E + Q_\beta,\end{aligned}$$

where Q_β is a quadratic form of rank p in linear combinations of the Y's. The rank on the left is n, which does not exceed the rank of Q_E plus p, so the rank of Q_E is at least $n - p$. Being at most $n - p$ it must be exactly $n - p$, and so by Cochran's theorem, Q_E/σ^2 and Q_β/σ^2 have independent chi-square distributions with $n - p$ and p degrees of freedom, respectively.

The testing of hypothesis can be developed in this general linear formulation to include as special cases all of the results given earlier. The development involves linear subspaces of the parameter space and the partitioning of matrices; it will not be presented here. (See Graybill [13] or Scheffé [20].)

Problems

12–13. Using the data of Problem 12–9, assuming a linear regression function,

(a) Construct a 90 per cent prediction interval for Y at $x = 6$.

(b) Construct a 70 per cent confidence interval for x_0, the value of x at which data 3, 7, and 5 are obtained.

12–14. Show that the estimator $T = \sum c_i \bar{Y}_i$, with c_i given and var $\bar{Y}_i = \sigma^2/n_i$, has minimum variance when the sample sizes n_i at x_i are chosen to be proportional to $|c_i|$. (Minimize var T subject to $\sum n_i = n$ using Lagrange's multiplier method.)

12–15. Use the calculus method of maximization to obtain equations (as the necessary condition) for a maximum and express them in matrix terms, in the general linear model. Show that when $\mathbf{X}'\mathbf{X}$ is nonsingular the equations are satisfied by the solutions obtained by completing the square.

12–16. Verify the expression for $(\mathbf{Y} - \mathbf{X}\boldsymbol{\beta})'(\mathbf{Y} - \mathbf{X}\boldsymbol{\beta})$, given early in Section 12.2.9, that "completes the square" and leads to the formula for $\hat{\boldsymbol{\beta}}$.

12–17. Verify that $\mathbf{A} \equiv \mathbf{I} - \mathbf{X}\mathbf{S}^{-1}\mathbf{X}'$ is symmetric and idempotent as claimed.

12–18. Verify the decomposition of $(\mathbf{Y} - \mathbf{X}\boldsymbol{\beta})'(\mathbf{Y} - \mathbf{X}\boldsymbol{\beta})$ into Q_E and $Q_{\boldsymbol{\beta}}$ at the end of Section 12.2.9, showing in particular that the cross-product term does indeed disappear.

12.3 Models for Designed Experiments

Like the preceding section on regression models, this section deals with inference about the means of several populations. There, the populations were indexed by one or more real parameters, and the interest lay partly in the functional dependence of the population means on those parameters. Here, the indexing or independent variables are categorical, and the basic question is whether or not a given variable has an effect. This section could have been titled "Classification Models," since it deals with models for populations that are classified and cross-classified according to factors that may be causing population differences. Another name, often used, is "Analysis of Variance Models," a name that is somewhat misleading, inasmuch as the ANOVA technique is also applied to regression models. Indeed, even the term *designed experiment* is used (Section 12.2.6) in regression analysis. As a matter of fact, there is a close relationship between regression models and classification models, for there is a unifying and encompassing general theory.

12.3.1 A Single Classification

Consider k independent populations, classified according to some factor or treatment whose categories or levels define the different populations. The population variable is a "response" or result of carrying out the experiment under the conditions defined by the corresponding treatment or factor category. Some examples are the yield of wheat as determined (along with unidentifiable factors) by seed type, student achievement in a course as affected by the particular instructor, the purity of a product as affected by the type of solvent used in its preparation, and the strength of a spot weld as determined by the machine used.

The models to be studied, here as in the case of the regression problem, make the basic assumption that the response variables are independent, normal, and *homoscedastic*—having constant variance σ^2. The notation will be as follows:

Population	Sample Size	Observations	Sample Means	Population Means	Population Variances
Y_1	n_1	$Y_{11}, \ldots, Y_{1n_1}$	$\bar{Y}_1$	μ_1	σ^2
$\vdots$	$\vdots$		$\vdots$	$\vdots$	$\vdots$
Y_k	n_k	$Y_{k1}, \ldots, Y_{kn_k}$	$\bar{Y}_k$	μ_k	σ^2

The likelihood function, given observations Y_{ij}, is a function of the k means and of σ^2:

$$L(\mu_1, \ldots, \mu_k; \sigma^2) = (2\pi\sigma^2)^{-n/2} \exp\left\{-\frac{1}{2\sigma^2}\sum\sum(Y_{ij} - \mu_i)^2\right\},$$

with maximum attained at the m.l.e.'s:

$$\hat{\mu}_i = \bar{Y}_i, \qquad \hat{\sigma}^2 = \frac{1}{n}\sum\sum(Y_{ij} - \bar{Y}_i)^2.$$

And, just as in the lemma of Section 12.2.4, the maximum value of L is given by

$$\sup L = (2\pi\hat{\sigma}^2)^{-n/2}e^{-n/2}.$$

The usual hypothesis to be tested is that there is *no* treatment or factor effect—that the k populations are identical, a condition equivalent to equality of means:

$$H_0\colon\ \mu_1 = \cdots = \mu_k \equiv \mu.$$

Under this hypothesis, the likelihood depends only on μ and σ^2:

$$L(\mu, \sigma^2) = (2\pi\sigma^2)^{-n/2} \exp\left\{-\frac{1}{2\sigma^2}\sum\sum(Y_{ij} - \mu)^2\right\},$$

with maximum value $(2\pi\hat{\sigma}_0^2)^{-n/2}e^{-n/2}$ achieved at

$$\hat{\mu} = \frac{1}{n}\sum\sum Y_{ij} = \frac{1}{n}\sum n_i\bar{Y}_i,$$

where $n = \sum n_i$, and

$$\hat{\sigma}_0^2 = \frac{1}{n}\sum\sum(Y_{ij} - \bar{Y})^2.$$

The likelihood ratio test for this null hypothesis in the context of the more general model is given by

$$\left(\frac{\hat{\sigma}_0^2}{\hat{\sigma}^2}\right)^{-n/2} < \text{const},$$

or equivalently, by

$$\frac{\hat{\sigma}^2 + (\hat{\sigma}_0^2 - \hat{\sigma}^2)}{\hat{\sigma}^2} = 1 + \frac{\hat{\sigma}_0^2 - \hat{\sigma}^2}{\hat{\sigma}^2} > \text{const.}$$

The difference of variance estimates in the numerator is itself a variance estimate:

$$\hat{\sigma}_0^2 - \hat{\sigma}^2 = \frac{1}{n}\sum n_i(\bar{Y}_i - \bar{Y})^2,$$

an identity resulting from a partitioning of $\hat{\sigma}_0^2$ much as in the regression case:

$$\sum\sum(Y_{ij} - \bar{Y}_i + \bar{Y}_i - \bar{Y})^2 = \sum\sum(Y_{ij} - \bar{Y}_i)^2 + \sum\sum(\bar{Y}_i - \bar{Y})^2,$$

where the cross-product term has again conveniently disappeared, as a result of the relation $\sum(Y_{ij} - \bar{Y}_i) = 0$, for $i = 1, \ldots, k$.

The quantity $\bar{Y}_i - \bar{Y}$ that has emerged is the m.l.e. of the parameter θ_i:

$$\theta_i \equiv \mu_i - \mu, \qquad \text{where } \mu \equiv \frac{1}{n}\sum n_i\mu_i.$$

(This μ reduces to the μ of H_0 when the μ_i are all equal.) This suggests the embedding of H_0 in the more general model by writing the latter as

$$H_0 + H_A\colon \quad Y_{ij} = \mu + \theta_i + \epsilon_{ij}, \qquad \text{where } \sum n_i\theta_i = 0,$$

and where ϵ_{ij} is a normal random "error" with mean 0 and variance σ^2. The hypothesis H_0 is now defined by $\theta_i \equiv 0$. The parameter θ_i is a "treatment effect," indicating a contribution to mean response peculiar to category i of the treatment.

The test statistic for H_0 versus $H_0 + H_A$ will be taken to be a multiple of the variance ratio given above:

$$F \equiv \frac{(\hat{\sigma}_0^2 - \hat{\sigma}^2)/(k - 1)}{\hat{\sigma}^2/(n - k)},$$

the divisors and the "F" terminology stemming from the following application of Cochran's theorem. Consider the decomposition:

$$\begin{aligned}\sum\sum(Y_{ij} - \mu_i)^2 &= \sum\sum[(Y_{ij} - \bar{Y}_i) + (\bar{Y}_i - \bar{Y} - \theta_i) + (\bar{Y} - \mu)]^2 \\ &= \sum\sum(Y_{ij} - \bar{Y}_i)^2 + \sum n_i(\bar{Y}_i - \bar{Y} - \theta_i)^2 + n(\bar{Y} - \mu)^2.\end{aligned}$$

The three cross-product terms have disappeared upon summation because of the relations

$$\sum n_i(\bar{Y}_i - \bar{Y} - \theta_i) = 0, \qquad \sum(Y_{ij} - \bar{Y}_i) = 0.$$

The ranks of the quadratic forms in the decomposition are readily seen to be $n - k$, $k - 1$, and 1, respectively, so that upon division by σ^2, their distributions are independent chi-square distributions with degrees of freedom corresponding to ranks. And then the variable termed F above has an F-distribution with $(k - 1, n - k)$ degrees of freedom when H_0 is true ($\theta_i \equiv 0$). When H_0 is not true, the distribution of F is noncentral F, and power would then be computed with the aid of noncentral F tables.

The test $F >$ const is eminently reasonable, since the numerator—a "between samples" estimate of σ^2 – will tend to be too large when there is an actual difference among the populations. The denominator—the "within samples" estimate —is unaffected by differences in population means, and so it can serve as a kind of standard against which to measure the numerator.

EXAMPLE 12–8. The following data are yields of a given process as determined by each of four analysts. It is desired to test whether the analysts differ significantly in their determinations. (A constant has been subtracted from the original data to simplify the calculations.)

Analyst	Yield	n_i	$\bar{X}_i$	$(X_{ij} - \bar{X}_i)^2$	$n_i(\bar{X}_i - \bar{X})^2$
1	8, 5, −1, 6, 5, 3	6	$\frac{13}{3}$	47.33	2.67
2	7, 12, 5, 3, 10	5	$\frac{37}{5}$	53.2	28.8
3	4, −2, 1	3	1	18	48
4	1, 6, 10, 7	4	6	42	4
Total	90	18	($\bar{X} = 5$)	160.53	83.47

An ANOVA table, giving the decomposition of the sum of squared deviations about $\bar{X}$, might be presented as follows:

Source	Sum of Squares	Degree of Freedom
Analysts	83.47	3
Error	160.53	14
Total	244	17

The F-ratio for testing the presence of an analyst effect is as follows:

$$F = \frac{83.47/3}{160.53/14} = 2.43.$$

The hypothesis of no difference among populations (i.e., among analysts) is accepted at the 5 per cent level, the critical value of F being $F_{.95}(3, 14) = 3.34$.

Among the situations that lend themselves to a single classification analysis are those in which subjects (people, animals, plots of ground, etc.) are given various categories of treatment. Each subject must be assigned a category;

and to minimize the introduction of any systematic effect of other factors that might give misleading results, it is generally recommended that assignments of subjects to categories be made at random. Such an experimental procedure is termed a *completely randomized* design. Fisher asserts [11] that this randomization must be carried out to make valid *any* meaningful analysis; but he also points out that it permits the construction of a permutation test (like that for the two-sample problem given in Example 11–18), which does not require the limiting assumption of normality. Carrying out such tests is rather laborious and not ordinarily done, because, happily, they are fairly well approximated by the normal model tests. (See also Scheffé [20], Chapter 9.)

The F-test for equality of k means is a generalization of the two-sided t-test given in Chapter 7 for the equality of two means (Problem 12–20). When there are more than two populations and when the means are not equal, it may be desired to pick out the more disparate populations or to estimate various relationships among the means. A method due to Scheffé of constructing simultaneous confidence intervals for linear combinations of means will be given next.

According to the application of Cochran's theorem given previously, the ratio

$$\frac{\sum n_i(\bar{Y}_i - \bar{Y} - \theta_i)^2}{(k - 1)S^2}$$

is $F(k - 1, n - k)$ when S^2 is the "pooled variance":

$$S^2 = \frac{1}{n - k}\sum\sum (Y_{ij} - \bar{Y}_i)^2.$$

For a given confidence coefficient η, let f denote the 100η percentile of this F-distribution: $P(F \le f) = \eta$. Then

$$P\left\{\sum n_i(\bar{Y}_i - \bar{Y} - \theta_i)^2 \le (k - 1)fS^2\right\} = \eta.$$

Now for any $\mathbf{a} = (a_1, \ldots, a_k)$, it follows from Schwarz' inequality that

$$\left[\sum a_i(\bar{Y}_i - \bar{Y} - \theta_i)\right]^2 \le \sum \frac{a_i^2}{n_i} \sum n_i(\bar{Y}_i - \bar{Y} - \theta_i)^2,$$

with equality for some $\mathbf{a}$, so that the right-hand side is actually the supremum of the left-hand side. Therefore,

$$P\left\{\sup_{\mathbf{a}} \left|\sum a_i(\bar{Y}_i - \bar{Y} - \theta_i)\right| \le S\sqrt{(k - 1)f\sum (a_i^2/n_i)}\right\} = \eta.$$

If the supremum on the left is taken over a finite set A of vectors $\mathbf{a}$, it is no larger, and the probability of the inequality is then at least η. Moreover, the supremum over A satisfies the inequality if and only if

$$\left|\sum a_i(\bar{Y}_i - \bar{Y} - \theta_i)\right| \le S\sqrt{(k - 1)f\sum (a_i^2/n_i)}$$

for *all* $\mathbf{a}$ in A; and so the probability of this is at least η. And then if one considers in particular, only a finite set A^* of *contrasts*—linear combinations $\sum a_i\mu_i$ such that $\sum a_i = 0$, which implies

$$\sum a_i(\bar{Y}_i - \bar{Y} - \theta_i) = \sum a_i\bar{Y}_i - \sum a_i\mu_i,$$

it may be asserted that the probability of

$$\sum a_i\bar{Y}_i - S\sqrt{b} \le \sum a_i\mu_i \le \sum a_i\bar{Y}_i + S\sqrt{b}$$

for all $\mathbf{a}$ in A^*, where $b = (k-1)f\sum(a_i^2/n_i)$, is at least η. These inequalities define confidence intervals, one for each contrast in A^*, with a probability of at least η that they simultaneously cover the corresponding contrasts.

EXAMPLE 12–9. Return to the situation and data of Example 12–8, suppose that it is desired to obtain simultaneous confidence intervals for $\mu_4 - \mu_3$ and for $\mu_2 - \mu_1$. If $\eta = .75$, the appropriate percentile of $F(3, 14)$ is $f = 1.53$. For the contrast $\mu_4 - \mu_3$, $a_4 = 1$ and $a_3 = -1$, so

$$\frac{a_4^2}{n_4} + \frac{a_3^2}{n_3} = \frac{7}{12},$$

and $b = 3 \times 1.53 \times \frac{7}{12}$. Then $S\sqrt{b} = 5.54$. Similarly, one finds, for the contrast $\mu_2 - \mu_1$, $S\sqrt{b} = 4.39$, so that 5 ± 5.54 and 3.067 ± 4.39 are simultaneous confidence intervals for $\mu_4 - \mu_3$ and $\mu_2 - \mu_1$, respectively, with confidence coefficient .75.

12.3.2 A Random Effects Model

In the model $Y_{ij} = \mu + \theta_i + \epsilon_{ij}$ for a single classification problem, the component θ_i is thought of as a "factor effect" or "treatment effect," and it was assumed to be a constant—a *fixed effect*, for each i. Conclusions drawn from the analysis, say in a problem in which the "factor" is the machine operator, are applicable only to those operators used in the experiment. However, it is sometimes desirable to be able to make a more sweeping inference, namely, an inference about the population of possible operators. This can be done in a *random effects* model:

$$Y_{ij} = \mu + Z_i + \epsilon_{ij},$$

where now the operators are thought of as drawn at random from a population of operators; the operator effect Z_i is a random variable.

Another instance in which such a model might be used is the experimental determination of a property of a product that comes in batches or sheets, with randomness from batch to batch or sheet to sheet (represented by the component Z_i) and also randomness in the experimental determination of the property of interest for a given batch or sheet, caused either by variations from point to point in the batch or sheet or by variations in the measurement process.

In the random effects model, it is assumed that the operator or batch component Z_i is normally distributed with mean 0 and variance ω^2; that the errors ϵ_{ij} are normally distributed with mean 0 and variance σ^2; and that all of the Z's and ϵ's are independent. With these assumptions, the observations are not generally independent. For instance,

$$\text{cov}\,(Y_{11}, Y_{12}) = \text{var}\, Z_1 + \text{cov}\,(\epsilon_{11}, \epsilon_{12}) = \omega^2.$$

On the other hand, if $\omega^2 = 0$, then the Y's *are* independent and $Z_i \equiv 0$ (with probability 1); so there is then no operator effect, and the structure of the model is precisely the same as the null structure in the fixed effect model. The distribution of the statistic used to test the hypothesis of no operator effect in that earlier model, namely,

$$F = \frac{\sum n_i(\bar{Y}_i - \bar{Y})^2/(k-1)}{\sum\sum (Y_{ij} - \bar{Y}_i)^2/(n-k)},$$

has the same distribution, $F(k-1, n-k)$, under the null hypothesis of either model.

The denominator of F, divided by σ^2, is chi-square with $n - k$ degrees of freedom, because

$$Y_{ij} - \bar{Y}_i = \epsilon_{ij} - \bar{\epsilon}_i,$$

from which it follows that the sums $\sum_j (Y_{ij} - \bar{Y}_i)^2/\sigma^2$ are $\chi^2(n_i - 1)$, respectively, and independent. The numerator of F has mean value

$$\sigma^2 + \frac{n\omega^2}{k-1}\left[1 - \sum \left(\frac{n_i}{n}\right)^2\right],$$

which tends to overestimate σ^2 when $\omega^2 \neq 0$. Thus, $F >$ const appears to be a reasonable critical region for testing $\omega^2 = 0$, with size α if the constant used is the $100(1 - \alpha)$ percentile of $F(k - 1, n - k)$. This test is precisely the test used in the fixed effects case, but the interpretation is now in terms of the population of operators.

Problems

12–19. A tensile test measures the quality of a spot-weld of an aluminum clad material. To determine whether there is a "machine effect," five welds from each of three machines are tested, with these results:

Machine A: 3.2, 4.1, 3.5, 3.0, 3.1.
Machine B: 4.9, 4.5, 4.5, 4.0, 4.2.
Machine C: 3.0, 2.9, 3.7, 3.5, 4.2.

(a) Test the hypothesis of no effect at the 5 per cent level.
(b) Construct a 90 per cent confidence interval for $\mu_B - \mu_C$, using data from all three machines in estimating the common variance.
(c) Construct simultaneous confidence intervals for $\mu_B - \mu_A$, $\mu_B - \mu_C$, and $\mu_A - \mu_B$, using $\eta = .95$.

12–20. Verify that the F-test for equality of means is equivalent to the two-sample t-test when $k = 2$.

12–21. Verify the ranks of the terms in the decomposition of $\sum \sum (Y_{ij} - \mu_i)^2$ under fixed effect model.

12–22. Verify the formula given for the mean value of the between-samples estimate of variance (i.e., of the numerator of the test ratio) in the random effects model.

12–23. Construct a likelihood ratio test for the equality of variances of several normal populations.

12.3.3 Two-Way Classifications

In studying the effect of a treatment applied in several categories, one usually finds that there are other factors that might be affecting the response. When these are not taken into account, as in the randomized design of Section 12.3.1, they enter the observations as contributions to random error, possibly in a way that invalidates the assumptions needed for the normal model tests. If the experiment is properly designed to take these factors into account, one finds that not only is the random error component reduced but he also now has the opportunity to study the effects of these factors.

The pairing of subjects in making comparisons (Section 7.3.3) was an example of such design, aimed at reducing variability. In each "matched pair" those factors thought to have an effect on the response but not under study would be matched as well as possible for the two subjects in the pair, and the two categories of the factor of main interest (e.g., "treatment" and "control") would be assigned randomly to the members of a pair.

More generally, when there are more than two treatment categories, the subjects may be grouped into matched *blocks* and randomly assigned treatment categories within each block. This is termed a *randomized block design.* When the matching or blocking is done according to the categories of a certain variable, that variable is called the *blocking variable*. This variable may also be of interest; and when it is, there is no particular reason to call one variable the "treatment" variable and the other the "blocking" variable. Both are factors in the response, and the purpose of the experiment is to sort out their effects.

EXAMPLE 12–10. Suppose that it is desired to study the differences among seed brands as they affect the yield of grain, and that for each brand the yield is to be measured from each of three plots of ground. Realizing that soil composition varies, one may divide an available piece of land into three blocks, each thought to be more homogeneous than the whole, and plant the four seed types in four subplots of each block, with the seed type assigned at random within the block. Results (say, in bu/acre) might look like this:

		Seed Type			
		A	*B*	*C*	*D*
	1	22	25	20	17
Block	2	29	26	23	22
	3	36	24	26	18

The factor of interest is seed type, although a look at the other factor might verify that blocking was indeed warranted.

Some experiments do not involve "subjects" to which treatments are applied but yet result in responses that are affected by one or more factors. In the case of two factors, when one observation Y_{ij} is obtained for each combination of a category i of one factor and a category j of the other factor, the data will appear to be of the same nature as the data in a randomized block design. The same normal model will be used for both situations.

EXAMPLE 12–11. A machine for testing wear has four weighted brushes, under which fabric samples are fixed. Resistance to abrasion is measured according to the loss of weight of the material after a given number of cycles. Tests on four fabrics yielded the following data:

		Fabric			
		A	*B*	*C*	*D*
	1	1.93	2.55	2.40	2.33
Brush	2	2.38	2.72	2.68	2.40
Position	3	2.20	2.75	2.31	2.38
	4	2.25	2.70	2.28	2.25

Here it might be desired to test the hypothesis that there is no difference among fabrics with regard to abrasion resistance. Brush position is a factor that is taken into account, in case it affects the response, but is likely not of primary interest.

Sometimes the term "factor" is reserved for a variable that has *levels*, or ordered categories—for example, a dosage that is low, medium, or high. Here it will simply denote an independent variable that may be affecting a response. This section will present the analysis of a two-factor experiment in which just one observation is obtained for each *cell*—each combination of a category of one factor with a category of the other. (An experiment with the same number of observations for each combination of factor categories is called a *factorial experiment*.)

As in the case of a one-way classification, such problems can be treated using a permutation test, or in the setting of a normal random effects model (when the categories of a factor can be thought of as randomly selected from a population). However, only the normal, fixed effects model will be considered.

The basic assumption will be, then, that the response Y_{ij} includes a random error:

$$Y_{ij} = \mu_{ij} + \epsilon_{ij},$$

and that these errors ϵ_{ij} are independently and normally distributed with constant variance σ^2. The mean μ_{ij} can always be written (this was observed also in Chapter 9) in the form

$$\mu_{ij} = \mu + \theta_i + \phi_j + \xi_{ij},$$

where

$$\sum \theta_i = \sum \phi_j = \sum_i \xi_{ij} = \sum_j \xi_{ij} = 0,$$

simply by defining

$$\mu = \frac{1}{n}\sum\sum \mu_{ij}, \qquad \mu_{i\cdot} = \frac{1}{c}\sum_j \mu_{ij}, \qquad \mu_{\cdot j} = \frac{1}{r}\sum_i \mu_{ij},$$

and

$$\theta_i = \mu_{i\cdot} - \mu, \qquad \phi_j = \mu_{\cdot j} - \mu, \qquad \xi_{ij} = \mu_{ij} - \mu_{i\cdot} - \mu_{\cdot j} + \mu.$$

This representation of μ_{ij}, in terms of which

$$Y_{ij} = \mu + \theta_i + \phi_j + \xi_{ij} + \epsilon_{ij},$$

isolates the effect θ_i of factor A ($i = 1, \ldots, r$), the effect ϕ_j of factor B ($j = 1, \ldots, c$), and the interaction (ξ_{ij}) that may be present in a response owing to a particular ingredient that is present when category i of A is combined with category j of B.

The above model is, of course, overly specific for the amount of data at hand, for one need simply choose $\mu_{ij} = Y_{ij}$ to have a perfect fit—and yet this model is almost certainly not correct. With one observation per cell one might hope to find an additive model—one in which the factor effects combine additively:

$$\mu_{ij} = \mu + \theta_i + \phi_j,$$

with no interaction term. Under this hypothesis the likelihood function is

$$(2\pi\sigma^2)^{-n/2} \exp\left\{-\frac{1}{2\sigma^2}\sum\sum (Y_{ij} - \mu - \theta_i - \phi_j)^2\right\},$$

with maximum value $(2\pi\hat{\sigma}^2)^{-n/2}e^{-n/2}$ when the parameters have the values of their m.l.e.'s:

$$\hat{\mu} = \bar{Y}, \qquad \hat{\theta}_i = \bar{Y}_{i\cdot} - \bar{Y}, \qquad \hat{\phi}_j = \bar{Y}_{\cdot j} - \bar{Y},$$

where $\bar{Y}_{i\cdot}$, $\bar{Y}_{\cdot j}$, and $\bar{Y}$ denote row means, column means, and overall mean, respectively, and

$$\begin{aligned}\hat{\sigma}^2 &= \frac{1}{n}\sum\sum(Y_{ij} - \bar{Y}_{i\cdot} - \bar{Y}_{\cdot j} + \bar{Y})^2 \\ &= \frac{1}{n}\sum\sum(Y_{ij} - \hat{\theta}_i - \hat{\phi}_j - \hat{\mu})^2.\end{aligned}$$

Along the lines of the Theorem in Section 12.2.2, giving a similar result for a regression problem, it is readily seen that $n\hat{\sigma}^2/\sigma^2$ is $\chi^2[(r-1)(c-1)]$, no matter what values the components θ_i and ϕ_j may have.

In the setting of this additive model, one may want to test the hypothesis that one factor or the other has no effect, say:

H_0: $\theta_i \equiv 0, \quad i = 1, \ldots, r.$

Under this hypothesis the m.l.e.'s of the remaining parameters are

$$\hat{\mu} = \bar{Y}, \qquad \hat{\phi}_j = \bar{Y}_{\cdot j} - \bar{Y},$$

and

$$\begin{aligned}n\hat{\sigma}_0^2 &= \sum\sum(Y_{ij} - \bar{Y}_{\cdot j})^2 \\ &= n\hat{\sigma}^2 + \sum_i c(\bar{Y}_{i\cdot} - \bar{Y})^2,\end{aligned}$$

and (still under H_0) $n\hat{\sigma}_0^2/\sigma^2$ is $\chi^2(r-1)$. The likelihood ratio test for H_0 against the alternative that H_0 is not true in the additive model is $(\hat{\sigma}_0^2/\sigma^2) >$ const, equivalent to

$$F_A \equiv \frac{\sum c(\bar{Y}_{i\cdot} - \bar{Y})^2/(r-1)}{\sum\sum(Y_{ij} - \bar{Y}_{i\cdot} - \bar{Y}_{\cdot j} + \bar{Y})^2/(r-1)(c-1)} > \text{const.}$$

Under H_0, the statistic F_A is distributed as $F[(r-1), (r-1)(c-1)]$, from which distribution the critical value is determined according to a given level.

EXAMPLE 12–12. Consider two factors with three levels or categories each, and assume that one observation per cell is obtained, as follows:

	Factor B (1)	(2)	(3)	$\bar{Y}_{i\cdot}$	$\bar{Y}_{i\cdot} - \bar{Y}$
Factor A (1)	3	5	4	4	−7
(2)	11	10	12	11	0
(3)	16	21	17	18	7
$\bar{Y}_{\cdot j}$	10	12	11	$11 = \bar{Y}$	
$\bar{Y}_{\cdot j} - \bar{Y}$	−1	1	0		

The sum of squares about the overall mean is 312, decomposed as follows:

Source	Sum of Squares	Degrees of Freedom
Factor A	294	2
Factor B	6	2
Error	12	4
Total	312	$8 = rc - 1$

(The error sum of squares, 12, is the sum of squares in the denominator of the test ratio for A and also of the test ratio for B.) The test ratios are as follows:

$$F_A = \frac{294/2}{12/4} = 49, \qquad F_B = \frac{6/2}{12/4} = 1.$$

Clearly, B would be judged as having no effect, but the factor A effect is significant [compare 49 with the 95th percentile of $F(2, 4)$].

When there are more than rc observations—more than one observation per cell, one can test hypotheses in the more general model with the interaction term, essentially because there is then available an estimate of error variance from the variation within the cells. One can test the presence of a factor effect, even if there is interaction, or can test the presence of interaction. The data will be of the form Y_{ijk}, representing the kth observation ($k = 1, \ldots, p$) in the ij-cell. Denoting by $\bar{Y}_{ij\cdot}$ the mean of the observations in the ij-cell one finds maximum likelihood estimates of factor effect terms to be obtained from these just as they were obtained from Y_{ij} in the simpler case:

$$\hat{\mu} = \bar{Y}, \qquad \hat{\theta}_i = \bar{Y}_{i\cdot\cdot} - \bar{Y}, \qquad \hat{\phi}_j = \bar{Y}_{\cdot j\cdot} - \bar{Y},$$

and

$$\hat{\xi}_{ij} = \bar{Y}_{ij\cdot} - \hat{\theta}_i - \hat{\phi}_j - \hat{\mu} = \bar{Y}_{ij\cdot} - \bar{Y}_{i\cdot\cdot} - \bar{Y}_{\cdot j\cdot} + \bar{Y}.$$

The m.l.e. of σ^2 is a pooled estimate of variance based on variation within the cells:

$$\hat{\sigma}^2 = \frac{1}{rcp} \sum \sum \sum (Y_{ijk} - \bar{Y}_{ij\cdot})^2.$$

The ANOVA table in this situation is as follows:

Source	Sum of Squares	Degree of Freedom
Factor A	$\sum cp(\bar{Y}_{i\cdot\cdot} - \bar{Y})^2$	$r - 1$
Factor B	$\sum rp(\bar{Y}_{\cdot j\cdot} - \bar{Y})^2$	$c - 1$
Interaction	$\sum\sum p(\bar{Y}_{ij\cdot} - \bar{Y}_{i\cdot\cdot} - \bar{Y}_{\cdot j\cdot} + \bar{Y})^2$	$(r - 1)(c - 1)$
Error	$\sum\sum\sum(Y_{ijk} - \bar{Y}_{ij\cdot})^2$	$rc(p - 1)$
Total	$\sum\sum\sum(Y_{ijk} - \bar{Y})^2$	$rcp - 1$

(See Problem 12–29.) In case it can be assumed that there is no interaction, the "interaction" and "error" sums of squares can be combined to yield an error sum of squares with $rcp - r - c + 1$ degrees of freedom to serve as the denominator in a test for factor A or for factor B. Otherwise, a test for factor A, say, is based on the F-ratio with the factor A sum of squares divided by its degrees of freedom in the numerator (an unbiased estimate of σ^2 if $\theta_i = 0$) and the error sum of squares divided by its degrees of freedom (an unbiased estimate of σ^2, no matter what) in the denominator.

12.3.4 Other Designs

More than two factors can be considered. Indeed, additional factors must be considered if there is any chance that they are confounding the results. For the case of three factors, at r, c, and p levels, with q observations per treatment, a linear model would be of the form

$$E(Y_{ijk}) = \mu + \alpha_i + \beta_j + \gamma_k + \zeta_{ij} + \xi_{jk} + \eta_{ki} + \delta_{ijk},$$

with the usual normalizing conditions that sums of these quantities (except μ) with respect to any one of the pertinent subscripts are zero. The α_i, β_j, and γ_k are factor effects, the ζ_{ij}, η_{ki}, and ξ_{jk} are second-order interactions, and δ_{ijk} is a third-order interaction, a contribution to the cell mean that is attributed to the peculiarity of the combination of levels i, j, and k of factors A, B, and C, respectively. With several observations per cell, the response variable would be Y_{ijkm}, with $m = 1, \ldots, q$. To carry out a test for the various factors and interactions, it would be necessary to have $rcpq$ observations. These could be more observations than are easily obtainable, and it is both interesting and useful that it is possible to test for one of the three factors in the presence of the other two with even fewer than the rcp observations that would be needed for one observation per cell. (With fewer observations, the sensitivity of the test is reduced, but the point is that there *is* a test.)

Suppose that the three factors are machines, operators, and time periods, and that there are four levels of each. Instead of obtaining a performance figure for each of the 64 combinations of machine, operator, and time period, one obtains the 16 observations as indicated in the accompanying table, each operator working once in each time period and once on each machine. The letter in the array indicates the machine to be used. An array (Latin) of the letters A, B, C, D such as is shown, with the property that no letter appears more than once in any column or in any row, is called a *Latin square*; these have been extensively studied, computed, and tabulated. (See Kempthorne [14].)

	Time Period			
	1	2	3	4
Operator 1	A	B	C	D
Operator 2	B	C	D	A
Operator 3	C	D	A	B
Operator 4	D	A	B	C

(Clearly, other Latin squares are obtained simply by permuting the letters A, B, C, and D. However, there is another set of squares based on:

$$\begin{matrix} A & B & C & D \\ B & D & A & C \\ C & A & D & B \\ D & C & B & A \end{matrix}$$

From all of these, one would select one at random for the experiment at hand.)

The model that can be tested with such data is of the form

$$Y_{ij(k)} = \mu + \alpha_i + \beta_j + \gamma_k + \epsilon_{ij(k)},$$

where $i = 1, \ldots, s$, $j = 1, \ldots, s$, and k is determined by i, j, and the particular Latin square used. Thus there are just s^2 observations.

The ANOVA table is as follows:

Source	Sum of Squares	Degrees of Freedom
Rows	$s \sum (\bar{Y}_{i\cdot} - \bar{Y})^2$	$s - 1$
Columns	$s \sum (\bar{Y}_{\cdot j} - \bar{Y})^2$	$s - 1$
"Letters"	$s \sum (\bar{Y}_{(k)} - \bar{Y})^2$	$s - 1$
Error	$\sum \sum (Y_{ij(k)} - \bar{Y}_{i\cdot} - \bar{Y}_{\cdot j} - \bar{Y}_{(k)} + 2\bar{Y})^2$	$(s - 2)(s - 1)$
Total	$\sum \sum (Y_{ij(k)} - \bar{Y})^2$	$s^2 - 1$

The numbers in the rightmost column denote the number of degrees of freedom of the corresponding chi-square distribution (when the term in question is divided by σ^2). The mean $\bar{Y}_{(k)}$ denotes the average of all entries in the table that are taken at the kth level of the third factor (letters).

Four factors can be considered simultaneously using a "Latin-Greco square," in which the levels of factor 3 are entered as Latin letters and the levels of factor 4 are entered as Greek letters. Each level of each factor appears once with each level of each other factor. A 3 × 3 square of this type (corresponding to three levels of each of four factors) follows.

		Factor 2	
	(1)	(2)	(3)
Factor 1 (1)	$A\alpha$	$B\beta$	$C\gamma$
Factor 1 (2)	$B\gamma$	$C\alpha$	$A\beta$
Factor 1 (3)	$C\beta$	$A\gamma$	$B\alpha$

The study of designs of experiments to handle many factors while permitting the testing of one of them has been very extensive but will not be developed further here. These methods are characteristically different from methods in which a factor is studied by keeping all other factors at fixed levels. In the statistical approach, these other factors are allowed to vary, and the results can be exploited to make inferences concerning these other factors as well as the one of interest.

12.3.5 The Matrix Formulation

The normal models used in this section to analyze data with regard to factor effects are again special cases of the general linear model introduced in Section 12.2.9 for regression problems. Observations are strung out in a vector, assumed to have a structure linear in p parameters:

$$\mathbf{Y} = \mathbf{X}\boldsymbol{\beta} + \boldsymbol{\epsilon},$$

where $\boldsymbol{\beta}$ is a $p \times 1$ vector of parameters, $\mathbf{X}$ is an $n \times p$ matrix defining the design of the experiment, and $\boldsymbol{\epsilon}$ is an $n \times 1$ vector of errors assumed, in the normal model, to have a multivariate normal distribution with mean vector $\mathbf{0}$ and covariance matrix $\sigma^2\mathbf{I}$.

To illustrate: Consider the single classification problem with observations Y_{ij} and parameters $\mu, \theta_1, \ldots, \theta_k$ put into vectors as follows:

$$\mathbf{Y}' = (Y_{11}, \ldots, Y_{1n_1}, \ldots, Y_{k1}, \ldots, Y_{kn_k}).$$
$$\boldsymbol{\beta} = (\mu, \theta_1, \ldots, \theta_k).$$

Let $\mathbf{X}$ be defined as the following $n \times p$ matrix:

$$\mathbf{X} = \begin{bmatrix} 1 & 1 & 0 & \cdots & 0 \\ 1 & 1 & 0 & \cdots & 0 \\ \vdots & & & & \\ 1 & 1 & 0 & \cdots & 0 \\ 1 & 0 & 1 & \cdots & 0 \\ \vdots & & & & \\ 1 & 0 & 1 & \cdots & 0 \\ \vdots & & & & \\ 1 & 0 & & \cdots & 1 \\ \vdots & & & & \\ 1 & 0 & & \cdots & 1 \end{bmatrix}.$$

It is readily verified that $\mathbf{Y} = \mathbf{X}\boldsymbol{\beta} + \boldsymbol{\epsilon}$ defines the previously given single classification model.

It is typical of classification or design models that the design matrix $\mathbf{X}$ in their matrix formulation contains 0's and 1's as it does above.

A least squares estimate of $\boldsymbol{\beta}$ would be a vector $\hat{\boldsymbol{\beta}}$ that minimizes the squared length of the error vector:

$$\boldsymbol{\epsilon}'\boldsymbol{\epsilon} = (\mathbf{Y} - \mathbf{X}\mathbf{b})'(\mathbf{Y} - \mathbf{X}\mathbf{b}) = \sum \left\{ Y_i - \sum x_{ij} b_j \right\}^2$$

when $\mathbf{b} = \hat{\boldsymbol{\beta}}$. Necessary conditions for this are that the partial derivatives with respect to the components of $\mathbf{b}$ vanish:

$$\left.\frac{\partial \boldsymbol{\epsilon}'\boldsymbol{\epsilon}}{\partial b_\nu}\right|_{\mathbf{b}=\hat{\boldsymbol{\beta}}} -2 \sum \left[Y_i - \sum x_{ij} \hat{\beta}_j \right] x_i = 0$$

for $\nu = 1, \ldots, p$. This condition can be written in matrix form:

$$\mathbf{X}'\mathbf{X}\boldsymbol{\beta} = \mathbf{X}'\mathbf{Y},$$

called the *normal equation* of this least squares process. If $\mathbf{X}'\mathbf{X}$ were of full rank (i.e., of rank p) as in the regression case, there would be a unique solution $(\mathbf{X}'\mathbf{X})^{-1}\mathbf{X}'\mathbf{Y}$. In classification problems, however, the typical situation is that $\mathbf{X}'\mathbf{X}$ is of less than full rank and that there are infinitely many solutions.

In the case of the one-way classification, with $\mathbf{X}$ as given above, the matrix $\mathbf{X'X}$ is easily computed:

$$\mathbf{X'X} = \begin{pmatrix} n & n_1 & n_2 & \cdots & n_k \\ n_1 & n_1 & 0 & & 0 \\ n_2 & 0 & n_2 & & 0 \\ \vdots & & & & \\ n_k & 0 & & \cdots & n_k \end{pmatrix}.$$

This is of rank $k = p - 1$, since rows 2 through p add up to the first row, whereas there is a $k \times k$ nonsingular submatrix. The normal equations are then:

$$\begin{cases} n\mu + \sum n_i\theta_i = \sum\sum Y_{ij} = n\bar{Y} \\ n_1\mu + n_1\theta_1 = \sum Y_{ij} = n_1\bar{Y}_1 \\ \quad\vdots \\ n_k\mu + n_k\theta_k = \sum Y_{kj} = n_k\bar{Y}_k. \end{cases}$$

Adding to these the condition $\sum n_i\theta_i = 0$ used in defining the θ's, one obtains the solution:

$$\hat{\mu} = \bar{Y}, \qquad \hat{\theta}_1 = \bar{Y}_1 - \bar{Y}, \ldots, \qquad \hat{\theta}_k = \bar{Y}_k - \bar{Y},$$

obtained earlier by maximizing the likelihood under the assumption of normality.

Again, further development of this approach, including the testing of hypotheses, will be left to treatises on linear models (Scheffé [20] and Graybill [13]).

Problems

12–24. Test the hypothesis of no difference in seed brands given the data of Example 12–10.

12–25. Test the hypothesis of no difference among fabric types, given the data of Example 12–11.

12–26. Show that the paired comparison t-test of Section 7.3.2 is a special case of the F-test for a treatment effect in a randomized block design, in which the "block" is a matched pair.

12–27. Obtain the m.l.e.'s for the parameters in the two-way classification model with one observation per cell. Use Cochran's theorem to show that $n\hat{\sigma}^2/\sigma^2$ has the distribution claimed, and that the numerator and denominator of F_A (used in testing for factor A) are independent.

12–28. Using the three observations per cell given, test for no interaction between factors. Test also the hypotheses that the factors have no effect.

		Machine			
		A	*B*	*C*	*D*
Operator	1	59	61	48	47
		43	49	47	40
		63	52	58	51
	2	60	49	40	38
		51	52	48	50
		48	55	56	41
	3	63	58	46	48
		60	48	53	50
		57	56	51	50

12–29. Verify the ANOVA table for the two-way classification model with p observations per cell. (Use Cochran's theorem to decompose $\sum\sum\sum(Y_{ijk} - \mu_{ij})^2$ and note in particular under what hypotheses the various sums of squares are chi-square when divided by σ^2.)

REFERENCES

1. Anderson, T. W. *An Introduction to Multivariate Statistical Analysis.* New York: John Wiley & Sons, Inc., 1958.
2. Bishop, Y. A., S. E. Fienberg, and P. W. Holland. *Discrete Multivariate Analysis: Theory and Practice.* Cambridge, Mass: M.I.T. Press, 1975.
3. Box, G. E. P., and G. C. Tiao. *Bayesian Inference in Statistical Analysis.* Reading, Mass.: Addison-Wesley Publishing Co., Inc., 1973.
4. Cox, D. R., and D. V. Hinkley. *Theoretical Statistics.* London: Chapman & Hall, Ltd., 1974.
5. Cramér, H. *Mathematical Methods of Statistics.* Princeton, N.J.: Princeton University Press, 1946.
6. DeGroot, M. H. *Optimal Statistical Decisions.* New York: McGraw-Hill Book Company, 1970.
7. Draper, N. R., and H. Smith. *Applied Regression Analysis.* New York, John Wiley & Sons, Inc., 1966.
8. Feller, W. *An Introduction to Probability Theory and Its Applications*, Vol. 1, 3rd ed. New York: John Wiley & Sons, Inc., 1968.
9. Ferguson, T. S. *Mathematical Statistics.* New York: Academic Press, Inc., 1967.
10. Fisher, R. A. *Statistical Methods and Scientific Inference*, 3rd ed. New York: Hafner Press, 1973.
11. Fisher, R. A. *The Design of Experiments.* Edinburgh, Scotland: Oliver and Boyd, Ltd., 1951.
12. Gibbons, J. D. *Nonparametric Statistical Inference.* New York: McGraw-Hill Book Company, 1971.
13. Graybill, F. A. *An Introduction to Linear Statistical Models*, Vol. 1. New York: McGraw-Hill Book Company, 1961.
14. Kempthorne, O. *The Design and Analysis of Experiments.* New York: John Wiley & Sons, Inc., 1952.
15. Kendall, M. G., and A. Stuart. *The Advanced Theory of Statistics*, Vol. 1, 3rd ed., 1969; Vol. II, 3rd ed., 1973. New York: Hafner Press.
16. Lehmann, E. *Testing Statistical Hypotheses.* New York: John Wiley & Sons, Inc., 1952.
17. Lindley, D. V. *Introduction to Probability and Statistics.* Cambridge, England: Cambridge University Press, 1965.
18. Loève, M. *Probability Theory*, 3rd ed. New York: Van Nostrand Reinhold Company, 1963.
19. Rao, C. R. *Linear Statistical Inference and Its Applications*, 2nd ed. New York: John Wiley & Sons, Inc., 1973,

20. Scheffé, H. *The Analysis of Variance*. New York: John Wiley & Sons, Inc., 1959.
21. Wald, A. *Sequential Analysis*. New York: John Wiley & Sons, Inc., 1950.
22. Wilks, S. S. *Mathematical Statistics*. New York: John Wiley & Sons, Inc., 1962.
23. Zacks, S. *The Theory of Statistical Inference*. New York: John Wiley & Sons, Inc., 1973.

APPENDIX TABLES

TABLE I Values of the Standard Normal Distribution Function

$$\Phi(z) = \int_{-\infty}^{z} \frac{1}{\sqrt{2\pi}} \exp\left(-\tfrac{1}{2}u^2\right) du = P(Z \leq z)$$

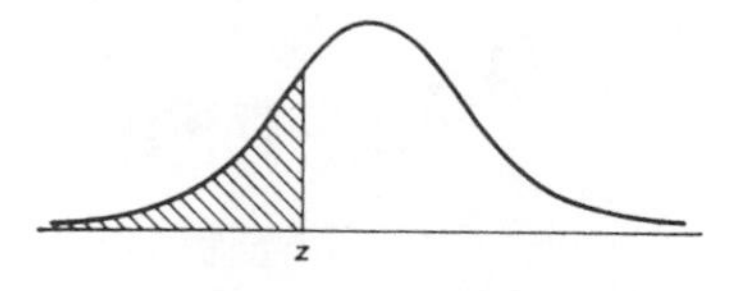

z	0	1	2	3	4	5	6	7	8	9
−3.	.0013	.0010	.0007	.0005	.0003	.0002	.0002	.0001	.0001	.0000
−2.9	.0019	.0018	.0017	.0017	.0016	.0016	.0015	.0015	.0014	.0014
−2.8	.0026	.0025	.0024	.0023	.0023	.0022	.0021	.0021	.0020	.0019
−2.7	.0035	.0034	.0033	.0032	.0031	.0030	.0029	.0028	.0027	.0026
−2.6	.0047	.0045	.0044	.0043	.0041	.0040	.0039	.0038	.0037	.0036
−2.5	.0062	.0060	.0059	.0057	.0055	.0054	.0052	.0051	.0049	.0048
−2.4	.0082	.0080	.0078	.0075	.0073	.0071	.0069	.0068	.0066	.0064
−2.3	.0107	.0104	.0102	.0099	.0096	.0094	.0091	.0089	.0087	.0084
−2.2	.0139	.0136	.0132	.0129	.0126	.0122	.0119	.0116	.0113	.0110
−2.1	.0179	.0174	.0170	.0166	.0162	.0158	.0154	.0150	.0146	.0143
−2.0	.0228	.0222	.0217	.0212	.0207	.0202	.0197	.0192	.0188	.0183
−1.9	.0287	.0281	.0274	.0268	.0262	.0256	.0250	.0244	.0238	.0233
−1.8	.0359	.0352	.0344	.0336	.0329	.0322	.0314	.0307	.0300	.0294
−1.7	.0446	.0436	.0427	.0418	.0409	.0401	.0392	.0384	.0375	.0367
−1.6	.0548	.0537	.0526	.0516	.0505	.0495	.0485	.0475	.0465	.0455
−1.5	.0668	.0655	.0643	.0630	.0618	.0606	.0594	.0582	.0570	.0559
−1.4	.0808	.0793	.0778	.0764	.0749	.0735	.0722	.0708	.0694	.0681
−1.3	.0968	.0951	.0934	.0918	.0901	.0885	.0869	.0853	.0838	.0823
−1.2	.1151	.1131	.1112	.1093	.1075	.1056	.1038	.1020	.1003	.0985
−1.1	.1357	.1335	.1314	.1292	.1271	.1251	.1230	.1210	.1190	.1170
−1.0	.1587	.1562	.1539	.1515	.1492	.1469	.1446	.1423	.1401	.1379
− .9	.1841	.1814	.1788	.1762	.1736	.1711	.1685	.1660	.1635	.1611
− .8	.2119	.2090	.2061	.2033	.2005	.1977	.1949	.1922	.1894	.1867
− .7	.2420	.2389	.2358	.2327	.2297	.2266	.2236	.2206	.2177	.2148
− .6	.2743	.2709	.2676	.2643	.2611	.2578	.2546	.2514	.2483	.2451
− .5	.3085	.3050	.3015	.2981	.2946	.2912	.2877	.2843	.2810	.2776
− .4	.3446	.3409	.3372	.3336	.3300	.3264	.3228	.3192	.3156	.3121
− .3	.3821	.3783	.3745	.3707	.3669	.3632	.3594	.3557	.3520	.3483
− .2	.4207	.4168	.4129	.4090	.4052	.4013	.3974	.3936	.3897	.3859
− .1	.4602	.4562	.4522	.4483	.4443	.4404	.4364	.4325	.4286	.4247
− 0	.5000	.4960	.4920	.4880	.4840	.4801	.4761	.4721	.4681	.4641

TABLE 1 Values of the Standard Normal Distribution Function (Continued)

z	0	1	2	3	4	5	6	7	8	9
.0	.5000	.5040	.5080	.5120	.5160	.5199	.5239	.5279	.5319	.5359
.1	.5398	.5438	.5478	.5517	.5557	.5596	.5636	.5675	.5714	.5753
.2	.5793	.5832	.5871	.5910	.5948	.5987	.6026	.6064	.6103	.6141
.3	.6179	.6217	.6255	.6293	.6331	.6368	.6406	.6443	.6480	.6517
.4	.6554	.6591	.6628	.6664	.6700	.6736	.6772	.6808	.6844	.6879
.5	.6915	.6950	.6985	.7019	.7054	.7088	.7123	.7157	.7190	.7224
.6	.7257	.7291	.7324	.7357	.7389	.7422	.7454	.7486	.7517	.7549
.7	.7580	.7611	.7642	.7673	.7703	.7734	.7764	.7794	.7823	.7852
.8	.7881	.7910	.7939	.7967	.7995	.8023	.8051	.8078	.8106	.8133
.9	.8159	.8186	.8212	.8238	.8264	.8289	.8315	.8340	.8365	.8389
1.0	.8413	.8438	.8461	.8485	.8508	.8531	.8554	.8577	.8599	.8621
1.1	.8643	.8665	.8686	.8708	.8729	.8749	.8770	.8790	.8810	.8830
1.2	.8849	.8869	.8888	.8907	.8925	.8944	.8962	.8980	.8997	.9015
1.3	.9032	.9049	.9066	.9082	.9099	.9115	.9131	.9147	.9162	.9177
1.4	.9192	.9207	.9222	.9236	.9251	.9265	.9278	.9292	.9306	.9319
1.5	.9332	.9345	.9357	.9370	.9382	.9394	.9406	.9418	.9430	.9441
1.6	.9452	.9463	.9474	.9484	.9495	.9505	.9515	.9525	.9535	.9545
1.7	.9554	.9564	.9573	.9582	.9591	.9599	.9608	.9616	.9625	.9633
1.8	.9641	.9648	.9656	.9664	.9671	.9678	.9686	.9693	.9700	.9706
1.9	.9713	.9719	.9726	.9732	.9738	.9744	.9750	.9756	.9762	.9767
2.0	.9772	.9778	.9783	.9788	.9793	.9798	.9803	.9808	.9812	.9817
2.1	.9821	.9826	.9830	.9834	.9838	.9842	.9846	.9850	.9854	.9857
2.2	.9861	.9864	.9868	.9871	.9874	.9878	.9881	.9884	.9887	.9890
2.3	.9893	.9896	.9898	.9901	.9904	.9906	.9909	.9911	.9913	.9916
2.4	.9918	.9920	.9922	.9925	.9927	.9929	.9931	.9932	.9934	.9936
2.5	.9938	.9940	.9941	.9943	.9945	.9946	.9948	.9949	.9951	.9952
2.6	.9953	.9955	.9956	.9957	.9959	.9960	.9961	.9962	.9963	.9964
2.7	.9965	.9966	.9967	.9968	.9969	.9970	.9971	.9972	.9973	.9974
2.8	.9974	.9975	.9976	.9977	.9977	.9978	.9979	.9979	.9980	.9981
2.9	.9981	.9982	.9982	.9983	.9984	.9984	.9985	.9985	.9986	.9986
3.	.9987	.9990	.9993	.9995	.9997	.9998	.9998	.9999	.9999	1.0000

Note 1: If a normal variable X is not "standard," its values must be "standardized": $Z = (X - \mu)/\sigma$, i.e., $P(X \leq x) = \Phi[(x - \mu)/\sigma]$.

Note 2: For "two-tail" probabilities, see Table Ib.

Note 3: For $z \geq 4$, $\Phi(z) = 1$ to four decimal places; for $z \leq -4$, $\Phi(z) = 0$ to four decimal places.

Note 4: Entries opposite 3 and -3 are for 3.0, 3.1, 3.2, etc., and -3.0, -3.1, etc., respectively.

TABLE Ia Percentiles of the Standard Normal Distribution

$P(Z \leq z)$	z
.001	−3.0902
.005	−2.5758
.01	−2.3263
.02	−2.0537
.03	−1.8808
.04	−1.7507
.05	−1.6449
.10	−1.2816
.15	−1.0364
.20	−.8416
.30	−.5244
.40	−.2533
.50	0
.60	.2533
.70	.5244
.80	.8416
.85	1.0364
.90	1.2816
.95	1.6449
.96	1.7507
.97	1.8808
.98	2.0537
.99	2.3263
.995	2.5758
.999	3.0902

TABLE Ib Two-Tail Probabilities for the Standard Normal Distribution

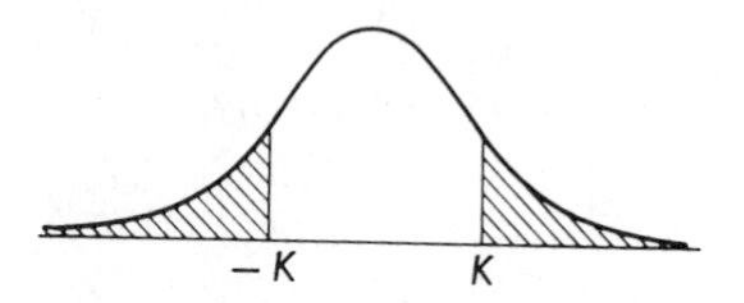

$P(\|Z\| > K)$	K
.001	3.2905
.002	3.0902
.005	2.80703
.01	2.5758
.02	2.3263
.03	2.1701
.04	2.0537
.05	1.9600
.06	1.8808
.08	1.7507
.10	1.6449
.15	1.4395
.20	1.2816
.30	1.0364

TABLE II Percentiles of the Chi-Square Distribution

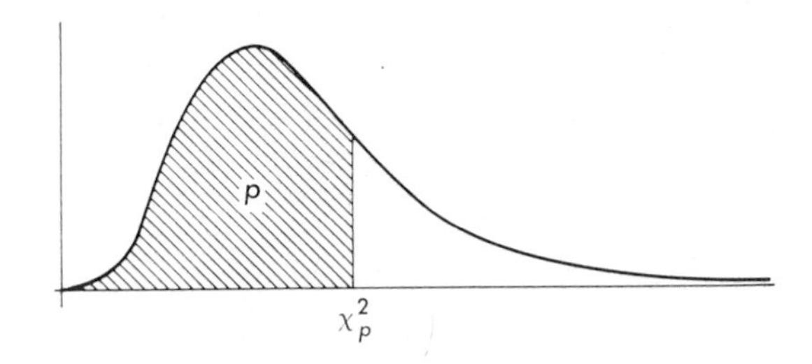

Degrees of freedom	$\chi^2_{.005}$	$\chi^2_{.01}$	$\chi^2_{.025}$	$\chi^2_{.05}$	$\chi^2_{.10}$	$\chi^2_{.20}$	$\chi^2_{.30}$	$\chi^2_{.50}$	$\chi^2_{.70}$	$\chi^2_{.80}$	$\chi^2_{.90}$	$\chi^2_{.95}$	$\chi^2_{.975}$	$\chi^2_{.99}$	$\chi^2_{.995}$
1	.000	.000	.001	.004	.016	.064	.148	.455	1.07	1.64	2.71	3.84	5.02	6.63	7.88
2	.010	.020	.051	.103	.211	.446	.713	1.39	2.41	3.22	4.61	5.99	7.38	9.21	10.6
3	.072	.115	.216	.352	.584	1.00	1.42	2.37	3.66	4.64	6.25	7.81	9.35.	11.3	12.8
4	.207	.297	.484	.711	1.06	1.65	2.20	3.36	4.88	5.99	7.78	9.49	11.1	13.3	14.9
5	.412	.554	.831	1.15	1.61	2.34	3.00	4.35	6.06	7.29	9.24	11.1	12.8	15.1	16.7
6	.676	.872	1.24	1.64	2.20	3.07	3.83	5.35	7.23	8.56	10.6	12.6	14.4	16.8	18.5
7	.989	1.24	1.69	2.17	2.83	3.82	4.67	6.35	8.38	9.80	12.0	14.1	16.0	18.5	20.3
8	1.34	1.65	2.18	2.73	3.49	4.59	5.53	7.34	9.52	11.0	13.4	15.5	17.5	20.1	22.0
9	1.73	2.09	2.70	3.33	4.17	5.38	6.39	8.34	10.7	12.2	14.7	16.9	19.0	21.7	23.6
10	2.16	2.56	3.25	3.94	4.87	6.18	7.27	9.34	11.8	13.4	16.0	18.3	20.5	23.2	25.2
11	2.60	3.05	3.82	4.57	5.58	6.99	8.15	10.3	12.9	14.6	17.3	19.7	21.9	24.7	26.8
12	3.07	3.57	4.40	5.23	6.30	7.81	9.03	11.3	14.0	15.8	18.5	21.0	23.3	26.2	28.3
13	3.57	4.11	5.01	5.89	7.04	8.63	9.93	12.3	15.1	17.0	19.8	22.4	24.7	27.7	29.8
14	4.07	4.66	5.63	6.57	7.79	9.47	10.8	13.3	16.2	18.2	21.1	23.7	26.1	29.1	31.3
15	4.60	5.23	6.26	7.26	8.55	10.3	11.7	14.3	17.3	19.3	22.3	25.0	27.5	30.6	32.8
16	5.14	5.81	6.91	7.96	9.31	11.2	12.6	15.3	18.4	20.5	23.5	26.3	28.8	32.0	34.3
17	5.70	6.41	7.56	8.67	10.1	12.0	13.5	16.3	19.5	21.6	24.8	27.6	30.2	33.4	35.7
18	6.26	7.01	8.23	9.39	10.9	12.9	14.4	17.3	20.6	22.8	26.0	28.9	31.5	34.8	37.2
19	6.83	7.63	8.91	10.1	11.7	13.7	15.4	18.3	21.7	23.9	27.2	30.1	32.9	36.2	38.6
20	7.43	8.26	9.59	10.9	12.4	14.6	16.3	19.3	22.8	25.0	28.4	31.4	34.2	37.6	40.0
21	8.03	8.90	10.3	11.6	13.2	15.4	17.2	20.3	23.9	26.2	29.6	32.7	35.5	38.9	41.4
22	8.64	9.54	11.0	12.3	14.0	16.3	18.1	21.3	24.9	27.3	30.8	33.9	36.8	40.3	42.8
23	9.26	10.2	11.7	13.1	14.8	17.2	19.0	22.3	26.0	28.4	32.0	35.2	38.1	41.6	44.2
24	9.89	10.9	12.4	13.8	15.7	18.1	19.9	23.3	27.1	29.6	33.2	36.4	39.4	43.0	45.6
25	10.5	11.5	13.1	14.6	16.5	18.9	20.9	24.3	28.2	30.7	34.4	37.7	40.6	44.3	46.9
26	11.2	12.2	13.8	15.4	17.3	19.8	21.8	25.3	29.2	31.8	35.6	38.9	41.9	45.6	48.3
27	11.8	12.9	14.6	16.2	18.1	20.7	22.7	26.3	30.3	32.9	36.7	40.1	43.2	47.0	49.6
28	12.5	13.6	15.3	16.9	18.9	21.6	23.6	27.3	31.4	34.0	37.9	41.3	44.5	48.3	51.0
29	13.1	14.3	16.0	17.7	19.8	22.5	24.6	28.3	32.5	35.1	39.1	42.6	45.7	49.6	52.3
30	13.8	15.0	16.8	18.5	20.6	23.4	25.5	29.3	33.5	36.2	40.3	43.8	47.0	50.9	53.7
40	20.7	22.1	24.4	26.5	29.0	32.3	34.9	39.3	44.2	47.3	51.8	55.8	59.3	63.7	66.8
50	28.0	29.7	32.3	34.8	37.7	41.3	44.3	49.3	54.7	58.2	63.2	67.5	71.4	76.2	79.5
60	35.5	37.5	40.5	43.2	46.5	50.6	53.8	59.3	65.2	69.0	74.4	79.1	83.3	88.4	92.0

Note: For degrees of freedom $k > 30$, use $\chi_p^2 = \frac{1}{2}(z_p + \sqrt{2k - 1})^2$, where z_p is the corresponding percentile of the standard normal distribution.

This table is adapted from Table VIII of *Biometrika Tables for Statisticians*, Vol. 1, 1954, by E. S. Pearson and H. O. Hartley, originally prepared by Catherine M. Thompson, with the kind permission of the editor of *Biometrika*.

TABLE III Percentiles of the t-Distribution

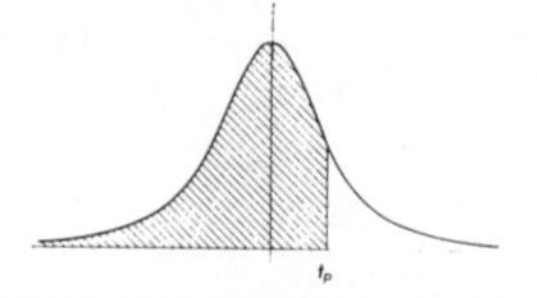

Degrees of Freedom	$t_{.55}$	$t_{.60}$	$t_{.65}$	$t_{.70}$	$t_{.75}$	$t_{.80}$	$t_{.85}$	$t_{.90}$	$t_{.95}$	$t_{.975}$	$t_{.99}$	$t_{.995}$	$t_{.9995}$
1	.158	.325	.510	.727	1.00	1.38	1.96	3.08	6.31	12.7	31.8	63.7	637
2	.142	.289	.445	.617	.816	1.06	1.39	1.89	2.92	4.30	6.96	9.92	31.6
3	.137	.277	.424	.584	.765	.978	1.25	1.64	2.35	3.18	4.54	5.84	12.9
4	.134	.271	.414	.569	.741	.941	1.19	1.53	2.13	2.78	3.75	4.60	8.61
5	.132	.267	.408	.559	.727	.920	1.16	1.48	2.01	2.57	3.36	4.03	6.86
6	.131	.265	.404	.553	.718	.906	1.13	1.44	1.94	2.45	3.14	3.71	5.96
7	.130	.263	.402	.549	.711	.896	1.12	1.42	1.90	2.36	3.00	3.50	5.40
8	.130	.262	.399	.546	.706	.889	1.11	1.40	1.86	2.31	2.90	3.36	5.04
9	.129	.261	.398	.543	.703	.883	1.10	1.38	1.83	2.26	2.82	3.25	4.78
10	.129	.260	.397	.542	.700	.879	1.09	1.37	1.81	2.23	2.76	3.17	4.59
11	.129	.260	.396	.540	.697	.876	1.09	1.36	1.80	2.20	2.72	3.11	4.44
12	.128	.259	.395	.539	.695	.873	1.08	1.36	1.78	2.18	2.68	3.06	4.32
13	.128	.259	.394	.538	.694	.870	1.08	1.35	1.77	2.16	2.65	3.01	4.22
14	.128	.258	.393	.537	.692	.868	1.08	1.34	1.76	2.14	2.62	2.98	4.14
15	.128	.258	.393	.536	.691	.866	1.07	1.34	1.75	2.13	2.60	2.95	4.07
16	.128	.258	.392	.535	.690	.865	1.07	1.34	1.75	2.12	2.58	2.92	4.02
17	.128	.257	.392	.534	.689	.863	1.07	1.33	1.74	2.11	2.57	2.90	3.96
18	.127	.257	.392	.534	.688	.862	1.07	1.33	1.73	2.10	2.55	2.88	3.92
19	.127	.257	.391	.533	.688	.861	1.07	1.33	1.73	2.09	2.54	2.86	3.88
20	.127	.257	.391	.533	.687	.860	1.06	1.32	1.72	2.09	2.53	2.84	3.85
21	.127	.257	.391	.532	.686	.859	1.06	1.32	1.72	2.08	2.52	2.83	3.82
22	.127	.256	.390	.532	.686	.858	1.06	1.32	1.72	2.07	2.51	2.82	3.79
23	.127	.256	.390	.532	.685	.858	1.06	1.32	1.71	2.07	2.50	2.81	3.77
24	.127	.256	.390	.531	.685	.857	1.06	1.32	1.71	2.06	2.49	2.80	3.74
25	.127	.256	.390	.531	.684	.856	1.06	1.32	1.71	2.06	2.48	2.79	3.72
26	.127	.256	.390	.531	.684	.856	1.06	1.32	1.70	2.06	2.48	2.78	3.71
27	.127	.256	.389	.531	.684	.855	1.06	1.31	1.70	2.05	2.47	2.77	3.69
28	.127	.256	.389	.530	.683	.855	1.06	1.31	1.70	2.05	2.47	2.76	3.67
29	.127	.256	.389	.530	.683	.854	1.05	1.31	1.70	2.04	2.46	2.76	3.66
30	.127	.256	.389	.530	.683	.854	1.05	1.31	1.70	2.04	2.46	2.75	3.65
∞	.126	.253	.385	.524	.674	.842	1.04	1.28	1.64	1.96	2.33	2.58	3.29

Note: For the lower percentiles, use the relation $t_\alpha = -t_{1-\alpha}$. In particular, $t_{.50} = -t_{.50} = 0$. For example, for 6 degrees of freedom, $t_{.35} = -t_{.65} = -.404$.

This table is abridged from Table II of Fisher and Yates, *Statistical Tables for Biological, Agricultural, and Medical Research* (5th ed.) and Fisher, *Statistical Methods for Research Workers*, published by Oliver and Boyd, Ltd., Edinburgh, by permission of the authors and publishers.

TABLE IVa $F_{.95}$, Ninety-Fifth Percentiles of the F-Distribution

Denominator Degrees of Freedom	Numerator Degrees of Freedom 1	2	3	4	5	6	8	10	12	15	20	24	30
1	161	200	216	225	230	234	239	242	244	246	248	249	250
2	18.5	19.0	19.2	19.2	19.3	19.3	19.4	19.4	19.4	19.4	19.4	19.5	19.5
3	10.1	9.55	9.28	9.12	9.01	8.94	8.85	8.79	8.74	8.70	8.66	8.64	8.62
4	7.71	6.94	6.59	6.39	6.26	6.16	6.04	5.96	5.91	5.86	5.80	5.77	5.75
5	6.61	5.79	5.41	5.19	5.05	4.95	4.82	4.74	4.68	4.62	4.56	4.53	4.50
6	5.99	5.14	4.76	4.53	4.39	4.28	4.15	4.06	4.00	3.94	3.87	3.84	3.81
7	5.59	4.74	4.35	4.12	3.97	3.87	3.73	3.64	3.57	3.51	3.44	3.41	3.38
8	5.32	4.46	4.07	3.84	3.69	3.58	3.44	3.35	3.28	3.22	3.15	3.12	3.08
9	5.12	4.26	3.86	3.63	3.48	3.37	3.23	3.14	3.07	3.01	2.94	2.90	2.86
10	4.96	4.10	3.71	3.48	3.33	3.22	3.07	2.98	2.91	2.85	2.77	2.74	2.70
11	4.84	3.98	3.59	3.36	3.20	3.09	2.95	2.85	2.79	2.72	2.65	2.61	2.57
12	4.75	3.89	3.49	3.26	3.11	3.00	2.85	2.75	2.69	2.62	2.54	2.51	2.47
13	4.67	3.81	3.41	3.18	3.03	2.92	2.77	2.67	2.60	2.53	2.46	2.42	2.38
14	4.60	3.74	3.34	3.11	2.96	2.85	2.70	2.60	2.53	2.46	2.39	2.35	2.31
15	4.54	3.68	3.29	3.06	2.90	2.79	2.64	2.54	2.48	2.40	2.33	2.29	2.25
16	4.49	3.63	3.24	3.01	2.85	2.74	2.59	2.49	2.42	2.35	2.28	2.24	2.19
17	4.45	3.59	3.20	2.96	2.81	2.70	2.55	2.45	2.38	2.31	2.23	2.19	2.15
18	4.41	3.55	3.16	2.93	2.77	2.66	2.51	2.41	2.34	2.27	2.19	2.15	2.11
19	4.38	3.52	3.13	2.90	2.74	2.63	2.48	2.38	2.31	2.23	2.16	2.11	2.07
20	4.35	3.49	3.10	2.87	2.71	2.60	2.45	2.35	2.28	2.20	2.12	2.08	2.04
21	4.32	3.47	3.07	2.84	2.68	2.57	2.42	2.32	2.25	2.18	2.10	2.05	2.01
22	4.30	3.44	3.05	2.82	2.66	2.55	2.40	2.30	2.23	2.15	2.07	2.03	1.98
23	4.28	3.42	3.03	2.80	2.64	2.53	2.37	2.27	2.20	2.13	2.05	2.01	1.96
24	4.26	3.40	3.01	2.78	2.62	2.51	2.36	2.25	2.18	2.11	2.03	1.98	1.94
25	4.24	3.39	2.99	2.76	2.60	2.49	2.34	2.24	2.16	2.09	2.01	1.96	1.92
30	4.17	3.32	2.92	2.69	2.53	2.42	2.27	2.16	2.09	2.01	1.93	1.89	1.84
40	4.08	3.23	2.84	2.61	2.45	2.34	2.18	2.08	2.00	1.92	1.84	1.79	1.74
60	4.00	3.15	2.76	2.53	2.37	2.25	2.10	1.99	1.92	1.84	1.75	1.70	1.65

This table is adapted from Table XVIII in *Biometrika Tables for Statisticians*, Vol. I, 1954, by E. S. Pearson and H. O. Hartley, originally prepared by M. Merrington and C. M. Thompson, with the kind permission of the editor of *Biometrika*.

TABLE IVb $F_{.99}$, Ninety-Ninth Percentiles of the F-Distribution

Denominator Degrees of Freedom	Numerator Degrees of Freedom 1	2	3	4	5	6	8	10	12	15	20	24	30
1	4050	5000	5400	5620	5760	5860	5980	6060	6110	6160	6210	6235	6260
2	98.5	99.0	99.2	99.2	99.3	99.3	99.4	99.4	99.4	99.4	99.4	99.5	99.5
3	34.1	30.8	29.5	28.7	28.2	27.9	27.5	27.3	27.1	26.9	26.7	26.6	26.5
4	21.2	18.0	16.7	16.0	15.5	15.2	14.8	14.5	14.4	14.2	14.0	13.9	13.8
5	16.3	13.3	12 1.	11.4	11.0	10.7	10.3	10.1	9.89	9.72	9.55	9.47	9.38
6	13.7	10.9	9.78	9.15	8.75	8.47	8.10	7.87	7.72	7.56	7.40	7.31	7.23
7	12.2	9.55	8.45	7.85	7.46	7.19	6.84	6.62	6.47	6.31	6.16	6.07	5.99
8	11.3	8.65	7.59	7.01	6.63	6.37	6.03	5.81	5.67	5.52	5.36	5.28	5.20
9	10.6	8.02	6.99	6.42	6.06	5.80	5.47	5.26	5.11	4.96	4.81	4.73	4.65
10	10.0	7.56	6.55	5.99	5.64	5.39	5.06	4.85	4.71	4.56	4.41	4.33	4.25
11	9.65	7.21	6.22	5.67	5.32	5.07	4.74	4.54	4.40	4.25	4.10	4.02	3.94
12	9.33	6.93	5.95	5.41	5.06	4.82	4.50	4.30	4.16	4.01	3.86	3.78	3.70
13	9.07	6.70	5.74	5.21	4.86	4.62	4.30	4.10	3.96	3.82	3.66	3.59	3.51
14	8.86	6.51	5.56	5.04	4.69	4.46	4.14	3.94	3.80	3.66	3.51	3.43	3.35
15	8.68	6.36	5.42	4.89	4.56	4.32	4.00	3.80	3.67	3.52	3.37	3.29	3.21
16	8.53	6.23	5.29	4.77	4.44	4.20	3.89	3.69	3.55	3.41	3.26	3.18	3.10
17	8.40	6.11	5.18	4.67	4.34	4.10	3.79	3.59	3.46	3.31	3.16	3.08	3.00
18	8.29	6.01	5.09	4.58	4.25	4.01	3.71	3.51	3.37	3.23	3.08	3.00	2.92
19	8.18	5.93	5.01	4.50	4.17	3.94	3.63	3.43	3.30	3.15	3.00	2.92	2.84
20	8.10	5.85	4.94	4.43	4.10	3.87	3.56	3.37	3.23	3.09	2.94	2.86	2.78
21	8.02	5.78	4.87	4.37	4.04	3.81	3.51	3.31	3.17	3.03	2.88	2.80	2.72
22	7.95	5.72	4.82	4.31	3.99	3.76	3.45	3.26	3.12	2.98	2.83	2.75	2.67
23	7.88	5.66	4.76	4.26	3.94	3.71	3.41	3.21	3.07	2.93	2.78	2.70	2.62
24	7.82	5.61	4.72	4.22	3.90	3.67	3.36	3.17	3.03	2.89	2.74	2.66	2.58
25	7.77	5.57	4.68	4.18	3.86	3.63	3.32	3.13	2.99	2.85	2.70	2.62	2.54
30	7.56	5.39	4.51	4.02	3.70	3.47	3.17	2.98	2.84	2.70	2.55	2.47	2.39
40	7.31	5.18	4.31	3.83	3.51	3.29	2.99	2.80	2.66	2.52	2.37	2.29	2.20
60	7.08	4.98	4.13	3.65	3.34	3.12	2.82	2.63	2.50	2.35	2.20	2.12	2.03

This table is adapted from Table XVIII in *Biometrika Tables for Statisticians*, Vol. I, 1954, by E. S. Pearson and H. O. Hartley, originally prepared by M. Merrington and C. M. Thompson, with the kind permission of the editor of *Biometrika*.

TABLE V Distribution of the Standardized Range $W = R/\sigma$ (assuming a normal population)

	Sample Size										
	2	3	4	5	6	7	8	9	10	12	15
$E(W)$	1.128	1.693	2.059	2.326	2.534	2.704	2.847	2.970	3.078	3.258	3.472
σ_W	.853	.888	.880	.864	.848	.833	.820	.808	.797	.778	.755
$W_{.005}$	.01	.13	.34	.55	.75	.92	1.08	1.21	1.33	1.55	1.80
$W_{.01}$	.02	.19	.43	.66	.87	1.05	1.20	1.34	1.47	1.68	1.93
$W_{.025}$	.04	.30	.59	.85	1.06	1.25	1.41	1.55	1.67	1.88	2.14
$W_{.05}$	.09	.43	.76	1.03	1.25	1.44	1.60	1.74	1.86	2.07	2.32
$W_{.1}$	.18	.62	.98	1.26	1.49	1.68	1.83	1.97	2.09	2.30	2.54
$W_{.2}$	.36	.90	1.29	1.57	1.80	1.99	2.14	2.28	2.39	2.59	2.83
$W_{.3}$	.55	1.14	1.53	1.82	2.04	2.22	2.38	2.51	2.62	2.82	3.04
$W_{.4}$	.74	1.36	1.76	2.04	2.26	2.44	2.59	2.71	2.83	3.01	3.23
$W_{.5}$	.95	1.59	1.98	2.26	2.47	2.65	2.79	2.92	3.02	3.21	3.42
$W_{.6}$	1.20	1.83	2.21	2.48	2.69	2.86	3.00	3.12	3.23	3.41	3.62
$W_{.7}$	1.47	2.09	2.47	2.73	2.94	3.10	3.24	3.35	3.46	3.63	3.83
$W_{.8}$	1.81	2.42	2.78	3.04	3.23	3.39	3.52	3.63	3.73	3.90	4.09
$W_{.9}$	2.33	2.90	3.24	3.48	3.66	3.81	3.93	4.04	4.13	4.29	4.47
$W_{.95}$	2.77	3.31	3.63	3.86	4.03	4.17	4.29	4.39	4.47	4.62	4.80
$W_{.975}$	3.17	3.68	3.98	4.20	4.36	4.49	4.61	4.70	4.79	4.92	5.09
$W_{.99}$	3.64	4.12	4.40	4.60	4.76	4.88	4.99	5.08	5.16	5.29	5.45
$W_{.995}$	3.97	4.42	4.69	4.89	5.03	5.15	5.26	5.34	5.42	5.54	5.70

This table is adapted from Tables XX and XXII in *Biometrika Tables for Statisticians*, Vol. I, 1954, by E. S. Pearson and H. O. Hartley, with the kind permission of the editor of *Biometrika*.

TABLE VI Acceptance Limits for the Kolmogorov–Smirnov Test of Goodness of Fit

Sample size (n)	Significance level				
	.20	.15	.10	.05	.01
1	.900	.925	.950	.975	.995
2	.684	.726	.776	.842	.929
3	.565	.597	.642	.708	.829
4	.494	.525	.564	.624	.734
5	.446	.474	.510	.563	.669
6	.410	.436	.470	.521	.618
7	.381	.405	.438	.486	.577
8	.358	.381	.411	.457	.543
9	.339	.360	.388	.432	.514
10	.322	.342	.368	.409	.486
11	.307	.326	.352	.391	.468
12	.295	.313	.338	.375	.450
13	.284	.302	.325	.361	.433
14	.274	.292	.314	.349	.418
15	.266	.283	.304	.338	.404
16	.258	.274	.295	.328	.391
17	.250	.266	.286	.318	.380
18	.244	.259	.278	.309	.370
19	.237	.252	.272	.301	.361
20	.231	.246	.264	.294	.352
25	.21	.22	.24	.264	.32
30	.19	.20	.22	.242	.29
35	.18	.19	.21	.23	.27
40				.21	.25
50				.19	.23
60				.17	.21
70				.16	.19
80				.15	.18
90				.14	
100				.14	
Asymptotic Formula:	$\frac{1.07}{\sqrt{n}}$	$\frac{1.14}{\sqrt{n}}$	$\frac{1.22}{\sqrt{n}}$	$\frac{1.36}{\sqrt{n}}$	$\frac{1.63}{\sqrt{n}}$

Reject the hypothetical distribution $F(x)$ if $D_n = \max |F_n(x) - F(x)|$ exceeds the tabulated value.

(For $\alpha = .01$ and .05, asymptotic formulas give values which are too high—by 1.5 per cent for $n = 80$.)

This table is taken from F. J. Massey, Jr., "The Kolmogorov–Smirnov Test for Goodness of Fit," *J. Am. Stat. Assn.* **46**, 68–78 (1951), except that certain corrections and additional entries are from Z. W. Birnbaum, "Numerical Tabulation of the Distribution of Kolmogorov's Statistic for Finite Sample Size," *J. Am. Stat. Assn.* **47**, 425–41 (1952), with the kind permission of the authors and the *J. Am. Stat. Assn.*

TABLE VII Acceptance Limits for the Two-Sample Kolmogorov–Smirnov Test

Sample Size n_2	Sample Size n_1: 1	2	3	4	5	6	7	8	9	10	12	15
1	* *	* *	* *	* *	* *	* *	* *	* *	* *	* *		
2		* *	* *	* *	* *	* *	* *	7/8 *	16/18 *	9/10 *		
3			* *	* *	12/15 *	5/6 *	18/21 *	18/24 *	7/9 8/9		9/12 11/12	
4				3/4 *	16/20 *	9/12 10/12	21/28 24/28	6/8 7/8	27/36 32/36	14/20 16/20	8/12 10/12	
5					4/5 4/5	20/30 25/30	25/35 30/35	27/40 32/40	31/45 36/45	7/10 8/10		10/15 11/15
6						4/6 5/6	29/42 35/42	16/24 18/24	12/18 14/18	19/30 22/30	7/12 9/12	
7							5/7 5/7	35/56 42/56	40/63 47/63	43/70 53/70		
8								5/8 6/8	45/72 54/72	23/40 28/40	14/24 16/24	
9									5/9 6/9	52/90 62/90	20/36 24/36	
10										6/10 7/10		15/30 19/30
12											6/12 7/12	30/60 35/60
15												7/15 8/15

Note 1: Reject H_0 if $D = \max |F_{n_2}(x) - F_{n_1}(x)|$ exceeds the tabulated value. The upper value gives a level at most .05 and the lower at most .01.

Note 2: Where * appears, do not reject H_0 at the given level.

Note 3: For large values of n_1 and n_2, the following approximate formulas may be used:

$$\alpha = .05: \quad 1.36\sqrt{\frac{n_1 + n_2}{n_1 n_2}}.$$

$$\alpha = .01: \quad 1.63\sqrt{\frac{n_1 + n_2}{n_1 n_2}}.$$

This table is derived from F. J. Massey, "Distribution Table for the Deviation Between Two Sample Cumulatives," *Ann. Math. Stat.* **23**, 435–41 (1952). Adapted with the kind permission of the author and the *Ann. Math. Stat.* Formulas for large-sample sizes were given by N. Smirnov, "Tables for Estimating the Goodness of Fit of Empirical Distributions," *Ann. Math. Stat.* **19**, 280–81 (1948).

TABLE VIII Cumulative Distribution of the Rank-Sum Statistic

c \ m, n	3, 3	3, 4	3, 5	3, 6	3, 7	3, 8	3, 9	3, 10
6	.050	.028	.018	.012	.008	.006		
7	.100	.057	.036	.024	.017	.012	.009	.007
8			.071	.048	.033	.024	.018	.014
9				.083	.058	.042	.032	.025
10					.092	.067	.050	.039
11						.097	.073	.056
12								.080
M	21	24	27	30	33	36	39	42

c \ m, n	4, 4	4, 5	4, 6	4, 7	4, 8	4, 9	4, 10
10	.014	.008					
11	.029	.016	.010	.006			
12	.057	.032	.019	.012	.008		
13	.100	.056	.033	.021	.014	.010	.007
14		.095	.057	.036	.024	.017	.012
15			.086	.055	.036	.025	.018
16				.082	.055	.038	.027
17					.077	.053	.038
18						.074	.053
19						.099	.071
20							.094
M	36	40	44	48	52	56	60

TABLE VIII Cumulative Distribution of the Rank-Sum Statistic (Continued)

c \ m, n	5, 5	5, 6	5, 7	5, 8	5, 9	5, 10
15	.004					
16	.008					
17	.016	.009				
18	.028	.015	.009			
19	.048	.026	.015	.009		
20	.075	.041	.024	.014	.009	
21		.063	.037	.023	.014	.010
22		.089	.053	.033	.021	.014
23			.074	.047	.030	.020
24				.064	.041	.028
25				.085	.056	.038
26					.073	.050
27					.095	.065
28						.082
M	55	60	65	70	75	80

c \ m, n	6, 6	6, 7	6, 8	6, 9	6, 10
24	.008				
25	.013	.007			
26	.021	.011			
27	.032	.017	.010		
28	.047	.026	.015	.009	
29	.066	.037	.021	.013	.008
30	.090	.051	.030	.018	.011
31		.069	.041	.025	.016
32		.090	.054	.033	.021
33			.071	.044	.028
34			.091	.057	.036
35				.072	.047
36				.091	.059
37					.074
38					.090
M	78	84	90	96	102

c \ m, n	7, 7	7, 8	7, 9	7, 10
34	.009			
35	.013	.007		
36	.019	.010		
37	.027	.014	.008	
38	.036	.020	.011	
39	.049	.027	.016	.009
40	.064	.036	.021	.012
41	.082	.047	.027	.017
42		.060	.036	.022
43		.076	.045	.028
44		.095	.057	.035
45			.071	.044
46			.087	.054
47				.067
48				.081
49				.097
M	105	112	119	126

TABLE VIII Cumulative Distribution of the Rank-Sum Statistic (Continued)

c \ m, n	$m=8$, $n=8$	$m=8$, $n=9$	$m=8$, $n=10$
46	.010		
47	.014	.008	
48	.019	.010	
49	.025	.014	.008
50	.032	.018	.010
51	.041	.023	.013
52	.052	.030	.017
53	.065	.037	.022
54	.080	.046	.027
55	.097	.057	.034
56		.069	.042
57		.084	.051
58			.061
59			.073
60			.086
M	136	144	152

c \ m, n	$m=9$, $n=9$	$m=9$, $n=10$
59	.009	
60	.012	
61	.016	.009
62	.020	.011
63	.025	.014
64	.031	.017
65	.039	.022
66	.047	.027
67	.057	.033
68	.068	.039
69	.081	.047
70	.095	.056
71		.067
72		.078
73		.091
M	171	180

c \ m, n	$m=10$, $n=10$
74	.009
75	.012
76	.014
77	.018
78	.022
79	.026
80	.032
81	.038
82	.045
83	.053
84	.062
85	.072
86	.083
87	.095
M	210

Note 1: m is the size of the smaller sample.
Note 2: Entry opposite M is the sum of the max and min rank sums.
Note 3: Entry opposite c is the cumulative probability in a tail:

$$P(R \leq c) = P(R \geq M - c).$$

Example: For a two-tail test at $\alpha = .05$, reject the null hypothesis if $R \leq 14$ or $R \geq 52 = 14 = 38$. (Level is then $.024 + .024$.)

TABLE IX The Wilcoxon Signed-Rank Statistic

$c \backslash n$	3	4	5	6	7	8	9	10	11	12	13	14	15
0	.125	.062	.031	.016	.008								
1	.250	.125	.062	.031	.016	.008							
2		.188	.094	.047	.023	.012							
3			.156	.078	.039	.020	.010						
4				.109	.055	.027	.014	.007					
5				.156	.078	.039	.020	.010					
6					.109	.055	.027	.014					
7					.148	.074	.037	.019	.009				
8					.188	.098	.049	.024	.012				
9						.125	.064	.032	.016				
10						.156	.082	.042	.021	.010			
11							.102	.053	.027	.013			
12							.125	.065	.034	.017			
13							.150	.080	.042	.021	.009		
14								.097	.051	.026	.011		
15									.062	.032	.013	.008	
16									.074	.039	.016	.010	
17									.087	.046	.020	.012	
18										.055	.024	.015	
19										.065	.029	.018	.009
20										.076	.034	.021	.011
21										.088	.040	.025	.013
22											.047	.029	.015
23											.055	.034	.018
24											.064	.039	.021
25											.073	.045	.024
26											.084	.052	.028
27											.095	.059	.032
28												.068	.036
29												.077	.042
30												.086	.047
31												.097	.053
32													.060
33													.068
34													.076
35													.084
36													.094

Table entry is $P(S \leq c) = P\left(S \geq \frac{n(n+1)}{2} - c\right)$.

Example: For $n = 9$, a two-sided test at $\alpha = .10$ would allow up to .05 in each tail. Since

$$P(S \leq 8) = P\left(S \geq \frac{9 \cdot 10}{2} - 8\right) = .049,$$

test is to reject null hypothesis if $S \leq 8$ or $S \geq 37$.

TABLE X Expected Values of Order Statistics from a Standard Normal Population

n	$E[X_{(n)}]$	$E[X_{(n-1)}]$	$E[X_{(n-2)}]$	$E[X_{(n-3)}]$	$E[X_{(n-4)}]$	$E[X_{(n-5)}]$	$E[X_{(n-6)}]$	$E[X_{(n-7)}]$	$E[X_{(n-8)}]$	$E[X_{(n-9)}]$
2	.564									
3	.846									
4	1.029	.297								
5	1.163	.495								
6	1.267	.642	.202							
7	1.352	.757	.353							
8	1.424	.852	.473	.153						
9	1.485	.932	.572	.275						
10	1.539	1.001	.656	.376	.123					
11	1.586	1.062	.729	.462	.225					
12	1.629	1.116	.793	.537	.312	.103				
13	1.668	1.164	.850	.603	.388	.190				
14	1.703	1.208	.901	.662	.456	.267	.088			
15	1.736	1.248	.948	.715	.516	.335	.165			
16	1.766	1.285	.990	.763	.570	.396	.234	.077		
17	1.794	1.319	1.030	.807	.619	.451	.295	.146		
18	1.820	1.350	1.066	.848	.665	.502	.351	.208	.069	
19	1.844	1.380	1.099	.886	.707	.548	.402	.264	.131	
20	1.867	1.408	1.131	.921	.745	.590	.448	.315	.187	.062

Adapted from Table XXVIII of *Biometrika Tables for Statisticians*, Vol. I, 1954, by E. S. Pearson and H. O. Hartley, with the kind permission of the editor of *Biometrika*.

TABLE XI Binomial Coefficients $\binom{n}{k}$

n \ k	2	3	4	5	6	7	8	9	10
2	1								
3	3	1							
4	6	4	1						
5	10	10	5	1					
6	15	20	15	6	1				
7	21	35	35	21	7	1			
8	28	56	70	56	28	8	1		
9	36	84	126	126	84	36	9	1	
10	45	120	210	252	210	120	45	10	1
11	55	165	330	462	462	330	165	55	11
12	66	220	495	792	924	792	495	220	66
13	78	286	715	1287	1716	1716	1287	715	286
14	91	364	1001	2002	3003	3432	3003	2002	1001
15	105	455	1365	3003	5005	6435	6435	5005	3003
16	120	560	1820	4368	8008	11440	12870	11440	8008
17	136	680	2380	6188	12376	19448	24310	24310	19448
18	153	816	3060	8568	18564	31824	43758	48620	43758
19	171	969	3876	11628	27132	50388	75582	92378	92378
20	190	1140	4845	15504	38760	77520	125970	167960	184756

TABLE XII Exponential Functions

x	e^{-x}	e^{x}	$\log_{10}e^{x}$
.01	.9900	1.0101	.00434
.02	.9802	1.0202	.00869
.03	.9704	1.0305	.01303
.04	.9608	1.0408	.01737
.05	.9512	1.0513	.02171
.06	.9418	1.0618	.02606
.07	.9324	1.0725	.03040
.08	.9231	1.0833	.03474
.09	.9139	1.0942	.03909
.10	.9048	1.1052	.04343
.20	.8187	1.2214	.08686
.30	.7408	1.3499	.13029
.40	.6703	1.4918	.17372
.50	.6065	1.6487	.21715
.60	.5488	1.8221	.26058
.70	.4966	2.0138	.30401
.80	.4493	2.2255	.34744
.90	.4066	2.4596	.39087
1.00	.3679	2.7183	.43429
2.00	.1353	7.3891	.86859
3.00	.04979	20.0886	1.30288
4.00	.01832	54.598	1.73718
5.00	.00674	148.41	2.17147
6.00	.00248	403.43	2.60577
7.00	.000912	1096.6	3.04006
8.00	.000335	2981.0	3.47536
9.00	.000123	8103.1	3.90865
10.00	.000045	22026.0	4.34294

TABLE XIII Poisson Distribution Function:

$$F(c) = \sum_{k=0}^{c} \frac{m^k e^{-m}}{k!} \quad m \text{ (expected value)}$$

c \ m	.02	.04	.06	.08	.10	.15	.20	.25	.30	.35	.40
0	.980	.961	.942	.923	.905	.861	.819	.779	.741	.705	.670
1	1.000	.999	.998	.997	.995	.990	.982	.974	.963	.951	.938
2		1.000	1.000	1.000	1.000	.999	.999	.998	.996	.994	.992
3						1.000	1.000	1.000	1.000	1.000	.999
4											1.000

c	.45	.50	.55	.60	.65	.70	.75	.80	.85	.90	.95
0	.638	.607	.577	.549	.522	.497	.472	.449	.427	.407	.387
1	.925	.910	.894	.878	.861	.844	.827	.809	.791	.772	.754
2	.989	.986	.982	.977	.972	.966	.959	.953	.945	.937	.929
3	.999	.998	.998	.997	.996	.994	.993	.991	.989	.987	.984
4	1.000	1.000	1.000	1.000	.999	.999	.999	.999	.998	.998	.997
5					1.000	1.000	1.000	1.000	1.000	1.000	1.000

c	1.0	1.1	1.2	1.3	1.4	1.5	1.6	1.7	1.8	1.9	2.0
0	.368	.333	.301	.273	.247	.223	.202	.183	.165	.150	.135
1	.736	.699	.663	.627	.592	.558	.525	.493	.463	.434	.406
2	.920	.900	.879	.857	.833	.809	.783	.757	.731	.704	.677
3	.981	.974	.966	.957	.946	.934	.921	.907	.891	.875	.857
4	.996	.995	.992	.989	.986	.981	.976	.970	.964	.956	.947
5	.999	.999	.998	.998	.997	.996	.994	.992	.990	.987	.983
6	1.000	1.000	1.000	1.000	.999	.999	.999	.998	.997	.997	.995
7					1.000	1.000	1.000	1.000	.999	.999	.999
8									1.000	1.000	1.000

c	2.2	2.4	2.6	2.8	3.0	3.2	3.4	3.6	3.8	4.0	4.2
0	.111	.091	.074	.061	.050	.041	.033	.027	.022	.018	.015
1	.355	.308	.267	.231	.199	.171	.147	.126	.107	.092	.078
2	.623	.570	.518	.469	.423	.380	.340	.303	.269	.238	.210
3	.819	.779	.736	.692	.647	.603	.558	.515	.473	.433	.395
4	.928	.904	.877	.848	.815	.781	.744	.706	.668	.629	.590
5	.975	.964	.951	.935	.916	.895	.871	.844	.816	.785	.753
6	.993	.988	.983	.976	.966	.955	.942	.927	.909	.889	.867
7	.998	.997	.995	.992	.988	.983	.977	.969	.960	.949	.936
8	1.000	.999	.999	.998	.996	.994	.992	.988	.984	.979	.972
9		1.000	1.000	.999	.999	.998	.997	.996	.994	.992	.989
10				1.000	1.000	1.000	.999	.999	.998	.997	.996
11							1.000	1.000	.999	.999	.999
12									1.000	1.000	1.000

TABLE XIII Poisson Distribution Function (Continued)

c	4.4	4.6	4.8	5.0	5.2	5.4	5.6	5.8	6.0	6.2	6.4
0	.012	.010	.008	.007	.006	.005	.004	.003	.002	.002	.002
1	.066	.056	.048	.040	.034	.029	.024	.021	.017	.015	.012
2	.185	.163	.143	.125	.109	.095	.082	.072	.062	.054	.046
3	.359	.326	.294	.265	.238	.213	.191	.170	.151	.134	.119
4	.551	.513	.476	.440	.406	.373	.342	.313	.285	.259	.235
5	.720	.686	.651	.616	.581	.546	.512	.478	.446	.414	.384
6	.844	.818	.791	.762	.732	.702	.670	.638	.606	.574	.542
7	.921	.905	.887	.867	.845	.822	.797	.771	.744	.716	.687
8	.964	.955	.944	.932	.918	.903	.886	.867	.847	.826	.803
9	.985	.980	.975	.968	.960	.951	.941	.929	.916	.902	.886
10	.994	.992	.990	.986	.982	.977	.972	.965	.957	.949	.939
11	.998	.997	.996	.995	.993	.990	.988	.984	.980	.975	.969
12	.999	.999	.999	.998	.997	.996	.995	.993	.991	.989	.986
13	1.000	1.000	1.000	.999	.999	.999	.998	.997	.996	.995	.994
14				1.000	1.000	1.000	.999	.999	.999	.998	.997
15							1.000	1.000	.999	.999	.999
16									1.000	1.000	1.000

c	6.6	6.8	7.0	7.2	7.4	7.6	7.8	8.0	8.5	9.0	9.5
0	.001	.001	.001	.001	.001	.001	.000	.000	.000	.000	.000
1	.010	.009	.007	.006	.005	.004	.004	.003	.002	.001	.001
2	.040	.034	.030	.025	.022	.019	.016	.014	.009	.006	.004
3	.105	.093	.082	.072	.063	.055	.048	.042	.030	.021	.015
4	.213	.192	.173	.156	.140	.125	.112	.100	.074	.055	.040
5	.355	.327	.301	.276	.253	.231	.210	.191	.150	.116	.089
6	.511	.480	.450	.420	.392	.365	.338	.313	.256	.207	.165
7	.658	.628	.599	.569	.539	.510	.481	.453	.386	.324	.269
8	.780	.755	.729	.703	.676	.648	.620	.593	.523	.456	.392
9	.869	.850	.830	.810	.788	.765	.741	.717	.653	.587	.522
10	.927	.915	.901	.887	.871	.854	.835	.816	.763	.706	.645
11	.963	.955	.947	.937	.926	.915	.902	.888	.849	.803	.752
12	.982	.978	.973	.967	.961	.954	.945	.936	.909	.876	.836
13	.992	.990	.987	.984	.980	.976	.971	.966	.949	.926	.898
14	.997	.996	.994	.993	.991	.989	.986	.983	.973	.959	.940
15	.999	.998	.998	.997	.996	.995	.993	.992	.986	.978	.967
16	.999	.999	.999	.999	.998	.998	.997	.996	.993	.989	.982
17	1.000	1.000	1.000	.999	.999	.999	.999	.998	.997	.995	.991
18				1.000	1.000	1.000	1.000	.999	.999	.998	.996
19								1.000	.999	.999	.998
20									1.000	1.000	.999
21											1.000

TABLE XIII Poisson Distribution Function (Continued)

c	10.0	10.5	11.0	11.5	12.0	12.5	13.0	13.5	14.0	14.5	15.0
2	.003	.002	.001	.001	.001	.000					
3	.010	.007	.005	.003	.002	.002	.001	.001	.000		
4	.029	.021	.015	.011	.008	.005	.004	.003	.002	.001	.001
5	.067	.050	.038	.028	.020	.015	.011	.008	.006	.004	.003
6	.130	.102	.079	.060	.046	.035	.026	.019	.014	.010	.008
7	.220	.179	.143	.114	.090	.070	.054	.041	.032	.024	.018
8	.333	.279	.232	.191	.155	.125	.100	.079	.062	.048	.037
9	.458	.397	.341	.289	.242	.201	.166	.135	.109	.088	.070
10	.583	.521	.460	.402	.347	.297	.252	.211	.176	.145	.118
11	.697	.639	.579	.520	.462	.406	.353	.304	.260	.220	.185
12	.792	.742	.689	.633	.576	.519	.463	.409	.358	.311	.268
13	.864	.825	.781	.733	.682	.628	.573	.518	.464	.413	.363
14	.917	.888	.854	.815	.772	.725	.675	.623	.570	.518	.466
15	.951	.932	.907	.878	.844	.806	.764	.718	.669	.619	.568
16	.973	.960	.944	.924	.899	.869	.835	.798	.756	.711	.664
17	.986	.978	.968	.954	.937	.916	.890	.861	.827	.790	.749
18	.993	.988	.982	.974	.963	.948	.930	.908	.883	.853	.819
19	.997	.994	.991	.986	.979	.969	.957	.942	.923	.901	.875
20	.998	.997	.995	.992	.988	.983	.975	.965	.952	.936	.917
21	.999	.999	.998	.996	.994	.991	.986	.980	.971	.960	.947
22	1.000	.999	.999	.998	.997	.995	.992	.989	.983	.976	.967
23		1.000	1.000	.999	.999	.998	.996	.994	.991	.986	.981
24				1.000	.999	.999	.998	.997	.995	.992	.989
25					1.000	.999	.999	.998	.997	.996	.994
26						1.000	1.000	.999	.999	.998	.997
27								1.000	.999	.999	.998
28									1.000	.999	.999
29										1.000	1.000

TABLE XIV Logarithms of Factorials

	0	1	2	3	4	5	6	7	8	9
00	0.0000	0.0000	0.3010	0.7782	1.3802	2.0792	2.8573	3.7024	4.6055	5.5598
10	6.5598	7.6012	8.6803	9.7943	10.9404	12.1165	13.3206	14.5511	15.8063	17.0851
20	18.3861	19.7083	21.0508	22.4125	23.7927	25.1906	26.6056	28.0370	29.4841	30.9465
30	32.4237	33.9150	35.4202	36.9387	38.4702	40.0142	41.5705	43.1387	44.7185	46.3096
40	47.9116	49.5244	51.1477	52.7811	54.4264	56.0778	57.7406	59.4127	61.0939	62.7841
50	64.4831	66.1906	67.9066	69.6309	71.3633	73.1037	74.8519	76.6077	78.3712	80.1420
60	81.9202	83.7055	85.4979	87.2972	89.1034	90.9163	92.7359	94.5619	96.3945	98.2333
70	100.0784	101.9297	103.7870	105.6503	107.5196	109.3946	111.2754	113.1619	115.0540	116.9516
80	118.8547	120.7632	122.6770	124.5961	126.5204	128.4498	130.3843	132.3238	134.2683	136.2177
90	138.1719	140.1310	142.0948	144.0632	146.0364	148.0141	149.9964	151.9831	153.9744	155.9700
100	157.9700	159.9743	161.9829	163.9958	166.0128	168.0340	170.0593	172.0887	174.1221	176.1595
110	178.2009	180.2462	182.2955	184.3485	186.4054	188.4661	190.5306	192.5988	194.6707	196.7462
120	198.8254	200.9082	202.9945	205.0844	207.1779	209.2748	211.3751	213.4790	215.5862	217.6967
130	219.8107	221.9280	224.0485	226.1724	228.2995	230.4298	232.5634	234.7001	236.8400	238.9830
140	241.1291	243.2783	245.4306	247.5860	249.7443	251.9057	254.0700	256.2374	258.4076	260.5808
150	262.7569	264.9359	267.1177	269.3024	271.4899	273.6803	275.8734	278.0693	280.2679	282.4693
160	284.6735	286.8803	289.0898	291.3020	293.5168	295.7343	297.9544	300.1771	302.4024	304.6303
170	306.8608	309.0938	311.3293	313.5674	315.8079	318.0509	320.2965	322.5444	324.7948	327.0477
180	329.3030	331.5606	333.8207	336.0832	338.3480	340.6152	342.8847	345.1565	347.4307	349.7071
190	351.9859	354.2669	356.5502	358.8358	361.1236	363.4136	365.7059	368.0003	370.2970	372.5959

TABLE XV Four-Place Common Logarithms

N	0	1	2	3	4	5	6	7	8	9
10	0000	0043	0086	0128	0170	0212	0253	0294	0334	0374
11	0414	0453	0492	0531	0569	0607	0645	0682	0719	0755
12	0792	0828	0864	0899	0934	0969	1004	1038	1072	1106
13	1139	1173	1206	1239	1271	1303	1335	1367	1399	1430
14	1461	1492	1523	1553	1584	1614	1644	1673	1703	1732
15	1761	1790	1818	1847	1875	1903	1931	1959	1987	2014
16	2041	2068	2095	2122	2148	2175	2201	2227	2253	2279
17	2304	2330	2355	2380	2405	2430	2455	2480	2504	2529
18	2553	2577	2601	2625	2648	2672	2695	2718	2742	2765
19	2788	2810	2833	2856	2878	2900	2923	2945	2967	2989
20	3010	3032	3054	3075	3096	3118	3139	3160	3181	3201
21	3222	3243	3263	3284	3304	3324	3345	3365	3385	3404
22	3424	3444	3464	3483	3502	3522	3541	3560	3579	3598
23	3617	3636	3655	3674	3692	3711	3729	3747	3766	3784
24	3802	3820	3838	3856	3874	3892	3909	3927	3945	3962
25	3979	3997	4014	4031	4048	4065	4082	4099	4116	4133
26	4150	4166	4183	4200	4216	4232	4249	4265	4281	4298
27	4314	4330	4346	4362	4378	4393	4409	4425	4440	4456
28	4472	4487	4502	4518	4533	4548	4564	4579	4594	4609
29	4624	4639	4654	4669	4683	4698	4713	4728	4742	4757
30	4771	4786	4800	4814	4829	4843	4857	4871	4886	4900
31	4914	4928	4942	4955	4969	4983	4997	5011	5024	5038
32	5051	5065	5079	5092	5105	5119	5132	5145	5159	5172
33	5185	5198	5211	5224	5237	5250	5263	5276	5289	5302
34	5315	5328	5340	5353	5366	5378	5391	5403	5416	5428
35	5441	5453	5465	5478	5490	5502	5514	5527	5539	5551
36	5563	5575	5587	5599	5611	5623	5635	5647	5658	5670
37	5682	5694	5705	5717	5729	5740	5752	5763	5775	5786
38	5798	5809	5821	5832	5843	5855	5866	5877	5888	5899
39	5911	5922	5933	5944	5955	5966	5977	5988	5999	6010
40	6021	6031	6042	6053	6064	6075	6085	6096	6107	6117
41	6128	6138	6149	6160	6170	6180	6191	6201	6212	6222
42	6232	6243	6253	6263	6274	6284	6294	6304	6314	6325
43	6335	6345	6355	6365	6375	6385	6395	6405	6415	6425
44	6435	6444	6454	6464	6474	6484	6493	6503	6513	6522
45	6532	6542	6551	6561	6571	6580	6590	6599	6609	6618
46	6628	6637	6646	6656	6665	6675	6684	6693	6702	6712
47	6721	6730	6739	6749	6758	6767	6776	6785	6794	6803
48	6812	6821	6830	6839	6848	6857	6866	6875	6884	6893
49	6902	6911	6920	6928	6937	6946	6955	6964	6972	6981
50	6990	6998	7007	7016	7024	7033	7042	7050	7059	7067
51	7076	7084	7093	7101	7110	7118	7126	7135	7143	7152
52	7160	7168	7177	7185	7193	7202	7210	7218	7226	7235
53	7243	7251	7259	7267	7275	7284	7292	7300	7308	7316
54	7324	7332	7340	7348	7356	7364	7372	7380	7388	7396
N	**0**	**1**	**2**	**3**	**4**	**5**	**6**	**7**	**8**	**9**

TABLE XV Four-Place Common Logarithms (Continued)

N	0	1	2	3	4	5	6	7	8	9
55	7404	7412	7419	7427	7435	7443	7451	7459	7466	7474
56	7482	7490	7497	7505	7513	7520	7528	7536	7543	7551
57	7559	7566	7574	7582	7589	7597	7604	7612	7619	7627
58	7634	7642	7649	7657	7664	7672	7679	7686	7694	7701
59	7709	7716	7723	7731	7738	7745	7752	7760	7767	7774
60	7782	7789	7796	7803	7810	7818	7825	7832	7839	7846
61	7853	7860	7868	7875	7882	7889	7896	7903	7910	7917
62	7924	7931	7938	7945	7952	7959	7966	7973	7980	7987
63	7993	8000	8007	8014	8021	8028	8035	8041	8048	8055
64	8062	8069	8075	8082	8089	8096	8102	8109	8116	8122
65	8129	8136	8142	8149	8156	8162	8169	8176	8182	8189
66	8195	8202	8209	8215	8222	8228	8235	8241	8248	8254
67	8261	8267	8274	8280	8287	8293	8299	8306	8312	8319
68	8325	8331	8338	8344	8351	8357	8363	8370	8376	8382
69	8388	8395	8401	8407	8414	8420	8426	8432	8439	8445
70	8451	8457	8463	8470	8476	8482	8488	8494	8500	8506
71	8513	8519	8525	8531	8537	8543	8549	8555	8561	8567
72	8573	8579	8585	8591	8597	8603	8609	8615	8621	8627
73	8633	8639	8645	8651	8657	8663	8669	8675	8681	8686
74	8692	8698	8704	8710	8716	8722	8727	8733	8739	8745
75	8751	8756	8762	8768	8774	8779	8785	8791	8797	8802
76	8808	8814	8820	8825	8831	8837	8842	8848	8854	8859
77	8865	8871	8876	8882	8887	8893	8899	8904	8910	8915
78	8921	8927	8932	8938	8943	8949	8954	8960	8965	8971
79	8976	8982	8987	8993	8998	9004	9009	9015	9020	9025
80	9031	9036	9042	9047	9053	9058	9063	9069	9074	9079
81	9085	9090	9096	9101	9106	9112	9117	9122	9128	9133
82	9138	9143	9149	9154	9159	9165	9170	9175	9180	9186
83	9191	9196	9201	9206	9212	9217	9222	9227	9232	9238
84	9243	9248	9253	9258	9263	9269	9274	9279	9284	9289
85	9294	9299	9304	9309	9315	9320	9325	9330	9335	9340
86	9345	9350	9355	9360	9365	9370	9375	9380	9385	9390
87	9395	9400	9405	9410	9415	9420	9425	9430	9435	9440
88	9445	9450	9455	9460	9465	9469	9474	9479	9484	9489
89	9494	9499	9504	9509	9513	9518	9523	9528	9533	9538
90	9542	9547	9552	9557	9562	9566	9571	9576	9581	9586
91	9590	9595	9600	9605	9609	9614	9619	9624	9628	9633
92	9638	9643	9647	9652	9657	9661	9666	9671	9675	9680
93	9685	9689	9694	9699	9703	9708	9713	9717	9722	9727
94	9731	9736	9741	9745	9750	9754	9759	9763	9768	9773
95	9777	9782	9786	9791	9795	9800	9805	9809	9814	9818
96	9823	9827	9832	9836	9841	9845	9850	9854	9859	9863
97	9868	9872	9877	9881	9886	9890	9894	9899	9903	9908
98	9912	9917	9921	9926	9930	9934	9939	9943	9948	9952
99	9956	9961	9965	9969	9974	9978	9983	9987	9991	9996
N	0	1	2	3	4	5	6	7	8	9

TABLE XVI Natural Logarithms

N	0	1	2	3	4	5	6	7	8	9
0.0		5.395	6.088	6.493	6.781	7.004	7.187	7.341	7.474	7.592
0.1	7.697	7.793	7.880	7.960	8.034	8.103	8.167	8.228	8.285	8.339
0.2	8.391	8.439	8.486	8.530	8.573	8.614	8.653	8.691	8.727	8.762
0.3	8.796	8.829	8.861	8.891	8.921	8.950	8.978	9.006	9.032	9.058
0.4	9.084	9.108	9.132	9.156	9.179	9.201	9.223	9.245	9.266	9.287
0.5	9.307	9.327	9.346	9.365	9.384	9.402	9.420	9.438	9.455	9.472
0.6	9.489	9.506	9.522	9.538	9.554	9.569	9.584	9.600	9.614	9.629
0.7	9.643	9.658	9.671	9.685	9.699	9.712	9.726	9.739	9.752	9.764
0.8	9.777	9.789	9.802	9.814	9.826	9.837	9.849	9.861	9.872	9.883
0.9	9.895	9.906	9.917	9.927	9.938	9.949	9.959	9.970	9.980	9.990
1.0	0.00000	0995	1980	2956	3922	4879	5827	6766	7696	8618
1.1	9531	*0436	*1333	*2222	*3103	*3976	*4842	*5700	*6511	*7395
1.2	0.1 8232	9062	9885	*0701	*1511	*2314	*3111	*3902	*4686	*5464
1.3	0.2 6236	7003	7763	8518	9267	*0010	*0748	*1481	*2208	*2930
1.4	0.3 3647	4359	5066	5767	6464	7156	7844	8526	9204	9878
1.5	0.4 0547	1211	1871	2527	3178	3825	4469	5108	5742	6373
1.6	7000	7623	8243	8858	9470	*0078	*0682	*1282	*1879	*2473
1.7	0.5 3063	3649	4232	4812	5389	5962	6531	7098	7661	8222
1.8	8779	9333	9884	*0432	*0977	*1519	*2058	*2594	*3127	*3658
1.9	0.6 4185	4710	5233	5752	6269	6783	7294	7803	8310	8813
2.0	9315	9813	*0310	*0804	*1295	*1784	*2271	*2755	*3237	*3716
2.1	0.7 4194	4669	5142	5612	6081	6547	7011	7473	7932	8390
2.2	8846	9299	9751	*0200	*0648	*1093	*1536	*1978	*2418	*2855
2.3	0.8 3291	3725	4157	4587	5015	5442	5866	6289	6710	7129
2.4	7547	7963	8377	8789	9200	9609	*0016	*0422	*0826	*1228
2.5	0.9 1629	2028	2426	2822	3216	3609	4001	4391	4779	5166
2.6	5551	5935	6317	6698	7078	7456	7833	8208	8582	8954
2.7	9325	9695	*0063	*0430	*0796	*1160	*1523	*1885	*2245	*2604
2.8	1.0 2962	3318	3674	4028	4380	4732	5082	5431	5779	6126
2.9	6471	6815	7158	7500	7841	8181	8519	8856	9192	9527
3.0	9861	*0194	*0526	*0856	*1186	*1514	*1841	*2168	*2493	*2817
3.1	1.1 3140	3462	3783	4103	4422	4740	5057	5373	5688	6002
3.2	6315	6627	6938	7248	7557	7865	8173	8479	8784	9089
3.3	9392	9695	9996	*0297	*0597	*0896	*1194	*1491	*1788	*2083
3.4	1.2 2378	2671	2964	3256	3547	3837	4127	4415	4703	4990
3.5	5276	5562	5846	6130	6413	6695	6976	7257	7536	7815
3.6	8093	8371	8647	8923	9198	9473	9746	*0019	*0291	*0563
3.7	1.3 0833	1103	1372	1641	1909	2176	2442	2708	2972	3237
3.8	3500	3763	4025	4286	4547	4807	5067	5325	5584	5841
3.9	6098	6354	6609	6864	7118	7372	7624	7877	8128	8379
4.0	8629	8879	9128	9377	9624	9872	*0118	*0364	*0610	*0854
N	**0**	**1**	**2**	**3**	**4**	**5**	**6**	**7**	**8**	**9**

Take tabular value—10

TABLE XVI Natural Logarithms (Continued)

N	0	1	2	3	4	5	6	7	8	9
4.0	8629	8879	9128	9377	9624	9872	*0118	*0364	*0610	*0854
4.1	1.4 1099	1342	1585	1828	2070	2311	2552	2792	3031	3270
4.2	3508	3746	3984	4220	4456	4692	4927	5161	5395	5629
4.3	5862	6094	6326	6557	6787	7018	7247	7476	7705	7933
4.4	8160	8387	8614	8840	9065	9290	9515	9739	9962	*0185
4.5	1.5 0408	0630	0851	1072	1293	1513	1732	1951	2170	2388
4.6	2606	2823	3039	3256	3471	3687	3902	4116	4330	4543
4.7	4756	4969	5181	5393	5604	5814	6025	6235	6444	6653
4.8	6862	7070	7277	7485	7691	7898	8104	8309	8515	8719
4.9	8924	9127	9331	9534	9737	9939	*0141	*0342	*0543	*0744
5.0	1.6 0944	1144	1343	1542	1741	1939	2137	2334	2531	2728
5.1	2924	3120	3315	3511	3705	3900	4094	4287	4481	4673
5.2	4866	5058	5250	5441	5632	5823	6013	6203	6393	6582
5.3	6771	6959	7147	7335	7523	7710	7896	8083	8269	8455
5.4	8640	8825	9010	9194	9378	9562	9745	9928	*0111	*0293
5.5	1.7 0475	0656	0838	1019	1199	1380	1560	1740	1919	2098
5.6	2277	2455	2633	2811	2988	3166	3342	3519	3695	3871
5.7	4047	4222	4397	4572	4746	4920	5094	5267	5440	5613
5.8	5786	5958	6130	6302	6473	6644	6815	6985	7156	7326
5.9	7495	7665	7834	8002	8171	8339	8507	8675	8842	9009
6.0	9176	9342	9509	9675	9840	*0006	*0171	*0336	*0500	*0665
6.1	1.8 0829	0993	1156	1319	1482	1645	1808	1970	2132	2294
6.2	2455	2616	2777	2938	3098	3258	3418	3578	3737	3896
6.3	4055	4214	4372	4530	4688	4845	5003	5160	5317	5473
6.4	5630	5786	5942	6097	6253	6408	6563	6718	6872	7026
6.5	7180	7334	7487	7641	7794	7947	8099	8251	8403	8555
6.6	8707	8858	9010	9160	9311	9462	9612	9762	9912	*0061
6.7	1.9 0211	0360	0509	0658	0806	0954	1102	1250	1398	1545
6.8	1692	1839	1986	2132	2279	2425	2571	2716	2862	3007
6.9	3152	3297	3442	3586	3730	3874	4018	4162	4305	4448
7.0	4591	4734	4876	5019	5161	5303	5445	5586	5727	5869
7.1	6009	6150	6291	6431	6571	6711	6851	6991	7130	7269
7.2	7408	7547	7685	7824	7962	8100	8238	8376	8513	8650
7.3	8787	8924	9061	9198	9334	9470	9606	9742	9877	*0013
7.4	2.0 0148	0283	0418	0553	0687	0821	0956	1089	1223	1357
7.5	1490	1624	1757	1890	2022	2155	2287	2419	2551	2683
7.6	2815	2946	3078	3209	3340	3471	3601	3732	3862	3992
7.7	4122	4252	4381	4511	4640	4769	4898	5027	5156	5284
7.8	5412	5540	5668	5796	5924	6051	6179	6306	6433	6560
7.9	6686	6813	6939	7065	7191	7317	7443	7568	7694	7819
8.0	7944	8069	8194	8318	8443	8567	8691	8815	8939	9063
N	0	1	2	3	4	5	6	7	8	9

TABLE XVI Natural Logarithms (Continued)

N	0	1	2	3	4	5	6	7	8	9
8.0	7944	8069	8194	8318	8443	8567	8691	8815	8939	9063
8.1	9186	9310	9433	9556	9679	9802	9924	*0047	*0169	*0291
8.2	2.1 0413	0535	0657	0779	0900	1021	1142	1263	1384	1505
8.3	1626	1746	1866	1986	2106	2226	2346	2465	2585	2704
8.4	2823	2942	3061	3180	3298	3417	3535	3653	3771	3889
8.5	4007	4124	4242	4359	4476	4593	4710	4827	4943	5060
8.6	5176	5292	5409	5524	5640	5756	5871	5987	6102	6217
8.7	6332	6447	6562	6677	6791	6905	7020	7134	7248	7361
8.8	7475	7589	7702	7816	7929	8042	8155	8267	8380	8493
8.9	8605	8717	8830	8942	9054	9165	9277	9389	9500	9611
9.0	9722	9834	9944	*0055	*0166	*0276	*0387	*0497	*0607	*0717
9.1	2.2 0827	0937	1047	1157	1266	1375	1485	1594	1703	1812
9.2	1920	2029	2138	2246	2354	2462	2570	2678	2786	2894
9.3	3001	3109	3216	3324	3431	3538	3645	3751	3858	3965
9.4	4071	4177	4284	4390	4496	4601	4707	4813	4918	5024
9.5	5129	5234	5339	5444	5549	5654	5759	5863	5968	6072
9.6	6176	6280	6384	6488	6592	6696	6799	6903	7006	7109
9.7	7213	7316	7419	7521	7624	7727	7829	7932	8034	8136
9.8	8238	8340	8442	8544	8646	8747	8849	8950	9051	9152
9.9	9253	9354	9455	9556	9657	9757	9858	9958	*0058	*0158
10.0	2.3 0259	0358	0458	0558	0658	0757	0857	0956	1055	1154
N	0	1	2	3	4	5	6	7	8	9

ANSWERS TO PROBLEMS

Chapter 1

1–1. 30

1–2. (a) 11,880
(b) 495
(c) 161,700
(d) n

1–3. k^m

1–4. 81

1–5. (a) 1024
(b) 56
(c) 252
(d) 36

1–6. (a) 216
(b) 2^n
(c) 210
(d) 5040

1–7. (a) 52
(b) $\binom{52}{5}$
(c) 26

1–11. (b) 2^n

1–12. (a) 31, 22, 13, 62, 53, 44, 35, 26, 66
(b) 22, 24, 26, 42, 44, 46, 62, 64, 66
(c) 11, 22, 33, 44, 55, 66
(d) 16, 15, 26, 61, 51, 62
(e) 46, 55, 64

1–15. (a) 1,098,240
(b) 123,552
(c) 3744
(d) 54,912
(e) 5148

1–17. ϕ, $\{a\}$, $\{b\}$, $\{c\}$, $\{a, b\}$, $\{c, a\}$, $\{b, c\}$, $\{a, b, c\}$

1–19. (a) Sets of form $A_X = \{\omega \mid X(\omega) = x\}$, $x = 10, 11, \ldots, 100$
(b) A_b = set of black students
A_b^c = set of nonblack students

1–20. (a) $\{HHH\}, \{HHT, HTH, THH\}, \{HTT, THT, TTH\}, \{TTT\}$
(b) $\{HHH, TTT\}$, $\{HHT, HTH, THH, HTT, THT, TTH\}$

1–21. 51, 61, 62, 15, 16, 26

1–22. (a) $\{a, b\}$, $\{c, d\}$, $\{e\}$
(b) Same [no]

1–23. Each partition set is a pair of lines: $x - y = \pm k$

1–24. (a) $\{(x, y, z) \mid x < y < z\}$
(b) Each set consists of the six permutations of a set of three numbers (a, b, c)

1–25. $\{R, W, Y, V\}$, $\{G, B\}$

1–26. (a) $\{\omega \mid -1 < \omega < 1\}$
(b) The interval $[0, 1)$

1–28. ϕ, $\{a, b\}$, $\{c, d\}$, Ω

1–30. (A finite interval is the intersection of half-infinite ones)

1–32. $P(E) + P(F) + P(G) - P(EF) - P(FG) - P(EG) + P(EFG)$

1–33. (a) $\frac{1}{4}$
(b) $\frac{3}{13}$
(c) $\frac{11}{26}$
(d) $\frac{4}{13}$
(e) $\frac{9}{13}$
(f) $\frac{7}{13}$
(g) $\frac{8}{13}$

1–34. (a) $\frac{1}{3}$
(b) $\frac{1}{3}$
(c) $\frac{7}{12}$
(d) $\frac{2}{3}$

(e) $\frac{2}{3}$
(f) $\frac{1}{3}$
1–35. (a) $\frac{1}{6}$
(b) $\frac{5}{6}$
(c) $\frac{11}{12}$
1–36. $\frac{1}{20}$
1–37. $\frac{3}{95}$
1–38. (a) $\frac{1}{4}$
(b) $\frac{1}{4}$
(c) $\frac{1}{6}$
(d) $\frac{1}{6}$
(e) $\frac{1}{12}$
1–39. Divide corresponding answers in Problem 1–15 by $\binom{52}{5}$
(a) .423
(b) .0475
(c) .00144
(d) .0211
(e) .00198
1–40. (a) $1/e^2$
(b) $1 - 1/e$
(c) $(e - 1)/e^2$
1–41. (a) $\frac{1}{2}$
(b) $\frac{1}{8}$
(c) $\frac{3}{4}$
(d) $1 - \pi/16$
(e) $\frac{3}{8}$
(f) $\frac{3}{4}$
(g) $\frac{3}{4}$
(h) $\frac{7}{8}$
(i) 0
1–42. $\frac{1}{6}$
1–43. $1 - 1/\sqrt{2}$
1–44. (a) $\frac{1}{3}$
(b) $\frac{1}{2}$
1–45. (a) $P(ade) = P(bcde) = \frac{3}{4}$, $P(de) = \frac{1}{2}, P(bc) = \frac{1}{4}$
(b) $\frac{1}{2}$
(c) $\frac{1}{2}$
1–46. (a) $\frac{1}{33}$
(b) $\frac{1}{33}$
(c) $\frac{1}{11}$
(d) $\frac{1}{3}$
(e) $2^6 6!/12!$
1–47. Not if the jailer has named Mark by chance. $\left(\frac{p}{1 + p}\right)$
1–48. (a) $\frac{1}{26}$
(b) $\frac{11}{221}$
(c) $\frac{6}{25}$
1–49. (a) $\frac{1}{2}$
(b) $\frac{1}{2}$
1–51. $\frac{1}{46}$
1–52. $\frac{7}{19}$
1–53. (a) .12
(b) .4
(c) .3
1–57. (a) $\frac{1}{3}, \frac{1}{3}$; No
(b) Yes
1–61. (a), (b) $(\frac{1}{6})^3(\frac{5}{6})^5$
(c) $(\frac{1}{6})^6(\frac{5}{6})^2$
1–62. 0.97, 0.34
1–63. 0.6561
1–64. 0.1225, 0.2450, 0.4350

Chapter 2

In the case of answers that are distribution functions, a particular function will be specified only for the range of values in which it changes, it being understood that the function is 0 to the left and 1 to the right of that range.

In the case of answers that are density functions, only the formula for the range of nonzero values will be given, it being then understood that the density vanishes outside the specified range.

2–1. $P(X = 0) = P(X = 4) = \frac{1}{16}$
$P(X = 1) = P(X = 3) = \frac{1}{4}$
$P(X = 2) = \frac{3}{8}$
2–2. (a) $\frac{1}{2}$
(b) $\frac{1}{32}$
(c) 0
(d) $\frac{239}{288}$
(e) $\frac{1}{4}$
2–3. (a) $\frac{1}{4}$
(b) $\frac{1}{4}$
(c) $\frac{1}{64}$
(d) $\frac{3}{32}$
2–5. .9192, .0099, .3314, .5371, .0228, .0026, .8400, .8426
2–6. $F(\lambda) = (1 + 2\lambda)/4, 0 \leq \lambda < 1$
2–7. (a) 0
(b) $\frac{1}{4}$
(c) $\frac{1}{2}$

(d) $\frac{3}{4}$
(e) $\frac{1}{4}$
(f) $\frac{3}{4}$

2–8. (a) $3e^{-2}/4$
(b) $\frac{1}{4}$
(c) $\frac{1}{4}$

2–9. $\frac{1}{3}$

2–10. $F_\theta(\lambda) = 1 - \cos\lambda, 0 \le \lambda \le \pi/2$
$F_\varphi(\lambda) = \frac{\lambda}{2\pi}, 0 \le \lambda \le 2\pi$

2–11. 3.7, 0.16, 0.48, 0.30

2–12. 0, 0.014, 2.4, 0.74

2–13. (b) One example:
$$Y(\omega) = \begin{cases} 0, \text{ for } \omega = a, \\ 1, \text{ for } \omega = b, c, d \end{cases}$$

2–14.

x_i	0	1
p_i	.8	.2

y_i	0	1	2
p_i	$\frac{1}{3}$	$\frac{8}{15}$	$\frac{2}{15}$

2–15. $p(1) = p(2) = p(3) = \frac{1}{5}$, $p(4) = \frac{2}{5}$

2–16.

x_i	0	1	3
p_i	$\frac{1}{3}$	$\frac{1}{2}$	$\frac{1}{6}$

2–17. (a)

y_i	1	2	3
p_i	.6	.3	.1

r_i	2	3	4
p_i	.3	.4	.3

(b) .9, .7

2–18. $f(k) = \frac{1}{36}(6 - |k - 7|)$

2–19. $f_X(x) = 1, 0 < x < 1$
$f_Y(\theta) = \frac{1}{2\pi}, 0 < \theta < 2\pi$

2–20. $f(x) = 2x, 0 < x < 1$

2–21. $[\pi(1 + x^2)]^{-1}$, 1, 0.5

2–22. $F(y) = 1 - e^{-2y}, y > 0$; $P(Y > 1) = e^{-2}$

2–23. (a) $F(x) = \frac{1}{2} + x - \frac{1}{2}x|x|$, $-1 < x < 1$
(b) $\frac{1}{2}, \frac{1}{4}$

2–24. $.2/\sqrt{\pi}$

2–26. (a) $k = \frac{1}{2}$
(b) $k = 6$
(c) No
(d) $k = \frac{1}{2}$

2–27. (a) No
(b) $k = 1 - p$

2–28. $f(x) = 1/(2A), -A < x < A$

2–31. (a) $f(x) = 1, 0 < x < 1$
(b) $f(x) = 1/(2\sqrt{x}), 0 < x < 1$

2–32. One way: $Y = 2X - 1$;
another: $Y_k = (sgn\ X_{2k-1})X_{2k}$

2–33.

y_k	−2	0	4
p_k	$\frac{1}{3}$	$\frac{1}{3}$	$\frac{1}{3}$

2–34. $F_Y(y) = y, 0 \le y \le 1$

2–35. $F_Y(y) = y, 0 \le y \le 1$

2–37. (a) $\pi/4$
(b) $\frac{1}{2}$
(c) 1
(d) $\frac{7}{16}$

2–40. (a) $\frac{5}{9}$
(b) $f(0) = \frac{25}{36}, f(1) = \frac{10}{36}$, $f(2) = \frac{1}{36}$ (for X or Y)

2–41. (a) $f_Y(y) = \frac{1}{3}, y = 2, 3, 4$
$f_X(x) = \frac{1}{4}$ for $x = 1, 3$,
$f_X(2) = \frac{1}{2}$
(b) $\frac{1}{3}$
(c) $\frac{1}{2}$
(d) $\frac{5}{12}$

2–42. (a) $\frac{1}{2}$
(b) $(4 - x^2)^{1/2}/(2\pi)$
(c) $\frac{3}{4}$
(d) $\lambda/4, 0 < \lambda < 4$
(e) $\lambda^2/4, 0 < \lambda < 2$

2–44. (a) $f_X(\lambda) = f_Y(\lambda) = e^{-\lambda}, \lambda > 0$
(b) $(1 - e^{-x})(1 - e^{-y})$, $x, y > 0$
(c) 0
(d) $1 - 5e^{-4}$
(e) $1 - (1 + \lambda)e^{-\lambda}, \lambda > 0$

2–45. (a) $f(\lambda) = 1, 0 < \lambda < 1$ (both X, Y)
(b) $w^2, 0 < w < 1$

2–46. $f(x, y) = 2, x > 0, y > 0$, $x + y < 1$ (if the altitude of the triangle is taken as a unit)

2–47. (a) $\frac{1}{2}$
(b) $\frac{1}{2}$
(c) $\frac{1}{3}$
(d)

y	2	3	4
$f(y \mid 2)$	$\frac{1}{3}$	0	$\frac{2}{3}$

(e)

x	1	2	3
$f(x \mid 2)$	$\frac{1}{4}$	$\frac{1}{2}$	$\frac{1}{4}$
$f(x \mid 3)$	$\frac{1}{2}$	0	$\frac{1}{2}$
$f(x \mid 4)$	0	1	0

2–48. (a) $f(x \mid 1) = 1/(2\sqrt{3})$, $-\sqrt{3} < x < \sqrt{3}$

(b) $2/\sqrt{15}$

2–49. $f(x \mid y) = \dfrac{2(1 - x - y)}{(1 - y)^2}$, $0 < x < 1 - y$

2–56. $f(x, y) = 1, 0 < x < 1, 0 < y < 1$

2–59. (a)

x	0	1	2
$f(x)$	$\frac{1}{2}$	$\frac{1}{6}$	$\frac{1}{3}$

(same for Y)

z	0	1
$f(z)$	$\frac{1}{2}$	$\frac{1}{2}$

(b) $f(0, 0) = \frac{1}{3}; f(2, 1) = f(2, 0) = f(1, 2) = f(0, 2) = \frac{1}{6}$

(c) $f(0, 0) = f(1, 2) = f(2, 0) = \frac{1}{3}$

2–60. $f(r) = 3r^2, 0 < r < 1$

2–61. (a) 6
(b) $6(1 - x - y)$, $0 < x < 1 - y < 1$
(c) $3(1 - x)^2, 0 < x < 1$
(d) 8, for $0 < x < \frac{1}{2} - y < \frac{1}{2}$

2–62. (a) $\exp(-\sum x_i), x_i > 0$
(b) $p^{\Sigma x_i}(1 - p)^{n - \Sigma x_i}, x_i = 0, 1$

2–65. (a) $\frac{4}{5}$
(b) $\frac{14}{5}$
(c) 1
(d) $\frac{10}{13}$

2–66. $EY = \frac{3}{2}, EZ = \frac{9}{2}, ER = 3$

2–67. 7

2–68. 10

2–69. 1

2–70. 0

2–71. 0

2–72. $\frac{1}{2}$

2–73. 0

2–74. 1

2–76. $\frac{1}{2}(1 - x)$

2–77. $3 - k/2, k = 1, 2, 3$

2–78. (a) $\frac{1}{3}$
(b) $\frac{1}{2}$

2–79. (a) $\frac{1}{2}$
(b) $\frac{1}{3}$

2–80. The sum is divergent

2–81. $E(Y \mid x) = x/3, EY = \frac{1}{6}$

2–82. $\frac{1}{20}$

2–83. (a) $y/2$
(b) $\frac{1}{3}$
(c) $\frac{1}{6}$

2–84. (a) 0
(b) $f_{X|y}(x)$ depends on y

2–85. Converse is false

2–86. (a) $\exp(-x - y - z)$, $x, y, z > 0$
(b) $E(X + Y) = 2EX = 2$
(c) $EZ = 1$
(d) 12
(e) $\frac{1}{3}$

2–87. (a) $\frac{3}{2}$
(b) 1

2–88. (a) $\frac{7}{2}$
(b) $\frac{91}{6}$

2–89. (a) .8
(b) $.2 + \dfrac{.8}{1 - t}$

2–92. $\frac{34}{25}$

2–93. 1

2–94. $1/\sqrt{18}$

2–95. .96

2–96. 0, 1

2–97. $\mu_3' - 3\mu_1'\mu_2' + 2\mu_1'^3$

2–99. $\frac{1}{4}$ (equal to the Chebyshev bound)

2–100. .02, $\frac{1}{8}$

2–101. $\frac{1}{3}$ versus $\frac{2}{9}$ when $b = 3$
$1/\sqrt{2}$ versus 1 when $b = \sqrt{2}$

2–106. Both equal to $\frac{8}{3}$

2–107. (a) 0
(b) $\frac{3}{4}$

2–108. $-10/147$

2–109. Dependent

2–110. $-\frac{1}{2}$

2–117. (a) 10
(b) 290/17

2–118. 4.58×10^{-3}

2–119. 1.12 miles

2–120. $B_0^2 \operatorname{var} \alpha + A_0^2 \operatorname{var} \beta$

2–121. (a) $e^{tb}, EX^r = b^r$
(b) $\mu_k = 0$

2–122. Divergent

2–123. $e^{bt}\psi_X(at)$

2–124. 2

2–125. $(2 - t)^{-1}, 1, 2$

2–126. $\psi(t) = (e^t - 1)/t = 1 + t/2 + t^2/6 + \cdots$, $\mu = \frac{1}{2}, \sigma^2 = \frac{1}{12}$

2–127. $\exp(t^2/2)$

2–128. $(-1)^m\binom{-n}{m}2^{-n-m} = \binom{m+n-1}{m}2^{-m-n}$

2–129. $\psi_Y(t) = (1-t)^{-n}$, $f_Y(y) = y^{n-1}e^{-y}/(n-1)!$

2–130. $\exp(nt^2/2)$; n

2–131. 25/216

2–132. .063

2–133. $(1 - it/\lambda)^{-1}$

2–134. $\exp(-t^2/2)$

2–135. $\exp(itm - |at|)$

2–136. (a) $1 + i\mu t + (\sigma^2 + \mu^2)\frac{1}{2}(it)^2 + o(t^2)$
(b) $1 - \frac{1}{2}\sigma^2 t^2 + o(t^2)$

2–138. (a) $\left(1 - \dfrac{it}{n\lambda}\right)^{-n}$
(b) $\exp(-\frac{1}{2}t^2/n)$
(c) $\exp(-|t|)$

2–139. $\Pi\psi_X(t_j - \bar{t}), \exp\left[\frac{1}{2}\sum(t_j - \bar{t})^2\right]$

2–142. .159

2–143. .92

2–144. 146.24

Chapter 3

3–1. $\binom{100}{50}\left(\frac{1}{2}\right)^{100} \doteq .079$

3–2. $\frac{1}{16}, \frac{7}{128}, \frac{7}{128}$

3–3. Bernoulli, $p = \frac{1}{4}$

3–4. $1 - (\frac{1}{12})^4$

3–5. $(.9)^4$

3–6. $\frac{16}{2187}$

3–7. .59, .92, 5

3–8. $(n)_r p^r$

3–9. 40.2

3–10. (a) 6
(b) 6, $\frac{9}{2}$

3–11. 3 and 4

3–15. Binomial, $p = F(x)$

3–16.

0	1	2	3	4
$\frac{1}{22}$	$\frac{10}{33}$	$\frac{5}{11}$	$\frac{2}{11}$	$\frac{1}{66}$

3–17. .0064, .0081

3–18. 1, 3

3–19. $\sum_{6}^{10}\binom{10}{k}(.45)^k(.55)^{10-k} \doteq .26$

3–21. $\binom{5}{k}\binom{15}{5-k}\Big/\binom{20}{5}$

3–22. (a) $\frac{1}{4}$
(b) Hypergeometric, $N = 10$, $M = 5$, $n = 3$
(c) 2.4

3–23. $2\binom{12}{6}$

3–24. .46, .08, .94

3–25. .19

3–26. .18

3–27. (a) $p(k) = .3(.7)^k$, $k = 0, 1, \ldots$
(b) $\left[\binom{7}{k}\Big/\binom{10}{k}\right]\dfrac{3}{10-k}$, $k = 0, 1, \ldots, 7$

3–30. .57, .30, .035

3–31. .184

3–32. (a) .147
(b) $e^{-2T/3}(2T/3)^k/k!$

3–33. .08

3–34. .018, .092

3–36. $9e^3$

3–38. .713

3–39. $\dfrac{1}{\lambda}\log 2$

3–40. (a) e^{-3}
(b) .423

3–41. Exponential, $\lambda = \frac{1}{5}$

3–44. .974, .074

3–45. .151

3–46. .049

3–47. (a) .074
(b) .265
(c) .0455

3–48. $\frac{3}{2}$

3–49. Same as initial distribution

3–52. .1336

3–53. .3830

3–54. .7257, .2881

3–56. 8.65, 11.35

3–57.

Interval	Probability
$(-\infty, 6), (14, \infty)$	.0228
(6, 7), (13, 14)	.0440
(7, 8), (12, 13)	.0919
(8, 9), (11, 12)	.1498
(9, 10), (10, 11)	.1915

3–59. $(2/\sigma)f(y/\sigma)$, $y > 0$, where f is standard normal; $(2\sigma^2/\pi)^{1/2}$

3–60. (b) e^{μ}

3–61. (a) 6.1 per cent, 2.6 per cent
(b) $3.8k$

3–63. $(\pi\sigma^2/2)^{1/2}$, $2\sigma^2(1 - \pi/4)$

3–64. $(2\pi u)^{-1/2}e^{-u/2}$, $u > 0$;
mean = 1, variance = 2

3–66. 69.03, .16

3–67. $(2^5 4!\,\sigma^2)^{-1}(y/\sigma^2)^4 e^{-y/(2\sigma^2)}$
$y > 0$; $10\sigma^2$, $20\sigma^4$

3–71.

y \ x	0	1	2
0	$\frac{1}{36}$	$\frac{1}{9}$	$\frac{1}{9}$
1	$\frac{1}{6}$	$\frac{1}{3}$	0
2	$\frac{1}{4}$	0	0

3–72. 81/1024

3–73. 1000/12,597

3–74. $8!/2^{20}$

Chapter 4

4–1. 0, 0, 9, 3.28

4–2. ± 2

4–3. (a) 75, 13
(b) 75, 13.15
(c) 80, 72.5

4–4. 29, 19, 6

4–5. (a) $(b - a)^{-n}$, $a < x_i < b$
$(i = 1, \ldots, n)$
(b) $(2\pi\sigma^2)^{-n/2} \times$
$\exp\{-\frac{1}{2}\sum(x_i - \mu)^2/\sigma^2\}$
(c) $p^{\Sigma x_i}(1 - p)^{n - \Sigma x_i}$, $x_i = 0, 1$
(d) $e^{-nm}m^{\Sigma x_i}/\prod x_i!$, $x_i = 0, 1, \ldots$
(e) $p^n(1 - p)^{\Sigma x_i}$, $x_i = 0, 1, \ldots$

4–6. $$\frac{\binom{M}{\sum x_i}\binom{N - M}{n - \sum x_i}}{\binom{n}{\sum x_i}\binom{N}{n}}$$

4–8. 13.28, 1.39

4–9. (a) $f(i, j) = \frac{1}{6}$, for
$i, j = 1, 2, 3, 4, i < j$
(b) $f(1) = \frac{1}{2}, f(2) = \frac{1}{3}, f(3) = \frac{1}{6}$
(c) $f(2.5) = \frac{1}{3}; f(x) = \frac{1}{6}$ for
$x = 1.5, 2, 3, 3.5$

4–10. .0228

4–11. (a) 0
(b) 4.004
(c) .0456

4–12. μ_k', $[\mu_{2k'} - (\mu_k')^2]/n$

4–15. $EX_{(1)} = \dfrac{1}{n + 1}$, $E(X)_{(n)} = \dfrac{n}{n + 1}$,
$\text{var } X_{(1)} = \text{var } X_{(n)} = \dfrac{n}{(n+2)(n+1)^2}$

4–16. $b - \dfrac{b - a}{n + 1}$, $\dfrac{n(b - a)^2}{(n + 2)(n + 1)^2}$

4–17. θ, $[2(n + 2)(n + 1)]^{-1}$

4–18. (a) $1 - [1 - R(x)]^n$
(b) $1 - \prod[1 - R_i(x)]$

4–19. (a) $[R(x)]^n$
(b) $\prod R_i(x)$

4–20. $\dfrac{n - 1}{n + 1}(b - a)$, $\dfrac{2(n - 1)(b - a)^2}{(n + 2)(n + 1)^2}$

4–21. $n(n - 1)[1 - F(v)]^{n-2}f(u)f(v)$

4–22. $\mathcal{N}[x_{.5}, \pi^2/(4n)]$

4–23. (a) $f(z_3 \mid A) = 1$;
$f(z_1 \mid B) = \frac{2}{3}, f(z_2 \mid B) = \frac{1}{3}$
(b) Yes
(c) No

4–24. U is a reduction of T

4–26. (b) Each has probability $\frac{1}{10}$
(c) (iv) {123, 234, 345},
{124, 134, 235, 245},
{125, 135, 145}

4–27. $1\Big/\binom{n}{k}$; yes

4–28. (a) No
(b) Yes
(c) No

4–30. If $N > 2$, conditional probabilities are $\frac{1}{3}$ for (1, 3), (2, 2), (3, 1); if $N = 2$, conditional probability is 1 for (2, 2)

4–31. $\sum X_i$ in each case

4–32. $\sum X_i$

4–33. $\sum Z_i^2$

4–34. $(\sum X_i, \sum X_i^2)$

4–35. $(f_1, \ldots, f_k)$, where f_i = frequency of x_i

4–36. (a), (b) $\sum X_i$
(c) $\sum(X_i - \mu)^2$
(d) $\sum X_i$
(e) $\sum Z_i^2$
(f) $\sum X_i^2$

4–37. $\{z_1, z_2\}$, $\{z_3\}$

4–38. $\sum X_i$

4–41. $(X_{(1)}, X_{(n)})$ is sufficient

4–42. (a) $(2\pi\sigma^2)^{-n/2} \times \exp[-\frac{1}{2}\sum (X_i - \mu)^2/\sigma^2]$
(b) $m^{\Sigma X_i} e^{-nm}/\prod X_i!$
(c) $p^{\Sigma X_i}(1 - p)^{n - \Sigma X_i}$
(d) $\lambda^n \exp\{-\lambda \sum X_i\}$
(e) $\prod \{\pi[1 + (X_i - \theta)^2]\}^{-1}$

4–43.

Y	Likelihood
0	$(5 - M)(4 - M)$
1	$M(5 - M)$
2	$M(M - 1)$

4–44. θ^{-n}, for $\theta > X_{(n)}$; 0 elsewhere
4–45. Common likelihood is $\lambda^k e^{-\lambda t_0}$
4–46. (a) $\hat{\mu} = \bar{X}, \hat{\sigma}^2 = S^2$
(b) $\hat{\theta} = \bar{X}$
(c) $\hat{\theta} = X_{(n)}$
4–47. (a) $\hat{\theta} = \bar{X}$
(b) $\hat{\lambda} = 1/\bar{X}$
4–51. (a) n
(b) $\dfrac{n}{pq}$
(c) n/λ^2
4–52. $\frac{1}{2}$

Chapter 5

5–2. $\sum a_i = 1$
5–3. Only in the trivial case $\sigma_T = 0$
5–4. (a) $EX_{(n)} = \dfrac{n}{n + 1}\,\theta \to \theta$
(b) $2\bar{X}, \theta^2/(3n)$
(c) $\dfrac{n + 1}{n} X_{(n)}, \dfrac{\theta^2}{n(n + 2)}$
5–5. $[\bar{X}(1 - \bar{X})/n]^{1/2}$
5–6. $\bar{X}/\sqrt{n}$
5–8. $pq/n, (n + 2)^{-2}[npq + (1 - 2p)^2]$
5–9. (a) $\dfrac{n\lambda}{n - 1}, \dfrac{(n - 1)}{n\bar{X}}$
5–12. $8/\pi^2$
5–14. $h(Y) = 2^{-Y}, e^{-3\lambda} - e^{-4\lambda}$, $\lambda/(e^{\lambda} - 1)$
5–16. $\Phi\left(\dfrac{\lambda - \bar{X}}{\sqrt{1 - 1/n}}\right)$
5–17. $\dfrac{n + 1}{n} X_{(n)}$
5–19. X^2; X^2
5–20. $\bar{X}, S^2$
5–21. $2\bar{X}$
5–22. By Property 2
5–23. $\hat{M} = 2$
5–24. $\hat{b} = X_{(1)}$
5–26. e^{-2X} (both)
5–27. $(X_{(1)}, X_{(n)}), X_{(n)} - X_{(1)}$
5–28. $\frac{1}{2}X_{(n)}, \frac{1}{2}[X_{(n)} + X_{(1)}]$
5–29. 664
5–30. (2.248, 2.352), (2.261, 2.339)
5–31. (2.290, 2.310)
5–32. $\dfrac{\bar{X}}{1 \pm 2/\sqrt{n}}$
5–33. (1.82, 2.43)

Chapter 6

6–1. S, C, C, S, C, C, S, C
6–2.

Number white	0	1	2	3	4	5
Type I error	0	.4				
Type II error			.3	.1	0	0

6–3. $K = \frac{1}{2}$
6–4. $\alpha = .506, \beta = .494$
6–5. (a) $n = 86, K = \frac{1}{2}$
(b) $n = 53, K = .64$
6–6. Partial answer: $\alpha = \frac{1}{2}, \beta = 1$ for the test that rejects H_0 for exactly one head
6–7. $\alpha = .208, \beta = .268$
6–8. $1 - 3p + 3p^2, \alpha = \frac{1}{4}$; $1 - 3p^2(1 - p), \alpha = \frac{5}{8}$
6–9. $n = 14, K = 104.4$
6–10. $(10 - M)(9 - M)(8 - M) \times (7 + 3M)/5040$
6–11. $K = .0351$, $\pi(\sigma^2) = \Phi\left(\dfrac{.99 - .0351/\sigma^2}{.141}\right)$
6–12. $K = .392, \pi(\mu) = 1 + \Phi\left(\dfrac{1 - \mu - K}{.2}\right) - \Phi\left(\dfrac{1 - \mu + K}{.2}\right)$
6–13. .067
6–14. (a) 337.30
(b) $Z > 9$ (reject)
(c) $1 - \Phi\left(\dfrac{267 - \mu}{13.74}\right)$
6–15. (a) $\Phi(\sqrt{n}\mu - K)$
(b) $n = 271, K = 0$
6–16. (a) $\frac{1}{2}[\pi_{C_1}(\theta) + \pi_{C_2}(\theta)]$
(b) $\phi(\mathbf{x}) = \frac{1}{2}$ in $C_1C_2^c + C_2C_1^c$, 1 in C_1C_2, 0 elsewhere

6–17. $\alpha = \frac{1}{2}f_0(2) + \sum_{t<2} f_0(t)$

6–18. $\phi(\mathbf{x}) = \begin{cases} 1, \sum x_i > 2, \\ .107, \sum x_i = 2, \\ 0, \sum x_i < 2 \end{cases}$

6–19. (a) z_3, z_1, z_5, z_4, z_2
(c) NP: $C = \{z_3, z_1\}$, $\beta = .4$ (compared with $\beta = .8$ for $\{z_4\}$)

6–21. $\bar{X} <$ const

6–22. (a) $\sum (X_i - \mu)^2 > K$ if $\sigma_1 > \sigma_0$
(b) $n = 73$, $K = 406$

6–23. Reject H_0 if $T > 5$; reject H_0 with probability .16 if $T = 5$

6–24. (a) $X > 1$; $\alpha = \beta = \frac{1}{4}$
(b) $X < -1$ or $X > 0$; $\alpha = \frac{3}{4}$, $\beta = .045$ ($X > -1$ has $\beta \doteq .10$)
(c) $\frac{7}{5} < X < 6$

6–25. p_0 = minimum of power function

6–27. C: $\sum X_i^2 > 37.7$

6–29. No

6–30. $S > 58$

6–31. $\sum (X_i - 1)^2 / \sum X_i^2 <$ const

6–32. $|X| >$ const; no

6–33.

C	α
ϕ	0
$\{z_3\}$	.1
$\{z_3, z_1\}$	.3
$\{z_3, z_1, z_4\}$	.7
Ω	1

6–34. Reject $v = v_0$ if $re^{-r} < K_1$, where $r = \hat{v}/v_0$

6–36. $\Lambda = \frac{1}{8}$, probability $\frac{1}{4}$; $\Lambda = \frac{27}{32}$, probability $\frac{3}{4}$

6–37. $S^2 > K$;
$\pi(\sigma^2) = 1 - F_{\chi^2(n-1)}(nK/\sigma^2)$

6–38. Reject $p = \frac{1}{2}$ if $\sum X_i \le .415n - 3$; accept $p = \frac{1}{2}$ if $\sum X_i \ge .415n + 2.17$; otherwise continue sampling

6–39. Continue sampling if $\sum X_i$ falls between

$$\tfrac{1}{2}n(\mu_0 + \mu_1) - \frac{\sigma^2}{\mu_1 - \mu_0} \log_e B$$

and

$$\tfrac{1}{2}n(\mu_0 + \mu_1) - \frac{\sigma^2}{\mu_1 - \mu_0} \log_e A$$

6–40. Continue sampling if $(\sum X_i) \log (m_1/m_0)$ falls between $-\log A + n(m_1 - m_0)$ and $-\log B + n(m_1 - m_0)$

6–42. $\mu = (h + 1)/2$,

$$\pi(\mu) = \frac{99^{2\mu-1}}{99^{2\mu-1} + 1}$$

6–43. $E(N) \doteq \dfrac{2\log_e 99}{2\mu - 1} \dfrac{99^{2\mu-1} - 1}{99^{2\mu-1} + 1}$
[with value $(\log 99)^2$ when $\mu = \frac{1}{2}$]

6–45.

θ	0	1	1.44	2	∞
$OC(\theta)$	0	.1	.56	.95	1
$E(N \mid \theta)$	0	7.74	13.5	10.3	3.25

Chapter 7

7–1. (a) 120
(b) $\frac{945}{32}\sqrt{\pi}$
(c) $\frac{4}{15}$
(d) $\frac{3}{32}\sqrt{2\pi}$
(e) $\frac{8}{81}$
(f) $\frac{1}{30030}$
(g) $3\pi/512$
(h) $\frac{1}{12}$

7–2. $EX = \alpha/\lambda$, var $X = \alpha/\lambda^2$

7–3. $\dfrac{rs}{(r + s)^2(r + s + 1)}$

7–4. $EW = \dfrac{r}{s - 1}$,

$$EW^2 = \frac{r(r + 1)}{(s - 1)(s - 2)}$$

7–7. $(k - 2)^{-1}$

7–8. $\sigma^2(1 - 1/n)$, $2\sigma^4(n - 1)/n^2$

7–11. 0

7–12. $\dfrac{2n^2(m + n - 2)}{m(n - 2)^2(n - 4)}$

7–13. .25

7–17. $k + 2\lambda$, $2k + 8\lambda$

7–18. (a) $K_1 = 3.317$, $K_2 = 10.9$
(b) Approx. 86 per cent (using S) versus 83 per cent

7–19. (a) $33.34 \times 10^{-8} < \sigma^2 < 65.24 \times 10^{-8}$
(b) $1.09302 < \mu < 1.09334$
(c) $70(\mu - 1.0932)^2 < 3.84\sigma^2$ together with (a)

7–20. $\sigma^2[1 - n\alpha_n^2/(n - 1)]$

7–21. $\alpha_{10} = .9227$; $.0571\sigma^2$, $.0546\sigma^2$, $.0540\sigma^2$, $.067\sigma^2$

7–22. 4.325, .0338 (standard error = .0081)

7–23. $1 - \Phi\left(\sqrt{\frac{2}{\pi}}\frac{K-\mu}{\sigma/\sqrt{n}}\right)$

7–25. $F = 2.85 < 6.39$ (accept H_0)

7–26. (a) $\frac{\tilde{S}_2^2}{\tilde{S}_1^2}F_{.05} < \frac{\sigma_2^2}{\sigma_1^2} < \frac{\tilde{S}_2^2}{\tilde{S}_1^2}F_{.95}$
(b) $.099 < \sigma_2^2/\sigma_1^2 < .857$ (11, 9 d.f.)

7–27. $1 - F(\theta F_{.95}) + F(\theta F_{.05})$, where F is the c.d.f. of $F(n_1 - 1, n_2 - 1)$ and $\theta = \sigma_2^2/\sigma_1^2$

7–28. $Z = 2.15$, reject H_0

7–29. (a) $T \doteq 2 < 2.18$, accept H_0 at $\alpha = .05$
(b) Becomes marginal at $\alpha = .05$ (using suggested test)

7–30. $t = 3.74$ (reject H_0)

7–31. $\rho = \frac{3}{4}$

Chapter 8

8–1. (a), (c), (g), (h), (i), (j)

8–3. $(\frac{2}{9}, \frac{4}{9}, \frac{1}{3}, 0)$

8–5. He should play

8–6. $m = b + 1$

8–7. (a) p (probability of winning)
(b) No
(c) Do not take if $p < \frac{1}{9}$

8–8. (a) $p = \frac{1}{10}$
(b) 1.90

8–9. a_3, a_1

8–10. Pure: a_1 (loss), a_2 (regret)
Mixed: (1, 0) (loss), $(\frac{1}{3}, \frac{2}{3})$ (regret)

8–11. (a) $(0, \alpha, 0, 0, 1 - \alpha)$, where $\frac{3}{4} \le \alpha \le 1$
(b) $(\frac{2}{3}, \frac{1}{3}, 0, 0, 0)$

8–12. Minimax at $a = \frac{1}{2}$

8–13. Pass with probability $\frac{1}{3}$, reject with probability $\frac{2}{3}$

8–14. a_1

8–15. a_2 if $w < \frac{2}{3}$; a_1 if $w > \frac{2}{3}$ (either if $w = \frac{2}{3}$)

8–16. (a) Reject lot
(b) Pass lot
(c) Pass if $p \le \frac{1}{3}$; reject if $p \ge \frac{1}{3}$

8–17. (a) $a = \frac{1}{2}$
(b) $a = \frac{1}{2}$

8–20. (a) One decision function has zero expected regrets
(b) The set of risk points is the same as without data

8–22. (a) $\theta(1 - \theta)/2$
(b) $(2\theta^2 - 2\theta + 1)/16$; neither is always better; for θ uniform d_2 is better

8–23. (a) If z_1 take a_3, if z_2 take a_1
(b), (c) Same as (a)

8–24. .175

8–25. $X > \frac{1}{2} + \ln 3$, $\alpha = .055$, $\beta = .725$

8–26. $R(p, C) = \begin{cases} 10p^4 - 8p^5 \text{ if } p \le \frac{1}{2}, \\ 1 - 5p^4 + 4p^5 \\ \quad \text{if } p > \frac{1}{2} \end{cases}$

8–27. No

8–28. (a) No
(c) No
(d) $X > 1$

8–29. (a) If $Z = z_1$, take a_3; if $Z = z_2$, take a_3 with probability $\frac{7}{17}$, a_2 with probability $\frac{10}{17}$
(b) (d_8, d_9) with probability $(\frac{1}{4}, \frac{3}{4})$

8–30. Partial answer: $R(\theta_1, d_1) = R(\theta_3, d_1) = .1600$, $R(\theta_2, d_1) = .3174$

8–31. $(\frac{48}{51}, \frac{3}{51})$

8–32. k/n

8–34. Distribution of $M - 1$ is binomial $(N - 1, p)$

8–35. $\mathcal{N}(\bar{X}, \sigma^2/n)$

8–36. 1.97

8–37. Gamma (α, λ); $E_h(\Theta) = \frac{1 + 2\alpha/n}{\overline{X^2} + 2\lambda/n}$

8–38. (a) 1981/2048
(b) $\frac{1}{2}$

8–39. (a) For gamma (α, λ), $\frac{\alpha + n}{\lambda + \sum X_i}$
(b) Improper priors; (2) yields m.l.e.
(c) Estimator tends to $1/\bar{X}$

Chapter 9

9–2. $\chi^2 = 5.64 < 7.81$; accept at $\alpha = .05$

9–4. (a), (b)

χ^2	Λ	Probability
12	1/729	3/729
7	64/3125	36/729
4	1/16	90/729
3	1/16	150/729
1	64/729	360/729
0	1	90/729

(d) $\frac{11}{16}$

9–6. (a) $.045 < p < .215$ (crude: $.017 < p < .183$)
(b) $.404 < p < .596$

9–7. Standard deviation of f/n is .0112

9–8. $Z = 1.76$

9–9. Distribution of Λ under H_0:

Λ	$\frac{1}{4}$	$\frac{1}{16}$
Probability	$\frac{2}{3}$	$\frac{1}{3}$

9–11. $\frac{n_{i+}n_{+j}}{n}$

9–12. Degrees of freedom: $(rc - 1) - (r - 1) - (c - 1)$

9–14. $\chi^2 = 4 > 3.84 = \chi^2_{.95}(1)$ (reject at $\alpha = .05$)

9–15. (a) $C = 1$
(b) 899.7 (800)

9–16. .225, .07

9–17. $\chi^2 = 46.29 > 16.8$ (reject at $\alpha = .01$)

9–18. Reject for either

4	0
5	3

or

1	3
8	0

9–19. Distribution of success patterns:

Pattern	Probability
1 3 0	96/495
2 2 0	108/495
1 1 2	288/495
4 0 0	3/495

Chapter 10

10–2. $\frac{1}{6\pi} \exp\left[-\frac{1}{18}(2u^2 - 2uv + 5v^2)\right]$

10–4. $f(z) = -\log z, \quad 0 < z < 1$

10–6. $f(u) = [B(\frac{1}{2}k, \frac{1}{2}m)]^{-1}u^{k/2-1} \times (1 + u)^{-(k+m)/2}$

10–7. $U = (X - Y)/\sqrt{2}$, $V = (X + Y - 2)/\sqrt{6}$

10–8. $f(\mathbf{x}) = n!$ for $\sum x_i < 1$ and $x_i > 0$ (all i)

10–9. $f(\mathbf{z}) = z_2 z_3^2 \cdots z_n^{n-1} e^{-z_n}$, for $z_n > 0$ and $0 < z_i < 1$, $i = 1, \ldots, n - 1$

10–10. $\frac{1}{\sqrt{10\pi}} \exp\{-\frac{1}{10}[9x^2 - 4x(y - 4) + (y - 4)^2]\}$

10–11. $\mu_X = \mu_Y = 0$, $\sigma_X^2 = \frac{4}{7}$, $\sigma_Y^2 = \frac{2}{7}$, $\sigma_{X,Y} = \frac{1}{7}$

10–12. (a) Circles
(c) Uniform on [0, 1]
(d) Uniform on $[0, 2\pi]$
(e) Uniform on $0 < u < 1$, $0 < v < 2\pi$

10–13. (a) Use $\theta = \frac{1}{2}\cot^{-1} 2$
(b) $U = 9X - 2Y$, $V = Y$

10–19. (b) $\frac{1}{5}\sqrt{10}$
(c) $\begin{pmatrix} 8 \\ 0 \\ 2 \end{pmatrix}, \begin{pmatrix} 13 & 9 & 11 \\ 9 & 13 & 11 \\ 11 & 11 & 11 \end{pmatrix}$
(d) $\phi(t_1, t_2) = \exp[8it_1 - -\frac{1}{2}(13t_1^2 + 18t_1t_2 + 13t_2^2)]$
$f(y_1, y_2) = \frac{1}{2\pi\sqrt{88}} \exp\{-\frac{1}{176}[13(y_1 - 8)^2 - 18y_2(y_1 - 8) + 13y_2^2]\}$

10–20. $\frac{1}{8\pi}\exp\{-\frac{1}{10}[x_1^2 + (2x_1 + \frac{5}{3}x_2)^2]\}$

10–23. (a) .80
(b) $1.03B - 1.94$ (reduction by factor $\doteq .6$)

10–24. Test statistic $= -1.77 < -1.645$ (reject H_0)

Chapter 11

11–1. $\frac{1}{6}$

11–2. $f(\mathbf{y}) = n!$ for $y_i > 0$, $\sum y_i < 1$

11–3. $12(1 - v_2)^2$, $0 < v_1 < v_2 < 1$

11–4. $\frac{(n - 1)!}{y_n^{n-1}}$, $0 < y_1 < \cdots < y_{n-1} < y_n$

11–6. (a) $D_n = .575 > .32$ (reject at $\alpha = .01$)

11–7. $D_n = .05 < .0957$ (accept at $\alpha = .10$)

11–8. .85 (approx. lower bound = .4)

11–9. $F_n(x) \pm d_\alpha$, where $P(D_n > d_\alpha) = \alpha$

11–11. Reject H_0 if $D_n^+ > .245$

11–12. (c) $D = \frac{24}{30} > \frac{15}{30}$ (reject H_0 at $\alpha = .05$)

11–14. (a) $r = 19 < 19.9 = r_{.95}$ (accept at $\alpha = .10$)

(b) $\frac{s - \mu_s}{\sigma_s} = \sqrt{\frac{451}{90}}$ (accept)

11–15. (a) $r \doteq 7$ (reject)

(b) $s \doteq 13$ (reject)

(c) $r = .61$ (reject: $Z \doteq 2$)

11–17. Approximately $\frac{2}{3}$ (large n)

11–19. First few: 5.64, 4.97, 5.15, 3.81, . . .

11–20. $Z = 2.12 > 1.96$ (reject at $\alpha = .05$)

11–21. Reject if 2 or fewer +'s ($\alpha = .088$)

$\pi(p) = (1 - p)^7(28p^2 + 7p + 1)$

(randomize; reject with probability .074 if 3 +'s)

11–22. No ($Z \doteq 1$)

11–23. .95

11–24. (a) $W_- = 1$ (reject)

(b) $T = 2.26 > 1.53$ (reject)

11–25. 255 to 278 (90.8%)

11–29. (a) $R_Y = 17$ (accept H_0)

(b) $c_1 = .191$

11–30. $R_X = 27$ (accept H_0)

11–31. $L = 2$, $R_Y = 17 > 15$, $c_1 \doteq .26$ (accept H_0)

11–34. 1

Chapter 12

12–1. $34y = 53 + 6x$

12–2. (a) $\hat{\gamma} = \sum x^2y/\sum x^4$

(b) $\hat{\alpha} = \frac{1}{\Delta}[\sum xy \sum z^2 - \sum xz \sum yz]$,

$\hat{\beta} = \frac{1}{\Delta}[\sum x^2 \sum yz - \sum xy \sum xz]$,

$\Delta = \sum x^2 \sum z^2 - (\sum xz)^2$

12–3. (a) $y = 2 + \frac{3}{17}(x - \frac{5}{2})$

(b) .303, .147

(c) $.245 < \sigma^2 < 14.28$

12–4. $\hat{\alpha} = \bar{Y} - \hat{\beta}\bar{x}$, $\text{cov}(\hat{\alpha}, \hat{\beta}) = -\frac{\bar{x}\sigma^2}{nS_X^2}$

12–5. $(\sum Y_i, \sum Y_i^2, \sum x_i Y_i)$

12–6. $\text{var}\,\hat{\beta} = n\sigma^2/(n - 1) \to \sigma^2$ ($\hat{\beta}$ not consistent)

12–7. $.122 + .0105x$

12–8. $.859 \pm .0617$

12–9. (a) $F = 0$ (accept at any level)

(b) $\hat{\alpha} = 5$, $\hat{\beta} = 1$, $\hat{\sigma}^2 = \frac{16}{9}$

(c) $\frac{(\alpha - 5)^2}{2.41} + \frac{(\beta - 1)^2}{3.61} < 1$

12–12.

Source	Sum of Squares	Degrees of Freedom
β after γ	.91	1
α after β, γ	.063	1
Error	.021	1
Fitting γ	1	3

12–13. 9 ± 5.58

12–19. (a) $F = 8.5 > 3.89$ (reject at $\alpha = .05$)

(b) $.96 \pm .51$

(c) $\Delta\bar{X} \pm .78$ for each case

12–23. $S^n >$ (const) $\prod \hat{\sigma}_i^{n_i}$, where S^2 = pooled variance and $\hat{\sigma}_i^2$ = variance of ith sample

12–24. $F = 4.1 < 4.76$ (accept H_0)

12–25. $F = 12.74 > 3.86$ (reject H_0)

12–28. No interaction ($F = .22 < 2.51$)

No Factor A effect ($F = 1.62 < 3.40$)

Factor B effect ($F = 4.77 > 3.01$)

INDEX